# FAR FC 2003

## Federal Aviation Regulations for Flight Crew

**Includes Parts 1, 25, 63, 65, and 121**

AF865136

U.S. Department of Transportation:
From Title 14 of the Code of
Federal Regulations (14 CFR)

Updated and Published by

Aviation Supplies & Academics, Inc.
7005 132nd Place SE
Newcastle, WA 98059-3153

**ASA-03-FAR-FC**

*FAR Flight Crew*
*(Federal Aviation Regulations for Flight Crew)*
2003 Edition

Aviation Supplies & Academics, Inc.
7005 132nd Place SE
Newcastle, WA 98059-3153


This publication contains current regulations as of September 10, 2002.
None of the material in this publication supersedes any documents, procedures, or regulations issued by the Federal Aviation Administration.

Visit the FAA's website to review changes to the regulations:
**http://av-info.faa.gov/dst/reference.htm#Regulations**



**ASA-03-FAR-FC**
ISBN 1-56027-479-4

Printed in the United States of America

03 02 9 8 7 6 5 4 3 2 1

# ASA's 2003 FAR and AIM Series

*FAR/AIM* *FAR Flight Crew* *FAR AMT*

ASA has been supplying the standard reference of the industry, the FAR/AIM series, for more than two decades. The 2003 series continues to provide information directly from the Federal Aviation Regulations and the *Aeronautical Information Manual*, along with these important features:

- Regulation and AIM changes are posted on the ASA website as a free download
- All changes since last printing are clearly identified
- Includes suggested study list of AIM paragraphs and regulations pertinent to specific pilot certificates and ratings
- Indexes included for both FAR and AIM provide an alphabetized listing of subject matter for quick look-up
- AIM produced with full-color graphics

Each regulation Part is preceded by a table of contents. Changes since last year's printing are identified on Page v and in the table of contents for each regulation Part (in bold and marked with an asterisk), as well as within the text for quick reference (changed text is indicated with a bold line in the margin). In the AIM, changes are explained in lists at the beginning, and the effective date of the AIM change is printed at the top of each updated page. It is recommended you familiarize yourself with all the changes to identify those that affect your aviation activities.

Changes affecting the regulations take place daily. The *Aeronautical Information Manual* changes twice a year. **ASA tracks all changes and posts them as Updates on the ASA website (see the "Product Updates" page), from the print date of this book through October 31, 2003**. Visit the ASA website for details on the **Sport Pilot** regulations, due for release in early 2003. You may also visit the FAA website at http://av-info.faa.gov/dst/reference.htm#Regulations to review advisory circulars, Notice of Proposed Regulations (NPRMs), current regulations, and FAA Orders and publications.

You'll find a Reader Response page in the back of this book which you can fill out and mail or fax to us. We welcome your suggestions and comments, as they help generate further improvements to future editions.

Although ASA is not a government agency, and we do not write the regulations or the AIM, we do work closely with the FAA. Questions or concerns can be forwarded to our attention, and we will in turn pass the comments on to the FAA. They are interested in user-feedback and your comments could foster improvements in the regulations which affect the whole industry.

**FAR/AIM Comments**
ASA, Inc.
7005 132nd Place SE
Newcastle, WA 98059-3153

**Internet:** www.asa2fly.com
**Fax:** 425.235.0128
**E-mail:** asa@asa2fly.com

# What's Changed Since Last Year?

**Changes since last year's printing** of the book are noted in the table of contents of each Part with an asterisk and bold title:

Example:

> ***61.5 Certificates and rating issued under this part.**

**The updated text** within the context of the regulation is indicated by a bold line in the margin:

> **(a)** The following certificates are issued under this part to an applicant who satisfactorily accomplishes the training and certification requirements for the certificate sought:
> (1) Pilot certificates—
> (i) Student pilot.
> (ii) Recreational pilot.
> (iii) Private pilot.
> (iv) Commercial pilot.
> (v) Airline transport pilot.
> (2) Flight instructor certificates.
> (3) Ground instructor certificates.

---

## How to Identify the Currency of the Regulations

**In each Part following the Table of Contents is a Source,** with the date of origin for that regulation.

Example:

SOURCE: Docket No. 11802, 38 FR 3161, Feb. 1, 1973, unless otherwise noted.

- **Docket number within the Federal Registers, where original text can be found**
- **Federal Register volume 38, page 3161**
- **Date regulation was written**

**If a change has taken place since the original Regulation was written,** it is noted at the end of the regulation.

Example:

[Doc. No. 11802, 38, FR 3161, Feb. 1, 1973, as amended by Amdt. 61-77, 51 FR 40703, Nov. 7, 1986]

- **Docket number within the Federal Registers**
- **Federal Register volume 38, page 3161, where the original text of regulation is**
- **Date original regulation was written**
- **Amendment number, the change to regulation; in the Federal Register**
- **Federal Register Volume 51, page number 40703, where Amendment can be found**
- **Date change took place**

# List of Changes Since September 2001

**Part 25**

§25.2 *Special retroactive requirements.* Compliance date is extended.

§25.731 *Wheels* is amended.

§25.735 *Brakes and braking systems* is amended.

§25.772 *Pilot Compartment doors is amended.*

§25.795 *Security considerations* is added.

**Part 63**

Special Federal Aviation Regulation No. 63 is removed.

SFAR No. 93 — *Temporary Extension of Time to Allow for Certain Training and Testing* is added.

SFAR No. 96 — *Relief for Participants in Operation Enduring Freedom* is added.

**Part 65**

Special Federal Aviation Regulation No. 63 is removed.

SFAR No. 96 — *Relief for Participants in Operation Enduring Freedom* is added.

**Part 121**

Special Federal Aviation Regulation No. 38–2 is removed.

Special Federal Aviation Regulation No. 52 is removed.

SFAR No. 92–4 — *Flightcrew Compartment Access and Door Designs* is added.

SFAR No. 93 — *Temporary Extension of Time to Allow for Certain Training and Testing* is added.

§121.313 *Miscellaneous equipment* is revised.

§121.538 *Aircraft security* is amended.

§121.547 *Admission to flight deck* is amended.

§121.575 *Alcoholic beverages* is revised.

§121.581 *Observer's seat; en route inspections* is revised.

§121.587 *Closing and locking of flightcrew compartment door* is revised.

§121.703 *Mechanical reliability reports.* Compliance date is extended.

§121.704 *Service difficulty reports (structural).* Compliance date is extended.

§121.705 *Mechanical interruption summary report.* Compliance date is extended.

Appendix I — *Drug Testing* is amended.

Appendix J — *Alcohol Misuse Prevention Program* is amended.

Appendix M — *Airplane Flight Recorder Specifications* is amended.

**For additional information on extended compliance dates and to view the rules currently in effect, visit ASA's website and click on "Product Updates."**

## Midyear Update Coupon

Mail or fax this completed coupon to ASA to receive the 1999 FAR-FC Midyear Update. The update will include information from the publication date of this book through May 15, 1999.

Please send me ASA's 1999 FAR-FC Midyear Update.

Name ______________________________

Address ______________________________

City ______________________________

State ______________________________

Zip ______________________________

Visit the ASA website for instant access to regulatory and procedural changes: www.asa2fly.com

**Note:**
It is **no longer necessary** to mail in a Midyear Update Coupon to stay abreast of regulatory and procedural changes. All changes made after the books are printed will be posted on the ASA website. Changes affecting the regulations take place daily. The AIM changes twice a year. ASA tracks all changes and posts them on the ASA website so you always have access to the most current information.

# FAR Parts Listed in Title 14 of the Code of Federal Regulations

**Location Key**

ALL .................. Published in ASA's *FAR/AIM*, *FAR-FC*, and *FAR-AMT* books
FAR/AIM .......... *FAR/AIM* combination book
FAR-FC ........... *FAR for Flight Crew* book
FAR-AMT ........ *FAR for Aviation Maintenance Technicians* book
GPO ................ Available from the Government Printing Office

# FAR for Flight Crew Contents

# SUBCHAPTER A DEFINITIONS

## PART 1
## DEFINITIONS AND ABBREVIATIONS

Sec.


**Authority:** 49 U.S.C. 106(g), 40113, 44701.

### §1.1 General definitions.

As used in Subchapters A through K of this chapter, unless the context requires otherwise:

***Administrator*** means the Federal Aviation Administrator or any person to whom he has delegated his authority in the matter concerned.

***Aerodynamic coefficients*** means nondimensional coefficients for aerodynamic forces and moments.

***Air carrier*** means a person who undertakes directly by lease, or other arrangement, to engage in air transportation.

***Air commerce*** means interstate, overseas, or foreign air commerce or the transportation of mail by aircraft or any operation or navigation of aircraft within the limits of any Federal airway or any operation or navigation of aircraft which directly affects, or which may endanger safety in, interstate, overseas, or foreign air commerce.

***Aircraft*** means a device that is used or intended to be used for flight in the air.

***Aircraft engine*** means an engine that is used or intended to be used for propelling aircraft. It includes turbosuperchargers, appurtenances, and accessories necessary for its functioning, but does not include propellers.

***Airframe*** means the fuselage, booms, nacelles, cowlings, fairings, airfoil surfaces (including rotors but excluding propellers and rotating airfoils of engines), and landing gear of an aircraft and their accessories and controls.

***Airplane*** means an engine-driven fixed-wing aircraft heavier than air, that is supported in flight by the dynamic reaction of the air against its wings.

***Airport*** means an area of land or water that is used or intended to be used for the landing and takeoff of aircraft, and includes its buildings and facilities, if any.

***Airship*** means an engine-driven lighter-than-air aircraft that can be steered.

***Air traffic*** means aircraft operating in the air or on an airport surface, exclusive of loading ramps and parking areas.

***Air traffic clearance*** means an authorization by air traffic control, for the purpose of preventing collision between known aircraft, for an aircraft to proceed under specified traffic conditions within controlled airspace.

***Air traffic control*** means a service operated by appropriate authority to promote the safe, orderly, and expeditious flow of air traffic.

***Air transportation*** means interstate, overseas, or foreign air transportation or the transportation of mail by aircraft.

***Alert Area.*** An alert area is established to inform pilots of a specific area wherein a high volume of pilot training or an unusual type of aeronautical activity is conducted.

***Alternate airport*** means an airport at which an aircraft may land if a landing at the intended airport becomes inadvisable.

***Altitude engine*** means a reciprocating aircraft engine having a rated takeoff power that is producible from sea level to an established higher altitude.

***Appliance*** means any instrument, mechanism, equipment, part, apparatus, appurtenance, or accessory, including communications equipment, that is used or intended to be used in operating or controlling an aircraft in flight, is installed in or attached to the aircraft, and is not part of an airframe, engine, or propeller.

***Approved***, unless used with reference to another person, means approved by the Administrator.

***Area navigation (RNAV)*** means a method of navigation that permits aircraft operations on any desired course within the coverage of station-referenced navigation signals or within the limits of self-contained system capability.

***Area navigation low route*** means an area navigation route within the airspace extending upward from 1,200 feet above the surface of the earth to, but not including, 18,000 feet MSL.

***Area navigation high route*** means an area navigation route within the airspace extending upward from, and including, 18,000 feet MSL to flight level 450.

***Armed Forces*** means the Army, Navy, Air Force, Marine Corps, and Coast Guard, including their regular and reserve components and members serving without component status.

***Autorotation*** means a rotorcraft flight condition in which the lifting rotor is driven entirely by action of the air when the rotorcraft is in motion.

***Auxiliary rotor*** means a rotor that serves either to counteract the effect of the main rotor torque on a rotorcraft or to maneuver the rotorcraft about one or more of its three principal axes.

***Balloon*** means a lighter-than-air aircraft that is not engine driven, and that sustains flight through the use of either gas buoyancy or an airborne heater.

***Brake horsepower*** means the power delivered at the propeller shaft (main drive or main output) of an aircraft engine.

***Calibrated airspeed*** means the indicated airspeed of an aircraft, corrected for position and instrument error. Calibrated airspeed is equal to true airspeed in standard atmosphere at sea level.

***Canard*** means the forward wing of a canard configuration and may be a fixed, movable, or variable geometry surface, with or without control surfaces.

***Canard configuration*** means a configuration in which the span of the forward wing is substantially less than that of the main wing.

***Category:***

(1) As used with respect to the certification, ratings, privileges, and limitations of airmen, means a broad classification of aircraft. Examples include: airplane; rotorcraft; glider; and lighter-than-air; and

(2) As used with respect to the certification of aircraft, means a grouping of aircraft based upon intended use or operating limitations. Examples include: transport, normal, utility, acrobatic, limited, restricted, and provisional.

***Category A***, with respect to transport category rotorcraft, means multiengine rotorcraft designed with engine and system isolation features specified in Part 29 and utilizing scheduled takeoff and landing operations under a critical engine failure concept which assures adequate designated surface area and adequate performance capability for continued safe flight in the event of engine failure.

***Category B***, with respect to transport category rotorcraft, means single-engine or multiengine rotorcraft which do not fully meet all Category A standards. Category B rotorcraft have no guaranteed stay-up ability in the event of engine failure and unscheduled landing is assumed.

***Category II operations***, with respect to the operation of aircraft, means a straight-in ILS approach to the runway of an airport under a Category II ILS instrument approach procedure issued by the Administrator or other appropriate authority.

***Category III operations***, with respect to the operation of aircraft, means an ILS approach to, and landing on, the runway of an airport using a Category III ILS instrument approach procedure issued by the Administrator or other appropriate authority.

***Category IIIa operations***, an ILS approach and landing with no decision height (DH), or a DH below 100 feet (30 meters), and controlling runway visual range not less than 700 feet (200 meters).

***Category IIIb operations***, an ILS approach and landing with no DH, or with a DH below 50 feet (15 meters), and controlling runway visual range less than 700 feet (200 meters), but not less than 150 feet (50 meters).

***Category IIIc operations***, an ILS approach and landing with no DH and no runway visual range limitation.

***Ceiling*** means the height above the earth's surface of the lowest layer of clouds or obscuring phenomena that is reported as "broken," "overcast," or "obscuration," and not classified as "thin" or "partial."

***Civil aircraft*** means aircraft other than public aircraft.

***Class:***

(1) As used with respect to the certification, ratings, privileges, and limitations of airmen, means a classification of aircraft within a category having similar operating characteristics. Examples include: single engine; multiengine; land; water; gyroplane; helicopter; airship; and free balloon; and

(2) As used with respect to the certification of aircraft, means a broad grouping of aircraft having similar characteristics of propulsion, flight, or landing. Examples include: airplane; rotorcraft; glider; balloon; landplane; and seaplane.

***Clearway*** means:

(1) For turbine engine powered airplanes certificated after August 29, 1959, an area beyond the runway, not less than 500 feet wide, centrally located about the extended centerline of the runway, and under the control of the airport authorities. The clearway is expressed in terms of a clearway plane, extending from the end of the runway with an upward slope not exceeding 1.25 percent, above which no object nor any terrain protrudes. However, threshold lights may protrude above the plane if their height above the end of the runway is 26 inches or less and if they are located to each side of the runway.

(2) For turbine engine powered airplanes certificated after September 30, 1958, but before August 30, 1959, an area beyond the takeoff runway extending no less than 300 feet on either side of the extended centerline of the runway, at an elevation no higher than the elevation of the end of the runway, clear of all fixed obstacles, and under the control of the airport authorities.

***Climbout speed***, with respect to rotorcraft, means a referenced airspeed which results in a flight path clear of the height-velocity envelope during initial climbout.

***Commercial operator*** means a person who, for compensation or hire, engages in the carriage by aircraft in air commerce of persons or property, other than as an air carrier or foreign air carrier or under the authority of Part 375 of this title. Where it is doubtful that an operation is for "compensation or hire", the test applied is whether the carriage by air is merely incidental to the person's other business or is, in itself, a major enterprise for profit.

***Controlled airspace*** means an airspace of defined dimensions within which air traffic control service is provided to IFR flights and to VFR flights in accordance with the airspace classification.

NOTE—Controlled airspace is a generic term that covers Class A, Class B, Class C, Class D, and Class E airspace.

***Controlled Firing Area.*** A controlled firing area is established to contain activities, which if not conducted in a controlled environment, would be hazardous to nonparticipating aircraft.

***Crewmember*** means a person assigned to perform duty in an aircraft during flight time.

***Critical altitude*** means the maximum altitude at which, in standard atmosphere, it is possible to maintain, at a specified rotational speed, a specified power or a specified manifold pressure. Unless otherwise stated, the critical altitude is the maximum altitude at which it is possible to maintain, at the maximum continuous rotational speed, one of the following:

(1) The maximum continuous power, in the case of engines for which this power rating is the same at sea level and at the rated altitude.

(2) The maximum continuous rated manifold pressure, in the case of engines, the maximum continuous power of which is governed by a constant manifold pressure.

***Critical engine*** means the engine whose failure would most adversely affect the performance or handling qualities of an aircraft.

***Decision height***, with respect to the operation of aircraft, means the height at which a decision must be made, during an ILS or PAR instrument approach, to either continue the approach or to execute a missed approach.

***Equivalent airspeed*** means the calibrated airspeed of an aircraft corrected for adiabatic compressible flow for the particular altitude. Equivalent airspeed is equal to calibrated airspeed in standard atmosphere at sea level.

***Extended over-water operation*** means—

(1) With respect to aircraft other than helicopters, an operation over water at a horizontal distance of more than 50 nautical miles from the nearest shoreline; and

(2) With respect to helicopters, an operation over water at a horizontal distance of more than 50 nautical miles from the nearest shoreline and more than 50 nautical miles from an off-shore heliport structure.

***External load*** means a load that is carried, or extends, outside of the aircraft fuselage.

***External-load attaching*** means the structural components used to attach an external load to an aircraft, including external-load containers, the backup structure at the attachment points, and any quick-release device used to jettison the external load.

***Fireproof***—

(1) With respect to materials and parts used to confine fire in a designated fire zone, means the capacity to withstand at least as well as steel in dimensions appropriate for the purpose for which they are used, the heat produced when there is a severe fire of extended duration in that zone; and

(2) With respect to other materials and parts, means the capacity to withstand the heat associated with fire at least as well as steel in dimensions appropriate for the purpose for which they are used.

***Fire resistant***—

(1) With respect to sheet or structural members means the capacity to withstand the heat associated with fire at least as well as aluminum alloy in dimensions appropriate for the purpose for which they are used; and

(2) With respect to fluid-carrying lines, fluid system parts, wiring, air ducts, fittings, and powerplant controls, means the capacity to perform the intended functions under the heat and other conditions likely to occur when there is a fire at the place concerned.

***Flame resistant*** means not susceptible to combustion to the point of propagating a flame, beyond safe limits, after the ignition source is removed.

***Flammable***, with respect to a fluid or gas, means susceptible to igniting readily or to exploding.

***Flap extended speed*** means the highest speed permissible with wing flaps in a prescribed extended position.

***Flash resistant*** means not susceptible to burning violently when ignited.

***Flight crewmember*** means a pilot, flight engineer, or flight navigator assigned to duty in an aircraft during flight time.

***Flight level*** means a level of constant atmospheric pressure related to a reference datum of 29.92 inches of mercury. Each is stated in three digits that represent hundreds of feet. For example, flight level 250 represents a barometric altimeter indication of 25,000 feet; flight level 255, an indication of 25,500 feet.

***Flight plan*** means specified information, relating to the intended flight of an aircraft, that is filed orally or in writing with air traffic control.

***Flight time*** means:

(1) Pilot time that commences when an aircraft moves under its own power for the purpose of flight and ends when the aircraft comes to rest after landing; or

(2) For a glider without self-launch capability, pilot times that commences when the glider is towed for the purpose of flight and ends when the glider comes to rest after landing.

***Flight visibility*** means the average forward horizontal distance, from the cockpit of an aircraft in flight, at which prominent unlighted objects may be seen and identified by day and prominent lighted objects may be seen and identified by night.

***Foreign air carrier*** means any person other than a citizen of the United States, who undertakes directly, by lease or other arrangement, to engage in air transportation.

***Foreign air commerce*** means the carriage by aircraft of persons or property for compensation or hire, or the carriage of mail by aircraft, or the operation or navigation of aircraft in the conduct or furtherance of a business or vocation, in commerce between a place in the United States and any place outside thereof; whether such commerce moves wholly by aircraft or partly by aircraft and partly by other forms of transportation.

***Foreign air transportation*** means the carriage by aircraft of persons or property as a common carrier for compensation or hire, or the carriage of mail by aircraft, in commerce between a place in the United States and any place outside of the United States, whether that commerce moves wholly by aircraft or partly by aircraft and partly by other forms of transportation.

***Forward wing*** means a forward lifting surface of a canard configuration or tandem-wing configuration airplane. The surface may be a fixed, movable, or variable geometry surface, with or without control surfaces.

***Glider*** means a heavier-than-air aircraft, that is supported in flight by the dynamic reaction of the air against its lifting surfaces and whose free flight does not depend principally on an engine.

***Go-around power or thrust setting*** means the maximum allowable in-flight power or thrust setting identified in the performance data.

***Ground visibility*** means prevailing horizontal visibility near the earth's surface as reported by the United States National Weather Service or an accredited observer.

***Gyrodyne*** means a rotorcraft whose rotors are normally engine-driven for takeoff, hovering, and landing, and for forward flight through part of its speed range, and whose means of propulsion, consisting usually of conventional propellers, is independent of the rotor system.

***Gyroplane*** means a rotorcraft whose rotors are not engine-driven, except for initial starting, but are made to rotate by action of the air when the rotorcraft is moving; and whose means of propulsion, consisting usually of conventional propellers, is independent of the rotor system.

***Helicopter*** means a rotorcraft that, for its horizontal motion, depends principally on its engine-driven rotors.

***Heliport*** means an area of land, water, or structure used or intended to be used for the landing and takeoff of helicopters.

***Idle thrust*** means the jet thrust obtained with the engine power control level set at the stop for the least thrust position at which it can be placed.

***IFR conditions*** means weather conditions below the minimum for flight under visual flight rules.

***IFR over-the-top***, with respect to the operation of aircraft, means the operation of an aircraft over-the-top on an IFR flight plan when cleared by air traffic control to maintain "VFR conditions" or "VFR conditions on top".

***Indicated airspeed*** means the speed of an aircraft as shown on its pitot static airspeed indicator calibrated to reflect standard atmosphere adiabatic compressible flow at sea level uncorrected for airspeed system errors.

***Instrument*** means a device using an internal mechanism to show visually or aurally the attitude, altitude, or operation of an aircraft or aircraft part. It includes electronic devices for automatically controlling an aircraft in flight.

***Interstate air commerce*** means the carriage by aircraft of persons or property for compensation or hire, or the carriage of mail by aircraft, or the operation or navigation of aircraft in the conduct or furtherance of a business or vocation, in commerce between a place in any State of the United States, or the District of Columbia, and a place in any other State of the United States, or the District of Columbia; or between places in the same State of the United States through the airspace over any place outside thereof; or between places in the same territory or possession of the United States, or the District of Columbia.

***Interstate air transportation*** means the carriage by aircraft of persons or property as a common carrier for compensation or hire, or the carriage of mail by aircraft in commerce:

(1) Between a place in a State or the District of Columbia and another place in another State or the District of Columbia;

(2) Between places in the same State through the airspace over any place outside that State; or

(3) Between places in the same possession of the United States;

Whether that commerce moves wholly by aircraft of partly by aircraft and partly by other forms of transportation.

***Intrastate air transportation*** means the carriage of persons or property as a common carrier for compensation or hire, by turbojet-powered aircraft capable of carrying thirty or more persons, wholly within the same State of the United States.

***Kite*** means a framework, covered with paper, cloth, metal, or other material, intended to be flown at the end of a rope or cable, and having as its only support the force of the wind moving past its surfaces.

***Landing gear extended speed*** means the maximum speed at which an aircraft can be safely flown with the landing gear extended.

***Landing gear operating speed*** means the maximum speed at which the landing gear can be safely extended or retracted.

***Large aircraft*** means aircraft of more than 12,500 pounds, maximum certificated takeoff weight.

***Lighter-than-air aircraft*** means aircraft that can rise and remain suspended by using contained gas weighing less than the air that is displaced by the gas.

***Load factor*** means the ratio of a specified load to the total weight of the aircraft. The specified load is expressed in terms of any of the following: aerodynamic forces, inertia forces, or ground or water reactions.

***Long-range communication system (LRCS).*** A system that uses satellite relay, data link, high frequency, or another approved communication system which extends beyond line of sight.

***Long-range navigation system (LRNS).*** An electronic navigation unit that is approved for use under instrument flight rules as a primary means of navigation, and has at least one source of navigational input, such as inertial navigation system, global positioning system, Omega/very low frequency, or Loran C.

***Mach number*** means the ratio of true airspeed to the speed of sound.

***Main rotor*** means the rotor that supplies the principal lift to a rotorcraft.

***Maintenance*** means inspection, overhaul, repair, preservation, and the replacement of parts, but excludes preventive maintenance.

***Major alteration*** means an alteration not listed in the aircraft, aircraft engine, or propeller specifications—

(1) That might appreciably affect weight, balance, structural strength, performance, powerplant operation, flight characteristics, or other qualities affecting airworthiness; or

(2) That is not done according to accepted practices or cannot be done by elementary operations.

***Major repair*** means a repair:

(1) That, if improperly done, might appreciably affect weight, balance, structural strength, performance, powerplant operation, flight characteristics, or other qualities affecting airworthiness; or

(2) That is not done according to accepted practices or cannot be done by elementary operations.

***Manifold pressure*** means absolute pressure as measured at the appropriate point in the induction system and usually expressed in inches of mercury.

***Maximum speed for stability characteristics, $V_{FC}/M_{FC}$*** means a speed that may not be less than a speed midway between maximum operating limit speed ($V_{MO}/M_{MO}$) and demonstrated flight diving speed ($V_{DF}/M_{DF}$), except that, for altitudes where the Mach number is the limiting factor, $M_{FC}$ need not exceed the Mach number at which effective speed warning occurs.

***Medical certificate*** means acceptable evidence of physical fitness on a form prescribed by the Administrator.

***Military operations area.*** A military operations area (MOA) is airspace established outside Class A airspace to separate or segregate certain nonhazardous military activities from IFR Traffic and to identify for VFR traffic where theses activities are conducted.

***Minimum descent altitude*** means the lowest altitude, expressed in feet above mean sea level, to which descent is authorized on final approach or during circle-to-land maneuvering in execution of a standard instrument approach procedure, where no electronic glide slope is provided.

***Minor alteration*** means an alteration other than a major alteration.

***Minor repair*** means a repair other than a major repair.

***Navigable airspace*** means airspace at and above the minimum flight altitudes prescribed by or under this chapter, including airspace needed for safe takeoff and landing.

***Night*** means the time between the end of evening civil twilight and the beginning of morning civil twilight, as published in the American Air Almanac, converted to local time.

***Nonprecision approach procedure*** means a standard instrument approach procedure in which no electronic glide slope is provided.

***Operate,*** with respect to aircraft, means use, cause to use or authorize to use aircraft, for the purpose (except as provided in §91.13 of this chapter) of air navigation including the piloting of aircraft, with or without the right of legal control (as owner, lessee, or otherwise).

***Operational control,*** with respect to a flight, means the exercise of authority over initiating, conducting or terminating a flight.

***Overseas air commerce*** means the carriage by aircraft of persons or property for compensation or hire, or the carriage of mail by aircraft, or the operation or navigation of aircraft in the conduct or furtherance of a business or vocation, in commerce between a place in any State of the United States, or the District of Columbia, and any place in a territory or possession of the United States; or between a place in a territory or possession of the United States, and a place in any other territory or possession of the United States.

***Overseas air transportation*** means the carriage by aircraft of persons or property as a common carrier for compensation or hire, or the carriage of mail by aircraft, in commerce:

(1) Between a place in a State or the District of Columbia and a place in a possession of the United States; or

(2) Between a place in a possession of the United States and a place in another possession of the United States; whether that commerce moves wholly by aircraft or partly by aircraft and partly by other forms of transportation.

***Over-the-top*** means above the layer of clouds or other obscuring phenomena forming the ceiling.

***Parachute*** means a device used or intended to be used to retard the fall of a body or object through the air.

***Person*** means an individual, firm, partnership, corporation, company, association, joint-stock association, or governmental entity. It includes a trustee, receiver, assignee, or similar representative of any of them.

***Pilotage*** means navigation by visual reference to landmarks.

***Pilot in command*** means the person who:

(1) Has final authority and responsibility for the operation and safety of the flight;

(2) Has been designated as pilot in command before or during the flight; and

(3) Holds the appropriate category, class, and type rating, if appropriate, for the conduct of the flight.

***Pitch setting*** means the propeller blade setting as determined by the blade angle measured in a manner, and at a radius, specified by the instruction manual for the propeller.

***Positive control*** means control of all air traffic, within designated airspace, by air traffic control.

***Powered-lift*** means a heavier-than-air aircraft capable of vertical takeoff, vertical landing, and low speed flight that depends principally on engine-driven lift devices or engine thrust for lift during these flight regimes and on nonrotating airfoil(s) for lift during horizontal flight.

***Precision approach procedure*** means a standard instrument approach procedure in which an electronic glide slope is provided, such as ILS and PAR.

***Preventive maintenance*** means simple or minor preservation operations and the replacement of small standard parts not involving complex assembly operations.

***Prohibited area.*** A prohibited area is airspace designated under part 73 within which no person may operate an aircraft without the permission of the using agency.

***Propeller*** means a device for propelling an aircraft that has blades on an engine-driven shaft and that, when rotated, produces by its action on the air, a thrust approximately perpendicular to its plane of rotation. It includes control components normally supplied by its manufacturer, but does not include main and auxiliary rotors or rotating airfoils of engines.

***Public aircraft*** means an aircraft used only for the United States Government, or owned and operated (except for commercial purposes), or exclusively leased for at least 90 continuous days, by a government (except the United States Government), including a State, the District of Columbia, or a territory or possession of the United States, or political subdivision of that government; but does not include a government-owned aircraft transporting property for commercial purposes, or transporting passengers other than transporting (for other than commercial purposes) crewmembers or other persons aboard the aircraft whose presence is required to perform, or is associated with the performance of, a governmental function such as firefighting, search and rescue, law enforcement, aeronautical research, or biological or geological resource management; or transporting (for other than commercial purposes) persons aboard the aircraft if the aircraft is operated by the Armed Forces or an intelligence agency of the United States. An aircraft described in the preceding sentence shall, notwithstanding any limitation relating to use of the aircraft for commercial purposes, be considered to be a public aircraft for the purposes of this Chapter without regard to whether the aircraft is operated by a unit of government on behalf of another unit of government, pursuant to a cost reimbursement agreement between such units of government, if the unit of government on whose behalf the operation is conducted certifies to the Administrator of the Federal Aviation Administration that the operation was necessary to respond to a significant and imminent threat to life or property (including natural resources) and that no service by a private operator was reasonably available to meet the threat.

***Rated continuous OEI power,*** with respect to rotorcraft turbine engines, means the approved brake horsepower developed under static conditions at specified altitudes and temperatures within the operating limitations established for the engine under Part 33 of this chapter, and limited in use to the time required to complete the flight after the failure of one engine of a multiengine rotorcraft.

***Rated maximum continuous augmented thrust,*** with respect to turbojet engine type certification, means the approved jet thrust that is developed statically or in flight, in standard atmosphere at a specified altitude, with fluid injection or with the burning of fuel in a separate combustion chamber, within the engine operating limitations established under Part 33 of this chapter, and approved for unrestricted periods of use.

***Rated maximum continuous power,*** with respect to reciprocating, turbopropeller, and turboshaft engines, means the approved brake horsepower that is developed statically or in flight, in standard atmosphere at a specified altitude, within the engine operating limitations established under Part 33, and approved for unrestricted periods of use.

***Rated maximum continuous thrust,*** with respect to turbojet engine type certification, means the approved jet thrust that is developed statically or in flight, in standard atmosphere at a specified altitude, without fluid injection and without the burning of fuel in a separate combustion chamber, within the engine operating limitations established under Part 33 of this chapter, and approved for unrestricted periods of use.

***Rated takeoff augmented thrust,*** with respect to turbojet engine type certification, means the approved jet thrust that is developed statically under standard sea level conditions, with fluid injection or with the burning of fuel in a separate combustion chamber, within the engine operating limitations established under Part 33 of this chapter, and limited in use to periods of not over 5 minutes for takeoff operation.

***Rated takeoff power,*** with respect to reciprocating, turbopropeller, and turboshaft engine type certification, means the approved brake horsepower that is developed statically under standard sea level conditions, within the engine operating limitations established under Part 33, and limited in use to periods of not over 5 minutes for takeoff operation.

***Rated takeoff thrust,*** with respect to turbojet engine type certification, means the approved jet thrust that is developed statically under standard sea level conditions, without fluid injection and without the burning of fuel in a separate combustion chamber, within the engine operating limitations established under Part 33 of this chapter, and limited in use to periods of not over 5 minutes for takeoff operation.

***Rated 30-minute OEI power,*** with respect to rotorcraft turbine engines, means the approved brake horsepower developed under static conditions at specified altitudes and temperatures within the operating limitations established for the engine under Part 33 of this chapter, and limited in use to a period of not more than 30 minutes after the failure of one engine of a multiengine rotorcraft.

***Rated 30-second OEI power,*** with respect to rotorcraft turbine engines, means the approved brake horsepower developed under static conditions at specified altitudes and temperatures within the operating limitations established for the engine under Part 33 of this chapter, for continued one-flight operation after the failure of one engine in multiengine rotorcraft, limited to three periods of use no longer than 30 seconds each in any one flight, and followed by mandatory inspection and prescribed maintenance action.

***Rated 2-minute OEI power,*** with respect to rotorcraft turbine engines, means the approved brake horsepower developed under static conditions at specified altitudes and temperatures within the operating limitations established for the engine under Part 33 of this chapter, for continued one-flight operation after the failure of one engine in multiengine rotorcraft, limited to three periods of use no longer than 2 minutes each in any one flight, and followed by mandatory inspection and prescribed maintenance action.

***Rated 2½-minute OEI power,*** with respect to rotorcraft turbine engines, means the approved brake horsepower developed under static conditions at specified altitudes and temperatures within the operating limitations established for the engine under Part 33 of this chapter, and limited in use to a period of not more than 2½ minutes after the failure of one engine of a multiengine rotorcraft.

***Rating*** means a statement that, as a part of a certificate, sets forth special conditions, privileges, or limitations.

***Reporting point*** means a geographical location in relation to which the position of an aircraft is reported.

***Restricted area.*** A restricted area is airspace designated under Part 73 within which the flight of aircraft, while not wholly prohibited, is subject to restriction.

***RNAV way point (W/P)*** means a predetermined geographical position used for route or instrument approach definition or progress reporting purposes that is defined relative to a VORTAC station position.

***Rocket*** means an aircraft propelled by ejected expanding gases generated in the engine from self-contained propellants and not dependent on the intake of outside substances. It includes any part which becomes separated during the operation.

***Rotorcraft*** means a heavier-than-air aircraft that depends principally for its support in flight on the lift generated by one or more rotors.

***Rotorcraft-load combination*** means the combination of a rotorcraft and an external-load, including the external-load attaching means. Rotorcraft-load combinations are designated as Class A, Class B, Class C, and Class D, as follows:

(1) Class A rotorcraft-load combination means one in which the external load cannot move freely, cannot be jettisoned, and does not extend below the landing gear.

(2) Class B rotorcraft-load combination means one in which the external load is jettisonable and is lifted free of land or water during the rotorcraft operation.

(3) Class C rotorcraft-load combination means one in which the external load is jettisonable and remains in contact with land or water during the rotorcraft operation.

(4) Class D rotorcraft-load combination means one in which the external-load is other than a Class A, B, or C and has been specifically approved by the Administrator for that operation.

***Route segment*** means a part of a route. Each end of that part is identified by:

(1) A continental or insular geographical location; or

(2) A point at which a definite radio fix can be established.

***Sea level engine*** means a reciprocating aircraft engine having a rated takeoff power that is producible only at sea level.

***Second in command*** means a pilot who is designated to be second in command of an aircraft during flight time.

***Show,*** unless the context otherwise requires, means to show to the satisfaction of the Administrator.

***Small aircraft*** means aircraft of 12,500 pounds or less, maximum certificated takeoff weight.

***Special VFR conditions*** mean meteorological conditions that are less than those required for basic VFR flight in controlled airspace and in which some aircraft are permitted flight under visual flight rules.

***Special VFR operations*** means aircraft operating in accordance with clearances within controlled airspace in meteorological conditions less than the basic VFR weather minima. Such operations must be requested by the pilot and approved by ATC.

***Standard atmosphere*** means the atmosphere defined in U.S. Standard Atmosphere, 1962 (Geopotential altitude tables).

***Stopway*** means an area beyond the takeoff runway, no less wide than the runway and centered upon the extended centerline of the runway, able to support the airplane during an aborted takeoff, without causing structural damage to the airplane, and designated by the airport authorities for use in decelerating the airplane during an aborted takeoff.

***Takeoff power:***

(1) With respect to reciprocating engines, means the brake horsepower that is developed under standard sea level conditions, and under the maximum conditions of crankshaft rotational speed and engine manifold pressure approved for the normal takeoff, and limited in continuous use to the period of time shown in the approved engine specification; and

(2) With respect to turbine engines, means the brake horsepower that is developed under static conditions at a specified altitude and atmospheric temperature, and under the maximum conditions of rotor shaft rotational speed and gas temperature approved for the normal takeoff, and limited in continuous use to the period of time shown in the approved engine specification.

***Takeoff safety speed*** means a referenced airspeed obtained after lift-off at which the required one-engine-inoperative climb performance can be achieved.

***Takeoff thrust,*** with respect to turbine engines, means the jet thrust that is developed under static conditions at a specific altitude and atmospheric temperature under the maximum conditions of rotorshaft rotational speed and gas temperature approved for the normal takeoff, and limited in continuous use to the period of time shown in the approved engine specification.

***Tandem wing configuration*** means a configuration having two wings of similar span, mounted in tandem.

***TCAS I*** means a TCAS that utilizes interrogations of, and replies from, airborne radar beacon transponders and provides traffic advisories to the pilot.

***TCAS II*** means a TCAS that utilizes interrogations of, and replies from airborne radar beacon transponders and provides traffic advisories and resolution advisories in the vertical plane.

***TCAS III*** means a TCAS that utilizes interrogation of, and replies from, airborne radar beacon transponders and provides traffic advisories and resolution advisories in the vertical and horizontal planes to the pilot.

***Time in service,*** with respect to maintenance time records, means the time from the moment an aircraft leaves the surface of the earth until it touches it at the next point of landing.

***True airspeed*** means the airspeed of an aircraft relative to undisturbed air. True airspeed is equal to equivalent airspeed multiplied by $(\rho 0/\rho)^{1/2}$.

***Traffic pattern*** means the traffic flow that is prescribed for aircraft landing at, taxiing on, or taking off from, an airport.

***Type:***

(1) As used with respect to the certification, ratings, privileges, and limitations of airmen, means a specific make and basic model of aircraft, including modifications thereto that do not change its handling or flight characteristics. Examples include: DC-7, 1049, and F-27; and

(2) As used with respect to the certification of aircraft, means those aircraft which are similar in design. Examples include: DC-7 and DC-7C; 1049G and 1049H; and F-27 and F-27F.

(3) As used with respect to the certification of aircraft engines means those engines which are similar in design. For example, JT8D and JT8D-7 are engines of the same type, and JT9D-3A and JT9D-7 are engines of the same type.

***United States,*** in a geographical sense, means (1) the States, the District of Columbia, Puerto Rico, and the possessions, including the territorial waters, and (2) the airspace of those areas.

***United States air carrier*** means a citizen of the United States who undertakes directly by lease, or other arrangement, to engage in air transportation.

***VFR over-the-top,*** with respect to the operation of aircraft, means the operation of an aircraft over-the-top under VFR when it is not being operated on an IFR flight plan.

***Warning area.*** A warning area is airspace of defined dimensions, extending from 3 nautical miles outward from the coast of the United States, that contains activity that may be hazardous to nonparticipating aircraft. The purpose of such warning areas is to warn nonparticipating pilots of the potential danger. A warning area may be located over domestic or international waters or both.

***Winglet or tip fin*** means an out-of-plane surface extending from a lifting surface. The surface may or may not have control surfaces.

[Docket No. 1150, 27 FR 4588, May 15, 1962; as amended by Amdt. 1–39, 60 FR 5075, Jan. 25, 1995; Amdt. 1–40, 60 FR 30749, June 9, 1995; Amdt. 1–42, 61 FR 2081, Jan. 24, 1996; Amdt. 1–43, 61 FR 5183, Feb. 9, 1996; Amdt. 1–44, 61 FR 7190, Feb. 26, 1996; Amdt. 1–46, 61 FR 31328, June 19, 1996; Amdt. 1–45, 61 FR 34547, July 2, 1996; Amdt. 1–47, 62 FR 16298, April 4, 1997]

## §1.2 Abbreviations and symbols.

In Subchapters A through K of this chapter:

"AGL" means above ground level.
"ALS" means approach light system.
"ASR" means airport surveillance radar.
"ATC" means air traffic control.
"CAS" means calibrated airspeed.
"CAT II" means Category II.
"CONSOL or CONSOLAN" means a kind of low or medium frequency long range navigational aid.
"DH" means decision height.
"DME" means distance measuring equipment compatible with TACAN.
"EAS" means equivalent airspeed.
"FAA" means Federal Aviation Administration.
"FM" means fan marker.
"GS" means glide slope.
"HIRL" means high-intensity runway light system.
"IAS" means indicated airspeed.
"ICAO" means International Civil Aviation Organization.
"IFR" means instrument flight rules.
"ILS" means instrument landing system.
"IM" means ILS inner marker.
"INT" means intersection.
"LDA" means localizer-type directional aid.
"LFR" means low-frequency radio range.
"LMM" means compass locator at middle marker.
"LOC" means ILS localizer.
"LOM" means compass locator at outer marker.
"M" means mach number.
"MAA" means maximum authorized IFR altitude.
"MALS" means medium intensity approach light system.
"MALSR" means medium intensity approach light system with runway alignment indicator lights.
"MCA" means minimum crossing altitude.
"MDA" means minimum descent altitude.
"MEA" means minimum en route IFR altitude.
"MM" means ILS middle marker.
"MOCA" means minimum obstruction clearance altitude.
"MRA" means minimum reception altitude.
"MSL" means mean sea level.
"NDB (ADF)" means nondirectional beacon (automatic direction finder).
"NOPT" means no procedure turn required.
"OEI" means one engine inoperative.
"OM" means ILS outer marker.
"PAR" means precision approach radar.
"RAIL" means runway alignment indicator light system.
"RBN" means radio beacon.
"RCLM" means runway centerline marking.
"RCLS" means runway centerline light system.
"REIL" means runway end identification lights.
"RR" means low or medium frequency radio range station.
"RVR" means runway visual range as measured in the touchdown zone area.
"SALS" means short approach light system.
"SSALS" means simplified short approach light system.
"SSALSR" means simplified short approach light system with runway alignment indicator lights.
"TACAN" means ultra-high frequency tactical air navigational aid.
"TAS" means true airspeed.
"TCAS" means a traffic alert and collision avoidance system.
"TDZL" means touchdown zone lights.
"TVOR" means very high frequency terminal omnirange station.
$V_A$ means design maneuvering speed.

$V_B$ means design speed for maximum gust intensity.

$V_C$ means design cruising speed.

$V_D$ means design diving speed.

$V_{DF}/M_{DF}$ means demonstrated flight diving speed.

$V_{EF}$ means the speed at which the critical engine is assumed to fail during takeoff.

$V_F$ means design flap speed.

$V_{FC}/M_{FC}$ means maximum speed for stability characteristics.

$V_{FE}$ means maximum flap extended speed.

$V_H$ means maximum speed in level flight with maximum continuous power.

$V_{LE}$ means maximum landing gear extended speed.

$V_{LO}$ means maximum landing gear operating speed.

$V_{LOF}$ means lift-off speed.

$V_{MC}$ means minimum control speed with the critical engine inoperative.

$V_{MO}/M_{MO}$ means maximum operating limit speed.

$V_{MU}$ means minimum unstick speed.

$V_{NE}$ means never-exceed speed.

$V_{NO}$ means maximum structural cruising speed.

$V_R$ means rotation speed.

$V_S$ means the stalling speed or the minimum steady flight speed at which the airplane is controllable.

$V_{S0}$ means the stalling speed or the minimum steady flight speed in the landing configuration.

$V_{S1}$ means the stalling speed or the minimum steady flight speed obtained in a specific configuration.

$V_{TOSS}$ means takeoff safety speed for Category A rotorcraft.

$V_X$ means speed for best angle of climb.

$V_Y$ means speed for best rate of climb.

$V_1$ means the maximum speed in the takeoff at which the pilot must take the first action (e.g., apply brakes, reduce thrust, deploy speed brakes) to stop the airplane within the accelerate-stop distance. $V_1$ also means the minimum speed in the takeoff, following a failure of the critical engine at $V_{EF}$, at which the pilot can continue the takeoff and achieve the required height above the takeoff surface within the takeoff distance.

$V_2$ means takeoff safety speed.

$V_{2min}$ means minimum takeoff safety speed.

"VFR" means visual flight rules.

"VHF" means very high frequency.

"VOR" means very high frequency omnirange station.

"VORTAC" means collocated VOR and TACAN.

[Docket No. 1150, 27 FR 4590, May 15, 1962; as amended by Amdt. 1–48, 63 FR 8318, Feb. 18, 1998]

## §1.3 Rules of construction.

(a) In Subchapters A through K of this chapter, unless the context requires otherwise:

(1) Words importing the singular include the plural;

(2) Words importing the plural include the singular; and

(3) Words importing the masculine gender include the feminine.

(b) In Subchapters A through K of this chapter, the word:

(1) "Shall" is used in an imperative sense;

(2) "May" is used in a permissive sense to state authority or permission to do the act prescribed, and the words "no person may * * *" or "a person may not * * *" mean that no person is required, authorized, or permitted to do the act prescribed; and

(3) "Includes" means "includes but is not limited to".

[Docket No. 1150, 27 FR 4590, May 15, 1962; as amended by Amdt. 1–10, 31 FR 5055, March 29, 1966]

# PART 25
## AIRWORTHINESS STANDARDS: TRANSPORT CATEGORY AIRPLANES

SPECIAL FEDERAL AVIATION REGULATION

SFAR No. 13

### Subpart A — General



### Subpart B — Flight

*GENERAL*



*PERFORMANCE*



*CONTROLLABILITY AND MANEUVERABILITY*



*TRIM*



*STABILITY*



*STALLS*



*GROUND AND WATER HANDLING CHARACTERISTICS*



*MISCELLANEOUS FLIGHT REQUIREMENTS*



### Subpart C — Structure

*GENERAL*



*FLIGHT LOADS*



*FLIGHT MANEUVER AND GUST CONDITIONS*



*SUPPLEMENTARY CONDITIONS*

**Authority**: 49 U.S.C 106(g), 40113, 44701, 44702 and 44704.

**Source:** Docket No. 5066, 29 FR 18291, Dec. 24, 1964, unless otherwise noted.

## SFAR No. 13 to Part 25

**1.** *Applicability.* Contrary provisions of the Civil Air Regulations regarding certification notwithstanding,[1] this regulation shall provide the basis for approval by the Administrator of modifications of individual Douglas DC-3 and Lockheed L-18 airplanes subsequent to the effective date of this regulation.

[1] It is not intended to waive compliance with such airworthiness requirements as are included in the operating parts of the Civil Air Regulations for specific types of operation.

**2.** *General modifications.* Except as modified in sections 3 and 4 of this regulation, an applicant for approval of modifications to a DC-3 or L-18 airplane which result in changes in design or in changes to approved limitations shall show that the modifications were accomplished in accordance with the rules of either Part 4a or Part 4b in effect on September 1, 1953, which are applicable to the modification being made: *Provided,* That an applicant may elect to accomplish a modification in accordance with the rules of Part 4b in effect on the date of application for the modification in lieu of Part 4a or Part 4b as in effect on September 1, 1953: *And provided further,* That each specific modification must be accomplished in accordance with all of the provisions contained in the elected rules relating to the particular modification.

**3.** *Specific conditions for approval.* An applicant for any approval of the following specific changes shall comply with section 2 of this regulation as modified by the applicable provisions of this section.

(a) *Increase in take-off power limitation—1,200 to 1,350 horsepower.* The engine take-off power limitation for the airplane may be increased to more than 1,200 horsepower but not to more than 1,350 horsepower per engine if the increase in power does not adversely affect the flight characteristics of the airplane.

(b) *Increase in take-off power limitation to more than 1,350 horsepower.* The engine take-off power limitation for the airplane may be increased to more than 1,350 horsepower per engine if compliance is shown with the flight characteristics and ground handling requirements of Part 4b.

(c) *Installation of engines of not more than 1,830 cubic inches displacement and not having a certificated take-off rating of more than 1,350 horsepower.* Engines of not more than 1,830 cubic inches displacement and not having a certificated take-off rating of more than 1,350 horsepower which necessitate a major modification of redesign of the engine installation may be installed, if the engine fire prevention and fire protection are equivalent to that on the prior engine installation.

(d) *Installation of engines of more than 1,830 cubic inches displacement or having certificated take-off rating of more than 1,350 horsepower.* Engines of more than 1,830 cubic inches displacement or having certificated take-off rating of more than 1,350 horsepower may be installed if compliance is shown with the engine installation requirements of Part 4b: *Provided,* That where literal compliance with the engine installation requirements of Part 4b is extremely difficult to accomplish and would not contribute materially to the objective sought, and the Administrator finds that the experience with the DC-3 or L-18 airplanes justifies it, he is authorized to accept such measures of compliance as he finds will effectively accomplish the basic objective.

**4.** *Establishment of new maximum certificated weights.* An applicant for approval of new maximum certificated weights shall apply for an amendment of the airworthiness certificate of the airplane and shall show that the weights sought have been established, and the appropriate manual material obtained, as provided in this section.

Note: Transport category performance requirements result in the establishment of maximum certificated weights for various altitudes.

(a) *Weights—25,200 to 26,900 for the DC-3 and 18,500 to 19,500 for the L-18.* New maximum certificated weights of more than 25,200 but not more than 26,900 pounds for DC-3 and more than 18,500 but not more than 19,500 pounds for L-18 airplanes may be established in accordance with the transport category performance requirements of either Part 4a or Part 4b, if the airplane at the new maximum weights can meet the structural requirements of the elected part.

(b) *Weights of more than 26,900 for the DC-3 and 19,500 for the L-18.* New maximum certificated weights of more than 26,900 pounds for DC-3 and 19,500 pounds for L-18 airplanes shall be established in accordance with the structural performance, flight characteristics, and ground handling requirements of Part 4b: *Provided,* That where literal compliance with the structural requirements of Part 4b is extremely difficult to accomplish and would not contribute materially to the objective sought, and the Administrator finds that the experience with the DC-3 or L-18 airplanes justifies it, he is authorized to accept such measures of compliance as he finds will effectively accomplish the basic objective.

(c) *Airplane flight manual-performance operating information.* An approved airplane flight manual shall be provided for each DC-3 and L-18 airplane which has had new maximum certificated weights established under this section. The airplane flight manual shall contain the applicable performance information prescribed in that part of the regulations under which the new certificated weights were established and such additional information as may be necessary to enable the application of the take-off, en route, and landing limitations prescribed for transport category airplanes in the operating parts of the Civil Air Regulations.

(d) *Performance operating limitations.* Each airplane for which new maximum certificated weights are established in accordance with paragraphs (a) or (b) of this section shall be considered a transport category airplane for the purpose of complying with the performance operating limitations applicable to the operations in which it is utilized.

**5.** *Reference.* Unless otherwise provided, all references in this regulation to Part 4a and Part 4b are those parts of the Civil Air Regulations in effect on September 1, 1953.

This regulation supersedes Special Civil Air Regulation SR-398 and shall remain effective until superseded or rescinded by the Board.

[19 FR 5039, Aug. 11, 1954. Redesignated at 29 FR 19099, Dec. 30, 1964]

# Subpart A—General

## §25.1 Applicability.

**(a)** This part prescribes airworthiness standards for the issue of type certificates, and changes to those certificates, for transport category airplanes.

**(b)** Each person who applies under Part 21 for such a certificate or change must show compliance with the applicable requirements in this part.

## §25.2 Special retroactive requirements.

The following special retroactive requirements are applicable to an airplane for which the regulations referenced in the type certificate predate the sections specified below—

**(a)** Irrespective of the date of application, each applicant for a supplemental type certificate (or an amendment to a type certificate) involving an increase in passenger seating capacity to a total greater than that for which the airplane has been type certificated must show that the airplane concerned meets the requirements of:

(1) Sections 25.721(d), 25.783(g), 25.785(c), 25.803(c) (2) through (9), 25.803 (d) and (e), 25.807 (a), (c), and (d), 25.809 (f) and (h), 25.811, 25.812, 25.813 (a), (b), and (c), 25.815, 25.817, 25.853 (a) and (b), 25.855(a), 25.993(f), and 25.1359(c) in effect on October 24, 1967, and

(2) Sections 25.803(b) and 25.803(c)(1) in effect on April 23, 1969.

**(b)** Irrespective of the date of application, each applicant for a supplemental type certificate (or an amendment to a type certificate) for an airplane manufactured after October 16, 1987, must show that the airplane meets the requirements of §25.807(c)(7) in effect on July 24, 1989.

**(c)** Compliance with subsequent revisions to the sections specified in paragraph (a) or (b) of this section may be elected or may be required in accordance with §21.101(a) of this chapter.

***Editorial Note:** Per the November 14, 2001 Federal Register, mandatory compliance was changed to June 10, 2003. To view the rule currently in effect, go to ASA's website and click on "Product Updates":*

www.asa2fly.com

[Docket No. 5066, 29 FR 18291, Dec. 24, 1964; as amended by Amdt. 25–72, 55 FR 29773, July 20, 1990; Amdt. No. 25–99, 65 FR 36266, June 7, 2000; Amdt. 25–99, 66 FR 56989, Nov. 14, 2001]

# Subpart B—Flight

General

## §25.21 Proof of compliance.

**(a)** Each requirement of this subpart must be met at each appropriate combination of weight and center of gravity within the range of loading conditions for which certification is requested. This must be shown—

(1) By tests upon an airplane of the type for which certification is requested, or by calculations based on, and equal in accuracy to, the results of testing; and

(2) By systematic investigation of each probable combination of weight and center of gravity, if compliance cannot be reasonably inferred from combinations investigated.

**(b)** [Reserved]

**(c)** The controllability, stability, trim, and stalling characteristics of the airplane must be shown for each altitude up to the maximum expected in operation.

**(d)** Parameters critical for the test being conducted, such as weight, loading (center of gravity and inertia), airspeed, power, and wind, must be maintained within acceptable tolerances of the critical values during flight testing.

**(e)** If compliance with the flight characteristics requirements is dependent upon a stability augmentation system or upon any other automatic or power-operated system, compliance must be shown with §§25.671 and 25.672.

**(f)** In meeting the requirements of §§25.105(d), 25.125, 25.233, and 25.237, the wind velocity must be measured at a height of 10 meters above the surface, or corrected for the difference between the height at which the wind velocity is measured and the 10-meter height.

[Docket No. 5066, 29 FR 18291, Dec. 24, 1964; as amended by Amdt. 25–23, 35 FR 5671, April 8, 1970; Amdt. 25–42, 43 FR 2320, Jan. 16, 1978; Amdt. 25–72, 55 FR 29774, July 20, 1990]

## §25.23 Load distribution limits.

**(a)** Ranges of weights and centers of gravity within which the airplane may be safely operated must be established. If a weight and center of gravity combination is allowable only within certain load distribution limits (such as spanwise) that could be inadvertently exceeded, these limits and the corresponding weight and center of gravity combinations must be established.

**(b)** The load distribution limits may not exceed—

(1) The selected limits;

(2) The limits at which the structure is proven; or

(3) The limits at which compliance with each applicable flight requirement of this subpart is shown.

## §25.25 Weight limits.

**(a)** *Maximum weights.* Maximum weights corresponding to the airplane operating conditions (such as ramp, ground or water taxi, takeoff, en route, and landing), environmental conditions (such as altitude and temperature), and loading conditions (such as zero fuel weight, center of gravity position and weight distribution) must be established so that they are not more than—

(1) The highest weight selected by the applicant for the particular conditions; or

(2) The highest weight at which compliance with each applicable structural loading and flight requirement is shown, except that for airplanes equipped with standby power rocket engines the maximum weight must not be more than the highest weight established in accordance with Appendix E of this part; or

(3) The highest weight at which compliance is shown with the certification requirements of Part 36 of this chapter.

**(b)** *Minimum weight.* The minimum weight (the lowest weight at which compliance with each applicable requirement of this part is shown) must be established so that it is not less than—

(1) The lowest weight selected by the applicant;

(2) The design minimum weight (the lowest weight at which compliance with each structural loading condition of this part is shown); or

(3) The lowest weight at which compliance with each applicable flight requirement is shown.

[Docket No. 5066, 29 FR 18291, Dec. 24, 1964; as amended by Amdt. 25–23, 35 FR 5671, April 8, 1970; Amdt. 25–63, 53 FR 16365, May 6, 1988]

## §25.27 Center of gravity limits.

The extreme forward and the extreme aft center of gravity limitations must be established for each practicably separable operating condition. No such limit may lie beyond—

**(a)** The extremes selected by the applicant;

**(b)** The extremes within which the structure is proven; or

**(c)** The extremes within which compliance with each applicable flight requirement is shown.

## §25.29 Empty weight and corresponding center of gravity.

**(a)** The empty weight and corresponding center of gravity must be determined by weighing the airplane with—

(1) Fixed ballast;

(2) Unusable fuel determined under §25.959; and

(3) Full operating fluids, including—

(i) Oil;

(ii) Hydraulic fluid; and

(iii) Other fluids required for normal operation of airplane systems, except potable water, lavatory precharge water, and fluids intended for injection in the engine.

**(b)** The condition of the airplane at the time of determining empty weight must be one that is well defined and can be easily repeated.

[Docket No. 5066, 29 FR 18291, Dec. 24, 1964; as amended by Amdt. 25–42, 43 FR 2320, Jan. 16, 1978; Amdt. 25–72, 55 FR 29774, July 20, 1990]

## §25.31 Removable ballast.

Removable ballast may be used on showing compliance with the flight requirements of this subpart.

## §25.33 Propeller speed and pitch limits.

**(a)** The propeller speed and pitch must be limited to values that will ensure—

(1) Safe operation under normal operating conditions; and

(2) Compliance with the performance requirements of §§25.101 through 25.125.

**(b)** There must be a propeller speed limiting means at the governor. It must limit the maximum possible governed engine speed to a value not exceeding the maximum allowable r.p.m.

**(c)** The means used to limit the low pitch position of the propeller blades must be set so that the engine does not exceed 103 percent of the maximum allowable engine rpm or 99 percent of an approved maximum overspeed, whichever is greater, with—

(1) The propeller blades at the low pitch limit and governor inoperative;

(2) The airplane stationary under standard atmospheric conditions with no wind; and

(3) The engines operating at the takeoff manifold pressure limit for reciprocating engine powered airplanes or the maximum takeoff torque limit for turbopropeller engine-powered airplanes.

[Docket No. 5066, 29 FR 18291, Dec. 24, 1964; as amended by Amdt. 25–57, 49 FR 6848, Feb. 23, 1984; Amdt. 25–72, 55 FR 29774, July 20, 1990]

PERFORMANCE

## §25.101 General.

**(a)** Unless otherwise prescribed, airplanes must meet the applicable performance requirements of this subpart for ambient atmospheric conditions and still air.

**(b)** The performance, as affected by engine power or thrust, must be based on the following relative humidities;

(1) For turbine engine powered airplanes, a relative humidity of—

(i) 80 percent, at and below standard temperatures; and

(ii) 34 percent, at and above standard temperatures plus 50°F.

Between these two temperatures, the relative humidity must vary linearly.

(2) For reciprocating engine powered airplanes, a relative humidity of 80 percent in a standard atmosphere. Engine power corrections for vapor pressure must be made in accordance with the following table:

| Altitude *H* (ft.) | Vapor pressure e (In. Hg.) | Specific humidity *w* (Lb. moisture per lb. dry air) | Density ratio $\rho/\sigma$ = 0.0023769 |
|---|---|---|---|
| 0 | 0.403 | 0.00849 | 0.99508 |
| 1,000 | .354 | .00773 | .96672 |
| 2,000 | .311 | .00703 | .93895 |
| 3,000 | .272 | .00638 | .91178 |
| 4,000 | .238 | .00578 | .88514 |
| 5,000 | .207 | .00523 | .85910 |
| 6,000 | .1805 | .00472 | .83361 |
| 7,000 | .1566 | .00425 | .80870 |
| 8,000 | .1356 | .00382 | .78434 |
| 9,000 | .1172 | .00343 | .76053 |
| 10,000 | .1010 | .00307 | .73722 |
| 15,000 | .0463 | .001710 | .62868 |
| 20,000 | .01978 | .000896 | .53263 |
| 25,000 | .00778 | .000436 | .44806 |

**(c)** The performance must correspond to the propulsive thrust available under the particular ambient atmospheric conditions, the particular flight condition, and the relative humidity specified in paragraph (b) of this section. The available propulsive thrust must correspond to engine power or thrust, not exceeding the approved power or thrust less—

(1) Installation losses; and

(2) The power or equivalent thrust absorbed by the accessories and services appropriate to the particular ambient atmospheric conditions and the particular flight condition.

**(d)** Unless otherwise prescribed, the applicant must select the takeoff, en route, approach, and landing configurations for the airplane.

**(e)** The airplane configurations may vary with weight, altitude, and temperature, to the extent they are compatible with the operating procedures required by paragraph (f) of this section.

**(f)** Unless otherwise prescribed, in determining the accelerate-stop distances, takeoff flight paths, takeoff distances, and landing distances, changes in the airplane's configuration, speed, power, and thrust, must be made in accordance with procedures established by the applicant for operation in service.

**(g)** Procedures for the execution of balked landings and missed approaches associated with the conditions prescribed in §§25.119 and 25.121(d) must be established.

**(h)** The procedures established under paragraphs (f) and (g) of this section must—

(1) Be able to be consistently executed in service by crews of average skill;

(2) Use methods or devices that are safe and reliable; and

(3) Include allowance for any time delays, in the execution of the procedures, that may reasonably be expected in service.

**(i)** The accelerate-stop and landing distances prescribed in §§25.109 and 25.125, respectively, must be determined with all the airplane wheel brake assemblies at the fully worn limit of their allowable wear range.

[Docket No. 5066, 29 FR 18291, Dec. 24, 1964; as amended by Amdt. 25–38, 41 FR 55466, Dec. 20, 1976; Amdt. 25–92, 63 FR 8318, Feb. 18, 1998]

## §25.103 Stalling speed.

**(a)** $V_S$ is the calibrated stalling speed, or the minimum steady flight speed, in knots, at which the airplane is controllable, with—

(1) Zero thrust at the stalling speed, or, if the resultant thrust has no appreciable effect on the stalling speed, with engines idling and throttles closed;

(2) Propeller pitch controls (if applicable) in the position necessary for compliance with paragraph (a)(1) of this section and the airplane in other respects (such as flaps and landing gear) in the condition existing in the test in which $V_S$ is being used;

25

(3) The weight used when $V_S$ is being used as a factor to determine compliance with a required performance standard; and

(4) The most unfavorable center of gravity allowable.

**(b)** The stalling speed $V_S$ is the minimum speed obtained as follows:

(1) Trim the airplane for straight flight at any speed not less than 1.2 $V_S$ or more than 1.4 $V_S$. At a speed sufficiently above the stall speed to ensure steady conditions, apply the elevator control at a rate so that the airplane speed reduction does not exceed one knot per second.

(2) Meet the flight characteristics provisions of §25.203.

## §25.105 Takeoff.

**(a)** The takeoff speeds described in §25.107, the accelerate-stop distance described in §25.109, the takeoff path described in §25.111, and the takeoff distance and takeoff run described in §25.113, must be determined—

(1) At each weight, altitude, and ambient temperature within the operational limits selected by the applicant; and

(2) In the selected configuration for takeoff.

**(b)** No takeoff made to determine the data required by this section may require exceptional piloting skill or alertness.

**(c)** The takeoff data must be based on—

(1) In the case of land planes and amphibians:

(i) Smooth, dry and wet, hard-surfaced runways; and

(ii) At the option of the applicant, grooved or porous friction course wet, hard-surfaced runways.

**(d)** The takeoff data must include, within the established operational limits of the airplane, the following operational correction factors:

(1) Not more than 50 percent of nominal wind components along the takeoff path opposite to the direction of takeoff, and not less than 150 percent of nominal wind components along the takeoff path in the direction of takeoff.

(2) Effective runway gradients.

[Docket No. 5066, 29 FR 18291, Dec. 24, 1964; as amended by Amdt. 25–92, 63 FR 8318, Feb. 18, 1998]

## §25.107 Takeoff speeds.

**(a)** $V_1$ must be established in relation to $V_{EF}$ as follows:

(1) $V_{EF}$ is the calibrated airspeed at which the critical engine is assumed to fail. $V_{EF}$ must be selected by the applicant, but may not be less than $V_{MCG}$ determined under §25.149(e).

(2) $V_1$, in terms of calibrated airspeed, is selected by the applicant; however, $V_1$ may not be less than $V_{EF}$ plus the speed gained with critical engine inoperative during the time interval between the instant at which the critical engine is failed, and the instant at which the pilot recognizes and reacts to the engine failure, as indicated by the pilot's initiation of the first action (e.g., applying brakes, reducing thrust, deploying speed brakes) to stop the airplane during accelerate-stop tests.

**(b)** $V_{2MIN}$, in terms of calibrated airspeed, may not be less than—

(1) 1.2 $V_S$ for—

(i) Two-engine and three-engine turbopropeller and reciprocating engine powered airplanes; and

(ii) Turbojet powered airplanes without provisions for obtaining a significant reduction in the one-engine-inoperative power-on stalling speed;

(2) 1.15 $V_S$ for—

(i) Turbopropeller and reciprocating engine powered airplanes with more than three engines; and

(ii) Turbojet powered airplanes with provisions for obtaining a significant reduction in the one-engine-inoperative power-on stalling speed; and

(3) 1.10 times $V_{MC}$ established under §25.149.

**(c)** $V_2$, in terms of calibrated airspeed, must be selected by the applicant to provide at least the gradient of climb required by §25.121(b) but may not be less than—

(1) $V_{2MIN}$, and

(2) $V_R$ plus the speed increment attained (in accordance with §25.111 (c)(2)) before reaching a height of 35 feet above the takeoff surface.

**(d)** $V_{MU}$ is the calibrated airspeed at and above which the airplane can safely lift off the ground, and continue the takeoff. $V_{MU}$ speeds must be selected by the applicant throughout the range of thrust-to-weight ratios to be certificated. These speeds may be established from free air data if these data are verified by ground takeoff tests.

**(e)** $V_R$, in terms of calibrated airspeed, must be selected in accordance with the conditions of paragraphs (e) (1) through (4) of this section:

(1) $V_R$ may not be less than—

(i) $V_1$

(ii) 105 percent of $V_{MC}$;

(iii) The speed (determined in accordance with §25.111(c)(2)) that allows reaching $V_2$ before reaching a height of 35 feet above the takeoff surface; or

(iv) A speed that, if the airplane is rotated at its maximum practicable rate, will result in a $V_{LOF}$ of not less than 110 percent of $V_{MU}$ in the all-engines-operating condition and not less than 105 percent of $V_{MU}$ determined at the thrust-to-weight ratio corresponding to the one-engine-inoperative condition.

(2) For any given set of conditions (such as weight, configuration, and temperature), a single value of $V_R$, obtained in accordance with this paragraph, must be used to show compliance

with both the one-engine-inoperative and the all-engines-operating takeoff provisions.

(3) It must be shown that the one-engine-inoperative takeoff distance, using a rotation speed of 5 knots less than $V_R$ established in accordance with paragraphs (e)(1) and (2) of this section, does not exceed the corresponding one-engine-inoperative takeoff distance using the established $V_R$. The takeoff distances must be determined in accordance with §25.113(a)(1).

(4) Reasonably expected variations in service from the established takeoff procedures for the operation of the airplane (such as over-rotation of the airplane and out-of-trim conditions) may not result in unsafe flight characteristics or in marked increases in the scheduled takeoff distances established in accordance with §25.113(a).

**(f)** $V_{LOF}$ is the calibrated airspeed at which the airplane first becomes airborne.

[Docket No. 5066, 29 FR 18291, Dec. 24, 1964; as amended by Amdt. 25–38, 41 FR 55466, Dec. 20, 1976; Amdt. 25–42, 43 FR 2320, Jan. 16, 1978; Amdt. 25–92, 63 FR 8318, Feb. 18, 1998; Amdt. 25–94, 63 FR 8848, Feb. 23, 1998]

## §25.109 Accelerate-stop distance.

**(a)** The accelerate-stop distance on a dry runway is the greater of the following distances:

(1) The sum of the distances necessary to—

(i) Accelerate the airplane from a standing start with all engines operating to $V_{EF}$ for takeoff from a dry runway;

(ii) Allow the airplane to accelerate from $V_{EF}$ to the highest speed reached during the rejected takeoff, assuming the critical engine fails at $V_{EF}$ and the pilot takes the first action to reject the takeoff at the $V_1$ for takeoff from a dry runway; and

(iii) Come to a full stop on a dry runway from the speed reached as prescribed in paragraph (a)(1)(ii) of this section; plus

(iv) A distance equivalent to 2 seconds at the $V_1$ for takeoff from a dry runway.

(2) The sum of the distances necessary to—

(i) Accelerate the airplane from a standing start with all engines operating to the highest speed reached during the rejected takeoff, assuming the pilot takes the first action to reject the takeoff at the $V_1$ for takeoff from a dry runway; and

(ii) With all engines still operating, come to a full stop on dry runway from the speed reached as prescribed in paragraph (a)(2)(i) of this section; plus

(iii) A distance equivalent to 2 seconds at the $V_1$ for takeoff from a dry runway.

**(b)** The accelerate-stop distance on a wet runway is the greater of the following distances:

(1) The accelerate-stop distance on a dry runway determined in accordance with paragraph (a) of this section; or

(2) The accelerate-stop distance determined in accordance with paragraph (a) of this section, except that the runway is wet and the corresponding wet runway values of $V_{EF}$ and $V_1$ are used. In determining the wet runway accelerate-stop distance, the stopping force from the wheel brakes may never exceed:

(i) The wheel brakes stopping force determined in meeting the requirements of §25.101(i) and paragraph (a) of this section; and

(ii) The force resulting from the wet runway braking coefficient of friction determined in accordance with paragraphs (c) or (d) of this section, as applicable, taking into account the distribution of the normal load between braked and unbraked wheels at the most adverse center-of-gravity position approved for takeoff.

**(c)** The wet runway braking coefficient of friction for a smooth wet runway is defined as a curve of friction coefficient versus ground speed and must be computed as follows:

(1) The maximum tire-to-ground wet runway braking coefficient of friction is defined as:

| Tire Pressure (psi) | Maximum Braking Coefficient (tire-to-ground) |
|---|---|
| 50 | $\mu_{t/g_{MAX}} = -0.0350\left(\frac{V}{100}\right)^3 + 0.306\left(\frac{V}{100}\right)^2 - 0.851\left(\frac{V}{100}\right) + 0.883$ |
| 100 | $\mu_{t/g_{MAX}} = -0.0437\left(\frac{V}{100}\right)^3 + 0.320\left(\frac{V}{100}\right)^2 - 0.805\left(\frac{V}{100}\right) + 0.804$ |
| 200 | $\mu_{t/g_{MAX}} = -0.0331\left(\frac{V}{100}\right)^3 + 0.252\left(\frac{V}{100}\right)^2 - 0.658\left(\frac{V}{100}\right) + 0.692$ |
| 300 | $\mu_{t/g_{MAX}} = -0.0401\left(\frac{V}{100}\right)^3 + 0.263\left(\frac{V}{100}\right)^2 - 0.611\left(\frac{V}{100}\right) + 0.614$ |

Where—

Tire Pressure = maximum airplane operating tire pressure (psi);

$\mu_{t/g_{MAX}}$ = maximum tire-to-ground braking coefficient;

V = airplane true ground speed (knots); and

Linear interpolation may be used for tire pressures other than those listed.

(2) The maximum tire-to-ground wet runway braking coefficient of friction must be adjusted to take into account the efficiency of the anti-skid system on a wet runway. Anti-skid system operation must be demonstrated by flight testing on a smooth wet runway, and its efficiency must be determined. Unless a specific anti-skid system efficiency is determined from a quantitative analysis of the flight testing on a smooth wet runway, the maximum tire-to-ground wet runway braking coefficient of friction determined in paragraph (c)(1) of this section must be multiplied by the efficiency value associated with the type of anti-skid system installed on the airplane:

| Type of anti-skid system | Efficiency value |
|---|---|
| On-Off | 0.30 |
| Quasi-Modulating | 0.50 |
| Full Modulating | 0.80 |

**(d)** At the option of the applicant, a higher wet runway braking coefficient of friction may be used for runway surfaces that have been grooved or treated with a porous friction course material. For grooved and porous friction course runways, the wet runway braking coefficient of friction is defined as either:

(1) 70 percent of the dry runway braking coefficient of friction used to determine the dry runway accelerate-stop distance; or

(2) The wet runway braking coefficient defined in paragraph (c) of this section, except that a specific anti-skid system efficiency, if determined, is appropriate for a grooved or porous friction course wet runway, and the maximum tire-to-ground wet runway braking coefficient of friction is defined as:

| Tire Pressure (psi) | Maximum Braking Coefficient (tire-to-ground) |
|---|---|
| 50 | $\mu_{t/g_{MAX}} = 0.1470\left(\frac{V}{100}\right)^5 - 1.050\left(\frac{V}{100}\right)^4 + 2.673\left(\frac{V}{100}\right)^3 - 2.683\left(\frac{V}{100}\right)^2 + 0.403\left(\frac{V}{100}\right) + 0.859$ |
| 100 | $\mu_{t/g_{MAX}} = 0.1106\left(\frac{V}{100}\right)^5 - 0.813\left(\frac{V}{100}\right)^4 + 2.130\left(\frac{V}{100}\right)^3 - 2.200\left(\frac{V}{100}\right)^2 + 0.317\left(\frac{V}{100}\right) + 0.807$ |
| 200 | $\mu_{t/g_{MAX}} = 0.0498\left(\frac{V}{100}\right)^5 - 0.398\left(\frac{V}{100}\right)^4 + 1.140\left(\frac{V}{100}\right)^3 - 1.285\left(\frac{V}{100}\right)^2 + 0.140\left(\frac{V}{100}\right) + 0.701$ |
| 300 | $\mu_{t/g_{MAX}} = 0.0314\left(\frac{V}{100}\right)^5 - 0.247\left(\frac{V}{100}\right)^4 + 0.703\left(\frac{V}{100}\right)^3 - 0.779\left(\frac{V}{100}\right)^2 + 0.00954\left(\frac{V}{100}\right) + 0.614$ |

Where—

Tire Pressure = maximum airplane operating tire pressure (psi);

$\mu_{t/g_{MAX}}$ = maximum tire-to-ground braking coefficient;

V = airplane true ground speed (knots); and

Linear interpolation may be used for tire pressures other than those listed.

**(e)** Except as provided in paragraph (f)(1) of this section, means other than wheel brakes may be used to determine the accelerate-stop distance if that means—

(1) Is safe and reliable;

(2) Is used so that consistent results can be expected under normal operating conditions; and

(3) Is such that exceptional skill is not required to control the airplane.

**(f)** The effects of available reverse thrust—

(1) Shall not be included as an additional means of deceleration when determining the accelerate-stop distance on a dry runway; and

(2) May be included as an additional means of deceleration using recommended reverse thrust procedures when determining the accelerate-stop distance on a wet runway, provided the requirements of paragraph (e) of this section are met.

**(g)** The landing gear must remain extended throughout the accelerate-stop distance.

**(h)** If the accelerate-stop distance includes a stopway with surface characteristics substantially different from those on the runway, the takeoff data must include operational correction factors for the accelerate-stop distance. The correction factors must account for the particular surface characteristics of the stopway and the variations in these characteristics with seasonal weather conditions (such as temperature, rain, snow, and ice) within the established operational limits.

**(i)** A flight test demonstration of the maximum brake kinetic energy accelerate-stop distance must be conducted with not more than 10 percent of the allowable brake wear range remaining on each of the airplane wheel brakes.

[Docket No. 5066, 29 FR 18291, Dec. 24, 1964; as amended by Amdt. 25–42, 43 FR 2321, Jan. 16, 1978; Amdt. 25–92, 63 FR 8318, Feb. 18, 1998]

## §25.111 Takeoff path.

**(a)** The takeoff path extends from a standing start to a point in the takeoff at which the airplane is 1,500 feet above the takeoff surface, or at which the transition from the takeoff to the en route configuration is completed and a speed is reached at which compliance with §25.121(c) is shown, whichever point is higher. In addition—

(1) The takeoff path must be based on the procedures prescribed in §25.101(f);

(2) The airplane must be accelerated on the ground to $V_{EF}$, at which point the critical engine must be made inoperative and remain inoperative for the rest of the takeoff; and

(3) After reaching $V_{EF}$, the airplane must be accelerated to $V_2$.

**(b)** During the acceleration to speed $V_2$, the nose gear may be raised off the ground at a speed not less than $V_R$. However, landing gear retraction may not be begun until the airplane is airborne.

**(c)** During the takeoff path determination in accordance with paragraphs (a) and (b) of this section—

(1) The slope of the airborne part of the takeoff path must be positive at each point;

(2) The airplane must reach $V_2$ before it is 35 feet above the takeoff surface and must continue at a speed as close as practical to, but not less than $V_2$, until it is 400 feet above the takeoff surface;

(3) At each point along the takeoff path, starting at the point at which the airplane reaches 400 feet above the takeoff surface, the available gradient of climb may not be less than—

(i) 1.2 percent for two-engine airplanes;

(ii) 1.5 percent for three-engine airplanes; and

(iii) 1.7 percent for four-engine airplanes; and

(4) Except for gear retraction and propeller feathering, the airplane configuration may not be changed, and no change in power or thrust that requires action by the pilot may be made, until the airplane is 400 feet above the takeoff surface.

**(d)** The takeoff path must be determined by a continuous demonstrated takeoff or by synthesis from segments. If the takeoff path is determined by the segmental method—

(1) The segments must be clearly defined and must be related to the distinct changes in the configuration, power or thrust, and speed;

(2) The weight of the airplane, the configuration, and the power or thrust must be constant throughout each segment and must correspond to the most critical condition prevailing in the segment;

(3) The flight path must be based on the airplane's performance without ground effect; and

(4) The takeoff path data must be checked by continuous demonstrated takeoffs up to the point at which the airplane is out of ground effect and its speed is stabilized, to ensure that the path is conservative relative to the continuous path.

The airplane is considered to be out of the ground effect when it reaches a height equal to its wing span.

**(e)** For airplanes equipped with standby power rocket engines, the takeoff path may be determined in accordance with section II of Appendix E.

[Docket No. 5066, 29 FR 18291, Dec. 24, 1964; as amended by Amdt. 25–6, 30 FR 8468, July 2, 1965; Amdt. 25–42, 43 FR 2321, Jan. 16, 1978; Amdt. 25–54, 45 FR 60172, Sept. 11, 1980; Amdt. 25–72, 55 FR 29774, July 20, 1990; Amdt. 25–94, 63 FR 8848, Feb. 23, 1998]

## §25.113 Takeoff distance and takeoff run.

**(a)** Takeoff distance on a dry runway is the greater of—

(1) The horizontal distance along the takeoff path from the start of the takeoff to the point at which the airplane is 35 feet above the takeoff surface, determined under §25.111 for a dry runway; or

(2) 115 percent of the horizontal distance along the takeoff path, with all engines operating, from the start of the takeoff to the point at which the airplane is 35 feet above the takeoff surface, as determined by a procedure consistent with §25.111.

**(b)** Takeoff distance on a wet runway is the greater of—

(1) The takeoff distance on a dry runway determined in accordance with paragraph (a) of this section; or

(2) The horizontal distance along the takeoff path from the start of the takeoff to the point at which the airplane is 15 feet above the takeoff surface, achieved in a manner consistent with the achievement of $V_2$ before reaching 35 feet above the takeoff surface, determined under §25.111 for a wet runway.

**(c)** If the takeoff distance does not include a clearway, the takeoff run is equal to the takeoff distance. If the takeoff distance includes a clearway—

(1) The takeoff run on a dry runway is the greater of—

(i) The horizontal distance along the takeoff path from the start of the takeoff to a point equidistant between the point at which $V_{LOF}$ is reached and the point at which the airplane is 35 feet above the takeoff surface, as determined under §25.111 for a dry runway; or

(ii) 115 percent of the horizontal distance along the takeoff path, with all engines operating, from the start of the takeoff to a point equidistant between the point at which $V_{LOF}$ is reached and the point at which the airplane is 35 feet above the takeoff surface, determined by a procedure consistent with §25.111.

(2) The takeoff run on a wet runway is the greater of—

(i) The horizontal distance along the takeoff path from the start of the takeoff to the point at which the airplane is 15 feet above the takeoff surface, achieved in a manner consistent with the achievement of $V_2$ before reaching 35 feet above the takeoff surface, as determined under §25.111 for a wet runway; or

(ii) 115 percent of the horizontal distance along the takeoff path, with all engines operating, from the start of the takeoff to a point equidistant between the point at which $V_{LOF}$ is reached and the point at which the airplane is 35 feet above the takeoff surface, determined by a procedure consistent with §25.111.

[Docket No. 5066, 29 FR 18291, Dec. 24, 1964; as amended by Amdt. 25–23, 35 FR 5671, April 8, 1970; Amdt. 25–92, 63 FR 8320, Feb. 18, 1998]

## §25.115 Takeoff flight path.

**(a)** The takeoff flight path shall be considered to begin 35 feet above the takeoff surface at the end of the takeoff distance determined in accordance with §25.113(a) or (b), as appropriate for the runway surface condition.

**(b)** The net takeoff flight path data must be determined so that they represent the actual takeoff flight paths (determined in accordance with §25.111 and with paragraph (a) of this section) reduced at each point by a gradient of climb equal to—

(1) 0.8 percent for two-engine airplanes;

(2) 0.9 percent for three-engine airplanes; and

(3) 1.0 percent for four-engine airplanes.

**(c)** The prescribed reduction in climb gradient may be applied as an equivalent reduction in acceleration along that part of the takeoff flight path at which the airplane is accelerated in level flight.

[Docket No. 5066, 29 FR 18291, Dec. 24, 1964; as amended by Amdt. 25–92, 63 FR 8320, Feb. 18, 1998]

## §25.117 Climb: general.

Compliance with the requirements of §§25.119 and 25.121 must be shown at each weight, altitude, and ambient temperature within the operational limits established for the airplane and with the most unfavorable center of gravity for each configuration.

## §25.119 Landing climb: All-engines-operating.

In the landing configuration, the steady gradient of climb may not be less than 3.2 percent, with—

**(a)** The engines at the power or thrust that is available eight seconds after initiation of movement of the power or thrust controls from the minimum flight idle to the go-around power or thrust setting; and

**(b)** A climb speed of not more than 1.3 $V_S$.

[Docket No. 5066, 29 FR 18291, Dec. 24, 1964; as amended by Amdt. 25–84, 60 FR 30749, June 9, 1995; Amdt. 25–94, 63 FR 8848, Feb. 23, 1998]

## §25.121 Climb: One-engine-inoperative.

**(a)** *Takeoff; landing gear extended.* In the critical takeoff configuration existing along the flight path (between the points at which the airplane reaches $V_{LOF}$ and at which the landing gear is fully retracted) and in the configuration used in §25.111 but without ground effect, the steady gradient of climb must be positive for two-engine airplanes, and not less than 0.3 percent for three-engine airplanes or 0.5 percent for four-engine airplanes, at $V_{LOF}$ and with—

(1) The critical engine inoperative and the remaining engines at the power or thrust available when retraction of the landing gear is begun in accordance with §25.111 unless there is a more critical power operating condition existing later along the flight path but before the point at which the landing gear is fully retracted; and

(2) The weight equal to the weight existing when retraction of the landing gear is begun, determined under §25.111.

**(b)** *Takeoff; landing gear retracted.* In the takeoff configuration existing at the point of the flight path at which the landing gear is fully retracted, and in the configuration used in §25.111 but without ground effect, the steady gradient of climb may not be less than 2.4 percent for two-engine airplanes, 2.7 percent for three-engine airplanes, and 3.0 percent for four-engine airplanes, at $V_2$ and with—

(1) The critical engine inoperative, the remaining engines at the takeoff power or thrust available at the time the landing gear is fully retracted, determined under §25.111, unless there is a more critical power operating condition existing

later along the flight path but before the point where the airplane reaches a height of 400 feet above the takeoff surface; and

(2) The weight equal to the weight existing when the airplane's landing gear is fully retracted, determined under §25.111.

**(c)** *Final takeoff.* In the en route configuration at the end of the takeoff path determined in accordance with §25.111, the steady gradient of climb may not be less than 1.2 percent for two-engine airplanes, 1.5 percent for three-engine airplanes, and 1.7 percent for four-engine airplanes, at not less than 1.25 $V_S$ and with—

(1) The critical engine inoperative and the remaining engines at the available maximum continuous power or thrust; and

(2) The weight equal to the weight existing at the end of the takeoff path, determined under §25.111.

**(d)** *Approach.* In the approach configuration corresponding to the normal all-engines-operating procedure in which $V_S$ for this configuration does not exceed 110 percent of the $V_S$ for the related landing configuration, the steady gradient of climb may not be less than 2.1 percent for two-engine airplanes, 2.4 percent for three-engine airplanes, and 2.7 percent for four-engine airplanes, with—

(1) The critical engine inoperative, the remaining engines at the go-around power or thrust setting;

(2) The maximum landing weight; and

(3) A climb speed established in connection with normal landing procedures, but not exceeding 1.5 $V_S$.

[Docket No. 5066, 29 FR 18291, Dec. 24, 1964; as amended by Amdt. 25–84, 60 FR 30749, June 9, 1995]

## §25.123 En route flight paths.

**(a)** For the en route configuration, the flight paths prescribed in paragraphs (b) and (c) of this section must be determined at each weight, altitude, and ambient temperature, within the operating limits established for the airplane. The variation of weight along the flight path, accounting for the progressive consumption of fuel and oil by the operating engines, may be included in the computation. The flight paths must be determined at any selected speed, with—

(1) The most unfavorable center of gravity;

(2) The critical engines inoperative;

(3) The remaining engines at the available maximum continuous power or thrust; and

(4)The means for controlling the engine-cooling air supply in the position that provides adequate cooling in the hot-day condition.

**(b)** The one-engine-inoperative net flight path data must represent the actual climb performance diminished by a gradient of climb of 1.1 percent for two-engine airplanes, 1.4 percent for three-engine airplanes, and 1.6 percent for four-engine airplanes.

**(c)** For three- or four-engine airplanes, the two-engine-inoperative net flight path data must represent the actual climb performance diminished by a gradient of climb of 0.3 percent for three-engine airplanes and 0.5 percent for four-engine airplanes.

## §25.125 Landing.

**(a)** The horizontal distance necessary to land and to come to a complete stop (or to a speed of approximately 3 knots for water landings) from a point 50 feet above the landing surface must be determined (for standard temperatures, at each weight, altitude, and wind within the operational limits established by the applicant for the airplane) as follows:

(1) The airplane must be in the landing configuration.

(2) A stabilized approach, with a calibrated airspeed of not less than 1.3 $V_S$ or $V_{MCL}$, whichever is greater, must be maintained down to the 50 foot height.

(3) Changes in configuration, power or thrust, and speed, must be made in accordance with the established procedures for service operation.

(4) The landing must be made without excessive vertical acceleration, tendency to bounce, nose over, ground loop, porpoise, or water loop.

(5) The landings may not require exceptional piloting skill or alertness.

**(b)** For landplanes and amphibians, the landing distance on land must be determined on a level, smooth, dry, hard-surfaced runway. In addition—

(1) The pressures on the wheel braking systems may not exceed those specified by the brake manufacturer;

(2) The brakes may not be used so as to cause excessive wear of brakes or tires; and

(3) Means other than wheel brakes may be used if that means—

(i) Is safe and reliable;

(ii) Is used so that consistent results can be expected in service; and

(iii) Is such that exceptional skill is not required to control the airplane.

**(c)** For seaplanes and amphibians, the landing distance on water must be determined on smooth water.

**(d)** For skiplanes, the landing distance on snow must be determined on smooth, dry, snow.

**(e)** The landing distance data must include correction factors for not more than 50 percent of the nominal wind components along the landing path opposite to the direction of landing, and not less than 150 percent of the nominal wind components along the landing path in the direction of landing.

**(f)** If any device is used that depends on the operation of any engine, and if the landing distance would be noticeably increased when a landing is made with that engine inoperative, the landing distance must be determined with that engine inoperative unless the use of compensating means will result in a landing distance not more than that with each engine operating.

[Docket No. 5066, 29 FR 18291, Dec. 24, 1964; as amended by Amdt. 25–72, 55 FR 29774, July 20, 1990; Amdt. 25–84, 60 FR 30749, June 9, 1995]

CONTROLLABILITY AND MANEUVERABILITY

## §25.143 General.

**(a)** The airplane must be safely controllable and maneuverable during—

(1) Takeoff;

(2) Climb;

(3) Level flight;

(4) Descent; and

(5) Landing.

**(b)** It must be possible to make a smooth transition from one flight condition to any other flight condition without exceptional piloting skill, alertness, or strength, and without danger of exceeding the airplane limit-load factor under any probable operating conditions, including—

(1) The sudden failure of the critical engine;

(2) For airplanes with three or more engines, the sudden failure of the second critical engine when the airplane is in the en route, approach, or landing configuration and is trimmed with the critical engine inoperative; and

(3) Configuration changes, including deployment or retraction of deceleration devices.

**(c)** The following table prescribes, for conventional wheel type controls, the maximum control forces permitted during the testing required by paragraphs (a) and (b) of this section:

| Force, in pounds, applied to the control wheel or rudder pedals | Pitch | Roll | Yaw |
|---|---|---|---|
| For short term application for pitch and roll control — two hands available for control | 75 | 50 | — |
| For short term application for pitch and roll control — one hand available for control | 50 | 25 | — |
| For short term application for yaw control | — | — | 150 |
| For long term application | 10 | 5 | 20 |

**(d)** Approved operating procedures or conventional operating practices must be followed when demonstrating compliance with the control force limitations for short term application that are prescribed in paragraph (c) of this section. The airplane must be in trim, or as near to being in trim as practical, in the immediately preceding steady flight condition. For the takeoff condition, the airplane must be trimmed according to the approved operating procedures.

**(e)** When demonstrating compliance with the control force limitations for long term application that are prescribed in paragraph (c) of this section, the airplane must be in trim, or as near to being in trim as practical.

**(f)** When maneuvering at a constant airspeed or Mach number (up to $V_{FC}/M_{FC}$), the stick forces and the gradient of the stick force versus maneuvering load factor must lie within satisfactory limits. The stick forces must not be so great as to make excessive demands on the pilot's strength when maneuvering the airplane, and must not be so low that the airplane can easily be overstressed inadvertently. Changes of gradient that occur with changes of load factor must not cause undue difficulty in maintaining control of the airplane, and local gradients must not be so low as to result in a danger of overcontrolling.

[Docket No. 5066, 29 FR 18291, Dec. 24, 1964; as amended by Amdt. 25–42, 43 FR 2321, Jan. 16, 1978; Amdt. 25–84, 60 FR 30749, June 9, 1995]

## §25.145 Longitudinal control.

**(a)** It must be possible at any speed between the trim speed prescribed in §25.103(b)(1) and $V_S$, to pitch the nose downward so that the acceleration to this selected trim speed is prompt with—

(1) The airplane trimmed at the trim speed prescribed in §25.103(b)(1).

(2) The landing gear extended;

(3) The wing flaps (i) retracted and (ii) extended; and

(4) Power (i) off and (ii) at maximum continuous power on the engines.

**(b)** With the landing gear extended, no change in trim control, or exertion of more than 50 pounds control force (representative of the maximum short term force that can be applied readily by one hand) may be required for the following maneuvers:

(1) With power off, flaps retracted, and the airplane trimmed at 1.4 $V_{S1}$, extend the flaps as rapidly as possible while maintaining the airspeed at approximately 40 percent above the stalling speed existing at each instant throughout the maneuver.

(2) Repeat paragraph (b)(1) except initially extend the flaps and then retract them as rapidly as possible.

(3) Repeat paragraph (b)(2), except at the go-around power or thrust setting.

(4) With power off, flaps retracted, and the airplane trimmed at 1.4 $V_{S1}$, rapidly set go-around power or thrust while maintaining the same airspeed.

(5) Repeat paragraph (b)(4) except with flaps extended.

(6) With power off, flaps extended, and the airplane trimmed at 1.4 $V_{S1}$, obtain and maintain airspeeds between 1.1 $V_{S1}$, and either 1.7 $V_{S1}$, or $V_{FE}$, whichever is lower.

**(c)** It must be possible, without exceptional piloting skill, to prevent loss of altitude when complete retraction of the high-lift devices from any position is begun during steady, straight, level flight at 1.1 $V_{S1}$ for propeller powered airplanes, or 1.2 $V_{S1}$ for turbojet powered airplanes, with—

(1) Simultaneous movement of the power or thrust controls to the go-around power or thrust setting;

(2) The landing gear extended; and

(3) The critical combinations of landing weights and altitudes.

**(d)** If gated high-lift device control positions are provided, paragraph (c) of this section applies to retractions of the high-lift devices from any position from the maximum landing position to the first gated position, between gated positions, and from the last gated position to the fully retracted position. The requirements of paragraph (c) of this section also apply to retractions from each approval landing position to the control position(s) associated with the high-lift device configuration(s) used to establish the go-around procedure(s) from that landing position. In addition, the first gated control position from the maximum landing position must correspond with a configuration of the high-lift devices used to establish a go-around procedure from a landing configuration. Each gated control position must require a separate and distinct motion of the control to pass through the gated position and must have features to prevent inadvertent movement of the control through the gated position. It must only be possible to make this separate and distinct motion once the control has reached the gated position.

[Docket No. 5066, 29 FR 18291, Dec. 24, 1964; as amended by Amdt. 25–23, 35 FR 5671, April 8, 1970; Amdt. 25–72, 55 FR 29774, July 20, 1990; Amdt. 25–84, 60 FR 30749, June 9, 1995; Amdt. 25–98, 64 FR 6164, Feb. 8, 1999]

## §25.147 Directional and lateral control.

**(a)** *Directional control; general.* It must be possible, with the wings level, to yaw into the operative engine and to safely make a reasonably sudden change in heading of up to 15 degrees in the direction of the critical inoperative engine. This must be shown at 1.4 $V_{S1}$ for heading changes up to 15 degrees (except that the heading change at which the rudder pedal force is 150 pounds need not be exceeded), and with—

(1) The critical engine inoperative and its propeller in the minimum drag position;

(2) The power required for level flight at 1.4 $V_{S1}$, but not more than maximum continuous power;

(3) The most unfavorable center of gravity;

(4) Landing gear retracted;

(5) Flaps in the approach position; and

(6) Maximum landing weight.

**(b)** *Directional control; airplanes with four or more engines.* Airplanes with four or more engines must meet the requirements of paragraph (a) of this section except that—

(1) The two critical engines must be inoperative with their propellers (if applicable) in the minimum drag position;

(2) [Reserved]

(3) The flaps must be in the most favorable climb position.

**(c)** *Lateral control; general.* It must be possible to make 20° banked turns, with and against the inoperative engine, from steady flight at a speed equal to 1.4 $V_{S1}$, with—

(1) The critical engine inoperative and its propeller (if applicable) in the minimum drag position;

(2) The remaining engines at maximum continuous power;

(3) The most unfavorable center of gravity;

(4) Landing gear (i) retracted and (ii) extended;

(5) Flaps in the most favorable climb position; and

(6) Maximum takeoff weight.

**(d)** *Lateral control; airplanes with four or more engines.* Airplanes with four or more engines must be able to make 20° banked turns, with and against the inoperative engines, from steady flight at a speed equal to 1.4 $V_{S1}$, with maximum continuous power, and with the airplane in the configuration prescribed by paragraph (b) of this section.

**(e)** *Lateral control; all engines operating.* With the engines operating, roll response must allow normal maneuvers (such as recovery from upsets produced by gusts and the initiation of evasive maneuvers). There must be enough excess lateral control in sideslips (up to sideslip angles that might be required in normal operation), to allow a limited amount of maneuvering and to correct for gusts. Lateral control must be enough at any speed up to $V_{FC}/M_{FC}$ to provide a peak roll rate necessary for safety, without excessive control forces or travel.

[Docket No. 5066, 29 FR 18291, Dec. 24, 1964; as amended by Amdt. 25–42, 43 FR 2321, Jan. 16, 1978; Amdt. 25–72, 55 FR 29774, July 20, 1990]

25

## §25.149 Minimum control speed.

**(a)** In establishing the minimum control speeds required by this section, the method used to simulate critical engine failure must represent the most critical mode of powerplant failure with respect to controllability expected in service.

**(b)** $V_{MC}$ is the calibrated airspeed at which, when the critical engine is suddenly made inoperative, it is possible to maintain control of the airplane with that engine still inoperative and maintain straight flight with an angle of bank of not more than 5 degrees.

**(c)** $V_{MC}$ may not exceed 1.2 $V_S$ with—

(1) Maximum available takeoff power or thrust on the engines;

(2) The most unfavorable center of gravity;

(3) The airplane trimmed for takeoff;

(4) The maximum sea level takeoff weight (or any lesser weight necessary to show $V_{MC}$);

(5) The airplane in the most critical takeoff configuration existing along the flight path after the airplane becomes airborne, except with the landing gear retracted;

(6) The airplane airborne and the ground effect negligible; and

(7) If applicable, the propeller of the inoperative engine—

(i) Windmilling;

(ii) In the most probable position for the specific design of the propeller control; or

(iii) Feathered, if the airplane has an automatic feathering device acceptable for showing compliance with the climb requirements of §25.121.

**(d)** The rudder forces required to maintain control at $V_{MC}$ may not exceed 150 pounds nor may it be necessary to reduce power or thrust of the operative engines. During recovery, the airplane may not assume any dangerous attitude or require exceptional piloting skill, alertness, or strength to prevent a heading change of more than 20 degrees.

**(e)** $V_{MCG}$, the minimum control speed on the ground, is the calibrated airspeed during the takeoff run at which, when the critical engine is suddenly made inoperative, it is possible to maintain control of the airplane using the rudder control alone (without the use of nosewheel steering), as limited by 150 pounds of force, and the lateral control to the extent of keeping the wings level to enable the takeoff to be safely continued using normal piloting skill. In the determination of $V_{MCG}$, assuming that the path of the airplane accelerating with all engines operating is along the centerline of the runway, its path from the point at which the critical engine is made inoperative to the point at which recovery to a direction parallel to the centerline is completed may not deviate more than 30 feet laterally from the centerline at any point. $V_{MCG}$ must be established with—

(1) The airplane in each takeoff configuration or, at the option of the applicant, in the most critical takeoff configuration;

(2) Maximum available takeoff power or thrust on the operating engines;

(3) The most unfavorable center of gravity;

(4) The airplane trimmed for takeoff; and

(5) The most unfavorable weight in the range of takeoff weights.

**(f)** $V_{MCL}$, the minimum control speed during approach and landing with all engines operating, is the calibrated airspeed at which, when the critical engine is suddenly made inoperative, it is possible to maintain control of the airplane with that engine still inoperative, and maintain straight flight with an angle of bank of not more than 5 degrees. $V_{MCL}$ must be established with—

(1) The airplane in the most critical configuration (or, at the option of the applicant, each configuration) for approach and landing with all engines operating;

(2) The most unfavorable center of gravity;

(3) The airplane trimmed for approach with all engines operating;

(4) The most favorable weight, or, at the option of the applicant, as a function of weight;

(5) For propeller airplanes, the propeller of the inoperative engine in the position it achieves without pilot action, assuming the engine fails while at the power or thrust necessary to maintain a three degree approach path angle; and

(6) Go-around power or thrust setting on the operating engine(s).

**(g)** For airplanes with three or more engines, $V_{MCL-2}$, the minimum control speed during approach and landing with one critical engine inoperative, is the calibrated airspeed at which, when a second critical engine is suddenly made inoperative, it is possible to maintain control of the airplane with both engines still inoperative, and maintain straight flight with an angle of bank of not more than 5 degrees. $V_{MCL-2}$ must be established with—

(1) The airplane in the most critical configuration (or, at the option of the applicant, each configuration) for approach and landing with one critical engine inoperative;

(2) The most unfavorable center of gravity;

(3) The airplane trimmed for approach with one critical engine inoperative;

(4) The most unfavorable weight, or, at the option of the applicant, as a function of weight;

(5) For propeller airplanes, the propeller of the more critical inoperative engine in the position it achieves without pilot action, assuming the engine fails while at the power or thrust necessary to maintain a three degree approach path angle, and the propeller of the other inoperative engine feathered;

(6) The power or thrust on the operating engine(s) necessary to maintain an approach path angle of three degrees when one critical engine is inoperative; and

(7) The power or thrust on the operating engine(s) rapidly changed, immediately after the second critical engine is made inoperative, from the power or thrust prescribed in paragraph (g)(6) of this section to—

(i) Minimum power or thrust; and

(ii) Go-around power or thrust setting.

**(h)** In demonstration of $V_{MCL}$ and $V_{MCL\text{-}2}$ —

(1) The rudder force may not exceed 150 pounds;

(2) The airplane may not exhibit hazardous flight characteristics or require exceptional piloting skill, alertness, or strength;

(3) Lateral control must be sufficient to roll the airplane, from an initial condition of steady flight, through an angle of 20 degrees in the direction necessary to initiate a turn away from the inoperative engine(s), in not more than 5 seconds; and

(4) For propeller airplanes, hazardous flight characteristics must not be exhibited due to any propeller position achieved when the engine fails or during any likely subsequent movements of the engine or propeller controls.

[Docket No. 5066, 29 FR 18291, Dec. 24, 1964; as amended by Amdt. 25–42, 43 FR 2321, Jan. 16, 1978; Amdt. 25–72, 55 FR 29774, July 20, 1990; 55 FR 37607, Sept. 12, 1990; Amdt. 25–84, 60 FR 30750, June 9, 1995]

### TRIM

## §25.161 Trim.

**(a)** *General.* Each airplane must meet the trim requirements of this section after being trimmed, and without further pressure upon, or movement of, either the primary controls or their corresponding trim controls by the pilot or the automatic pilot.

**(b)** *Lateral and directional trim.* The airplane must maintain lateral and directional trim with the most adverse lateral displacement of the center of gravity within the relevant operating limitations, during normally expected conditions of operation (including operation at any speed from 1.4 $V_{S1}$ to $V_{MO}/M_{MO}$).

**(c)** *Longitudinal trim.* The airplane must maintain longitudinal trim during—

(1) A climb with maximum continuous power at a speed not more than 1.4 $V_{S1}$, with the landing gear retracted, and the flaps (i) retracted and (ii) in the takeoff position;

(2) A glide with power off at a speed not more than 1.4 $V_{S1}$, with the landing gear extended, the wing flaps (i) retracted and (ii) extended, the most unfavorable center of gravity position approved for landing with the maximum landing weight, and with the most unfavorable center of gravity position approved for landing regardless of weight; and

(3) Level flight at any speed from 1.4 $V_{S1}$, to $V_{MO}/M_{MO}$, with the landing gear and flaps retracted, and from 1.4 $V_{S1}$ to $V_{LE}$ with the landing gear extended.

**(d)** *Longitudinal, directional, and lateral trim.* The airplane must maintain longitudinal, directional, and lateral trim (and for the lateral trim, the angle of bank may not exceed five degrees) at 1.4 $V_{S1}$ during climbing flight with—

(1) The critical engine inoperative;

(2) The remaining engines at maximum continuous power; and

(3) The landing gear and flaps retracted.

**(e)** *Airplanes with four or more engines.* Each airplane with four or more engines must maintain trim in rectilinear flight—

(1) At the climb speed, configuration, and power required by §25.123(a) for the purpose of establishing the rate of climb;

(2) With the most unfavorable center of gravity position; and

(3) At the weight at which the two-engine-inoperative climb is equal to at least 0.013 $V_{S0}{}^2$ at an altitude of 5,000 feet.

[Docket No. 5066, 29 FR 18291, Dec. 24, 1964; as amended by Amdt. 25–23, 35 FR 5671, April 8, 1970; Amdt. 25–38, 41 FR 55466, Dec. 20, 1976]

### STABILITY

## §25.171 General.

The airplane must be longitudinally, directionally, and laterally stable in accordance with the provisions of §§25.173 through 25.177. In addition, suitable stability and control feel (static stability) is required in any condition normally encountered in service, if flight tests show it is necessary for safe operation.

[Docket No. 5066, 29 FR 18291, Dec. 24, 1964; as amended by Amdt. 25–7, 30 FR 13117, Oct. 15, 1965]

## §25.173 Static longitudinal stability.

Under the conditions specified in §25.175, the characteristics of the elevator control forces (including friction) must be as follows:

**(a)** A pull must be required to obtain and maintain speeds below the specified trim speed, and a push must be required to obtain and maintain speeds above the specified trim speed. This must be shown at any speed that can be obtained except speeds higher than the landing gear or wing flap operating limit speeds or $V_{FC}/M_{FC}$, whichever is appropriate, or lower than the minimum speed for steady unstalled flight.

**(b)** The airspeed must return to within 10 percent of the original trim speed for the climb, approach, and landing conditions specified in

§25.175 (a), (c), and (d), and must return to within 7.5 percent of the original trim speed for the cruising condition specified in §25.175(b), when the control force is slowly released from any speed within the range specified in paragraph (a) of this section.

**(c)** The average gradient of the stable slope of the stick force versus speed curve may not be less than 1 pound for each 6 knots.

**(d)** Within the free return speed range specified in paragraph (b) of this section, it is permissible for the airplane, without control forces, to stabilize on speeds above or below the desired trim speeds if exceptional attention on the part of the pilot is not required to return to and maintain the desired trim speed and altitude.

[Docket No. 5066, 29 FR 18291, Dec. 24, 1964; as amended by Amdt. 25–7, 30 FR 13117, Oct. 15, 1965]

## §25.175 Demonstration of static longitudinal stability.

Static longitudinal stability must be shown as follows:

**(a)** *Climb.* The stick force curve must have a stable slope at speeds between 85 and 115 percent of the speed at which the airplane—

(1) Is trimmed, with—

(i) Wing flaps retracted;

(ii) Landing gear retracted;

(iii) Maximum takeoff weight; and

(iv) 75 percent of maximum continuous power for reciprocating engines or the maximum power or thrust selected by the applicant as an operating limitation for use during climb for turbine engines; and

(2) Is trimmed at the speed for best rate-of-climb except that the speed need not be less than 1.4 $V_{S1}$.

**(b)** *Cruise.* Static longitudinal stability must be shown in the cruise condition as follows:

(1) With the landing gear retracted at high speed, the stick force curve must have a stable slope at all speeds within a range which is the greater of 15 percent of the trim speed plus the resulting free return speed range, or 50 knots plus the resulting free return speed range, above and below the trim speed (except that the speed range need not include speeds less than 1.4 $V_{S1}$, nor speeds greater than $V_{FC}/M_{FC}$, nor speeds that require a stick force of more than 50 pounds), with—

(i) The wing flaps retracted;

(ii) The center of gravity in the most adverse position (see §25.27);

(iii) The most critical weight between the maximum takeoff and maximum landing weights;

(iv) 75 percent of maximum continuous power for reciprocating engines or for turbine engines, the maximum cruising power selected by the applicant as an operating limitation (see §25.1521), except that the power need not exceed that required at $V_{MO}/M_{MO}$; and

(v) The airplane trimmed for level flight with the power required in paragraph (b)(1)(iv) of this section.

(2) With the landing gear retracted at low speed, the stick force curve must have a stable slope at all speeds within a range which is the greater of 15 percent of the trim speed plus the resulting free return speed range, or 50 knots plus the resulting free return speed range, above and below the trim speed (except that the speed range need not include speeds less than 1.4 $V_{S1}$, nor speeds greater than the minimum speed of the applicable speed range prescribed in paragraph (b)(1), nor speeds that require a stick force of more than 50 pounds), with—

(i) Wing flaps, center of gravity position, and weight as specified in paragraph (b)(1) of this section;

(ii) Power required for level flight at a speed equal to $V_{MO}$ + 1.4 $V_{S1}$ /2; and

(iii) The airplane trimmed for level flight with the power required in paragraph (b)(2)(ii) of this section.

(3) With the landing gear extended, the stick force curve must have a stable slope at all speeds within a range which is the greater of 15 percent of the trim speed plus the resulting free return speed range, or 50 knots plus the resulting free return speed range, above and below the trim speed (except that the speed range need not include speeds less than 1.4 $V_{S1}$, nor speeds greater than $V_{LE}$, nor speeds that require a stick force of more than 50 pounds), with—

(i) Wing flap, center of gravity position, and weight as specified in paragraph (b)(1) of this section;

(ii) 75 percent of maximum continuous power for reciprocating engines or, for turbine engines, the maximum cruising power selected by the applicant as an operating limitation, except that the power need not exceed that required for level flight at $V_{LE}$; and

(iii) The aircraft trimmed for level flight with the power required in paragraph (b)(3)(ii) of this section.

**(c)** *Approach.* The stick force curve must have a stable slope at speeds between 1.1 $V_{S1}$ and 1.8 $V_{S1}$, with—

(1) Wing flaps in the approach position;

(2) Landing gear retracted;

(3) Maximum landing weight; and

(4) The airplane trimmed at 1.4 $V_{S1}$ with enough power to maintain level flight at this speed.

**(d)** *Landing.* The stick force curve must have a stable slope, and the stick force may not exceed

80 pounds, at speeds between 1.1 $V_{S0}$ and 1.3 $V_{S0}$ with—

(1) Wing flaps in the landing position;

(2) Landing gear extended;

(3) Maximum landing weight;

(4) Power or thrust off on the engines; and

(5) The airplane trimmed at 1.4 $V_{S0}$ with power or thrust off.

[Docket No. 5066, 29 FR 18291, Dec. 24, 1964; as amended by Amdt. 25–7, 30 FR 13117, Oct. 15, 1965]

## §25.177 Static lateral-directional stability.

**(a)** [Reserved]

**(b)** [Reserved]

**(c)** In straight, steady sideslips, the aileron and rudder control movements and forces must be substantially proportional to the angle of sideslip in a stable sense; and the factor of proportionality must lie between limits found necessary for safe operation throughout the range of sideslip angles appropriate to the operation of the airplane. At greater angles, up to the angle at which full rudder is used or a rudder force of 180 pounds is obtained, the rudder pedal forces may not reverse; and increased rudder deflection must be needed for increased angles of sideslip. Compliance with this paragraph must be demonstrated for all landing gear and flap positions and symmetrical power conditions at speeds from 1.2 $V_{S1}$ to $V_{FE}$, $V_{LE}$, or $V_{FC}/M_{FC}$, as appropriate.

**(d)** The rudder gradients must meet the requirements of paragraph (c) at speeds between $V_{MO}/M_{MO}$ and $V_{FC}/M_{FC}$ except that the dihedral effect (aileron deflection opposite the corresponding rudder input) may be negative provided the divergence is gradual, easily recognized, and easily controlled by the pilot.

[Docket No. 5066, 29 FR 18291, Dec. 24, 1964; as amended by Amdt. 25–72, 55 FR 29774, July 20, 1990; 55 FR 37607, Sept. 12, 1990]

## §25.181 Dynamic stability.

**(a)** Any short period oscillation, not including combined lateral-directional oscillations, occurring between 1.2 $V_S$ and maximum allowable speed appropriate to the configuration of the airplane must be heavily damped with the primary controls—

(1) Free; and

(2) In a fixed position.

**(b)** Any combined lateral-directional oscillations ("Dutch roll") occurring between 1.2 $V_S$ and maximum allowable speed appropriate to the configuration of the airplane must be positively damped with controls free, and must be controllable with normal use of the primary controls without requiring exceptional pilot skill.

[Docket No. 5066, 29 FR 18291, Dec. 24, 1964; as amended by Amdt. 25–42, 43 FR 2322, Jan. 16, 1978; Amdt. 25–72, 55 FR 29775, July 20, 1990; 55 FR 37607, Sept. 12, 1990]

### STALLS

## §25.201 Stall demonstration.

**(a)** Stalls must be shown in straight flight and in 30 degree banked turns with—

(1) Power off; and

(2) The power necessary to maintain level flight at 1.6 $V_{S1}$ (where $V_{S1}$ corresponds to the stalling speed with flaps in the approach position, the landing gear retracted, and maximum landing weight).

**(b)** In each condition required by paragraph (a) of this section, it must be possible to meet the applicable requirements of §25.203 with—

(1) Flaps, landing gear, and deceleration devices in any likely combination of positions approved for operation;

(2) Representative weights within the range for which certification is requested;

(3) The most adverse center of gravity for recovery; and

(4) The airplane trimmed for straight flight at the speed prescribed in §25.103(b)(1).

**(c)** The following procedures must be used to show compliance with §25.203;

(1) Starting at a speed sufficiently above the stalling speed to ensure that a steady rate of speed reduction can be established, apply the longitudinal control so that the speed reduction does not exceed one knot per second until the airplane is stalled.

(2) In addition, for turning flight stalls, apply the longitudinal control to achieve airspeed deceleration rates up to 3 knots per second.

(3) As soon as the airplane is stalled, recover by normal recovery techniques.

**(d)** The airplane is considered stalled when the behavior of the airplane gives the pilot a clear and distinctive indication of an acceptable nature that the airplane is stalled. Acceptable indications of a stall, occurring either individually or in combination, are—

(1) A nose-down pitch that cannot be readily arrested;

(2) Buffeting, of a magnitude and severity that is a strong and effective deterrent to further speed reduction; or

(3) The pitch control reaches the aft stop and no further increase in pitch attitude occurs when the control is held full aft for a short time before recovery is initiated.

[Docket No. 5066, 29 FR 18291, Dec. 24, 1964; as amended by Amdt. 25–38, 41 FR 55466, Dec. 20, 1976; Amdt. 25–42, 43 FR 2322, Jan. 16, 1978; Amdt. 25–84, 60 FR 30750, June 9, 1995]

## §25.203 Stall characteristics.

**(a)** It must be possible to produce and to correct roll and yaw by unreversed use of the aileron and rudder controls, up to the time the airplane is stalled. No abnormal nose-up pitching may occur. The longitudinal control force must be positive up to and throughout the stall. In addition, it must be possible to promptly prevent stalling and to recover from a stall by normal use of the controls.

**(b)** For level wing stalls, the roll occurring between the stall and the completion of the recovery may not exceed approximately 20 degrees.

**(c)** For turning flight stalls, the action of the airplane after the stall may not be so violent or extreme as to make it difficult, with normal piloting skill, to effect a prompt recovery and to regain control of the airplane. The maximum bank angle that occurs during the recovery may not exceed—

(1) Approximately 60 degrees in the original direction of the turn, or 30 degrees in the opposite direction, for deceleration rates up to 1 knot per second; and

(2) Approximately 90 degrees in the original direction of the turn, or 60 degrees in the opposite direction, for deceleration rates in excess of 1 knot per second.

[Docket No. 5066, 29 FR 18291, Dec. 24, 1964; as amended by Amdt. 25–84, 60 FR 30750, June 9, 1995]

## §25.207 Stall warning.

**(a)** Stall warning with sufficient margin to prevent inadvertent stalling with the flaps and landing gear in any normal position must be clear and distinctive to the pilot in straight and turning flight.

**(b)** The warning may be furnished either through the inherent aerodynamic qualities of the airplane or by a device that will give clearly distinguishable indications under expected conditions of flight. However, a visual stall warning device that requires the attention of the crew within the cockpit is not acceptable by itself. If a warning device is used, it must provide a warning in each of the airplane configurations prescribed in paragraph (a) of this section at the speed prescribed in paragraph (c) of this section.

**(c)** The stall warning must begin at a speed exceeding the stalling speed (i.e., the speed at which the airplane stalls or the minimum speed demonstrated, whichever is applicable under the provisions of §25.201(d)) by seven percent or at any lesser margin if the stall warning has enough clarity, duration, distinctiveness, or similar properties.

[Docket No. 5066, 29 FR 18291, Dec. 24, 1964; as amended by Amdt. 25–7, 30 FR 13118, Oct. 15, 1965; Amdt. 25–42, 43 FR 2322, Jan. 16, 1978]

### Ground and Water Handling Characteristics

## §25.231 Longitudinal stability and control.

**(a)** Landplanes may have no uncontrollable tendency to nose over in any reasonably expected operating condition or when rebound occurs during landing or takeoff. In addition—

(1) Wheel brakes must operate smoothly and may not cause any undue tendency to nose over; and

(2) If a tail-wheel landing gear is used, it must be possible, during the takeoff ground run on concrete, to maintain any altitude up to thrust line level, at 80 percent of $V_{S1}$.

**(b)** For seaplanes and amphibians, the most adverse water conditions safe for takeoff, taxiing, and landing, must be established.

## §25.233 Directional stability and control.

**(a)** There may be no uncontrollable ground-looping tendency in 90° cross winds, up to a wind velocity of 20 knots or 0.2 $V_{S0}$, whichever is greater, except that the wind velocity need not exceed 25 knots at any speed at which the airplane may be expected to be operated on the ground. This may be shown while establishing the 90° cross component of wind velocity required by §25.237.

**(b)** Landplanes must be satisfactorily controllable, without exceptional piloting skill or alertness, in power-off landings at normal landing speed, without using brakes or engine power to maintain a straight path. This may be shown during power-off landings made in conjunction with other tests.

**(c)** The airplane must have adequate directional control during taxiing. This may be shown during taxiing prior to takeoffs made in conjunction with other tests.

[Docket No. 5066, 29 FR 18291, Dec. 24, 1964; as amended by Amdt. 25–23, 35 FR 5671, April 8, 1970; Amdt. 25–42, 43 FR 2322, Jan. 16, 1978; Amdt. 25–94, 63 FR 8848, Feb. 23, 1998]

## §25.235 Taxiing condition.

The shock absorbing mechanism may not damage the structure of the airplane when the airplane is taxied on the roughest ground that may reasonably be expected in normal operation.

## §25.237 Wind velocities.

**(a)** For landplanes and amphibians, a 90-degree cross component of wind velocity, demonstrated to be safe for takeoff and landing, must be established for dry runways and must be at least

20 knots or 0.2 $V_{S0}$, whichever is greater, except that it need not exceed 25 knots.

**(b)** For seaplanes and amphibians, the following applies:

(1) A 90-degree cross component of wind velocity, up to which takeoff and landing is safe under all water conditions that may reasonably be expected in normal operation, must be established and must be at least 20 knots or 0.2 $V_{S0}$, whichever is greater, except that it need not exceed 25 knots.

(2) A wind velocity, for which taxiing is safe in any direction under all water conditions that may reasonably be expected in normal operation, must be established and must be at least 20 knots or 0.2 $V_{S0}$, whichever is greater, except that it need not exceed 25 knots.

[Docket No. 5066, 29 FR 18291, Dec. 24, 1964; as amended by Amdt. 25–42, 43 FR 2322, Jan. 16, 1978]

## §25.239 Spray characteristics, control, and stability on water.

**(a)** For seaplanes and amphibians, during take-off, taxiing, and landing, and in the conditions set forth in paragraph (b) of this section, there may be no—

(1) Spray characteristics that would impair the pilot's view, cause damage, or result in the taking in of an undue quantity of water;

(2) Dangerously uncontrollable porpoising, bounding, or swinging tendency; or

(3) Immersion of auxiliary floats or sponsons, wing tips, propeller blades, or other parts not designed to withstand the resulting water loads.

**(b)** Compliance with the requirements of paragraph (a) of this section must be shown—

(1) In water conditions, from smooth to the most adverse condition established in accordance with §25.231;

(2) In wind and cross-wind velocities, water currents, and associated waves and swells that may reasonably be expected in operation on water;

(3) At speeds that may reasonably be expected in operation on water;

(4) With sudden failure of the critical engine at any time while on water; and

(5) At each weight and center of gravity position, relevant to each operating condition, within the range of loading conditions for which certification is requested.

**(c)** In the water conditions of paragraph (b) of this section, and in the corresponding wind conditions, the seaplane or amphibian must be able to drift for five minutes with engines inoperative, aided, if necessary, by a sea anchor.

MISCELLANEOUS FLIGHT REQUIREMENTS

## §25.251 Vibration and buffeting.

**(a)** The airplane must be demonstrated in flight to be free from any vibration and buffeting that would prevent continued safe flight in any likely operating condition.

**(b)** Each part of the airplane must be demonstrated in flight to be free from excessive vibration under any appropriate speed and power conditions up to $V_{DF}/M_{DF}$. The maximum speeds shown must be used in establishing the operating limitations of the airplane in accordance with §25.1505.

**(c)** Except as provided in paragraph (d) of this section, there may be no buffeting condition, in normal flight, including configuration changes during cruise, severe enough to interfere with the control of the airplane, to cause excessive fatigue to the crew, or to cause structural damage. Stall warning buffeting within these limits is allowable.

**(d)** There may be no perceptible buffeting condition in the cruise configuration in straight flight at any speed up to $V_{MO}/M_{MO}$, except that stall warning buffeting is allowable.

**(e)** For an airplane with $M_D$ greater than .6 or with a maximum operating altitude greater than 25,000 feet, the positive maneuvering load factors at which the onset of perceptible buffeting occurs must be determined with the airplane in the cruise configuration for the ranges of airspeed or Mach number, weight, and altitude for which the airplane is to be certificated. The envelopes of load factor, speed, altitude, and weight must provide a sufficient range of speeds and load factors for normal operations. Probable inadvertent excursions beyond the boundaries of the buffet onset envelopes may not result in unsafe conditions.

[Docket No. 5066, 29 FR 18291, Dec. 24, 1964; as amended by Amdt. 25–23, 35 FR 5671, April 8, 1970; Amdt. 25–72, 55 FR 29775, July 20, 1990; Amdt. 25–77, 57 FR 28949, June 29, 1992]

## §25.253 High-speed characteristics.

**(a)** *Speed increase and recovery characteristics.* The following speed increase and recovery characteristics must be met:

(1) Operating conditions and characteristics likely to cause inadvertent speed increases (including upsets in pitch and roll) must be simulated with the airplane trimmed at any likely cruise speed up to $V_{MO}/M_{MO}$. These conditions and characteristics include gust upsets, inadvertent control movements, low stick force gradient in relation to control friction, passenger movement, leveling off from climb, and descent from Mach to airspeed limit altitudes.

(2) Allowing for pilot reaction time after effective inherent or artificial speed warning occurs, it must be shown that the airplane can be recovered to a

normal attitude and its speed reduced to $V_{MO}/M_{MO}$, without—

(i) Exceptional piloting strength or skill;

(ii) Exceeding $V_D/M_D$, $V_{DF}/M_{DF}$, or the structural limitations; and

(iii) Buffeting that would impair the pilot's ability to read the instruments or control the airplane for recovery.

(3) With the airplane trimmed at any speed up to $V_{MO}/M_{MO}$, there must be no reversal of the response to control input about any axis at any speed up to $V_{DF}/M_{DF}$. Any tendency to pitch, roll, or yaw must be mild and readily controllable, using normal piloting techniques. When the airplane is trimmed at $V_{MO}/M_{MO}$, the slope of the elevator control force versus speed curve need not be stable at speeds greater than $V_{FC}/M_{FC}$, but there must be a push force at all speeds up to $V_{DF}/M_{DF}$ and there must be no sudden or excessive reduction of elevator control force as $V_{DF}/M_{DF}$ is reached.

**(b)** *Maximum speed for stability characteristics, $V_{FC}/M_{FC}$.* $V_{FC}/M_{FC}$ is the maximum speed at which the requirements of §§25.143(f), 25.147(e), 25.175(b)(1), 25.177, and 25.181 must be met with flaps and landing gear retracted. It may not be less than a speed midway between $V_{MO}/M_{MO}$ and $V_{DF}/M_{DF}$, except that for altitudes where Mach number is the limiting factor, $M_{FC}$ need not exceed the Mach number at which effective speed warning occurs.

[Docket No. 5066, 29 FR 18291, Dec. 24, 1964; as amended by Amdt. 25–23, 35 FR 5671, April 8, 1970; Amdt. 25–54, 45 FR 60172, Sept. 11, 1980; Amdt. 25–72, 55 FR 29775, July 20, 1990; Amdt. 25–84, 60 FR 30750, June 9, 1995]

## §25.255 Out-of-trim characteristics.

**(a)** From an initial condition with the airplane trimmed at cruise speeds up to $V_{MO}/M_{MO}$, the airplane must have satisfactory maneuvering stability and controllability with the degree of out-of-trim in both the airplane nose-up and nose-down directions, which results from the greater of—

(1) A three-second movement of the longitudinal trim system at its normal rate for the particular flight condition with no aerodynamic load (or an equivalent degree of trim for airplanes that do not have a power-operated trim system), except as limited by stops in the trim system, including those required by §25.655(b) for adjustable stabilizers; or

(2) The maximum mistrim that can be sustained by the autopilot while maintaining level flight in the high speed cruising condition.

**(b)** In the out-of-trim condition specified in paragraph (a) of this section, when the normal acceleration is varied from +1 g to the positive and negative values specified in paragraph (c) of this section—

(1) The stick force vs. g curve must have a positive slope at any speed up to and including $V_{FC}/M_{FC}$; and

(2) At speeds between $V_{FC}/M_{FC}$ and $V_{DF}/M_{DF}$ the direction of the primary longitudinal control force may not reverse.

**(c)** Except as provided in paragraphs (d) and (e) of this section, compliance with the provisions of paragraph (a) of this section must be demonstrated in flight over the acceleration range—

(1) -1 g to +2.5 g; or

(2) 0 g to 2.0 g, and extrapolating by an acceptable method to -1 g and +2.5 g.

**(d)** If the procedure set forth in paragraph (c)(2) of this section is used to demonstrate compliance and marginal conditions exist during flight test with regard to reversal of primary longitudinal control force, flight tests must be accomplished from the normal acceleration at which a marginal condition is found to exist to the applicable limit specified in paragraph (b)(1) of this section.

**(e)** During flight tests required by paragraph (a) of this section, the limit maneuvering load factors prescribed in §§25.333(b) and 25.337, and the maneuvering load factors associated with probable inadvertent excursions beyond the boundaries of the buffet onset envelopes determined under §25.251(e), need not be exceeded. In addition, the entry speeds for flight test demonstrations at normal acceleration values less than 1 g must be limited to the extent necessary to accomplish a recovery without exceeding $V_{DF}/M_{DF}$.

**(f)** In the out-of-trim condition specified in paragraph (a) of this section, it must be possible from an overspeed condition at $V_{DF}/M_{DF}$ to produce at least 1.5 g for recovery by applying not more than 125 pounds of longitudinal control force using either the primary longitudinal control alone or the primary longitudinal control and the longitudinal trim system. If the longitudinal trim is used to assist in producing the required load factor, it must be shown at $V_{DF}/M_{DF}$ that the longitudinal trim can be actuated in the airplane nose-up direction with the primary surface loaded to correspond to the least of the following airplane nose-up control forces:

(1) The maximum control forces expected in service as specified in §§25.301 and 25.397.

(2) The control force required to produce 1.5 g.

(3) The control force corresponding to buffeting or other phenomena of such intensity that it is a strong deterrent to further application of primary longitudinal control force.

[Docket No. 5066, 29 FR 18291, Dec. 24, 1964; as amended by Amdt. 25–42, 43 FR 2322, Jan. 16, 1978]

# Subpart C—Structure

GENERAL

## §25.301 Loads.

**(a)** Strength requirements are specified in terms of limit loads (the maximum loads to be expected in service) and ultimate loads (limit loads multiplied by prescribed factors of safety). Unless otherwise provided, prescribed loads are limit loads.

**(b)** Unless otherwise provided, the specified air, ground, and water loads must be placed in equilibrium with inertia forces, considering each item of mass in the airplane. These loads must be distributed to conservatively approximate or closely represent actual conditions. Methods used to determine load intensities and distribution must be validated by flight load measurement unless the methods used for determining those loading conditions are shown to be reliable.

**(c)** If deflections under load would significantly change the distribution of external or internal loads, this redistribution must be taken into account.

[Docket No. 5066, 29 FR 18291, Dec. 24, 1964; as amended by Amdt. 25–23, 35 FR 5672, April 8, 1970]

## §25.303 Factor of safety.

Unless otherwise specified, a factor of safety of 1.5 must be applied to the prescribed limit load which are considered external loads on the structure. When a loading condition is prescribed in terms of ultimate loads, a factor of safety need not be applied unless otherwise specified.

[Docket No. 5066, 29 FR 18291, Dec. 24, 1964; as amended by Amdt. 25–23, 35 FR 5672, April 8, 1970]

## §25.305 Strength and deformation.

**(a)** The structure must be able to support limit loads without detrimental permanent deformation. At any load up to limit loads, the deformation may not interfere with safe operation.

**(b)** The structure must be able to support ultimate loads without failure for at least 3 seconds. However, when proof of strength is shown by dynamic tests simulating actual load conditions, the 3-second limit does not apply. Static tests conducted to ultimate load must include the ultimate deflections and ultimate deformation induced by the loading. When analytical methods are used to show compliance with the ultimate load strength requirements, it must be shown that—

(1) The effects of deformation are not significant;

(2) The deformations involved are fully accounted for in the analysis; or

(3) The methods and assumptions used are sufficient to cover the effects of these deformations.

**(c)** Where structural flexibility is such that any rate of load application likely to occur in the operating conditions might produce transient stresses appreciably higher than those corresponding to static loads, the effects of this rate of application must be considered.

**(d)** [Reserved]

**(e)** The airplane must be designed to withstand any vibration and buffeting that might occur in any likely operating condition up to $V_D/M_D$, including stall and probable inadvertent excursions beyond the boundaries of the buffet onset envelope. This must be shown by analysis, flight tests, or other tests found necessary by the Administrator.

**(f)** Unless shown to be extremely improbable, the airplane must be designed to withstand any forced structural vibration resulting from any failure, malfunction or adverse condition in the flight control system. These must be considered limit loads and must be investigated at airspeeds up to $V_C/M_C$.

[Docket No. 5066, 29 FR 18291, Dec. 24, 1964; as amended by Amdt. 25–23, 35 FR 5672, April 8, 1970; Amdt. 25–54, 45 FR 60172, Sept. 11, 1980; Amdt. 25–77, 57 FR 28949, June 29, 1992; Amdt. 25–86, 61 FR 5220, Feb. 9, 1996]

## §25.307 Proof of structure.

**(a)** Compliance with the strength and deformation requirements of this subpart must be shown for each critical loading condition. Structural analysis may be used only if the structure conforms to that for which experience has shown this method to be reliable. The Administrator may require ultimate load tests in cases where limit load tests may be inadequate.

**(b) – (c)** [Reserved]

**(d)** When static or dynamic tests are used to show compliance with the requirements of §25.305(b) for flight structures, appropriate material correction factors must be applied to the test results, unless the structure, or part thereof, being tested has features such that a number of elements contribute to the total strength of the structure and the failure of one element results in the redistribution of the load through alternate load paths.

[Docket No. 5066, 29 FR 18291, Dec. 24, 1964; as amended by Amdt. 25–23, 35 FR 5672, April 8, 1970; Amdt. 25–54, 45 FR 60172, Sept. 11, 1980; Amdt. 25–72, 55 FR 29775, July 20, 1990]

FLIGHT LOADS

## §25.321 General.

**(a)** Flight load factors represent the ratio of the aerodynamic force component (acting normal to the assumed longitudinal axis of the airplane) to the weight of the airplane. A positive load factor is one in which the aerodynamic force acts upward with respect to the airplane.

**(b)** Considering compressibility effects at each speed, compliance with the flight load requirements of this subpart must be shown—

(1) At each critical altitude within the range of altitudes selected by the applicant;

(2) At each weight from the design minimum weight to the design maximum weight appropriate to each particular flight load condition; and

(3) For each required altitude and weight, for any practicable distribution of disposable load within the operating limitations recorded in the Airplane Flight Manual.

**(c)** Enough points on and within the boundaries of the design envelope must be investigated to ensure that the maximum load for each part of the airplane structure is obtained.

**(d)** The significant forces acting on the airplane must be placed in equilibrium in a rational or conservative manner. The linear inertia forces must be considered in equilibrium with the thrust and all aerodynamic loads, while the angular (pitching) inertia forces must be considered in equilibrium with thrust and all aerodynamic moments, including moments due to loads on components such as tail surfaces and nacelles. Critical thrust values in the range from zero to maximum continuous thrust must be considered.

[Docket No. 5066, 29 FR 18291, Dec. 24, 1964; as amended by Amdt. 25–23, 35 FR 5672, April 8, 1970; Amdt. 25–86, 61 FR 5220, Feb. 9, 1996]

FLIGHT MANEUVER AND GUST CONDITIONS

## §25.331 Symmetric maneuvering conditions.

**(a)** *Procedure.* For the analysis of the maneuvering flight conditions specified in paragraphs (b) and (c) of this section, the following provisions apply:

(1) Where sudden displacement of a control is specified, the assumed rate of control surface displacement may not be less than the rate that could be applied by the pilot through the control system.

(2) In determining elevator angles and chordwise load distribution in the maneuvering conditions of paragraphs (b) and (c) of this section, the effect of corresponding pitching velocities must be taken into account. The in-trim and out-of-trim flight conditions specified in §25.255 must be considered.

**(b)** *Maneuvering balanced conditions.* Assuming the airplane to be in equilibrium with zero pitching acceleration, the maneuvering conditions A through I on the maneuvering envelope in §25.333(b) must be investigated.

**(c)** *Pitch maneuver conditions.* The conditions specified in paragraphs (c)(1) and (2) of this section must be investigated. The movement of the pitch control surfaces may be adjusted to take into account limitations imposed by the maximum pilot effort specified by §25.397(b), control system stops and any indirect effect imposed by limitations in the output side of the control system (for example, stalling torque or maximum rate obtainable by a power control system.)

(1) *Maximum pitch control displacement at $V_A$.* The airplane is assumed to be flying in steady level flight (point A1, §25.333(b)) and the cockpit pitch control is suddenly moved to obtain extreme nose up pitching acceleration. In defining the tail load, the response of the airplane must be taken into account. Airplane loads that occur subsequent to the time when normal acceleration at the c.g. exceeds the positive limit maneuvering load factor (at point $A_2$ in §25.333(b)), or the resulting tailplane normal load reaches its maximum, whichever occurs first, need not be considered.

(2) *Specified control displacement.* A checked maneuver, based on a rational pitching control motion vs. time profile, must be established in which the design limit load factor specified in §25.337 will not be exceeded. Unless lesser values cannot be exceeded, the airplane response must result in pitching accelerations not less than the following:

(i) A positive pitching acceleration (nose up) is assumed to be reached concurrently with the airplane load factor of 1.0 (Points $A_1$ to $D_1$, §25.333(b)). The positive acceleration must be equal to at least

$$\frac{39n}{v}(n-1.5), \text{(Radians/sec.}^2\text{)}$$

where—

n is the positive load factor at the speed under consideration, and V is the airplane equivalent speed in knots.

(ii) A negative pitching acceleration (nose down) is assumed to be reached concurrently with the positive maneuvering load factor (points $A_2$ to $D_2$, §25.333(b)). This negative pitching acceleration must be equal to at least

$$\frac{-26n}{v}(n-1.5), \text{(Radians/sec.}^2\text{)}$$

where—

n is the positive load factor at the speed under consideration; and V is the airplane equivalent speed in knots.

[Docket No. 5066, 29 FR 18291, Dec. 24, 1964; as amended by Amdt. 25–23, 35 FR 5672, April 8, 1970; Amdt. 25–46, 43 FR 50594, Oct. 30, 1978; 43 FR 52495, Nov. 13, 1978; 43 FR 54082, Nov. 20, 1978; Amdt. 25–72, 55 FR 29775, July 20, 1990; 55 FR 37607, Sept. 12, 1990; Amdt. 25–86, 61 FR 5220, Feb. 9, 1996; Amdt. 25–91, 62 FR 40704, July 29, 1997]

## §25.333 Flight maneuvering envelope.

**(a)** *General.* The strength requirements must be met at each combination of airspeed and load factor on and within the boundaries of the representative maneuvering envelope (*V-n* diagram) of paragraph (b) of this section. This envelope must also be used in determining the airplane structural operating limitations as specified in §25.1501.

**(b)** *Maneuvering envelope.*

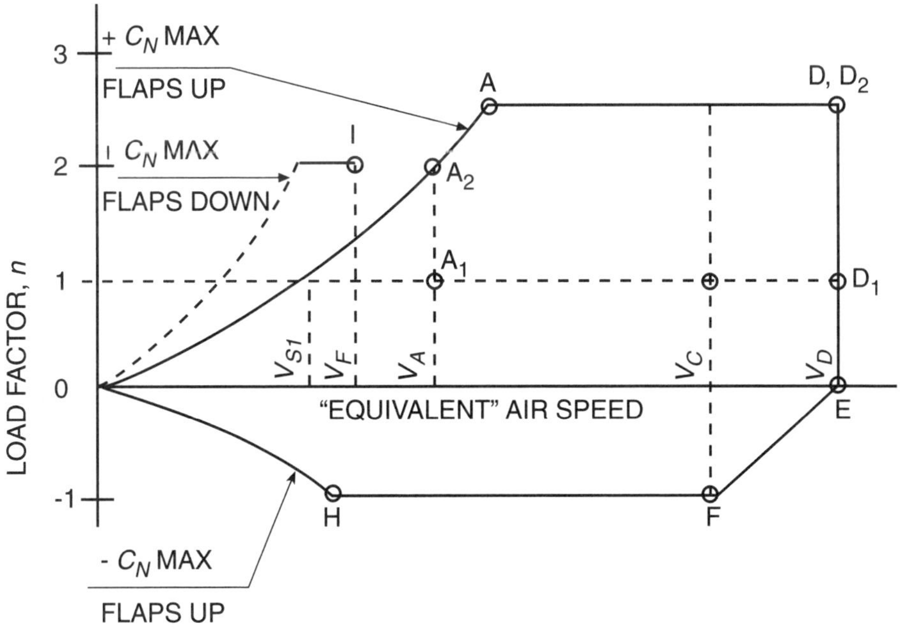


[Docket No. 5066, 29 FR 18291, Dec. 24, 1964; as amended by Amdt. 25–86, 61 FR 5220, Feb. 9, 1996]

## §25.335 Design airspeeds.

The selected design airspeeds are equivalent airspeeds (EAS). Estimated values of $V_{S0}$ and $V_{S1}$ must be conservative.

**(a)** *Design cruising speed,* $V_C$. For $V_C$, the following apply:

(1) The minimum value of $V_C$ must be sufficiently greater than $V_B$ to provide for inadvertent speed increases likely to occur as a result of severe atmospheric turbulence.

(2) Except as provided in §25.335(d)(2), $V_C$ may not be less than $V_B$ + 1.32 $U_{REF}$ (with $U_{REF}$ as specified in §25.341(a)(5)(i)). However $V_C$ need not exceed the maximum speed in level flight at maximum continuous power for the corresponding altitude.

(3) At altitudes where $V_D$ is limited by Mach number, $V_C$ may be limited to a selected Mach number.

**(b)** *Design dive speed,* $V_D$. $V_D$ must be selected so that $V_C/M_C$ is not greater than 0.8 $V_D/M_D$, or so that the minimum speed margin between $V_C/M_C$ and $V_D/M_D$ is the greater of the following values:

(1) From an initial condition of stabilized flight at $V_C/M_C$, the airplane is upset, flown for 20 seconds along a flight path 7.5° below the initial path, and then pulled up at a load factor of 1.5 g (0.5 g acceleration increment). The speed increase occurring in this maneuver may be calculated if reliable or conservative aerodynamic data is used. Power as specified in §25.175(b)(1)(iv) is assumed until the pullup is initiated, at which time power reduc-

tion and the use of pilot controlled drag devices may be assumed;

(2) The minimum speed margin must be enough to provide for atmospheric variations (such as horizontal gusts, and penetration of jet streams and cold fronts) and for instrument errors and airframe production variations. These factors may be considered on a probability basis. The margin at altitude where $M_C$ is limited by compressibility effects must not be less than 0.07M unless a lower margin is determined using a rational analysis that includes the effects of any automatic systems. In any case, the margin may not be reduced to less than 0.05M.

**(c)** *Design maneuvering speed $V_A$*. For $V_A$, the following apply:

(1) $V_A$ may not be less than $V_{S1} \sqrt{n}$ where—

(i) *n* is the limit positive maneuvering load factor at $V_C$; and

(ii) $V_{S1}$ is the stalling speed with flaps retracted.

(2) $V_A$ and $V_S$ must be evaluated at the design weight and altitude under consideration.

(3) $V_A$ need not be more than $V_C$ or the speed at which the positive $C_{Nmax}$ curve intersects the positive maneuver load factor line, whichever is less.

**(d)** *Design speed for maximum gust intensity, $V_B$.*

(1) $V_B$ may not be less than

$$V_{S1}\left[1+\frac{K_g U_{ref} V_c a}{498w}\right]^{1/2}$$

where—

$V_{S1}$ = the 1-g stalling speed based on $C_{NAmax}$ with the flaps retracted at the particular weight under consideration;

$V_C$ = design cruise speed (knots equivalent airspeed);

$U_{ref}$ = the reference gust velocity (feet per second equivalent airspeed) from §25.341(a)(5)(i);

w = average wing loading (pounds per square foot) at the particular weight under consideration.

$$K_g = \frac{.88\mu}{5.3+\mu}$$

$$\mu = \frac{2w}{\rho c a g}$$

ρ = density of air (slugs/ft$^3$);

c = mean geometric chord of the wing (feet);

g = acceleration due to gravity (ft/sec$^2$);

a = slope of the airplane normal force coefficient curve, $C_{NA}$ per radian;

(2) At altitudes where $V_C$ is limited by Mach number—

(i) $V_B$ may be chosen to provide an optimum margin between low and high speed buffet boundaries; and,

(ii) $V_B$ need not be greater than $V_C$.

**(e)** *Design flap speeds, $V_F$*. For $V_F$, the following apply:

(1) The design flap speed for each flap position (established in accordance with §25.697(a)) must be sufficiently greater than the operating speed recommended for the corresponding stage of flight (including balked landings) to allow for probable variations in control of airspeed and for transition from one flap position to another.

(2) If an automatic flap positioning or load limiting device is used, the speeds and corresponding flap positions programmed or allowed by the device may be used.

(3) $V_F$ may not be less than—

(i) 1.6 $V_{S1}$ with the flaps in takeoff position at maximum takeoff weight;

(ii) 1.8 $V_{S1}$ with the flaps in approach position at maximum landing weight, and

(iii) 1.8 $V_{S0}$ with the flaps in landing position at maximum landing weight.

**(f)** *Design drag device speeds, $V_{DD}$*. The selected design speed for each drag device must be sufficiently greater than the speed recommended for the operation of the device to allow for probable variations in speed control. For drag devices intended for use in high speed descents, $V_{DD}$ may not be less than $V_D$. When an automatic drag device positioning or load limiting means is used, the speeds and corresponding drag device positions programmed or allowed by the automatic means must be used for design.

[Docket No. 5066, 29 FR 18291, Dec. 24, 1964; as amended by Amdt. 25–23, 35 FR 5672, April 8, 1970; Amdt. 25–86, 61 FR 5220, Feb. 9, 1996; 25–91, 62 FR 40704, July 29, 1997]

## §25.337 Limit maneuvering load factors.

**(a)** Except where limited by maximum (static) lift coefficients, the airplane is assumed to be subjected to symmetrical maneuvers resulting in the limit maneuvering load factors prescribed in this section. Pitching velocities appropriate to the corresponding pull-up and steady turn maneuvers must be taken into account.

**(b)** The positive limit maneuvering load factor "n" for any speed up to $V_N$ may not be less than 2.1+24,000/ (W +10,000) except that "n" may not be less than 2.5 and need not be greater than 3.8—where "W" is the design maximum takeoff weight.

**(c)** The negative limit maneuvering load factor—

(1) May not be less than –1.0 at speeds up to $V_C$; and

(2) Must vary linearly with speed from the value at $V_C$ to zero at $V_D$.

**(d)** Maneuvering load factors lower than those specified in this section may be used if the airplane has design features that make it impossible to exceed these values in flight.

[Docket No. 5066, 29 FR 18291, Dec. 24, 1964; as amended by Amdt. 25–23, 35 FR 5672, April 8, 1970]

## §25.341 Gust and turbulence loads.

**(a)** *Discrete Gust Design Criteria.* The airplane is assumed to be subjected to symmetrical vertical and lateral gusts in level flight. Limit gust loads must be determined in accordance with the provisions:

(1) Loads on each part of the structure must be determined by dynamic analysis. The analysis must take into account unsteady aerodynamic characteristics and all significant structural degrees of freedom including rigid body motions.

(2) The shape of the gust must be:

$$U = \frac{U_{ds}}{2}\left[1 - Cos\left(\frac{\pi s}{H}\right)\right]$$

for $0 \le s \le 2H$

where—

s = distance penetrated into the gust (feet);

$U_{ds}$ = the design gust velocity in equivalent airspeed specified in paragraph (a)(4) of this section; and

H = the gust gradient which is the distance (feet) parallel to the airplane's flight path for the gust to reach its peak velocity.

(3) A sufficient number of gust gradient distances in the range 30 feet to 350 feet must be investigated to find the critical response for each load quantity.

(4) The design gust velocity must be:

$$U_{ds} = U_{ref}F_g\left(H/350\right)^{1/6}$$

where—

$U_{ref}$ = the reference gust velocity in equivalent airspeed defined in paragraph (a)(5) of this section.

$F_g$ = the flight profile alleviation factor defined in paragraph (a)(6) of this section.

(5) The following reference gust velocities apply:

(i) At the airplane design speed $V_C$: Positive and negative gusts with reference gust velocities of 56.0 ft/sec EAS must be considered at sea level. The reference gust velocity may be reduced linearly from 56.0 ft/sec EAS at sea level to 44.0 ft/sec EAS at 15000 feet. The reference gust velocity may be further reduced linearly from 44.0 ft/sec EAS at 15000 feet to 26.0 ft/sec EAS at 50000 feet.

(ii) At the airplane design speed $V_D$: The reference gust velocity must be 0.5 times the value obtained under §25.341(a)(5)(i).

(6) The flight profile alleviation factor, Fg, must be increased linearly from the sea level value to a value of 1.0 at the maximum operating altitude defined in §25.1527. At sea level, the flight profile alleviation factor is determined by the following equation:

$$F_g = 0.5\left(F_{gz} + F_{gm}\right)$$

Where:

$$F_{gz} = 1 - \frac{Z_{mo}}{250000}$$

$$F_{gm} = \sqrt{R_2 Tan\,(\pi R_1/4)}$$

$$R_1 = \frac{\text{Maximum Landing Weight}}{\text{Maximum Take-off Weight}}$$

$$R_2 = \frac{\text{Maximum Zero Fuel Weight}}{\text{Maximum Take-off Weight}}$$

$Z_{mo}$ = Maximum operating altitude defined in §25.1527.

(7) When a stability augmentation system is included in the analysis, the effect of any significant system nonlinearities should be accounted for when deriving limit loads from limit gust conditions.

**(b)** *Continuous Gust Design Criteria.* The dynamic response of the airplane to vertical and lateral continuous turbulence must be taken into account. The continuous gust design criteria of Appendix G of this part must be used to establish the dynamic response unless more rational criteria are shown.

[Docket No. 27902, 61 FR 5221, Feb. 9, 1996; as amended by Amdt. 25–86, 61 FR 9533, March 8, 1996]

## §25.343 Design fuel and oil loads.

**(a)** The disposable load combinations must include each fuel and oil load in the range from zero fuel and oil to the selected maximum fuel and oil load. A structural reserve fuel condition, not exceeding 45 minutes of fuel under the operating conditions in §25.1001(e) and (f), as applicable, may be selected.

25

**(b)** If a structural reserve fuel condition is selected, it must be used as the minimum fuel weight condition for showing compliance with the flight load requirements as prescribed in this subpart. In addition—

(1) The structure must be designed for a condition of zero fuel and oil in the wing at limit loads corresponding to—

(i) A maneuvering load factor of +2.25; and

(ii) The gust conditions of §25.341(a) but assuming 85% of the design velocities prescribed in §25.341(a)(4).

(2) Fatigue evaluation of the structure must account for any increase in operating stresses resulting from the design condition of paragraph (b)(1) of this section; and

(3) The flutter, deformation, and vibration requirements must also be met with zero fuel.

[Docket No. 5066, 29 FR 18291, Dec. 24, 1964; as amended by Amdt. 25–18, 33 FR 12226, Aug. 30, 1968; Amdt. 25–72, 55 FR 37607, Sept. 12, 1990; Amdt. 25–86, 61 FR 5221, Feb. 9, 1996]

## §25.345 High lift devices.

**(a)** If wing flaps are to be used during takeoff, approach, or landing, at the design flap speeds established for these stages of flight under §25.335(e) and with the wing flaps in the corresponding positions, the airplane is assumed to be subjected to symmetrical maneuvers and gusts. The resulting limit loads must correspond to the conditions determined as follows:

(1) Maneuvering to a positive limit load factor of 2.0; and

(2) Positive and negative gusts of 25 ft/sec EAS acting normal to the flight path in level flight. Gust loads resulting on each part of the structure must be determined by rational analysis. The analysis must take into account the unsteady aerodynamic characteristics and rigid body motions of the aircraft. The shape of the gust must be as described in §25.341(a)(2) except that—

$U_{ds}$ = 25 ft/sec EAS;

H = 12.5 c; and

c = mean geometric chord of the wing (feet).

**(b)** The airplane must be designed for the conditions prescribed in paragraph (a) of this section, except that the airplane load factor need not exceed 1.0, taking into account, as separate conditions, the effects of—

(1) Propeller slipstream corresponding to maximum continuous power at the design flap speeds $V_F$, and with takeoff power at not less than 1.4 times the stalling speed for the particular flap position and associated maximum weight; and

(2) A head-on gust of 25 feet per second velocity (EAS).

**(c)** If flaps or other high lift devices are to be used in en route conditions, and with flaps in the appropriate position at speeds up to the flap design speed chosen for these conditions, the airplane is assumed to be subjected to symmetrical maneuvers and gusts within the range determined by—

(1) Maneuvering to a positive limit load factor as prescribed in §25.337(b); and

(2) The discrete vertical gust criteria in §25.341(a).

**(d)** The airplane must be designed for a maneuvering load factor of 1.5g at the maximum take-off weight with the wing-flaps and similar high lift devices in the landing configurations.

[Docket No. 5066, 29 FR 18291, Dec. 24, 1964; as amended by Amdt. 25–46, 43 FR 50595, Oct. 30, 1978; Amdt. 25–72, 55 FR 37607, Sept. 17, 1990; Amdt. 25–86, 61 FR 5221, Feb. 9, 1996; Amdt. 25–91, 62 FR 40704, July 29, 1997]

## §25.349 Rolling conditions.

The airplane must be designed for loads resulting from the rolling conditions specified in paragraphs (a) and (b) of this section. Unbalanced aerodynamic moments about the center of gravity must be reacted in a rational or conservative manner, considering the principal masses furnishing the reacting inertia forces.

**(a)** *Maneuvering.* The following conditions, speeds, and aileron deflections (except as the deflections may be limited by pilot effort) must be considered in combination with an airplane load factor of zero and of two-thirds of the positive maneuvering factor used in design. In determining the required aileron deflections, the torsional flexibility of the wing must be considered in accordance with §25.301(b):

(1) Conditions corresponding to steady rolling velocities must be investigated. In addition, conditions corresponding to maximum angular acceleration must be investigated for airplanes with engines or other weight concentrations outboard of the fuselage. For the angular acceleration conditions, zero rolling velocity may be assumed in the absence of a rational time history investigation of the maneuver.

(2) At $V_A$, a sudden deflection of the aileron to the stop is assumed.

(3) At $V_C$, the aileron deflection must be that required to produce a rate of roll not less than that obtained in paragraph (a)(2) of this section.

(4) At $V_D$, the aileron deflection must be that required to produce a rate of roll not less than one-third of that in paragraph (a)(2) of this section.

**(b)** *Unsymmetrical gusts.* The airplane is assumed to be subjected to unsymmetrical vertical gusts in level flight. The resulting limit loads must be determined from either the wing maximum airload derived directly from §25.341(a), or the wing maximum airload derived indirectly from the vertical load factor calculated from §25.341(a). It must be assumed that 100 percent of the wing air load acts on one side of the airplane and 80 percent of the wing air load acts on the other side.

[Docket No. 5066, 29 FR 18291, Dec. 24, 1964; as amended by Amdt. 25–23, 35 FR 5672, April 8, 1970; Amdt. 25–86, 61 FR 5222, Feb. 9, 1996; Amdt. 25–94, 63 FR 8848, Feb. 23, 1998]

## §25.351 Yaw maneuver conditions.

The airplane must be designed for loads resulting from the yaw maneuver conditions specified in paragraphs (a) through (d) of this section at speeds from $V_{MC}$ to $V_D$. Unbalanced aerodynamic moments about the center of gravity must be reacted in a rational or conservative manner considering the airplane inertia forces. In computing the tail loads the yawing velocity may be assumed to be zero.

**(a)** With the airplane in unaccelerated flight at zero yaw, it is assumed that the cockpit rudder control is suddenly displaced to achieve the resulting rudder deflection, as limited by:

(1) The control system on control surface stops; or

(2) A limit pilot force of 300 pounds from $V_{MC}$ to $V_A$ and 200 pounds from $V_C/M_C$ to $V_D/M_D$, with a linear variation between $V_A$ and $V_C/M_C$.

**(b)** With the cockpit rudder control deflected so as always to maintain the maximum rudder deflection available within the limitations specified in paragraph (a) of this section, it is assumed that the airplane yaws to the overswing sideslip angle.

**(c)** With the airplane yawed to the static equilibrium sideslip angle, it is assumed that the cockpit rudder control is held so as to achieve the maximum rudder deflection available within the limitations specified in paragraph (a) of this section.

**(d)** With the airplane yawed to the static equilibrium sideslip angle of paragraph (c) of this section, it is assumed that the cockpit rudder control is suddenly returned to neutral.

[Docket No. 5066, 29 FR 18291, Dec. 24, 1964; as amended by Amdt. 25–23, 35 FR 5672, April 8, 1970; Amdt. 25–46, 43 FR 50595, Oct. 30, 1978; Amdt. 25–72, 55 FR 29775, July 20, 1990; 55 FR 37607, Sept. 12, 1990; 55 FR 41415, Oct. 11, 1990; Amdt. 25–86, 61 FR 5222, Feb. 9, 1996; Amdt. 25–91, 62 FR 40704, July 29, 1997]

### Supplementary Conditions

## §25.361 Engine torque.

**(a)** Each engine mount and its supporting structure must be designed for the effects of—

(1) A limit engine torque corresponding to takeoff power and propeller speed acting simultaneously with 75 percent of the limit loads from flight condition A of §25.333(b);

(2) A limit torque corresponding to the maximum continuous power and propeller speed, acting simultaneously with the limit loads from flight condition A of §25.333(b); and

(3) For turbopropeller installations, in addition to the conditions specified in paragraphs (a)(1) and (2) of this section, a limit engine torque corresponding to takeoff power and propeller speed, multiplied by a factor accounting for propeller control system malfunction, including quick feathering, acting simultaneously with 1g level flight loads. In the absence of a rational analysis, a factor of 1.6 must be used.

**(b)** For turbine engine installations, the engine mounts and supporting structure must be designed to withstand each of the following:

(1) A limit engine torque load imposed by sudden engine stoppage due to malfunction or structural failure (such as compressor jamming).

(2) A limit engine torque load imposed by the maximum acceleration of the engine.

**(c)** The limit engine torque to be considered under paragraph (a) of this section must be obtained by multiplying mean torque for the specified power and speed by a factor of—

(1) 1.25 for turbopropeller installations;

(2) 1.33 for reciprocating engines with five or more cylinders; or

(3) Two, three, or four, for engines with four, three, or two cylinders, respectively.

[Docket No. 5066, 29 FR 18291, Dec. 24, 1964; as amended by Amdt. 25–23, 35 FR 5672, April 8, 1970; Amdt. 25–46, 43 FR 50595, Oct. 30, 1978; Amdt. 25–72, 55 FR 29776, July 20, 1990]

## §25.363 Side load on engine and auxiliary power unit mounts.

**(a)** Each engine and auxiliary power unit mount and its supporting structure must be designed for a limit load factor in lateral direction, for the side load on the engine and auxiliary power unit mount, at least equal to the maximum load factor obtained in the yawing conditions but not less than—

(1) 1.33; or

(2) One-third of the limit load factor for flight condition A as prescribed in §25.333(b).

**(b)** The side load prescribed in paragraph (a) of this section may be assumed to be independent of other flight conditions.

[Docket No. 5066, 29 FR 18291, Dec. 24, 1964; as amended by Amdt. 25–23, 35 FR 5672, April 8, 1970; Amdt. 25–91, 61 FR 40704, July 29, 1997]

25

## §25.365 Pressurized compartment loads.

For airplanes with one or more pressurized compartments the following apply:

**(a)** The airplane structure must be strong enough to withstand the flight loads combined with pressure differential loads from zero up to the maximum relief valve setting.

**(b)** The external pressure distribution in flight, and stress concentrations and fatigue effects must be accounted for.

**(c)** If landings may be made with the compartment pressurized, landing loads must be combined with pressure differential loads from zero up to the maximum allowed during landing.

**(d)** The airplane structure must be designed to be able to withstand the pressure differential loads corresponding to the maximum relief valve setting multiplied by a factor of 1.33 for airplanes to be approved for operation to 45,000 feet or by a factor of 1.67 for airplanes to be approved for operation above 45,000 feet, omitting other loads.

**(e)** Any structure, component or part, inside or outside a pressurized compartment, the failure of which could interfere with continued safe flight and landing, must be designed to withstand the effects of a sudden release of pressure through an opening in any compartment at any operating altitude resulting from each of the following conditions:

(1) The penetration of the compartment by a portion of an engine following an engine disintegration;

(2) Any opening in any pressurized compartment up to the size $H_o$ in square feet; however, small compartments may be combined with an adjacent pressurized compartment and both considered as a single compartment for openings that cannot reasonably be expected to be confined to the small compartment. The size $H_o$ must be computed by the following formula:

$H_o = PA_S$

where,

$H_o$ = Maximum opening in square feet, need not exceed 20 square feet.

$$P = \frac{A_S}{6240} + .024$$

$A_S$ = Maximum cross-sectional area of the pressurized shell normal to the longitudinal axis, in square feet; and

(3) The maximum opening caused by airplane or equipment failures not shown to be extremely improbable.

**(f)** In complying with paragraph (e) of this section, the fail-safe features of the design may be considered in determining the probability of failure or penetration and probable size of openings, provided that possible improper operation of closure devices and inadvertent door openings are also considered. Furthermore, the resulting differential pressure loads must be combined in a rational and conservative manner with 1-g level flight loads and any loads arising from emergency depressurization conditions. These loads may be considered as ultimate conditions; however, any deformations associated with these conditions must not interfere with continued safe flight and landing. The pressure relief provided by intercompartment venting may also be considered.

**(g)** Bulkheads, floors, and partitions in pressurized compartments for occupants must be designed to withstand the conditions specified in paragraph (e) of this section. In addition, reasonable design precautions must be taken to minimize the probability of parts becoming detached and injuring occupants while in their seats.

[Docket No. 5066, 29 FR 18291, Dec. 24, 1964; as amended by Amdt. 25–54, 45 FR 60172, Sept. 11, 1980; Amdt. 25–71, 55 FR 13477, April 10, 1990; Amdt. 25–72, 55 FR 29776, July 20, 1990; Amdt. 25–87, 61 FR 28695, June 5, 1996]

## §25.367 Unsymmetrical loads due to engine failure.

**(a)** The airplane must be designed for the unsymmetrical loads resulting from the failure of the critical engine. Turbopropeller airplanes must be designed for the following conditions in combination with a single malfunction of the propeller drag limiting system, considering the probable pilot corrective action on the flight controls:

(1) At speeds between $V_{MC}$ and $V_D$, the loads resulting from power failure because of fuel flow interruption are considered to be limit loads.

(2) At speeds between $V_{MC}$ and $V_C$, the loads resulting from the disconnection of the engine compressor from the turbine or from loss of the turbine blades are considered to be ultimate loads.

(3) The time history of the thrust decay and drag build-up occurring as a result of the prescribed engine failures must be substantiated by test or other data applicable to the particular engine-propeller combination.

(4) The timing and magnitude of the probable pilot corrective action must be conservatively estimated, considering the characteristics of the particular engine-propeller-airplane combination.

**(b)** Pilot corrective action may be assumed to be initiated at the time maximum yawing velocity is reached, but not earlier than two seconds after the engine failure. The magnitude of the corrective action may be based on the control forces specified in §25.397(b) except that lower forces may be assumed where it is shown by analysis or test that these forces can control the yaw and roll resulting from the prescribed engine failure conditions.

## §25.371 Gyroscopic loads.

The structure supporting any engine or auxiliary power unit must be designed for the loads including the gyroscopic loads arising from the conditions specified in §§25.331, 25.341(a), 25.349, 25.351, 25.473, 25.479, and 25.481, with the engine or auxiliary power unit at maximum rpm appropriate to the condition. For the purposes of compliance with this section, the pitch maneuver in §25.331(c)(1) must be carried out until the positive limit maneuvering load factor (point $A_2$ in §25.333(b)) is reached.

[Docket No. 5066, 29 FR 18291, Dec. 24, 1964; as amended by Amdt. 25–86, 61 FR 5222, Feb. 9, 1996; Amdt. 25–91, 62 FR 40704, July 29, 1997]

## §25.373 Speed control devices.

If speed control devices (such as spoilers and drag flaps) are installed for use in en route conditions—

**(a)** The airplane must be designed for the symmetrical maneuvers prescribed in §25.333 and §25.337, the yawing maneuvers prescribed in §25.351, and the vertical and later gust conditions prescribed in §25.341(a), at each setting and the maximum speed associated with that setting; and

**(b)** If the device has automatic operating or load limiting features, the airplane must be designed for the maneuver and gust conditions prescribed in paragraph (a) of this section, at the speeds and corresponding device positions that the mechanism allows.

[Docket No. 5066, 29 FR 18291, Dec. 24, 1964; as amended by Amdt. 25–72, 55 FR 29776, July 20, 1990; Amdt. 25–86, 61 FR 5222, Feb. 9, 1996]

### Control Surface and System Loads

## §25.391 Control surface loads: general.

The control surfaces must be designed for the limit loads resulting from the flight conditions in §§25.331, 25.341(a), 25.349 and 25.351 and the ground gust conditions in §25.415, considering the requirements for—

**(a)** Loads parallel to hinge line, in §25.393;
**(b)** Pilot effort effects, in §25.397;
**(c)** Trim tab effects, in §25.407;
**(d)** Unsymmetrical loads, in §25.427; and
**(e)** Auxiliary aerodynamic surfaces, in §25.445.

[Docket No. 5066, 29 FR 18291, Dec. 24, 1964; as amended by Amdt. 25–86, 61 FR 5222, Feb. 9, 1996]

## §25.393 Loads parallel to hinge line.

**(a)** Control surfaces and supporting hinge brackets must be designed for inertia loads acting parallel to the hinge line.

**(b)** In the absence of more rational data, the inertia loads may be assumed to be equal to *KW*, where—

(1) K = 24 for vertical surfaces;
(2) K = 12 for horizontal surfaces; and
(3) W = weight of the movable surfaces.

## §25.395 Control system.

**(a)** Longitudinal, lateral, directional, and drag control system and their supporting structures must be designed for loads corresponding to 125 percent of the computed hinge moments of the movable control surface in the conditions prescribed in §25.391.

**(b)** The system limit loads, except the loads resulting from ground gusts, need not exceed the loads that can be produced by the pilot (or pilots) and by automatic or power devices operating the controls.

**(c)** The loads must not be less than those resulting from application of the minimum forces prescribed in §25.397(c).

[Docket No. 5066, 29 FR 18291, Dec. 24, 1964; as amended by Amdt. 25–23, 35 FR 5672, April 8, 1970; Amdt. 25–72, 55 FR 29776, July 20, 1990]

## §25.397 Control system loads.

**(a)** *General.* The maximum and minimum pilot forces, specified in paragraph (c) of this section, are assumed to act at the appropriate control grips or pads (in a manner simulating flight conditions) and to be reacted at the attachment of the control system to the control surface horn.

**(b)** *Pilot effort effects.* In the control surface flight loading condition, the air loads on movable surfaces and the corresponding deflections need not exceed those that would result in flight from the application of any pilot force within the ranges specified in paragraph (c) of this section. Two-thirds of the maximum values specified for the aileron and elevator may be used if control surface hinge moments are based on reliable data. In applying this criterion, the effects of servo mechanisms, tabs, and automatic pilot systems, must be considered.

**(c)** *Limit pilot forces and torques.* The limit pilot forces and torques are as follows:

| Control | Maximum forces or torques | Minimum forces or torques |
|---|---|---|
| **Aileron:** | | |
| Stick | 100 lbs. | 40 lbs. |
| Wheel[1] | 80 D in.-lbs[2] | 40 D in.-lbs. |
| **Elevator:** | | |
| Stick | 250 lbs. | 100 lbs. |
| Wheel (symmetrical) | 300 lbs. | 100 lbs. |
| Wheel (unsymmetrical)[3] | — | 100 lbs. |
| **Rudder** | 300 lbs. | 130 lbs. |

[1] The critical parts of the aileron control system must be designed for a single tangential force with a limit value equal to 1.25 times the couple force determined from these criteria.

[2] D = wheel diameter (inches).

[3] The unsymmetrical forces must be applied at one of the normal handgrip points on the periphery of the control wheel.

[Docket 5066, 29 FR 18291, Dec. 24, 1964; as amended by Amdt. 25–38, 41 FR 55466, Dec. 20, 1976; Amdt. 25–72, 55 FR 29776, July 20, 1990]

## §25.399 Dual control system.

**(a)** Each dual control system must be designed for the pilots operating in opposition, using individual pilot forces not less than—

(1) 0.75 times those obtained under §25.395; or

(2) The minimum forces specified in §25.397(c).

**(b)** The control system must be designed for pilot forces applied in the same direction, using individual pilot forces not less than 0.75 times those obtained under §25.395.

## §25.405 Secondary control system.

Secondary controls, such as wheel brake, spoiler, and tab controls, must be designed for the maximum forces that a pilot is likely to apply to those controls. The following values may be used:

Pilot Control Force Limits (Secondary Controls)

| Control | Limit pilot forces |
|---|---|
| Miscellaneous:<br>* Crank, wheel or lever | $\left(\frac{1+R}{3}\right) \times 50$ lbs.<br>but not less than 50 lbs. nor more than 150 lbs. (R = radius). (Applicable to any angle within 20° of plane of control). |
| Twist | 133 in.-lbs. |
| Push-pull | To be chosen by applicant |

*Limited to flap, tab, stabilizer, spoiler, and landing gear operation controls.

## §25.407 Trim tab effects.

The effects of trim tabs on the control surface design conditions must be accounted for only where the surface loads are limited by maximum pilot effort. In these cases, the tabs are considered to be deflected in the direction that would assist the pilot, and the deflections are—

**(a)** For elevator trim tabs, those required to trim the airplane at any point within the positive portion of the pertinent flight envelope in §25.333(b), except as limited by the stops; and

**(b)** For aileron and rudder trim tabs, those required to trim the airplane in the critical unsymmetrical power and loading conditions, with appropriate allowance for rigging tolerances.

## §25.409 Tabs.

**(a)** *Trim tabs.* Trim tabs must be designed to withstand loads arising from all likely combinations of tab setting, primary control position, and airplane speed (obtainable without exceeding the flight load conditions prescribed for the airplane as a whole), when the effect of the tab is opposed by pilot effort forces up to those specified in §25.397(b).

**(b)** *Balancing tabs.* Balancing tabs must be designed for deflections consistent with the primary control surface loading conditions.

**(c)** *Servo tabs.* Servo tabs must be designed for deflections consistent with the primary control surface loading conditions obtainable within the pilot maneuvering effort, considering possible opposition from the trim tabs.

## §25.415 Ground gust conditions.

**(a)** The control system must be designed as follows for control surface loads due to ground gusts and taxiing downwind:

(1) The control system between the stops nearest the surfaces and the cockpit controls must be designed for loads corresponding to the limit hinge moments H of paragraph (a)(2) of this section. These loads need not exceed—

(i) The loads corresponding to the maximum pilot loads in §25.397(c) for each pilot alone; or

(ii) 0.75 times these maximum loads for each pilot when the pilot forces are applied in the same direction.

(2) The control system stops nearest the surfaces, the control system locks, and the parts of the systems (if any) between these stops and locks and the control surface horns, must be designed for limit hinge moments H, in foot pounds, obtained from the formula, $H = .0034KV^2cS$, where—

V = 65 (wind speed in knots)

K = limit hinge moment factor for ground gusts derived in paragraph (b) of this section.

c = mean chord of the control surface aft of the hinge line (ft);

S = area of the control surface aft of the hinge line (sq ft).

**(b)** The limit hinge moment factor K for ground gusts must be derived as follows:

| Surface | K | Position of controls |
|---|---|---|
| (a) Aileron | 0.75 | Control column locked or lashed in mid-position |
| (b) do | [11]±0.50 | Ailerons at full throw |
| (c) Elevator | [11]±0.75 | (c) Elevator full down |
| (d) do | [11]±0.75 | (d) Elevator full up |
| (e) Rudder | 0.75 | (e) Rudder in neutral |
| (f) do | 0.75 | (f) Rudder at full throw |

[1] A positive value of K indicates a moment tending to depress the surface, while a negative value of K indicates a moment tending to raise the surface.

[Docket No. 5066, 29 FR 18291, Dec. 24, 1964; as amended by Amdt. 25–72, 55 FR 29776, July 20, 1990; Amdt. 25–91, 62 FR 40705, July 29, 1997]

## §25.427 Unsymmetrical loads.

**(a)** In designing the airplane for lateral gust, yaw maneuver and roll maneuver conditions, account must be taken of unsymmetrical loads on the empennage arising from effects such as slipstream and aerodynamic interference with the wing, vertical fin and other aerodynamic surfaces.

**(b)** The horizontal tail must be assumed to be subjected to unsymmetrical loading conditions determined as follows:

(1) 100 percent of the maximum loading from the symmetrical maneuver conditions of §25.331 and the vertical gust conditions of §25.341(a) acting separately on the surface on one side of the plane of symmetry; and

(2) 80 percent of these loadings acting on the other side.

**(c)** For empennage arrangements where the horizontal tail surfaces have dihedral angles greater than plus or minus 10 degrees, or are supported by the vertical tail surfaces, the surfaces and the supporting structure must be designed for gust velocities specified in §25.341(a) acting in any orientation at right angles to the flight path.

**(d)** Unsymmetrical loading on the empennage arising from buffet conditions of §25.305(e) must be taken into account.

[Docket No. 5066, 29 FR 18291, Dec. 24, 1964; as amended by Amdt. 25–23, 35 FR 5673, April 8, 1970 Amdt. 25–86, 61 FR 5222, Feb. 9, 1996]

## §25.445 Auxiliary aerodynamic surfaces.

**(a)** When significant, the aerodynamic influence between auxiliary aerodynamic surfaces, such as outboard fins and winglets, and their supporting aerodynamic surfaces, must be taken into account for all loading conditions including pitch, roll, and yaw maneuvers, and gusts as specified in §25.341(a) acting at any orientation at right angles to the flight path.

**(b)** To provide for unsymmetrical loading when outboard fins extend above and below the horizontal surface, the critical vertical surface loading (load per unit area) determined under §25.391 must also be applied as follows:

(1) 100 percent to the area of the vertical surfaces above (or below) the horizontal surface.

(2) 80 percent to the area below (or above) the horizontal surface.

[Docket No. 5066, 29 FR 18291, Dec. 24, 1964; as amended by Amdt. 25–86, 61 FR 5222, Feb. 9, 1996]

## §25.457 Wing flaps.

Wing flaps, their operating mechanisms, and their supporting structures must be designed for critical loads occurring in the conditions prescribed in §25.345, accounting for the loads occurring during transition from one flap position and airspeed to another.

## §25.459 Special devices.

The loading for special devices using aerodynamic surfaces (such as slots, slats and spoilers) must be determined from test data.

[Docket No. 5066, 29 FR 18291, Dec. 24, 1964; as amended by Amdt. 25–72, 55 FR 29776, July 20, 1990]

### Ground Loads

## §25.471 General.

**(a)** *Loads and equilibrium.* For limit ground loads—

(1) Limit ground loads obtained under this subpart are considered to be external forces applied to the airplane structure; and

(2) In each specified ground load condition, the external loads must be placed in equilibrium with the linear and angular inertia loads in a rational or conservative manner.

**(b)** *Critical centers of gravity.* The critical centers of gravity within the range for which certification is requested must be selected so that the maximum design loads are obtained in each landing gear element. Fore and aft, vertical, and lateral airplane centers of gravity must be considered. Lateral displacements of the c.g. from the airplane centerline which would result in main gear loads not greater than 103 percent of the critical design load for symmetrical loading conditions may be selected without considering the effects of these lateral c.g. displacements on the loading of the main gear elements, or on the airplane structure provided—

(1) The lateral displacement of the c.g. results from random passenger or cargo disposition within the fuselage or from random unsymmetrical fuel loading or fuel usage; and

(2) Appropriate loading instructions for random disposable loads are included under the provisions of §25.1583(c)(1) to ensure that the lateral displacement of the center of gravity is maintained within these limits.

**(c)** *Landing gear dimension data.* Figure 1 of Appendix A contains the basic landing gear dimension data.

[Docket No. 5066, 29 FR 18291, Dec. 24, 1964; as amended by Amdt. 25–23, 35 FR 5673, April 8, 1970]

## §25.473 Landing load conditions and assumptions.

**(a)** For the landing conditions specified in §25.479 to §25.485 the airplane is assumed to contact the ground—

(1) In the attitudes defined in §25.479 and §25.481;

(2) With a limit descent velocity of 10 fps at the design landing weight (the maximum weight for landing conditions at the maximum descent velocity); and

(3) With a limit descent velocity of 6 fps at the design takeoff weight (the maximum weight for landing conditions at a reduced descent velocity).

(4) The prescribed descent velocities may be modified if it is shown that the airplane has design features that make it impossible to develop these velocities.

**(b)** Airplane lift, not exceeding airplane weight, may be assumed unless the presence of systems or procedures significantly affects the lift.

**(c)** The method of analysis of airplane and landing gear loads must take into account at least the following elements:

(1) Landing gear dynamic characteristics.

(2) Spin-up and springback.

(3) Rigid body response.

(4) Structural dynamic response of the airframe, if significant.

**(d)** The landing gear dynamic characteristics must be validated by tests as defined in §25.723(a).

**(e)** The coefficient of friction between the tires and the ground may be established by considering the effects of skidding velocity and tire pressure. However, this coefficient of friction need not be more than 0.8.

[Docket No. 5066, 29 FR 18291, Dec. 24, 1964; as amended by Amdt. 25–23, 35 FR 5673, April 8, 1970; Amdt. 25–91, 62 FR 40705, July 29, 1997; Amdt. 25–103, 66 FR 27394, May 16, 2001]

## §25.477 Landing gear arrangement.

Sections 25.479 through 25.485 apply to airplanes with conventional arrangements of main and nose gears, or main and tail gears, when normal operating techniques are used.

## §25.479 Level landing conditions.

**(a)** In the level attitude, the airplane is assumed to contact the ground at forward velocity components, ranging from $V_{L1}$ to 1.25 $V_{L2}$ parallel to the ground under the conditions prescribed in §25.473 with—

(1) $V_{L1}$ equal to $V_{S0}$ (TAS) at the appropriate landing weight and in standard sea level conditions; and

(2) $V_{L2}$ equal to $V_{S0}$ (TAS) at the appropriate landing weight and altitudes in a hot day temperature of 41 degrees F above standard.

(3) The effects of increased contact speed must be investigated if approval of downwind landings exceeding 10 knots is requested.

**(b)** For the level landing attitude for airplanes with tail wheels, the conditions specified in this section must be investigated with the airplane horizontal reference line horizontal in accordance with Figure 2 of Appendix A of this part.

**(c)** For the level landing attitude for airplanes with nose wheels, shown in Figure 2 of Appendix A of this part, the conditions specified in this section must be investigated assuming the following attitudes:

(1) An attitude in which the main wheels are assumed to contact the ground with the nose wheel just clear of the ground; and

(2) If reasonably attainable at the specified descent and forward velocities, an attitude in which the nose and main wheels are assumed to contact the ground simultaneously.

**(d)** In addition to the loading conditions prescribed in paragraph (a) of this section, but with maximum vertical ground reactions calculated from paragraph (a), the following apply:

(1) The landing gear and directly affected attaching structure must be designed for the maximum vertical ground reaction combined with an aft acting drag component of not less than 25% of this maximum vertical ground reaction.

(2) The most severe combination of loads that are likely to arise during a lateral drift landing must be taken into account. In absence of a more rational analysis of this condition, the following must be investigated:

(i) A vertical load equal to 75% of the maximum ground reaction of §25.473 must be considered in combination with a drag and side load of 40% and 25% respectively of that vertical load.

(ii) The shock absorber and tire deflections must be assumed to be 75% of the deflection corresponding to the maximum ground reaction of §25.473(a)(2). This load case need not be considered in combination with flat tires.

(3) The combination of vertical and drag components is considered to be acting at the wheel axle centerline.

[Docket No. 5066, 29 FR 18291, Dec. 24, 1964; as amended by Amdt. 25–23, 35 FR 5673, April 8, 1970; Amdt. 25–91, 62 FR 40705, July 29, 1997; Amdt. 25–91, 62 FR 45481, Aug. 27, 1997]

## §25.481 Tail-down landing conditions.

**(a)** In the tail-down attitude, the airplane is assumed to contact the ground at forward velocity components, ranging from $V_{L1}$ to $V_{L2}$ parallel to the ground under the conditions prescribed in §25.473 with—

(1) $V_{L1}$ equal to $V_{S0}$ (TAS) at the appropriate landing weight and in standard sea level conditions; and

(2) $V_{L2}$ equal to $V_{S0}$ (TAS) at the appropriate landing weight and altitudes in a hot day temperature of 41 degrees F above standard.

(3) The combination of vertical and drag components considered to be acting at the main wheel axle centerline.

**(b)** For the tail-down landing condition for airplanes with tail wheels, the main and tail wheels are assumed to contact the ground simultaneously, in accordance with figure 3 of Appendix A. Ground reaction conditions on the tail wheel are assumed to act—

(1) Vertically; and

(2) Up and aft through the axle at 45 degrees to the ground line.

**(c)** For the tail-down landing condition for airplanes with nose wheels, the airplane is assumed to be at an attitude corresponding to either the stalling angle or the maximum angle allowing clearance with the ground by each part of the airplane other than the main wheels, in accordance with figure 3 of Appendix A, whichever is less.

[Docket No. 5066, 29 FR 18291, Dec. 24, 1964; as amended by Amdt. 25–91, 62 FR 40705, July 29, 1997; Amdt. 25–94, 63 FR 8848, Feb. 23, 1998]

## §25.483 One-gear landing conditions.

For the one-gear landing conditions, the airplane is assumed to be in the level attitude and to contact the ground on one main landing gear, in accordance with Figure 4 of Appendix A of this part. In this attitude—

**(a)** The ground reactions must be the same as those obtained on that side under §25.479(d)(1); and

**(b)** Each unbalanced external load must be reacted by airplane inertia in a rational or conservative manner.

[Docket No. 5066, 29 FR 18291, Dec. 24, 1964; as amended by Amdt. 25–91, 62 FR 40705, July 29, 1997]

## §25.485 Side load conditions.

In addition to §25.479(d)(2) the following conditions must be considered:

**(a)** For the side load condition, the airplane is assumed to be in the level attitude with only the main wheels contacting the ground, in accordance with figure 5 of Appendix A.

**(b)** Side loads of 0.8 of the vertical reaction (on one side) acting inward and 0.6 of the vertical reaction (on the other side) acting outward must be combined with one-half of the maximum vertical ground reactions obtained in the level landing conditions. These loads are assumed to be applied at the ground contact point and to be resisted by the inertia of the airplane. The drag loads may be assumed to be zero.

[Docket No. 5066, 29 FR 18291, Dec. 24, 1964; as amended by Amdt. 25–91, 62 FR 40705, July 29, 1997]

## §25.487 Rebound landing condition.

**(a)** The landing gear and its supporting structure must be investigated for the loads occurring during rebound of the airplane from the landing surface.

**(b)** With the landing gear fully extended and not in contact with the ground, a load factor of 20.0 must act on the unsprung weights of the landing gear. This load factor must act in the direction of motion of the unsprung weights as they reach their limiting positions in extending with relation to the sprung parts of the landing gear.

## §25.489 Ground handling conditions.

Unless otherwise prescribed, the landing gear and airplane structure must be investigated for the conditions in §§25.491 through 25.509 with the airplane at the design ramp weight (the maximum weight for ground handling conditions). No wing lift may be considered. The shock absorbers and tires may be assumed to be in their static position.

[Docket No. 5066, 29 FR 18291, Dec. 24, 1964; as amended by Amdt. 25–23, 35 FR 5673, April 8, 1970]

## §25.491 Taxi, takeoff and landing roll.

Within the range of appropriate ground speeds and approved weights, the airplane structure and landing gear are assumed to be subjected to loads not less than those obtained when the aircraft is operating over the roughest ground that may reasonably be expected in normal operation.

[Docket No. 5066, 29 FR 18291, Dec. 24, 1964; as amended by Amdt. 25–91, 62 FR 40705, July 29, 1997]

## §25.493 Braked roll conditions.

**(a)** An airplane with a tail wheel is assumed to be in the level attitude with the load on the main wheels, in accordance with figure 6 of Appendix A. The limit vertical load factor is 1.2 at the design landing weight and 1.0 at the design ramp weight. A drag reaction equal to the vertical reaction multiplied by a coefficient of friction of 0.8, must be combined with the vertical ground reaction and applied at the ground contact point.

**(b)** For an airplane with a nose wheel the limit vertical load factor is 1.2 at the design landing weight, and 1.0 at the design ramp weight. A drag reaction equal to the vertical reaction, multiplied by a coefficient of friction of 0.8, must be combined with the vertical reaction and applied at the ground contact point of each wheel with brakes. The following two attitudes, in accordance with figure 6 of Appendix A, must be considered:

(1) The level attitude with the wheels contacting the ground and the loads distributed between the main and nose gear. Zero pitching acceleration is assumed.

(2) The level attitude with only the main gear contacting the ground and with the pitching moment resisted by angular acceleration.

**(c)** A drag reaction lower than that prescribed in this section may be used if it is substantiated that an effective drag force of 0.8 times the vertical reaction cannot be attained under any likely loading condition.

**(d)** An airplane equipped with a nose gear must be designed to withstand the loads arising from the dynamic pitching motion of the airplane due to sudden application of maximum braking force. The airplane is considered to be at design takeoff weight with the nose and main gears in contact with the ground, and with a steady-state vertical load factor of 1.0. The steady-state nose gear reaction must be combined with the maximum incremental nose gear vertical reaction caused by the sudden application of maximum braking force as described in paragraphs (b) and (c) of this section.

**(e)** In the absence of a more rational analysis, the nose gear vertical reaction prescribed in paragraph (d) of this section must be calculated according to the following formula:

$$V_N = \frac{W_T}{A+B}\left[B + \frac{f\mu AE}{A+B+\mu E}\right]$$

Where:

$V_N$ = Nose gear vertical reaction.

$W_T$ = Design takeoff weight.

A = Horizontal distance between the c.g. of the airplane and the nose wheel.

B = Horizontal distance between the c.g. of the airplane and the line joining the centers of the main wheels.

E = Vertical height of the c.g. of the airplane above the ground in the 1.0 g static condition.

μ = Coefficient of friction of 0.80.

f = Dynamic response factor; 2.0 is to be used unless a lower factor is substantiated. In the absence of other information, the dynamic response factor f may be defined by the equation:

$$f = 1 + \exp\left(\frac{-\pi\xi}{\sqrt{1-\xi^2}}\right)$$

Where:

$\xi$ is the effective critical damping ratio of the rigid body pitching mode about the main landing gear effective ground contact point.

[Docket No. 5066, 29 FR 18291, Dec. 24, 1964; as amended by Amdt. 25–23, 35 FR 5673, April 8, 1970; Amdt. 25–97, 63 FR 29072, May 27, 1998]

## §25.495 Turning.

In the static position, in accordance with figure 7 of Appendix A, the airplane is assumed to execute a steady turn by nose gear steering, or by application of sufficient differential power, so that the limit load factors applied at the center of gravity are 1.0 vertically and 0.5 laterally. The side ground reaction of each wheel must be 0.5 of the vertical reaction.

## §25.497 Tail-wheel yawing.

**(a)** A vertical ground reaction equal to the static load on the tail wheel, in combination with a side component of equal magnitude, is assumed.

**(b)** If there is a swivel, the tail wheel is assumed to be swiveled 90° to the airplane longitudinal axis with the resultant load passing through the axle.

**(c)** If there is a lock, steering device, or shimmy damper the tail wheel is also assumed to be in the trailing position with the side load acting at the ground contact point.

## §25.499 Nose-wheel yaw and steering.

**(a)** A vertical load factor of 1.0 at the airplane center of gravity, and a side component at the nose wheel ground contact equal to 0.8 of the vertical ground reaction at that point are assumed.

**(b)** With the airplane assumed to be in static equilibrium with the loads resulting from the use of brakes on one side of the main landing gear, the nose gear, its attaching structure, and the fuselage structure forward of the center of gravity must be designed for the following loads:

(1) A vertical load factor at the center of gravity of 1.0.

(2) A forward acting load at the airplane center of gravity of 0.8 times the vertical load on one main gear.

(3) Side and vertical loads at the ground contact point on the nose gear that are required for static equilibrium.

(4) A side load factor at the airplane center of gravity of zero.

**(c)** If the loads prescribed in paragraph (b) of this section result in a nose gear side load higher than 0.8 times the vertical nose gear load, the design nose gear side load may be limited to 0.8 times the vertical load, with unbalanced yawing moments assumed to be resisted by airplane inertia forces.

**(d)** For other than the nose gear, its attaching structure, and the forward fuselage structure, the loading conditions are those prescribed in paragraph (b) of this section, except that—

(1) A lower drag reaction may be used if an effective drag force of 0.8 times the vertical reaction cannot be reached under any likely loading condition; and

(2) The forward acting load at the center of gravity need not exceed the maximum drag reaction on one main gear, determined in accordance with §25.493(b).

**(e)** With the airplane at design ramp weight, and the nose gear in any steerable position, the combined application of full normal steering torque and vertical force equal to 1.33 times the maximum static reaction on the nose gear must be considered in designing the nose gear, its attaching structure, and the forward fuselage structure.

[Docket No. 5066, 29 FR 18291, Dec. 24, 1964; as amended by Amdt. 25–23, 35 FR 5673, April 8, 1970; Amdt. 25–46, 43 FR 50595, Oct. 30, 1978; Amdt. 25–91, 62 FR 40705, July 29, 1997]

## §25.503 Pivoting.

**(a)** The airplane is assumed to pivot about one side of the main gear with the brakes on that side locked. The limit vertical load factor must be 1.0 and the coefficient of friction 0.8.

**(b)** The airplane is assumed to be in static equilibrium, with the loads being applied at the ground contact points, in accordance with figure 8 of Appendix A.



## §25.507 Reversed braking.

**(a)** The airplane must be in a three point static ground attitude. Horizontal reactions parallel to the ground and directed forward must be applied at the ground contact point of each wheel with brakes. The limit loads must be equal to 0.55 times the vertical load at each wheel or to the load developed by 1.2 times the nominal maximum static brake torque, whichever is less.

**(b)** For airplanes with nose wheels, the pitching moment must be balanced by rotational inertia.

**(c)** For airplanes with tail wheels, the resultant of the ground reactions must pass through the center of gravity of the airplane.

## §25.509 Towing loads.

**(a)** The towing loads specified in paragraph (d) of this section must be considered separately. These loads must be applied at the towing fittings and must act parallel to the ground. In addition—

(1) A vertical load factor equal to 1.0 must be considered acting at the center of gravity;

(2) The shock struts and tires must be in their static positions; and

(3) With $W_T$ as the design ramp weight, the towing load, $F_{TOW}$, is—

(i) 0.3 $W_T$ for $W_T$ less than 30,000 pounds;

(ii) (6 $W_T$ + 450,000)/7 for $W_T$ between 30,000 and 100,000 pounds; and

(iii) 0.15 $W_T$ for $W_T$ over 100,000 pounds.

**(b)** For towing points not on the landing gear but near the plane of symmetry of the airplane, the drag and side tow load components specified for the auxiliary gear apply. For towing points located outboard of the main gear, the drag and side tow load components specified for the main gear apply. Where the specified angle of swivel cannot be reached, the maximum obtainable angle must be used.

**(c)** The towing loads specified in paragraph (d) of this section must be reacted as follows:

(1) The side component of the towing load at the main gear must be reacted by a side force at the static ground line of the wheel to which the load is applied.

(2) The towing loads at the auxiliary gear and the drag components of the towing loads at the main gear must be reacted as follows:

(i) A reaction with a maximum value equal to the vertical reaction must be applied at the axle of the wheel to which the load is applied. Enough airplane inertia to achieve equilibrium must be applied.

(ii) The loads must be reacted by airplane inertia.

**(d)** The prescribed towing loads are as follows:

| Tow point | Position | Load | | |
|---|---|---|---|---|
| | | Magnitude | No. | Direction |
| Main gear | | 0.75 $F_{Tow}$ per main gear unit | 1 | Forward, parallel to drag axis |
| | | | 2 | Forward, at 30° to drag axis |
| | | | 3 | Aft, parallel to drag axis |
| | | | 4 | Aft, at 30° to drag axis |
| Auxiliary gear | Swiveled forward | 1.0 $F_{Tow}$ | 5 | Forward |
| | | | 6 | Aft |
| | Swiveled aft | do | 7 | Forward |
| | | | 8 | Aft |
| | Swiveled 45° from forward | 0.5 $F_{Tow}$ | 9 | Forward, in plane of wheel |
| | | | 10 | Aft, in plane of wheel |
| | Swiveled 45° from aft | do | 11 | Forward, in plane of wheel |
| | | | 12 | Aft, in plane of wheel |

[Docket No. 5066, 29 FR 18291, Dec. 24, 1964; as amended by Amdt. 25–23, 35 FR 5673, April 8, 1970]

## §25.511 Ground load: unsymmetrical loads on multiple-wheel units.

**(a)** *General.* Multiple-wheel landing gear units are assumed to be subjected to the limit ground loads prescribed in this subpart under paragraphs (b) through (f) of this section. In addition—

(1) A tandem strut gear arrangement is a multiple-wheel unit; and

(2) In determining the total load on a gear unit with respect to the provisions of paragraphs (b) through (f) of this section, the transverse shift in the load centroid, due to unsymmetrical load distribution on the wheels, may be neglected.

**(b)** *Distribution of limit loads to wheels*; tires inflated. The distribution of the limit loads among the wheels of the landing gear must be established for each landing, taxiing, and ground handling condition, taking into account the effects of the following factors:

(1) The number of wheels and their physical arrangements. For truck type landing gear units, the effects of any seesaw motion of the truck during the landing impact must be considered in determining the maximum design loads for the fore and aft wheel pairs.

(2) Any differentials in tire diameters resulting from a combination of manufacturing tolerances, tire growth, and tire wear. A maximum tire-diameter differential equal to 2/3 of the most unfavorable combination of diameter variations that is obtained when taking into account manufacturing tolerances, tire growth, and tire wear, may be assumed.

(3) Any unequal tire inflation pressure, assuming the maximum variation to be ±5 percent of the nominal tire inflation pressure.

(4) A runway crown of zero and a runway crown having a convex upward shape that may be approximated by a slope of $1\frac{1}{2}$ percent with the horizontal. Runway crown effects must be considered with the nose gear unit on either slope of the crown.

(5) The airplane attitude.

(6) Any structural deflections.

**(c)** *Deflated tires.* The effect of deflated tires on the structure must be considered with respect to the loading conditions specified in paragraphs (d) through (f) of this section, taking into account the physical arrangement of the gear components. In addition—

(1) The deflation of any one tire for each multiple wheel landing gear unit, and the deflation of any two critical tires for each landing gear unit using four or more wheels per unit, must be considered; and

(2) The ground reactions must be applied to the wheels with inflated tires except that, for multiple-wheel gear units with more than one shock strut, a rational distribution of the ground reactions between the deflated and inflated tires, accounting for the differences in shock strut extensions resulting from a deflated tire, may be used.

**(d)** *Landing conditions.* For one and for two deflated tires, the applied load to each gear unit is assumed to be 60 percent and 50 percent, respectively, of the limit load applied to each gear for each of the prescribed landing conditions. However, for the drift landing condition of §25.485, 100 percent of the vertical load must be applied.

**(e)** *Taxiing and ground handling conditions.* For one and for two deflated tires—

(1) The applied side or drag load factor, or both factors, at the center of gravity must be the most critical value up to 50 percent and 40 percent, respectively, of the limit side or drag load factors, or both factors, corresponding to the most severe condition resulting from consideration of the prescribed taxiing and ground handling conditions;

(2) For the braked roll conditions of §25.493 (a) and (b)(2), the drag loads on each inflated tire may not be less than those at each tire for the symmetrical load distribution with no deflated tires;

(3) The vertical load factor at the center of gravity must be 60 percent and 50 percent, respectively, of the factor with no deflated tires, except that it may not be less than 1g; and

(4) Pivoting need not be considered.

**(f)** *Towing conditions.* For one and for two deflated tires, the towing load, $F_{TOW}$, must be 60 percent and 50 percent, respectively, of the load prescribed.

## §25.519 Jacking and tie-down provisions.

**(a)** *General.* The airplane must be designed to withstand the limit load conditions resulting from the static ground load conditions of paragraph (b) of this section and, if applicable, paragraph (c) of this section at the most critical combinations of airplane weight and center of gravity. The maximum allowable load at each jack pad must be specified.

**(b)** *Jacking.* The airplane must have provisions for jacking and must withstand the following limit loads when the airplane is supported on jacks—

(1) For jacking by the landing gear at the maximum ramp weight of the airplane, the airplane structure must be designed for a vertical load of 1.33 times the vertical static reaction at each jacking point acting singly and in combination with a horizontal load of 0.33 times the vertical static reaction applied in any direction.

(2) For jacking by other airplane structure at maximum approved jacking weight:

(i) The airplane structure must be designed for a vertical load of 1.33 times the vertical reaction at each jacking point acting singly and in combination with a horizontal load of 0.33 times the vertical static reaction applied in any direction.

(ii) The jacking pads and local structure must be designed for a vertical load of 2.0 times the vertical static reaction at each jacking point, acting singly and in combination with a horizontal load of 0.33 times the vertical static reaction applied in any direction.

**(c)** *Tie-down.* If tie-down points are provided, the main tie-down points and local structure must withstand the limit loads resulting from a 65-knot horizontal wind from any direction.

[Docket No. 26129, 59 FR 22102, April 28, 1994]

### Water Loads

## §25.521 General.

**(a)** Seaplanes must be designed for the water loads developed during takeoff and landing, with the seaplane in any attitude likely to occur in normal operation, and at the appropriate forward and sinking velocities under the most severe sea conditions likely to be encountered.

**(b)** Unless a more rational analysis of the water loads is made, or the standards in ANC-3 are used, §§25.523 through 25.537 apply.

**(c)** The requirements of this section and §§25.523 through 25.537 apply also to amphibians.

## §25.523 Design weights and center of gravity positions.

**(a)** *Design weights.* The water load requirements must be met at each operating weight up to the design landing weight except that, for the takeoff condition prescribed in §25.531, the design water takeoff weight (the maximum weight for water taxi and takeoff run) must be used.

**(b)** *Center of gravity positions.* The critical centers of gravity within the limits for which certification is requested must be considered to reach maximum design loads for each part of the seaplane structure.

[Docket No. 5066, 29 FR 18291, Dec. 24, 1964; as amended by Amdt. 25–23, 35 FR 5673, April 8, 1970]

## §25.525 Application of loads.

**(a)** Unless otherwise prescribed, the seaplane as a whole is assumed to be subjected to the loads corresponding to the load factors specified in §25.527.

**(b)** In applying the loads resulting from the load factors prescribed in §25.527, the loads may be distributed over the hull or main float bottom (in order to avoid excessive local shear loads and bending moments at the location of water load application) using pressures not less than those prescribed in §25.533(b).

**(c)** For twin float seaplanes, each float must be treated as an equivalent hull on a fictitious seaplane with a weight equal to one-half the weight of the twin float seaplane.

**(d)** Except in the takeoff condition of §25.531, the aerodynamic lift on the seaplane during the impact is assumed to be $2/3$ of the weight of the seaplane.

## §25.527 Hull and main float load factors.

**(a)** Water reaction load factors $n_W$ must be computed in the following manner:

(1) For the step landing case

$$n_w = \frac{C_1 V_{SO}{}^2}{\left(\mathrm{Tan}^{2/3}\beta\right)W^{1/3}}$$

(2) For the bow and stern landing cases

$$n_w = \frac{C_1 V_{SO}{}^2}{\left(\mathrm{Tan}^{2/3}\beta\right)W^{1/3}} \times \frac{K_1}{\left(1 + r_x{}^2\right)^{2/3}}$$

**(b)** The following values are used:

(1) $n_W$ = water reaction load factor (that is, the water reaction divided by seaplane weight).

(2) $C_1$ = empirical seaplane operations factor equal to 0.012 (except that this factor may not be less than that necessary to obtain the minimum value of step load factor of 2.33).

(3) $V_{SO}$ = seaplane stalling speed in knots with flaps extended in the appropriate landing position and with no slipstream effect.

(4) $\beta$ = angle of dead rise at the longitudinal station at which the load factor is being determined in accordance with figure 1 of Appendix B.

(5) W = seaplane design landing weight in pounds.

(6) $K_1$ = empirical hull station weighing factor, in accordance with figure 2 of Appendix B.

(7) $r_x$ = ratio of distance, measured parallel to hull reference axis, from the center of gravity of the seaplane to the hull longitudinal station at which the load factor is being computed to the radius of gyration in pitch of the seaplane, the hull reference axis being a straight line, in the plane of symmetry, tangential to the keel at the main step.

**(c)** For a twin float seaplane, because of the effect of flexibility of the attachment of the floats to the seaplane, the factor $K_1$ may be reduced at the bow and stern to 0.8 of the value shown in figure 2 of Appendix B. This reduction applies only to the design of the carrythrough and seaplane structure.

[Docket No. 5066, 29 FR 18291, Dec. 24, 1964; as amended by Amdt. 25–23, 35 FR 5673, April 8, 1970]

## §25.529 Hull and main float landing conditions.

**(a)** *Symmetrical step, bow, and stern landing.* For symmetrical step, bow, and stern landings, the limit water reaction load factors are those computed under §25.527. In addition—

(1) For symmetrical step landings, the resultant water load must be applied at the keel, through the center of gravity, and must be directed perpendicularly to the keel line;

(2) For symmetrical bow landings, the resultant water load must be applied at the keel, one-fifth of the longitudinal distance from the bow to the step, and must be directed perpendicularly to the keel line; and

(3) For symmetrical stern landings, the resultant water load must be applied at the keel, at a point 85 percent of the longitudinal distance from the step to the stern post, and must be directed perpendicularly to the keel line.

**(b)** *Unsymmetrical landing for hull and single float seaplanes.* Unsymmetrical step, bow, and stern landing conditions must be investigated. In addition—

(1) The loading for each condition consists of an upward component and a side component equal, respectively, to 0.75 and 0.25 tan $\beta$ times the resultant load in the corresponding symmetrical landing condition; and

(2) The point of application and direction of the upward component of the load is the same as that in the symmetrical condition, and the point of application of the side component is at the same longitudinal station as the upward component but is directed inward perpendicularly to the plane of symmetry at a point midway between the keel and chine lines.

**(c)** *Unsymmetrical landing; twin float seaplanes.* The unsymmetrical loading consists of an upward load at the step of each float of 0.75 and a side load of 0.25 tan $\beta$ at one float times the step landing load reached under §25.527. The side load is directed inboard, perpendicularly to the plane of symmetry midway between the keel and chine lines of the float, at the same longitudinal station as the upward load.

## §25.531 Hull and main float takeoff condition.

For the wing and its attachment to the hull or main float—

**(a)** The aerodynamic wing lift is assumed to be zero; and

**(b)** A downward inertia load, corresponding to a load factor computed from the following formula, must be applied:

$$n = \frac{C_{TO} V_{S1}{}^2}{\left(\tan^{2/3}\beta\right)W^{1/3}}$$

where—

n = inertia load factor;

$C_{TO}$ = empirical seaplane operations factor equal to 0.004;

$V_{S1}$ = seaplane stalling speed (knots) at the design takeoff weight with the flaps extended in the appropriate takeoff position;

$\beta$ = angle of dead rise at the main step (degrees); and

W = design water takeoff weight in pounds.

[Docket No. 5066, 29 FR 18291, Dec. 24, 1964; as amended by Amdt. 25–23, 35 FR 5673, April 8, 1970]

## §25.533 Hull and main float bottom pressures.

**(a)** *General.* The hull and main float structure, including frames and bulkheads, stringers, and bottom plating, must be designed under this section.

**(b)** *Local pressures.* For the design of the bottom plating and stringers and their attachments to the supporting structure, the following pressure distributions must be applied:

(1) For an unflared bottom, the pressure at the chine is 0.75 times the pressure at the keel, and the pressures between the keel and chine vary linearly, in accordance with figure 3 of Appendix B. The pressure at the keel (psi) is computed as follows:

$$P_k = C_2 \times \frac{K_2 V_{S1}^{\ 2}}{\tan \beta_k}$$

where—

$P_k$ = pressure (p.s.i.) at the keel;

$C_2$ = 0.00213;

$K_2$ = hull station weighing factor, in accordance with figure 2 of Appendix B;

$V_{S1}$ = seaplane stalling speed (Knots) at the design water takeoff weight with flaps extended in the appropriate takeoff position; and

$\beta_k$ = angle of dead rise at keel, in accordance with figure 1 of Appendix B.

(2) For a flared bottom, the pressure at the beginning of the flare is the same as that for an unflared bottom, and the pressure between the chine and the beginning of the flare varies linearly, in accordance with figure 3 of Appendix B. The pressure distribution is the same as that prescribed in paragraph (b)(1) of this section for an unflared bottom except that the pressure at the chine is computed as follows:

$$P_{ch} = C_3 \times \frac{K_2 V_{S1}^{\ 2}}{\tan \beta}$$

where—

$P_{ch}$ = pressure (p.s.i.) at the chine;

$C_3$ = 0.0016;

$K_2$ = hull station weighing factor, in accordance with figure 2 of Appendix B;

$V_{S1}$ = seaplane stalling speed at the design water takeoff weight with flaps extended in the appropriate takeoff position; and

$\beta$ = angle of dead rise at appropriate station.

The area over which these pressures are applied must simulate pressures occurring during high localized impacts on the hull or float, but need not extend over an area that would induce critical stresses in the frames or in the overall structure.

**(c)** *Distributed pressures.* For the design of the frames, keel, and chine structure, the following pressure distributions apply:

(1) Symmetrical pressures are computed as follows:

$$P = C_4 \times \frac{K_2 V_{S0}^{\ 2}}{\tan \beta}$$

where—

P = pressure (p.s.i.);

$C_4$ = 0.078 $C_1$ (with $C_1$ computed under §25.527);

$K_2$ = hull station weighing factor, determined in accordance with figure 2 of Appendix B;

$V_{S0}$ = seaplane stalling speed (Knots) with landing flaps extended in the appropriate position and with no slipstream effect; and

$V_{S0}$ = seaplane stalling speed with landing flaps extended in the appropriate position and with no slipstream effect; and

$\beta$ = angle of dead rise at appropriate station.

(2) The unsymmetrical pressure distribution consists of the pressures prescribed in paragraph (c)(1) of this section on one side of the hull or main float centerline and one-half of that pressure on the other side of the hull or main float centerline, in accordance with figure 3 of Appendix B.

These pressures are uniform and must be applied simultaneously over the entire hull or main float bottom. The loads obtained must be carried into the sidewall structure of the hull proper, but need not be transmitted in a fore and aft direction as shear and bending loads.

[Docket No. 5066, 29 FR 18291, Dec. 24, 1964; as amended by Amdt. 25–23, 35 FR 5673, April 8, 1970]

25

## §25.535 Auxiliary float loads.

**(a)** *General.* Auxiliary floats and their attachments and supporting structures must be designed for the conditions prescribed in this section. In the cases specified in paragraphs (b) through (e) of this section, the prescribed water loads may be distributed over the float bottom to avoid excessive local loads, using bottom pressures not less than those prescribed in paragraph (g) of this section.

**(b)** *Step loading.* The resultant water load must be applied in the plane of symmetry of the float at a point three-fourths of the distance from the bow to the step and must be perpendicular to the keel. The resultant limit load is computed as follows, except that the value of L need not exceed three times the weight of the displaced water when the float is completely submerged:

$$L = \frac{C_5 V_{S0}{}^2 W^{2/3}}{\tan^{2/3} \beta_s \left(1 + r_y^2\right)^{2/3}}$$

where—

L = limit load (lbs.);

$C_5$ = 0.0053;

$V_{S0}$ = seaplane stalling speed (knots) with landing flaps extended in the appropriate position and with no slipstream effect;

W = seaplane design landing weight in pounds;

$\beta_S$ = angle of dead rise at a station ¾ of the distance from the bow to the step, but need not be less than 15 degrees; and

$r_y$ = ratio of the lateral distance between the center of gravity and the plane of symmetry of the float to the radius of gyration in roll.

**(c)** *Bow loading.* The resultant limit load must be applied in the plane of symmetry of the float at a point one-fourth of the distance from the bow to the step and must be perpendicular to the tangent to the keel line at that point. The magnitude of the resultant load is that specified in paragraph (b) of this section.

**(d)** *Unsymmetrical step loading.* The resultant water load consists of a component equal to 0.75 times the load specified in paragraph (a) of this section and a side component equal to 3.25 tan β times the load specified in paragraph (b) of this section. The side load must be applied perpendicularly to the plane of symmetry of the float at a point midway between the keel and the chine.

**(e)** *Unsymmetrical bow loading.* The resultant water load consists of a component equal to 0.75 times the load specified in paragraph (b) of this section and a side component equal to 0.25 tan β times the load specified in paragraph (c) of this section. The side load must be applied perpendicularly to the plane of symmetry at a point midway between the keel and the chine.

**(f)** *Immersed float condition.* The resultant load must be applied at the centroid of the cross section of the float at a point one-third of the distance from the bow to the step. The limit load components are as follows:

$$vertical = \rho g V_{S0}$$

$$aft = C_x \frac{\rho}{2} V^{2/3} (K V_{S0})^2$$

$$side = C_y \frac{\rho}{2} V^{2/3} (K V_{S0})^2$$

where—

ρ = mass density of water (slugs/ft.$^2$);

V = volume of float (ft.$^2$);

$C_x$ = coefficient of drag force, equal to 0.133;

$C_y$ = coefficient of side force, equal to 0.106;

K = 0.8, except that lower values may be used if it is shown that the floats are incapable of submerging at a speed of 0.8 $V_{S0}$ in normal operations;

$V_{S0}$ = seaplane stalling speed (knots) with landing flaps extended in the appropriate position and with no slipstream effect; and

g = acceleration due to gravity (ft./sec.$^2$).

**(g)** *Float bottom pressures.* The float bottom pressures must be established under §25.533, except that the value of $K_2$ in the formulae may be taken as 1.0. The angle of dead rise to be used in determining the float bottom pressures is set forth in paragraph (b) of this section.

[Docket No. 5066, 29 FR 18291, Dec. 24, 1964; as amended by Amdt. 25–23, 35 FR 5673, April 8, 1970]

## §25.537 Seawing loads.

Seawing design loads must be based on applicable test data.

### Emergency Landing Conditions

## §25.561 General.

**(a)** The airplane, although it may be damaged in emergency landing conditions on land or water, must be designed as prescribed in this section to protect each occupant under those conditions.

**(b)** The structure must be designed to give each occupant every reasonable chance of escaping serious injury in a minor crash landing when—

(1) Proper use is made of seats, belts, and all other safety design provisions;

(2) The wheels are retracted (where applicable); and

(3) The occupant experiences the following ultimate inertia forces acting separately relative to the surrounding structure:

(i) Upward, 3.0g

(ii) Forward, 9.0g

(iii) Sideward, 3.0g on the airframe; and 4.0g on the seats and their attachments.

(iv) Downward, 6.0g

(v) Rearward, 1.5g

**(c)** For equipment, cargo in the passenger compartments and any other large masses, the following apply:

(1) Except as provided in paragraph (c)(2) of this section, these items must be positioned so that if they break loose they will be unlikely to:

(i) Cause direct injury to occupants;

(ii) Penetrate fuel tanks or lines or cause fire or explosion hazard by damage to adjacent systems; or

(iii) Nullify any of the escape facilities provided for use after an emergency landing.

(2) When such positioning is not practical (e.g. fuselage mounted engines or auxiliary power units) each such item of mass shall be restrained under all loads up to those specified in paragraph (b)(3) of this section. The local attachments for these items should be designed to withstand 1.33 times the specified loads if these items are subject to severe wear and tear through frequent removal (e.g. quick change interior items).

**(d)** Seats and items of mass (and their supporting structure) must not deform under any loads up to those specified in paragraph (b)(3) of this section in any manner that would impede subsequent rapid evacuation of occupants.

[Docket No. 5066, 29 FR 18291, Dec. 24, 1964; as amended by Amdt. 25–23, 35 FR 5673, April 8, 1970; Amdt. 25–64, 53 FR 17646, May 17, 1988; Amdt. 25–91, 62 FR 40706, July 29, 1997]

## §25.562 Emergency landing dynamic conditions.

**(a)** The seat and restraint system in the airplane must be designed as prescribed in this section to protect each occupant during an emergency landing condition when—

(1) Proper use is made of seats, safety belts, and shoulder harnesses provided for in the design; and

(2) The occupant is exposed to loads resulting from the conditions prescribed in this section.

**(b)** Each seat type design approved for crew or passenger occupancy during takeoff and landing must successfully complete dynamic tests or be demonstrated by rational analysis based on dynamic tests of a similar type seat, in accordance with each of the following emergency landing conditions. The tests must be conducted with an occupant simulated by a 170-pound anthropomorphic test dummy, as defined by 49 CFR Part 572, Subpart B, or its equivalent, sitting in the normal upright position.

(1) A change in downward vertical velocity ($\Delta$v) of not less than 35 feet per second, with the airplane's longitudinal axis canted downward 30 degrees with respect to the horizontal plane and with the wings level. Peak floor deceleration must occur in not more than 0.08 seconds after impact and must reach a minimum of 14g.

(2) A change in forward longitudinal velocity ($\Delta$v) of not less than 44 feet per second, with the airplane's longitudinal axis horizontal and yawed 10 degrees either right or left, whichever would cause the greatest likelihood of the upper torso restraint system (where installed) moving off the occupant's shoulder, and with the wings level. Peak floor deceleration must occur in not more than 0.09 seconds after impact and must reach a minimum of 16g. Where floor rails or floor fittings are used to attach the seating devices to the test fixture, the rails or fittings must be misaligned with respect to the adjacent set of rails or fittings by at least 10 degrees vertically (i.e., out of Parallel) with one rolled 10 degrees.

**(c)** The following performance measures must not be exceeded during the dynamic tests conducted in accordance with paragraph (b) of this section:

(1) Where upper torso straps are used for crewmembers, tension loads in individual straps must not exceed 1,750 pounds. If dual straps are used for restraining the upper torso, the total strap tension loads must not exceed 2,000 pounds.

(2) The maximum compressive load measured between the pelvis and the lumbar column of the anthropomorphic dummy must not exceed 1,500 pounds.

(3) The upper torso restraint straps (where installed) must remain on the occupant's shoulder during the impact.

(4) The lap safety belt must remain on the occupant's pelvis during the impact.

(5) Each occupant must be protected from serious head injury under the conditions prescribed in paragraph (b) of this section. Where head contact with seats or other structure can occur, protection must be provided so that the head impact does not exceed a Head Injury Criterion (HIC) of 1,000 units. The level of HIC is defined by the equation:

$$HIC = \left\{ (t_2 - t_1) \left[ \frac{1}{(t_2 - t_1)} \int_{t_1}^{t_2} a(t)dt \right]^{2.5} \right\}_{max}$$

Where:

$t_1$ is the initial integration time,

$t_2$ is the final integration time, and

a(t) is the total acceleration vs. time curve for the head strike, and where

(t) is in seconds, and (a) is in units of gravity (g).

(6) Where leg injuries may result from contact with seats or other structure, protection must be provided to prevent axially compressive loads exceeding 2,250 pounds in each femur.

(7) The seat must remain attached at all points of attachment, although the structure may have yielded.

(8) Seats must not yield under the tests specified in paragraphs (b)(1) and (b)(2) of this section to the extent they would impede rapid evacuation of the airplane occupants.

[Docket No. 5066, 29 FR 18291, Dec. 24, 1964; as amended by Amdt. 25–64, 53 FR 17646, May 17, 1988]

## §25.563 Structural ditching provisions.

Structural strength considerations of ditching provisions must be in accordance with §25.801(e).

FATIGUE EVALUATION

## §25.571 Damage-tolerance and fatigue evaluation of structure.

**(a)** *General.* An evaluation of the strength, detail design, and fabrication must show that catastrophic failure due to fatigue, corrosion, manufacturing defects, or accidental damage, will be avoided throughout the operational life of the airplane. This evaluation must be conducted in accordance with the provisions of paragraphs (b) and (e) of this section, except as specified in paragraph (c) of this section, for each part of the structure that could contribute to a catastrophic failure (such as wing, empennage, control surfaces and their systems, the fuselage, engine mounting, landing gear, and their related primary attachments). For turbojet powered airplanes, those parts that could contribute to a catastrophic failure must also be evaluated under paragraph (d) of this section. In addition, the following apply:

(1) Each evaluation required by this section must include—

(i) The typical loading spectra, temperatures, and humidities expected in service;

(ii) The identification of principal structural elements and detail design points, the failure of which could cause catastrophic failure of the airplane; and

(iii) An analysis, supported by test evidence, of the principal structural elements and detail design points identified in paragraph (a)(1)(ii) of this section.

(2) The service history of airplanes of similar structural design, taking due account of differences in operating conditions and procedures, may be used in the evaluations required by this section.

(3) Based on the evaluations required by this section, inspections or other procedures must be established, as necessary, to prevent catastrophic failure, and must be included in the Airworthiness Limitations Section of the Instructions for Continued Airworthiness required by §25.1529. Inspection thresholds for the following types of structure must be established based on crack growth analyses and/or tests, assuming the structure contains an initial flaw of the maximum probable size that could exist as a result of manufacturing or service-induced damage:

(i) Single load path structure, and

(ii) Multiple load path "fail-safe" structure and crack arrest "fail-safe" structure, where it cannot be demonstrated that load path failure, partial failure, or crack arrest will be detected and repaired during normal maintenance, inspection, or operation of an airplane prior to failure of the remaining structure.

**(b)** *Damage-tolerance evaluation.* The evaluation must include a determination of the probable locations and modes of damage due to fatigue, corrosion, or accidental damage. Repeated load and static analyses supported by test evidence and (if available) service experience must also be incorporated in the evaluation. Special consideration for widespread fatigue damage must be included where the design is such that this type of damage could occur. It must be demonstrated with sufficient full-scale fatigue test evidence that widespread fatigue damage will not occur within the design service goal of the airplane. The type certificate may be issued prior to completion of full-scale fatigue testing, provided the Administrator has approved a plan for completing the required tests, and the airworthiness limitations section of the instructions for continued airworthiness required by §25.1529 of this part specifies that no airplane may be operated beyond a number of cycles equal to $\frac{1}{2}$ the number of cycles accumulated on the fatigue test article, until such testing is completed. The extent of damage for residual strength evaluation at any time within the operational life of the airplane must be consistent with the initial detectability and subsequent growth under repeated loads. The residual strength evaluation must show that the remaining structure is able to withstand loads (considered as static ultimate loads) corresponding to the following conditions:

(1) The limit symmetrical maneuvering conditions specified in §25.337 at all speeds up to $V_C$ and in §25.345.

(2) The limit gust conditions specified in §25.341 at the specified speeds up to $V_C$ and in §25.345.

(3) The limit rolling conditions specified in §25.349 and the limit unsymmetrical conditions specified in §§25.367 and 25.427 (a) through (c), at speeds up to $V_C$.

(4) The limit yaw maneuvering conditions specified in §25.351(a) at the specified speeds up to $V_C$.

(5) For pressurized cabins, the following conditions:

(i) The normal operating differential pressure combined with the expected external aerodynamic pressures applied simultaneously with the flight loading conditions specified in paragraphs (b) (1) through (4) of this section, if they have a significant effect.

(ii) The maximum value of normal operating differential pressure (including the expected external aerodynamic pressures during 1 g level flight) multiplied by a factor of 1.15, omitting other loads.

(6) For landing gear and directly-affected airframe structure, the limit ground loading conditions specified in §§25.473, 25.491, and 25.493.

If significant changes in structural stiffness or geometry, or both, follow from a structural failure, or partial failure, the effect on damage tolerance must be further investigated.

**(c)** *Fatigue (safe-life) evaluation.* Compliance with the damage-tolerance requirements of paragraph (b) of this section is not required if the applicant establishes that their application for particular structure is impractical. This structure must be shown by analysis, supported by test evidence, to be able to withstand the repeated loads of variable magnitude expected during its service life without detectable cracks. Appropriate safe-life scatter factors must be applied.

**(d)** *Sonic fatigue strength.* It must be shown by analysis, supported by test evidence, or by the service history of airplanes of similar structural design and sonic excitation environment, that—

(1) Sonic fatigue cracks are not probable in any part of the flight structure subject to sonic excitation; or

(2) Catastrophic failure caused by sonic cracks is not probable assuming that the loads prescribed in paragraph (b) of this section are applied to all areas affected by those cracks.

**(e)** *Damage-tolerance (discrete source) evaluation.* The airplane must be capable of successfully completing a flight during which likely structural damage occurs as a result of—

(1) Impact with a 4-pound bird when the velocity of the airplane relative to the bird along the airplane's flight path is equal to $V_C$ at sea level or $0.85V_C$ at 8,000 feet, whichever is more critical;

(2) Uncontained fan blade impact;

(3) Uncontained engine failure; or

(4) Uncontained high energy rotating machinery failure.

The damaged structure must be able to withstand the static loads (considered as ultimate loads) which are reasonably expected to occur on the flight. Dynamic effects on these static loads need not be considered. Corrective action to be taken by the pilot following the incident, such as limiting maneuvers, avoiding turbulence, and reducing speed, must be considered. If significant changes in structural stiffness or geometry, or both, follow from a structural failure or partial failure, the effect on damage tolerance must be further investigated.

[Docket No. 5066, 29 FR 18291, Dec. 24, 1964; as amended by Amdt. 25–45, 43 FR 46242, Oct. 5, 1978; Amdt. 25–54, 45 FR 60173, Sept. 11, 1980; Amdt. 25–72, 55 FR 29776, July 20, 1990; Amdt. 25–86, 61 FR 5222, Feb. 9, 1996; Amdt. 25–96, 63 FR 15714, March 31, 1998; Amdt. 25–92, 63 FR 23338, April 28, 1998]

LIGHTNING PROTECTION

## §25.581 Lightning protection.

**(a)** The airplane must be protected against catastrophic effects from lightning.

**(b)** For metallic components, compliance with paragraph (a) of this section may be shown by—

(1) Bonding the components properly to the airframe; or

(2) Designing the components so that a strike will not endanger the airplane.

**(c)** For nonmetallic components, compliance with paragraph (a) of this section may be shown by—

(1) Designing the components to minimize the effect of a strike; or

(2) Incorporating acceptable means of diverting the resulting electrical current so as not to endanger the airplane.

[Docket No. 5066, 29 FR 18291, Dec. 24, 1964; as amended by Amdt. 25–23, 35 FR 5674, April 8, 1970]

# Subpart D— Design and Construction

GENERAL

## §25.601 General.

The airplane may not have design features or details that experience has shown to be hazardous or unreliable. The suitability of each questionable design detail and part must be established by tests.

## §25.603 Materials.

The suitability and durability of materials used for parts, the failure of which could adversely affect safety, must—

**(a)** Be established on the basis of experience or tests;

**(b)** Conform to approved specifications (such as industry or military specifications, or Technical Standard Orders) that ensure their having the strength and other properties assumed in the design data; and

**(c)** Take into account the effects of environmental conditions, such as temperature and humidity, expected in service.

[Docket No. 5066, 29 FR 18291, Dec. 24, 1964; as amended by Amdt. 25–38, 41 FR 55466, Dec. 20 1976; Amdt. 25–46, 43 FR 50595, Oct. 30, 1978]

## §25.605 Fabrication methods.

**(a)** The methods of fabrication used must produce a consistently sound structure. If a fabrication process (such as gluing, spot welding, or heat treating) requires close control to reach this objective, the process must be performed under an approved process specification.

**(b)** Each new aircraft fabrication method must be substantiated by a test program.

[Docket No. 5066, 29 FR 18291, Dec. 24, 1964; as amended by Amdt. 25–46, 43 FR 50595, Oct. 30, 1978]

## §25.607 Fasteners.

**(a)** Each removable bolt, screw, nut, pin, or other removable fastener must incorporate two separate locking devices if—

(1) Its loss could preclude continued flight and landing within the design limitations of the airplane using normal pilot skill and strength; or

(2) Its loss could result in reduction in pitch, yaw, or roll control capability or response below that required by Subpart B of this chapter.

**(b)** The fasteners specified in paragraph (a) of this section and their locking devices may not be adversely affected by the environmental conditions associated with the particular installation.

**(c)** No self-locking nut may be used on any bolt subject to rotation in operation unless a nonfriction locking device is used in addition to the self-locking device.

[Docket No. 5066, 29 FR 18291, Dec. 24, 1964; as amended by Amdt. 25–23, 35 FR 5674, April 8, 1970]

## §25.609 Protection of structure.

Each part of the structure must—

**(a)** Be suitably protected against deterioration or loss of strength in service due to any cause, including—

(1) Weathering;

(2) Corrosion; and

(3) Abrasion; and

**(b)** Have provisions for ventilation and drainage where necessary for protection.

## §25.611 Accessibility provisions.

Means must be provided to allow inspection (including inspection of principal structural elements and control systems), replacement of parts normally requiring replacement, adjustment, and lubrication as necessary for continued airworthiness. The inspection means for each item must be practicable for the inspection interval for the item. Nondestructive inspection aids may be used to inspect structural elements where it is impracticable to provide means for direct visual inspection if it is shown that the inspection is effective and the inspection procedures are specified in the maintenance manual required by §25.1529.

[Docket No. 5066, 29 FR 18291, Dec. 24, 1964; as amended by Amdt. 25–23, 35 FR 5674, April 8, 1970]

## §25.613 Material strength properties and design values.

**(a)** Material strength properties must be based on enough tests of material meeting approved specifications to establish design values on a statistical basis.

**(b)** Design values must be chosen to minimize the probability of structural failures due to material variability. Except as provided in paragraph (e) of this section, compliance with this paragraph must be shown by selecting design values which assure material strength with the following probability:

(1) Where applied loads are eventually distributed through a single member within an assembly, the failure of which would result in loss of structural integrity of the component, 99 percent probability with 95 percent confidence.

(2) For redundant structure, in which the failure of individual elements would result in applied loads being safely distributed to other load carrying members, 90 percent probability with 95 percent confidence.

**(c)** The effects of temperature on allowable stresses used for design in an essential component or structure must be considered where thermal effects are significant under normal operating conditions.

**(d)** The strength, detail design, and fabrication of the structure must minimize the probability of disastrous fatigue failure, particularly at points of stress concentration.

**(e)** Greater design values may be used if a "premium selection" of the material is made in which a specimen of each individual item is tested before use to determine that the actual strength properties of that particular item will equal or exceed those used in design.

[Docket No. 5066, 29 FR 18291, Dec. 24, 1964; as amended by Amdt. 25–46, 43 FR 50595, Oct. 30, 1978; Amdt. 25–72, 55 FR 29776, July 20, 1990]

## §25.619 Special factors.

The factor of safety prescribed in §25.303 must be multiplied by the highest pertinent special factor of safety prescribed in §§25.621 through 25.625 for each part of the structure whose strength is—

**(a)** Uncertain;

**(b)** Likely to deteriorate in service before normal replacement; or

**(c)** Subject to appreciable variability because of uncertainties in manufacturing processes or inspection methods.

[Docket No. 5066, 29 FR 18291, Dec. 24, 1964; as amended by Amdt. 25–23, 35 FR 5674, April 8, 1970]

## §25.621 Casting factors.

**(a)** *General.* The factors, tests, and inspections specified in paragraphs (b) through (d) of this section must be applied in addition to those necessary to establish foundry quality control. The inspections must meet approved specifications. Paragraphs (c) and (d) of this section apply to any structural castings except castings that are pressure tested as parts of hydraulic or other fluid systems and do not support structural loads.

**(b)** *Bearing stresses and surfaces.* The casting factors specified in paragraphs (c) and (d) of this section—

(1) Need not exceed 1.25 with respect to bearing stresses regardless of the method of inspection used; and

(2) Need not be used with respect to the bearing surfaces of a part whose bearing factor is larger than the applicable casting factor.

**(c)** *Critical castings.* For each casting whose failure would preclude continued safe flight and landing of the airplane or result in serious injury to occupants, the following apply:

(1) Each critical casting must—

(i) Have a casting factor of not less than 1.25; and

(ii) Receive 100 percent inspection by visual, radiographic, and magnetic particle or penetrant inspection methods or approved equivalent nondestructive inspection methods.

(2) For each critical casting with a casting factor less than 1.50, three sample castings must be static tested and shown to meet—

(i) The strength requirements of §25.305 at an ultimate load corresponding to a casting factor of 1.25; and

(ii) The deformation requirements of §25.305 at a load of 1.15 times the limit load.

(3) Examples of these castings are structural attachment fittings, parts of flight control systems, control surface hinges and balance weight attachments, seat, berth, safety belt, and fuel and oil tank supports and attachments, and cabin pressure valves.

**(d)** *Noncritical castings.* For each casting other than those specified in paragraph (c) of this section, the following apply:

(1) Except as provided in paragraphs (d)(2) and (3) of this section, the casting factors and corresponding inspections must meet the following table:

| Casting factor | Inspection |
|---|---|
| 2.0 or more | 100 percent visual |
| Less than 2.0 but more than 1.5 | 100 percent visual, and magnetic particle or penetrant or equivalent nondestructive inspection methods |
| 1.25 through 1.50 | 100 percent visual, magnetic particle or penetrant, and radiographic, or approved equivalent nondestructive inspection methods |

(2) The percentage of castings inspected by nonvisual methods may be reduced below that specified in paragraph (d)(1) of this section when an approved quality control procedure is established.

(3) For castings procured to a specification that guarantees the mechanical properties of the material in the casting and provides for demonstration of these properties by test of coupons cut from the castings on a sampling basis—

(i) A casting factor of 1.0 may be used; and

(ii) The castings must be inspected as provided in paragraph (d)(1) of this section for casting factors of "1.25 through 1.50" and tested under paragraph (c)(2) of this section.

## §25.623 Bearing factors.

**(a)** Except as provided in paragraph (b) of this section, each part that has clearance (free fit), and that is subject to pounding or vibration, must have a bearing factor large enough to provide for the effects of normal relative motion.

**(b)** No bearing factor need be used for a part for which any larger special factor is prescribed.

## §25.625 Fitting factors.

For each fitting (a part or terminal used to join one structural member to another), the following apply:

**(a)** For each fitting whose strength is not proven by limit and ultimate load tests in which actual stress conditions are simulated in the fitting and surrounding structures, a fitting factor of at least 1.15 must be applied to each part of—

(1) The fitting;

(2) The means of attachment; and

(3) The bearing on the joined members.

**(b)** No fitting factor need be used—

(1) For joints made under approved practices and based on comprehensive test data (such as continuous joints in metal plating, welded joints, and scarf joints in wood); or

(2) With respect to any bearing surface for which a larger special factor is used.

**(c)** For each integral fitting, the part must be treated as a fitting up to the point at which the section properties become typical of the member.

**(d)** For each seat, berth, safety belt, and harness, the fitting factor specified in §25.785(f)(3) applies.

[Docket No. 5066, 29 FR 18291, Dec. 24, 1964; as amended by Amdt. 25–23, 35 FR 5674, April 8, 1970; Amdt. 25–72, 55 FR 29776, July 20, 1990]

## §25.629 Aeroelastic stability requirements.

**(a)** *General.* The aeroelastic stability evaluations required under this section include flutter, divergence, control reversal and any undue loss of stability and control as a result of structural deformation. The aeroelastic evaluation must include whirl modes associated with any propeller or rotating device that contributes significant dynamic forces. Compliance with this section must be shown by analyses, wind tunnel tests, ground vibration tests, flight tests, or other means found necessary by the Administrator.

**(b)** *Aeroelastic stability envelopes.* The airplane must be designed to be free from aeroelastic instability for all configurations and design conditions within the aeroelastic stability envelopes as follows:

(1) For normal conditions without failures, malfunctions, or adverse conditions, all combinations of altitudes and speeds encompassed by the $V_D/M_D$ versus altitude envelope enlarged at all points by an increase of 15 percent in equivalent airspeed at both constant Mach number and constant altitude. In addition, a proper margin of stability must exist at all speeds up to $V_D/M_D$ and, there must be no large and rapid reduction in stability as $V_D/M_D$ is approached. The enlarged envelope may be limited to Mach 1.0 when $M_D$ is less than 1.0 at all design altitudes, and

(2) For the conditions described in §25.629(d) below, for all approved altitudes, any airspeed up to the greater airspeed defined by;

(i) The $V_D/M_D$ envelope determined by §25.335(b); or

(ii) An altitude-airspeed envelope defined by a 15-percent increase in equivalent airspeed above $V_C$ at constant altitude, from sea level to the altitude of the intersection of 1.15 $V_C$ with the extension of the constant cruise Mach number line, $M_C$, then a linear variation in equivalent airspeed to $M_C$+.05 at the altitude of the lowest $V_C/M_C$ intersection; then, at higher altitudes, up to the maximum flight altitude, the boundary defined by a .05 Mach increase in $M_C$ at constant altitude.

**(c)** *Balance weights.* If concentrated balance weights are used, their effectiveness and strength, including supporting structure, must be substantiated.

**(d)** *Failures, malfunctions, and adverse conditions.* The failures, malfunctions, and adverse conditions which must be considered in showing compliance with this section are:

(1) Any critical fuel loading conditions, not shown to be extremely improbable, which may result from mismanagement of fuel.

(2) Any single failure in any flutter damper system.

(3) For airplanes not approved for operation in icing conditions, the maximum likely ice accumulation expected as a result of an inadvertent encounter.

(4) Failure of any single element of the structure supporting any engine, independently mounted propeller shaft, large auxiliary power unit, or large externally mounted aerodynamic body (such as an external fuel tank).

(5) For airplanes with engines that have propellers or large rotating devices capable of significant dynamic forces, any single failure of the engine structure that would reduce the rigidity of the rotational axis.

(6) The absence of aerodynamic or gyroscopic forces resulting from the most adverse combination of feathered propellers or other rotating devices capable of significant dynamic forces. In addition, the effect of a single feathered propeller or rotating device must be coupled with the failures of paragraphs (d)(4) and (d)(5) of this section.

(7) Any single propeller or rotating device capable of significant dynamic forces rotating at the highest likely overspeed.

(8) Any damage or failure condition, required or selected for investigation by §25.571. The single structural failures described in paragraphs (d)(4) and (d)(5) of this section need not be considered in showing compliance with this section if:

(i) The structural element could not fail due to discrete source damage resulting from the conditions described in §25.571(e), and

(ii) A damage tolerance investigation in accordance with §25.571(b) shows that the maximum extent of damage assumed for the purpose of residual strength evaluation does not involve complete failure of the structural element.

(9) Any damage, failure, or malfunction considered under §§25.631, 25.671, 25.672, and 25.1309.

(10) Any other combination of failures, malfunctions, or adverse conditions not shown to be extremely improbable.

**(e)** *Flight flutter testing.* Full scale flight flutter tests at speeds up to $V_{DF}/M_{DF}$ must be conducted for new type designs and for modifications to a type design unless the modifications have been shown to have an insignificant effect on the aeroelastic stability. These tests must demonstrate that the airplane has a proper margin of damping at all speeds up to $V_{DF}/M_{DF}$, and that there is no large and rapid reduction in damping as $V_{DF}/M_{DF}$, is approached. If a failure, malfunction, or adverse condition is simulated during flight test in showing compliance with paragraph (d) of this section, the maximum speed investigated need not exceed $V_{FC}/M_{FC}$ if it is shown, by correlation of the flight test data with other test data or analyses, that the airplane is free from any aeroelastic instability at all speeds within the altitude-airspeed envelope described in paragraph (b)(2) of this section.

[Docket No. 26007, 57 FR 28949, June 29, 1992]

## §25.631 Bird strike damage.

The empennage structure must be designed to assure capability of continued safe flight and landing of the airplane after impact with an 8-pound bird when the velocity of the airplane (relative to the bird along the airplane's flight path) is equal to $V_C$ at sea level, selected under §25.335(a). Compliance with this section by provision of redundant structure and protected location of control system elements or protective devices such as splitter plates or energy absorbing material is acceptable. Where compliance is shown by analysis, tests, or both, use of data on airplanes having similar structural design is acceptable.

[Docket No. 5066, 29 FR 18291, Dec. 24, 1964; as amended by Amdt. 25–23, 35 FR 5674, April 8, 1970]

### CONTROL SURFACES

## §25.651 Proof of strength.

**(a)** Limit load tests of control surfaces are required. These tests must include the horn or fitting to which the control system is attached.

**(b)** Compliance with the special factors requirements of §§25.619 through 25.625 and 25.657 for control surface hinges must be shown by analysis or individual load tests.

## §25.655 Installation.

**(a)** Movable tail surfaces must be installed so that there is no interference between any surfaces when one is held in its extreme position and the others are operated through their full angular movement.

**(b)** If an adjustable stabilizer is used, it must have stops that will limit its range of travel to the maximum for which the airplane is shown to meet the trim requirements of §25.161.

## §25.657 Hinges.

**(a)** For control surface hinges, including ball, roller, and self-lubricated bearing hinges, the approved rating of the bearing may not be exceeded. For nonstandard bearing hinge configurations, the rating must be established on the basis of experience or tests and, in the absence of a rational investigation, a factor of safety of not less than 6.67 must be used with respect to the ultimate bearing strength of the softest material used as a bearing.

**(b)** Hinges must have enough strength and rigidity for loads parallel to the hinge line.

[Docket No. 5066, 29 FR 18291, Dec. 24, 1964; as amended by Amdt. 25–23, 35 FR 5674, April 8, 1970]

### CONTROL SYSTEMS

## §25.671 General.

**(a)** Each control and control system must operate with the ease, smoothness, and positiveness appropriate to its function.

**(b)** Each element of each flight control system must be designed, or distinctively and permanently marked, to minimize the probability of incorrect assembly that could result in the malfunctioning of the system.

**(c)** The airplane must be shown by analysis, tests, or both, to be capable of continued safe flight and landing after any of the following failures or jamming in the flight control system and surfaces (including trim, lift, drag, and feel systems), within the normal flight envelope, without requiring exceptional piloting skill or strength. Probable malfunctions must have only minor effects on control system operation and must be capable of being readily counteracted by the pilot.

(1) Any single failure, excluding jamming (for example, disconnection or failure of mechanical elements, or structural failure of hydraulic components, such as actuators, control spool housing, and valves).

(2) Any combination of failures not shown to be extremely improbable, excluding jamming (for example, dual electrical or hydraulic system failures, or any single failure in combination with any probable hydraulic or electrical failure).

(3) Any jam in a control position normally encountered during takeoff, climb, cruise, normal turns, descent, and landing unless the jam is shown to be extremely improbable, or can be alleviated. A runaway of a flight control to an adverse position and jam must be accounted for if such runaway and subsequent jamming is not extremely improbable.

**(d)** The airplane must be designed so that it is controllable if all engines fail. Compliance with this requirement may be shown by analysis where that method has been shown to be reliable.

[Docket No. 5066, 29 FR 18291, Dec. 24, 1964; as amended by Amdt. 25–23, 35 FR 5674, April 8, 1970]

## §25.672 Stability augmentation and automatic and power-operated systems.

If the functioning of stability augmentation or other automatic or power-operated systems is necessary to show compliance with the flight characteristics requirements of this part, such systems must comply with §25.671 and the following:

**(a)** A warning which is clearly distinguishable to the pilot under expected flight conditions without requiring his attention must be provided for any failure in the stability augmentation system or in any other automatic or power-operated system which could result in an unsafe condition if the pilot were not aware of the failure. Warning systems must not activate the control systems.

**(b)** The design of the stability augmentation system or of any other automatic or power-operated system must permit initial counteraction of failures of the type specified in §25.671(c) without requiring exceptional pilot skill or strength, by either the deactivation of the system, or a failed portion thereof, or by overriding the failure by movement of the flight controls in the normal sense.

**(c)** It must be shown that after any single failure of the stability augmentation system or any other automatic or power-operated system—

(1) The airplane is safely controllable when the failure or malfunction occurs at any speed or altitude within the approved operating limitations that is critical for the type of failure being considered;

(2) The controllability and maneuverability requirements of this part are met within a practical operational flight envelope (for example, speed, altitude, normal acceleration, and airplane configurations) which is described in the Airplane Flight Manual; and

(3) The trim, stability, and stall characteristics are not impaired below a level needed to permit continued safe flight and landing.

[Docket No. 5066, 29 FR 18291, Dec. 24, 1964; as amended by Amdt. 25–23, 35 FR 5675 April 8, 1970]

## §25.675 Stops.

**(a)** Each control system must have stops that positively limit the range of motion of each movable aerodynamic surface controlled by the system.

**(b)** Each stop must be located so that wear, slackness, or take-up adjustments will not adversely affect the control characteristics of the airplane because of a change in the range of surface travel.

**(c)** Each stop must be able to withstand any loads corresponding to the design conditions for the control system.

[Docket No. 5066, 29 FR 18291, Dec. 24, 1964; as amended by Amdt. 25–38, 41 FR 55466, Dec. 20, 1976]

## §25.677 Trim systems.

**(a)** Trim controls must be designed to prevent inadvertent or abrupt operation and to operate in the plane, and with the sense of motion, of the airplane.

**(b)** There must be means adjacent to the trim control to indicate the direction of the control movement relative to the airplane motion. In addition, there must be clearly visible means to indicate the position of the trim device with respect to the range of adjustment.

**(c)** Trim control systems must be designed to prevent creeping in flight. Trim tab controls must be irreversible unless the tab is appropriately balanced and shown to be free from flutter.

**(d)** If an irreversible tab control system is used, the part from the tab to the attachment of the irreversible unit to the airplane structure must consist of a rigid connection.

[Docket No. 5066, 29 FR 18291, Dec. 24, 1964; as amended by Amdt. 25–23, 35 FR 5675, April 8, 1970]

## §25.679 Control system gust locks.

**(a)** There must be a device to prevent damage to the control surfaces (including tabs), and to the control system, from gusts striking the airplane while it is on the ground or water. If the device, when engaged, prevents normal operation of the control surfaces by the pilot, it must—

(1) Automatically disengage when the pilot operates the primary flight controls in a normal manner; or

(2) Limit the operation of the airplane so that the pilot receives unmistakable warning at the start of takeoff.

**(b)** The device must have means to preclude the possibility of it becoming inadvertently engaged in flight.

## §25.681 Limit load static tests.

**(a)** Compliance with the limit load requirements of this Part must be shown by tests in which—

(1) The direction of the test loads produces the most severe loading in the control system; and

(2) Each fitting, pulley, and bracket used in attaching the system to the main structure is included.

**(b)** Compliance must be shown (by analyses or individual load tests) with the special factor requirements for control system joints subject to angular motion.

## §25.683 Operation tests.

It must be shown by operation tests that when portions of the control system subject to pilot effort loads are loaded to 80 percent of the limit load specified for the system and the powered portions of the control system are loaded to the maximum load expected in normal operation, the system is free from—

**(a)** Jamming;

**(b)** Excessive friction; and

**(c)** Excessive deflection.

[Docket No. 5066, 29 FR 18291, Dec. 24, 1964; as amended by Amdt. 25–23, 35 FR 5675, April 8, 1970]

## §25.685 Control system details.

**(a)** Each detail of each control system must be designed and installed to prevent jamming, chafing, and interference from cargo, passengers, loose objects, or the freezing of moisture.

**(b)** There must be means in the cockpit to prevent the entry of foreign objects into places where they would jam the system.

**(c)** There must be means to prevent the slapping of cables or tubes against other parts.

**(d)** Sections 25.689 and 25.693 apply to cable systems and joints.

[Docket No. 5066, 29 FR 18291, Dec. 24, 1964; as amended by Amdt. 25–38, 41 FR 55466, Dec. 20, 1976]

## §25.689 Cable systems.

**(a)** Each cable, cable fitting, turnbuckle, splice, and pulley must be approved. In addition—

(1) No cable smaller than $\frac{1}{8}$ inch in diameter may be used in the aileron, elevator, or rudder systems; and

(2) Each cable system must be designed so that there will be no hazardous change in cable tension throughout the range of travel under operating conditions and temperature variations.

**(b)** Each kind and size of pulley must correspond to the cable with which it is used. Pulleys and sprockets must have closely fitted guards to prevent the cables and chains from being displaced or fouled. Each pulley must lie in the plane passing through the cable so that the cable does not rub against the pulley flange.

**(c)** Fairleads must be installed so that they do not cause a change in cable direction of more than three degrees.

**(d)** Clevis pins subject to load or motion and retained only by cotter pins may not be used in the control system.

**(e)** Turnbuckles must be attached to parts having angular motion in a manner that will positively prevent binding throughout the range of travel.

**(f)** There must be provisions for visual inspection of fairleads, pulleys, terminals, and turnbuckles.



## §25.693 Joints.

Control system joints (in push-pull systems) that are subject to angular motion, except those in ball and roller bearing systems, must have a special factor of safety of not less than 3.33 with respect to the ultimate bearing strength of the softest material used as a bearing. This factor may be reduced to 2.0 for joints in cable control systems. For ball or roller bearings, the approved ratings may not be exceeded.

[Docket No. 5066, 29 FR 18291, Dec. 24, 1964; as amended by Amdt. 25–72, 55 FR 29777, July 20, 1990]

## §25.697 Lift and drag devices, controls.

**(a)** Each lift device control must be designed so that the pilots can place the device in any takeoff, en route, approach, or landing position established under §25.101(d). Lift and drag devices must maintain the selected positions, except for movement produced by an automatic positioning or load limiting device, without further attention by the pilots.

**(b)** Each lift and drag device control must be designed and located to make inadvertent operation improbable. Lift and drag devices intended for ground operation only must have means to prevent the inadvertent operation of their controls in flight if that operation could be hazardous.

(c) The rate of motion of the surfaces in response to the operation of the control and the characteristics of the automatic positioning or load limiting device must give satisfactory flight and performance characteristics under steady or changing conditions of airspeed, engine power, and airplane attitude.

(d) The lift device control must be designed to retract the surfaces from the fully extended position, during steady flight at maximum continuous engine power at any speed below $V_F$ +9.0 (knots).

[Docket No. 5066, 29 FR 18291, Dec. 24, 1964; as amended by Amdt. 25–23, 35 FR 5675, April 8, 1970; Amdt. 25–46, 43 FR 50595, Oct. 30, 1978; Amdt. 25–57, 49 FR 6848, Feb. 23, 1984]

## §25.699 Lift and drag device indicator.

**(a)** There must be means to indicate to the pilots the position of each lift or drag device having a separate control in the cockpit to adjust its position. In addition, an indication of unsymmetrical operation or other malfunction in the lift or drag device systems must be provided when such indication is necessary to enable the pilots to prevent or counteract an unsafe flight or ground condition, considering the effects on flight characteristics and performance.

**(b)** There must be means to indicate to the pilots the takeoff, en route, approach, and landing lift device positions.

**(c)** If any extension of the lift and drag devices beyond the landing position is possible, the controls must be clearly marked to identify this range of extension.

[Docket No. 5066, 29 FR 18291, Dec. 24, 1964; as amended by Amdt. 25–23, 35 FR 5675, April 8, 1970]

## §25.701 Flap and slat interconnection.

**(a)** Unless the airplane has safe flight characteristics with the flaps or slats retracted on one side and extended on the other, the motion of flaps or slats on opposite sides of the plane of symmetry must be synchronized by a mechanical interconnection or approved equivalent means.

**(b)** If a wing flap or slat interconnection or equivalent means is used, it must be designed to account for the applicable unsymmetrical loads, including those resulting from flight with the engines on one side of the plane of symmetry inoperative and the remaining engines at takeoff power.

**(c)** For airplanes with flaps or slats that are not subjected to slipstream conditions, the structure must be designed for the loads imposed when the wing flaps or slats on one side are carrying the most severe load occurring in the prescribed symmetrical conditions and those on the other side are carrying not more than 80 percent of that load.

**(d)** The interconnection must be designed for the loads resulting when interconnected flap or slat surfaces on one side of the plane of symmetry are jammed and immovable while the surfaces on the other side are free to move and the full power of the surface actuating system is applied.

[Docket No. 5066, 29 FR 18291, Dec. 24, 1964; as amended by Amdt. 25–72, 55 FR 29777, July 20, 1990]

## §25.703 Takeoff warning system.

A takeoff warning system must be installed and must meet the following requirements:

**(a)** The system must provide to the pilots an aural warning that is automatically activated during the initial portion of the takeoff roll if the airplane is in a configuration, including any of the following, that would not allow a safe takeoff:

(1) The wing flaps or leading edge devices are not within the approved range of takeoff positions.

(2) Wing spoilers (except lateral control spoilers meeting the requirements of §25.671), speed brakes, or longitudinal trim devices are in a position that would not allow a safe takeoff.

**(b)** The warning required by paragraph (a) of this section must continue until—

(1) The configuration is changed to allow a safe takeoff;

(2) Action is taken by the pilot to terminate the takeoff roll;

(3) The airplane is rotated for takeoff; or

(4) The warning is manually deactivated by the pilot.

**(c)** The means used to activate the system must function properly throughout the ranges of takeoff weights, altitudes, and temperatures for which certification is requested.

[Docket No. 5066, 29 FR 18291, Dec. 24, 1964; as amended by Amdt. 25–42, 43 FR 2323, Jan. 16, 1978]

### LANDING GEAR

## §25.721 General.

**(a)** The main landing gear system must be designed so that if it fails due to overloads during takeoff and landing (assuming the overloads to act in the upward and aft directions), the failure mode is not likely to cause—

(1) For airplanes that have passenger seating configuration, excluding pilots seats, of nine seats or less, the spillage of enough fuel from any fuel system in the fuselage to constitute a fire hazard; and

(2) For airplanes that have a passenger seating configuration, excluding pilots seats, of 10 seats or more, the spillage of enough fuel from any part of the fuel system to constitute a fire hazard.

**(b)** Each airplane that has a passenger seating configuration excluding pilots seats, of 10 seats or more must be designed so that with the airplane under control it can be landed on a paved runway with any one or more landing gear legs not extended without sustaining a structural component failure that is likely to cause the spillage of enough fuel to constitute a fire hazard.

**(c)** Compliance with the provisions of this section may be shown by analysis or tests, or both.

[Docket No. 5066, 29 FR 18291, Dec. 24, 1964; as amended by Amdt. 25–32, 37 FR 3969, Feb. 24, 1972]

## §25.723 Shock absorption tests.

**(a)** The analytical representation of the landing gear dynamic characteristics that is used in determining the landing loads must be validated by energy absorption tests. A range of tests must be conducted to ensure that the analytical representation is valid for the design conditions specified in §25.473.

(1) The configurations subjected to energy absorption tests at limit design conditions must include at least the design landing weight or the design takeoff weight, whichever produces the greater value of landing impact energy.

(2) The test attitude of the landing gear unit and the application of appropriate drag loads during the test must simulate the airplane landing conditions in a manner consistent with the development of rational or conservative limit loads.

**(b)** The landing gear may not fail in a test, demonstrating its reserve energy absorption capacity, simulating a descent velocity of 12 f.p.s. at design landing weight, assuming airplane lift not greater than airplane weight acting during the landing impact.

**(c)** In lieu of the tests prescribed in this section, changes in previously approved design weights and minor changes in design may be substantiated by analyses based on previous tests conducted on the same basic landing gear system that has similar energy absorption characteristics.

[Docket No. FAA–1999–5835, 66 FR 27394, May 16, 2001]

## §25.725 [Reserved]

## §25.727 [Reserved]

## §25.729 Retracting mechanism.

**(a)** *General.* For airplanes with retractable landing gear, the following apply:

(1) The landing gear retracting mechanism, wheel well doors, and supporting structure, must be designed for—

(i) The loads occurring in the flight conditions when the gear is in the retracted position,

(ii) The combination of friction loads, inertia loads, brake torque loads, air loads, and gyroscopic loads resulting from the wheels rotating at a peripheral speed equal to 1.3 $V_S$ (with the flaps in takeoff position at design takeoff weight), occurring during retraction and extension at any airspeed up to 1.6 $V_{S1}$ (with the flaps in the approach position at design landing weight), and

(iii) Any load factor up to those specified in §25.345(a) for the flaps extended condition.

(2) Unless there are other means to decelerate the airplane in flight at this speed, the landing gear, the retracting mechanism, and the airplane structure (including wheel well doors) must be designed to withstand the flight loads occurring with the landing gear in the extended position at any speed up to 0.67 $V_C$.

(3) Landing gear doors, their operating mechanism, and their supporting structures must be designed for the yawing maneuvers prescribed for the airplane in addition to the conditions of airspeed and load factor prescribed in paragraphs (a)(1) and (2) of this section.

**(b)** *Landing gear lock.* There must be positive means to keep the landing gear extended, in flight and on the ground.

**(c)** *Emergency operation.* There must be an emergency means for extending the landing gear in the event of—

(1) Any reasonably probable failure in the normal retraction system; or

(2) The failure of any single source of hydraulic, electric, or equivalent energy supply.

**(d)** *Operation test.* The proper functioning of the retracting mechanism must be shown by operation tests.

**(e)** *Position indicator and warning device.* If a retractable landing gear is used, there must be a landing gear position indicator (as well as necessary switches to actuate the indicator) or other means to inform the pilot that the gear is secured in the extended (or retracted) position. This means must be designed as follows:

(1) If switches are used, they must be located and coupled to the landing gear mechanical systems in a manner that prevents an erroneous indication of "down and locked" if the landing gear is not in a fully extended position, or of "up and locked" if the landing gear is not in the fully retracted position. The switches may be located where they are operated by the actual landing gear locking latch or device.

(2) The flightcrew must be given an aural warning that functions continuously, or is periodically repeated, if a landing is attempted when the landing gear is not locked down.

(3) The warning must be given in sufficient time to allow the landing gear to be locked down or a go-around to be made.

(4) There must be a manual shut-off means readily available to the flightcrew for the warning required by paragraph (e)(2) of this section such that it could be operated instinctively, inadvertently, or by habitual reflexive action.

(5) The system used to generate the aural warning must be designed to eliminate false or inappropriate alerts.

(6) Failures of systems used to inhibit the landing gear aural warning, that would prevent the warning system from operating, must be improbable.

**(f)** *Protection of equipment in wheel wells.* Equipment that is essential to safe operation of the airplane and that is located in wheel wells must be protected from the damaging effects of—

(1) A bursting tire, unless it is shown that a tire cannot burst from overheat; and

(2) A loose tire tread, unless it is shown that a loose tire tread cannot cause damage.

[Docket No. 5066, 29 FR 18291, Dec. 24, 1964; as amended by Amdt. 25–23, 35 FR 5676, April 8, 1970; Amdt. 25–42, 43 FR 2323, Jan. 16, 1978; Amdt. 25–72, 55 FR 29777, July 20, 1990; Amdt. 25–75, 56 FR 63762, Dec. 5, 1991]

## §25.731 Wheels.

**(a)** Each main and nose wheel must be approved.

**(b)** The maximum static load rating of each wheel may not be less than the corresponding static ground reaction with—

(1) Design maximum weight; and

(2) Critical center of gravity.

**(c)** The maximum limit load rating of each wheel must equal or exceed the maximum radial limit load determined under the applicable ground load requirements of this part.

**(d)** *Overpressure burst prevention.* Means must be provided in each wheel to prevent wheel failure and tire burst that may result from excessive pressurization of the wheel and tire assembly.

**(e)** *Braked wheels.* Each braked wheel must meet the applicable requirements of §25.735.

[Docket No. 5066, 29 FR 18291, Dec. 24, 1964; as amended by Amdt. 25–72, 55 FR 29777, July 20, 1990; Amdt. 25–107, 67 FR 20420, April 24, 2002]

## §25.733 Tires.

**(a)** When a landing gear axle is fitted with a single wheel and tire assembly, the wheel must be fitted with a suitable tire of proper fit with a speed rating approved by the Administrator that is not exceeded under critical conditions and with a load rating approved by the Administrator that is not exceeded under—

(1) The loads on the main wheel tire, corresponding to the most critical combination of airplane weight (up to maximum weight) and center of gravity position, and

(2) The loads corresponding to the ground reactions in paragraph (b) of this section, on the nose wheel tire, except as provided in paragraphs (b)(2) and (b)(3) of this section.

**(b)** The applicable ground reactions for nose wheel tires are as follows:

(1) The static ground reaction for the tire corresponding to the most critical combination of airplane weight (up to maximum ramp weight) and center of gravity position with a force of 1.0g acting downward at the center of gravity. This load may not exceed the load rating of the tire.

(2) The ground reaction of the tire corresponding to the most critical combination of airplane weight (up to maximum landing weight) and center of gravity position combined with forces of 1.0g downward and 0.31g forward acting at the center of gravity. The reactions in this case must be distributed to the nose and main wheels by the principles of statics with a drag reaction equal to 0.31 times the vertical load at each wheel with brakes capable of producing this ground reaction. This nose tire load may not exceed 1.5 times the load rating of the tire.

(3) The ground reaction of the tire corresponding to the most critical combination of airplane weight (up to maximum ramp weight) and center of gravity position combined with forces of 1.0g downward and 0.20g forward acting at the center of gravity. The reactions in this case must be distributed to the nose and main wheels by the principles of statics with a drag reaction equal to 0.20 times the vertical load at each wheel with brakes capable of producing this ground reaction. This nose tire load may not exceed 1.5 times the load rating of the tire.

**(c)** When a landing gear axle is fitted with more than one wheel and tire assembly, such as dual or dual-tandem, each wheel must be fitted with a suitable tire of proper fit with a speed rating approved by the Administrator that is not exceeded under critical conditions, and with a load rating approved by the Administrator that is not exceeded by—

(1) The loads on each main wheel tire, corresponding to the most critical combination of airplane weight (up to maximum weight) and center of gravity position, when multiplied by a factor of 1.07; and

(2) Loads specified in paragraphs (a)(2), (b)(1), (b)(2), and (b)(3) of this section on each nose wheel tire.

**(d)** Each tire installed on a retractable landing gear system must, at the maximum size of the tire type expected in service, have a clearance to surrounding structure and systems that is adequate to prevent unintended contact between the tire and any part of the structure or systems.

**(e)** For an airplane with a maximum certificated takeoff weight of more than 75,000 pounds, tires mounted on braked wheels must be inflated with dry nitrogen or other gases shown to be inert so that the gas mixture in the tire does not contain oxygen in excess of 5 percent by volume, unless it can be shown that the tire liner material will not produce a volatile gas when heated or that means are provided to prevent tire temperatures from reaching unsafe levels.

[Docket No. 5066, 29 FR 18291, Dec. 24, 1964; as amended by Amdt. 25–48, 44 FR 68752, Nov. 29, 1979; Amdt. 25–72, 55 FR 29777, July 20, 1990; Amdt. 25–78, 58 FR 11781, Feb. 26, 1993]

## §25.735 Brakes and braking systems.

**(a)** *Approval.* Each assembly consisting of a wheel(s) and brake(s) must be approved.

**(b)** *Brake system capability.* The brake system, associated systems and components must be designed and constructed so that:

(1) If any electrical, pneumatic, hydraulic, or mechanical connecting or transmitting element fails, or if any single source of hydraulic or other brake operating energy supply is lost, it is possible to bring the airplane to rest with a braked roll stopping distance of not more than two times that obtained in determining the landing distance as prescribed in §25.125.

(2) Fluid lost from a brake hydraulic system following a failure in, or in the vicinity of, the brakes is insufficient to cause or support a hazardous fire on the ground or in flight.

**(c)** *Brake controls.* The brake controls must be designed and constructed so that:

(1) Excessive control force is not required for their operation.

(2) If an automatic braking system is installed, means are provided to:

(i) Arm and disarm the system, and

(ii) Allow the pilot(s) to override the system by use of manual braking.

**(d)** *Parking brake.* The airplane must have a parking brake control that, when selected on, will, without further attention, prevent the airplane from rolling on a dry and level paved runway when the most adverse combination of maximum thrust on one engine and up to maximum ground idle thrust on any, or all, other engine(s) is applied. The control must be suitably located or be adequately protected to prevent inadvertent operation. There must be indication in the cockpit when the parking brake is not fully released.

**(e)** *Antiskid system.* If an antiskid system is installed:

(1) It must operate satisfactorily over the range of expected runway conditions, without external adjustment.

(2) It must, at all times, have priority over the automatic braking system, if installed.

**(f)** *Kinetic energy capacity—*

(1) Design landing stop. The design landing stop is an operational landing stop at maximum landing weight. The design landing stop brake kinetic energy absorption requirement of each wheel, brake, and tire assembly must be determined. It must be substantiated by dynamometer testing that the wheel, brake and tire assembly is capable of absorbing not less than this level of kinetic energy throughout the defined wear range of the brake. The energy absorption rate derived from the airplane manufacturer's braking requirements must be achieved. The mean deceleration must not be less than 10 fps$^2$.

(2) Maximum kinetic energy accelerate-stop. The maximum kinetic energy accelerate-stop is a rejected takeoff for the most critical combination of airplane takeoff weight and speed. The accelerate-stop brake kinetic energy absorption requirement of each wheel, brake, and tire assembly must be determined. It must be substantiated by dynamometer testing that the wheel, brake, and tire assembly is capable of absorbing not less than this level of kinetic energy throughout the defined wear range of the brake. The energy absorption rate derived from the airplane manufacturer's braking requirements must be achieved. The mean deceleration must not be less than 6 fps$^2$.

(3) Most severe landing stop. The most severe landing stop is a stop at the most critical combination of airplane landing weight and speed. The most severe landing stop brake kinetic energy absorption requirement of each wheel, brake, and tire assembly must be determined. It must be substantiated by dynamometer testing that, at the declared fully worn limit(s) of the brake heat sink, the wheel, brake and tire assembly is capable of absorbing not less than this level of kinetic energy. The most severe landing stop need not be considered for extremely improbable failure conditions or if the maximum kinetic energy accelerate-stop energy is more severe.

**(g)** *Brake condition after high kinetic energy dynamometer stop(s).* Following the high kinetic energy stop demonstration(s) required by paragraph (f) of this section, with the parking brake promptly and fully applied for at least 3 minutes, it must be demonstrated that for at least 5 minutes from application of the parking brake, no condition occurs (or has occurred during the stop), including fire associated with the tire or wheel and brake assembly, that could prejudice the safe and complete evacuation of the airplane.

**(h)** *Stored energy systems.* An indication to the flightcrew of the usable stored energy must be provided if a stored energy system is used to show compliance with paragraph (b)(1) of this section. The available stored energy must be sufficient for:

(1) At least 6 full applications of the brakes when an antiskid system is not operating; and

(2) Bringing the airplane to a complete stop when an antiskid system is operating, under all runway surface conditions for which the airplane is certificated.

**(i)** *Brake wear indicators.* Means must be provided for each brake assembly to indicate when the heat sink is worn to the permissible limit. The means must be reliable and readily visible.

**(j)** *Overtemperature burst prevention.* Means must be provided in each braked wheel to prevent a wheel failure, a tire burst, or both, that may result from elevated brake temperatures. Additionally, all wheels must meet the requirements of §25.731(d).

**(k)** *Compatibility.* Compatibility of the wheel and brake assemblies with the airplane and its systems must be substantiated.

[Docket No. FAA–1999–6063, 67 FR 20420, April 24, 2002]

## §25.737 Skis.

Each ski must be approved. The maximum limit load rating of each ski must equal or exceed the maximum limit load determined under the applicable ground load requirements of this part.

### Floats and Hulls

## §25.751 Main float buoyancy.

Each main float must have—

**(a)** A buoyancy of 80 percent in excess of that required to support the maximum weight of the seaplane or amphibian in fresh water; and

**(b)** Not less than five watertight compartments approximately equal in volume.

## §25.753 Main float design.

Each main float must be approved and must meet the requirements of §25.521.

## §25.755 Hulls.

**(a)** Each hull must have enough watertight compartments so that, with any two adjacent compartments flooded, the buoyancy of the hull and auxiliary floats (and wheel tires, if used) provides a margin of positive stability great enough to minimize the probability of capsizing in rough, fresh water.

**(b)** Bulkheads with watertight doors may be used for communication between compartments.

### Personnel and Cargo Accommodations

## §25.771 Pilot compartment.

**(a)** Each pilot compartment and its equipment must allow the minimum flight crew (established under §25.1523) to perform their duties without unreasonable concentration or fatigue.

**(b)** The primary controls listed in §25.779(a), excluding cables and control rods, must be located with respect to the propellers so that no member of the minimum flight crew (established under §25.1523), or part of the controls, lies in the region between the plane of rotation of any inboard propeller and the surface generated by a line passing through the center of the propeller hub making an angle of five degrees forward or aft of the plane of rotation of the propeller.

**(c)** If provision is made for a second pilot, the airplane must be controllable with equal safety from either pilot seat.

**(d)** The pilot compartment must be constructed so that, when flying in rain or snow, it will not leak in a manner that will distract the crew or harm the structure.

**(e)** Vibration and noise characteristics of cockpit equipment may not interfere with safe operation of the airplane.

[Docket No. 5066, 29 FR 18291, Dec. 24, 1964; as amended by Amdt. 25–4, 30 FR 6113, April 30, 1965]

## §25.772 Pilot compartment doors.

For an airplane that has a lockable door installed between the pilot compartment and the passenger compartment:

**(a)** For airplanes with a maximum passenger seating configuration of more than 20 seats, the emergency exit configuration must be designed so that neither crewmembers nor passengers require use of the flightdeck door in order to reach the emergency exits provided for them; and

**(b)** Means must be provided to enable flight crewmembers to directly enter the passenger compartment from the pilot compartment if the cockpit door becomes jammed.

**(c)** There must be an emergency means to enable a flight attendant to enter the pilot compartment in the event that the flightcrew becomes incapacitated.

[Docket No. 24344, 55 FR 29777, July 20, 1990; as amended by Amdt. 25–106, 67 FR 2127, Jan. 15, 2002]

## §25.773 Pilot compartment view.

**(a)** *Nonprecipitation conditions.* For nonprecipitation conditions, the following apply:

(1) Each pilot compartment must be arranged to give the pilots a sufficiently extensive, clear, and undistorted view, to enable them to safely perform any maneuvers within the operating limitations of the airplane, including taxiing takeoff, approach, and landing.

(2) Each pilot compartment must be free of glare and reflection that could interfere with the normal duties of the minimum flight crew (established under §25.1523). This must be shown in day and night flight tests under nonprecipitation conditions.

**(b)** *Precipitation conditions.* For precipitation conditions, the following apply:

(1) The airplane must have a means to maintain a clear portion of the windshield, during precipitation conditions, sufficient for both pilots to have a sufficiently extensive view along the flight path in normal flight attitudes of the airplane. This means must be designed to function, without continuous attention on the part of the crew, in—

(i) Heavy rain at speeds up to 1.6 $V_{S1}$ with lift and drag devices retracted; and

(ii) The icing conditions specified in §25.1419 if certification with ice protection provisions is requested.

(2) The first pilot must have—

(i) A window that is openable under the conditions prescribed in paragraph (b)(1) of this section when the cabin is not pressurized, provides the view specified in that paragraph, and gives sufficient protection from the elements against impairment of the pilot's vision; or

(ii) An alternate means to maintain a clear view under the conditions specified in paragraph (b)(1) of this section, considering the probable damage due to a severe hail encounter.

**(c)** *Internal windshield and window fogging.* The airplane must have a means to prevent fogging of the internal portions of the windshield and window panels over an area which would provide the visibility specified in paragraph (a) of this section under all internal and external ambient conditions, including precipitation conditions, in which the airplane is intended to be operated.

**(d)** Fixed markers or other guides must be installed at each pilot station to enable the pilots to position themselves in their seats for an optimum combination of outside visibility and instrument scan. If lighted markers or guides are used they must comply with the requirements specified in §25.1381.

[Docket No. 5066, 29 FR 18291, Dec. 24, 1964; as amended by Amdt. 25–23, 35 FR 5676, April 8, 1970; Amdt. 25–46, 43 FR 50595, Oct. 30, 1978; Amdt. 25–72, 55 FR 29778, July 20, 1990]

## §25.775 Windshields and windows.

**(a)** Internal panes must be made of nonsplintering material.

**(b)** Windshield panes directly in front of the pilots in the normal conduct of their duties, and the supporting structures for these panes, must withstand, without penetration, the impact of a four-pound bird when the velocity of the airplane (relative to the bird along the airplane's flight path) is equal to the value of $V_C$, at sea level, selected under §25.335(a).

**(c)** Unless it can be shown by analysis or tests that the probability of occurrence of a critical windshield fragmentation condition is of a low order, the airplane must have a means to minimize the danger to the pilots from flying windshield fragments due to bird impact. This must be shown for each transparent pane in the cockpit that—

(1) Appears in the front view of the airplane;

(2) Is inclined 15 degrees or more to the longitudinal axis of the airplane; and

(3) Has any part of the pane located where its fragmentation will constitute a hazard to the pilots.

**(d)** The design of windshields and windows in pressurized airplanes must be based on factors peculiar to high altitude operation, including the effects of continuous and cyclic pressurization loadings, the inherent characteristics of the material used, and the effects of temperatures and temperature differentials. The windshield and window panels must be capable of withstanding the maximum cabin pressure differential loads combined with critical aerodynamic pressure and temperature effects after any single failure in the installation or associated systems. It may be assumed that, after a single failure that is obvious to the flight crew (established under §25.1523), the cabin pressure differential is reduced from the maximum, in accordance with appropriate operating limitations, to allow continued safe flight of the airplane with a cabin pressure altitude of not more than 15,000 feet.

**(e)** The windshield panels in front of the pilots must be arranged so that, assuming the loss of vision through any one panel, one or more panels remain available for use by a pilot seated at a pilot station to permit continued safe flight and landing.

[Docket No. 5066, 29 FR 18291, Dec. 24, 1964; as amended by Amdt. 25–23, 35 FR 5676, April 8, 1970; Amdt. 25–38, 41 FR 55466, Dec. 20, 1976]

## §25.777 Cockpit controls.

**(a)** Each cockpit control must be located to provide convenient operation and to prevent confusion and inadvertent operation.

**(b)** The direction of movement of cockpit controls must meet the requirements of §25.779. Wherever practicable, the sense of motion involved in the operation of other controls must correspond to the sense of the effect of the operation upon the airplane or upon the part operated. Controls of a variable nature using a rotary motion must move clockwise from the off position, through an increasing range, to the full on position.

**(c)** The controls must be located and arranged, with respect to the pilots' seats, so that there is full and unrestricted movement of each control without interference from the cockpit structure or the clothing of the minimum flight crew (established under §25.1523) when any member of this flight crew, from 5'2" to 6'3" in height, is seated with the seat belt and shoulder harness (if provided) fastened.

**(d)** Identical powerplant controls for each engine must be located to prevent confusion as to the engines they control.

**(e)** Wing flap controls and other auxiliary lift device controls must be located on top of the pedestal, aft of the throttles, centrally or to the right of the pedestal centerline, and not less than 10 inches aft of the landing gear control.

**(f)** The landing gear control must be located forward of the throttles and must be operable by each pilot when seated with seat belt and shoulder harness (if provided) fastened.

**(g)** Control knobs must be shaped in accordance with §25.781. In addition, the knobs must be of the same color, and this color must contrast with the color of control knobs for other purposes and the surrounding cockpit.

**(h)** If a flight engineer is required as part of the minimum flight crew (established under §25.1523), the airplane must have a flight engineer station located and arranged so that the flight crewmembers can perform their functions efficiently and without interfering with each other.

[Docket No. 5066, 29 FR 18291, Dec. 24, 1964; as amended by Amdt. 25–46, 43 FR 50596, Oct. 30, 1978]

## §25.779 Motion and effect of cockpit controls.

Cockpit controls must be designed so that they operate in accordance with the following movement and actuation:

**(a)** Aerodynamic controls:

(1) *Primary.*

| Controls | Motion and effect |
|---|---|
| Aileron | Right (clockwise) for right wing down |
| Elevator | Rearward for nose up |
| Rudder | Right pedal forward for nose right |

(2) *Secondary.*

| Controls | Motion and effect |
|---|---|
| Flaps (or auxiliary lift devices) | Forward for flaps up; rearward for flaps down |
| Trim tabs (or equivalent) | Rotate to produce similar rotation of the airplane about an axis parallel to the axis of the control |

**(b)** Powerplant and auxiliary controls:

(1) *Powerplant.*

| Controls | Motion and effect |
|---|---|
| Power or thrust | Forward to increase forward thrust and rearward to increase rearward thrust |
| Propellers | Forward to increase rpm |
| Mixture | Forward or upward for rich |
| Carburetor air heat | Forward or upward for cold |
| Supercharger | Forward or upward for low blower For turbosuperchargers, forward, upward, or clockwise, to increase pressure |

(2) *Auxiliary.*

| Controls | Motion and effect |
|---|---|
| Landing gear | Down to extend |

[Docket No. 5066, 29 FR 18291, Dec. 24, 1964; as amended by Amdt. 25–72, 55 FR 29778, July 20, 1990]

## §25.781 Cockpit control knob shape.

Cockpit control knobs must conform to the general shapes (but not necessarily the exact sizes or specific proportions) in the following figure:

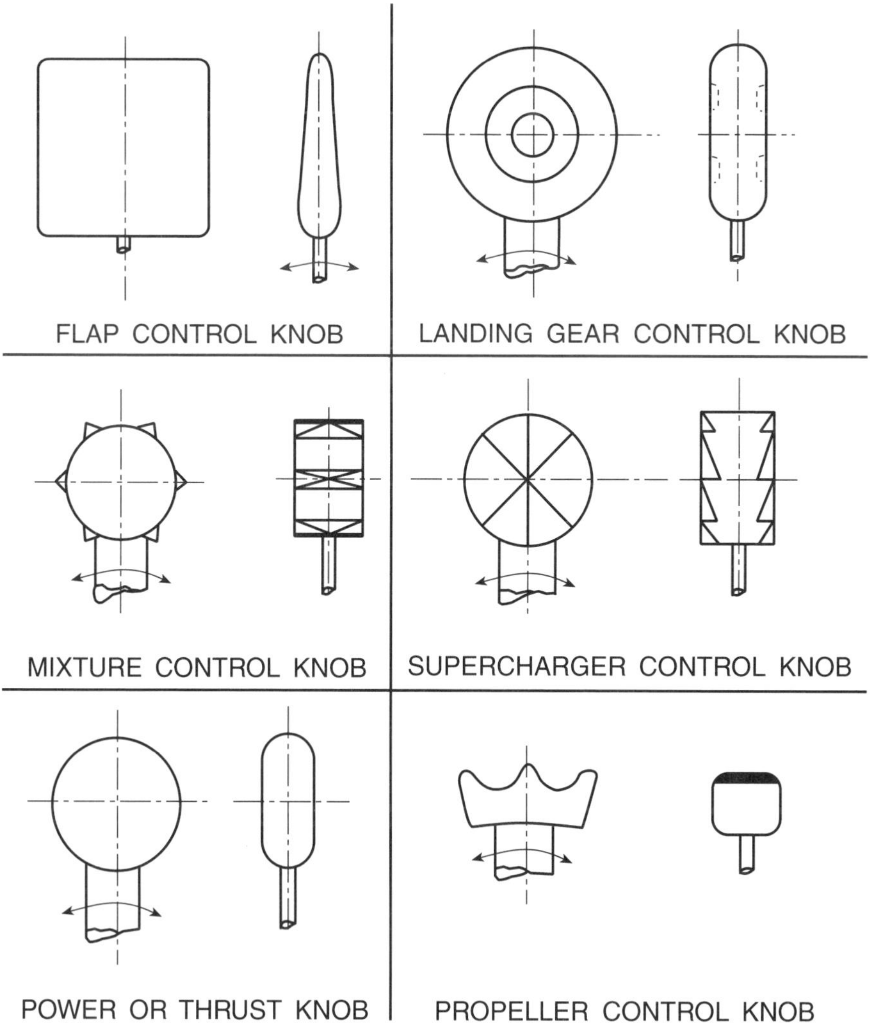


[Docket No. 5066, 29 FR 18291, Dec. 24, 1964; as amended by Amdt. 25–72, 55 FR 29779, July 20, 1990]

## §25.783 Doors.

**(a)** Each cabin must have at least one easily accessible external door.

**(b)** There must be a means to lock and safeguard each external door against opening in flight (either inadvertently by persons or as a result of mechanical failure or failure of a single structural element either during or after closure). Each external door must be openable from both the inside and the outside, even though persons may be crowded against the door on the inside of the airplane. Inward opening doors may be used if there are means to prevent occupants from crowding against the door to an extent that would interfere with the opening of the door. The means of opening must be simple and obvious and must be arranged and marked so that it can be readily located and operated, even in darkness. Auxiliary locking devices may be used.

**(c)** Each external door must be reasonably free from jamming as a result of fuselage deformation in a minor crash.

**(d)** Each external door must be located where persons using them will not be endangered by the propellers when appropriate operating procedures are used.

**(e)** There must be a provision for direct visual inspection of the locking mechanism to determine if external doors, for which the initial opening movement is not inward (including passenger, crew, service, and cargo doors), are fully closed and locked. The provision must be discernible under operational lighting conditions by appropriate crewmembers using a flashlight or equivalent lighting source. In addition, there must be a visual warning means to signal the appropriate flight crewmembers if any external door is not fully closed and locked. The means must be designed such that any failure or combination of failures that would result in an erroneous closed and locked indication is improbable for doors for which the initial opening movement is not inward.

**(f)** External doors must have provisions to prevent the initiation of pressurization of the airplane to an unsafe level if the door is not fully closed and locked. In addition, it must be shown by safety analysis that inadvertent opening is extremely improbable.

**(g)** Cargo and service doors not suitable for use as emergency exits need only meet paragraphs (e) and (f) of this section and be safeguarded against opening in flight as a result of mechanical failure or failure of a single structural element.

**(h)** Each passenger entry door in the side of the fuselage must meet the applicable requirements of §§25.807 through 25.813 for a Type II or larger passenger emergency exit.

**(i)** If an integral stair is installed in a passenger entry door that is qualified as a passenger emergency exit, the stair must be designed so that under the following conditions the effectiveness of passenger emergency egress will not be impaired:

(1) The door, integral stair, and operating mechanism have been subjected to the inertia forces specified in §25.561(b)(3), acting separately relative to the surrounding structure.

(2) The airplane is in the normal ground attitude and in each of the attitudes corresponding to collapse of one or more legs of the landing gear.

**(j)** All lavatory doors must be designed to preclude anyone from becoming trapped inside the lavatory, and if a locking mechanism is installed, it be capable of being unlocked from the outside without the aid of special tools.

[Docket No. 5066, 29 FR 18291, Dec. 24, 1964; as amended by Amdt. 25–15, 32 FR 13262, Sept. 20, 1967; Amdt. 25–23, 35 FR 5676, April 8, 1970; Amdt. 25–54, 45 FR 60173, Sept. 11, 1980; Amdt. 25–72, 55 FR 29780, July 20, 1990; Amdt. 25–88, 61 FR 57956, Nov. 8, 1996]

## §25.785 Seats, berths, safety belts, and harnesses.

**(a)** A seat (or berth for a nonambulant person) must be provided for each occupant who has reached his or her second birthday.

**(b)** Each seat, berth, safety belt, harness, and adjacent part of the airplane at each station designated as occupiable during takeoff and landing must be designed so that a person making proper use of these facilities will not suffer serious injury in an emergency landing as a result of the inertia forces specified in §§25.561 and 25.562.

**(c)** Each seat or berth must be approved.

**(d)** Each occupant of a seat that makes more than an 18-degree angle with the vertical plane containing the airplane centerline must be protected from head injury by a safety belt and an energy absorbing rest that will support the arms, shoulders, head, and spine, or by a safety belt and shoulder harness that will prevent the head from contacting any injurious object. Each occupant of any other seat must be protected from head injury by a safety belt and, as appropriate to the type, location, and angle of facing of each seat, by one or more of the following:

(1) A shoulder harness that will prevent the head from contacting any injurious object.

(2) The elimination of any injurious object within striking radius of the head.

(3) An energy absorbing rest that will support the arms, shoulders, head, and spine.

**(e)** Each berth must be designed so that the forward part has a padded end board, canvas diaphragm, or equivalent means, that can withstand the static load reaction of the occupant when subjected to the forward inertia force specified in §25.561. Berths must be free from corners and protuberances likely to cause injury to a person occupying the berth during emergency conditions.

**(f)** Each seat or berth, and its supporting structure, and each safety belt or harness and its anchorage must be designed for an occupant weight of 170 pounds, considering the maximum load factors, inertia forces, and reactions among the occupant, seat, safety belt, and harness for each relevant flight and ground load condition (including the emergency landing conditions prescribed in §25.561). In addition—

(1) The structural analysis and testing of the seats, berths, and their supporting structures may be determined by assuming that the critical load in the forward, sideward, downward, upward, and rearward directions (as determined from the prescribed flight, ground, and emergency landing conditions) acts separately or using selected combinations of loads if the required strength in each specified direction is substantiated. The forward load factor need not be applied to safety belts for berths.

(2) Each pilot seat must be designed for the reactions resulting from the application of the pilot forces prescribed in §25.395.

(3) The inertia forces specified in §25.561 must be multiplied by a factor of 1.33 (instead of the fitting factor prescribed in §25.625) in determining the strength of the attachment of each seat to the structure and each belt or harness to the seat or structure.

**(g)** Each seat at a flight deck station must have a restraint system consisting of a combined safety belt and shoulder harness with a single-point release that permits the flight deck occupant, when seated with the restraint system fastened, to perform all of the occupant's necessary flight deck functions. There must be a means to secure each combined restraint system when not in use to prevent interference with the operation of the airplane and with rapid egress in an emergency.

**(h)** Each seat located in the passenger compartment and designated for use during takeoff and landing by a flight attendant required by the operating rules of this chapter must be:

(1) Near a required floor level emergency exit, except that another location is acceptable if the emergency egress of passengers would be enhanced with that location. A flight attendant seat must be located adjacent to each Type A or B emergency exit. Other flight attendant seats must be evenly distributed among the required floor-level emergency exits to the extent feasible.

(2) To the extent possible, without compromising proximity to a required floor level emergency exit, located to provide a direct view of the cabin area for which the flight attendant is responsible.

(3) Positioned so that the seat will not interfere with the use of a passageway or exit when the seat is not in use.

(4) Located to minimize the probability that occupants would suffer injury by being struck by items dislodged from service areas, stowage compartments, or service equipment.

(5) Either forward or rearward facing with an energy absorbing rest that is designed to support the arms, shoulders, head, and spine.

(6) Equipped with a restraint system consisting of a combined safety belt and shoulder harness unit with a single point release. There must be means to secure each restraint system when not in use to prevent interference with rapid egress in an emergency.

**(i)** Each safety belt must be equipped with a metal to metal latching device.

**(j)** If the seat backs do not provide a firm handhold, there must be a handgrip or rail along each aisle to enable persons to steady themselves while using the aisles in moderately rough air.

**(k)** Each projecting object that would injure persons seated or moving about the airplane in normal flight must be padded.

**(l)** Each forward observer's seat required by the operating rules must be shown to be suitable for use in conducting the necessary enroute inspection.

[Docket No. 5066, 29 FR 18291, Dec. 24, 1964; as amended by Amdt. 25–72, 55 FR 29780, July 20, 1990; Amdt. 25–88, 61 FR 57956, Nov. 8, 1996]

## §25.787 Stowage compartments.

**(a)** Each compartment for the stowage of cargo, baggage, carry-on articles, and equipment (such as life rafts), and any other stowage compartment must be designed for its placarded maximum weight of contents and for the critical load distribution at the appropriate maximum load factors corresponding to the specified flight and ground load conditions, and to the emergency landing conditions of §25.561(b), except that the forces specified in the emergency landing conditions need not be applied to compartments located below, or forward, of all occupants in the airplane. If the airplane has a passenger seating configuration, excluding pilots seats, of 10 seats or more, each stowage compartment in the passenger cabin, except for underseat and overhead compartments for passenger convenience, must be completely enclosed.

**(b)** There must be a means to prevent the contents in the compartments from becoming a hazard by shifting, under the loads specified in paragraph (a) of this section. For stowage compartments in the passenger and crew cabin, if the means used is a latched door, the design must take into consideration the wear and deterioration expected in service.

**(c)** If cargo compartment lamps are installed, each lamp must be installed so as to prevent contact between lamp bulb and cargo.

[Docket No. 5066, 29 FR 18291, Dec. 24, 1964; as amended by Amdt. 25–32, 37 FR 3969, Feb. 24, 1972; Amdt. 25–38, 41 FR 55466, Dec. 20, 1976; Amdt. 25–51, 45 FR 7755, Feb. 4, 1980]

## §25.789 Retention of items of mass in passenger and crew compartments and galleys.

**(a)** Means must be provided to prevent each item of mass (that is part of the airplane type design) in a passenger or crew compartment or galley from becoming a hazard by shifting under the appropriate maximum load factors corresponding to the specified flight and ground load conditions, and to the emergency landing conditions of §25.561(b).

**(b)** Each interphone restraint system must be designed so that when subjected to the load factors specified in §25.561(b)(3), the interphone will remain in its stowed position.

[Docket No. 5066, 29 FR 18291, Dec. 24, 1964; as amended by Amdt. 25–32, 37 FR 3969, Feb. 24, 1972; Amdt. 25–46, 43 FR 50596, Oct. 30, 1978]

## §25.791 Passenger information signs and placards.

**(a)** If smoking is to be prohibited, there must be at least one placard so stating that is legible to each person seated in the cabin. If smoking is to be allowed, and if the crew compartment is separated from the passenger compartment, there must be at least one sign notifying when smoking is prohibited. Signs which notify when smoking is prohibited must be operable by a member of the flightcrew and, when illuminated, must be legible under all probable conditions of cabin illumination to each person seated in the cabin.

**(b)** Signs that notify when seat belts should be fastened and that are installed to comply with the operating rules of this chapter must be operable by a member of the flightcrew and, when illuminated, must be legible under all probable conditions of cabin illumination to each person seated in the cabin.

**(c)** A placard must be located on or adjacent to the door of each receptacle used for the disposal of flammable waste materials to indicate that use of the receptacle for disposal of cigarettes, etc., is prohibited.

**(d)** Lavatories must have "No Smoking" or "No Smoking in Lavatory" placards conspicuously located on or adjacent to each side of the entry door.

**(e)** Symbols that clearly express the intent of the sign or placard may be used in lieu of letters.

[Docket No. 5066, 29 FR 18291, Dec. 24, 1964; as amended by Amdt. 25–72, 55 FR 29780, July 20, 1990]

## §25.793 Floor surfaces.

The floor surface of all areas which are likely to become wet in service must have slip resistant properties.

[Docket No. 5066, 29 FR 18291, Dec. 24, 1964; as amended by Amdt. 25–51, 45 FR 7755, Feb. 4, 1980]

## §25.795 Security considerations.

**(a)** *Protection of flightdeck.* If a flightdeck door is required by operating rules, the door installation must be designed to:

(1) Resist forcible intrusion by unauthorized persons and be capable of withstanding impacts of 300 Joules (221.3 foot-pounds) at the critical locations on the door, as well as a 250 pound (1113 Newtons) constant tensile load on the knob or handle, and

(2) Resist penetration by small arms fire and fragmentation devices to a level equivalent to level IIIa of the National Institute of Justice Standard (NIJ) 0101.04.

**(b)** [Reserved]

[Docket No. FAA–2001–11032, 67 FR 2127, Jan. 15, 2002]

### Emergency Provisions

## §25.801 Ditching.

**(a)** If certification with ditching provisions is requested, the airplane must meet the requirements of this section and §§25.807(e), 25.1411, and 25.1415(a).

**(b)** Each practicable design measure, compatible with the general characteristics of the airplane, must be taken to minimize the probability that in an emergency landing on water, the behavior of the airplane would cause immediate injury to the occupants or would make it impossible for them to escape.

**(c)** The probable behavior of the airplane in a water landing must be investigated by model tests or by comparison with airplanes of similar configuration for which the ditching characteristics are known. Scoops, flaps, projections, and any other factor likely to affect the hydrodynamic characteristics of the airplane, must be considered.

**(d)** It must be shown that, under reasonably probable water conditions, the flotation time and trim of the airplane will allow the occupants to leave the airplane and enter the liferafts required by §25.1415. If compliance with this provision is shown by buoyancy and trim computations, appropriate allowances must be made for probable structural damage and leakage. If the airplane has fuel tanks (with fuel jettisoning provisions) that can reasonably be expected to withstand a ditching without leakage, the jettisonable volume of fuel may be considered as buoyancy volume.

**(e)** Unless the effects of the collapse of external doors and windows are accounted for in the investigation of the probable behavior of the airplane in a water landing (as prescribed in paragraphs (c) and (d) of this section), the external doors and windows must be designed to withstand the probable maximum local pressures.

[Docket No. 5066, 29 FR 18291, Dec. 24, 1964; as amended by Amdt. 25–72, 55 FR 29781, July 20, 1990]

## §25.803 Emergency evacuation.

**(a)** Each crew and passenger area must have emergency means to allow rapid evacuation in crash landings, with the landing gear extended as well as with the landing gear retracted, considering the possibility of the airplane being on fire.

**(b)** [Reserved]

**(c)** For airplanes having a seating capacity of more than 44 passengers, it must be shown that the maximum seating capacity, including the number of crewmembers required by the operating rules for which certification is requested, can be evacuated from the airplane to the ground under simulated emergency conditions within 90 seconds. Compliance with this requirement must be shown by actual demonstration using the test criteria outlined in Appendix J of this part unless the Administrator finds that a combination of analysis and testing will provide data equivalent to that which would be obtained by actual demonstration.

**(d)–(e)** [Reserved]

[Docket No. 24344, 55 FR 29781, July 20, 1990]

## §25.807 Emergency exits.

**(a)** *Type.* For the purpose of this part, the types of exits are defined as follows:

(1) *Type I.* This type is a floor-level exit with a rectangular opening of not less than 24 inches wide by 48 inches high, with corner radii not greater than eight inches.

(2) *Type II.* This type is a rectangular opening of not less than 20 inches wide by 44 inches high, with corner radii not greater than seven inches. Type II exits must be floor-level exits unless located over the wing, in which case they must not have a step-up inside the airplane of more than 10 inches nor a step-down outside the airplane of more than 17 inches.

(3) *Type III.* This type is a rectangular opening of not less than 20 inches wide by 36 inches high with corner radii not greater than seven inches, and with a step-up inside the airplane of not more than 20 inches. If the exit is located over the wing, the step-down outside the airplane may not exceed 27 inches.

(4) *Type IV.* This type is a rectangular opening of not less than 19 inches wide by 26 inches high, with corner radii not greater than 6.3 inches, located over the wing, with a step-up inside the airplane of not more than 29 inches and a step-down outside the airplane of not more than 36 inches.

(5) *Ventral.* This type is an exit from the passenger compartment through the pressure shell and the bottom fuselage skin. The dimensions and physical configuration of this type of exit must allow at least the same rate of egress as a Type I exit with the airplane in the normal ground attitude, with landing gear extended.

(6) *Tailcone.* This type is an aft exit from the passenger compartment through the pressure shell and through an openable cone of the fuselage aft of the pressure shell. The means of opening the tailcone must be simple and obvious and must employ a single operation.

(7) *Type A.* This type is a floor-level exit with a rectangular opening of not less than 42 inches wide by 72 inches high, with corner radii not greater than seven inches.

(8) *Type B.* This type is a floor-level exit with a rectangular opening of not less than 32 inches wide by 72 inches high, with corner radii not greater than six inches.

(9) *Type C.* This type is a floor-level exit with a rectangular opening of not less than 30 inches wide by 48 inches high, with corner radii not greater than 10 inches.

**(b)** *Step down distance.* Step down distance, as used in this section, means the actual distance between the bottom of the required opening and a usable foot hold, extending out from the fuselage, that is large enough to be effective without searching by sight or feel.

**(c)** *Over-sized exits.* Openings larger than those specified in this section, whether or not of rectangular shape, may be used if the specified rectangular opening can be inscribed within the opening and the base of the inscribed rectangular opening meets the specified step-up and step-down heights.

**(d)** *Asymmetry.* Exits of an exit pair need not be diametrically opposite each other nor of the same size; however, the number of passenger seats permitted under paragraph (g) of this section is based on the smaller of the two exits.

**(e)** *Uniformity.* Exits must be distributed as uniformly as practical, taking into account passenger seat distribution.

**(f)** *Location.*

(1) Each required passenger emergency exit must be accessible to the passengers and located where it will afford the most effective means of passenger evacuation.

(2) If only one floor-level exit per side is prescribed, and the airplane does not have a tailcone or ventral emergency exit, the floor-level exits must be in the rearward part of the passenger compartment unless another location affords a more effective means of passenger evacuation.

(3) If more than one floor-level exit per side is prescribed, and the airplane does not have a combination cargo and passenger configuration, at least one floor-level exit must be located in each side near each end of the cabin.

(4) For an airplane that is required to have more than one passenger emergency exit for each side of the fuselage, no passenger emergency exit shall be more than 60 feet from any adjacent passenger emergency exit on the same side of the same deck of the fuselage, as measured parallel to the airplane's longitudinal axis between the nearest exit edges.

**(g)** *Type and number required.* The maximum number of passenger seats permitted depends on the type and number of exits installed in each side of the fuselage. Except as further restricted in paragraphs (g)(1) through (g)(9) of this section, the maximum number of passenger seats permitted for each exit of a specific type installed in each side of the fuselage is as follows:

| | |
|---|---|
| Type A | 110 |
| Type B | 75 |
| Type C | 55 |
| Type I | 45 |
| Type II | 40 |
| Type III | 35 |
| Type IV | 9 |

(1) For a passenger seating configuration of 1 to 9 seats, there must be at least one Type IV or larger overwing exit in each side of the fuselage or, if overwing exits are not provided, at least one exit in each side that meets the minimum dimensions of a Type III exit.

(2) For a passenger seating configuration of more than 9 seats, each exit must be a Type III or larger exit.

(3) For a passenger seating configuration of 10 to 19 seats, there must be at least one Type III or larger exit in each side of the fuselage.

(4) For a passenger seating configuration of 20 to 40 seats, there must be at least two exits, one of which must be a Type II or larger exit, in each side of the fuselage.

(5) For a passenger seating configuration of 41 to 110 seats, there must be at least two exits, one of which must be a Type I or larger exit, in each side of the fuselage.

(6) For a passenger seating configuration of more than 110 seats, the emergency exits in each side of the fuselage must include at least two Type I or larger exits.

(7) The combined maximum number of passenger seats permitted for all Type III exits is 70, and the combined maximum number of passenger seats permitted for two Type III exits in each side of the fuselage that are separated by fewer than three passenger seat rows is 65.

(8) If a Type A, Type B, or Type C exit is installed, there must be at least two Type C or larger exits in each side of the fuselage.

(9) If a passenger ventral or tail cone exit is installed and that exit provides at least the same rate of egress as a Type III exit with the airplane in the most adverse exit opening condition that would result from the collapse of one or more legs of the landing gear, an increase in the passenger seating configuration is permitted as follows:

(i) For a ventral exit, 12 additional passenger seats.

(ii) For a tail cone exit incorporating a floor level opening of not less than 20 inches wide by 60 inches high, with corner radii not greater than seven inches, in the pressure shell and incorporating an approved assist means in accordance with §25.810(a), 25 additional passenger seats.

(iii) For a tail cone exit incorporating an opening in the pressure shell which is at least equivalent to a Type III emergency exit with respect to dimensions, step-up and step-down distance, and with the top of the opening not less than 56 inches from the passenger compartment floor, 15 additional passenger seats.

**(h)** *Excess exits.* Each emergency exit in the passenger compartment in excess of the minimum number of required emergency exits must meet the applicable requirements of §25.809 through §25.812, and must be readily accessible.

**(i)** *Ditching emergency exits for passengers.* Whether or not ditching certification is requested, ditching emergency exits must be provided in accordance with the following requirements, unless the emergency exits required by paragraph (g) of this section already meet them:

(1) For airplanes that have a passenger seating configuration of nine or fewer seats, excluding pilot seats, one exit above the waterline in each side of the airplane, meeting at least the dimensions of a Type IV exit.

(2) For airplanes that have a passenger seating configuration of 10 of more seats, excluding pilot seats, one exit above the waterline in a side of the airplane, meeting at least the dimensions of a Type III exit for each unit (or part of a unit) of 35 passenger seats, but no less than two such exits in the passenger cabin, with one on each side of the airplane. The passenger seat/exit ratio may be increased through the use of larger exits, or other means, provided it is shown that the evacuation

capability during ditching has been improved accordingly.

(3) If it is impractical to locate side exits above the waterline, the side exits must be replaced by an equal number of readily accessible overhead hatches of not less than the dimensions of a Type III exit, except that for airplanes with a passenger configuration of 35 or fewer seats, excluding pilot seats, the two required Type III side exits need be replaced by only one overhead hatch.

**(j)** *Flightcrew emergency exits.* For airplanes in which the proximity of passenger emergency exits to the flightcrew area does not offer a convenient and readily accessible means of evacuation of the flightcrew, and for all airplanes having a passenger seating capacity greater than 20, flightcrew exits shall be located in the flightcrew area. Such exits shall be of sufficient size and so located as to permit rapid evacuation by the crew. One exit shall be provided on each side of the airplane; or, alternatively, a top hatch shall be provided. Each exit must encompass an unobstructed rectangular opening of at least 19 by 20 inches unless satisfactory exit utility can be demonstrated by a typical crewmember.

[Docket No. 5066, 29 FR 18291, Dec. 24, 1964; as amended by Amdt. 25–72, 55 FR 29781, July 20, 1990; Amdt. 25–88, 61 FR 57957, Nov. 8, 1996; Amdt. 25–88, 62 FR 1817, Jan. 13, 1997; Amdt. 25–94, 63 FR 8848, Feb. 23, 1998; Amdt. 25–94, 63 FR 12862, March 16, 1998]

## §25.809 Emergency exit arrangement.

**(a)** Each emergency exit, including a flight crew emergency exit, must be a movable door or hatch in the external walls of the fuselage, allowing unobstructed opening to the outside.

**(b)** Each emergency exit must be openable from the inside and the outside except that sliding window emergency exits in the flight crew area need not be openable from the outside if other approved exits are convenient and readily accessible to the flight crew area. Each emergency exit must be capable of being opened, when there is no fuselage deformation—

(1) With the airplane in the normal ground attitude and in each of the attitudes corresponding to collapse of one or more legs of the landing gear; and

(2) Within 10 seconds measured from the time when the opening means is actuated to the time when the exit is fully opened.

**(c)** The means of opening emergency exits must be simple and obvious and may not require exceptional effort. Internal exit-opening means involving sequence operations (such as operation of two handles or latches or the release of safety catches) may be used for flight crew emergency exits if it can be reasonably established that these means are simple and obvious to crewmembers trained in their use.

**(d)** If a single power-boost or single power-operated system is the primary system for operating more than one exit in an emergency, each exit must be capable of meeting the requirements of paragraph (b) of this section in the event of failure of the primary system. Manual operation of the exit (after failure of the primary system) is acceptable.

**(e)** Each emergency exit must be shown by tests, or by a combination of analysis and tests, to meet the requirements of paragraphs (b) and (c) of this section.

**(f)** There must be a means to lock each emergency exit and to safeguard against its opening in flight, either inadvertently by persons or as a result of mechanical failure. In addition, there must be a means for direct visual inspection of the locking mechanism by crewmembers to determine that each emergency exit, for which the initial opening movement is outward, is fully locked.

**(g)** There must be provisions to minimize the probability of jamming of the emergency exits resulting from fuselage deformation in a minor crash landing.

**(h)** When required by the operating rules for any large passenger-carrying turbojet-powered airplane, each ventral exit and tailcone exit must be—

(1) Designed and constructed so that it cannot be opened during flight; and

(2) Marked with a placard readable from a distance of 30 inches and installed at a conspicuous location near the means of opening the exit, stating that the exit has been designed and constructed so that it cannot be opened during flight.

[Docket No. 5066, 29 FR 18291, Dec. 24, 1964; as amended by Amdt. 25–15, 32 FR 13264, Sept. 20, 1967; Amdt. 25–32, 37 FR 3970, Feb. 24, 1972; Amdt. 25–34, 37 FR 25355, Nov. 30, 1972; Amdt. 25–46, 43 FR 50597, Oct. 30, 1978; Amdt. 25–47, 44 FR 61325, Oct. 25, 1979; Amdt. 25–72, 55 FR 29782, July 20, 1990]

## §25.810 Emergency egress assist means and escape routes.

**(a)** Each non over-wing Type A, Type B or Type C exit, and any other non over-wing landplane emergency exit more than 6 feet from the ground with the airplane on the ground and the landing gear extended, must have an approved means to assist the occupants in descending to the ground.

(1) The assisting means for each passenger emergency exit must be a self-supporting slide or equivalent; and, in the case of a Type A or Type B exits, it must be capable of carrying simultaneously two parallel lines of evacuees. In addition, the assisting means must be designed to meet the following requirements—

25

(i) It must be automatically deployed and deployment must begin during the interval between the time the exit opening means is actuated from inside the airplane and the time the exit is fully opened. However, each passenger emergency exit which is also a passenger entrance door or a service door must be provided with means to prevent deployment of the assisting means when it is opened from either the inside or the outside under nonemergency conditions for normal use.

(ii) Except for assisting means installed at Type C exits, it must be automatically erected within 6 seconds after deployment is begun. Assisting means installed at Type C exits must be automatically erected within 10 seconds from the time the opening means of the exit is actuated.

(iii) It must be of such length after full deployment that the lower end is self-supporting on the ground and provides safe evacuation of occupants to the ground after collapse of one or more legs of the landing gear.

(iv) It must have the capability, in 25-knot winds directed from the most critical angle, to deploy and, with the assistance of only one person, to remain usable after full deployment to evacuate occupants safely to the ground.

(v) For each system installation (mockup or airplane installed), five consecutive deployment and inflation tests must be conducted (per exit) without failure, and at least three tests of each such five-test series must be conducted using a single representative sample of the device. The sample devices must be deployed and inflated by the system's primary means after being subjected to the inertia forces specified in §25.561(b). If any part of the system fails or does not function properly during the required tests, the cause of the failure or malfunction must be corrected by positive means and after that, the full series of five consecutive deployment and inflation tests must be conducted without failure.

(2) The assisting means for flightcrew emergency exits may be a rope or any other means demonstrated to be suitable for the purpose. If the assisting means is a rope, or an approved device equivalent to a rope, it must be—

(i) Attached to the fuselage structure at or above the top of the emergency exit opening, or, for a device at a pilot's emergency exit window, at another approved location if the stowed device, or its attachment, would reduce the pilot's view in flight;

(ii) Able (with its attachment) to withstand a 400-pound static load.

**(b)** Assist means from the cabin to the wing are required for each Type A or Type B exit located above the wing and having a stepdown unless the exit without an assist-means can be shown to have a rate of passenger egress at least equal to that of the same type of nonoverwing exit. If an assist means is required, it must be automatically deployed and automatically erected concurrent with the opening of the exit. In the case of assist means installed at Type C exits, it must be self-supporting within 10 seconds from the time the opening means of the exits is actuated. For all other exit types, it must be self-supporting 6 seconds after deployment is begun.

**(c)** An escape route must be established from each overwing emergency exit, and (except for flap surfaces suitable as slides) covered with a slip resistant surface. Except where a means for channeling the flow of evacuees is provided—

(1) The escape route from each Type A or Type B passenger emergency exit, or any common escape route from two Type III passenger emergency exits, must be at least 42 inches wide; that from any other passenger emergency exit must be at least 24 inches wide; and

(2) The escape route surface must have a reflectance of at least 80 percent, and must be defined by markings with a surface-to-marking contrast ratio of at least 5:1.

**(d)** Means must be provided to assist evacuees to reach the ground for all Type C exits located over the wing and, if the place on the airplane structure at which the escape route required in paragraph (c) of this section terminates is more than 6 feet from the ground with the airplane on the ground and the landing gear extended, for all other exit types.

(1) If the escape route is over the flap, the height of the terminal edge must be measured with the flap in the takeoff or landing position, whichever is higher from the ground.

(2) The assisting means must be usable and self-supporting with one or more landing gear legs collapsed and under a 25-knot wind directed from the most critical angle.

(3) The assisting means provided for each escape route leading from a Type A or B emergency exit must be capable of carrying simultaneously two parallel lines of evacuees; and, the assisting means leading from any other exit type must be capable of carrying as many parallel lines of evacuees as there are required escape routes.

(4) The assisting means provided for each escape route leading from a Type C exit must be automatically erected within 10 seconds from the time the opening means of the exit is actuated, and that provided for the escape route leading from any other exit type must be automatically erected within 10 seconds after actuation of the erection system.

[Docket No. 5066, 29 FR 18291, Dec. 24, 1964; as amended by Amdt. 25–72, 55 FR 29782, July 20, 1990; Amdt. 25–88, 61 FR 57958, Nov. 8, 1996; Amdt. 25–88, 62 FR 1817, Jan. 13, 1997]

## §25.811 Emergency exit marking.

**(a)** Each passenger emergency exit, its means of access, and its means of opening must be conspicuously marked.

**(b)** The identity and location of each passenger emergency exit must be recognizable from a distance equal to the width of the cabin.

**(c)** Means must be provided to assist the occupants in locating the exits in conditions of dense smoke.

**(d)** The location of each passenger emergency exit must be indicated by a sign visible to occupants approaching along the main passenger aisle (or aisles). There must be—

(1) A passenger emergency exit locator sign above the aisle (or aisles) near each passenger emergency exit, or at another overhead location if it is more practical because of low headroom, except that one sign may serve more than one exit if each exit can be seen readily from the sign;

(2) A passenger emergency exit marking sign next to each passenger emergency exit, except that one sign may serve two such exits if they both can be seen readily from the sign; and

(3) A sign on each bulkhead or divider that prevents fore and aft vision along the passenger cabin to indicate emergency exits beyond and obscured by the bulkhead or divider, except that if this is not possible the sign may be placed at another appropriate location.

**(e)** The location of the operating handle and instructions for opening exits from the inside of the airplane must be shown in the following manner—

(1) Each passenger emergency exit must have, on or near the exit, a marking that is readable from a distance of 30 inches.

(2) Each Type A, Type B, Type C or Type I passenger emergency exit operating handle must—

(i) Be self-illuminated with an initial brightness of at least 160 microlamberts; or

(ii) Be conspicuously located and well illuminated by the emergency lighting even in conditions of occupant crowding at the exit.

(3) [Reserved]

(4) Each Type A, Type B, Type C, Type I, or Type II passenger emergency exit with a locking mechanism released by rotary motion of the handle must be marked—

(i) With a red arrow, with a shaft at least three-fourths of an inch wide and a head twice the width of the shaft, extending along at least 70 degrees of arc at a radius approximately equal to three-fourths of the handle length.

(ii) So that the centerline of the exit handle is within ±1 inch of the projected point of the arrow when the handle has reached full travel and has released the locking mechanism, and

(iii) With the word "open" in red letters 1 inch high, placed horizontally near the head of the arrow.

**(f)** Each emergency exit that is required to be openable from the outside, and its means of opening, must be marked on the outside of the airplane. In addition, the following apply:

(1) The outside marking for each passenger emergency exit in the side of the fuselage must include a 2-inch colored band outlining the exit.

(2) Each outside marking including the band, must have color contrast to be readily distinguishable from the surrounding fuselage surface. The contrast must be such that if the reflectance of the darker color is 15 percent or less, the reflectance of the lighter color must be at least 45 percent. "Reflectance" is the ratio of the luminous flux reflected by a body to the luminous flux it receives. When the reflectance of the darker color is greater than 15 percent, at least a 30-percent difference between its reflectance and the reflectance of the lighter color must be provided.

(3) In the case of exists other than those in the side of the fuselage, such as ventral or tail cone exists, the external means of opening, including instructions if applicable, must be conspicuously marked in red, or bright chrome yellow if the background color is such that red is inconspicuous. When the opening means is located on only one side of the fuselage, a conspicuous marking to that effect must be provided on the other side.

**(g)** Each sign required by paragraph (d) of this section may use the word "exit" in its legend in place of the term "emergency exit."

[Docket No. 5066, 29 FR 18291, Dec. 24, 1964; as amended by Amdt. 25–15, 32 FR 13264, Sept. 20, 1967; Amdt. 25–32, 37 FR 3970, Feb. 24, 1972; Amdt. 25–46, 43 FR 50597, Oct. 30, 1978; 43 FR 52495, Nov. 13, 1978; Amdt. 25–79, 58 FR 45229, Aug. 26, 1993; Amdt. 25–88, 61 FR 57958, Nov. 8, 1996]

## §25.812 Emergency lighting.

**(a)** An emergency lighting system, independent of the main lighting system, must be installed. However, the sources of general cabin illumination may be common to both the emergency and the main lighting systems if the power supply to the emergency lighting system is independent of the power supply to the main lighting system. The emergency lighting system must include:

(1) Illuminated emergency exit marking and locating signs, sources of general cabin illumination, interior lighting in emergency exit areas, and floor proximity escape path marking.

(2) Exterior emergency lighting.

**(b)** Emergency exit signs—

(1) For airplanes that have a passenger seating configuration, excluding pilot seats, of 10 seats or more must meet the following requirements:

25

(i) Each passenger emergency exit locator sign required by §25.811(d)(1) and each passenger emergency exit marking sign required by §25.811(d)(2) must have red letters at least 1½ inches high on an illuminated white background, and must have an area of at least 21 square inches excluding the letters. The lighted background-to-letter contrast must be at least 10:1. The letter height to stroke-width ratio may not be more than 7:1 nor less than 6:1. These signs must be internally electrically illuminated with a background brightness of at least 25 foot-lamberts and a high-to-low background contrast no greater than 3:1.

(ii) Each passenger emergency exit sign required by §25.811(d)(3) must have red letters at least 1½ inches high on a white background having an area of at least 21 square inches excluding the letters. These signs must be internally electrically illuminated or self-illuminated by other than electrical means and must have an initial brightness of at least 400 microlamberts. The colors may be reversed in the case of a sign that is self-illuminated by other than electrical means.

(2) For airplanes that have a passenger seating configuration, excluding pilot seats, of nine seats or less, that are required by §25.811(d) (1), (2), and (3) must have red letters at least 1 inch high on a white background at least 2 inches high. These signs may be internally electrically illuminated, or self-illuminated by other than electrical means, with an initial brightness of at least 160 microlamberts. The colors may be reversed in the case of a sign that is self-illuminated by other than electrical means.

**(c)** General illumination in the passenger cabin must be provided so that when measured along the centerline of main passenger aisle(s), and cross aisle(s) between main aisles, at seat armrest height and at 40-inch intervals, the average illumination is not less than 0.05 foot-candle and the illumination at each 40-inch interval is not less than 0.01 foot-candle. A main passenger aisle(s) is considered to extend along the fuselage from the most forward passenger emergency exit or cabin occupant seat, whichever is farther forward, to the most rearward passenger emergency exit or cabin occupant seat, whichever is farther aft.

**(d)** The floor of the passageway leading to each floor-level passenger emergency exit, between the main aisles and the exit openings, must be provided with illumination that is not less than 0.02 foot-candle measured along a line that is within 6 inches of and parallel to the floor and is centered on the passenger evacuation path.

**(e)** Floor proximity emergency escape path marking must provide emergency evacuation guidance for passengers when all sources of illumination more than 4 feet above the cabin aisle floor are totally obscured. In the dark of the night, the floor proximity emergency escape path marking must enable each passenger to—

(1) After leaving the passenger seat, visually identify the emergency escape path along the cabin aisle floor to the first exits or pair of exits forward and aft of the seat; and

(2) Readily identify each exit from the emergency escape path by reference only to markings and visual features not more than 4 feet above the cabin floor.

**(f)** Except for subsystems provided in accordance with paragraph (h) of this section that serve no more than one assist means, are independent of the airplane's main emergency lighting system, and are automatically activated when the assist means is erected, the emergency lighting system must be designed as follows.

(1) The lights must be operable manually from the flight crew station and from a point in the passenger compartment that is readily accessible to a normal flight attendant seat.

(2) There must be a flight crew warning light which illuminates when power is on in the airplane and the emergency lighting control device is not armed.

(3) The cockpit control device must have an "on," "off," and "armed" position so that when armed in the cockpit or turned on at either the cockpit or flight attendant station the lights will either light or remain lighted upon interruption (except an interruption caused by a transverse vertical separation of the fuselage during crash landing) of the airplane's normal electric power. There must be a means to safeguard against inadvertent operation of the control device from the "armed" or "on" positions.

**(g)** Exterior emergency lighting must be provided as follows:

(1) At each overwing emergency exit the illumination must be—

(i) Not less than 0.03 foot-candle (measured normal to the direction of the incident light) on a 2-square-foot area where an evacuee is likely to make his first step outside the cabin;

(ii) Not less than 0.05 foot-candle (measured normal to the direction of incident light) along the 30 percent of the slip-resistant portion of the escape route required in §25.810(c) that is farthest from the exit for the minimum required width of the escape route; and

(iii) Not less than 0.03 foot-candle on the ground surface with the landing gear extended (measured normal to the direction of the incident light) where an evacuee using the established escape route would normally make first contact with the ground.

(2) At each non-overwing emergency exit not required by §25.809(f) to have descent assist means the illumination must be not less than 0.03 foot-candle (measured normal to the direction of

the incident light) on the ground surface with the landing gear extended where an evacuee is likely to make his first contact with the ground outside the cabin.

**(h)** The means required in §25.809 (f)(1) and (h) to assist the occupants in descending to the ground must be illuminated so that the erected assist means is visible from the airplane.

(1) If the assist means is illuminated by exterior emergency lighting, it must provide illumination of not less than 0.03 foot-candle (measured normal to the direction of the incident light) at the ground end of the erected assist means where an evacuee using the established escape route would normally make first contact with the ground, with the airplane in each of the attitudes corresponding to the collapse of one or more legs of the landing gear.

(2) If the emergency lighting subsystem illuminating the assist means serves no other assist means, is independent of the airplane's main emergency lighting system, and is automatically activated when the assist means is erected, the lighting provisions—

(i) May not be adversely affected by stowage; and

(ii) Must provide illumination of not less than 0.03 foot-candle (measured normal to the direction of incident light) at the ground and of the erected assist means where an evacuee would normally make first contact with the ground, with the airplane in each of the attitudes corresponding to the collapse of one or more legs of the landing gear.

**(i)** The energy supply to each emergency lighting unit must provide the required level of illumination for at least 10 minutes at the critical ambient conditions after emergency landing.

**(j)** If storage batteries are used as the energy supply for the emergency lighting system, they may be recharged from the airplane's main electric power system: *Provided*, That, the charging circuit is designed to preclude inadvertent battery discharge into charging circuit faults.

**(k)** Components of the emergency lighting system, including batteries, wiring relays, lamps, and switches must be capable of normal operation after having been subjected to the inertia forces listed in §25.561(b).

**(l)** The emergency lighting system must be designed so that after any single transverse vertical separation of the fuselage during crash landing—

(1) Not more than 25 percent of all electrically illuminated emergency lights required by this section are rendered inoperative, in addition to the lights that are directly damaged by the separation;

(2) Each electrically illuminated exit sign required under §25.811(d)(2) remains operative exclusive of those that are directly damaged by the separation; and

(3) At least one required exterior emergency light for each side of the airplane remains operative exclusive of those that are directly damaged by the separation.

[Docket No. 5066, 29 FR 18291, Dec. 24, 1964; as amended by Amdt. 25–15, 32 FR 13265, Sept. 20, 1967; Amdt. 25–28, 36 FR 16899, Aug. 26, 1971; Amdt. 25–32, 37 FR 3971, Feb. 24, 1972; Amdt. 25–46, 43 FR 50597, Oct. 30, 1978; Amdt. 25–58, 49 FR 43186, Oct. 26, 1984; Amdt. 25–88, 61 FR 57958, Nov. 8, 1996]

## §25.813 Emergency exit access.

Each required emergency exit must be accessible to the passengers and located where it will afford an effective means of evacuation. Emergency exit distribution must be as uniform as practical, taking passenger distribution into account; however, the size and location of exits on both sides of the cabin need not be symmetrical. If only one floor level exit per side is prescribed, and the airplane does not have a tail cone or ventral emergency exit, the floor level exit must be in the rearward part of the passenger compartment, unless another location affords a more effective means of passenger evacuation. Where more than one floor level exit per side is prescribed, at least one floor level exit per side must be located near each end of the cabin, except that this provision does not apply to combination cargo/passenger configurations. In addition—

**(a)** There must be a passageway leading from the nearest main aisle to each Type A, Type B, Type C, Type I, or Type II emergency exit and between individual passenger areas. Each passageway leading to a Type A or Type B exit must be unobstructed and at least 36 inches wide. Passageways between individual passenger areas and those leading to Type I, Type II, or Type C emergency exits must be unobstructed and at least 20 inches wide. Unless there are two or more main aisles, each Type A or B exit must be located so that there is passenger flow along the main aisle to that exit from both the forward and aft directions. If two or more main aisles are provided, there must be unobstructed cross-aisles at least 20 inches wide between main aisles. There must be—

(1) A cross-aisle which leads directly to each passageway between the nearest main aisle and a Type A or B exit; and

(2) A cross-aisle which leads to the immediate vicinity of each passageway between the nearest main aisle and a Type I, Type II, or Type III exit; except that when two Type III exits are located within three passenger rows of each other, a single cross-aisle may be used if it leads to the vicinity between the passageways from the nearest main aisle to each exit.

**(b)** Adequate space to allow crewmember(s) to assist in the evacuation of passengers must be provided as follows:

(1) The assist space must not reduce the unobstructed width of the passageway below that required for the exit.

(2) For each Type A or Type B exit, assist space must be provided at each side of the exit regardless of whether a means is required by §25.810(a) to assist passengers in descending to the ground from that exit.

(3) Assist space must be provided at one side of any other type exit required by §25.810(a) to have a means to assist passengers in descending to the ground from that exit.

**(c)** The following must be provided for each Type III or Type IV exit—

(1) There must be access from the nearest to each exit. In addition, for each Type III exit in an airplane that has a passenger seating configuration of 60 or more—

(i) Except as provided in paragraph (c)(1)(ii), the access must be provided by an unobstructed passageway that is at least 10 inches in width for interior arrangements in which the adjacent seat rows on the exit side of the aisle contain no more than two seats, or 20 inches in width for interior arrangements in which those rows contain three seats. The width of the passageway must be measured with adjacent seats adjusted to their most adverse position. The centerline of the required passageway width must not be displaced more than 5 inches horizontally from that of the exit.

(ii) In lieu of one 10- or 20-inch passageway, there may be two passageways, between seat rows only, that must be at least 6 inches in width and lead to an unobstructed space adjacent to each exit. (Adjacent exits must not share a common passageway.) The width of the passageways must be measured with adjacent seats adjusted to their most adverse position. The unobstructed space adjacent to the exit must extend vertically from the floor to the ceiling (or bottom of sidewall stowage bins), inboard from the exit for a distance not less than the width of the narrowest passenger seat installed on the airplane, and from the forward edge of the forward passageway to the aft edge of the aft passageway. The exit opening must be totally within the fore and aft bounds of the unobstructed space.

(2) In addition to the access—

(i) For airplanes that have a passenger seating configuration of 20 or more, the projected opening of the exit provided must be obstructed and there must be no interference in opening the exit by seats, berths, or other protrusions (including any seatback in the most adverse position) for a distance from that exit not less than the width of the narrowest passenger seat installed on the airplane.

(ii) For airplanes that have a passenger seating configuration of 19 or fewer, there may be minor obstructions in this region, if there are compensating factors to maintain the effectiveness of the exit.

(3) For each Type III exit, regardless of the passenger capacity of the airplane in which it is installed, there must be placards that—

(i) Are readable by all persons seated adjacent to and facing a passageway to the exit;

(ii) Accurately state or illustrate that proper method of opening the exit including the use of handholds; and

(iii) If the exit is a removable hatch, state the weight of the hatch and indicate an appropriate location to place the hatch after removal.

**(d)** If it is necessary to pass through a passageway between passenger compartments to reach any required emergency exit from any seat in the passenger cabin, the passageway must be unobstructed. However, curtains may be used if they allow free entry through the passageway.

**(e)** No door may be installed in any partition between passenger compartments.

**(f)** If it is necessary to pass through a doorway separating the passenger cabin from other areas to reach any required emergency exit from any passenger seat, the door must have a means to latch it in open position. The latching means must be able to withstand the loads imposed upon it when the door is subjected to the ultimate inertia forces, relative to the surrounding structure, listed in §25.561(b).

[Docket No. 5066, 29 FR 18291, Dec. 24, 1964; as amended by Amdt. 25–1, 30 FR 3204, March 9, 1965; Amdt. 25–15, 32 FR 13265, Sept. 20, 1967; Amdt. 25–32, 37 FR 3971, Feb. 24, 1972; Amdt. 25–46, 43 FR 50597, Oct. 30, 1978; Amdt. 25–72, 55 FR 29783, July 20, 1990; Amdt. 25–76, 57 FR 19244, May 4, 1992; Amdt. 25–76, 57 FR 29120, June 30, 1992; Amdt. 25–88, 61 FR 57958, Nov. 8, 1996]

## §25.815 Width of aisle.

The passenger aisle width at any point between seats must equal or exceed the values in the following table:

| Passenger seating capacity | Minimum passenger aisle width (inches) | |
|---|---|---|
| | Less than 25 in. from floor | 25 in. and more from floor |
| 10 or less | [1]12 | 15 |
| 11 through 19 | 12 | 20 |
| 20 or more | 15 | 20 |

[1]A narrower width not less than 9 inches may be approved when substantiated by tests found necessary by the Administrator.

[Docket No. 5066, 29 FR 18291, Dec. 24, 1964; as amended by Amdt. 25–15, 32 FR 13265, Sept. 20, 1967; Amdt. 25–38, 41 FR 55466, Dec. 20, 1976]

## §25.817 Maximum number of seats abreast.

On airplanes having only one passenger aisle, no more than three seats abreast may be placed on each side of the aisle in any one row.

[Docket No. 5066, 29 FR 18291, Dec. 24, 1964; as amended by Amdt. 25–15, 32 FR 13265, Sept. 20, 1967]

## §25.819 Lower deck service compartments (including galleys).

For airplanes with a service compartment located below the main deck, which may be occupied during taxi or flight but not during takeoff or landing, the following apply:

**(a)** There must be at least two emergency evacuation routes, one at each end of each lower deck service compartment or two having sufficient separation within each compartment, which could be used by each occupant of the lower deck service compartment to rapidly evacuate to the main deck under normal and emergency lighting conditions. The routes must provide for the evacuation of incapacitated persons, with assistance. The use of the evacuation routes may not be dependent on any powered device. The routes must be designed to minimize the possibility of blockage which might result from fire, mechanical or structural failure, or persons standing on top of or against the escape routes. In the event the airplane's main power system or compartment main lighting system should fail, emergency illumination for each lower deck service compartment must be automatically provided.

**(b)** There must be a means for two-way voice communication between the flight deck and each lower deck service compartment.

**(c)** There must be an aural emergency alarm system, audible during normal and emergency conditions, to enable crewmembers on the flight deck and at each required floor level emergency exit to alert occupants of each lower deck service compartment of an emergency situation.

**(d)** There must be a means, readily detectable by occupants of each lower deck service compartment, that indicates when seat belts should be fastened.

**(e)** If a public address system is installed in the airplane, speakers must be provided in each lower deck service compartment.

**(f)** For each occupant permitted in a lower deck service compartment, there must be a forward or aft facing seat which meets the requirements of §25.785(c) and must be able to withstand maximum flight loads when occupied.

**(g)** For each powered lift system installed between a lower deck service compartment and the main deck for the carriage of persons or equipment, or both, the system must meet the following requirements:

(1) Each lift control switch outside the lift, except emergency stop buttons, must be designed to prevent the activation of the life if the lift door, or the hatch required by paragraph (g)(3) of this section, or both are open.

(2) An emergency stop button, that when activated will immediately stop the lift, must be installed within the lift and at each entrance to the lift.

(3) There must be a hatch capable of being used for evacuating persons from the lift that is openable from inside and outside the lift without tools, with the lift in any position.

[Docket No. 5066, 29 FR 18291, Dec. 24, 1964; as amended by Amdt. 25–53, 45 FR 41593, June 19, 1980; 45 FR 43154, June 26, 1980]

25

### Ventilation and Heating

## §25.831 Ventilation.

**(a)** Under normal operating conditions and in the event of any probable failure conditions of any system which would adversely affect the ventilating air, the ventilation system must be designed to provide a sufficient amount of uncontaminated air to enable the crewmembers to perform their duties without undue discomfort or fatigue and to provide reasonable passenger comfort. For normal operating conditions, the ventilation system must be designed to provide each occupant with an airflow containing at least 0.55 pounds of fresh air per minute.

**(b)** Crew and passenger compartment air must be free from harmful or hazardous concentrations of gases or vapors. In meeting this requirement, the following apply:

(1) Carbon monoxide concentrations in excess of 1 part in 20,000 parts of air are considered hazardous. For test purposes, any acceptable carbon monoxide detection method may be used.

(2) Carbon dioxide concentration during flight must be shown not to exceed 0.5 percent by volume (sea level equivalent) in compartments normally occupied by passengers or crewmembers.

**(c)** There must be provisions made to ensure that the conditions prescribed in paragraph (b) of this section are met after reasonably probable failures or malfunctioning of the ventilating, heating, pressurization, or other systems and equipment.

**(d)** If accumulation of hazardous quantities of smoke in the cockpit area is reasonably probable, smoke evacuation must be readily accomplished, starting with full pressurization and without depressurizing beyond safe limits.

**(e)** Except as provided in paragraph (f) of this section, means must be provided to enable the occupants of the following compartments and areas to control the temperature and quantity of ventilating air supplied to their compartment or area independently of the temperature and quantity of air supplied to other compartments and areas:

(1) The flight crew compartment.

(2) Crewmember compartments and areas other than the flight crew compartment unless the crewmember compartment or area is ventilated by air interchange with other compartments or areas under all operating conditions.

**(f)** Means to enable the flight crew to control the temperature and quantity of ventilating air supplied to the flight crew compartment independently of the temperature and quantity of ventilating air supplied to other compartments are not required if all of the following conditions are met:

(1) The total volume of the flight crew and passenger compartments is 800 cubic feet or less.

(2) The air inlets and passages for air to flow between flight crew and passenger compartments are arranged to provide compartment temperatures within 5 degrees F of each other and adequate ventilation to occupants in both compartments.

(3) The temperature and ventilation controls are accessible to the flight crew.

**(g)** The exposure time at any given temperature must not exceed the values shown in the following graph after any improbable failure condition.

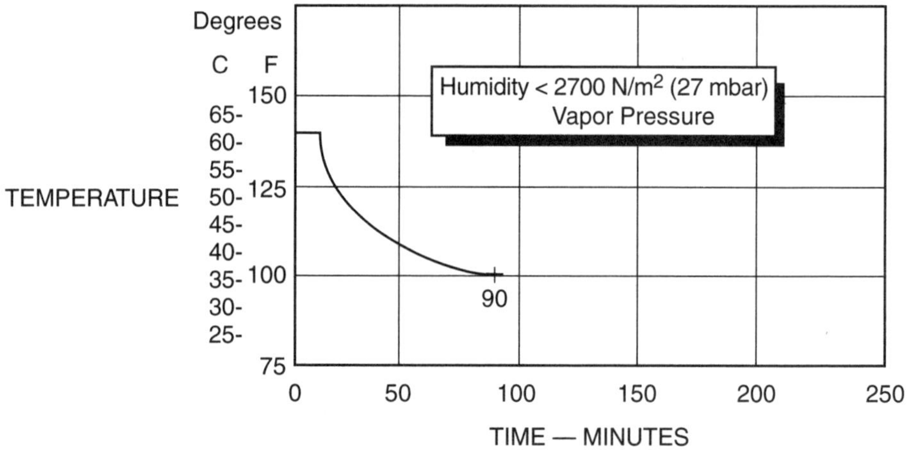


TIME — TEMPERATURE RELATIONSHIP

[Docket No. 5066, 29 FR 18291, Dec. 24, 1964; as amended by Amdt. 25–41, 42 FR 36970, July 18, 1977; Amdt. 25–87, 61 FR 28695, June 5, 1996; Amdt. 25–89, 61 FR 63956, Dec. 2, 1996]

## §25.832 Cabin ozone concentration.

**(a)** The airplane cabin ozone concentration during flight must be shown not to exceed—

(1) 0.25 parts per million by volume, sea level equivalent, at any time above flight level 320; and

(2) 0.1 parts per million by volume, sea level equivalent, time-weighted average during any 3-hour interval above flight level 270.

**(b)** For the purpose of this section, "sea level equivalent" refers to conditions of 25°C and 760 millimeters of mercury pressure.

**(c)** Compliance with this section must be shown by analysis or tests based on airplane operational procedures and performance limitations, that demonstrate that either—

(1) The airplane cannot be operated at an altitude which would result in cabin ozone concentrations exceeding the limits prescribed by paragraph (a) of this section; or

(2) The airplane ventilation system, including any ozone control equipment, will maintain cabin ozone concentrations at or below the limits prescribed by paragraph (a) of this section.

[Docket No. 5066, 29 FR 18291, Dec. 24, 1964; as amended by Amdt. 25–50, 45 FR 3883, Jan. 1, 1980; Amdt. 25–56, 47 FR 58489, Dec. 30, 1982; Amdt. 25–94, 63 FR 8848, Feb. 23, 1998]

## §25.833 Combustion heating systems.

Combustion heaters must be approved.

[Docket No. 5066, 29 FR 18291, Dec. 24, 1964; as amended by Amdt. 25–72, 55 FR 29783, July 20, 1990]

PRESSURIZATION

## §25.841 Pressurized cabins.

**(a)** Pressurized cabins and compartments to be occupied must be equipped to provide a cabin pressure altitude of not more than 8,000 feet at the maximum operating altitude of the airplane under normal operating conditions.

(1) If certification for operation above 25,000 feet is requested, the airplane must be designed so that occupants will not be exposed to cabin pressure altitudes in excess of 15,000 feet after any probable failure condition in the pressurization system.

(2) The airplane must be designed so that occupants will not be exposed to a cabin pressure altitude that exceeds the following after decompression from any failure condition not shown to be extremely improbable:

(i) Twenty-five thousand (25,000) feet for more than 2 minutes; or

(ii) Forty thousand (40,000) feet for any duration.

(3) Fuselage structure, engine and system failures are to be considered in evaluating the cabin decompression.

**(b)** Pressurized cabins must have at least the following valves, controls, and indicators for controlling cabin pressure:

(1) Two pressure relief valves to automatically limit the positive pressure differential to a predetermined value at the maximum rate of flow delivered by the pressure source. The combined capacity of the relief valves must be large enough so that the failure of any one valve would not cause an appreciable rise in the pressure differential. The pressure differential is positive when the internal pressure is greater than the external.

(2) Two reverse pressure differential relief valves (or their equivalents) to automatically prevent a negative pressure differential that would damage the structure. One valve is enough, however, if it is of a design that reasonably precludes its malfunctioning.

(3) A means by which the pressure differential can be rapidly equalized.

(4) An automatic or manual regulator for controlling the intake or exhaust airflow, or both, for maintaining the required internal pressures and airflow rates.

(5) Instruments at the pilot or flight engineer station to show the pressure differential, the cabin pressure altitude, and the rate of change of the cabin pressure altitude.

(6) Warning indication at the pilot or flight engineer station to indicate when the safe or preset pressure differential and cabin pressure altitude limits are exceeded. Appropriate warning markings on the cabin pressure differential indicator meet the warning requirement for pressure differential limits and an aural or visual signal (in addition to cabin altitude indicating means) meets the warning requirement for cabin pressure altitude limits if it warns the flight crew when the cabin pressure altitude exceeds 10,000 feet.

(7) A warning placard at the pilot or flight engineer station if the structure is not designed for pressure differentials up to the maximum relief valve setting in combination with landing loads.

(8) The pressure sensors necessary to meet the requirements of paragraphs (b)(5) and (b)(6) of this section and §25.1447(c), must be located and the sensing system designed so that, in the event of loss of cabin pressure in any passenger or crew compartment (including upper and lower lobe galleys), the warning and automatic presentation devices, required by those provisions, will be actuated without any delay that would significantly increase the hazards resulting from decompression.

[Docket No. 5066, 29 FR 18291, Dec. 24, 1964; as amended by Amdt. 25–38, 41 FR 55466, Dec. 20, 1976; Amdt. 25–87, 61 FR 28696, June 5, 1996]

## §25.843 Tests for pressurized cabins.

**(a)** *Strength test.* The complete pressurized cabin, including doors, windows, and valves, must be tested as a pressure vessel for the pressure differential specified in §25.365(d).

**(b)** *Functional tests.* The following functional tests must be performed:

(1) Tests of the functioning and capacity of the positive and negative pressure differential valves, and of the emergency release valve, to stimulate the effects of closed regulator valves.

(2) Tests of the pressurization system to show proper functioning under each possible condition of pressure, temperature, and moisture, up to the maximum altitude for which certification is requested.

(3) Flight tests, to show the performance of the pressure supply, pressure and flow regulators, indicators, and warning signals, in steady and stepped climbs and descents at rates corresponding to the maximum attainable within the operating limitations of the airplane, up to the maximum altitude for which certification is requested.

(4) Tests of each door and emergency exit, to show that they operate properly after being subjected to the flight tests prescribed in paragraph (b)(3) of this section.

25

FIRE PROTECTION

## §25.851 Fire extinguishers.

**(a)** *Hand fire extinguishers.*

(1) The following minimum number of hand fire extinguishers must be conveniently located and evenly distributed in passenger compartments:

| Passenger capacity | No. of extinguishers |
|---|---|
| 7 through 30 | 1 |
| 31 through 60 | 2 |
| 61 through 200 | 3 |
| 201 through 300 | 4 |
| 301 through 400 | 5 |
| 401 through 500 | 6 |
| 501 through 600 | 7 |
| 601 through 700 | 8 |

(2) At least one hand fire extinguisher must be conveniently located in the pilot compartment.

(3) At least one readily accessible hand fire extinguisher must be available for use in each Class A or Class B cargo or baggage compartment and in each Class E cargo or baggage compartment that is accessible to crewmembers in flight.

(4) At least one hand fire extinguisher must be located in, or readily accessible for use in, each galley located above or below the passenger compartment.

(5) Each hand fire extinguisher must be approved.

(6) At least one of the required fire extinguishers located in the passenger compartment of an airplane with a passenger capacity of at least 31 and not more than 60, and at least two of the fire extinguishers located in the passenger compartment of an airplane with a passenger capacity of 61 or more must contain Halon 1211 (bromochlorodifluoromethane $CBrC_1F_2$), or equivalent, as the extinguishing agent. The type of extinguishing agent used in any other extinguisher required by this section must be appropriate for the kinds of fires likely to occur where used.

(7) The quantity of extinguishing agent used in each extinguisher required by this section must be appropriate for the kinds of fires likely to occur where used.

(8) Each extinguisher intended for use in a personnel compartment must be designed to minimize the hazard of toxic gas concentration.

**(b)** Built-in fire extinguishers. If a built-in fire extinguisher is provided—

(1) Each built-in fire extinguishing system must be installed so that—

(i) No extinguishing agent likely to enter personnel compartments will be hazardous to the occupants; and

(ii) No discharge of the extinguisher can cause structural damage.

(2) The capacity of each required built-in fire extinguishing system must be adequate for any fire likely to occur in the compartment where used, considering the volume of the compartment and the ventilation rate.

[Docket No. 5066, 29 FR 18291, Dec. 24, 1964; as amended by Amdt. 25-74, 56 FR 15456, April 16, 1991]

## §25.853 Compartment interiors.

For each compartment occupied by the crew or passengers, the following apply:

**(a)** Materials (including finishes or decorative surfaces applied to the materials) must meet the applicable test criteria prescribed in Part I of Appendix F of this part, or other approved equivalent methods, regardless of the passenger capacity of the airplane.

**(b)** [Reserved]

**(c)** In addition to meeting the requirements of paragraph (a) of this section, seat cushions, except those on flight crewmember seats, must meet the test requirements of part II of Appendix F of this part, or other equivalent methods, regardless of the passenger capacity of the airplane.

**(d)** Except as provided in paragraph (e) of this section, the following interior components of airplanes with passenger capacities of 20 or more must also meet the test requirements of parts IV and V of Appendix F of this part, or other approved equivalent method, in addition to the flammability requirements prescribed in paragraph (a) of this section:

(1) Interior ceiling and wall panels, other than lighting lenses and windows;

(2) Partitions, other than transparent panels needed to enhance cabin safety;

(3) Galley structure, including exposed surfaces of stowed carts and standard containers and the cavity walls that are exposed when a full complement of such carts or containers is not carried; and

(4) Large cabinets and cabin stowage compartments, other than underseat stowage compartments for stowing small items such as magazines and maps.

**(e)** The interiors of compartments, such as pilot compartments, galleys, lavatories, crew rest quarters, cabinets and stowage compartments, need not meet the standards of paragraph (d) of this section, provided the interiors of such compartments are isolated from the main passenger cabin by doors or equivalent means that would normally be closed during an emergency landing condition.

**(f)** Smoking is not to be allowed in lavatories. If smoking is to be allowed in any other compartment occupied by the crew or passengers, an adequate number of self-contained, removable ashtrays must be provided for all seated occupants.

**(g)** Regardless of whether smoking is allowed in any other part of the airplane, lavatories must have self-contained, removable ashtrays located conspicuously on or near the entry side of each lavatory door, except that one ashtray may serve more than one lavatory door if the ashtray can be seen readily from the cabin side of each lavatory served.

**(h)** Each receptacle used for the disposal of flammable waste material must be fully enclosed, constructed of at least fire resistant materials, and must contain fires likely to occur in it under normal use. The capability of the receptacle to contain those fires under all probable conditions of wear, misalignment, and ventilation expected in service must be demonstrated by test.

[Docket No. 5066, 29 FR 18291, Dec. 24, 1964; as amended by Amdt. 25–72, 55 FR 29783, July 20, 1990; Amdt. 25–83, 60 FR 6623, Feb. 2, 1995]

## §25.854 Lavatory fire protection.

For airplanes with a passenger capacity of 20 or more:

**(a)** Each lavatory must be equipped with a smoke detector system or equivalent that provides a warning light in the cockpit, or provides a warning light or audible warning in the passenger cabin that would be readily detected by a flight attendant; and

**(b)** Each lavatory must be equipped with a built-in fire extinguisher for each disposal receptacle for towels, paper, or waste, located within the lavatory. The extinguisher must be designed to discharge automatically into each disposal receptacle upon occurrence of a fire in that receptacle.

[Docket No. 5066, 29 FR 18291, Dec. 24, 1964; as amended by Amdt. 25–74, 56 FR 15456, April 16, 1991]

## §25.855 Cargo or baggage compartments.

For each cargo and baggage compartment not occupied by crew or passengers, the following apply:

**(a)** The compartment must meet one of the class requirements of §25.857.

**(b)** Class B through Class E cargo or baggage compartments, as defined in §25.857, must have a liner, and the liner must be separate from (but may be attached to) the airplane structure.

**(c)** Ceiling and sidewall liner panels of Class C compartments must meet the test requirements of part III of Appendix F of this part or other approved equivalent methods.

**(d)** All other materials used in the construction of the cargo or baggage compartment must meet the applicable test criteria prescribed in part I of Appendix F of this part or other approved equivalent methods.

**(e)** No compartment may contain any controls, wiring, lines, equipment, or accessories whose damage or failure would affect safe operation, unless those items are protected so that—

(1) They cannot be damaged by the movement of cargo in the compartment, and

(2) Their breakage or failure will not create a fire hazard.

**(f)** There must be means to prevent cargo or baggage from interfering with the functioning of the fire protective features of the compartment.

**(g)** Sources of heat within the compartment must be shielded and insulated to prevent igniting the cargo or baggage.

**(h)** Flight tests must be conducted to show compliance with the provisions of §25.857 concerning—

(1) Compartment accessibility,

(2) The entries of hazardous quantities of smoke or extinguishing agent into compartments occupied by the crew or passengers, and

(3) The dissipation of the extinguishing agent in Class C compartments.

**(i)** During the above tests, it must be shown that no inadvertent operation of smoke or fire detectors in any compartment would occur as a result of fire contained in any other compartment, either during or after extinguishment, unless the extinguishing system floods each such compartment simultaneously.

[Docket No. 5066, 29 FR 18291, Dec. 24, 1964; as amended by Amdt. 25–72, 55 FR 29784, July 20, 1990; Amdt. 25–93, 63 FR 8048, Feb. 17, 1998]

## §25.857 Cargo compartment classification.

**(a)** *Class A.* A Class A cargo or baggage compartment is one in which—

(1) The presence of a fire would be easily discovered by a crewmember while at his station; and

(2) Each part of the compartment is easily accessible in flight.

**(b)** *Class B.* A Class B cargo or baggage compartment is one in which—

(1) There is sufficient access in flight to enable a crewmember to effectively reach any part of the compartment with the contents of a hand fire extinguisher;

(2) When the access provisions are being used, no hazardous quantity of smoke, flames, or extinguishing agent, will enter any compartment occupied by the crew or passengers;

(3) There is a separate approved smoke detector or fire detector system to give warning at the pilot or flight engineer station.

**(c)** *Class C* A Class C cargo or baggage compartment is one not meeting the requirements for either a Class A or B compartment but in which—

(1) There is a separate approved smoke detector or fire detector system to give warning at the pilot or flight engineer station;

(2) There is an approved built-in fire extinguishing or suppression system controllable from the cockpit;

(3) There are means to exclude hazardous quantities of smoke, flames, or extinguishing agent, from any compartment occupied by the crew or passengers;

(4) There are means to control ventilation and drafts within the compartment so that the extinguishing agent used can control any fire that may start within the compartment.

**(d)** [Reserved]

(4) [Reserved]

(5) Consideration is given to the effect of heat within the compartment on adjacent critical parts of the airplane. For compartments of 500 cu. ft. or less, an airflow of 1500 cu. ft. per hour is acceptable.

(6) The compartment volume does not exceed 1,000 cubic feet.

**(e)** *Class E.* A Class E cargo compartment is one on airplanes used only for the carriage of cargo and in which—

(1) [Reserved]

(2) There is a separate approved smoke or fire detector system to give warning at the pilot or flight engineer station;

(3) There are means to shut off the ventilating airflow to, or within, the compartment, and the controls for these means are accessible to the flight crew in the crew compartment;

(4) There are means to exclude hazardous quantities of smoke, flames, or noxious gases, from the flight crew compartment; and

(5) The required crew emergency exits are accessible under any cargo loading condition.

[Docket No. 5066, 29 FR 18291, Dec. 24, 1964; as amended by Amdt. 25–32, 37 FR 3972, Feb. 24, 1972; Amdt. 25–60, 51 FR 18243, May 16, 1986; Amdt. 25–93, 63 FR 8048, Feb. 17, 1998]

## §25.858 Cargo or baggage compartment smoke or fire detection systems.

If certification with cargo or baggage compartment smoke or fire detection provisions is requested, the following must be met for each cargo or baggage compartment with those provisions:

**(a)** The detection system must provide a visual indication to the flight crew within one minute after the start of a fire.

**(b)** The system must be capable of detecting a fire at a temperature significantly below that at which the structural integrity of the airplane is substantially decreased.

**(c)** There must be means to allow the crew to check in flight, the functioning of each fire detector circuit.

**(d)** The effectiveness of the detection system must be shown for all approved operating configurations and conditions.

[Docket No. 5066, 29 FR 18291, Dec. 24, 1964; as amended by Amdt. 25–54, 45 FR 60173, Sept. 11, 1980; Amdt. 25–93, 63 FR 8048, Feb. 17, 1998]

## §25.859 Combustion heater fire protection.

**(a)** *Combustion heater fire zones.* The following combustion heater fire zones must be protected from fire in accordance with the applicable provisions of §§25.1181 through 25.1191 and §§25.1195 through 25.1203;

(1) The region surrounding the heater, if this region contains any flammable fluid system components (excluding the heater fuel system), that could—

(i) Be damaged by heater malfunctioning; or

(ii) Allow flammable fluids or vapors to reach the heater in case of leakage.

(2) The region surrounding the heater, if the heater fuel system has fittings that, if they leaked, would allow fuel or vapors to enter this region.

(3) The part of the ventilating air passage that surrounds the combustion chamber. However, no fire extinguishment is required in cabin ventilating air passages.

**(b)** *Ventilating air ducts.* Each ventilating air duct passing through any fire zone must be fireproof. In addition—

(1) Unless isolation is provided by fireproof valves or by equally effective means, the ventilating air duct downstream of each heater must be fireproof for a distance great enough to ensure that any fire originating in the heater can be contained in the duct; and

(2) Each part of any ventilating duct passing through any region having a flammable fluid system must be constructed or isolated from that system so that the malfunctioning of any component of that system cannot introduce flammable fluids or vapors into the ventilating airstream.

**(c)** *Combustion air ducts.* Each combustion air duct must be fireproof for a distance great enough to prevent damage from backfiring or reverse flame propagation. In addition—

(1) No combustion air duct may have a common opening with the ventilating airstream unless flames from backfires or reverse burning cannot enter the ventilating airstream under any operating condition, including reverse flow or malfunctioning of the heater or its associated components; and

(2) No combustion air duct may restrict the prompt relief of any backfire that, if so restricted, could cause heater failure.

**(d)** *Heater controls; general.* Provision must be made to prevent the hazardous accumulation of water or ice on or in any heater control component, control system tubing, or safety control.

**(e)** *Heater safety controls.* For each combustion heater there must be the following safety control means:

(1) Means independent of the components provided for the normal continuous control of air temperature, airflow, and fuel flow must be provided, for each heater, to automatically shut off the ignition and fuel supply to that heater at a point remote from that heater when any of the following occurs:

(i) The heat exchanger temperature exceeds safe limits.

(ii) The ventilating air temperature exceeds safe limits.

(iii) The combustion airflow becomes inadequate for safe operation.

(iv) The ventilating airflow becomes inadequate for safe operation.

(2) The means of complying with paragraph (e)(1) of this section for any individual heater must—

(i) Be independent of components serving any other heater whose heat output is essential for safe operation; and

(ii) Keep the heater off until restarted by the crew.

(3) There must be means to warn the crew when any heater whose heat output is essential for safe operation has been shut off by the automatic means prescribed in paragraph (e)(1) of this section.

**(f)** *Air intakes.* Each combustion and ventilating air intake must be located so that no flammable fluids or vapors can enter the heater system under any operating condition—

(1) During normal operation; or

(2) As a result of the malfunctioning of any other component.

**(g)** *Heater exhaust.* Heater exhaust systems must meet the provisions of §§25.1121 and 25.1123. In addition, there must be provisions in the design of the heater exhaust system to safely expel the products of combustion to prevent the occurrence of—

(1) Fuel leakage from the exhaust to surrounding compartments;

(2) Exhaust gas impingement on surrounding equipment or structure;

(3) Ignition of flammable fluids by the exhaust, if the exhaust is in a compartment containing flammable fluid lines; and

(4) Restriction by the exhaust of the prompt relief of backfires that, if so restricted, could cause heater failure.

**(h)** *Heater fuel systems.* Each heater fuel system must meet each powerplant fuel system requirement affecting safe heater operation. Each heater fuel system component within the ventilating airstream must be protected by shrouds so that no leakage from those components can enter the ventilating airstream.

**(i)** *Drains.* There must be means to safely drain fuel that might accumulate within the combustion chamber or the heat exchanger. In addition—

(1) Each part of any drain that operates at high temperatures must be protected in the same manner as heater exhausts; and

(2) Each drain must be protected from hazardous ice accumulation under any operating condition.

[Docket No. 5066, 29 FR 18291, Dec. 24 1964; as amended by Amdt. 25–11, 32 FR 6912, May 5, 1967; Amdt. 25–23, 35 FR 5676, April 8, 1970]

## §25.863 Flammable fluid fire protection.

**(a)** In each area where flammable fluids or vapors might escape by leakage of a fluid system, there must be means to minimize the probability of ignition of the fluids and vapors, and the resultant hazards if ignition does occur.

**(b)** Compliance with paragraph (a) of this section must be shown by analysis or tests, and the following factors must be considered:

(1) Possible sources and paths of fluid leakage, and means of detecting leakage.

(2) Flammability characteristics of fluids, including effects of any combustible or absorbing materials.

(3) Possible ignition sources, including electrical faults, overheating of equipment, and malfunctioning of protective devices.

(4) Means available for controlling or extinguishing a fire, such as stopping flow of fluids, shutting down equipment, fireproof containment, or use of extinguishing agents.

(5) Ability of airplane components that are critical to safety of flight to withstand fire and heat.

**(c)** If action by the flight crew is required to prevent or counteract a fluid fire (e.g., equipment shutdown or actuation of a fire extinguisher) quick acting means must be provided to alert the crew.

**(d)** Each area where flammable fluids or vapors might escape by leakage of a fluid system must be identified and defined.

[Docket No. 5066, 29 FR 18291, Dec. 24, 1964; as amended by Amdt. 25–23, 35 FR 5676, April 8, 1970; Amdt. 25–46, 43 FR 50597, Oct. 30, 1978]

### §25.865 Fire protection of flight controls, engine mounts, and other flight structure.

Essential flight controls, engine mounts, and other flight structures located in designated fire zones or in adjacent areas which would be subjected to the effects of fire in the fire zone must be constructed of fireproof material or shielded so that they are capable of withstanding the effects of fire.

[Docket No. 5066, 29 FR 18291, Dec. 24, 1964; as amended by Amdt. 25–23, 35 FR 5676, April 8, 1970]

### §25.867 Fire protection: other components.

**(a)** Surfaces to the rear of the nacelles, within one nacelle diameter of the nacelle centerline, must be at least fire-resistant.

**(b)** Paragraph (a) of this section does not apply to tail surfaces to the rear of the nacelles that could not be readily affected by heat, flames, or sparks coming from a designated fire zone or engine compartment of any nacelle.

[Docket No. 5066, 29 FR 18291, Dec. 24, 1964; as amended by Amdt. 25–23, 35 FR 5676, April 8, 1970]

### §25.869 Fire protection: systems.

**(a)** Electrical system components:

(1) Components of the electrical system must meet the applicable fire and smoke protection requirements of §§25.831(c) and 25.863.

(2) Electrical cables, terminals, and equipment in designated fire zones, that are used during emergency procedures, must be at least fire resistant.

(3) Main power cables (including generator cables) in the fuselage must be designed to allow a reasonable degree of deformation and stretching without failure and must be—

(i) Isolated from flammable fluid lines; or

(ii) Shrouded by means of electrically insulated, flexible conduit, or equivalent, which is in addition to the normal cable insulation.

(4) Insulation on electrical wire and electrical cable installed in any area of the fuselage must be self-extinguishing when tested in accordance with the applicable portions of part I, Appendix F of this part.

**(b)** Each vacuum air system line and fitting on the discharge side of the pump that might contain flammable vapors or fluids must meet the requirements of §25.1183 if the line or fitting is in a designated fire zone. Other vacuum air systems components in designated fire zones must be at least fire resistant.

**(c)** Oxygen equipment and lines must—

(1) Not be located in any designated fire zone,

(2) Be protected from heat that may be generated in, or escape from, any designated fire zone, and

(3) Be installed so that escaping oxygen cannot cause ignition of grease, fluid, or vapor accumulations that are present in normal operation or as a result of failure or malfunction of any system.

[Docket No. 5066, 29 FR 18291, Dec. 24, 1964; as amended by Amdt. 25–72, 55 FR 29784, July 20, 1990]

MISCELLANEOUS

### §25.871 Leveling means.

There must be means for determining when the airplane is in a level position on the ground.

[Docket No. 5066, 29 FR 18291, Dec. 24, 1964; as amended by Amdt. 25–23, 35 FR 5676, April 8, 1970]

### §25.875 Reinforcement near propellers.

**(a)** Each part of the airplane near the propeller tips must be strong and stiff enough to withstand the effects of the induced vibration and of ice thrown from the propeller.

**(b)** No window may be near the propeller tips unless it can withstand the most severe ice impact likely to occur.

## Subpart E—Powerplant

GENERAL

### §25.901 Installation.

**(a)** For the purpose of this part, the airplane powerplant installation includes each component that—

(1) Is necessary for propulsion;

(2) Affects the control of the major propulsive units; or

(3) Affects the safety of the major propulsive units between normal inspections or overhauls.

**(b)** For each powerplant—

(1) The installation must comply with—

(i) The installation instructions provided under §33.5 of this chapter; and

(ii) The applicable provisions of this subpart;

(2) The components of the installation must be constructed, arranged, and installed so as to ensure their continued safe operation between normal inspections or overhauls;

(3) The installation must be accessible for necessary inspections and maintenance; and

(4) The major components of the installation must be electrically bonded to the other parts of the airplane.

**(c)** For each powerplant and auxiliary power unit installation, it must be established that no single failure or malfunction or probable combination of failures will jeopardize the safe operation of the airplane except that the failure of structural elements need not be considered if the probability of such failure is extremely remote.

**(d)** Each auxiliary power unit installation must meet the applicable provisions of this subpart.

[Docket No. 5066, 29 FR 18291, Dec. 24, 1964; as amended by Amdt. 25–23, 35 FR 5676, April 8, 1970; Amdt. 25–40, 42 FR 15042, March 17, 1977; Amdt. 25–46, 43 FR 50597, Oct. 30, 1978]

## §25.903 Engines.

**(a)** *Engine type certificate.*

(1) Each engine must have a type certificate and must meet the applicable requirements of part 34 of this chapter.

(2) Each turbine engine must comply with one of the following:

(i) Sections 33.76, 33.77 and 33.78 of this chapter in effect on December 13, 2000, or as subsequently amended; or

(ii) Sections 33.77 and 33.78 of this chapter in effect on April 30, 1998, or as subsequently amended before December 13, 2000; or

(iii) Comply with §33.77 of this chapter in effect on October 31, 1974, or as subsequently amended prior to April 30, 1998, unless that engine's foreign object ingestion service history has resulted in an unsafe condition; or

(iv) Be shown to have a foreign object ingestion service history in similar installation locations which has not resulted in any unsafe condition.

**(b)** *Engine isolation.* The powerplants must be arranged and isolated from each other to allow operation, in at least one configuration, so that the failure or malfunction of any engine, or of any system that can affect the engine, will not—

(1) Prevent the continued safe operation of the remaining engines; or

(2) Require immediate action by any crewmember for continued safe operation.

**(c)** *Control of engine rotation.* There must be means for stopping the rotation of any engine individually in flight, except that, for turbine engine installations, the means for stopping the rotation of any engine need be provided only where continued rotation could jeopardize the safety of the airplane. Each component of the stopping system on the engine side of the firewall that might be exposed to fire must be at least fire-resistant. If hydraulic propeller feathering systems are used for this purpose, the feathering lines must be at least fire resistant under the operating conditions that may be expected to exist during feathering.

**(d)** *Turbine engine installations.* For turbine engine installations—

(1) Design precautions must be taken to minimize the hazards to the airplane in the event of an engine rotor failure or of a fire originating within the engine which burns through the engine case.

(2) The powerplant systems associated with engine control devices, systems, and instrumentation, must be designed to give reasonable assurance that those engine operating limitations that adversely affect turbine rotor structural integrity will not be exceeded in service.

**(e)** *Restart capability.*

(1) Means to restart any engine in flight must be provided.

(2) An altitude and airspeed envelope must be established for in-flight engine restarting, and each engine must have a restart capability within that envelope.

(3) For turbine engine powered airplanes, if the minimum windmilling speed of the engines, following the inflight shutdown of all engines, is insufficient to provide the necessary electrical power for engine ignition, a power source independent of the engine-driven electrical power generating system must be provided to permit inflight engine ignition for restarting.

**(f)** *Auxiliary Power Unit.* Each auxiliary power unit must be approved or meet the requirements of the category for its intended use.

[Docket No. 5066, 29 FR 18291, Dec. 24, 1964; as amended by Amdt. 25–23, 35 FR 5676, April 8, 1970; Amdt. 25–40, 42 FR 15042, March 17, 1977; Amdt. 25–57, 49 FR 6848, Feb. 23, 1984; Amdt. 25–72, 55 FR 29784, July 20, 1990; Amdt. 25–73, 55 FR 32861, Aug. 10, 1990; Amdt. 25–94, 63 FR 8848, Feb. 23, 1998; Amdt. 25–95, 63 FR 14798, March 26, 1998; Amdt. 25–100, 65 FR 55854, Sept. 14, 2000]

## §25.904 Automatic takeoff thrust control system (ATTCS).

Each applicant seeking approval for installation of an engine power control system that automatically resets the power or thrust on the operating engine(s) when any engine fails during the takeoff must comply with the requirements of Appendix I of this part.

[Docket No. 5066, 29 FR 18291, Dec. 24, 1964; as amended by Amdt. 25–62, 52 FR 43156, Nov. 9, 1987]

## §25.905 Propellers.

**(a)** Each propeller must have a type certificate.

**(b)** Engine power and propeller shaft rotational speed may not exceed the limits for which the propeller is certificated.

**(c)** Each component of the propeller blade pitch control system must meet the requirements of §35.42 of this chapter.

**(d)** Design precautions must be taken to minimize the hazards to the airplane in the event a propeller blade fails or is released by a hub failure. The hazards which must be considered include damage to structure and vital systems due to impact of a failed or released blade and the unbalance created by such failure or release.

[Docket No. 5066, 29 FR 18291, Dec. 24, 1964; as amended by Amdt. 25–54, 45 FR 60173, Sept. 11, 1980; Amdt. 25–57, 49 FR 6848, Feb. 23, 1984; Amdt. 25–72, 55 FR 29784, July 20, 1990]

## §25.907 Propeller vibration.

**(a)** The magnitude of the propeller blade vibration stresses under any normal condition of operation must be determined by actual measurement or by comparison with similar installations for which these measurements have been made.

**(b)** The determined vibration stresses may not exceed values that have been shown to be safe for continuous operation.

## §25.925 Propeller clearance.

Unless smaller clearances are substantiated, propeller clearances with the airplane at maximum weight, with the most adverse center of gravity, and with the propeller in the most adverse pitch position, may not be less than the following:

**(a)** *Ground clearance.* There must be a clearance of at least seven inches (for each airplane with nose wheel landing gear) or nine inches (for each airplane with tail wheel landing gear) between each propeller and the ground with the landing gear statically deflected and in the level takeoff, or taxiing attitude, whichever is most critical. In addition, there must be positive clearance between the propeller and the ground when in the level takeoff attitude with the critical tire(s) completely deflated and the corresponding landing gear strut bottomed.

**(b)** *Water clearance.* There must be a clearance of at least 18 inches between each propeller and the water, unless compliance with §25.239(a) can be shown with a lesser clearance.

**(c)** *Structural clearance.* There must be—

(1) At least one inch radial clearance between the blade tips and the airplane structure, plus any additional radial clearance necessary to prevent harmful vibration;

(2) At least one-half inch longitudinal clearance between the propeller blades or cuffs and stationary parts of the airplane; and

(3) Positive clearance between other rotating parts of the propeller or spinner and stationary parts of the airplane.

[Docket No. 5066, 29 FR 18291, Dec. 24, 1964; as amended by Amdt. 25–72, 55 FR 29784, July 20, 1990]

## §25.929 Propeller deicing.

**(a)** For airplanes intended for use where icing may be expected, there must be a means to prevent or remove hazardous ice accumulation on propellers or on accessories where ice accumulation would jeopardize engine performance.

**(b)** If combustible fluid is used for propeller deicing, §§25.1181 through 25.1185 and 25.1189 apply.

## §25.933 Reversing systems.

**(a)** For turbojet reversing systems—

(1) Each system intended for ground operation only must be designed so that during any reversal in flight the engine will produce no more than flight idle thrust. In addition, it must be shown by analysis or test, or both, that—

(i) Each operable reverser can be restored to the forward thrust position; and

(ii) The airplane is capable of continued safe flight and landing under any possible position of the thrust reverser.

(2) Each system intended for inflight use must be designed so that no unsafe condition will result during normal operation of the system, or from any failure (or reasonably likely combination of failures) of the reversing system, under any anticipated condition of operation of the airplane including ground operation. Failure of structural elements need not be considered if the probability of this kind of failure is extremely remote.

(3) Each system must have means to prevent the engine from producing more than idle thrust when the reversing system malfunctions, except that it may produce any greater forward thrust that is shown to allow directional control to be maintained, with aerodynamic means alone, under the most critical reversing condition expected in operation.

**(b)** For propeller reversing systems—

(1) Each system intended for ground operation only must be designed so that no single failure (or reasonably likely combination of failures) or malfunction of the system will result in unwanted reverse thrust under any expected operating condition. Failure of structural elements need not be considered if this kind of failure is extremely remote.

(2) Compliance with this section may be shown by failure analysis or testing, or both, for propeller systems that allow propeller blades to move from the flight low-pitch position to a position that is substantially less than that at the normal flight low-pitch position. The analysis may include or be supported by the analysis made to show compliance with the requirements of §35.21 of this chapter for the propeller and associated installation components.

[Docket No. 5066, 29 FR 18291, Dec. 24, 1964; as amended by Amdt. 25–72, 55 FR 29784, July 20, 1990]

## §25.934 Turbojet engine thrust reverser system tests.

Thrust reversers installed on turbojet engines must meet the requirements of §33.97 of this chapter.

[Docket No. 5066, 29 FR 18291, Dec. 24, 1964; as amended by Amdt. 25–23, 35 FR 5677, April 8, 1970]

## §25.937 Turbopropeller-drag limiting systems.

Turbopropeller power airplane propeller-drag limiting systems must be designed so that no single failure or malfunction of any of the systems during normal or emergency operation results in propeller drag in excess of that for which the airplane was designed under §25.367. Failure of structural elements of the drag limiting systems need not be considered if the probability of this kind of failure is extremely remote.

## §25.939 Turbine engine operating characteristics.

**(a)** Turbine engine operating characteristics must be investigated in flight to determine that no adverse characteristics (such as stall, surge, or flameout) are present, to a hazardous degree, during normal and emergency operation within the range of operating limitations of the airplane and of the engine.

**(b)** [Reserved]

**(c)** The turbine engine air inlet system may not, as a result of air flow distortion during normal operation, cause vibration harmful to the engine.

[Docket No. 5066, 29 FR 18291, Dec. 24, 1964; as amended by Amdt. 25–11, 32 FR 6912, May 5, 1967; Amdt. 25–40, 42 FR 15043, March 17, 1977]

## §25.941 Inlet, engine, and exhaust compatibility.

For airplanes using variable inlet or exhaust system geometry, or both—

**(a)** The system comprised of the inlet, engine (including thrust augmentation systems, if incorporated), and exhaust must be shown to function properly under all operating conditions for which approval is sought, including all engine rotating speeds and power settings, and engine inlet and exhaust configurations;

**(b)** The dynamic effects of the operation of these (including consideration of probable malfunctions) upon the aerodynamic control of the airplane may not result in any condition that would require exceptional skill, alertness, or strength on the part of the pilot to avoid exceeding an operational or structural limitation of the airplane; and

**(c)** In showing compliance with paragraph (b) of this section, the pilot strength required may not exceed the limits set forth in §25.143(c), subject to the conditions set forth in paragraphs (d) and (e) of §25.143.

[Docket No. 5066, 29 FR 18291, Dec. 24, 1964; as amended by Amdt. 25–38, 41 FR 55467, Dec. 20, 1976]

## §25.943 Negative acceleration.

No hazardous malfunction of an engine, an auxiliary power unit approved for use in flight, or any component or system associated with the powerplant or auxiliary power unit may occur when the airplane is operated at the negative accelerations within the flight envelopes prescribed in §25.333. This must be shown for the greatest duration expected for the acceleration.

[Docket No. 5066, 29 FR 18291, Dec. 24, 1964; as amended by Amdt. 25–40, 42 FR 15043, March 17, 1977]

## §25.945 Thrust or power augmentation system.

**(a)** *General.* Each fluid injection system must provide a flow of fluid at the rate and pressure established for proper engine functioning under each intended operating condition. If the fluid can freeze, fluid freezing may not damage the airplane or adversely affect airplane performance.

**(b)** *Fluid tanks.* Each augmentation system fluid tank must meet the following requirements:

(1) Each tank must be able to withstand without failure the vibration, inertia, fluid, and structural loads that it may be subject to in operation.

(2) The tanks as mounted in the airplane must be able to withstand without failure or leakage an internal pressure 1.5 times the maximum operating pressure.

(3) If a vent is provided, the venting must be effective under all normal flight conditions.

(4) [Reserved]

**(c)** Augmentation system drains must be designed and located in accordance with §25.1455 if—

(1) The augmentation system fluid is subject to freezing; and

(2) The fluid may be drained in flight or during ground operation.

**(d)** The augmentation liquid tank capacity available for the use of each engine must be large enough to allow operation of the airplane under the approved procedures for the use of liquid-augmented power. The computation of liquid consumption must be based on the maximum approved rate appropriate for the desired engine output and must include the effect of temperature on engine performance as well as any other factors that might vary the amount of liquid required.

**(e)** This section does not apply to fuel injection systems.

25

[Docket No. 5066, 29 FR 18291, Dec. 24, 1964; as amended by Amdt. 25–40, 42 FR 15043, March 17, 1977; Amdt. 25–72, 55 FR 29785, July 20, 1990]

FUEL SYSTEM

## §25.951 General.

**(a)** Each fuel system must be constructed and arranged to ensure a flow of fuel at a rate and pressure established for proper engine and auxiliary power unit functioning under each likely operating condition, including any maneuver for which certification is requested and during which the engine or auxiliary power unit is permitted to be in operation.

**(b)** Each fuel system must be arranged so that any air which is introduced into the system will not result in—

(1) Power interruption for more than 20 seconds for reciprocating engines; or

(2) Flameout for turbine engines.

**(c)** Each fuel system for a turbine engine must be capable of sustained operation throughout its flow and pressure range with fuel initially saturated with water at 80°F and having 0.75cc of free water per gallon added and cooled to the most critical condition for icing likely to be encountered in operation.

**(d)** Each fuel system for a turbine engine powered airplane must meet the applicable fuel venting requirements of part 34 of this chapter.

[Docket No. 5066, 29 FR 18291, Dec. 24, 1964; as amended by Amdt. 25–23, 35 FR 5677, April 8, 1970; Amdt. 25–36, 39 FR 35460, Oct. 1, 1974; Amdt. 25–38, 41 FR 55467, Dec. 20, 1976; Amdt. 25–73, 55 FR 32861, Aug. 10, 1990]

## §25.952 Fuel system analysis and test.

**(a)** Proper fuel system functioning under all probable operating conditions must be shown by analysis and those tests found necessary by the Administrator. Tests, if required, must be made using the airplane fuel system or a test article that reproduces the operating characteristics of the portion of the fuel system to be tested.

**(b)** The likely failure of any heat exchanger using fuel as one of its fluids may not result in a hazardous condition.

[Docket No. 5066, 29 FR 18291, Dec. 24, 1964; as amended by Amdt. 25–40, 42 FR 15043, Mar. 17, 1977]

## §25.953 Fuel system independence.

Each fuel system must meet the requirements of §25.903(b) by—

**(a)** Allowing the supply of fuel to each engine through a system independent of each part of the system supplying fuel to any other engine; or

**(b)** Any other acceptable method.

## §25.954 Fuel system lightning protection.

The fuel system must be designed and arranged to prevent the ignition of fuel vapor within the system by—

**(a)** Direct lightning strikes to areas having a high probability of stroke attachment;

**(b)** Swept lightning strokes to areas where swept strokes are highly probable; and

**(c)** Corona and streamering at fuel vent outlets.

[Docket No. 5066, 29 FR 18291, Dec. 24, 1964; as amended by Amdt. 25–14, 32 FR 11629, Aug. 11, 1967]

## §25.955 Fuel flow.

**(a)** Each fuel system must provide at least 100 percent of the fuel flow required under each intended operating condition and maneuver. Compliance must be shown as follows:

(1) Fuel must be delivered to each engine at a pressure within the limits specified in the engine type certificate.

(2) The quantity of fuel in the tank may not exceed the amount established as the unusable fuel supply for that tank under the requirements of §25.959 plus that necessary to show compliance with this section.

(3) Each main pump must be used that is necessary for each operating condition and attitude for which compliance with this section is shown, and the appropriate emergency pump must be substituted for each main pump so used.

(4) If there is a fuel flowmeter, it must be blocked and the fuel must flow through the meter or its bypass.

**(b)** If an engine can be supplied with fuel from more than one tank, the fuel system must—

(1) For each reciprocating engine, supply the full fuel pressure to that engine in not more than 20 seconds after switching to any other fuel tank containing usable fuel when engine malfunctioning becomes apparent due to the depletion of the fuel supply in any tank from which the engine can be fed; and

(2) For each turbine engine, in addition to having appropriate manual switching capability, be designed to prevent interruption of fuel flow to that engine, without attention by the flight crew, when any tank supplying fuel to that engine is depleted of usable fuel during normal operation, and any other tank, that normally supplies fuel to that engine alone, contains usable fuel.

[Docket No. 5066, 29 FR 18291, Dec. 24, 1964; as amended by Amdt. 25–11, 32 FR 6912, May 5, 1967]

## §25.957 Flow between interconnected tanks.

If fuel can be pumped from one tank to another in flight, the fuel tank vents and the fuel transfer system must be designed so that no structural damage to the tanks can occur because of overfilling.

## §25.959 Unusable fuel supply.

The unusable fuel quantity for each fuel tank and its fuel system components must be established at not less than the quantity at which the first evidence of engine malfunction occurs under the most adverse fuel feed condition for all intended operations and flight maneuvers involving fuel feeding from that tank. Fuel system component failures need not be considered.

[Docket No. 5066, 29 FR 18291, Dec. 24, 1964; as amended by Amdt. 25–23, 35 FR 5677, April 8, 1970; Amdt. 25–40, 42 FR 15043, March 17, 1977]

## §25.961 Fuel system hot weather operation.

**(a)** The fuel system must perform satisfactorily in hot weather operation. This must be shown by showing that the fuel system from the tank outlets to each engine is pressurized, under all intended operations, so as to prevent vapor formation, or must be shown by climbing from the altitude of the airport elected by the applicant to the maximum altitude established as an operating limitation under §25.1527. If a climb test is elected, there may be no evidence of vapor lock or other malfunctioning during the climb test conducted under the following conditions:

(1) For reciprocating engine powered airplanes, the engines must operate at maximum continuous power, except that takeoff power must be used for the altitudes from 1,000 feet below the critical altitude through the critical altitude. The time interval during which takeoff power is used may not be less than the takeoff time limitation.

(2) For turbine engine powered airplanes, the engines must operate at takeoff power for the time interval selected for showing the takeoff flight path, and at maximum continuous power for the rest of the climb.

(3) The weight of the airplane must be the weight with full fuel tanks, minimum crew, and the ballast necessary to maintain the center of gravity within allowable limits.

(4) The climb airspeed may not exceed—

(i) For reciprocating engine powered airplanes, the maximum airspeed established for climbing from takeoff to the maximum operating altitude with the airplane in the following configuration:

(A) Landing gear retracted.

(B) Wing flaps in the most favorable position.

(C) Cowl flaps (or other means of controlling the engine cooling supply) in the position that provides adequate cooling in the hot-day condition.

(D) Engine operating within the maximum continuous power limitations.

(E) Maximum takeoff weight; and

(ii) For turbine engine powered airplanes, the maximum airspeed established for climbing from takeoff to the maximum operating altitude.

(5) The fuel temperature must be at least 110°F.

**(b)** The test prescribed in paragraph (a) of this section may be performed in flight or on the ground under closely simulated flight conditions. If a flight test is performed in weather cold enough to interfere with the proper conduct of the test, the fuel tank surfaces, fuel lines, and other fuel system parts subject to cold air must be insulated to simulate, insofar as practicable, flight in hot weather.

[Docket No. 5066, 29 FR 18291, Dec. 24, 1964; as amended by Amdt. 25–11, 32 FR 6912, May 5, 1967; Amdt. 25–57, 49 FR 6848, Feb. 23, 1984]

## §25.963 Fuel tanks: general.

**(a)** Each fuel tank must be able to withstand, without failure, the vibration, inertia, fluid, and structural loads that it may be subjected to in operation.

**(b)** Flexible fuel tank liners must be approved or must be shown to be suitable for the particular application.

**(c)** Integral fuel tanks must have facilities for interior inspection and repair.

**(d)** Fuel tanks within the fuselage contour must be able to resist rupture and to retain fuel, under the inertia forces prescribed for the emergency landing conditions in §25.561. In addition, these tanks must be in a protected position so that exposure of the tanks to scraping action with the ground is unlikely.

**(e)** Fuel tank access covers must comply with the following criteria in order to avoid loss of hazardous quantities of fuel:

(1) All covers located in an area where experience or analysis indicates a strike is likely must be shown by analysis or tests to minimize penetration and deformation by tire fragments, low energy engine debris, or other likely debris.

(2) All covers must be fire resistant as defined in part 1 of this chapter.

**(f)** For pressurized fuel tanks, a means with failsafe features must be provided to prevent the buildup of an excessive pressure difference between the inside and the outside of the tank.

[Docket No. 5066, 29 FR 18291, Dec. 24, 1964; as amended by Amdt. 25–40, 42 FR 15043, March 17, 1977; Amdt. 25–69, 54 FR 40354, Sept. 29, 1989]

## §25.965 Fuel tank tests.

**(a)** It must be shown by tests that the fuel tanks, as mounted in the airplane, can withstand, without failure or leakage, the more critical of the pressures resulting from the conditions specified in paragraphs (a)(1) and (2) of this section. In addition, it must be shown by either analysis or tests, that tank surfaces subjected to more critical pressures resulting from the condition of paragraphs (a)(3) and (4) of this section, are able to withstand the following pressures:

(1) An internal pressure of 3.5 psi.

(2) 125 percent of the maximum air pressure developed in the tank from ram effect.

(3) Fluid pressures developed during maximum limit accelerations, and deflections, of the airplane with a full tank.

(4) Fluid pressures developed during the most adverse combination of airplane roll and fuel load.

**(b)** Each metallic tank with large unsupported or unstiffened flat surfaces, whose failure or deformation could cause fuel leakage, must be able to withstand the following test, or its equivalent, without leakage or excessive deformation of the tank walls:

(1) Each complete tank assembly and its supports must be vibration tested while mounted to simulate the actual installation.

(2) Except as specified in paragraph (b)(4) of this section, the tank assembly must be vibrated for 25 hours at an amplitude of not less than $\frac{1}{32}$ of an inch (unless another amplitude is substantiated) while $\frac{2}{3}$ filled with water or other suitable test fluid.

(3) The test frequency of vibration must be as follows:

(i) If no frequency of vibration resulting from any r.p.m. within the normal operating range of engine speeds is critical, the test frequency of vibration must be 2,000 cycles per minute.

(ii) If only one frequency of vibration resulting from any r.p.m. within the normal operating range of engine speeds is critical, that frequency of vibration must be the test frequency.

(iii) If more than one frequency of vibration resulting from any r.p.m. within the normal operating range of engine speeds is critical, the most critical of these frequencies must be the test frequency.

(4) Under paragraphs (b)(3)(ii) and (iii) of this section, the time of test must be adjusted to accomplish the same number of vibration cycles that would be accomplished in 25 hours at the frequency specified in paragraph (b)(3)(i) of this section.

(5) During the test, the tank assembly must be rocked at the rate of 16 to 20 complete cycles per minute, through an angle of 15° on both sides of the horizontal (30 total), about the most critical axis, for 25 hours. If motion about more than one axis is likely to be critical, the tank must be rocked about each critical axis for $12\frac{1}{2}$ hours.

**(c)** Except where satisfactory operating experience with a similar tank in a similar installation is shown, nonmetallic tanks must withstand the test specified in paragraph (b)(5) of this section, with fuel at a temperature of 110°F. During this test, a representative specimen of the tank must be installed in a supporting structure simulating the installation in the airplane.

**(d)** For pressurized fuel tanks, it must be shown by analysis or tests that the fuel tanks can withstand the maximum pressure likely to occur on the ground or in flight.

[Docket No. 5066, 29 FR 18291, Dec. 24, 1964; as amended by Amdt. 25–11, 32 FR 6913, May 5, 1967; Amdt. 25–40, 42 FR 15043, March 17, 1977]

## §25.967 Fuel tank installations.

**(a)** Each fuel tank must be supported so that tank loads (resulting from the weight of the fuel in the tanks) are not concentrated on unsupported tank surfaces. In addition—

(1) There must be pads, if necessary, to prevent chafing between the tank and its supports;

(2) Padding must be nonabsorbent or treated to prevent the absorption of fluids;

(3) If a flexible tank liner is used, it must be supported so that it is not required to withstand fluid loads; and

(4) Each interior surface of the tank compartment must be smooth and free of projections that could cause wear of the liner unless—

(i) Provisions are made for protection of the liner at these points; or

(ii) The construction of the liner itself provides that protection.

**(b)** Spaces adjacent to tank surfaces must be ventilated to avoid fume accumulation due to minor leakage. If the tank is in a sealed compartment, ventilation may be limited to drain holes large enough to prevent excessive pressure resulting from altitude changes.

**(c)** The location of each tank must meet the requirements of §25.1185(a).

**(d)** No engine nacelle skin immediately behind a major air outlet from the engine compartment may act as the wall of an integral tank.

**(e)** Each fuel tank must be isolated from personnel compartments by a fumeproof and fuelproof enclosure.

## §25.969 Fuel tank expansion space.

Each fuel tank must have an expansion space of not less than 2 percent of the tank capacity. It must be impossible to fill the expansion space inadvertently with the airplane in the normal ground attitude. For pressure fueling systems, compliance with this section may be shown with the means provided to comply with §25.979(b).

[Docket No. 5066, 29 FR 18291, Dec. 24, 1964; as amended by Amdt. 25–11, 32 FR 6913, May 5, 1967]

## §25.971 Fuel tank sump.

**(a)** Each fuel tank must have a sump with an effective capacity, in the normal ground attitude, of not less than the greater of 0.10 percent of the tank capacity or one-sixteenth of a gallon unless operating limitations are established to ensure that the accumulation of water in service will not exceed the sump capacity.

**(b)** Each fuel tank must allow drainage of any hazardous quantity of water from any part of the tank to its sump with the airplane in the ground attitude.

**(c)** Each fuel tank sump must have an accessible drain that—

(1) Allows complete drainage of the sump on the ground;

(2) Discharges clear of each part of the airplane; and

(3) Has manual or automatic means for positive locking in the closed position.

## §25.973 Fuel tank filler connection.

Each fuel tank filler connection must prevent the entrance of fuel into any part of the airplane other than the tank itself. In addition—

**(a)** [Reserved]

**(b)** Each recessed filler connection that can retain any appreciable quantity of fuel must have a drain that discharges clear of each part of the airplane;

**(c)** Each filler cap must provide a fuel-tight seal; and

**(d)** Each fuel filling point, except pressure fueling connection points, must have a provision for electrically bonding the airplane to ground fueling equipment.

[Docket No. 5066, 29 FR 18291, Dec. 24, 1964; as amended by Amdt. 25–40, 42 FR 15043, March 17, 1977; Amdt. 25–72, 55 FR 29785, July 20, 1990]

## §25.975 Fuel tank vents and carburetor vapor vents.

**(a)** *Fuel tank vents.* Each fuel tank must be vented from the top part of the expansion space so that venting is effective under any normal flight condition. In addition—

(1) Each vent must be arranged to avoid stoppage by dirt or ice formation;

(2) The vent arrangement must prevent siphoning of fuel during normal operation;

(3) The venting capacity and vent pressure levels must maintain acceptable differences of pressure between the interior and exterior of the tank, during—

(i) Normal flight operation;

(ii) Maximum rate of ascent and descent; and

(iii) Refueling and defueling (where applicable);

(4) Airspaces of tanks with interconnected outlets must be interconnected;

(5) There may be no point in any vent line where moisture can accumulate with the airplane in the ground attitude or the level flight attitude, unless drainage is provided; and

(6) No vent or drainage provision may end at any point—

(i) Where the discharge of fuel from the vent outlet would constitute a fire hazard; or

(ii) From which fumes could enter personnel compartments.

**(b)** *Carburetor vapor vents.* Each carburetor with vapor elimination connections must have a vent line to lead vapors back to one of the fuel tanks. In addition—

(1) Each vent system must have means to avoid stoppage by ice; and

(2) If there is more than one fuel tank, and it is necessary to use the tanks in a definite sequence, each vapor vent return line must lead back to the fuel tank used for takeoff and landing.

## §25.977 Fuel tank outlet.

**(a)** There must be a fuel strainer for the fuel tank outlet or for the booster pump. This strainer must—

(1) For reciprocating engine powered airplanes, have 8 to 16 meshes per inch; and

(2) For turbine engine powered airplanes, prevent the passage of any object that could restrict fuel flow or damage any fuel system component.

**(b)** [Reserved]

**(c)** The clear area of each fuel tank outlet strainer must be at least five times the area of the outlet line.

**(d)** The diameter of each strainer must be at least that of the fuel tank outlet.

**(e)** Each finger strainer must be accessible for inspection and cleaning.

[Docket No. 5066, 29 FR 18291, Dec. 24, 1964; as amended by Amdt. 25–11, 32 FR 6913, May 5, 1967; Amdt. 25–36, 39 FR 35460, Oct. 1, 1974]

## §25.979 Pressure fueling system.

For pressure fueling systems, the following apply:

**(a)** Each pressure fueling system fuel manifold connection must have means to prevent the escape of hazardous quantities of fuel from the system if the fuel entry valve fails.

**(b)** An automatic shutoff means must be provided to prevent the quantity of fuel in each tank from exceeding the maximum quantity approved for that tank. This means must—

(1) Allow checking for proper shutoff operation before each fueling of the tank; and

(2) Provide indication at each fueling station of failure of the shutoff means to stop the fuel flow at the maximum quantity approved for that tank.

**(c)** A means must be provided to prevent damage to the fuel system in the event of failure of the automatic shutoff means prescribed in paragraph (b) of this section.

**(d)** The airplane pressure fueling system (not including fuel tanks and fuel tank vents) must withstand an ultimate load that is 2.0 times the load arising from the maximum pressures, including surge, that is likely to occur during fueling. The maximum surge pressure must be established with any combination of tank valves being either intentionally or inadvertently closed.

**(e)** The airplane defueling system (not including fuel tanks and fuel tank vents) must withstand an ultimate load that is 2.0 times the load arising from the maximum permissible defueling pressure (positive or negative) at the airplane fueling connection.

[Docket No. 5066, 29 FR 18291, Dec. 24, 1964; as amended by Amdt. 25–11, 32 FR 6913, May 5, 1967; Amdt. 25–38, 41 FR 55467, Dec. 20, 1976; Amdt. 25–72, 55 FR 29785, July 20, 1990]

## §25.981 Fuel tank ignition prevention.

**(a)** No ignition source may be present at each point in the fuel tank or fuel tank system where catastrophic failure could occur due to ignition of fuel or vapors. This must be shown by:

(1) Determining the highest temperature allowing a safe margin below the lowest expected autoignition temperature of the fuel in the fuel tanks.

(2) Demonstrating that no temperature at each place inside each fuel tank where fuel ignition is possible will exceed the temperature determined under paragraph (a)(1) of this section. This must be verified under all probable operating, failure, and malfunction conditions of each component whose operation, failure, or malfunction could increase the temperature inside the tank.

(3) Demonstrating that an ignition source could not result from each single failure, from each single failure in combination with each latent failure condition not shown to be extremely remote, and from all combinations of failures not shown to be extremely improbable. The effects of manufacturing variability, aging, wear, corrosion, and likely damage must be considered.

**(b)** Based on the evaluations required by this section, critical design configuration control limitations, inspections, or other procedures must be established, as necessary, to prevent development of ignition sources within the fuel tank system and must be included in the Airworthiness Limitations section of the Instructions for Continued Airworthiness required by §25.1529. Visible means to identify critical features of the design must be placed in areas of the airplane where maintenance actions, repairs, or alterations may be apt to violate the critical design configuration limitations (e.g., color-coding of wire to identify separation limitation).

**(c)** The fuel tank installation must include either—

(1) Means to minimize the development of flammable vapors in the fuel tanks (in the context of this rule, "minimize" means to incorporate practicable design methods to reduce the likelihood of flammable vapors); or

(2) Means to mitigate the effects of an ignition of fuel vapors within fuel tanks such that no damage caused by an ignition will prevent continued safe flight and landing.

[Docket No. FAA–1999–6411, 66 FR 23130, May 7, 2001]

### Fuel System Components

## §25.991 Fuel pumps.

**(a)** *Main pumps.* Each fuel pump required for proper engine operation, or required to meet the fuel system requirements of this subpart (other than those in paragraph (b) of this section, is a main pump. For each main pump, provision must be made to allow the bypass of each positive displacement fuel pump other than a fuel injection pump (a pump that supplies the proper flow and pressure for fuel injection when the injection is not accomplished in a carburetor) approved as part of the engine.

**(b)** *Emergency pumps.* There must be emergency pumps or another main pump to feed each engine immediately after failure of any main pump (other than a fuel injection pump approved as part of the engine).

## §25.993 Fuel system lines and fittings.

**(a)** Each fuel line must be installed and supported to prevent excessive vibration and to withstand loads due to fuel pressure and accelerated flight conditions.

**(b)** Each fuel line connected to components of the airplane between which relative motion could exist must have provisions for flexibility.

**(c)** Each flexible connection in fuel lines that may be under pressure and subjected to axial loading must use flexible hose assemblies.

**(d)** Flexible hose must be approved or must be shown to be suitable for the particular application.

**(e)** No flexible hose that might be adversely affected by exposure to high temperatures may be used where excessive temperatures will exist during operation or after engine shut-down.

**(f)** Each fuel line within the fuselage must be designed and installed to allow a reasonable degree of deformation and stretching without leakage.

[Docket No. 5066, 29 FR 18291, Dec. 24, 1964; as amended by Amdt. 25–15, 32 FR 13266, Sept. 20, 1967]

## §25.994 Fuel system components.

Fuel system components in an engine nacelle or in the fuselage must be protected from damage which could result in spillage of enough fuel to constitute a fire hazard as a result of a wheels-up landing on a paved runway.

[Docket No. 5066, 29 FR 18291, Dec. 24, 1964; as amended by Amdt. 25–57, 49 FR 6848, Feb. 23, 1984]

## §25.995 Fuel valves.

In addition to the requirements of §25.1189 for shutoff means, each fuel valve must—

**(a)** [Reserved]

**(b)** Be supported so that no loads resulting from their operation or from accelerated flight conditions are transmitted to the lines attached to the valve.

[Docket No. 5066, 29 FR 18291, Dec. 24, 1964; amended by Amdt. 25–40, 42 FR 15043, Mar. 17, 1977]

## §25.997 Fuel strainer or filter.

There must be a fuel strainer or filter between the fuel tank outlet and the inlet of either the fuel metering device or an engine driven positive displacement pump, whichever is nearer the fuel tank outlet. This fuel strainer or filter must—

**(a)** Be accessible for draining and cleaning and must incorporate a screen or element which is easily removable;

**(b)** Have a sediment trap and drain except that it need not have a drain if the strainer or filter is easily removable for drain purposes;

**(c)** Be mounted so that its weight is not supported by the connecting lines or by the inlet or outlet connections of the strainer or filter itself, unless adequate strength margins under all loading conditions are provided in the lines and connections; and

**(d)** Have the capacity (with respect to operating limitations established for the engine) to ensure that engine fuel system functioning is not impaired, with the fuel contaminated to a degree (with respect to particle size and density) that is greater than that established for the engine in Part 33 of this chapter.

[Docket No. 5066, 29 FR 18291, Dec. 24, 1964; as amended by Amdt. 25–36, 39 FR 35460, Oct. 1, 1974; Amdt. 25–57, 49 FR 6848, Feb. 23, 1984]

## §25.999 Fuel system drains.

**(a)** Drainage of the fuel system must be accomplished by the use of fuel strainer and fuel tank sump drains.

**(b)** Each drain required by paragraph (a) of this section must—

(1) Discharge clear of all parts of the airplane;

(2) Have manual or automatic means for positive locking in the closed position; and

(3) Have a drain valve—

(i) That is readily accessible and which can be easily opened and closed; and

(ii) That is either located or protected to prevent fuel spillage in the event of a landing with landing gear retracted.

[Docket No. 5066, 29 FR 18291, Dec. 24, 1964; as amended by Amdt. 25–38, 41 FR 55467, Dec. 20, 1976]

## §25.1001 Fuel jettisoning system.

**(a)** A fuel jettisoning system must be installed on each airplane unless it is shown that the airplane meets the climb requirements of §§25.119 and 25.121(d) at maximum takeoff weight, less the actual or computed weight of fuel necessary for a 15-minute flight comprised of a takeoff, go-around, and landing at the airport of departure with the airplane configuration, speed, power, and thrust the same as that used in meeting the applicable takeoff, approach, and landing climb performance requirements of this part.

**(b)** If a fuel jettisoning system is required it must be capable of jettisoning enough fuel within 15 minutes, starting with the weight given in paragraph (a) of this section, to enable the airplane to meet the climb requirements of §§25.119 and 25.121(d), assuming that the fuel is jettisoned under the conditions, except weight, found least favorable during the flight tests prescribed in paragraph (c) of this section.

**(c)** Fuel jettisoning must be demonstrated beginning at maximum takeoff weight with flaps and landing gear up and in—

(1) A power-off glide at 1.4 $V_{S1}$;

(2) A climb at the one-engine inoperative best rate-of-climb speed, with the critical engine inoperative and the remaining engines at maximum continuous power; and

(3) Level flight at 1.4 $V_{S1}$; if the results of the tests in the conditions specified in paragraphs (c)(1) and (2) of this section show that this condition could be critical.

**(d)** During the flight tests prescribed in paragraph (c) of this section, it must be shown that—

(1) The fuel jettisoning system and its operation are free from fire hazard;

(2) The fuel discharges clear of any part of the airplane;

25

(3) Fuel or fumes do not enter any parts of the airplane; and

(4) The jettisoning operation does not adversely affect the controllability of the airplane.

**(e)** For reciprocating engine powered airplanes, means must be provided to prevent jettisoning the fuel in the tanks used for takeoff and landing below the level allowing 45 minutes flight at 75 percent maximum continuous power. However, if there is an auxiliary control independent of the main jettisoning control, the system may be designed to jettison the remaining fuel by means of the auxiliary jettisoning control.

**(f)** For turbine engine powered airplanes, means must be provided to prevent jettisoning the fuel in the tanks used for takeoff and landing below the level allowing climb from sea level to 10,000 feet and thereafter allowing 45 minutes cruise at a speed for maximum range. However, if there is an auxiliary control independent of the main jettisoning control, the system may be designed to jettison the remaining fuel by means of the auxiliary jettisoning control.

**(g)** The fuel jettisoning valve must be designed to allow flight personnel to close the valve during any part of the jettisoning operation.

**(h)** Unless it is shown that using any means (including flaps, slots, and slats) for changing the airflow across or around the wings does not adversely affect fuel jettisoning, there must be a placard, adjacent to the jettisoning control, to warn flight crewmembers against jettisoning fuel while the means that change the airflow are being used.

**(i)** The fuel jettisoning system must be designed so that any reasonably probable single malfunction in the system will not result in a hazardous condition due to unsymmetrical jettisoning of, or inability to jettison, fuel.

[Docket No. 5066, 29 FR 18291, Dec. 24, 1964; as amended by Amdt. 25–18, 33 FR 12226, Aug. 30, 1968; Amdt. 25–57, 49 FR 6848, Feb. 23, 1984]

### OIL SYSTEM

## §25.1011 General.

**(a)** Each engine must have an independent oil system that can supply it with an appropriate quantity of oil at a temperature not above that safe for continuous operation.

**(b)** The usable oil capacity may not be less than the product of the endurance of the airplane under critical operating conditions and the approved maximum allowable oil consumption of the engine under the same conditions, plus a suitable margin to ensure system circulation. Instead of a rational analysis of airplane range for the purpose of computing oil requirements for reciprocating engine powered airplanes, the following fuel/oil ratios may be used:

(1) For airplanes without a reserve oil or oil transfer system, a fuel/oil ratio of 30:1 by volume.

(2) For airplanes with either a reserve oil or oil transfer system, a fuel/oil ratio of 40:1 by volume.

**(c)** Fuel/oil ratios higher than those prescribed in paragraphs (b) (1) and (2) of this section may be used if substantiated by data on actual engine oil consumption.

## §25.1013 Oil tanks.

**(a)** *Installation.* Each oil tank installation must meet the requirements of §25.967.

**(b)** *Expansion space.* Oil tank expansion space must be provided as follows:

(1) Each oil tank used with a reciprocating engine must have an expansion space of not less than the greater of 10 percent of the tank capacity or 0.5 gallon, and each oil tank used with a turbine engine must have an expansion space of not less than 10 percent of the tank capacity.

(2) Each reserve oil tank not directly connected to any engine may have an expansion space of not less than two percent of the tank capacity.

(3) It must be impossible to fill the expansion space inadvertently with the airplane in the normal ground attitude.

**(c)** *Filler connection.* Each recessed oil tank filler connection that can retain any appreciable quantity of oil must have a drain that discharges clear of each part of the airplane. In addition, each oil tank filler cap must provide an oil-tight seal.

**(d)** *Vent.* Oil tanks must be vented as follows:

(1) Each oil tank must be vented from the top part of the expansion space so that venting is effective under any normal flight condition.

(2) Oil tank vents must be arranged so that condensed water vapor that might freeze and obstruct the line cannot accumulate at any point.

**(e)** *Outlet.* There must be means to prevent entrance into the tank itself, or into the tank outlet, of any object that might obstruct the flow of oil through the system. No oil tank outlet may be enclosed by any screen or guard that would reduce the flow of oil below a safe value at any operating temperature. There must be a shutoff valve at the outlet of each oil tank used with a turbine engine, unless the external portion of the oil system (including the oil tank supports) is fireproof.

**(f)** *Flexible oil tank liners.* Each flexible oil tank liner must be approved or must be shown to be suitable for the particular application.

[Docket No. 5066, 29 FR 18291, Dec. 24; as amended by Amdt. 25–19, 33 FR 15410, Oct. 17, 1968; Amdt. 25–23, 35 FR 5677, April 8, 1970; Amdt. 25–36, 39 FR 35460, Oct. 1, 1974; Amdt. 25–57, 49 FR 6848, Feb. 23, 1984; Amdt. 25–72, 55 FR 29785, July 20, 1990]

## §25.1015 Oil tank tests.

Each oil tank must be designed and installed so that—

**(a)** It can withstand, without failure, each vibration, inertia, and fluid load that it may be subjected to in operation; and

**(b)** It meets the provisions of §25.965, except—

(1) The test pressure—

(i) For pressurized tanks used with a turbine engine, may not be less than 5 p.s.i. plus the maximum operating pressure of the tank instead of the pressure specified in §25.965(a); and

(ii) For all other tanks may not be less than 5 p.s.i. instead of the pressure specified in §25.965(a); and

(2) The test fluid must be oil at 250°F instead of the fluid specified in §25.965(c).

[Docket No. 5066, 29 FR 18291, Dec. 24, 1964; as amended by Amdt. 25–36, 39 FR 35461, Oct. 1, 1974]

## §25.1017 Oil lines and fittings.

**(a)** Each oil line must meet the requirements of §25.993 and each oil line and fitting in any designated fire zone must meet the requirements of §25.1183.

**(b)** Breather lines must be arranged so that—

(1) Condensed water vapor that might freeze and obstruct the line cannot accumulate at any point;

(2) The breather discharge does not constitute a fire hazard if foaming occurs or causes emitted oil to strike the pilot's windshield; and

(3) The breather does not discharge into the engine air induction system.

## §25.1019 Oil strainer or filter.

**(a)** Each turbine engine installation must incorporate an oil strainer or filter through which all of the engine oil flows and which meets the following requirements:

(1) Each oil strainer or filter that has a bypass must be constructed and installed so that oil will flow at the normal rate through the rest of the system with the strainer or filter completely blocked.

(2) The oil strainer or filter must have the capacity (with respect to operating limitations established for the engine) to ensure that engine oil system functioning is not impaired when the oil is contaminated to a degree (with respect to particle size and density) that is greater than that established for the engine under Part 33 of this chapter.

(3) The oil strainer or filter, unless it is installed at an oil tank outlet, must incorporate an indicator that will indicate contamination before it reaches the capacity established in accordance with paragraph (a)(2) of this section.

(4) The bypass of a strainer or filter must be constructed and installed so that the release of collected contaminants is minimized by appropriate location of the bypass to ensure that collected contaminants are not in the bypass flow path.

(5) An oil strainer or filter that has no bypass, except one that is installed at an oil tank outlet, must have a means to connect it to the warning system required in §25.1305(c)(7).

**(b)** Each oil strainer or filter in a powerplant installation using reciprocating engines must be constructed and installed so that oil will flow at the normal rate through the rest of the system with the strainer or filter element completely blocked.

[Docket No. 5066, 29 FR 18291, Dec. 24, 1964; as amended by Amdt. 25–36, 39 FR 35461, Oct. 1, 1974; Amdt. 25–57, 49 FR 6848, Feb. 23, 1984]



## §25.1021 Oil system drains.

A drain (or drains) must be provided to allow safe drainage of the oil system. Each drain must—

**(a)** Be accessible; and

**(b)** Have manual or automatic means for positive locking in the closed position.

[Docket No. 5066, 29 FR 18291, Dec. 24, 1964; as amended by Amdt. 25–57, 49 FR 6848, Feb. 23, 1984]

## §25.1023 Oil radiators.

**(a)** Each oil radiator must be able to withstand, without failure, any vibration, inertia, and oil pressure load to which it would be subjected in operation.

**(b)** Each oil radiator air duct must be located so that, in case of fire, flames coming from normal openings of the engine nacelle cannot impinge directly upon the radiator.

## §25.1025 Oil valves.

**(a)** Each oil shutoff must meet the requirements of §25.1189.

**(b)** The closing of oil shutoff means may not prevent propeller feathering.

**(c)** Each oil valve must have positive stops or suitable index provisions in the "on" and "off" positions and must be supported so that no loads resulting from its operation or from accelerated flight conditions are transmitted to the lines attached to the valve.

## §25.1027 Propeller feathering system.

**(a)** If the propeller feathering system depends on engine oil, there must be means to trap an amount of oil in the tank if the supply becomes depleted due to failure of any part of the lubricating system other than the tank itself.

**(b)** The amount of trapped oil must be enough to accomplish the feathering operation and must be available only to the feathering pump.

**(c)** The ability of the system to accomplish feathering with the trapped oil must be shown. This may be done on the ground using an auxiliary source of oil for lubricating the engine during operation.

**(d)** Provision must be made to prevent sludge or other foreign matter from affecting the safe operation of the propeller feathering system.

[Docket No. 5066, 29 FR 18291, Dec. 24, 1964; as amended by Amdt. 25–38, 41 FR 55467, Dec. 20, 1976]

COOLING

## §25.1041 General.

The powerplant and auxiliary power unit cooling provisions must be able to maintain the temperatures of powerplant components, engine fluids, and auxiliary power unit components and fluids within the temperature limits established for these components and fluids, under ground, water, and flight operating conditions, and after normal engine or auxiliary power unit shutdown, or both.

[Docket No. 5066, 29 FR 18291, Dec. 24, 1964; as amended by Amdt. 25–38, 41 FR 55467, Dec. 20, 1976]

## §25.1043 Cooling tests.

**(a)** *General.* Compliance with §25.1041 must be shown by tests, under critical ground, water, and flight operating conditions. For these tests, the following apply:

(1) If the tests are conducted under conditions deviating from the maximum ambient atmospheric temperature, the recorded powerplant temperatures must be corrected under paragraphs (c) and (d) of this section.

(2) No corrected temperatures determined under paragraph (a)(1) of this section may exceed established limits.

(3) For reciprocating engines, the fuel used during the cooling tests must be the minimum grade approved for the engines, and the mixture settings must be those normally used in the flight stages for which the cooling tests are conducted. The test procedures must be as prescribed in §25.1045.

**(b)** *Maximum ambient atmospheric temperature.* A maximum ambient atmospheric temperature corresponding to sea level conditions of at least 100 degrees F must be established. The assumed temperature lapse rate is 3.6 degrees F per thousand feet of altitude above sea level until a temperature of -69.7 degrees F is reached, above which altitude the temperature is considered constant at -69.7 degrees F. However, for winterization installations, the applicant may select a maximum ambient atmospheric temperature corresponding to sea level conditions of less than 100 degrees F.

**(c)** *Correction factor (except cylinder barrels).* Unless a more rational correction applies, temperatures of engine fluids and powerplant components (except cylinder barrels) for which temperature limits are established, must be corrected by adding to them the difference between the maximum ambient atmospheric temperature and the temperature of the ambient air at the time of the first occurrence of the maximum component or fluid temperature recorded during the cooling test.

**(d)** *Correction factor for cylinder barrel temperatures.* Unless a more rational correction applies, cylinder barrel temperatures must be corrected by adding to them 0.7 times the difference between the maximum ambient atmospheric temperature and the temperature of the ambient air at the time of the first occurrence of the maximum cylinder barrel temperature recorded during the cooling test.

[Docket No. 5066, 29 FR 18291, Dec. 24, 1964; as amended by Amdt. 25–42, 43 FR 2323, Jan. 16, 1978]

## §25.1045 Cooling test procedures.

**(a)** Compliance with §25.1041 must be shown for the takeoff, climb, en route, and landing stages of flight that correspond to the applicable performance requirements. The cooling tests must be conducted with the airplane in the configuration, and operating under the conditions, that are critical relative to cooling during each stage of flight. For the cooling tests, a temperature is "stabilized" when its rate of change is less than two degrees F per minute.

**(b)** Temperatures must be stabilized under the conditions from which entry is made into each stage of flight being investigated, unless the entry condition normally is not one during which component and the engine fluid temperatures would stabilize (in which case, operation through the full entry condition must be conducted before entry into the stage of flight being investigated in order to allow temperatures to reach their natural levels at the time of entry). The takeoff cooling test must be preceded by a period during which the powerplant component and engine fluid temperatures are stabilized with the engines at ground idle.

**(c)** Cooling tests for each stage of flight must be continued until—

(1) The component and engine fluid temperatures stabilize;

(2) The stage of flight is completed; or

(3) An operating limitation is reached.

**(d)** For reciprocating engine powered airplanes, it may be assumed, for cooling test purposes, that the takeoff stage of flight is complete when the airplane reaches an altitude of 1,500 feet above the takeoff surface or reaches a point in the takeoff where the transition from the takeoff to the en route configuration is completed and a speed is reached at which compliance with §25.121(c) is shown, whichever point is at a higher altitude. The airplane must be in the following configuration:

(1) Landing gear retracted.

(2) Wing flaps in the most favorable position.

(3) Cowl flaps (or other means of controlling the engine cooling supply) in the position that provides adequate cooling in the hot-day condition.

(4) Critical engine inoperative and its propeller stopped.

(5) Remaining engines at the maximum continuous power available for the altitude.

**(e)** For hull seaplanes and amphibians, cooling must be shown during taxiing downwind for 10 minutes, at five knots above step speed.

[Docket No. 5066, 29 FR 18291, Dec. 24, 1964; as amended by Amdt. 25–57, 49 FR 6848, Feb. 23, 1984]

### Induction System

## §25.1091 Air induction.

**(a)** The air induction system for each engine and auxiliary power unit must supply—

(1) The air required by that engine and auxiliary power unit under each operating condition for which certification is requested; and

(2) The air for proper fuel metering and mixture distribution with the induction system valves in any position.

**(b)** Each reciprocating engine must have an alternate air source that prevents the entry of rain, ice, or any other foreign matter.

**(c)** Air intakes may not open within the cowling, unless—

(1) That part of the cowling is isolated from the engine accessory section by means of a fireproof diaphragm; or

(2) For reciprocating engines, there are means to prevent the emergence of backfire flames.

**(d)** For turbine engine powered airplanes and airplanes incorporating auxiliary power units—

(1) There must be means to prevent hazardous quantities of fuel leakage or overflow from drains, vents, or other components of flammable fluid systems from entering the engine or auxiliary power unit intake system; and

(2) The airplane must be designed to prevent water or slush on the runway, taxiway, or other airport operating surfaces from being directed into the engine or auxiliary power unit air inlet ducts in hazardous quantities, and the air inlet ducts must be located or protected so as to minimize the ingestion of foreign matter during takeoff, landing, and taxiing.

**(e)** If the engine induction system contains parts or components that could be damaged by foreign objects entering the air inlet, it must be shown by tests or, if appropriate, by analysis that the induction system design can withstand the foreign object ingestion test conditions of §§33.76, 33.77 and 33.78(a)(1) of this chapter without failure of parts or components that could create a hazard.

[Docket No. 5066, 29 FR 18291, Dec. 24, 1964; as amended by Amdt. 25–38, 41 FR 55467, Dec. 20, 1976; Amdt. 25–40, 42 FR 15043, March 17, 1977; Amdt. 25–57, 49 FR 6849, Feb. 23, 1984; Amdt. 25–100, 65 FR 55854, Sept. 14, 2000]

## §25.1093 Induction system icing protection.



**(a)** *Reciprocating engines.* Each reciprocating engine air induction system must have means to prevent and eliminate icing. Unless this is done by other means, it must be shown that, in air free of visible moisture at a temperature of 30°F, each airplane with altitude engines using—

(1) Conventional venturi carburetors have a preheater that can provide a heat rise of 120°F with the engine at 60 percent of maximum continuous power; or

(2) Carburetors tending to reduce the probability of ice formation has a preheater that can provide a heat rise of 100°F with the engine at 60 percent of maximum continuous power.

**(b)** *Turbine engines.*

(1) Each turbine engine must operate throughout the flight power range of the engine (including idling), without the accumulation of ice on the engine, inlet system components, or airframe components that would adversely affect engine operation or cause a serious loss of power or thrust—

(i) Under the icing conditions specified in Appendix C, and

(ii) In falling and blowing snow within the limitations established for the airplane for such operation.

(2) Each turbine engine must idle for 30 minutes on the ground, with the air bleed available for engine icing protection at its critical condition, without adverse effect, in an atmosphere that is at a temperature between 15° and 30°F (between -9° and -1°C) and has a liquid water content not less than 0.3 grams per cubic meter in the form of drops having a mean effective diameter not less than 20 microns, followed by momentary operation at takeoff power or thrust. During the 30 minutes of idle operation, the engine may be run up periodically to a moderate power or thrust setting in a manner acceptable to the Administrator.

**(c)** *Supercharged reciprocating engines.* For each engine having a supercharger to pressurize the air before it enters the carburetor, the heat rise in the air caused by that supercharging at any altitude may be utilized in determining compliance with paragraph (a) of this section if the heat rise utilized is that which will be available, automatically, for the applicable altitude and operating condition because of supercharging.

[Docket No. 5066, 29 FR 18291, Dec. 24, 1964; as amended by Amdt. 25–38, 41 FR 55467, Dec. 20, 1976; Amdt. 25–40, 42 FR 15043, March 17, 1977; Amdt. 25–57, 49 FR 6849, Feb. 23, 1984; Amdt. 25–72, 55 FR 29785, July 20, 1990]

## §25.1101 Carburetor air preheater design.

Each carburetor air preheater must be designed and constructed to—

**(a)** Ensure ventilation of the preheater when the engine is operated in cold air;

**(b)** Allow inspection of the exhaust manifold parts that it surrounds; and

**(c)** Allow inspection of critical parts of the preheater itself.

## §25.1103 Induction system ducts and air duct systems.

**(a)** Each induction system duct upstream of the first stage of the engine supercharger and of the auxiliary power unit compressor must have a drain to prevent the hazardous accumulation of fuel and moisture in the ground attitude. No drain may discharge where it might cause a fire hazard.

**(b)** Each induction system duct must be—

(1) Strong enough to prevent induction system failures resulting from normal backfire conditions; and

(2) Fire-resistant if it is in any fire zone for which a fire-extinguishing system is required, except that ducts for auxiliary power units must be fireproof within the auxiliary power unit fire zone.

**(c)** Each duct connected to components between which relative motion could exist must have means for flexibility.

**(d)** For turbine engine and auxiliary power unit bleed air duct systems, no hazard may result if a duct failure occurs at any point between the air duct source and the airplane unit served by the air.

**(e)** Each auxiliary power unit induction system duct must be fireproof for a sufficient distance upstream of the auxiliary power unit compartment to prevent hot gas reverse flow from burning through auxiliary power unit ducts and entering any other compartment or area of the airplane in which a hazard would be created resulting from the entry of hot gases. The materials used to form the remainder of the induction system duct and plenum chamber of the auxiliary power unit must be capable of resisting the maximum heat conditions likely to occur.

**(f)** Each auxiliary power unit induction system duct must be constructed of materials that will not absorb or trap hazardous quantities of flammable fluids that could be ignited in the event of a surge or reverse flow condition.

[Docket No. 5066, 29 FR 18291, Dec. 24, 1964; as amended by Amdt. 25–46, 43 FR 50597, Oct. 30, 1978]

## §25.1105 Induction system screens.

If induction system screens are used—

**(a)** Each screen must be upstream of the carburetor;

**(b)** No screen may be in any part of the induction system that is the only passage through which air can reach the engine, unless it can be deiced by heated air;

**(c)** No screen may be deiced by alcohol alone; and

**(d)** It must be impossible for fuel to strike any screen.

## §25.1107 Inter-coolers and after-coolers.

Each inter-cooler and after-cooler must be able to withstand any vibration, inertia, and air pressure load to which it would be subjected in operation.

### Exhaust System

## §25.1121 General.

For powerplant and auxiliary power unit installations the following apply:

**(a)** Each exhaust system must ensure safe disposal of exhaust gases without fire hazard or carbon monoxide contamination in any personnel compartment. For test purposes, any acceptable carbon monoxide detection method may be used to show the absence of carbon monoxide.

**(b)** Each exhaust system part with a surface hot enough to ignite flammable fluids or vapors must be located or shielded so that leakage from any system carrying flammable fluids or vapors will not result in a fire caused by impingement of the fluids or vapors on any part of the exhaust system including shields for the exhaust system.

**(c)** Each component that hot exhaust gases could strike, or that could be subjected to high temperatures from exhaust system parts, must be fireproof. All exhaust system components must be separated by fireproof shields from adjacent parts of the airplane that are outside the engine and auxiliary power unit compartments.

**(d)** No exhaust gases may discharge so as to cause a fire hazard with respect to any flammable fluid vent or drain.

**(e)** No exhaust gases may discharge where they will cause a glare seriously affecting pilot vision at night.

(f) Each exhaust system component must be ventilated to prevent points of excessively high temperature.

(g) Each exhaust shroud must be ventilated or insulated to avoid, during normal operation, a temperature high enough to ignite any flammable fluids or vapors external to the shroud.

[Docket No. 5066, 29 FR 18291, Dec. 24, 1964; as amended by Amdt. 25–40, 42 FR 15043, March 17, 1977]

## §25.1123 Exhaust piping.

For powerplant and auxiliary power unit installations, the following apply:

(a) Exhaust piping must be heat and corrosion resistant, and must have provisions to prevent failure due to expansion by operating temperatures.

(b) Piping must be supported to withstand any vibration and inertia loads to which it would be subjected in operation; and

(c) Piping connected to components between which relative motion could exist must have means for flexibility.

[Docket No. 5066, 29 FR 18291, Dec. 24, 1964; as amended by Amdt. 25–40, 42 FR 15044, March 17, 1977]

## §25.1125 Exhaust heat exchangers.

For reciprocating engine powered airplanes, the following apply:

(a) Each exhaust heat exchanger must be constructed and installed to withstand each vibration, inertia, and other load to which it would be subjected in operation. In addition—

(1) Each exchanger must be suitable for continued operation at high temperatures and resistant to corrosion from exhaust gases;

(2) There must be means for the inspection of the critical parts of each exchanger;

(3) Each exchanger must have cooling provisions wherever it is subject to contact with exhaust gases; and

(4) No exhaust heat exchanger or muff may have any stagnant areas or liquid traps that would increase the probability of ignition of flammable fluids or vapors that might be present in case of the failure or malfunction of components carrying flammable fluids.

(b) If an exhaust heat exchanger is used for heating ventilating air—

(1) There must be a secondary heat exchanger between the primary exhaust gas heat exchanger and the ventilating air system; or

(2) Other means must be used to preclude the harmful contamination of the ventilating air.

[Docket No. 5066, 29 FR 18291, Dec. 24, 1964; as amended by Amdt. 25–38, 41 FR 55467, Dec. 20, 1976]

## §25.1127 Exhaust driven turbo-superchargers.

(a) Each exhaust driven turbo-supercharger must be approved or shown to be suitable for the particular application. It must be installed and supported to ensure safe operation between normal inspections and overhauls. In addition, there must be provisions for expansion and flexibility between exhaust conduits and the turbine.

(b) There must be provisions for lubricating the turbine and for cooling turbine parts where temperatures are critical.

(c) If the normal turbo-supercharger control system malfunctions, the turbine speed may not exceed its maximum allowable value. Except for the waste gate operating components, the components provided for meeting this requirement must be independent of the normal turbo-supercharger controls.

### POWERPLANT CONTROLS AND ACCESSORIES

## §25.1141 Powerplant controls: general.

Each powerplant control must be located, arranged, and designed under §§25.777 through 25.781 and marked under §25.1555. In addition, it must meet the following requirements:

(a) Each control must be located so that it cannot be inadvertently operated by persons entering, leaving, or moving normally in, the cockpit.

(b) Each flexible control must be approved or must be shown to be suitable for the particular application.

(c) Each control must have sufficient strength and rigidity to withstand operating loads without failure and without excessive deflection.

(d) Each control must be able to maintain any set position without constant attention by flight crewmembers and without creep due to control loads or vibration.

(e) The portion of each powerplant control located in a designated fire zone that is required to be operated in the event of fire must be at least fire resistant.

(f) Powerplant valve controls located in the cockpit must have—

(1) For manual valves, positive stops or in the case of fuel valves suitable index provisions, in the open and closed position; and

(2) For power-assisted valves, a means to indicate to the flight crew when the valve—

(i) Is in the fully open or fully closed position; or

(ii) Is moving between the fully open and fully closed position.

[Docket No. 5066, 29 FR 18291, Dec. 24, 1964; as amended by Amdt. 25–40, 42 FR 15044, March 17, 1977; Amdt. 25–72, 55 FR 29785, July 20, 1990]

## §25.1142 Auxiliary power unit controls.

Means must be provided on the flight deck for starting, stopping, and emergency shutdown of each installed auxiliary power unit.

[Docket No. 5066, 29 FR 18291, Dec. 24, 1964; as amended by Amdt. 25–46, 43 FR 50598, Oct. 30, 1978]

## §25.1143 Engine controls.

**(a)** There must be a separate power or thrust control for each engine.

**(b)** Power and thrust controls must be arranged to allow—

(1) Separate control of each engine; and

(2) Simultaneous control of all engines.

**(c)** Each power and thrust control must provide a positive and immediately responsive means of controlling its engine.

**(d)** For each fluid injection (other than fuel) system and its controls not provided and approved as part of the engine, the applicant must show that the flow of the injection fluid is adequately controlled.

**(e)** If a power or thrust control incorporates a fuel shutoff feature, the control must have a means to prevent the inadvertent movement of the control into the shutoff position. The means must—

(1) Have a positive lock or stop at the idle position; and

(2) Require a separate and distinct operation to place the control in the shutoff position.

[Docket No. 5066, 29 FR 18291, Dec. 24, 1964; as amended by Amdt. 25–23, 35 FR 5677, April 8, 1970; Amdt. 25–38, 41 FR 55467, Dec. 20, 1976; Amdt. 25–57, 49 FR 6849, Feb. 23, 1984]

## §25.1145 Ignition switches.

**(a)** Ignition switches must control each engine ignition circuit on each engine.

**(b)** There must be means to quickly shut off all ignition by the grouping of switches or by a master ignition control.

**(c)** Each group of ignition switches, except ignition switches for turbine engines for which continuous ignition is not required, and each master ignition control must have a means to prevent its inadvertent operation.

[Docket No. 5066, 29 FR 18291, Dec. 24, 1964; as amended by Amdt. 25–40, 42 FR 15044 March 17, 1977]

## §25.1147 Mixture controls.

**(a)** If there are mixture controls, each engine must have a separate control. The controls must be grouped and arranged to allow—

(1) Separate control of each engine; and

(2) Simultaneous control of all engines.

**(b)** Each intermediate position of the mixture controls that corresponds to a normal operating setting must be identifiable by feel and sight.

**(c)** The mixture controls must be accessible to both pilots. However, if there is a separate flight engineer station with a control panel, the controls need be accessible only to the flight engineer.

## §25.1149 Propeller speed and pitch controls.

**(a)** There must be a separate propeller speed and pitch control for each propeller.

**(b)** The controls must be grouped and arranged to allow—

(1) Separate control of each propeller; and

(2) Simultaneous control of all propellers.

**(c)** The controls must allow synchronization of all propellers.

**(d)** The propeller speed and pitch controls must be to the right of, and at least one inch below, the pilot's throttle controls.

## §25.1153 Propeller feathering controls.

**(a)** There must be a separate propeller feathering control for each propeller. The control must have means to prevent its inadvertent operation.

**(b)** If feathering is accomplished by movement of the propeller pitch or speed control lever, there must be means to prevent the inadvertent movement of this lever to the feathering position during normal operation.

[Docket No. 5066, 29 FR 18291, Dec. 24, 1964; as amended by Amdt. 25–11, 32 FR 6913, May 5, 1967]

## §25.1155 Reverse thrust and propeller pitch settings below the flight regime.

Each control for reverse thrust and for propeller pitch settings below the flight regime must have means to prevent its inadvertent operation. The means must have a positive lock or stop at the flight idle position and must require a separate and distinct operation by the crew to displace the control from the flight regime (forward thrust regime for turbojet powered airplanes).

[Docket No. 5066, 29 FR 18291, Dec. 24, 1964; as amended by Amdt. 25–11, 32 FR 6913, May 5, 1967]

## §25.1157 Carburetor air temperature controls.

There must be a separate carburetor air temperature control for each engine.

## §25.1159 Supercharger controls.

Each supercharger control must be accessible to the pilots or, if there is a separate flight engineer station with a control panel, to the flight engineer.

## §25.1161 Fuel jettisoning system controls.

Each fuel jettisoning system control must have guards to prevent inadvertent operation. No control may be near any fire extinguisher control or other control used to combat fire.

## §25.1163 Powerplant accessories.

**(a)** Each engine mounted accessory must—

(1) Be approved for mounting on the engine involved;

(2) Use the provisions on the engine for mounting; and

(3) Be sealed to prevent contamination of the engine oil system and the accessory system.

**(b)** Electrical equipment subject to arcing or sparking must be installed to minimize the probability of contact with any flammable fluids or vapors that might be present in a free state.

**(c)** If continued rotation of an engine-driven cabin supercharger or of any remote accessory driven by the engine is hazardous if malfunctioning occurs, there must be means to prevent rotation without interfering with the continued operation of the engine.

[Docket No. 5066, 29 FR 18291, Dec. 24, 1964; as amended by Amdt. 25–57, 49 FR 6849, Feb. 23, 1984]

## §25.1165 Engine ignition systems.

**(a)** Each battery ignition system must be supplemented by a generator that is automatically available as an alternate source of electrical energy to allow continued engine operation if any battery becomes depleted.

**(b)** The capacity of batteries and generators must be large enough to meet the simultaneous demands of the engine ignition system and the greatest demands of any electrical system components that draw electrical energy from the same source.

**(c)** The design of the engine ignition system must account for—

(1) The condition of an inoperative generator;

(2) The condition of a completely depleted battery with the generator running at its normal operating speed; and

(3) The condition of a completely depleted battery with the generator operating at idling speed, if there is only one battery.

**(d)** Magneto ground wiring (for separate ignition circuits) that lies on the engine side of the fire wall, must be installed, located, or protected, to minimize the probability of simultaneous failure of two or more wires as a result of mechanical damage, electrical faults, or other cause.

**(e)** No ground wire for any engine may be routed through a fire zone of another engine unless each part of that wire within that zone is fireproof.

**(f)** Each ignition system must be independent of any electrical circuit, not used for assisting, controlling, or analyzing the operation of that system.

**(g)** There must be means to warn appropriate flight crewmembers if the malfunctioning of any part of the electrical system is causing the continuous discharge of any battery necessary for engine ignition.

**(h)** Each engine ignition system of a turbine powered airplane must be considered an essential electrical load.

[Docket No. 5066, 29 FR 18291, Dec. 24, 1964; as amended by Amdt. 25–23, 35 FR 5677, April 8, 1970; Amdt. 25–72, 55 FR 29785, July 20, 1990]

25

## §25.1167 Accessory gearboxes.

For airplanes equipped with an accessory gearbox that is not certificated as part of an engine—

**(a)** The engine with gearbox and connecting transmissions and shafts attached must be subjected to the tests specified in §33.49 or §33.87 of this chapter, as applicable;

**(b)** The accessory gearbox must meet the requirements of §§33.25 and 33.53 or 33.91 of this chapter, as applicable; and

**(c)** Possible misalignments and torsional loadings of the gearbox, transmission, and shaft system, expected to result under normal operating conditions must be evaluated.

[Docket No. 5066, 29 FR 18291, Dec. 24, 1964; as amended by Amdt. 25–38, 41 FR 55467, Dec. 20, 1976]

### Powerplant Fire Protection

## §25.1181 Designated fire zones; regions included.

**(a)** Designated fire zones are—

(1) The engine power section;

(2) The engine accessory section;

(3) Except for reciprocating engines, any complete powerplant compartment in which no isolation is provided between the engine power section and the engine accessory section;

(4) Any auxiliary power unit compartment;

(5) Any fuel-burning heater and other combustion equipment installation described in §25.859;

(6) The compressor and accessory sections of turbine engines; and

(7) Combustor, turbine, and tailpipe sections of turbine engine installations that contain lines or components carrying flammable fluids or gases.

**(b)** Each designated fire zone must meet the requirements of §§25.867, and 25.1185 through 25.1203.

[Docket No. 5066, 29 FR 18291, Dec. 24, 1964; as amended by Amdt. 25–11, 32 FR 6913, May 5, 1967; Amdt. 25–23, 35 FR 5677, April 8, 1970; Amdt. 25–72, 55 FR 29785, July 20, 1990]

## §25.1182 Nacelle areas behind firewalls, and engine pod attaching structures containing flammable fluid lines.

**(a)** Each nacelle area immediately behind the firewall, and each portion of any engine pod attaching structure containing flammable fluid lines, must meet each requirement of §§25.1103(b), 25.1165 (d) and (e), 25.1183, 25.1185(c), 25.1187, 25.1189, and 25.1195 through 25.1203, including those concerning designated fire zones. However, engine pod attaching structures need not contain fire detection or extinguishing means.

**(b)** For each area covered by paragraph (a) of this section that contains a retractable landing gear, compliance with that paragraph need only be shown with the landing gear retracted.

[Docket No. 5066, 29 FR 18291, Dec. 24, 1964; as amended by Amdt. 25–11, 32 FR 6913, May 5, 1967]

## §25.1183 Flammable fluid-carrying components.

**(a)** Except as provided in paragraph (b) of this section, each line, fitting, and other component carrying flammable fluid in any area subject to engine fire conditions, and each component which conveys or contains flammable fluid in a designated fire zone must be fire resistant, except that flammable fluid tanks and supports in a designated fire zone must be fireproof or be enclosed by a fireproof shield unless damage by fire to any non-fireproof part will not cause leakage or spillage of flammable fluid. Components must be shielded or located to safeguard against the ignition of leaking flammable fluid. An integral oil sump of less than 25-quart capacity on a reciprocating engine need not be fireproof nor be enclosed by a fireproof shield.

**(b)** Paragraph (a) of this section does not apply to—

(1) Lines, fittings, and components which are already approved as part of a type certificated engine; and

(2) Vent and drain lines, and their fittings, whose failure will not result in, or add to, a fire hazard.

**(c)** All components, including ducts, within a designated fire zone must be fireproof if, when exposed to or damaged by fire, they could—

(1) Result in fire spreading to other regions of the airplane; or

(2) Cause unintentional operation of, or inability to operate, essential services or equipment.

[Docket No. 5066, 29 FR 18291, Dec. 24, 1964; as amended by Amdt. 25–11, 32 FR 6913, May 5, 1967; Amdt. 25–36, 39 FR 35461, Oct. 1, 1974; Amdt. 25–57, 49 FR 6849, Feb. 23, 1984; Amdt. 25–101, 65 FR 79710, Dec. 19, 2000]

## §25.1185 Flammable fluids.

**(a)** Except for the integral oil sumps specified in §25.1183 (a), no tank or reservoir that is a part of a system containing flammable fluids or gases may be in a designated fire zone unless the fluid contained, the design of the system, the materials used in the tank, the shut-off means, and all connections, lines, and control provide a degree of safety equal to that which would exist if the tank or reservoir were outside such a zone.

**(b)** There must be at least one-half inch of clear airspace between each tank or reservoir and each firewall or shroud isolating a designated fire zone.

**(c)** Absorbent materials close to flammable fluid system components that might leak must be covered or treated to prevent the absorption of hazardous quantities of fluids.

[Docket No. 5066, 29 FR 18291, Dec. 24, 1964; as amended by Amdt. 25–19, 33 FR 15410, Oct. 17, 1968; Amdt. 25–94, 63 FR 8848, Feb. 23, 1998]

## §25.1187 Drainage and ventilation of fire zones.

**(a)** There must be complete drainage of each part of each designated fire zone to minimize the hazards resulting from failure or malfunctioning of any component containing flammable fluids. The drainage means must be—

(1) Effective under conditions expected to prevail when drainage is needed; and

(2) Arranged so that no discharged fluid will cause an additional fire hazard.

**(b)** Each designated fire zone must be ventilated to prevent the accumulation of flammable vapors.

**(c)** No ventilation opening may be where it would allow the entry of flammable fluids, vapors, or flame from other zones.

**(d)** Each ventilation means must be arranged so that no discharged vapors will cause an additional fire hazard.

**(e)** Unless the extinguishing agent capacity and rate of discharge are based on maximum air flow through a zone, there must be means to allow the crew to shut off sources of forced ventilation to any fire zone except the engine power section of the nacelle and the combustion heater ventilating air ducts.

## §25.1189 Shutoff means.

(a) Each engine installation and each fire zone specified in §25.1181(a) (4) and (5) must have a means to shut off or otherwise prevent hazardous quantities of fuel, oil, deicer, and other flammable fluids, from flowing into, within, or through any designated fire zone, except that shutoff means are not required for—

(1) Lines, fittings, and components forming an integral part of an engine; and

(2) Oil systems for turbine engine installations in which all components of the system in a designated fire zone, including oil tanks, are fireproof or located in areas not subject to engine fire conditions.

(b) The closing of any fuel shutoff valve for any engine may not make fuel unavailable to the remaining engines.

(c) Operation of any shutoff may not interfere with the later emergency operation of other equipment, such as the means for feathering the propeller.

(d) Each flammable fluid shutoff means and control must be fireproof or must be located and protected so that any fire in a fire zone will not affect its operation.

(e) No hazardous quantity of flammable fluid may drain into any designated fire zone after shutoff.

(f) There must be means to guard against inadvertent operation of the shutoff means and to make it possible for the crew to reopen the shutoff means in flight after it has been closed.

(g) Each tank-to-engine shutoff valve must be located so that the operation of the valve will not be affected by powerplant or engine mount structural failure.

(h) Each shutoff valve must have a means to relieve excessive pressure accumulation unless a means for pressure relief is otherwise provided in the system.

[Docket No. 5066, 29 FR 18291, Dec. 24, 1964; as amended by Amdt. 25–23, 35 FR 5677, April 8, 1970; Amdt. 25–57, 49 FR 6849, Feb. 23, 1984]

## §25.1191 Firewalls.

(a) Each engine, auxiliary power unit, fuel-burning heater, other combustion equipment intended for operation in flight, and the combustion, turbine, and tailpipe sections of turbine engines, must be isolated from the rest of the airplane by firewalls, shrouds, or equivalent means.

(b) Each firewall and shroud must be—

(1) Fireproof;

(2) Constructed so that no hazardous quantity of air, fluid, or flame can pass from the compartment to other parts of the airplane;

(3) Constructed so that each opening is sealed with close fitting fireproof grommets, bushings, or firewall fittings; and

(4) Protected against corrosion.

## §25.1192 Engine accessory section diaphragm.

For reciprocating engines, the engine power section and all portions of the exhaust system must be isolated from the engine accessory compartment by a diaphragm that complies with the firewall requirements of §25.1191.

[Docket No. 5066, 29 FR 18291, Dec. 24, 1964; as amended by Amdt. 25–23, 35 FR 5678, April 8, 1970]

## §25.1193 Cowling and nacelle skin.

(a) Each cowling must be constructed and supported so that it can resist any vibration, inertia, and air load to which it may be subjected in operation.

(b) Cowling must meet the drainage and ventilation requirements of §25.1187.

(c) On airplanes with a diaphragm isolating the engine power section from the engine accessory section, each part of the accessory section cowling subject to flame in case of fire in the engine power section of the powerplant must—

(1) Be fireproof; and

(2) Meet the requirements of §25.1191.

(d) Each part of the cowling subject to high temperatures due to its nearness to exhaust system parts or exhaust gas impingement must be fireproof.

(e) Each airplane must—

(1) Be designed and constructed so that no fire originating in any fire zone can enter, either through openings or by burning through external skin, any other zone or region where it would create additional hazards;

(2) Meet paragraph (e)(1) of this section with the landing gear retracted (if applicable); and

(3) Have fireproof skin in areas subject to flame if a fire starts in the engine power or accessory sections.

## §25.1195 Fire extinguishing systems.

(a) Except for combustor, turbine, and tail pipe sections of turbine engine installations that contain lines or components carrying flammable fluids or gases for which it is shown that a fire originating in these sections can be controlled, there must be a fire extinguisher system serving each designated fire zone.

**(b)** The fire extinguishing system, the quantity of the extinguishing agent, the rate of discharge, and the discharge distribution must be adequate to extinguish fires. It must be shown by either actual or simulated flights tests that under critical airflow conditions in flight the discharge of the extinguishing agent in each designated fire zone specified in paragraph (a) of this section will provide an agent concentration capable of extinguishing fires in that zone and of minimizing the probability of reignition. An individual "one-shot" system may be used for auxiliary power units, fuel burning heaters, and other combustion equipment. For each other designated fire zone, two discharges must be provided each of which produces adequate agent concentration.

**(c)** The fire extinguishing system for a nacelle must be able to simultaneously protect each zone of the nacelle for which protection is provided.

[Docket No. 5066, 29 FR 18291, Dec. 24, 1964; as amended by Amdt. 25–46, 43 FR 50598, Oct. 30, 1978]

## §25.1197 Fire extinguishing agents.

**(a)** Fire extinguishing agents must—

(1) Be capable of extinguishing flames emanating from any burning of fluids or other combustible materials in the area protected by the fire extinguishing system; and

(2) Have thermal stability over the temperature range likely to be experienced in the compartment in which they are stored.

**(b)** If any toxic extinguishing agent is used, provisions must be made to prevent harmful concentrations of fluid or fluid vapors (from leakage during normal operation of the airplane or as a result of discharging the fire extinguisher on the ground or in flight) from entering any personnel compartment, even though a defect may exist in the extinguishing system. This must be shown by test except for built-in carbon dioxide fuselage compartment fire extinguishing systems for which—

(1) Five pounds or less of carbon dioxide will be discharged, under established fire control procedures, into any fuselage compartment; or

(2) There is protective breathing equipment for each flight crewmember on flight deck duty.

[Docket No. 5066, 29 FR 18291, Dec. 24, 1964; as amended by Amdt. 25–38, 41 FR 55467, Dec. 20, 1976; Amdt. 25–40, 42 FR 15044, March 17, 1977]

## §25.1199 Extinguishing agent containers.

**(a)** Each extinguishing agent container must have a pressure relief to prevent bursting of the container by excessive internal pressures.

**(b)** The discharge end of each discharge line from a pressure relief connection must be located so that discharge of the fire extinguishing agent would not damage the airplane. The line must also be located or protected to prevent clogging caused by ice or other foreign matter.

**(c)** There must be a means for each fire extinguishing agent container to indicate that the container has discharged or that the charging pressure is below the established minimum necessary for proper functioning.

**(d)** The temperature of each container must be maintained, under intended operating conditions, to prevent the pressure in the container from—

(1) Falling below that necessary to provide an adequate rate of discharge; or

(2) Rising high enough to cause premature discharge.

**(e)** If a pyrotechnic capsule is used to discharge the extinguishing agent, each container must be installed so that temperature conditions will not cause hazardous deterioration of the pyrotechnic capsule.

[Docket No. 5066, 29 FR 18291, Dec. 24, 1964; as amended by Amdt. 25–23, 35 FR 5678, April 8, 1970; Amdt. 25–40, 42 FR 15044, March 17, 1977]

## §25.1201 Fire extinguishing system materials.

**(a)** No material in any fire extinguishing system may react chemically with any extinguishing agent so as to create a hazard.

**(b)** Each system component in an engine compartment must be fireproof.

## §25.1203 Fire detector system.

**(a)** There must be approved, quick acting fire or overheat detectors in each designated fire zone, and in the combustion, turbine, and tailpipe sections of turbine engine installations, in numbers and locations ensuring prompt detection of fire in those zones.

**(b)** Each fire detector system must be constructed and installed so that—

(1) It will withstand the vibration, inertia, and other loads to which it may be subjected in operation;

(2) There is a means to warn the crew in the event that the sensor or associated wiring within a designated fire zone is severed at one point, unless the system continues to function as a satisfactory detection system after the severing; and

(3) There is a means to warn the crew in the event of a short circuit in the sensor or associated wiring within a designated fire zone, unless the system continues to function as a satisfactory detection system after the short circuit.

**(c)** No fire or overheat detector may be affected by any oil, water, other fluids or fumes that might be present.

(d) There must be means to allow the crew to check, in flight, the functioning of each fire or overheat detector electric circuit.

(e) Wiring and other components of each fire or overheat detector system in a fire zone must be at least fire-resistant.

(f) No fire or overheat detector system component for any fire zone may pass through another fire zone, unless—

(1) It is protected against the possibility of false warnings resulting from fires in zones through which it passes; or

(2) Each zone involved is simultaneously protected by the same detector and extinguishing system.

(g) Each fire detector system must be constructed so that when it is in the configuration for installation it will not exceed the alarm activation time approved for the detectors using the response time criteria specified in the appropriate Technical Standard Order for the detector.

[Docket No. 5066, 29 FR 18291, Dec. 24, 1964; as amended by Amdt. 25–23, 35 FR 5678, April 8, 1970; Amdt. 25–26, 36 FR 5493, March 24, 1971]

### §25.1207 Compliance.

Unless otherwise specified, compliance with the requirements of §§25.1181 through 25.1203 must be shown by a full scale fire test or by one or more of the following methods:

(a) Tests of similar powerplant configurations;

(b) Tests of components;

(c) Service experience of aircraft with similar powerplant configurations;

(d) Analysis.

[Docket No. 5066, 29 FR 18291, Dec. 24, 1964; as amended by Amdt. 25–46, 43 FR 50598, Oct. 30, 1978]

## Subpart F—Equipment

GENERAL

### §25.1301 Function and installation.

Each item of installed equipment must—

(a) Be of a kind and design appropriate to its intended function;

(b) Be labeled as to its identification, function, or operating limitations, or any applicable combination of these factors;

(c) Be installed according to limitations specified for that equipment; and

(d) Function properly when installed.

### §25.1303 Flight and navigation instruments.

(a) The following flight and navigation instruments must be installed so that the instrument is visible from each pilot station:

(1) A free air temperature indicator or an air-temperature indicator which provides indications that are convertible to free-air temperature.

(2) A clock displaying hours, minutes, and seconds with a sweep-second pointer or digital presentation.

(3) A direction indicator (nonstabilized magnetic compass).

(b) The following flight and navigation instruments must be installed at each pilot station:

(1) An airspeed indicator. If airspeed limitations vary with altitude, the indicator must have a maximum allowable airspeed indicator showing the variation of $V_{MO}$ with altitude.

(2) An altimeter (sensitive).

(3) A rate-of-climb indicator (vertical speed).

(4) A gyroscopic rate-of-turn indicator combined with an integral slip-skid indicator (turn-and-bank indicator) except that only a slip-skid indicator is required on large airplanes with a third attitude instrument system usable through flight attitudes of 360° of pitch and roll and installed in accordance with §121.305(k) of this title.

(5) A bank and pitch indicator (gyroscopically stabilized).

(6) A direction indicator (gyroscopically stabilized, magnetic or nonmagnetic).

(c) The following flight and navigation instruments are required as prescribed in this paragraph:

(1) A speed warning device is required for turbine engine powered airplanes and for airplanes with $V_{MO}/M_{MO}$ greater than 0.8 $V_{DF}/M_{DF}$ or 0.8 $V_D/M_D$. The speed warning device must give effective aural warning (differing distinctively from aural warnings used for other purposes) to the pilots, whenever the speed exceeds $V_{MO}$ plus 6 knots or $M_{MO}$ +0.01. The upper limit of the production tolerance for the warning device may not exceed the prescribed warning speed.

(2) A machmeter is required at each pilot station for airplanes with compressibility limitations not otherwise indicated to the pilot by the airspeed indicating system required under paragraph (b)(1) of this section.

[Docket No. 5066, 29 FR 18291, Dec. 24, 1964; as amended by Amdt. 25–23, 35 FR 5678, April 8, 1970; Amdt. 25–24, 35 FR 7108, May 6, 1970; Amdt. 25–38, 41 FR 55467, Dec. 20, 1976; Amdt. 25–90, 62 FR 13253, March 19, 1997]

## §25.1305 Powerplant instruments.

The following are required powerplant instruments:

**(a)** *For all airplanes.*

(1) A fuel pressure warning means for each engine, or a master warning means for all engines with provision for isolating the individual warning means from the master warning means.

(2) A fuel quantity indicator for each fuel tank.

(3) An oil quantity indicator for each oil tank.

(4) An oil pressure indicator for each independent pressure oil system of each engine.

(5) An oil pressure warning means for each engine, or a master warning means for all engines with provision for isolating the individual warning means from the master warning means.

(6) An oil temperature indicator for each engine.

(7) Fire-warning indicators.

(8) An augmentation liquid quantity indicator (appropriate for the manner in which the liquid is to be used in operation) for each tank.

**(b)** *For reciprocating engine-powered airplanes.* In addition to the powerplant instruments required by paragraph (a) of this section, the following powerplant instruments are required:

(1) A carburetor air temperature indicator for each engine.

(2) A cylinder head temperature indicator for each air-cooled engine.

(3) A manifold pressure indicator for each engine.

(4) A fuel pressure indicator (to indicate the pressure at which the fuel is supplied) for each engine.

(5) A fuel flowmeter, or fuel mixture indicator, for each engine without an automatic altitude mixture control.

(6) A tachometer for each engine.

(7) A device that indicates, to the flight crew (during flight), any change in the power output, for each engine with—

(i) An automatic propeller feathering system, whose operation is initiated by a power output measuring system; or

(ii) A total engine piston displacement of 2,000 cubic inches or more.

(8) A means to indicate to the pilot when the propeller is in reverse pitch, for each reversing propeller.

**(c)** *For turbine engine-powered airplanes.* In addition to the powerplant instruments required by paragraph (a) of this section, the following powerplant instruments are required:

(1) A gas temperature indicator for each engine.

(2) A fuel flowmeter indicator for each engine.

(3) A tachometer (to indicate the speed of the rotors with established limiting speeds) for each engine.

(4) A means to indicate, to the flight crew, the operation of each engine starter that can be operated continuously but that is neither designed for continuous operation nor designed to prevent hazard if it failed.

(5) An indicator to indicate the functioning of the powerplant ice protection system for each engine.

(6) An indicator for the fuel strainer or filter required by §25.997 to indicate the occurrence of contamination of the strainer or filter before it reaches the capacity established in accordance with §25.997(d).

(7) A warning means for the oil strainer or filter required by §25.1019, if it has no bypass, to warn the pilot of the occurrence of contamination of the strainer or filter screen before it reaches the capacity established in accordance with §25.1019(a)(2).

(8) An indicator to indicate the proper functioning of any heater used to prevent ice clogging of fuel system components.

**(d)** *For turbojet engine powered airplanes.* In addition to the powerplant instruments required by paragraphs (a) and (c) of this section, the following powerplant instruments are required:

(1) An indicator to indicate thrust, or a parameter that is directly related to thrust, to the pilot. The indication must be based on the direct measurement of thrust or of parameters that are directly related to thrust. The indicator must indicate a change in thrust resulting from any engine malfunction, damage, or deterioration.

(2) A position indicating means to indicate to the flight crew when the thrust reversing device is in the reverse thrust position, for each engine using a thrust reversing device.

(3) An indicator to indicate rotor system unbalance.

**(e)** *For turbopropeller-powered airplanes.* In addition to the powerplant instruments required by paragraphs (a) and (c) of this section, the following powerplant instruments are required:

(1) A torque indicator for each engine.

(2) Position indicating means to indicate to the flight crew when the propeller blade angle is below the flight low pitch position, for each propeller.

**(f)** For airplanes equipped with fluid systems (other than fuel) for thrust or power augmentation, an approved means must be provided to indicate the proper functioning of that system to the flight crew.

[Docket No. 5066, 29 FR 18291, Dec. 24, 1964; as amended by Amdt. 25–23, 35 FR 5678, April 8, 1970; Amdt. 25–35, 39 FR 1831, Jan. 15, 1974; Amdt. 25–36, 39 FR 35461, Oct. 1, 1974; Amdt. 25–38, 41 FR 55467, Dec. 20, 1976; Amdt. 25–54, 45 FR 60173, Sept. 11, 1980; Amdt. 25–72, 55 FR 29785, July 20, 1990]

## §25.1307 Miscellaneous equipment.

The following is required miscellaneous equipment:

**(a)** [Reserved]

**(b)** Two or more independent sources of electrical energy.

**(c)** Electrical protective devices, as prescribed in this part.

**(d)** Two systems for two-way radio communications, with controls for each accessible from each pilot station, designed and installed so that failure of one system will not preclude operation of the other system. The use of a common antenna system is acceptable if adequate reliability is shown.

**(e)** Two systems for radio navigation, with controls for each accessible from each pilot station, designed and installed so that failure of one system will not preclude operation of the other system. The use of a common antenna system is acceptable if adequate reliability is shown.

[Docket No. 5066, 29 FR 18291, Dec. 24, 1964; as amended by Amdt. 25–23, 35 FR 5678, April 8, 1970; Amdt. 25–46, 43 FR 50598, Oct. 30, 1978; Amdt. 25–54, 45 FR 60173, Sept. 11, 1980; Amdt. 25–72, 55 FR 29785, July 20, 1990]

## §25.1309 Equipment, systems, and installations.

**(a)** The equipment, systems, and installations whose functioning is required by this subchapter, must be designed to ensure that they perform their intended functions under any foreseeable operating condition.

**(b)** The airplane systems and associated components, considered separately and in relation to other systems, must be designed so that—

(1) The occurrence of any failure condition which would prevent the continued safe flight and landing of the airplane is extremely improbable, and

(2) The occurrence of any other failure conditions which would reduce the capability of the airplane or the ability of the crew to cope with adverse operating conditions is improbable.

**(c)** Warning information must be provided to alert the crew to unsafe system operating conditions, and to enable them to take appropriate corrective action. Systems, controls, and associated monitoring and warning means must be designed to minimize crew errors which could create additional hazards.

**(d)** Compliance with the requirements of paragraph (b) of this section must be shown by analysis, and where necessary, by appropriate ground, flight, or simulator tests. The analysis must consider—

(1) Possible modes of failure, including malfunctions and damage from external sources.

(2) The probability of multiple failures and undetected failures.

(3) The resulting effects on the airplane and occupants, considering the stage of flight and operating conditions, and

(4) The crew warning cues, corrective action required, and the capability of detecting faults.

**(e)** Each installation whose functioning is required by this subchapter, and that requires a power supply, is an "essential load" on the power supply. The power sources and the system must be able to supply the following power loads in probable operating combinations and for probable durations:

(1) Loads connected to the system with the system functioning normally.

(2) Essential loads, after failure of any one prime mover, power converter, or energy storage device.

(3) Essential loads after failure of—

(i) Any one engine on two-engine airplanes; and

(ii) Any two engines on three-or-more-engine airplanes.

(4) Essential loads for which an alternate source of power is required by this chapter, after any failure or malfunction in any one power supply system, distribution system, or other utilization system.

**(f)** In determining compliance with paragraphs (e) (2) and (3) of this section, the power loads may be assumed to be reduced under a monitoring procedure consistent with safety in the kinds of operation authorized. Loads not required in controlled flight need not be considered for the two-engine-inoperative condition on airplanes with three or more engines.

**(g)** In showing compliance with paragraphs (a) and (b) of this section with regard to the electrical system and equipment design and installation, critical environmental conditions must be considered. For electrical generation, distribution, and utilization equipment required by or used in complying with this chapter, except equipment covered by Technical Standard Orders containing environmental test procedures, the ability to provide continuous, safe service under foreseeable environmental conditions may be shown by environmental tests, design analysis, or reference to previous comparable service experience on other aircraft.

[Docket No. 5066, 29 FR 18291, Dec. 24, 1964; as amended by Amdt. 25–23, 35 FR 5679, April 8, 1970; Amdt. 25–38, 41 FR 55467, Dec. 20, 1976; Amdt. 25–41, 42 FR 36970, July 18, 1977]

## §25.1316 System lightning protection.

**(a)** For functions whose failure would contribute to or cause a condition that would prevent the continued safe flight and landing of the airplane, each electrical and electronic system that performs these functions must be designed and installed to ensure that the operation and operational capabilities of the systems to perform these functions are not adversely affected when the airplane is exposed to lightning.

**(b)** For functions whose failure would contribute to or cause a condition that would reduce the capability of the airplane or the ability of the flightcrew to cope with adverse operating conditions, each electrical and electronic system that performs these functions must be designed and installed to ensure that these functions can be recovered in a timely manner after the airplane is exposed to lightning.

**(c)** Compliance with the lightning protection criteria prescribed in paragraphs (a) and (b) of this section must be shown for exposure to a severe lightning environment. The applicant must design for and verify that aircraft electrical/electronic systems are protected against the effects of lightning by:

(1) Determining the lightning strike zones for the airplane;

(2) Establishing the external lightning environment for the zones;

(3) Establishing the internal environment;

(4) Identifying all the electrical and electronic systems that are subject to the requirements of this section, and their locations on or within the airplane;

(5) Establishing the susceptibility of the systems to the internal and external lightning environment;

(6) Designing protection; and

(7) Verifying that the protection is adequate.

[Docket No. 25912, 59 FR 22116, April 28, 1994]

INSTRUMENTS: INSTALLATION

## §25.1321 Arrangement and visibility.

**(a)** Each flight, navigation, and powerplant instrument for use by any pilot must be plainly visible to him from his station with the minimum practicable deviation from his normal position and line of vision when he is looking forward along the flight path.

**(b)** The flight instruments required by §25.1303 must be grouped on the instrument panel and centered as nearly as practicable about the vertical plane of the pilot's forward vision. In addition—

(1) The instrument that most effectively indicates attitude must be on the panel in the top center position;

(2) The instrument that most effectively indicates airspeed must be adjacent to and directly to the left of the instrument in the top center position:

(3) The instrument that most effectively indicates altitude must be adjacent to and directly to the right of the instrument in the top center position; and

(4) The instrument that most effectively indicates direction of flight must be adjacent to and directly below the instrument in the top center position.

**(c)** Required powerplant instruments must be closely grouped on the instrument panel. In addition—

(1) The location of identical powerplant instruments for the engines must prevent confusion as to which engine each instrument relates; and

(2) Powerplant instruments vital to the safe operation of the airplane must be plainly visible to the appropriate crewmembers.

**(d)** Instrument panel vibration may not damage or impair the accuracy of any instrument.

**(e)** If a visual indicator is provided to indicate malfunction of an instrument, it must be effective under all probable cockpit lighting conditions.

[Docket No. 5066, 29 FR 18291, Dec. 24, 1964; as amended by Amdt. 25–23, 35 FR 5679, April 8, 1970; Amdt. 25–41, 42 FR 36970, July 18, 1977]

## §25.1322 Warning, caution, and advisory lights.

If warning, caution or advisory lights are installed in the cockpit, they must, unless otherwise approved by the Administrator, be—

**(a)** Red, for warning lights (lights indicating a hazard which may require immediate corrective action);

**(b)** Amber, for caution lights (lights indicating the possible need for future corrective action);

**(c)** Green, for safe operation lights; and

**(d)** Any other color, including white, for lights not described in paragraphs (a) through (c) of this section, provided the color differs sufficiently from the colors prescribed in paragraphs (a) through (c) of this section to avoid possible confusion.

[Docket No. 5066, 29 FR 18291, Dec. 24, 1964; as amended by Amdt. 25–38, 41 FR 55467, Dec. 20, 1976]

## §25.1323 Airspeed indicating system.

For each airspeed indicating system, the following apply:

**(a)** Each airspeed indicating instrument must be approved and must be calibrated to indicate true airspeed (at sea level with a standard atmosphere) with a minimum practicable instrument calibration error when the corresponding pitot and static pressures are applied.

**(b)** Each system must be calibrated to determine the system error (that is, the relation between IAS and CAS) in flight and during the accelerated takeoff ground run. The ground run calibration must be determined—

(1) From 0.8 of the minimum value of $V_1$ to the maximum value of $V_2$, considering the approved ranges of altitude and weight; and

(2) With the flaps and power settings corresponding to the values determined in the establishment of the takeoff path under §25.111 assuming that the critical engine fails at the minimum value of $V_1$.

**(c)** The airspeed error of the installation, excluding the airspeed indicator instrument calibration error, may not exceed three percent or five knots, whichever is greater, throughout the speed range, from—

(1) $V_{MO}$ to 1.3 $V_{S1}$, with flaps retracted; and

(2) 1.3 $V_{S0}$ to $V_{FE}$ with flaps in the landing position.

**(d)** Each system must be arranged, so far as practicable, to prevent malfunction or serious error due to the entry of moisture, dirt, or other substances.

**(e)** Each system must have a heated pitot tube or an equivalent means of preventing malfunction due to icing.

**(f)** Where duplicate airspeed indicators are required, their respective pitot tubes must be far enough apart to avoid damage to both tubes in a collision with a bird.

[Docket No. 5066, 29 FR 18291, Dec. 24, 1964; as amended by Amdt. 25–57, 49 FR 6849, Feb. 23, 1984]

## §25.1325 Static pressure systems.

**(a)** Each instrument with static air case connections must be vented to the outside atmosphere through an appropriate piping system.

**(b)** Each static port must be designed and located in such manner that the static pressure system performance is least affected by airflow variation, or by moisture or other foreign matter, and that the correlation between air pressure in the static pressure system and true ambient atmospheric static pressure is not changed when the airplane is exposed to the continuous and intermittent maximum icing conditions defined in Appendix C of this part.

**(c)** The design and installation of the static pressure system must be such that—

(1) Positive drainage of moisture is provided; chafing of the tubing and excessive distortion or restriction at bends in the tubing is avoided; and the materials used are durable, suitable for the purpose intended, and protected against corrosion; and

(2) It is airtight except for the port into the atmosphere. A proof test must be conducted to demonstrate the integrity of the static pressure system in the following manner:

(i) *Unpressurized airplanes.* Evacuate the static pressure system to a pressure differential of approximately 1 inch of mercury or to a reading on the altimeter, 1,000 feet above the airplane elevation at the time of the test. Without additional pumping for a period of 1 minute, the loss of indicated altitude must not exceed 100 feet on the altimeter.

(ii) *Pressurized airplanes.* Evacuate the static pressure system until a pressure differential equivalent to the maximum cabin pressure differential for which the airplane is type certificated is achieved. Without additional pumping for a period of 1 minute, the loss of indicated altitude must not exceed 2 percent of the equivalent altitude of the maximum cabin differential pressure or 100 feet, whichever is greater.

**(d)** Each pressure altimeter must be approved and must be calibrated to indicate pressure altitude in a standard atmosphere, with a minimum practicable calibration error when the corresponding static pressures are applied.

**(e)** Each system must be designed and installed so that the error in indicated pressure altitude, at sea level, with a standard atmosphere, excluding instrument calibration error, does not result in an error of more than ±30 feet per 100 knots speed for the appropriate configuration in the speed range between 1.3 $V_{S0}$ with flaps extended and 1.8 $V_{S1}$ with flaps retracted. However, the error need not be less than ±30 feet.

**(f)** If an altimeter system is fitted with a device that provides corrections to the altimeter indication, the device must be designed and installed in such manner that it can be bypassed when it malfunctions, unless an alternate altimeter system is provided. Each correction device must be fitted with a means for indicating the occurrence of reasonably probable malfunctions, including power failure, to the flight crew. The indicating means must be effective for any cockpit lighting condition likely to occur.

**(g)** Except as provided in paragraph (h) of this section, if the static pressure system incorporates both a primary and an alternate static pressure source, the means for selecting one or the other source must be designed so that—

(1) When either source is selected, the other is blocked off; and

(2) Both sources cannot be blocked off simultaneously.

**(h)** For unpressurized airplanes, paragraph (g)(1) of this section does not apply if it can be demonstrated that the static pressure system calibration, when either static pressure source is selected, is not changed by the other static pressure source being open or blocked.

[Docket No. 5066, 29 FR 18291, Dec. 24, 1964; as amended by Amdt. 25–5, 30 FR 8261, June 29, 1965; Amdt. 25–12, 32 FR 7587, May 24, 1967; Amdt. 25–41, 42 FR 36970, July 18, 1977]

## §25.1326 Pitot heat indication systems.

If a flight instrument pitot heating system is installed, an indication system must be provided to indicate to the flight crew when that pitot heating system is not operating. The indication system must comply with the following requirements:

**(a)** The indication provided must incorporate an amber light that is in clear view of a flight crewmember.

**(b)** The indication provided must be designed to alert the flight crew if either of the following conditions exist:

(1) The pitot heating system is switched "off."

(2) The pitot heating system is switched "on" and any pitot tube heating element is inoperative.

[Docket No. 5066, 29 FR 18291, Dec. 24, 1964; as amended by Amdt. 25–43, 43 FR 10339, March 13, 1978]

## §25.1327 Magnetic direction indicator.

**(a)** Each magnetic direction indicator must be installed so that its accuracy is not excessively affected by the airplane's vibration or magnetic fields.

**(b)** The compensated installation may not have a deviation, in level flight, greater than 10 degrees on any heading.

## §25.1329 Automatic pilot system.

**(a)** Each automatic pilot system must be approved and must be designed so that the automatic pilot can be quickly and positively disengaged by the pilots to prevent it from interfering with their control of the airplane.

**(b)** Unless there is automatic synchronization, each system must have a means to readily indicate to the pilot the alignment of the actuating device in relation to the control system it operates.

**(c)** Each manually operated control for the system must be readily accessible to the pilots.

**(d)** Quick release (emergency) controls must be on both control wheels, on the side of each wheel opposite the throttles.

**(e)** Attitude controls must operate in the plane and sense of motion specified in §§25.777(b) and 25.779(a) for cockpit controls. The direction of motion must be plainly indicated on, or adjacent to, each control.

**(f)** The system must be designed and adjusted so that, within the range of adjustment available to the human pilot, it cannot produce hazardous loads on the airplane, or create hazardous deviations in the flight path, under any condition of flight appropriate to its use, either during normal operation or in the event of a malfunction, assuming that corrective action begins within a reasonable period of time.

**(g)** If the automatic pilot integrates signals from auxiliary controls or furnishes signals for operation of other equipment, there must be positive interlocks and sequencing of engagement to prevent improper operation. Protection against adverse interaction of integrated components, resulting from a malfunction, is also required.

**(h)** If the automatic pilot system can be coupled to airborne navigation equipment, means must be provided to indicate to the flight crew the current mode of operation. Selector switch position is not acceptable as a means of indication.

[Docket No. 5066, 29 FR 18291, Dec. 24, 1964; as amended by Amdt. 25–46, 43 FR 50598, Oct. 30, 1978]

## §25.1331 Instruments using a power supply.

**(a)** For each instrument required by §25.1303(b) that uses a power supply, the following apply:

(1) Each instrument must have a visual means integral with, the instrument, to indicate when power adequate to sustain proper instrument performance is not being supplied. The power must be measured at or near the point where it enters the instruments. For electric instruments, the power is considered to be adequate when the voltage is within approved limits.

(2) Each instrument must, in the event of the failure of one power source, be supplied by another power source. This may be accomplished automatically or by manual means.

(3) If an instrument presenting navigation data receives information from sources external to that instrument and loss of that information would render the presented data unreliable, the instrument must incorporate a visual means to warn the crew, when such loss of information occurs, that the presented data should not be relied upon.

**(b)** As used in this section, "instrument" includes devices that are physically contained in one unit, and devices that are composed of two or more physically separate units or components connected together (such as a remote indicating gyroscopic direction indicator that includes a magnetic sensing element, a gyroscopic unit, an amplifier and an indicator connected together).

[Docket No. 5066, 29 FR 18291, Dec. 24, 1964; as amended by Amdt. 25–41, 42 FR 36970, July 18, 1977]

## §25.1333 Instrument systems.

For systems that operate the instruments required by §25.1303(b) which are located at each pilot's station—

**(a)** Means must be provided to connect the required instruments at the first pilot's station to operating systems which are independent of the operating systems at other flight crew stations, or other equipment;

**(b)** The equipment, systems, and installations must be designed so that one display of the information essential to the safety of flight which is provided by the instruments, including attitude, direction, airspeed, and altitude will remain available to the pilots, without additional crewmember action, after any single failure or combination of failures that is not shown to be extremely improbable; and

**(c)** Additional instruments, systems, or equipment may not be connected to the operating systems for the required instruments, unless provisions are made to ensure the continued normal functioning of the required instruments in the event of any malfunction of the additional instruments, systems, or equipment which is not shown to be extremely improbable.

[Docket No. 5066, 29 FR 18291, Dec. 24, 1964; as amended by Amdt. 25–23, 35 FR 5679, April 8, 1970; Amdt. 25–41, 42 FR 36970, July 18, 1977]

## §25.1335 Flight director systems.

If a flight director system is installed, means must be provided to indicate to the flight crew its current mode of operation. Selector switch position is not acceptable as a means of indication.

[Docket No. 5066, 29 FR 18291, Dec. 24, 1964; as amended by Amdt. 25–41, 42 FR 36970, July 18, 1977]

## §25.1337 Powerplant instruments.

**(a)** *Instruments and instrument lines.*

(1) Each powerplant and auxiliary power unit instrument line must meet the requirements of §§25.993 and 25.1183.

(2) Each line carrying flammable fluids under pressure must—

(i) Have restricting orifices or other safety devices at the source of pressure to prevent the escape of excessive fluid if the line fails; and

(ii) Be installed and located so that the escape of fluids would not create a hazard.

(3) Each powerplant and auxiliary power unit instrument that utilizes flammable fluids must be installed and located so that the escape of fluid would not create a hazard.

**(b)** *Fuel quantity indicator.* There must be means to indicate to the flight crewmembers, the quantity, in gallons or equivalent units, of usable fuel in each tank during flight. In addition—

(1) Each fuel quantity indicator must be calibrated to read "zero" during level flight when the quantity of fuel remaining in the tank is equal to the unusable fuel supply determined under §25.959;

(2) Tanks with interconnected outlets and airspaces may be treated as one tank and need not have separate indicators; and

(3) Each exposed sight gauge, used as a fuel quantity indicator, must be protected against damage.

**(c)** *Fuel flowmeter system.* If a fuel flowmeter system is installed, each metering component must have a means for bypassing the fuel supply if malfunction of that component severely restricts fuel flow.

**(d)** *Oil quantity indicator.* There must be a stick gauge or equivalent means to indicate the quantity of oil in each tank. If an oil transfer or reserve oil supply system is installed, there must be a means to indicate to the flight crew, in flight, the quantity of oil in each tank.

**(e)** *Turbopropeller blade position indicator.* Required turbopropeller blade position indicators must begin indicating before the blade moves more than eight degrees below the flight low pitch stop. The source of indication must directly sense the blade position.

**(f)** *Fuel pressure indicator.* There must be means to measure fuel pressure, in each system supplying reciprocating engines, at a point downstream of any fuel pump except fuel injection pumps. In addition—

(1) If necessary for the maintenance of proper fuel delivery pressure, there must be a connection to transmit the carburetor air intake static pressure to the proper pump relief valve connection; and

(2) If a connection is required under paragraph (f)(1) of this section, the gauge balance lines must be independently connected to the carburetor inlet pressure to avoid erroneous readings.

[Docket No. 5066, 29 FR 18291, Dec. 24, 1964; as amended by Amdt. 25–40, 42 FR 15044, March 17, 1977]

### Electrical Systems and Equipment

## §25.1351 General.

**(a)** *Electrical system capacity.* The required generating capacity, and number and kinds of power sources must—

(1) Be determined by an electrical load analysis; and

(2) Meet the requirements of §25.1309.

**(b)** *Generating system.* The generating system includes electrical power sources, main power busses, transmission cables, and associated control, regulation, and protective devices. It must be designed so that—

(1) Power sources function properly when independent and when connected in combination;

25

(2) No failure or malfunction of any power source can create a hazard or impair the ability of remaining sources to supply essential loads;

(3) The system voltage and frequency (as applicable) at the terminals of all essential load equipment can be maintained within the limits for which the equipment is designed, during any probable operating condition; and

(4) System transients due to switching, fault clearing, or other causes do not make essential loads inoperative, and do not cause a smoke or fire hazard.

(5) There are means accessible, in flight, to appropriate crewmembers for the individual and collective disconnection of the electrical power sources from the system.

(6) There are means to indicate to appropriate crewmembers the generating system quantities essential for the safe operation of the system, such as the voltage and current supplied by each generator.

**(c)** *External power.* If provisions are made for connecting external power to the airplane, and that external power can be electrically connected to equipment other than that used for engine starting, means must be provided to ensure that no external power supply having a reverse polarity, or a reverse phase sequence, can supply power to the airplane's electrical system.

**(d)** *Operation without normal electrical power.* It must be shown by analysis, tests, or both, that the airplane can be operated safely in VFR conditions, for a period of not less than five minutes, with the normal electrical power (electrical power sources excluding the battery) inoperative, with critical type fuel (from the standpoint of flameout and restart capability), and with the airplane initially at the maximum certificated altitude. Parts of the electrical system may remain on if—

(1) A single malfunction, including a wire bundle or junction box fire, cannot result in loss of both the part turned off and the part turned on; and

(2) The parts turned on are electrically and mechanically isolated from the parts turned off.

[Docket No. 5066, 29 FR 18291, Dec. 24, 1964; as amended by Amdt. 25–41, 42 FR 36970, July 18, 1977; Amdt. 25–72, 55 FR 29785, July 20, 1990]

## §25.1353 Electrical equipment and installations.

**(a)** Electrical equipment, controls, and wiring must be installed so that operation of any one unit or system of units will not adversely affect the simultaneous operation of any other electrical unit or system essential to the safe operation.

**(b)** Cables must be grouped, routed, and spaced so that damage to essential circuits will be minimized if there are faults in heavy current-carrying cables.

**(c)** Storage batteries must be designed and installed as follows:

(1) Safe cell temperatures and pressures must be maintained during any probable charging or discharging condition. No uncontrolled increase in cell temperature may result when the battery is recharged (after previous complete discharge)—

(i) At maximum regulated voltage or power;

(ii) During a flight of maximum duration; and

(iii) Under the most adverse cooling condition likely to occur in service.

(2) Compliance with paragraph (c)(1) of this section must be shown by test unless experience with similar batteries and installations has shown that maintaining safe cell temperatures and pressures presents no problem.

(3) No explosive or toxic gases emitted by any battery in normal operation, or as the result of any probable malfunction in the charging system or battery installation, may accumulate in hazardous quantities within the airplane.

(4) No corrosive fluids or gases that may escape from the battery may damage surrounding airplane structures or adjacent essential equipment.

(5) Each nickel cadmium battery installation capable of being used to start an engine or auxiliary power unit must have provisions to prevent any hazardous effect on structure or essential systems that may be caused by the maximum amount of heat the battery can generate during a short circuit of the battery or of its individual cells.

(6) Nickel cadmium battery installations capable of being used to start an engine or auxiliary power unit must have—

(i) A system to control the charging rate of the battery automatically so as to prevent battery overheating;

(ii) A battery temperature sensing and over-temperature warning system with a means for disconnecting the battery from its charging source in the event of an over-temperature condition; or

(iii) A battery failure sensing and warning system with a means for disconnecting the battery from its charging source in the event of battery failure.

[Docket No. 5066, 29 FR 18291, Dec. 24, 1964; as amended by Amdt. 25–41, 42 FR 36970, July 18, 1977; Amdt. 25–42, 43 FR 2323, Jan. 16, 1978]

## §25.1355 Distribution system.

**(a)** The distribution system includes the distribution busses, their associated feeders, and each control and protective device.

**(b)** [Reserved]

**(c)** If two independent sources of electrical power for particular equipment or systems are required by this chapter, in the event of the failure of one power source for such equipment or system, another power source (including its separate feeder) must be automatically provided or be manually selectable to maintain equipment or system operation.

[Docket No. 5066, 29 FR 18291, Dec. 24, 1964; as amended by Amdt. 25–23, 35 FR 5679, April 8, 1970; Amdt. 25–38, 41 FR 55468, Dec. 20, 1976]

## §25.1357 Circuit protective devices.

**(a)** Automatic protective devices must be used to minimize distress to the electrical system and hazard to the airplane in the event of wiring faults or serious malfunction of the system or connected equipment.

**(b)** The protective and control devices in the generating system must be designed to de-energize and disconnect faulty power sources and power transmission equipment from their associated busses with sufficient rapidity to provide protection from hazardous over-voltage and other malfunctioning.

**(c)** Each resettable circuit protective device must be designed so that, when an overload or circuit fault exists, it will open the circuit irrespective of the position of the operating control.

**(d)** If the ability to reset a circuit breaker or replace a fuse is essential to safety in flight, that circuit breaker or fuse must be located and identified so that it can be readily reset or replaced in flight.

**(e)** Each circuit for essential loads must have individual circuit protection. However, individual protection for each circuit in an essential load system (such as each position light circuit in a system) is not required.

**(f)** If fuses are used, there must be spare fuses for use in flight equal to at least 50 percent of the number of fuses of each rating required for complete circuit protection.

**(g)** Automatic reset circuit breakers may be used as integral protectors for electrical equipment (such as thermal cut-outs) if there is circuit protection to protect the cable to the equipment.

## §25.1363 Electrical system tests.

**(a)** When laboratory tests of the electrical system are conducted—

(1) The tests must be performed on a mock-up using the same generating equipment used in the airplane;

(2) The equipment must simulate the electrical characteristics of the distribution wiring and connected loads to the extent necessary for valid test results; and

(3) Laboratory generator drives must simulate the actual prime movers on the airplane with respect to their reaction to generator loading, including loading due to faults.

**(b)** For each flight condition that cannot be simulated adequately in the laboratory or by ground tests on the airplane, flight tests must be made.

### Lights

## §25.1381 Instrument lights.

**(a)** The instrument lights must—

(1) Provide sufficient illumination to make each instrument, switch and other device necessary for safe operation easily readable unless sufficient illumination is available from another source; and

(2) Be installed so that—

(i) Their direct rays are shielded from the pilot's eyes; and

(ii) No objectionable reflections are visible to the pilot.

**(b)** Unless undimmed instrument lights are satisfactory under each expected flight condition, there must be a means to control the intensity of illumination.

[Docket No. 5066, 29 FR 18291, Dec. 24, 1964; as amended by Amdt. 25–72, 55 FR 29785, July 20, 1990]

## §25.1383 Landing lights.

**(a)** Each landing light must be approved, and must be installed so that—

(1) No objectionable glare is visible to the pilot;

(2) The pilot is not adversely affected by halation; and

(3) It provides enough light for night landing.

**(b)** Except when one switch is used for the lights of a multiple light installation at one location, there must be a separate switch for each light.

**(c)** There must be a means to indicate to the pilots when the landing lights are extended.

## §25.1385 Position light system installation.

**(a)** *General.* Each part of each position light system must meet the applicable requirements of this section and each system as a whole must meet the requirements of §§25.1387 through 25.1397.

**(b)** *Forward position lights.* Forward position lights must consist of a red and a green light spaced laterally as far apart as practicable and installed forward on the airplane so that, with the airplane in the normal flying position, the red light is on the left side and the green light is on the right side. Each light must be approved.

**(c)** *Rear position light.* The rear position light must be a white light mounted as far aft as practicable on the tail or on each wing tip, and must be approved.

**(d)** *Light covers and color filters.* Each light cover or color filter must be at least flame resistant and may not change color or shape or lose any appreciable light transmission during normal use.

[Docket No. 5066, 29 FR 18291, Dec. 24, 1964; as amended by Amdt. 25–38, 41 FR 55468, Dec. 20, 1976]

## §25.1387 Position light system dihedral angles.

**(a)** Except as provided in paragraph (e) of this section, each forward and rear position light must, as installed, show unbroken light within the dihedral angles described in this section.

**(b)** Dihedral angle *L* (left) is formed by two intersecting vertical planes, the first parallel to the longitudinal axis of the airplane, and the other at 110 degrees to the left of the first, as viewed when looking forward along the longitudinal axis.

**(c)** Dihedral angle *R* (right) is formed by two intersecting vertical planes, the first parallel to the longitudinal axis of the airplane, and the other at 110 degrees to the right of the first, as viewed when looking forward along the longitudinal axis.

**(d)** Dihedral angle *A* (aft) is formed by two intersecting vertical planes making angles of 70 degrees to the right and to the left, respectively, to a vertical plane passing through the longitudinal axis, as viewed when looking aft along the longitudinal axis.

**(e)** If the rear position light, when mounted as far aft as practicable in accordance with §25.1385(c), cannot show unbroken light within dihedral angle *A* (as defined in paragraph (d) of this section), a solid angle or angles of obstructed visibility totaling not more than 0.04 steradians is allowable within that dihedral angle, if such solid angle is within a cone whose apex is at the rear position light and whose elements make an angle of 30° with a vertical line passing through the rear position light.

[Docket No. 5066, 29 FR 18291, Dec. 24, 1964; as amended by Amdt. 25–30, 36 FR 21278, Nov. 5, 1971]

## §25.1389 Position light distribution and intensities.

**(a)** *General.* The intensities prescribed in this section must be provided by new equipment with light covers and color filters in place. Intensities must be determined with the light source operating at a steady value equal to the average luminous output of the source at the normal operating voltage of the airplane. The light distribution and intensity of each position light must meet the requirements of paragraph (b) of this section.

**(b)** *Forward and rear position lights.* The light distribution and intensities of forward and rear position lights must be expressed in terms of minimum intensities in the horizontal plane, minimum intensities in any vertical plane, and maximum intensities in overlapping beams, within dihedral angles *L*, *R*, and *A*, and must meet the following requirements:

(1) *Intensities in the horizontal plane.* Each intensity in the horizontal plane (the plane containing the longitudinal axis of the airplane and perpendicular to the plane of symmetry of the airplane) must equal or exceed the values in §25.1391.

(2) *Intensities in any vertical plane.* Each intensity in any vertical plane (the plane perpendicular to the horizontal plane) must equal or exceed the appropriate value in §25.1393, where *I* is the minimum intensity prescribed in §25.1391 for the corresponding angles in the horizontal plane.

(3) *Intensities in overlaps between adjacent signals.* No intensity in any overlap between adjacent signals may exceed the values given in §25.1395, except that higher intensities in overlaps may be used with main beam intensities substantially greater than the minima specified in §§25.1391 and 25.1393 if the overlap intensities in relation to the main beam intensities do not adversely affect signal clarity. When the peak intensity of the forward position lights is more than 100 candles, the maximum overlap intensities between them may exceed the values given in §25.1395 if the overlap intensity in Area A is not more than 10 percent of peak position light intensity and the overlap intensity in Area B is not greater than 2.5 percent of peak position light intensity.

## §25.1391 Minimum intensities in the horizontal plane of forward and rear position lights.

Each position light intensity must equal or exceed the applicable values in the following table:

| Dihedral angle (light included) | Angle from right or left of longitudinal axis, measured from dead ahead | Intensity (candles) |
|---|---|---|
| *L* and *R* (forward red and green) | 0° to 10° | 40 |
| | 10° to 20° | 30 |
| | 20° to 110° | 5 |
| A (rear white) | 110° to 180° | 20 |

## §25.1393 Minimum intensities in any vertical plane of forward and rear position lights.

Each position light intensity must equal or exceed the applicable values in the following table:

| Angle above or below the horizontal plane | Intensity, I |
|---|---|
| 0° | 1.00 |
| 0° to 5° | 0.90 |
| 5° to 10° | 0.80 |
| 10° to 15° | 0.70 |
| 15° to 20° | 0.50 |
| 20° to 30° | 0.30 |
| 30° to 40° | 0.10 |
| 40° to 90° | 0.05 |

## §25.1395 Maximum intensities in overlapping beams of forward and rear position lights.

No position light intensity may exceed the applicable values in the following table, except as provided in §25.1389(b)(3).

| Overlaps | Maximum Intensity | |
|---|---|---|
| | Area A (candles) | Area B (candles) |
| Green in dihedral angle *L* | 10 | 1 |
| Red in dihedral angle *R* | 10 | 1 |
| Green in dihedral angle *A* | 5 | 1 |
| Red in dihedral angle *A* | 5 | 1 |
| Rear white in dihedral angle *L* | 5 | 1 |
| Rear white in dihedral angle *R* | 5 | 1 |

Where—

**(a)** Area A includes all directions in the adjacent dihedral angle that pass through the light source and intersect the common boundary plane at more than 10 degrees but less than 20 degrees; and

**(b)** Area B includes all directions in the adjacent dihedral angle that pass through the light source and intersect the common boundary plane at more than 20 degrees.

## §25.1397 Color specifications.

Each position light color must have the applicable International Commission on Illumination chromaticity coordinates as follows:

**(a)** *Aviation red*—

"y" is not greater than 0.335; and
"z" is not greater than 0.002.

**(b)** *Aviation green*—

"x" is not greater than 0.440 – 0.320 y;
"x" is not greater than y – 0.170; and
"y" is not less than 0.390 – 0.170 x.

**(c)** *Aviation white*—

"x" is not less than 0.300 and not greater than 0.540;

"y" is not less than "x – 0.040" or "$y_0$ – 0.010", whichever is the smaller; and

"y" is not greater than "x+0.020" nor "0.636–0.400 x";

Where "$y_0$" is the "y" coordinate of the Planckian radiator for the value of "x" considered.

[Docket No. 5066, 29 FR 18291, Dec. 24, 1964; as amended by Amdt. 25–27, 36 FR 12972, July 10, 1971]

25

## §25.1399 Riding light.

**(a)** Each riding (anchor) light required for a seaplane or amphibian must be installed so that it can—

(1) Show a white light for at least 2 nautical miles at night under clear atmospheric conditions; and

(2) Show the maximum unbroken light practicable when the airplane is moored or drifting on the water.

**(b)** Externally hung lights may be used.

## §25.1401 Anticollision light system.

**(a)** *General.* The airplane must have an anticollision light system that—

(1) Consists of one or more approved anticollision lights located so that their light will not impair the crew's vision or detract from the conspicuity of the position lights; and

(2) Meets the requirements of paragraphs (b) through (f) of this section.

**(b)** *Field of coverage.* The system must consist of enough lights to illuminate the vital areas around the airplane considering the physical configuration and flight characteristics of the airplane. The field of coverage must extend in each direction within at least 75 degrees above and 75 degrees below the horizontal plane of the airplane, except that a solid angle or angles of obstructed visibility totaling not more than 0.03 steradians is allowable within a solid angle equal to 0.15 steradians centered about the longitudinal axis in the rearward direction.

**(c)** *Flashing characteristics.* The arrangement of the system, that is, the number of light sources, beam width, speed of rotation, and other characteristics, must give an effective flash frequency of not less than 40, nor more than 100 cycles per minute. The effective flash frequency is the frequency at which the airplane's complete anticollision light system is observed from a distance, and applies to each sector of light including any overlaps that exist when the system consists of more than one light source. In overlaps, flash frequencies may exceed 100, but not 180 cycles per minute.

**(d)** *Color.* Each anticollision light must be either aviation red or aviation white and must meet the applicable requirements of §25.1397.

**(e)** *Light intensity.* The minimum light intensities in all vertical planes, measured with the red filter (if used) and expressed in terms of "effective" intensities, must meet the requirements of paragraph (f) of this section. The following relation must be assumed:

$$I_e = \frac{\int_{t_1}^{t_2} I(t)dt}{0.2 + (t_2 - t_1)}$$

where:

$I_e$ = effective intensity (candles).

I(t) = instantaneous intensity as a function of time.

$t_2 - t_1$ = flash time interval (seconds).

Normally, the maximum value of effective intensity is obtained when $t_2$ and $t_1$ are chosen so that the effective intensity is equal to the instantaneous intensity at $t_2$ and $t_1$.

**(f)** *Minimum effective intensities for anticollision lights.* Each anticollision light effective intensity must equal or exceed the applicable values in the following table:

| Angle above or below the horizontal plane | Effective Intensity (candles) |
|---|---|
| 0° to 5° | 400 |
| 5° to 10° | 240 |
| 10° to 20° | 80 |
| 20° to 30° | 40 |
| 30° to 75° | 20 |

[Docket No. 5066, 29 FR 18291, Dec. 24, 1964; as amended by Amdt. 25–27, 36 FR 12972, July 10, 1971; Amdt. 25–41, 42 FR 36970, July 18, 1977]

## §25.1403 Wing icing detection lights.

Unless operations at night in known or forecast icing conditions are prohibited by an operating limitation, a means must be provided for illuminating or otherwise determining the formation of ice on the parts of the wings that are critical from the standpoint of ice accumulation. Any illumination that is used must be of a type that will not cause glare or reflection that would handicap crewmembers in the performance of their duties.

[Docket No. 5066, 29 FR 18291, Dec. 24, 1964; as amended by Amdt. 25–38, 41 FR 55468, Dec. 20, 1976]

### Safety Equipment

## §25.1411 General.

**(a)** *Accessibility.* Required safety equipment to be used by the crew in an emergency must be readily accessible.

**(b)** *Stowage provisions.* Stowage provisions for required emergency equipment must be furnished and must—

(1) Be arranged so that the equipment is directly accessible and its location is obvious; and

(2) Protect the safety equipment from inadvertent damage.

**(c)** *Emergency exit descent device.* The stowage provisions for the emergency exit descent device required by §25.809(f) must be at the exits for which they are intended.

**(d)** *Liferafts.*

(1) The stowage provisions for the liferafts described in §25.1415 must accommodate enough rafts for the maximum number of occupants for which certification for ditching is requested.

(2) Liferafts must be stowed near exits through which the rafts can be launched during an unplanned ditching.

(3) Rafts automatically or remotely released outside the airplane must be attached to the airplane by means of the static line prescribed in §25.1415.

(4) The stowage provisions for each portable liferaft must allow rapid detachment and removal of the raft for use at other than the intended exits.

**(e)** *Long-range signaling device.* The stowage provisions for the long-range signaling device required by §25.1415 must be near an exit available during an unplanned ditching.

**(f)** *Life preserver stowage provisions.* The stowage provisions for life preservers described in §25.1415 must accommodate one life preserver for each occupant for which certification for ditching is requested. Each life preserver must be within easy reach of each seated occupant.

**(g)** *Life line stowage provisions.* If certification for ditching under §25.801 is requested, there must be provisions to store life lines. These provisions must—

(1) Allow one life line to be attached to each side of the fuselage; and

(2) Be arranged to allow the life lines to be used to enable the occupants to stay on the wing after ditching.

[Docket No. 5066, 29 FR 18291, Dec. 24, 1964; as amended by Amdt. 25–32, 37 FR 3972, Feb. 24, 1972; Amdt. 25–46, 43 FR 50598, Oct. 30, 1978; Amdt. 25–53, 45 FR 41593, June 19, 1980; Amdt. 25–70, 54 FR 43925, Oct. 27, 1989; Amdt. 25–79, 58 FR 45229, Aug. 26, 1993]

## §25.1415 Ditching equipment.

**(a)** Ditching equipment used in airplanes to be certificated for ditching under §25.801, and required by the operating rules of this chapter, must meet the requirements of this section.

**(b)** Each liferaft and each life preserver must be approved. In addition—

(1) Unless excess rafts of enough capacity are provided, the buoyancy and seating capacity beyond the rated capacity of the rafts must accommodate all occupants of the airplane in the event of a loss of one raft of the largest rated capacity; and

(2) Each raft must have a trailing line, and must have a static line designed to hold the raft near the airplane but to release it if the airplane becomes totally submerged.

**(c)** Approved survival equipment must be attached to each liferaft.

**(d)** There must be an approved survival type emergency locator transmitter for use in one life raft.

**(e)** For airplanes not certificated for ditching under §25.801 and not having approved life preservers, there must be an approved flotation means for each occupant. This means must be within easy reach of each seated occupant and must be readily removable from the airplane.

[Docket No. 5066, 29 FR 18291, Dec. 24, 1964; as amended by Amdt. 25–29, 36 FR 18722, Sept. 21, 1971; Amdt 25–50, 45 FR 38348, June 9, 1980; Amdt. 25–72, 55 FR 29785, July 20, 1990; as Amdt. 25–82, 59 FR 32057, June 21, 1994]

## §25.1419 Ice protection.

If certification with ice protection provisions is desired, the airplane must be able to safely operate in the continuous maximum and intermittent maximum icing conditions of Appendix C. To establish that the airplane can operate within the continuous maximum and intermittent maximum conditions of Appendix C:

**(a)** An analysis must be performed to establish that the ice protection for the various components of the airplane is adequate, taking into account the various airplane operational configurations; and

**(b)** To verify the ice protection analysis, to check for icing anomalies, and to demonstrate that the ice protection system and its components are effective, the airplane or its components must be flight tested in the various operational configurations, in measured natural atmospheric icing conditions and, as found necessary, by one or more of the following means:

(1) Laboratory dry air or simulated icing tests, or a combination of both, of the components or models of the components.

(2) Flight dry air tests of the ice protection system as a whole, or of its individual components.

(3) Flight tests of the airplane or its components in measured simulated icing conditions.

**(c)** Caution information, such as an amber caution light or equivalent, must be provided to alert the flightcrew when the anti-ice or de-ice system is not functioning normally.

**(d)** For turbine engine powered airplanes, the ice protection provisions of this section are considered to be applicable primarily to the airframe. For the powerplant installation, certain additional provisions of subpart E of this part may be found applicable.

[Docket No. 5066, 29 FR 18291, Dec. 24, 1964; as amended by Amdt. 25–72, 55 FR 29785, July 20, 1990]

## §25.1421 Megaphones.

If a megaphone is installed, a restraining means must be provided that is capable of restraining the megaphone when it is subjected to the ultimate inertia forces specified in §25.561(b)(3).

[Docket No. 5066, 29 FR 18291, Dec. 24, 1964; as amended by Amdt. 25–41, 42 FR 36970, July 18, 1977]

### Miscellaneous Equipment

## §25.1423 Public address system.

A public address system required by this chapter must—

**(a)** Be powerable when the aircraft is in flight or stopped on the ground, after the shutdown or failure of all engines and auxiliary power units, or the disconnection or failure of all power sources dependent on their continued operation, for—

(1) A time duration of at least 10 minutes, including an aggregate time duration of at least 5 minutes of announcements made by flight and cabin crewmembers, considering all other loads which may remain powered by the same source when all other power sources are inoperative; and

(2) An additional time duration in its standby state appropriate or required for any other loads that are powered by the same source and that are essential to safety of flight or required during emergency conditions.

**(b)** Be capable of operation within 10 seconds by a flight attendant at those stations in the passenger compartment from which they system is accessible.

**(c)** Be intelligible at all passenger seats, lavatories, and flight attendant seats and work stations.

**(d)** Be designed so that no unused, unstowed microphone will render the system inoperative.

**(e)** Be capable of functioning independently of any required crewmember interphone system.

**(f)** Be accessible for immediate use from each of two flight crewmember stations in the pilot compartment.

**(g)** For each required floor-level passenger emergency exit which has an adjacent flight attendant seat, have a microphone which is readily accessible to the seated flight attendant, except that one microphone may serve more than one exit, provided the proximity of the exits allows unassisted verbal communication between seated flight attendants.

[Docket No. 26003, 58 FR 45229, Aug. 26, 1993]

## §25.1431 Electronic equipment.

**(a)** In showing compliance with §25.1309 (a) and (b) with respect to radio and electronic equipment and their installations, critical environmental conditions must be considered.

**(b)** Radio and electronic equipment must be supplied with power under the requirements of §25.1355(c).

**(c)** Radio and electronic equipment, controls, and wiring must be installed so that operation of any one unit or system of units will not adversely affect the simultaneous operation of any other radio or electronic unit, or system of units, required by this chapter.

## §25.1433 Vacuum systems.

There must be means, in addition to the normal pressure relief, to automatically relieve the pressure in the discharge lines from the vacuum air pump when the delivery temperature of the air becomes unsafe.

[Docket No. 5066, 29 FR 18291, Dec. 24, 1964; as amended by Amdt. 25–72, 55 FR 29785, July 20, 1990]

## §25.1435 Hydraulic systems.

**(a)** *Element design.* Each element of the hydraulic system must be designed to:

(1) Withstand the proof pressure without permanent deformation that would prevent it from performing its intended functions, and the ultimate pressure without rupture. The proof and ultimate pressures are defined in terms of the design operating pressure (DOP) as follows:

| Element | Proof (xDOP) | Ultimate (xDOP) |
|---|---|---|
| 1. Tubes and fittings | 1.5 | 3.0 |
| 2. Pressure vessels containing gas: | | |
| High pressure (e.g., accumulators) | 3.0 | 4.0 |
| Low pressure (e.g., reservoirs) | 1.5 | 3.0 |
| 3. Hoses | 2.0 | 4.0 |
| 4. All other elements | 1.5 | 2.0 |

(2) Withstand, without deformation that would prevent it from performing its intended function, the design operating pressure in combination with limit structural loads that may be imposed;

(3) Withstand, without rupture, the design operating pressure multiplied by a factor of 1.5 in combination with ultimate structural load that can reasonably occur simultaneously;

(4) Withstand the fatigue effects of all cyclic pressures, including transients, and associated externally induced loads, taking into account the consequences of element failure; and

(5) Perform as intended under all environmental conditions for which the airplane is certificated.

**(b)** *System design.* Each hydraulic system must:

(1) Have means located at a flightcrew station to indicate appropriate system parameters, if

(i) It performs a function necessary for continued safe flight and landing; or

(ii) In the event of hydraulic system malfunction, corrective action by the crew to ensure continued safe flight and landing is necessary;

(2) Have means to ensure that system pressures, including transient pressures and pressures from fluid volumetric changes in elements that are likely to remain closed long enough for such changes to occur, are within the design capabilities of each element, such that they meet the requirements defined in §25.1435(a)(1) through (a)(5);

(3) Have means to minimize the release of harmful or hazardous concentrations of hydraulic fluid or vapors into the crew and passenger compartments during flight;

(4) Meet the applicable requirements of §§25.863, 25.1183, 25.1185, and 25.1189 if a flammable hydraulic fluid is used; and

(5) Be designed to use any suitable hydraulic fluid specified by the airplane manufacturer, which must be identified by appropriate markings as required by §25.1541.

**(c)** *Tests.* Tests must be conducted on the hydraulic system(s), and/or subsystem(s) and elements, except that analysis may be used in place of or to supplement testing, where the analysis is shown to be reliable and appropriate. All internal and external influences must be taken into account to an extent necessary to evaluate their effects, and to assure reliable system and element functioning and integration. Failure or unacceptable deficiency of an element or system must be corrected and be sufficiently retested, where necessary.

(1) The system(s), subsystem(s), or element(s) must be subjected to performance, fatigue, and endurance tests representative of airplane ground and flight operations.

(2) The complete system must be tested to determine proper functional performance and rela-

tion to the other systems, including simulation of relevant failure conditions, and to support or validate element design.

(3) The complete hydraulic system(s) must be functionally tested on the airplane in normal operation over the range of motion of all associated user systems. The test must be conducted at the system relief pressure or 1.25 times the DOP if a system pressure relief device is not part of the system design. Clearances between hydraulic system elements and other systems or structural elements must remain adequate and there must be no detrimental effects.

[Docket No. 28617, 66 FR 27402, May 16, 2001]

## §25.1438 Pressurization and pneumatic systems.

**(a)** Pressurization system elements must be burst pressure tested to 2.0 times, and proof pressure tested to 1.5 times, the maximum normal operating pressure.

**(b)** Pneumatic system elements must be burst pressure tested to 3.0 times, and proof pressure tested to 1.5 times, the maximum normal operating pressure.

**(c)** An analysis, or a combination of analysis and test, may be substituted for any test required by paragraph (a) or (b) of this section if the Administrator finds it equivalent to the required test.

[Docket No. 5066, 29 FR 18291, Dec. 24, 1964; as amended by Amdt. 25–41, 42 FR 36971, July 18, 1977]

## §25.1439 Protective breathing equipment.

**(a)** If there is a class A, B, or E cargo compartment, protective breathing equipment must be installed for the use of appropriate crewmembers. In addition, protective breathing equipment must be installed in each isolated separate compartment in the airplane, including upper and lower lobe galleys, in which crewmember occupancy is permitted during flight for the maximum number of crewmembers expected to be in the area during any operation.

**(b)** For protective breathing equipment required by paragraph (a) of this section or by any operating rule of this chapter, the following apply:

(1) The equipment must be designed to protect the flight crew from smoke, carbon dioxide, and other harmful gases while on flight deck duty and while combating fires in cargo compartments.

(2) The equipment must include—

(i) Masks covering the eyes, nose, and mouth; or

(ii) Masks covering the nose and mouth, plus accessory equipment to cover the eyes.

(3) The equipment, while in use, must allow the flight crew to use the radio equipment and to communicate with each other, while at their assigned duty stations.

(4) The part of the equipment protecting the eyes may not cause any appreciable adverse effect on vision and must allow corrective glasses to be worn.

(5) The equipment must supply protective oxygen of 15 minutes duration per crewmember at a pressure altitude of 8,000 feet with a respiratory minute volume of 30 liters per minute BTPD. If a demand oxygen system is used, a supply of 300 liters of free oxygen at 70°F and 760 mm. Hg. pressure is considered to be of 15-minute duration at the prescribed altitude and minute volume. If a continuous flow protective breathing system is used (including a mask with a standard rebreather bag) a flow rate of 60 liters per minute at 8,000 feet (45 liters per minute at sea level) and a supply of 600 liters of free oxygen at 70°F and 760 mm. Hg. pressure is considered to be of 15-minute duration at the prescribed altitude and minute volume. BTPD refers to body temperature conditions (that is, 37°C, at ambient pressure, dry).

(6) The equipment must meet the requirements of paragraphs (b) and (c) of §25.1441.

[Docket No. 5066, 29 FR 18291, Dec. 24, 1964; as amended by Amdt. 25–38, 41 FR 55468, Dec. 20, 1976]

## §25.1441 Oxygen equipment and supply.

**(a)** If certification with supplemental oxygen equipment is requested, the equipment must meet the requirements of this section and §§25.1443 through 25.1453.

**(b)** The oxygen system must be free from hazards in itself, in its method of operation, and in its effect upon other components.

**(c)** There must be a means to allow the crew to readily determine, during flight, the quantity of oxygen available in each source of supply.

**(d)** The oxygen flow rate and the oxygen equipment for airplanes for which certification for operation above 40,000 feet is requested must be approved.

## §25.1443 Minimum mass flow of supplemental oxygen.

**(a)** If continuous flow equipment is installed for use by flight crewmembers, the minimum mass flow of supplemental oxygen required for each crewmember may not be less than the flow required to maintain, during inspiration, a mean tracheal oxygen partial pressure of 149 mm. Hg. when breathing 15 liters per minute, BTPS, and with a maximum tidal volume of 700 cc. with a constant time interval between respirations.

**(b)** If demand equipment is installed for use by flight crewmembers, the minimum mass flow of supplemental oxygen required for each crewmember may not be less than the flow required to maintain, during inspiration, a mean tracheal oxygen partial pressure of 122 mm. Hg., up to and including a cabin pressure altitude of 35,000 feet, and 95 percent oxygen between cabin pressure altitudes of 35,000 and 40,000 feet, when breathing 20 liters per minute BTPS. In addition, there must be means to allow the crew to use undiluted oxygen at their discretion.

**(c)** For passengers and cabin attendants, the minimum mass flow of supplemental oxygen required for each person at various cabin pressure altitudes may not be less than the flow required to maintain, during inspiration and while using the oxygen equipment (including masks) provided, the following mean tracheal oxygen partial pressures:

(1) At cabin pressure altitudes above 10,000 feet up to and including 18,500 feet, a mean tracheal oxygen partial pressure of 100 mm. Hg. when breathing 15 liters per minute, BTPS, and with a tidal volume of 700 cc. with a constant time interval between respirations.

(2) At cabin pressure altitudes above 18,500 feet up to and including 40,000 feet, a mean tracheal oxygen partial pressure of 83.8 mm. Hg. when breathing 30 liters per minute, BTPS, and with a tidal volume of 1,100 cc. with a constant time interval between respirations.

**(d)** If first-aid oxygen equipment is installed, the minimum mass flow of oxygen to each user may not be less than four liters per minute, STPD. However, there may be a means to decrease this flow to not less than two liters per minute, STPD, at any cabin altitude. The quantity of oxygen required is based upon an average flow rate of three liters per minute per person for whom first-aid oxygen is required.

**(e)** If portable oxygen equipment is installed for use by crewmembers, the minimum mass flow of supplemental oxygen is the same as specified in paragraph (a) or (b) of this section, whichever is applicable.

## §25.1445 Equipment standards for the oxygen distributing system.

**(a)** When oxygen is supplied to both crew and passengers, the distribution system must be designed for either—

(1) A source of supply for the flight crew on duty and a separate source for the passengers and other crewmembers; or

(2) A common source of supply with means to separately reserve the minimum supply required by the flight crew on duty.

**(b)** Portable walk-around oxygen units of the continuous flow, diluter-demand, and straight demand kinds may be used to meet the crew or passenger breathing requirements.

## §25.1447 Equipment standards for oxygen dispensing units.

If oxygen dispensing units are installed, the following apply:

**(a)** There must be an individual dispensing unit for each occupant for whom supplemental oxygen is to be supplied. Units must be designed to cover the nose and mouth and must be equipped with a suitable means to retain the unit in position on the face. Flight crew masks for supplemental oxygen must have provisions for the use of communication equipment.

**(b)** If certification for operation up to and including 25,000 feet is requested, an oxygen supply terminal and unit of oxygen dispensing equipment for the immediate use of oxygen by each crewmember must be within easy reach of that crewmember. For any other occupants, the supply terminals and dispensing equipment must be located to allow the use of oxygen as required by the operating rules in this chapter.

**(c)** If certification for operation above 25,000 feet is requested, there must be oxygen dispensing equipment meeting the following requirements:

(1) There must be an oxygen dispensing unit connected to oxygen supply terminals immediately available to each occupant, wherever seated, and at least two oxygen dispensing units connected to oxygen terminals in each lavatory. The total number of dispensing units and outlets in the cabin must exceed the number of seats by at least 10 percent. The extra units must be as uniformly distributed throughout the cabin as practicable. If certification for operation above 30,000 feet is requested, the dispensing units providing the required oxygen flow must be automatically presented to the occupants before the cabin pressure altitude exceeds 15,000 feet. The crew must be provided with a manual means of making the dispensing units immediately available in the event of failure of the automatic system.

(2) Each flight crewmember on flight deck duty must be provided with a quick-donning type oxygen dispensing unit connected to an oxygen supply terminal. This dispensing unit must be immediately available to the flight crewmember when seated at his station, and installed so that it:

(i) Can be placed on the face from its ready position, properly secured, sealed, and supplying oxygen upon demand, with one hand, within five seconds and without disturbing eyeglasses or causing delay in proceeding with emergency duties; and

(ii) Allows, while in place, the performance of normal communication functions.

(3) The oxygen dispensing equipment for the flight crewmembers must be:

(i) The diluter demand or pressure demand (pressure demand mask with a diluter demand pressure breathing regulator) type, or other approved oxygen equipment shown to provide the same degree of protection, for airplanes to be operated above 25,000 feet.

(ii) The pressure demand (pressure demand mask with a diluter demand pressure breathing regulator) type with mask-mounted regulator, or other approved oxygen equipment shown to provide the same degree of protection, for airplanes operated at altitudes where decompressions that are not extremely improbable may expose the flightcrew to cabin pressure altitudes in excess of 34,000 feet.

(4) Portable oxygen equipment must be immediately available for each cabin attendant.

[Docket No. 5066, 29 FR 18291, Dec. 24, 1964; as amended by Amdt. 25–41, 42 FR 36971, July 18, 1977; Amdt. 25–87, 61 FR 28696, June 5, 1996]

## §25.1449 Means for determining use of oxygen.

There must be a means to allow the crew to determine whether oxygen is being delivered to the dispensing equipment.

## §25.1450 Chemical oxygen generators.

**(a)** For the purpose of this section, a chemical oxygen generator is defined as a device which produces oxygen by chemical reaction.

**(b)** Each chemical oxygen generator must be designed and installed in accordance with the following requirements:

(1) Surface temperature developed by the generator during operation may not create a hazard to the airplane or to its occupants.

(2) Means must be provided to relieve any internal pressure that may be hazardous.

**(c)** In addition to meeting the requirements in paragraph (b) of this section, each portable chemical oxygen generator that is capable of sustained operation by successive replacement of a generator element must be placarded to show—

(1) The rate of oxygen flow, in liters per minute;

(2) The duration of oxygen flow, in minutes, for the replaceable generator element; and

(3) A warning that the replaceable generator element may be hot, unless the element construction is such that the surface temperature cannot exceed 100 degrees F.

[Docket No. 5066, 29 FR 18291, Dec. 24, 1964; as amended by Amdt. 25–41, 42 FR 36971, July 18, 1977]

## §25.1453 Protection of oxygen equipment from rupture.

Oxygen pressure tanks, and lines between tanks and the shutoff means, must be—

**(a)** Protected from unsafe temperatures; and

**(b)** Located where the probability and hazards of rupture in a crash landing are minimized.

## §25.1455 Draining of fluids subject to freezing.

If fluids subject to freezing may be drained overboard in flight or during ground operation, the drains must be designed and located to prevent the formation of hazardous quantities of ice on the airplane as a result of the drainage.

[Docket No. 5066, 29 FR 18291, Dec. 24, 1964; as amended by Amdt. 25–23, 35 FR 5680, April 8, 1970]

## §25.1457 Cockpit voice recorders.

**(a)** Each cockpit voice recorder required by the operating rules of this chapter must be approved and must be installed so that it will record the following:

(1) Voice communications transmitted from or received in the airplane by radio.

(2) Voice communications of flight crewmembers on the flight deck.

(3) Voice communications of flight crewmembers on the flight deck, using the airplane's interphone system.

(4) Voice or audio signals identifying navigation or approach aids introduced into a headset or speaker.

(5) Voice communications of flight crewmembers using the passenger loudspeaker system, if there is such a system and if the fourth channel is available in accordance with the requirements of paragraph (c)(4)(ii) of this section.

**(b)** The recording requirements of paragraph (a)(2) of this section must be met by installing a cockpit-mounted area microphone, located in the best position for recording voice communications originating at the first and second pilot stations and voice communications of other crewmembers on the flight deck when directed to those stations. The microphone must be so located and, if necessary, the preamplifiers and filters of the recorder must be so adjusted or supplemented, that the intelligibility of the recorded communications is as high as practicable when recorded under flight cockpit noise conditions and played back. Repeated aural or visual playback of the record may be used in evaluating intelligibility.

**(c)** Each cockpit voice recorder must be installed so that the part of the communication or audio signals specified in paragraph (a) of this section obtained from each of the following sources is recorded on a separate channel:

(1) For the first channel, from each boom, mask, or hand-held microphone, headset, or speaker used at the first pilot station.

25

(2) For the second channel from each boom, mask, or hand-held microphone, headset, or speaker used at the second pilot station.

(3) For the third channel—from the cockpit-mounted area microphone.

(4) For the fourth channel, from—

(i) Each boom, mask, or hand-held microphone, headset, or speaker used at the station for the third and fourth crew members; or

(ii) If the stations specified in paragraph (c)(4)(i) of this section are not required or if the signal at such a station is picked up by another channel, each microphone on the flight deck that is used with the passenger loudspeaker system, if its signals are not picked up by another channel.

(5) As far as is practicable all sounds received by the microphone listed in paragraphs (c)(1), (2), and (4) of this section must be recorded without interruption irrespective of the position of the interphone-transmitter key switch. The design shall ensure that sidetone for the flight crew is produced only when the interphone, public address system, or radio transmitters are in use.

**(d)** Each cockpit voice recorder must be installed so that—

(1) It receives its electric power from the bus that provides the maximum reliability for operation of the cockpit voice recorder without jeopardizing service to essential or emergency loads;

(2) There is an automatic means to simultaneously stop the recorder and prevent each erasure feature from functioning, within 10 minutes after crash impact; and

(3) There is an aural or visual means for preflight checking of the recorder for proper operation.

**(e)** The record container must be located and mounted to minimize the probability of rupture of the container as a result of crash impact and consequent heat damage to the record from fire. In meeting this requirement, the record container must be as far aft as practicable, but may not be where aft mounted engines may crush the container during impact. However, it need not be outside of the pressurized compartment.

**(f)** If the cockpit voice recorder has a bulk erasure device, the installation must be designed to minimize the probability of inadvertent operation and actuation of the device during crash impact.

**(g)** Each recorder container must—

(1) Be either bright orange or bright yellow;

(2) Have reflective tape affixed to its external surface to facilitate its location under water; and

(3) Have an underwater locating device, when required by the operating rules of this chapter, on or adjacent to the container which is secured in such manner that they are not likely to be separated during crash impact.

[Docket No. 5066, 29 FR 18291, Dec. 24, 1964; as amended by Amdt. 25–2, 30 FR 3932, March 26, 1965; Amdt. 25–16, 32 FR 13914, Oct. 6, 1967; Amdt. 25–41, 42 FR 36971, July 18, 1977; Amdt. 25–65, 53 FR 26143, July 11, 1988]

## §25.1459 Flight recorders.

**(a)** Each flight recorder required by the operating rules of this chapter must be installed so that—

(1) It is supplied with airspeed, altitude, and directional data obtained from sources that meet the accuracy requirements of §§25.1323, 25.1325, and 25.1327, as appropriate;

(2) The vertical acceleration sensor is rigidly attached, and located longitudinally either within the approved center of gravity limits of the airplane, or at a distance forward or aft of these limits that does not exceed 25 percent of the airplane's mean aerodynamic chord;

(3) It receives its electrical power from the bus that provides the maximum reliability for operation of the flight recorder without jeopardizing service to essential or emergency loads;

(4) There is an aural or visual means for preflight checking of the recorder for proper recording of data in the storage medium.

(5) Except for recorders powered solely by the engine-driven electrical generator system, there is an automatic means to simultaneously stop a recorder that has a data erasure feature and prevent each erasure feature from functioning, within 10 minutes after crash impact; and

(6) There is a means to record data from which the time of each radio transmission either to or from ATC can be determined.

**(b)** Each nonejectable record container must be located and mounted so as to minimize the probability of container rupture resulting from crash impact and subsequent damage to the record from fire. In meeting this requirement the record container must be located as far aft as practicable, but need not be aft of the pressurized compartment, and may not be where aft-mounted engines may crush the container upon impact.

**(c)** A correlation must be established between the flight recorder readings of airspeed, altitude, and heading and the corresponding readings (taking into account correction factors) of the first pilot's instruments. The correlation must cover the airspeed range over which the airplane is to be operated, the range of altitude to which the airplane is limited, and 360 degrees of heading. Correlation may be established on the ground as appropriate.

**(d)** Each recorder container must—

(1) Be either bright orange or bright yellow;

(2) Have reflective tape affixed to its external surface to facilitate its location under water; and

(3) Have an underwater locating device, when required by the operating rules of this chapter, on or adjacent to the container which is secured in such a manner that they are not likely to be separated during crash impact.

**(e)** Any novel or unique design or operational characteristics of the aircraft shall be evaluated to determine if any dedicated parameters must be recorded on flight recorders in addition to or in place of existing requirements.

[Docket No. 5066, 29 FR 18291, Dec. 24, 1964; as amended by Amdt. 25–8, 31 FR 127, Jan. 6, 1966; Amdt. 25–25, 35 FR 13192, Aug. 19, 1970; Amdt. 25–37, 40 FR 2577, Jan. 14, 1975; Amdt. 25–41, 42 FR 36971, July 18, 1977; Amdt. 25–65, 53 FR 26144, July 11, 1988]

### §25.1461 Equipment containing high energy rotors.

**(a)** Equipment containing high energy rotors must meet paragraph (b), (c), or (d) of this section.

**(b)** High energy rotors contained in equipment must be able to withstand damage caused by malfunctions, vibration, abnormal speeds, and abnormal temperatures. In addition—

(1) Auxiliary rotor cases must be able to contain damage caused by the failure of high energy rotor blades; and

(2) Equipment control devices, systems, and instrumentation must reasonably ensure that no operating limitations affecting the integrity of high energy rotors will be exceeded in service.

**(c)** It must be shown by test that equipment containing high energy rotors can contain any failure of a high energy rotor that occurs at the highest speed obtainable with the normal speed control devices inoperative.

**(d)** Equipment containing high energy rotors must be located where rotor failure will neither endanger the occupants nor adversely affect continued safe flight.

[Docket No. 5066, 29 FR 18291, Dec. 24, 1964; as amended by Amdt. 25–41, 42 FR 36971, July 18, 1977]

## Subpart G— Operating Limitations and Information

### §25.1501 General.

**(a)** Each operating limitation specified in §§25.1503 through 25.1533 and other limitations and information necessary for safe operation must be established.

**(b)** The operating limitations and other information necessary for safe operation must be made available to the crewmembers as prescribed in §§25.1541 through 25.1587.

[Docket No. 5066, 29 FR 18291, Dec. 24, 1964; as amended by Amdt. 25–42, 43 FR 2323, Jan. 16, 1978]

OPERATING LIMITATIONS

### §25.1503 Airspeed limitations: general.

When airspeed limitations are a function of weight, weight distribution, altitude, or Mach number, limitations corresponding to each critical combination of these factors must be established.

### §25.1505 Maximum operating limit speed.

The maximum operating limit speed ($V_{MO}/M_{MO}$ airspeed or Mach Number, whichever is critical at a particular altitude) is a speed that may not be deliberately exceeded in any regime of flight (climb, cruise, or descent), unless a higher speed is authorized for flight test or pilot training operations. $V_{MO}/M_{MO}$ must be established so that it is not greater than the design cruising speed $V_C$ and so that it is sufficiently below $V_D/M_D$ or $V_{DF}/M_{DF}$, to make it highly improbable that the latter speeds will be inadvertently exceeded in operations. The speed margin between $V_{MO}/M_{MO}$ and $V_D/M_D$ or $V_{DF}/M_{DF}$ may not be less than that determined under §25.335(b) or found necessary during the flight tests conducted under §25.253.

[Docket No. 5066, 29 FR 18291, Dec. 24, 1964; as amended by Amdt. 25–23, 35 FR 5680, April 8, 1970]

### §25.1507 Maneuvering speed.

The maneuvering speed must be established so that it does not exceed the design maneuvering speed $V_A$ determined under §25.335(c).

### §25.1511 Flap extended speed.

The established flap extended speed $V_{FE}$ must be established so that it does not exceed the design flap speed $V_F$ chosen under §§25.335(e) and 25.345, for the corresponding flap positions and engine powers.

### §25.1513 Minimum control speed.

The minimum control speed $V_{MC}$ determined under §25.149 must be established as an operating limitation.

### §25.1515 Landing gear speeds.

**(a)** The established landing gear operating speed or speeds, $V_{LO}$, may not exceed the speed at which it is safe both to extend and to retract the landing gear, as determined under §25.729 or by flight characteristics. If the extension speed is not the same as the retraction speed, the two speeds must be designated as $V_{LO(EXT)}$ and $V_{LO(RET)}$, respectively.

**(b)** The established landing gear extended speed $V_{LE}$ may not exceed the speed at which it is safe to fly with the landing gear secured in the fully extended position, and that determined under §25.729.

[Docket No. 5066, 29 FR 18291, Dec. 24, 1964; as amended by Amdt. 25–38, 41 FR 55468, Dec. 20, 1976]

## § 25.1516 Other speed limitations.

Any other limitation associated with speed must be established.

[Docket No. FAA–2000–8511, 66 FR 34024, June 26, 2001]

## §25.1517 Rough air speed, $V_{RA}$.

A rough air speed, $V_{RA}$, for use as the recommended turbulence penetration airspeed in §25.1585(a)(8), must be established, which—

(1) Is not greater than the design airspeed for maximum gust intensity, selected for $V_B$; and

(2) Is not less than the minimum value of $V_B$ specified in §25.335(d); and

(3) Is sufficiently less than $V_{MO}$ to ensure that likely speed variation during rough air encounters will not cause the overspeed warning to operate too frequently. In the absence of a rational investigation substantiating the use of other values, $V_{RA}$ must be less than $V_{MO}$ — 35 knots (TAS).

[Docket No. 27902, 61 FR 5222, Feb. 9, 1996]

## §25.1519 Weight, center of gravity, and weight distribution.

The airplane weight, center of gravity, and weight distribution limitations determined under §§25.23 through 25.27 must be established as operating limitations.

## §25.1521 Powerplant limitations.

**(a)** *General.* The powerplant limitations prescribed in this section must be established so that they do not exceed the corresponding limits for which the engines or propellers are type certificated and do not exceed the values on which compliance with any other requirement of this part is based.

**(b)** *Reciprocating engine installations.* Operating limitations relating to the following must be established for reciprocating engine installations:

(1) Horsepower or torque, r.p.m., manifold pressure, and time at critical pressure altitude and sea level pressure altitude for—

(i) Maximum continuous power (relating to unsupercharged operation or to operation in each supercharger mode as applicable); and

(ii) Takeoff power (relating to unsupercharged operation or to operation in each supercharger mode as applicable).

(2) Fuel grade or specification.

(3) Cylinder head and oil temperatures.

(4) Any other parameter for which a limitation has been established as part of the engine type certificate except that a limitation need not be established for a parameter that cannot be exceeded during normal operation due to the design of the installation or to another established limitation.

**(c)** *Turbine engine installations.* Operating limitations relating to the following must be established for turbine engine installations:

(1) Horsepower, torque or thrust, r.p.m., gas temperature, and time for—

(i) Maximum continuous power or thrust (relating to augmented or unaugmented operation as applicable).

(ii) Takeoff power or thrust (relating to augmented or unaugmented operation as applicable).

(2) Fuel designation or specification.

(3) Any other parameter for which a limitation has been established as part of the engine type certificate except that a limitation need not be established for a parameter that cannot be exceeded during normal operation due to the design of the installation or to another established limitation.

**(d)** *Ambient temperature.* An ambient temperature limitation (including limitations for winterization installations, if applicable) must be established as the maximum ambient atmospheric temperature established in accordance with §25.1043(b).

[Docket No. 5066, 29 FR 18291, Dec. 24, 1964; as amended by Amdt. 25–72, 55 FR 29786, July 20, 1990]

## §25.1522 Auxiliary power unit limitations.

If an auxiliary power unit is installed in the airplane, limitations established for the auxiliary power unit, including categories of operation, must be specified as operating limitations for the airplane.

[Docket No. 5066, 29 FR 18291, Dec. 24, 1964; as amended by Amdt. 25–72, 55 FR 29786, July 20, 1990]

## §25.1523 Minimum flight crew.

The minimum flight crew must be established so that it is sufficient for safe operation, considering—

**(a)** The workload on individual crewmembers;

**(b)** The accessibility and ease of operation of necessary controls by the appropriate crewmember; and

**(c)** The kind of operation authorized under §25.1525. The criteria used in making the determinations required by this section are set forth in Appendix D.

[Docket No. 5066, 29 FR 18291, Dec. 24, 1964; as amended by Amdt. 25–3, 30 FR 6067, April 29, 1965]

## §25.1525 Kinds of operation.

The kinds of operation to which the airplane is limited are established by the category in which it is eligible for certification and by the installed equipment.

## § 25.1527 Ambient air temperature and operating altitude.

The extremes of the ambient air temperature and operating altitude for which operation is allowed, as limited by flight, structural, powerplant, functional, or equipment characteristics, must be established.

[Docket No. FAA–2000–8511, 66 FR 34024, June 26, 2001]

## §25.1529 Instructions for Continued Airworthiness.

The applicant must prepare Instructions for Continued Airworthiness in accordance with Appendix H to this part that are acceptable to the Administrator. The instructions may be incomplete at type certification if a program exists to ensure their completion prior to delivery of the first airplane or issuance of a standard certificate of airworthiness, whichever occurs later.

[Docket No. 5066, 29 FR 18291, Dec. 24, 1964; as amended by Amdt. 25–54, 45 FR 60173, Sept. 11, 1980]

## §25.1531 Maneuvering flight load factors.

Load factor limitations, not exceeding the positive limit load factors determined from the maneuvering diagram in §25.333(b), must be established.

## §25.1533 Additional operating limitations.

**(a)** Additional operating limitations must be established as follows:

(1) The maximum takeoff weights must be established as the weights at which compliance is shown with the applicable provisions of this part (including the takeoff climb provisions of §25.121(a) through (c), for altitudes and ambient temperatures).

(2) The maximum landing weights must be established as the weights at which compliance is shown with the applicable provisions of this part (including the landing and approach climb provisions of §§25.119 and 25.121(d) for altitudes and ambient temperatures).

(3) The minimum takeoff distances must be established as the distances at which compliance is shown with the applicable provisions of this part (including the provisions of §§25.109 and 25.113, for weights, altitudes, temperatures, wind components, runway surface conditions (dry and wet), and runway gradients for smooth, hard-surfaced runways. Additionally, at the option of the applicant, wet runway takeoff distances may be established for runway surfaces that have been grooved or treated with a porous friction course, and may be approved for use on runways where such surfaces have been designed, constructed, and maintained in a manner acceptable to the Administrator.

**(b)** The extremes for variable factors (such as altitude, temperature, wind, and runway gradients) are those at which compliance with the applicable provisions of this part is shown.

[Docket No. 5066, 29 FR 18291, Dec. 24, 1964; as amended by Amdt. 25–38, 41 FR 55468, Dec. 20, 1976; Amdt. 25–72, 55 FR 29786, July 20, 1990; Amdt. 25–92, 63 FR 8321, Feb. 18, 1998]

25

### Markings and Placards

## §25.1541 General.

**(a)** The airplane must contain—

(1) The specified markings and placards; and

(2) Any additional information, instrument markings, and placards required for the safe operation if there are unusual design, operating, or handling characteristics.

**(b)** Each marking and placard prescribed in paragraph (a) of this section—

(1) Must be displayed in a conspicuous place; and

(2) May not be easily erased, disfigured, or obscured.

## §25.1543 Instrument markings: general.

For each instrument—

**(a)** When markings are on the cover glass of the instrument, there must be means to maintain the correct alignment of the glass cover with the face of the dial; and

**(b)** Each instrument marking must be clearly visible to the appropriate crewmember.

[Docket No. 5066, 29 FR 18291, Dec. 24, 1964; as amended by Amdt. 25–72, 55 FR 29786, July 20, 1990]

## §25.1545 Airspeed limitation information.

The airspeed limitations required by §25.1583 (a) must be easily read and understood by the flight crew.

## §25.1547 Magnetic direction indicator.

**(a)** A placard meeting the requirements of this section must be installed on, or near, the magnetic direction indicator.

**(b)** The placard must show the calibration of the instrument in level flight with the engines operating.

**(c)** The placard must state whether the calibration was made with radio receivers on or off.

**(d)** Each calibration reading must be in terms of magnetic heading in not more than 45 degree increments.

## §25.1549 Powerplant and auxiliary power unit instruments.

For each required powerplant and auxiliary power unit instrument, as appropriate to the type of instrument—

**(a)** Each maximum and, if applicable, minimum safe operating limit must be marked with a red radial or a red line;

**(b)** Each normal operating range must be marked with a green arc or green line, not extending beyond the maximum and minimum safe limits;

**(c)** Each takeoff and precautionary range must be marked with a yellow arc or a yellow line; and

**(d)** Each engine, auxiliary power unit, or propeller speed range that is restricted because of excessive vibration stresses must be marked with red arcs or red lines.

[Docket No. 5066, 29 FR 18291, Dec. 24, 1964; as amended by Amdt. 25–40, 42 FR 15044, March 17, 1977]

## §25.1551 Oil quantity indication.

Each oil quantity indicating means must be marked to indicate the quantity of oil readily and accurately.

[Docket No. 5066, 29 FR 18291, Dec. 24, 1964; as amended by Amdt. 25–72, 55 FR 29786, July 20, 1990]

## §25.1553 Fuel quantity indicator.

If the unusable fuel supply for any tank exceeds one gallon, or five percent of the tank capacity, whichever is greater, a red arc must be marked on its indicator extending from the calibrated zero reading to the lowest reading obtainable in level flight.

## §25.1555 Control markings.

**(a)** Each cockpit control, other than primary flight controls and controls whose function is obvious, must be plainly marked as to its function and method of operation.

**(b)** Each aerodynamic control must be marked under the requirements of §§25.677 and 25.699.

**(c)** For powerplant fuel controls—

(1) Each fuel tank selector control must be marked to indicate the position corresponding to each tank and to each existing cross feed position;

(2) If safe operation requires the use of any tanks in a specific sequence, that sequence must be marked on, or adjacent to, the selector for those tanks; and

(3) Each valve control for each engine must be marked to indicate the position corresponding to each engine controlled.

**(d)** For accessory, auxiliary, and emergency controls—

(1) Each emergency control (including each fuel jettisoning and fluid shutoff must be colored red; and

(2) Each visual indicator required by §25.729(e) must be marked so that the pilot can determine at any time when the wheels are locked in either extreme position, if retractable landing gear is used.

## §25.1557 Miscellaneous markings and placards.

**(a)** *Baggage and cargo compartments and ballast location.* Each baggage and cargo compartment, and each ballast location must have a placard stating any limitations on contents, including weight, that are necessary under the loading requirements. However, underseat compartments designed for the storage of carry-on articles weighing not more than 20 pounds need not have a loading limitation placard.

**(b)** *Powerplant fluid filler openings.* The following apply:

(1) Fuel filler openings must be marked at or near the filler cover with—

(i) The word "fuel";

(ii) For reciprocating engine powered airplanes, the minimum fuel grade;

(iii) For turbine engine powered airplanes, the permissible fuel designations; and

(iv) For pressure fueling systems, the maximum permissible fueling supply pressure and the maximum permissible defueling pressure.

(2) Oil filler openings must be marked at or near the filler cover with the word "oil."

(3) Augmentation fluid filler openings must be marked at or near the filler cover to identify the required fluid.

**(c)** *Emergency exit placards.* Each emergency exit placard must meet the requirements of §25.811.

**(d)** *Doors.* Each door that must be used in order to reach any required emergency exit must have a suitable placard stating that the door is to be latched in the open position during takeoff and landing.

[Docket No. 5066, 29 FR 18291, Dec. 24, 1964; as amended by Amdt. 25–32, 37 FR 3972, Feb. 24, 1972; Amdt. 25–38, 41 FR 55468, Dec. 20, 1976; Amdt. 25–72, 55 FR 29786, July 20, 1990]

## §25.1561 Safety equipment.

**(a)** Each safety equipment control to be operated by the crew in emergency, such as controls for automatic liferaft releases, must be plainly marked as to its method of operation.

**(b)** Each location, such as a locker or compartment, that carries any fire extinguishing, signaling, or other life saving equipment must be marked accordingly.

**(c)** Stowage provisions for required emergency equipment must be conspicuously marked to identify the contents and facilitate the easy removal of the equipment.

**(d)** Each liferaft must have obviously marked operating instructions.

**(e)** Approved survival equipment must be marked for identification and method of operation.

[Docket No. 5066, 29 FR 18291, Dec. 24, 1964; as amended by Amdt. 25–46, 43 FR 50598, Oct. 30, 1978]

## §25.1563 Airspeed placard.

A placard showing the maximum airspeeds for flap extension for the takeoff, approach, and landing positions must be installed in clear view of each pilot.

### AIRPLANE FLIGHT MANUAL

## §25.1581 General.

**(a)** *Furnishing information.* An Airplane Flight Manual must be furnished with each airplane, and it must contain the following:

(1) Information required by §§25.1583 through 25.1587.

(2) Other information that is necessary for safe operation because of design, operating, or handling characteristics.

(3) Any limitation, procedure, or other information established as a condition of compliance with the applicable noise standards of part 36 of this chapter.

**(b)** *Approved information.* Each part of the manual listed in §§25.1583 through 25.1587, that is appropriate to the airplane, must be furnished, verified, and approved, and must be segregated, identified, and clearly distinguished from each unapproved part of that manual.

**(c)** [Reserved]

**(d)** Each Airplane Flight Manual must include a table of contents if the complexity of the manual indicates a need for it.

[Docket No. 5066, 29 FR 18291, Dec. 24, 1964; as amended by Amdt. 25–42, 43 FR 2323, Jan. 16, 1978; Amdt. 25–72, 55 FR 29786, July 20, 1990]

## §25.1583 Operating limitations.

**(a)** *Airspeed limitations.* The following airspeed limitations and any other airspeed limitations necessary for safe operation must be furnished:

(1) The maximum operating limit speed $V_{MO}/M_{MO}$ and a statement that this speed limit may not be deliberately exceeded in any regime of flight (climb, cruise, or descent) unless a higher speed is authorized for flight test or pilot training.

(2) If an airspeed limitation is based upon compressibility effects, a statement to this effect and information as to any symptoms, the probable behavior of the airplane, and the recommended recovery procedures.

(3) The maneuvering speed $V_A$ and a statement that full application of rudder and aileron controls, as well as maneuvers that involve angles of attack near the stall, should be confined to speeds below this value.

(4) The flap extended speed $V_{FE}$ and the pertinent flap positions and engine powers.

(5) The landing gear operating speed or speeds, and a statement explaining the speeds as defined in §25.1515(a).

(6) The landing gear extended speed $V_{LE}$, if greater than $V_{LO}$, and a statement that this is the maximum speed at which the airplane can be safely flown with the landing gear extended.

**(b)** *Powerplant limitations.* The following information must be furnished:

(1) Limitations required by §25.1521 and §25.1522.

(2) Explanation of the limitations, when appropriate.

(3) Information necessary for marking the instruments required by §§25.1549 through 25.1553.

**(c)** *Weight and loading distribution.* The weight and center of gravity limitations established under §25.1519 must be furnished in the Airplane Flight Manual. All of the following information, including the weight distribution limitations established under §25.1519, must be presented either in the Airplane Flight Manual or in a separate weight and balance control and loading document that is incorporated by reference in the Airplane Flight Manual:

(1) The condition of the airplane and the items included in the empty weight as defined in accordance with §25.29.

(2) Loading instructions necessary to ensure loading of the airplane within the weight and center of gravity limits, and to maintain the loading within these limits in flight.

(3) If certification for more than one center of gravity range is requested, the appropriate limitations, with regard to weight and loading procedures, for each separate center of gravity range.

**(d)** *Flight crew.* The number and functions of the minimum flight crew determined under §25.1523 must be furnished.

**(e)** *Kinds of operation.* The kinds of operation approved under §25.1525 must be furnished.

**(f)** *Ambient air temperatures and operating altitudes.* The extremes of the ambient air temperatures and operating altitudes established under §25.1527 must be furnished.

**(g)** [Reserved]

**(h)** *Additional operating limitations.* The operating limitations established under §25.1533 must be furnished.

**(i)** *Maneuvering flight load factors.* The positive maneuvering limit load factors for which the structure is proven, described in terms of accelerations, must be furnished.

[Docket No. 5066, 29 FR 1891, Dec. 24, 1964; as amended by Amdt. 25–38, 41 FR 55468, Dec, 20, 1976; Amdt. 25–42, 43 FR 2323, Jan. 16, 1978; Amdt. 25–46, 43 FR 50598, Oct. 30, 1978; Amdt. 25–72, 55 FR 29787, July 20, 1990; Amdt. 25–105, 66 FR 34024, June 26, 2001]

## §25.1585 Operating procedures.

**(a)** Operating procedures must be furnished for—

(1) Normal procedures peculiar to the particular type or model encountered in connection with routine operations;

(2) Non-normal procedures for malfunction cases and failure conditions involving the use of special systems or the alternative use of regular systems; and

(3) Emergency procedures for foreseeable but unusual situations in which immediate and precise action by the crew may be expected to substantially reduce the risk of catastrophe.

**(b)** Information or procedures not directly related to airworthiness or not under the control of the crew, must not be included, nor must any procedure that is accepted as basic airmanship.

**(c)** Information identifying each operating condition in which the fuel system independence prescribed in §25.953 is necessary for safety must be furnished, together with instructions for placing the fuel system in a configuration used to show compliance with that section.

**(d)** The buffet onset envelopes, determined under §25.251 must be furnished. The buffet onset envelopes presented may reflect the center of gravity at which the airplane is normally loaded during cruise if corrections for the effect of different center of gravity locations are furnished.

**(e)** Information must be furnished that indicates that when the fuel quantity indicator reads "zero" in level flight, any fuel remaining in the fuel tank cannot be used safely in flight.

**(f)** Information on the total quantity of usable fuel for each fuel tank must be furnished.

[Docket No. FAA–2000–8511, 66 FR 34024, June 26, 2001]

## §25.1587 Performance information.

**(a)** Each Airplane Flight Manual must contain information to permit conversion of the indicated temperature to free air temperature if other than a free air temperature indicator is used to comply with the requirements of §25.1303(a)(1).

**(b)** Each Airplane Flight Manual must contain the performance information computed under the applicable provisions of this part (including §§25.115, 25.123, and 25.125 for the weights, altitudes, temperatures, wind components, and runway gradients, as applicable) within the operational limits of the airplane, and must contain the following:

(1) In each case, the conditions of power, configuration, and speeds, and the procedures for handling the airplane and any system having a significant effect on the performance information.

(2) $V_S$ determined in accordance with §25.103.

(3) The following performance information (determined by extrapolation and computed for the range of weights between the maximum landing weight and the maximum takeoff weight):

(i) Climb in the landing configuration.

(ii) Climb in the approach configuration.

(iii) Landing distance.

(4) Procedures established under §25.101(f) and (g) that are related to the limitations and information required by §25.1533 and by this paragraph (b) in the form of guidance material, including any relevant limitations or information.

(5) An explanation of significant or unusual flight or ground handling characteristics of the airplane.

(6) Corrections to indicated values of airspeed, altitude, and outside air temperature.

(7) An explanation of operational landing runway length factors included in the presentation of the landing distance, if appropriate.

[Docket No. FAA–2000–8511, 66 FR 34024, June 26, 2001]

# APPENDIX A TO PART 25

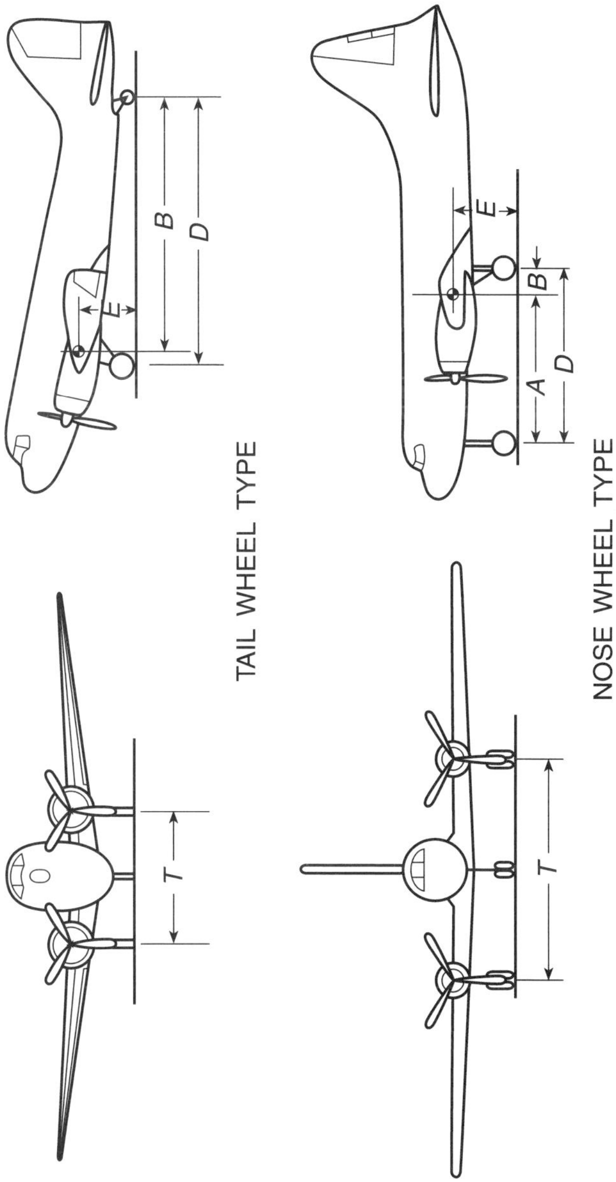


FIGURE 1—Basic landing gear dimension data.

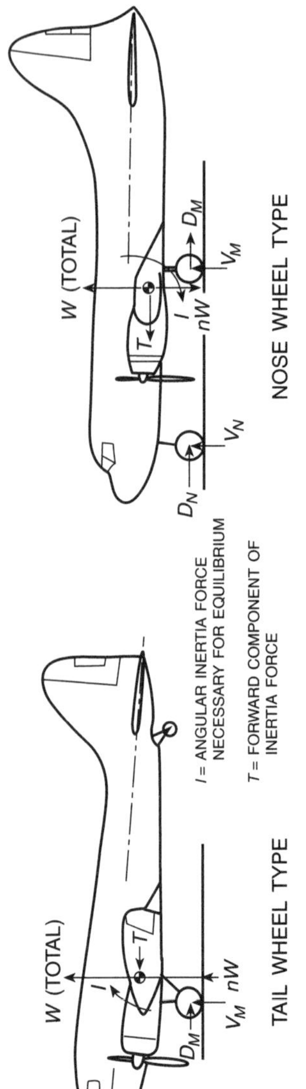


FIGURE 2—Level landing.

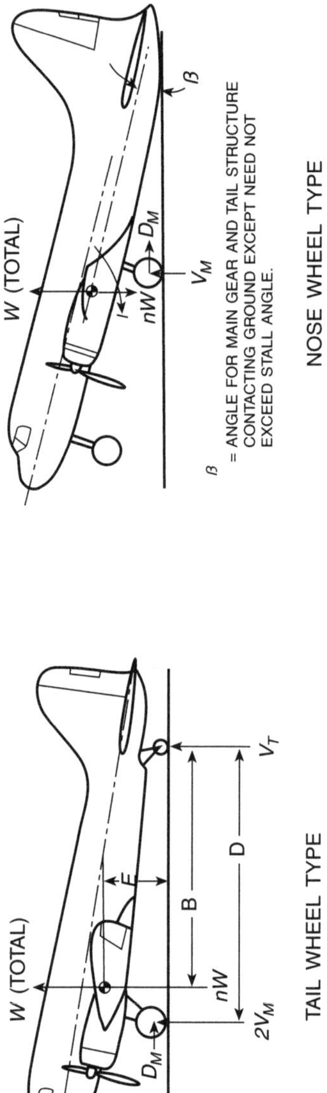


FIGURE 3—Tail-down landing

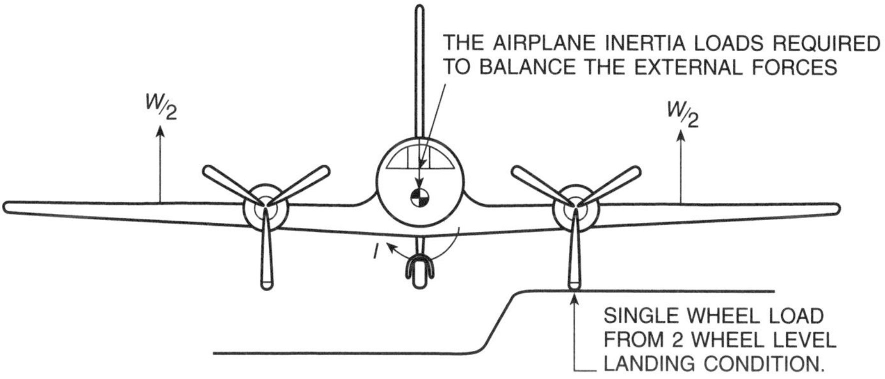


NOSE OR TAIL WHEEL TYPE

FIGURE 4

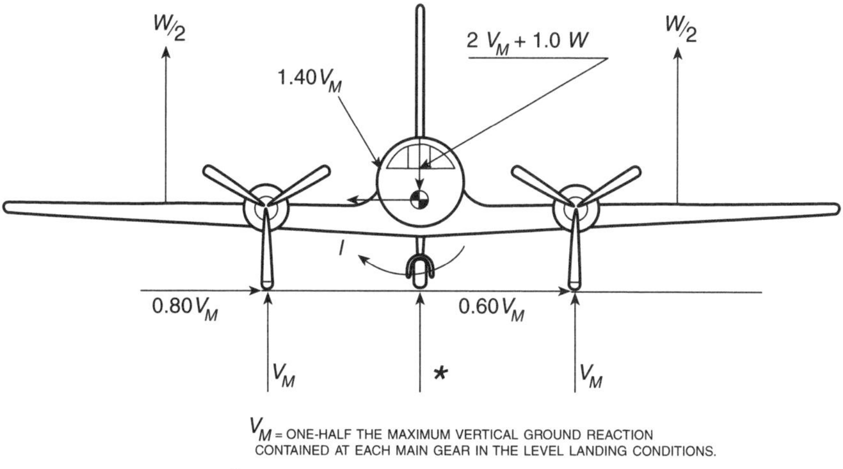


NOSE OR TAIL WHEEL TYPE AIRPLANE IN LEVEL ALTITUDE

FIGURE 5—Lateral drift landing.

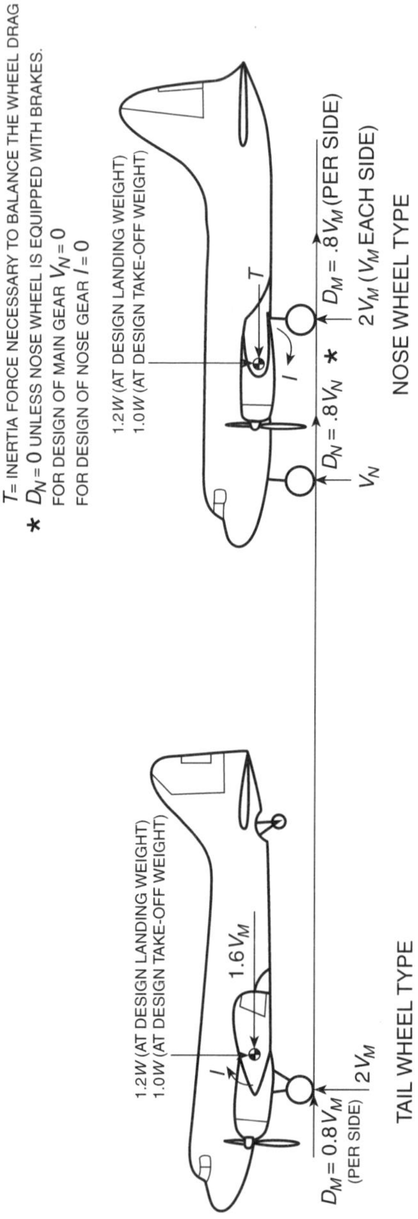


FIGURE 6—Braked roll.

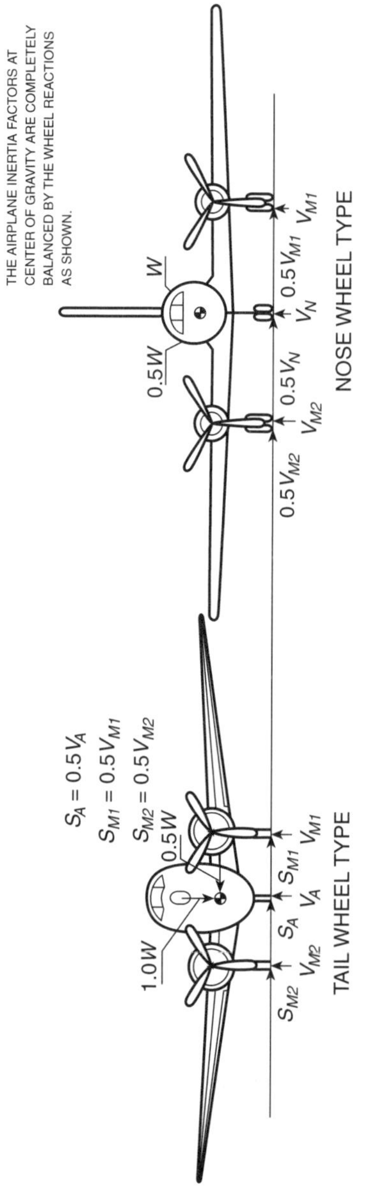


FIGURE 7—Ground turning.

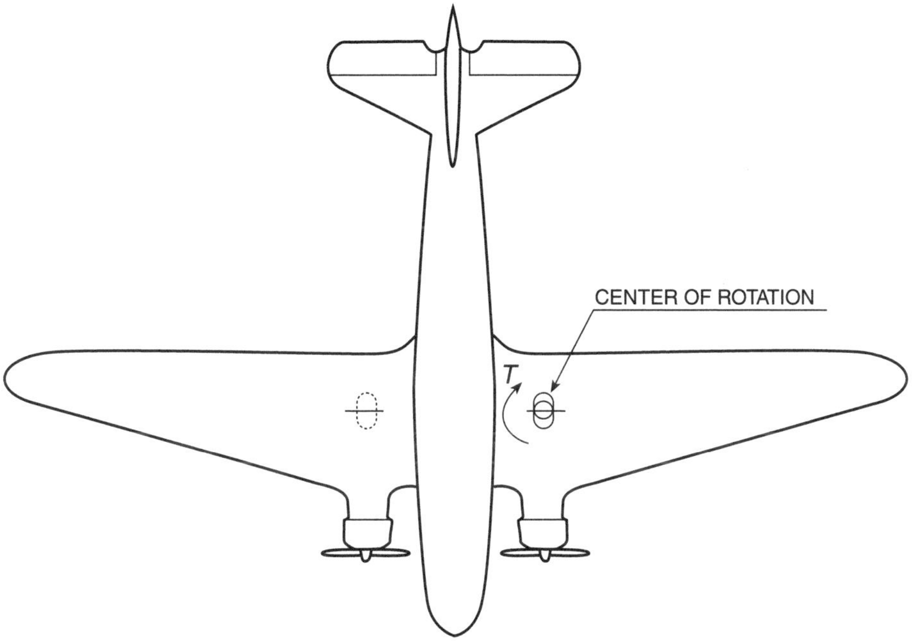


*$V_N$ and $V_M$ are static ground reactions. For tail wheel type the airplane is in the three point attitude. Pivoting is assumed to take place about one main landing gear unit.*

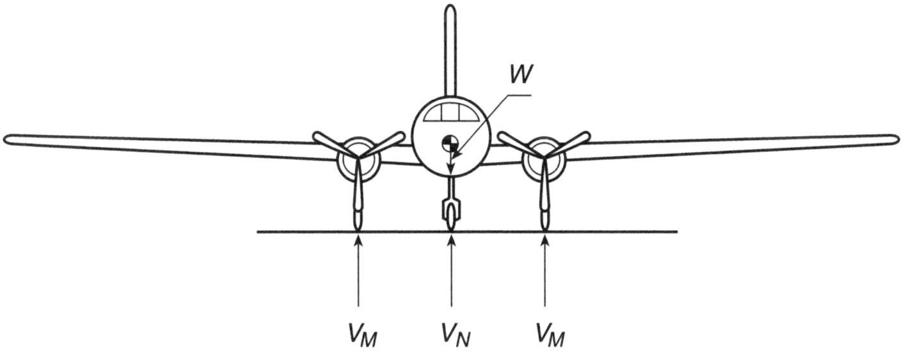


FIGURE 8—Pivoting, nose or tail wheel type.

## APPENDIX B TO PART 25

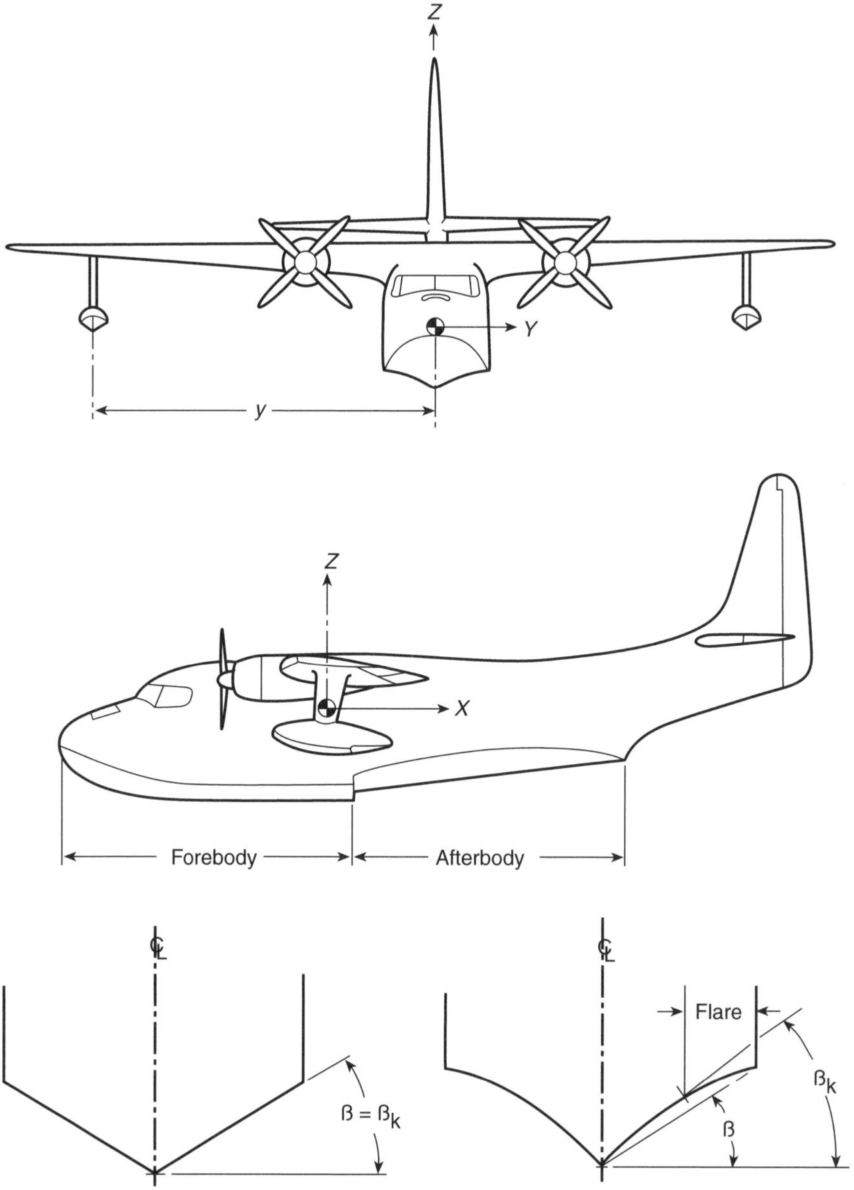


FIGURE 1 — Pictorial definition of angles, dimensions, and directions on a seaplane.

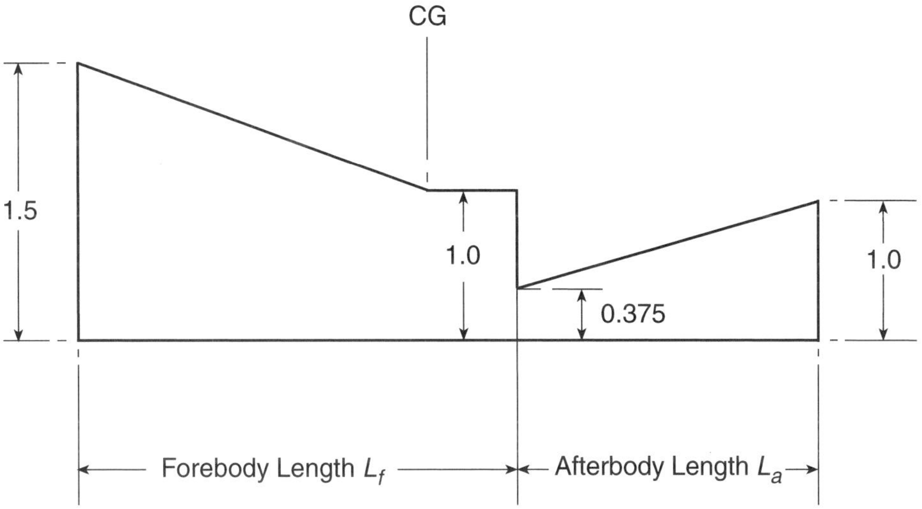


$K_1$ (Vertical Loads)

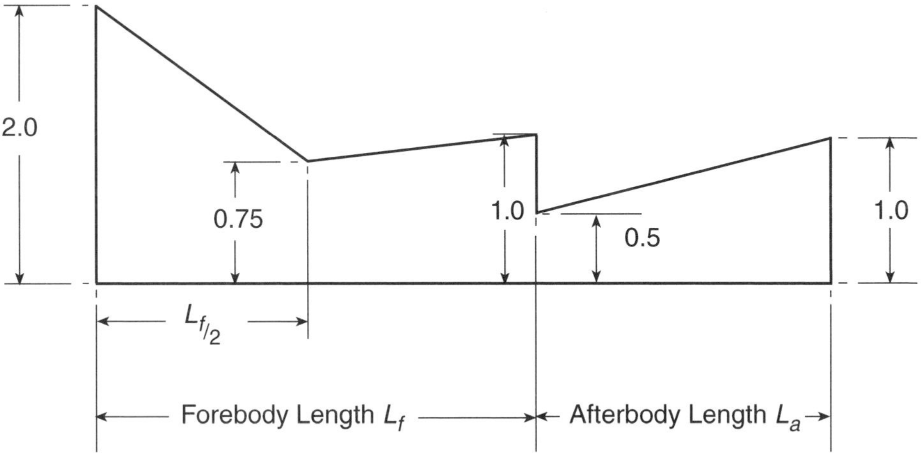


$K_2$ (Bottom Pressures)

FIGURE 2 — Hull station weighing factor.

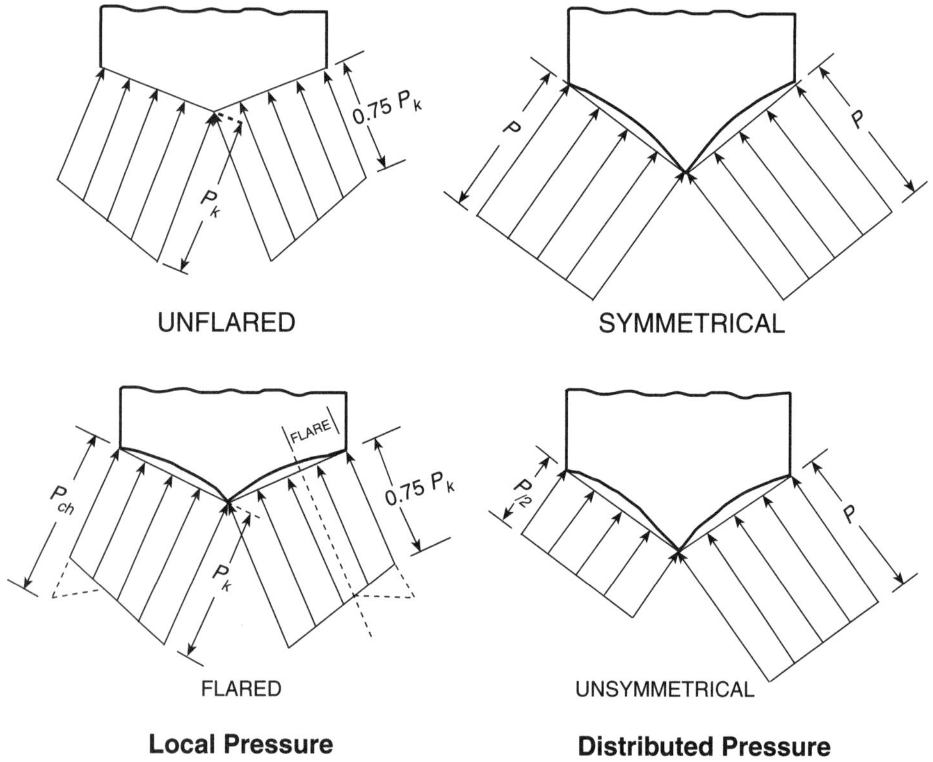


FIGURE 3 — Transverse pressure distributions.

## Appendix C to Part 25

**(a)** *Continuous maximum icing.* The maximum continuous intensity of atmospheric icing conditions (continuous maximum icing) is defined by the variables of the cloud liquid water content, the mean effective diameter of the cloud droplets, the ambient air temperature, and the interrelationship of these three variables as shown in figure 1 of this appendix. The limiting icing envelope in terms of altitude and temperature is given in figure 2 of this appendix. The inter-relationship of cloud liquid water content with drop diameter and altitude is determined from figures 1 and 2. The cloud liquid water content for continuous maximum icing conditions of a horizontal extent, other than 17.4 nautical miles, is determined by the value of liquid water content of figure 1, multiplied by the appropriate factor from figure 3 of this appendix.

**(b)** *Intermittent maximum icing.* The intermittent maximum intensity of atmospheric icing conditions (intermittent maximum icing) is defined by the variables of the cloud liquid water content, the mean effective diameter of the cloud droplets, the ambient air temperature, and the interrelationship of these three variables as shown in figure 4 of this appendix. The limiting icing envelope in terms of altitude and temperature is given in figure 5 of this appendix. The inter-relationship of cloud liquid water content with drop diameter and altitude is determined from figures 4 and 5. The cloud liquid water content for intermittent maximum icing conditions of a horizontal extent, other than 2.6 nautical miles, is determined by the value of cloud liquid water content of figure 4 multiplied by the appropriate factor in figure 6 of this appendix.

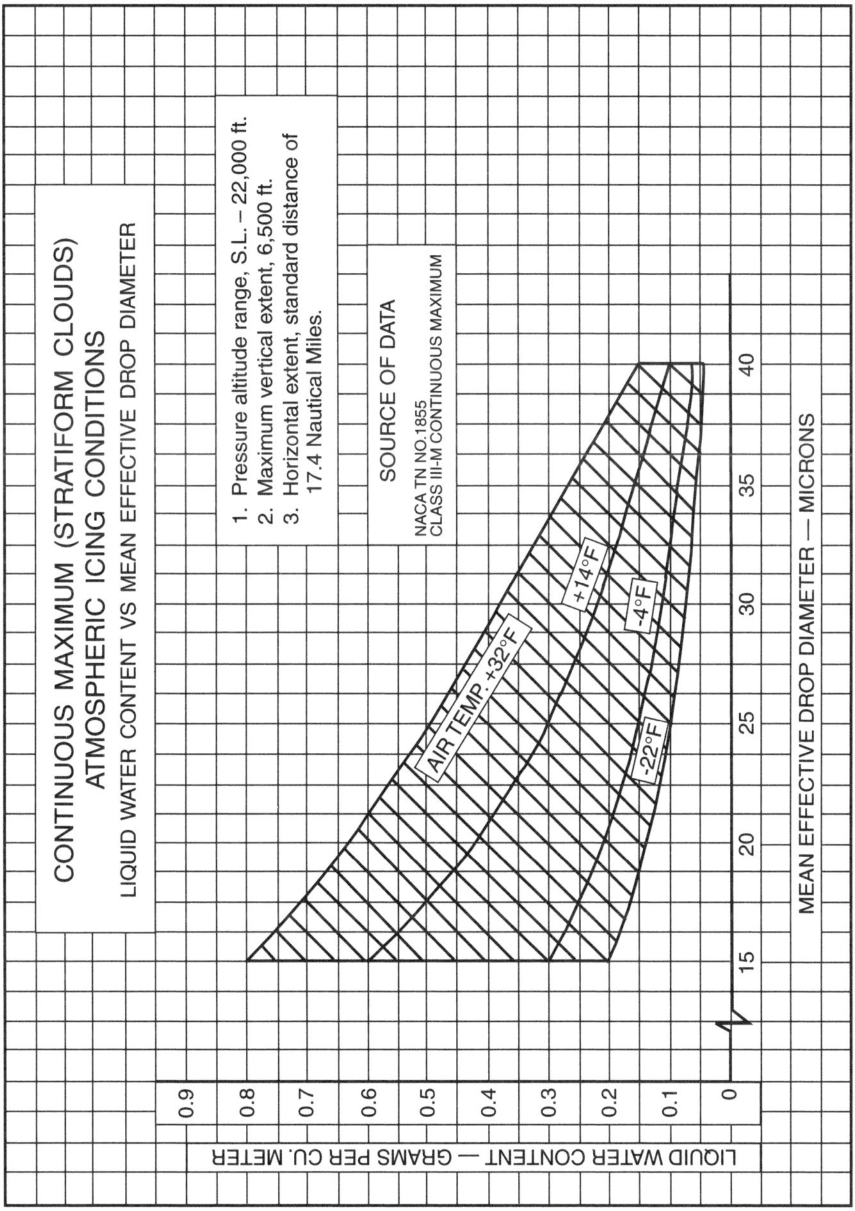
CONTINUOUS MAXIMUM (STRATIFORM CLOUDS)
ATMOSPHERIC ICING CONDITIONS
LIQUID WATER CONTENT VS MEAN EFFECTIVE DROP DIAMETER
1. Pressure altitude range, S.L. – 22,000 ft.
2. Maximum vertical extent, 6,500 ft.
3. Horizontal extent, standard distance of 17.4 Nautical Miles.
SOURCE OF DATA
NACA TN NO.1855
CLASS III-M CONTINUOUS MAXIMUM
AIR TEMP. +32°F
+14°F
-4°F
-22°F
LIQUID WATER CONTENT — GRAMS PER CU. METER
0.9
0.8
0.7
0.6
0.5
0.4
0.3
0.2
0.1
0
15
20
25
30
35
40
MEAN EFFECTIVE DROP DIAMETER — MICRONS

FIGURE 1

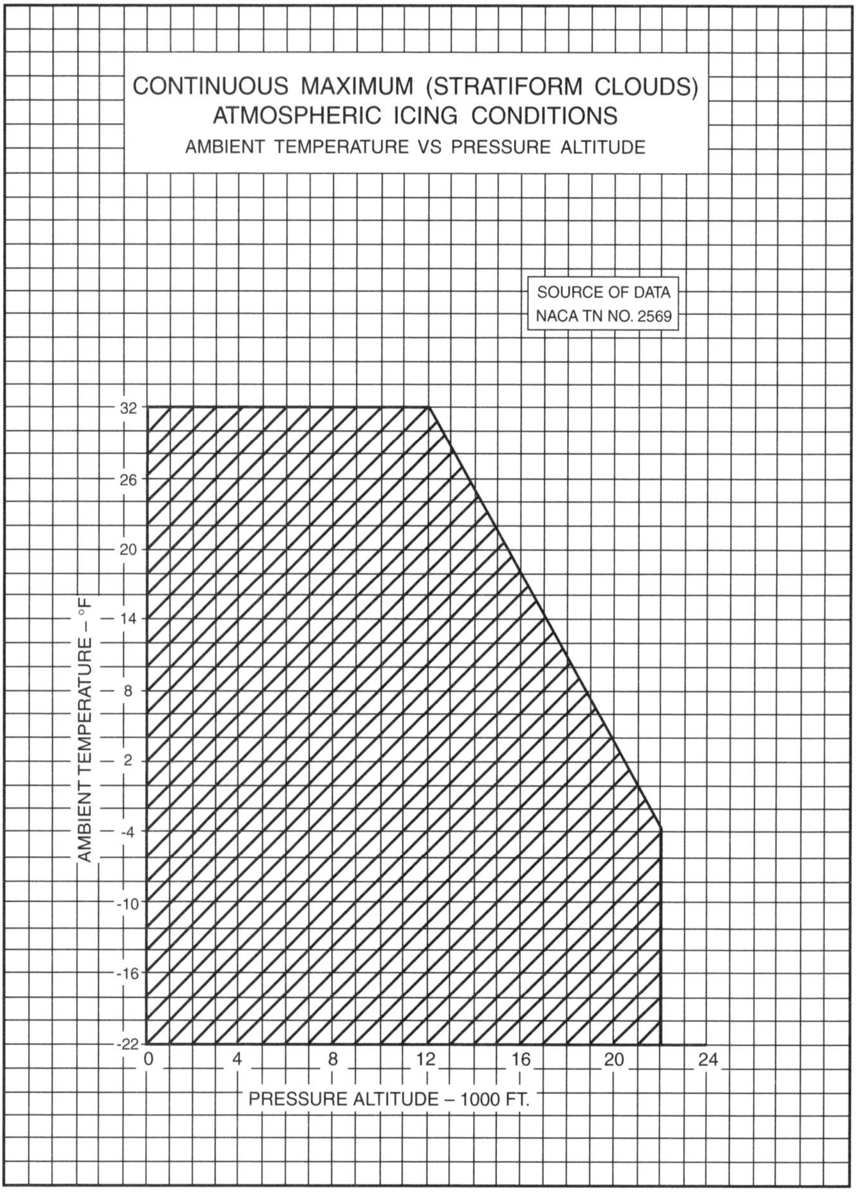
CONTINUOUS MAXIMUM (STRATIFORM CLOUDS)
ATMOSPHERIC ICING CONDITIONS
AMBIENT TEMPERATURE VS PRESSURE ALTITUDE
SOURCE OF DATA
NACA TN NO. 2569
AMBIENT TEMPERATURE – °F
32
26
20
14
8
2
-4
-10
-16
-22
0
4
8
12
16
20
24
PRESSURE ALTITUDE – 1000 FT.

FIGURE 2

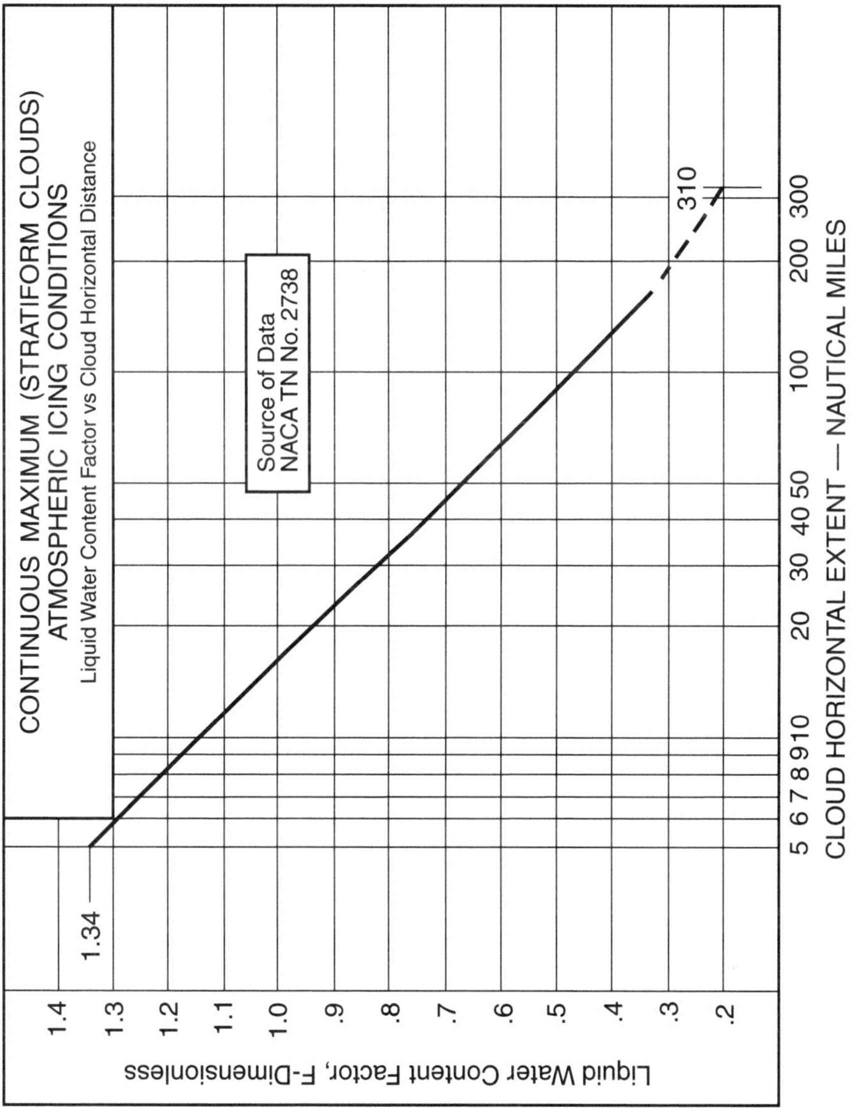
CONTINUOUS MAXIMUM (STRATIFORM CLOUDS)
ATMOSPHERIC ICING CONDITIONS
Liquid Water Content Factor vs Cloud Horizontal Distance
Source of Data
NACA TN No. 2738
Liquid Water Content Factor, F-Dimensionless
1.4
1.3
1.2
1.1
1.0
.9
.8
.7
.6
.5
.4
.3
.2
1.34
310
5 6 7 8 9 10
20
30
40 50
100
200
300
CLOUD HORIZONTAL EXTENT — NAUTICAL MILES

FIGURE 3

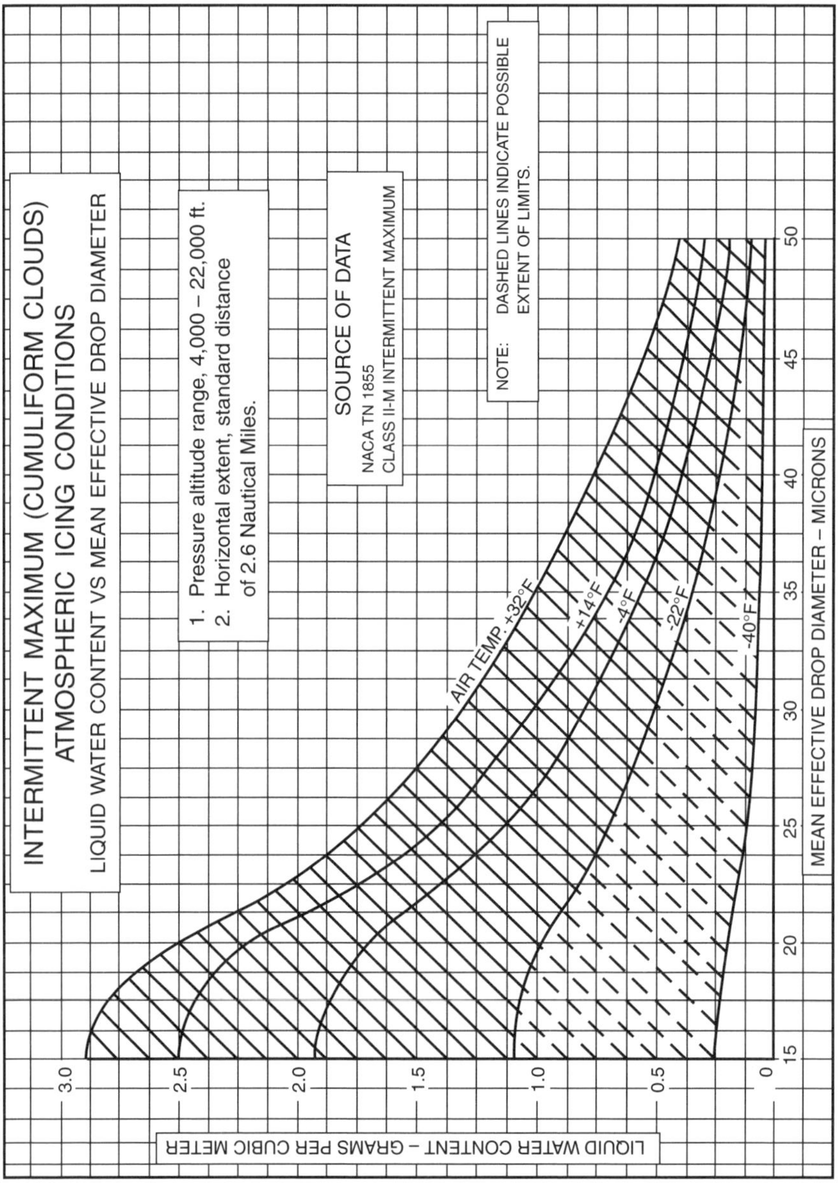
INTERMITTENT MAXIMUM (CUMULIFORM CLOUDS)
ATMOSPHERIC ICING CONDITIONS
LIQUID WATER CONTENT VS MEAN EFFECTIVE DROP DIAMETER
1. Pressure altitude range, 4,000 – 22,000 ft.
2. Horizontal extent, standard distance of 2.6 Nautical Miles.
SOURCE OF DATA
NACA TN 1855
CLASS II-M INTERMITTENT MAXIMUM
NOTE: DASHED LINES INDICATE POSSIBLE EXTENT OF LIMITS.
AIR TEMP. +32°F
+14°F
-4°F
-22°F
-40°F
LIQUID WATER CONTENT – GRAMS PER CUBIC METER
3.0
2.5
2.0
1.5
1.0
0.5
0
15
20
25
30
35
40
45
50
MEAN EFFECTIVE DROP DIAMETER – MICRONS

FIGURE 4

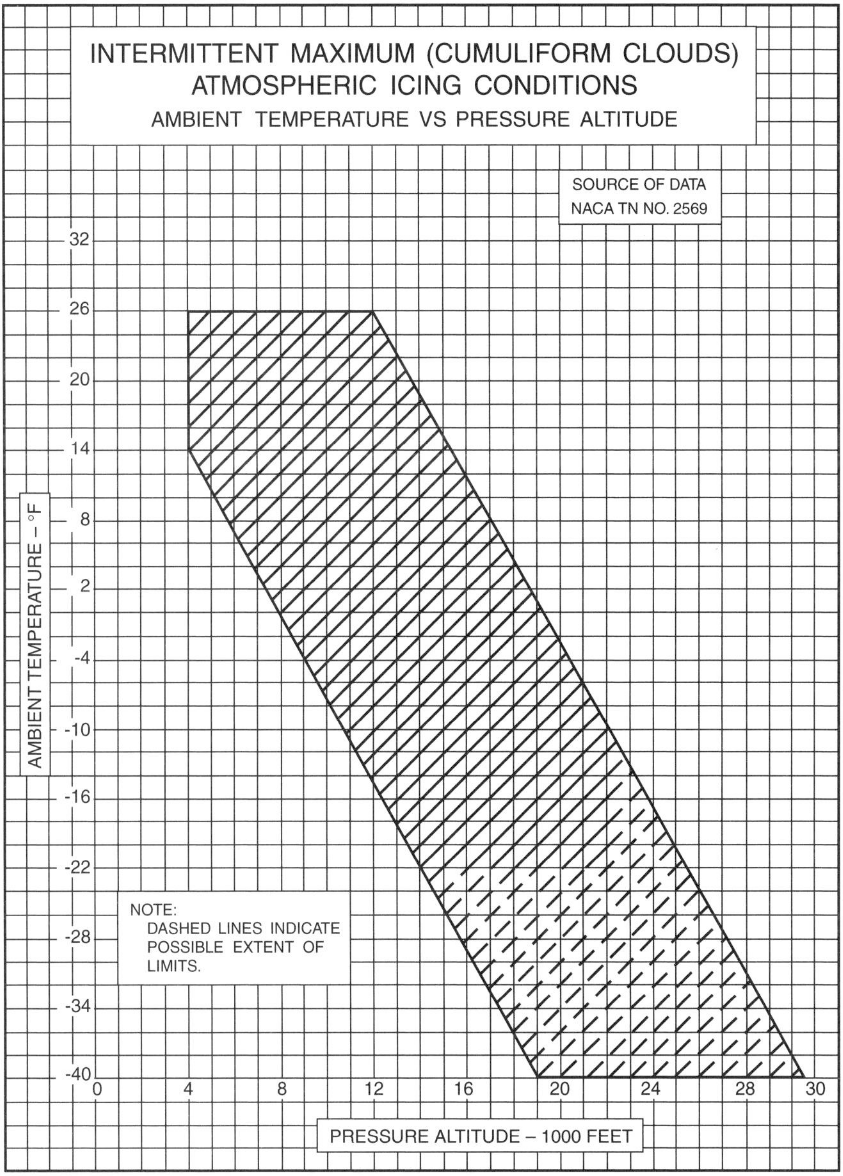
INTERMITTENT MAXIMUM (CUMULIFORM CLOUDS)
ATMOSPHERIC ICING CONDITIONS
AMBIENT TEMPERATURE VS PRESSURE ALTITUDE
SOURCE OF DATA
NACA TN NO. 2569
AMBIENT TEMPERATURE – °F
32
26
20
14
8
2
-4
-10
-16
-22
-28
-34
-40
NOTE:
DASHED LINES INDICATE POSSIBLE EXTENT OF LIMITS.
0
4
8
12
16
20
24
28
30
PRESSURE ALTITUDE – 1000 FEET

FIGURE 5

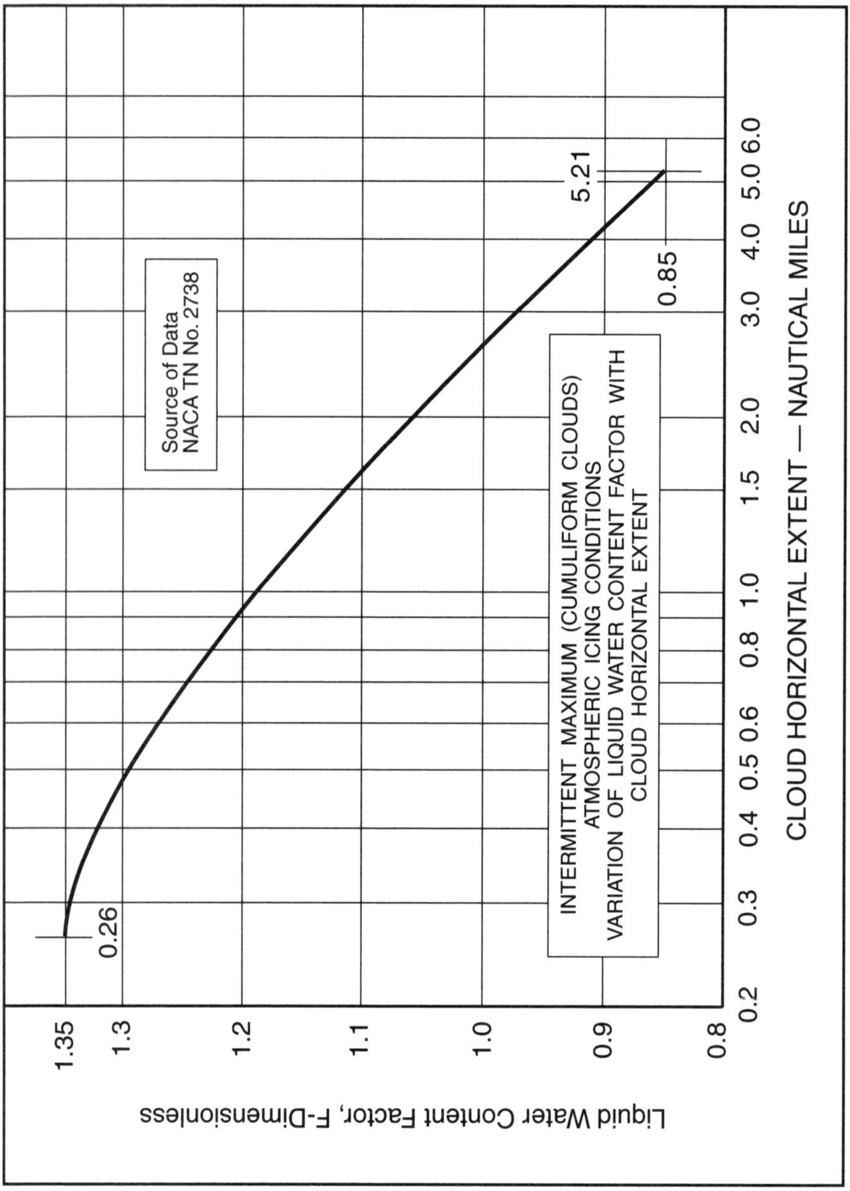

Source of Data
NACA TN No. 2738
INTERMITTENT MAXIMUM (CUMULIFORM CLOUDS)
ATMOSPHERIC ICING CONDITIONS
VARIATION OF LIQUID WATER CONTENT FACTOR WITH
CLOUD HORIZONTAL EXTENT
Liquid Water Content Factor, F-Dimensionless
1.35
1.3
1.2
1.1
1.0
0.9
0.8
0.26
5.21
0.85
0.2 0.3 0.4 0.5 0.6 0.8 1.0 1.5 2.0 3.0 4.0 5.0 6.0
CLOUD HORIZONTAL EXTENT — NAUTICAL MILES


FIGURE 6

## Appendix D to Part 25

*Criteria for determining minimum flight crew.* The following are considered by the Agency in determining the minimum flight crew under §25.1523:

**(a)** *Basic workload functions.* The following basic workload functions are considered:

(1) Flight path control.

(2) Collision avoidance.

(3) Navigation.

(4) Communications.

(5) Operation and monitoring of aircraft engines and systems.

(6) Command decisions.

**(b)** *Workload factors.* The following workload factors are considered significant when analyzing and demonstrating workload for minimum flight crew determination:

(1) The accessibility, ease, and simplicity of operation of all necessary flight, power, and equipment controls, including emergency fuel shutoff valves, electrical controls, electronic controls, pressurization system controls, and engine controls.

(2) The accessibility and conspicuity of all necessary instruments and failure warning devices such as fire warning, electrical system malfunction, and other failure or caution indicators. The extent to which such instruments or devices direct the proper corrective action is also considered.

(3) The number, urgency, and complexity of operating procedures with particular consideration given to the specific fuel management schedule imposed by center of gravity, structural or other considerations of an airworthiness nature, and to the ability of each engine to operate at all times from a single tank or source which is automatically replenished if fuel is also stored in other tanks.

(4) The degree and duration of concentrated mental and physical effort involved in normal operation and in diagnosing and coping with malfunctions and emergencies.

(5) The extent of required monitoring of the fuel, hydraulic, pressurization, electrical, electronic, deicing, and other systems while en route.

(6) The actions requiring a crewmember to be unavailable at his assigned duty station, including: observation of systems, emergency operation of any control, and emergencies in any compartment.

(7) The degree of automation provided in the aircraft systems to afford (after failures or malfunctions) automatic crossover or isolation of difficulties to minimize the need for flight crew action to guard against loss of hydraulic or electric power to flight controls or to other essential systems.

(8) The communications and navigation workload.

(9) The possibility of increased workload associated with any emergency that may lead to other emergencies.

(10) Incapacitation of a flight crewmember whenever the applicable operating rule requires a minimum flight crew of at least two pilots.

**(c)** *Kind of operation authorized.* The determination of the kind of operation authorized requires consideration of the operating rules under which the airplane will be operated. Unless an applicant desires approval for a more limited kind of operation. It is assumed that each airplane certificated under this Part will operate under IFR conditions.

[Docket No. 5066, 29 FR 18291, Dec. 24, 1964; as amended by Amdt. 25–3, 30 FR 6067, April 29, 1965]



## Appendix E to Part 25

### *I — Limited Weight Credit For Airplanes Equipped With Standby Power*

**(a)** Each applicant for an increase in the maximum certificated takeoff and landing weights of an airplane equipped with a type-certificated standby power rocket engine may obtain an increase as specified in paragraph (b) if—

(1) The installation of the rocket engine has been approved and it has been established by flight test that the rocket engine and its controls can be operated safely and reliably at the increase in maximum weight; and

(2) The Airplane Flight Manual, or the placard, markings or manuals required in place thereof, set forth in addition to any other operating limitations the Administrator may require, the increased weight approved under this regulation and a prohibition against the operation of the airplane at the approved increased weight when—

(i) The installed standby power rocket engines have been stored or installed in excess of the time limit established by the manufacturer of the rocket engine (usually stenciled on the engine casing); or

(ii) The rocket engine fuel has been expended or discharged.

**(b)** The currently approved maximum takeoff and landing weights at which an airplane is certificated without a standby power rocket engine installation may be increased by an amount that does not exceed any of the following:

(1) An amount equal in pounds to 0.014 IN, where I is the maximum usable impulse in pounds-seconds available from each standby power rocket engine and N is the number of rocket engines installed.

(2) An amount equal to 5 percent of the maximum certificated weight approved in accordance with the applicable airworthiness regulations without standby power rocket engines installed.

(3) An amount equal to the weight of the rocket engine installation.

(4) An amount that, together with the currently approved maximum weight, would equal the maximum structural weight established for the airplane without standby rocket engines installed.

### *II — Performance Credit for Transport Category Airplanes Equipped With Standby Power*

The Administrator may grant performance credit for the use of standby power on transport category airplanes. However, the performance credit applies only to the maximum certificated takeoff and landing weights, the takeoff distance, and the takeoff paths, and may not exceed that found by the Administrator to result in an overall level of safety in the takeoff, approach, and landing regimes of flight equivalent to that prescribed in the regulations under which the airplane was originally certificated without standby power. For the purposes of this Appendix, "standby power" is power or thrust, or both, obtained from rocket engines for a relatively short period and actuated only in cases of emergency. The following provisions apply:

**(1)** *Takeoff; general.* The takeoff data prescribed in paragraphs (2) and (3) of this Appendix must be determined at all weights and altitudes, and at ambient temperatures if applicable, at which performance credit is to be applied.

**(2)** *Takeoff path.*

(a) The one-engine-inoperative takeoff path with standby power in use must be determined in accordance with the performance requirements of the applicable airworthiness regulations.

(b) The one-engine-inoperative takeoff path (excluding that part where the airplane is on or just above the takeoff surface) determined in accordance with paragraph (a) of this section must lie above the one-engine-inoperative takeoff path without standby power at the maximum takeoff weight at which all of the applicable airworthiness requirements are met. For the purpose of this comparison, the flight path is considered to extend to at least a height of 400 feet above the takeoff surface.

(c) The takeoff path with all engines operating, but without the use of standby power, must reflect a conservatively greater overall level of performance than the one-engine-inoperative takeoff path established in accordance with paragraph (a) of this section. The margin must be established by the Administrator to insure safe day-to-day operations, but in no case may it be less than 15 percent. The all-engines-operating takeoff path must be determined by a procedure consistent with that established in complying with paragraph (a) of this section.

(d) For reciprocating-engine-powered airplanes, the takeoff path to be scheduled in the Airplane Flight Manual must represent the one-engine-operative takeoff path determined in accordance with paragraph (a) of this section and modified to reflect the procedure (see paragraph (6)) established by the applicant for flap retraction and attainment of the en route speed. The scheduled takeoff path must have a positive slope at all points of the airborne portion and at no point must it lie above the takeoff path specified in paragraph (a) of this section.

**(3)** *Takeoff distance.* The takeoff distance must be the horizontal distance along the one-engine-inoperative take off path determined in accordance with paragraph (2)(a) from the start of the takeoff to the point where the airplane attains a height of 50 feet above the takeoff surface for reciprocating-engine-powered airplanes and a height of 35 feet above the takeoff surface for turbine-powered airplanes.

**(4)** *Maximum certificated takeoff weights.* The maximum certificated takeoff weights must be determined at all altitudes, and at ambient temperatures, if applicable, at which performance credit is to be applied and may not exceed the weights established in compliance with paragraphs (a) and (b) of this section.

(a) The conditions of paragraphs (2) (b) through (d) must be met at the maximum certificated takeoff weight.

(b) Without the use of standby power, the airplane must meet all of the en route requirements of the applicable airworthiness regulations under which the airplane was originally certificated. In addition, turbine-powered airplanes without the use of standby power must meet the final takeoff climb requirements prescribed in the applicable airworthiness regulations.

**(5)** *Maximum certificated landing weights.*

(a) The maximum certificated landing weights (one-engine-inoperative approach and all-engine-operating landing climb) must be determined at all altitudes, and at ambient temperatures if applicable, at which performance credit is to be applied and must not exceed that established in compliance with paragraph (b) of this section.

(b) The flight path, with the engines operating at the power or thrust, or both, appropriate to the airplane configuration and with standby power in use, must lie above the flight path without standby power in use at the maximum weight at which all of the applicable airworthiness requirements are met. In addition, the flight paths must comply with subparagraphs (i) and (ii) of this paragraph.

(i) The flight paths must be established without changing the appropriate airplane configuration.

(ii) The flight paths must be carried out for a minimum height of 400 feet above the point where standby power is actuated.

**(6)** *Airplane configuration, speed, and power and thrust; general.* Any change in the airplane's configuration, speed, and power or thrust, or both, must be made in accordance with the procedures established by the applicant for the operation of the airplane in service and must comply with paragraphs (a) through (c) of this section. In addition, procedures must be established for the execution of balked landings and missed approaches.

(a) The Administrator must find that the procedure can be consistently executed in service by crews of average skill.

(b) The procedure may not involve methods or the use of devices which have not been proven to be safe and reliable.

(c) Allowances must be made for such time delays in the execution of the procedures as may be reasonably expected to occur during service.

**(7)** *Installation and operation; standby power.* The standby power unit and its installation must comply with paragraphs (a) and (b) of this section.

(a) The standby power unit and its installation must not adversely affect the safety of the airplane.

(b) The operation of the standby power unit and its control must have proven to be safe and reliable.

[Docket No. 5066, 29 FR 18291, Dec. 24, 1964; as amended by Amdt. 25–6, 30 FR 8468, July 2, 1965]

## APPENDIX F TO PART 25

### *PART I — TEST CRITERIA AND PROCEDURES FOR SHOWING COMPLIANCE WITH §25.853, OR 25.855.*

**(a)** Material test criteria—

(1) *Interior compartments occupied by crew or passengers.*

(i) Interior ceiling panels, interior wall panels, partitions, galley structure, large cabinet walls, structural flooring, and materials used in the construction of stowage compartments (other than underseat stowage compartments and compartments for stowing small items such as magazines and maps) must be self-extinguishing when tested vertically in accordance with the applicable portions of part I of this appendix. The average burn length may not exceed 6 inches and the average flame time after removal of the flame source may not exceed 15 seconds. Drippings from the test specimen may not continue to flame for more than an average of 3 seconds after falling.

(ii) Floor covering, textiles (including draperies and upholstery), seat cushions, padding, decorative and nondecorative coated fabrics, leather, trays and galley furnishings, electrical conduit, thermal and acoustical insulation and insulation covering, air ducting, joint and edge covering, liners of Class B and E cargo or baggage compartments, floor panels of Class B, C, D, or E cargo or baggage compartments, insulation blankets, cargo covers and transparencies, molded and thermoformed parts, air ducting joints, and trim strips (decorative and chafing), that are constructed of materials not covered in subparagraph (iv) below, must be self-extinguishing when tested vertically in accordance with the applicable portions of part I of this appendix or other approved equivalent means. The average burn length may not exceed 8 inches, and the average flame time after removal of the flame source may not exceed 15 seconds. Drippings from the test specimen may not continue to flame for more than an average of 5 seconds after falling.

(iii) Motion picture film must be safety film meeting the Standard Specifications for Safety Photographic Film PHI.25 (available from the American National Standards Institute, 1430 Broadway, New York, NY 10018). If the film travels through ducts, the ducts must meet the requirements of subparagraph (ii) of this paragraph.

(iv) Clear plastic windows and signs, parts constructed in whole or in part of elastomeric materials, edge lighted instrument assemblies consisting of two or more instruments in a common housing, seat belts, shoulder harnesses, and cargo and baggage tiedown equipment, including containers, bins, pallets, etc., used in passenger or crew compartments, may not have an average burn rate greater than 2.5 inches per minute when tested horizontally in accordance with the applicable portions of this appendix.

(v) Except for small parts (such as knobs, handles, rollers, fasteners, clips, grommets, rub strips, pulleys, and small electrical parts) that would not contribute significantly to the propagation of a fire and for electrical wire and cable insulation, materials in items not specified in paragraphs (a)(1)(i), (ii), (iii), or (iv) of part I of this appendix may not have a burn rate greater than 4.0 inches per minute when tested horizontally in accordance with the applicable portions of this appendix.

(2) *Cargo and baggage compartments not occupied by crew or passengers.*

(i) Thermal and acoustic insulation (including coverings) used in each cargo and baggage compartment must be constructed of materials that meet the requirements set forth in paragraph (a)(1)(ii) of part I of this appendix.

(ii) A cargo or baggage compartment defined in §25.857 as Class B or E must have a liner constructed of materials that meet the requirements of paragraph (a)(1)(ii) of part I of this appendix and separated from the airplane structure (except for attachments). In addition, such liners must be subjected to the 45 degree angle test. The flame may not penetrate (pass through) the material during application of the flame or subsequent to its removal. The average flame time after removal of the flame source may not exceed 15 seconds, and the average glow time may not exceed 10 seconds.

(iii) A cargo or baggage compartment defined in §25.857 as Class B, C, D, or E must have floor panels constructed of materials which meet the requirements of paragraph (a)(1)(ii) of part I of this appendix and which are separated from the airplane structure (except for attachments). Such panels must be subjected to the 45 degree angle test. The flame may not penetrate (pass through) the material during application of the flame or subsequent to its removal. The average flame time after removal of the flame source may not exceed 15 seconds, and the average glow time may not exceed 10 seconds.

(iv) Insulation blankets and covers used to protect cargo must be constructed of materials that meet the requirements of paragraph (a)(1)(ii) of part I of this appendix. Tiedown equipment (including containers, bins, and pallets) used in each cargo and baggage compartment must be constructed of materials that meet the requirements of paragraph (a)(1)(v) of part I of this appendix.

(3) *Electrical system components.* Insulation on electrical wire or cable installed in any area of the fuselage must be self-extinguishing when subjected to the 60 degree test specified in part I of this appendix. The average burn length may not exceed 3 inches, and the average flame time after removal of the flame source may not exceed 30 seconds. Drippings from the test specimen may not continue to flame for more than an average of 3 seconds after falling.

**(b)** *Test Procedures—*

(1) *Conditioning.* Specimens must be conditioned to 70 ±5 F, and at 50 percent ±5 percent relative humidity until moisture equilibrium is reached or for 24 hours. Each specimen must remain in the conditioning environment until it is subjected to the flame.

(2) *Specimen configuration.* Except for small parts and electrical wire and cable insulation, materials must be tested either as section cut from a fabricated part as installed in the airplane or as a specimen simulating a cut section, such as a specimen cut from a flat sheet of the material or a model of the fabricated part. The specimen may be cut from any location in a fabricated part; however, fabricated units, such as sandwich panels, may not be separated for test. Except as noted below, the specimen thickness must be no thicker than the minimum thickness to be qualified for use in the airplane. Test specimens of thick foam parts, such as seat cushions, must be ½-inch in thickness. Test specimens of materials that must meet the requirements of paragraph (a)(1)(v) of part I of this appendix must be no more than ⅛-inch in thickness. Electrical wire and cable specimens must be the same size as used in the airplane. In the case of fabrics, both the warp and fill direction of the weave must be tested to determine the most critical flammability condition. Specimens must be mounted in a metal frame so that the two long edges and the upper edge are held securely during the vertical test prescribed in subparagraph (4) of this paragraph and the two long edges and the edge away from the flame are held securely during the horizontal test prescribed in subparagraph (5) of this paragraph. The exposed area of the specimen must be at least 2 inches wide and 12 inches long, unless the actual size used in the airplane is smaller. The edge to which the burner flame is applied must not consist of the finished or protected edge of the specimen but must be representative of the actual cross-section of the material or part as installed in the airplane. The specimen must be mounted in a metal frame so that all four edges are held securely and the exposed area of the specimen is at least 8 inches by 8 inches during the 45° test prescribed in subparagraph (6) of this paragraph.

(3) *Apparatus.* Except as provided in subparagraph (7) of this paragraph, tests must be conducted in a draft-free cabinet in accordance with Federal Test Method Standard 191 Model 5903 (revised Method 5902) for the vertical test, or Method 5906 for horizontal test (available from the General Services Administration, Business Service Center, Region 3, Seventh & D Streets SW., Washington, DC 20407). Specimens which are too large for the cabinet must be tested in similar draft-free conditions.

(4) *Vertical test.* A minimum of three specimens must be tested and results averaged. For fabrics, the direction of weave corresponding to the most critical flammability conditions must be parallel to the longest dimension. Each specimen must be supported vertically. The specimen must be exposed to a Bunsen or Tirrill burner with a nominal ⅜-inch I.D. tube adjusted to give a flame of 1½ inches in height. The minimum flame temperature measured by a calibrated thermocouple pyrometer in the center of the flame must be 1550°F The lower edge of the specimen must be ¾-inch above the top edge of the burner. The flame must be applied to the center line of the lower edge of the specimen. For materials covered by paragraph (a)(1)(i) of part I of this appendix, the flame must be applied for 60 seconds and then removed. For materials covered by paragraph (a)(1)(ii) of part I of this appendix, the flame must be applied for 12 seconds and then removed. Flame time, burn length, and flaming time of drippings, if any, may be recorded. The burn length determined in accordance with subparagraph (7)

of this paragraph must be measured to the nearest tenth of an inch.

(5) *Horizontal test.* A minimum of three specimens must be tested and the results averaged. Each specimen must be supported horizontally. The exposed surface, when installed in the aircraft, must be face down for the test. The specimen must be exposed to a Bunsen or Tirrill burner with a nominal ⅜-inch I.D. tube adjusted to give a flame of 1½ inches in height. The minimum flame temperature measured by a calibrated thermocouple pyrometer in the center of the flame must be 1550°F. The specimen must be positioned so that the edge being tested is centered ¾-inch above the top of the burner. The flame must be applied for 15 seconds and then removed. A minimum of 10 inches of specimen must be used for timing purposes, approximately 1½ inches must burn before the burning front reaches the timing zone, and the average burn rate must be recorded.

(6) *Forty-five degree test.* A minimum of three specimens must be tested and the results averaged. The specimens must be supported at an angle of 45° to a horizontal surface. The exposed surface when installed in the aircraft must be face down for the test. The specimens must be exposed to a Bunsen or Tirrill burner with a nominal ⅜-inch I.D. tube adjusted to give a flame of 1½ inches in height. The minimum flame temperature measured by a calibrated thermocouple pyrometer in the center of the flame must be 1550°F. Suitable precautions must be taken to avoid drafts. The flame must be applied for 30 seconds with one-third contacting the material at the center of the specimen and then removed. Flame time, glow time, and whether the flame penetrates (passes through) the specimen must be recorded.

(7) *Sixty degree test.* A minimum of three specimens of each wire specification (make and size) must be tested. The specimen of wire or cable (including insulation) must be placed at an angle of 60° with the horizontal in the cabinet specified in subparagraph (3) of this paragraph with the cabinet door open during the test, or must be placed within a chamber approximately 2 feet high by 1 foot by 1 foot, open at the top and at one vertical side (front), and which allows sufficient flow of air for complete combustion, but which is free from drafts. The specimen must be parallel to and approximately 6 inches from the front of the chamber. The lower end of the specimen must be held rigidly clamped. The upper end of the specimen must pass over a pulley or rod and must have an appropriate weight attached to it so that the specimen is held tautly throughout the flammability test. The test specimen span between lower clamp and upper pulley or rod must be 24 inches and must be marked 8 inches from the lower end to indicate the central point for flame application. A flame from a Bunsen or Tirrill burner must be applied for 30 seconds at the test mark. The burner must be mounted underneath the test mark on the specimen, perpendicular to the specimen and at an angle of 30° to the vertical plane of the specimen. The burner must have a nominal bore of ⅜-inch and be adjusted to provide a 3-inch high flame with an inner cone approximately one-third of the flame height. The minimum temperature of the hottest portion of the flame, as measured with a calibrated thermocouple pyrometer, may not be less than 1750°F. The burner must be positioned so that the hottest portion of the flame is applied to the test mark on the wire. Flame time, burn length, and flaming time of drippings, if any, must be recorded. The burn length determined in accordance with paragraph (8) of this paragraph must be measured to the nearest tenth of an inch. Breaking of the wire specimens is not considered a failure.

(8) *Burn length.* Burn length is the distance from the original edge to the farthest evidence of damage to the test specimen due to flame impingement, including areas of partial or complete consumption, charring, or embrittlement, but not including areas sooted, stained, warped, or discolored, nor areas where material has shrunk or melted away from the heat source.

## *Part II — Flammability of Seat Cushions*

**(a)** *Criteria for Acceptance.* Each seat cushion must meet the following criteria:

(1) At least three sets of seat bottom and seat back cushion specimens must be tested.

(2) If the cushion is constructed with a fire blocking material, the fire blocking material must completely enclose the cushion foam core material.

(3) Each specimen tested must be fabricated using the principal components (i.e., foam core, flotation material, fire blocking material, if used, and dress covering) and assembly processes (representative seams and closures) intended for use in the production articles. If a different material combination is used for the back cushion than for the bottom cushion, both material combinations must be tested as complete specimen sets, each set consisting of a back cushion specimen and a bottom cushion specimen. If a cushion, including outer dress covering, is demonstrated to meet the requirements of this appendix using the oil burner test, the dress covering of that cushion may be replaced with a similar dress covering provided the burn length of the replacement covering, as determined by the test specified in §25.853(c), does not exceed the corresponding burn length of the dress covering used on the cushion subjected to the oil burner test.

25

(4) For at least two-thirds of the total number of specimen sets tested, the burn length from the burner must not reach the side of the cushion opposite the burner. The burn length must not exceed 17 inches. Burn length is the perpendicular distance from the inside edge of the seat frame closest to the burner to the farthest evidence of damage to the test specimen due to flame impingement, including areas of partial or complete consumption, charring, or embrittlement, but not including areas sooted, stained, warped, or discolored, or areas where material has shrunk or melted away from the heat source.

(5) The average percentage weight loss must not exceed 10 percent. Also, at least two-thirds of the total number of specimen sets tested must not exceed 10 percent weight loss. All droppings falling from the cushions and mounting stand are to be discarded before the after-test weight is determined. The percentage weight loss for a specimen set is the weight of the specimen set before testing less the weight of the specimen set after testing expressed as the percentage of the weight before testing.

**(b)** *Test Conditions.* Vertical air velocity should average 25 fpm ±10 fpm at the top of the back seat cushion. Horizontal air velocity should be below 10 fpm just above the bottom seat cushion. Air velocities should be measured with the ventilation hood operating and the burner motor off.

**(c)** *Test Specimens.*

(1) For each test, one set of cushion specimens representing a seat bottom and seat back cushion must be used.

(2) The seat bottom cushion specimen must be 18±⅛ inches (457±3 mm) wide by 20±⅛ inches (508±3 mm) deep by 4±⅛ inches (102±3 mm) thick, exclusive of fabric closures and seam overlap.

(3) The seat back cushion specimen must be 18±⅛ inches (432±3 mm) wide by 25±⅛ inches (635±3 mm) high by 2±⅛ inches (51±3 mm) thick, exclusive of fabric closures and seam overlap.

(4) The specimens must be conditioned at 70±5°F (21±2°C) 55%±10% relative humidity for at least 24 hours before testing.

**(d)** *Test Apparatus.* The arrangement of the test apparatus is shown in Figures 1 through 5 and must include the components described in this section. Minor details of the apparatus may vary, depending on the model burner used.

(1) *Specimen Mounting Stand.* The mounting stand for the test specimens consists of steel angles, as shown in Figure 1. The length of the mounting stand legs is 12 ±⅛ inches (305±3 mm). The mounting stand must be used for mounting the test specimen seat bottom and seat back, as shown in Figure 2. The mounting stand should also include a suitable drip pan lined with aluminum foil, dull side up.

(2) *Test Burner.* The burner to be used in testing must—

(i) Be a modified gun type;

(ii) Have an 80-degree spray angle nozzle nominally rated for 2.25 gallons/hour at 100 psi;

(iii) Have a 12-inch (305 mm) burner cone installed at the end of the draft tube, with an opening 6 inches (152 mm) high and 11 inches (280 mm) wide, as shown in Figure 3; and

(iv) Have a burner fuel pressure regulator that is adjusted to deliver a nominal 2.0 gallon/hour of # 2 Grade kerosene or equivalent required for the test.

Burner models which have been used successfully in testing are the Lennox Model OB-32, Carlin Model 200 CRD, and Park Model DPL 3400. FAA published reports pertinent to this type of burner are: (1) Powerplant Engineering Report No. 3A, Standard Fire Test Apparatus and Procedure for Flexible Hose Assemblies, dated March 1978; and (2) Report No. DOT/FAA/RD/76/213, Reevaluation of Burner Characteristics for Fire Resistance Tests, dated January 1977.

(3) *Calorimeter.*

(i) The calorimeter to be used in testing must be a (0-15.0 BTU/ft$^2$ -sec. 0-17.0 W/cm$^2$) calorimeter, accurate ±3%, mounted in a 6-inch by 12-inch (152 by 305 mm) by ¾-inch (19 mm) thick calcium silicate insulating board which is attached to a steel angle bracket for placement in the test stand during burner calibration, as shown in Figure 4.

(ii) Because crumbling of the insulating board with service can result in misalignment of the calorimeter, the calorimeter must be monitored and the mounting shimmed, as necessary, to ensure that the calorimeter face is flush with the exposed plane of the insulating board in a plane parallel to the exit of the test burner cone.

(4) *Thermocouples.* The seven thermocouples to be used for testing must be 1/16- to ⅛-inch metal sheathed, ceramic packed, type K, grounded thermocouples with a nominal 22 to 30 American wire gage (AWG)-size conductor. The seven thermocouples must be attached to a steel angle bracket to form a thermocouple rake for placement in the test stand during burner calibration, as shown in Figure 5.

(5) *Apparatus Arrangement.* The test burner must be mounted on a suitable stand to position the exit of the burner cone a distance of 4±⅛ inches (102±3 mm) from one side of the specimen mounting stand. The burner stand should have the capability of allowing the burner to be swung away from the specimen mounting stand during warmup periods.

(6) *Data Recording.* A recording potentiometer or other suitable calibrated instrument with an appropriate range must be used to measure and record the outputs of the calorimeter and the thermocouples.

(7) *Weight Scale.* Weighing Device—A device must be used that with proper procedures may determine the before and after test weights of each set of seat cushion specimens within 0.02 pound (9 grams). A continuous weighing system is preferred.

(8) *Timing Device.* A stopwatch or other device (calibrated to ±1 second) must be used to measure the time of application of the burner flame and self-extinguishing time or test duration.

**(e)** *Preparation of Apparatus.* Before calibration, all equipment must be turned on and the burner fuel must be adjusted as specified in paragraph (d)(2).

**(f)** *Calibration.* To ensure the proper thermal output of the burner, the following test must be made:

(1) Place the calorimeter on the test stand as shown in Figure 4 at a distance of $4 \pm 1/8$ inches (102 ±3 mm) from the exit of the burner cone.

(2) Turn on the burner, allow it to run for 2 minutes for warmup, and adjust the burner air intake damper to produce a reading of $10.5 \pm 0.5$ BTU/ft$^2$ -sec. ($11.9 \pm 0.6$ w/cm$^2$) on the calorimeter to ensure steady state conditions have been achieved. Turn off the burner.

(3) Replace the calorimeter with the thermocouple rake (Figure 5).

(4) Turn on the burner and ensure that the thermocouples are reading 1900 ±100°F (1038 ±38°C) to ensure steady state conditions have been achieved.

(5) If the calorimeter and thermocouples do not read within range, repeat steps in paragraphs 1 through 4 and adjust the burner air intake damper until the proper readings are obtained. The thermocouple rake and the calorimeter should be used frequently to maintain and record calibrated test parameters. Until the specific apparatus has demonstrated consistency, each test should be calibrated. After consistency has been confirmed, several tests may be conducted with the pre-test calibration before and a calibration check after the series.

**(g)** *Test Procedure.* The flammability of each set of specimens must be tested as follows:

(1) Record the weight of each set of seat bottom and seat back cushion specimens to be tested to the nearest 0.02 pound (9 grams).

(2) Mount the seat bottom and seat back cushion test specimens on the test stand as shown in Figure 2, securing the seat back cushion specimen to the test stand at the top.

(3) Swing the burner into position and ensure that the distance from the exit of the burner cone to the side of the seat bottom cushion specimen is $4 \pm 1/8$ inches ($102 \pm 3$ mm).

(4) Swing the burner away from the test position. Turn on the burner and allow it to run for 2 minutes to provide adequate warmup of the burner cone and flame stabilization.

(5) To begin the test, swing the burner into the test position and simultaneously start the timing device.

(6) Expose the seat bottom cushion specimen to the burner flame for 2 minutes and then turn off the burner. Immediately swing the burner away from the test position. Terminate test 7 minutes after initiating cushion exposure to the flame by use of a gaseous extinguishing agent (i.e., Halon or $CO_2$).

(7) Determine the weight of the remains of the seat cushion specimen set left on the mounting stand to the nearest 0.02 pound (9 grams) excluding all droppings.

**(h)** *Test Report.* With respect to all specimen sets tested for a particular seat cushion for which testing of compliance is performed, the following information must be recorded:

(1) An identification and description of the specimens being tested.

(2) The number of specimen sets tested.

(3) The initial weight and residual weight of each set, the calculated percentage weight loss of each set, and the calculated average percentage weight loss for the total number of sets tested.

(4) The burn length for each set tested.

**Figure 1**

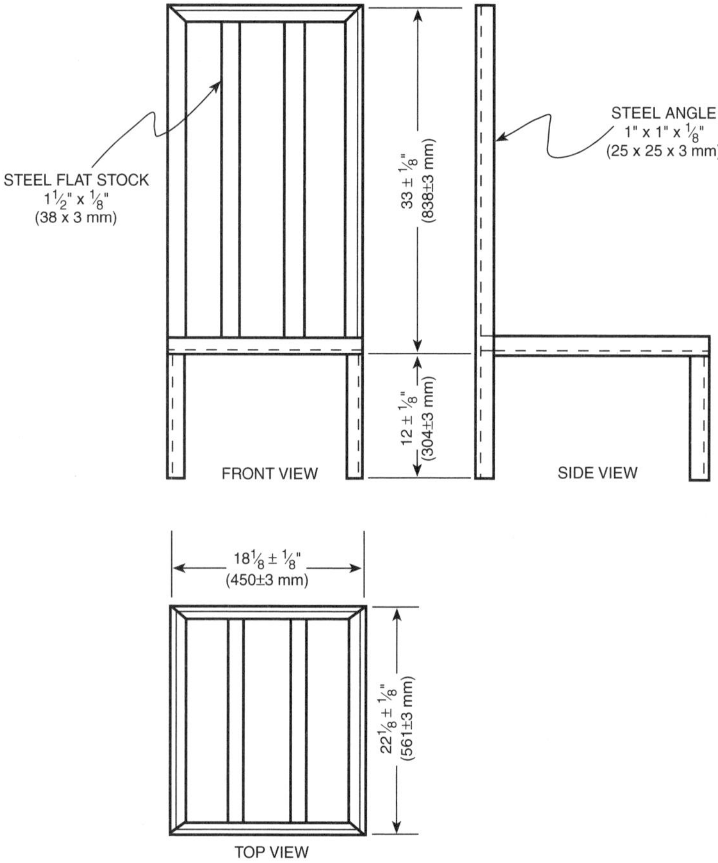


NOTE:

ALL JOINTS WELDED
FLAT STOCK BUTT WELDED
ALL MEASUREMENTS INSIDE

Figure 2

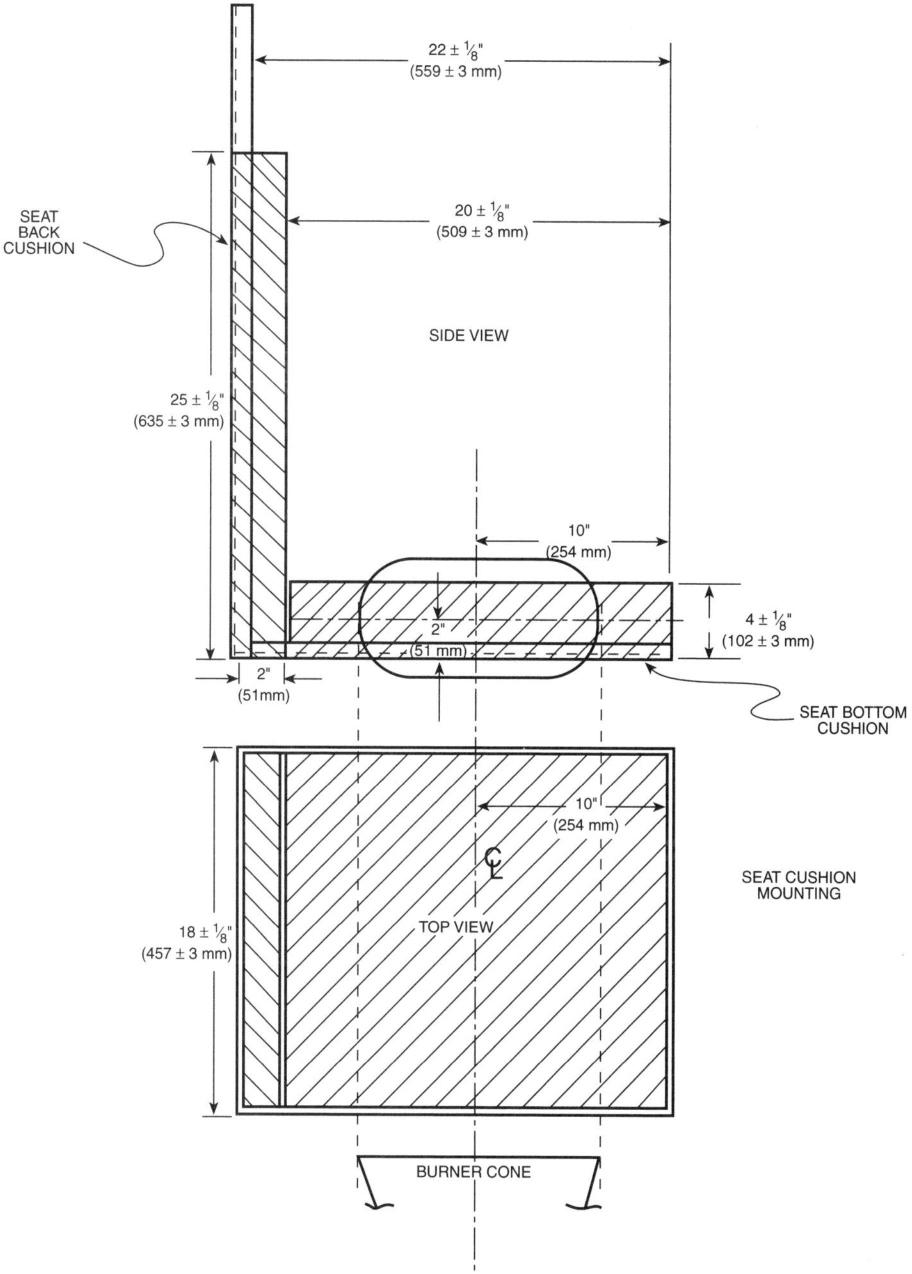
22 ± 1/8"
(559 ± 3 mm)
SEAT BACK CUSHION
20 ± 1/8"
(509 ± 3 mm)
SIDE VIEW
25 ± 1/8"
(635 ± 3 mm)
10"
(254 mm)
2"
(51 mm)
4 ± 1/8"
(102 ± 3 mm)
2"
(51mm)
SEAT BOTTOM CUSHION
10"
(254 mm)
SEAT CUSHION MOUNTING
18 ± 1/8"
(457 ± 3 mm)
TOP VIEW
BURNER CONE

Figure 3

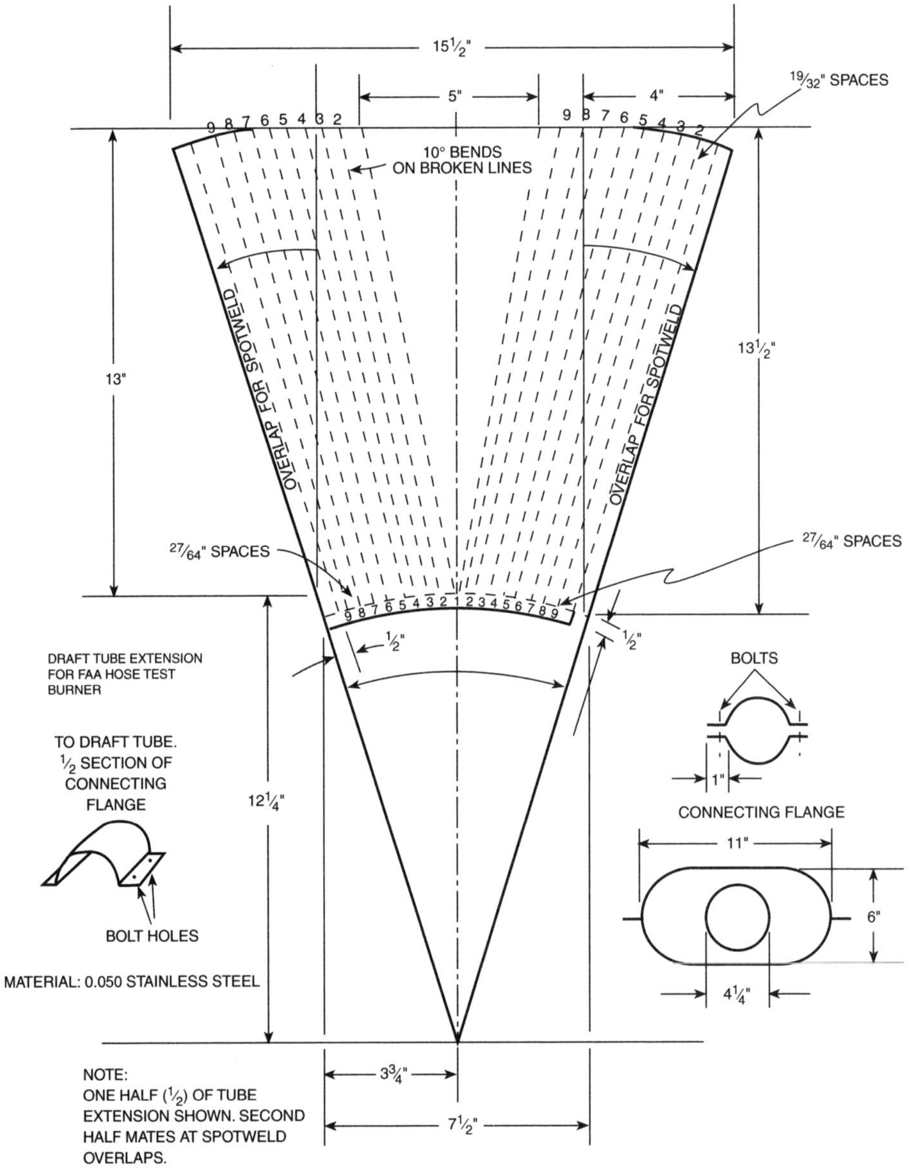
15½"
5"
4"
19/32" SPACES
9 8 7 6 5 4 3 2
9 8 7 6 5 4 3 2
10° BENDS
ON BROKEN LINES
OVERLAP FOR SPOTWELD
OVERLAP FOR SPOTWELD
13"
13½"
27/64" SPACES
27/64" SPACES
9 8 7 6 5 4 3 2 1 2 3 4 5 6 7 8 9
½"
½"
DRAFT TUBE EXTENSION
FOR FAA HOSE TEST
BURNER
TO DRAFT TUBE.
½ SECTION OF
CONNECTING
FLANGE
BOLT HOLES
12¼"
BOLTS
1"
CONNECTING FLANGE
11"
6"
4¼"
MATERIAL: 0.050 STAINLESS STEEL
3¾"
7½"
NOTE:
ONE HALF (½) OF TUBE
EXTENSION SHOWN. SECOND
HALF MATES AT SPOTWELD
OVERLAPS.

Figure 4

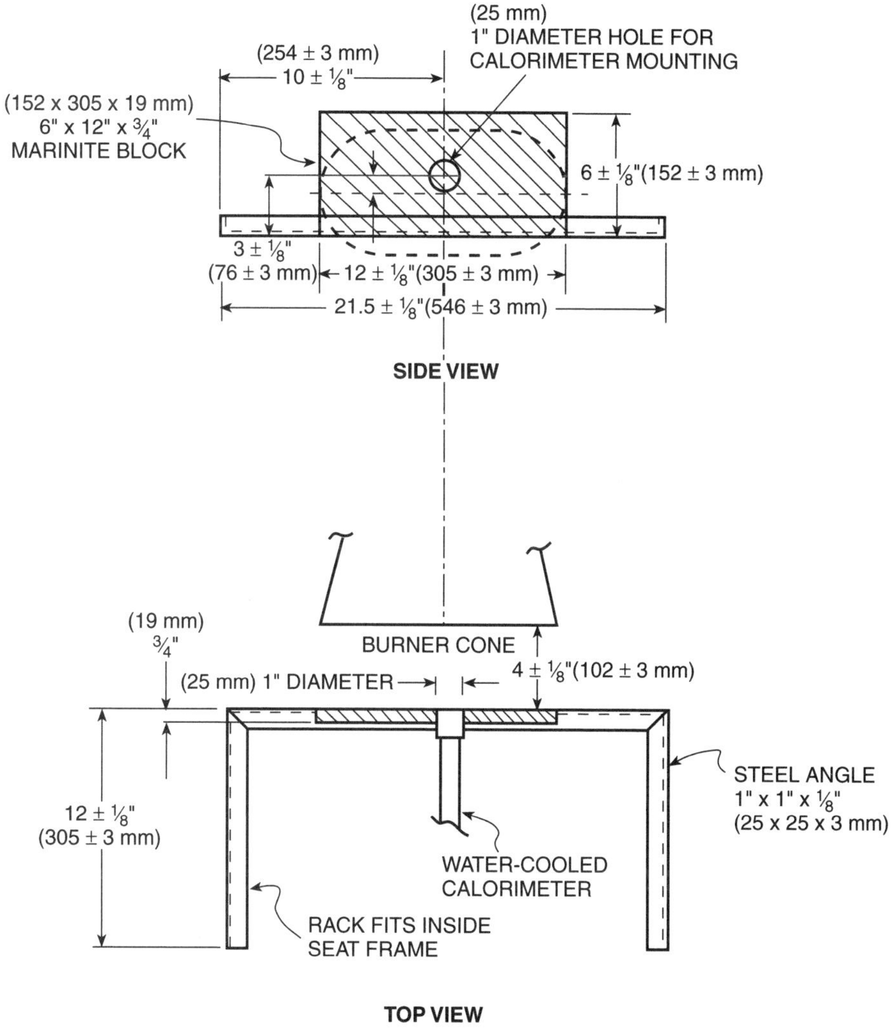


CALORIMETER BRACKET

**Figure 5**

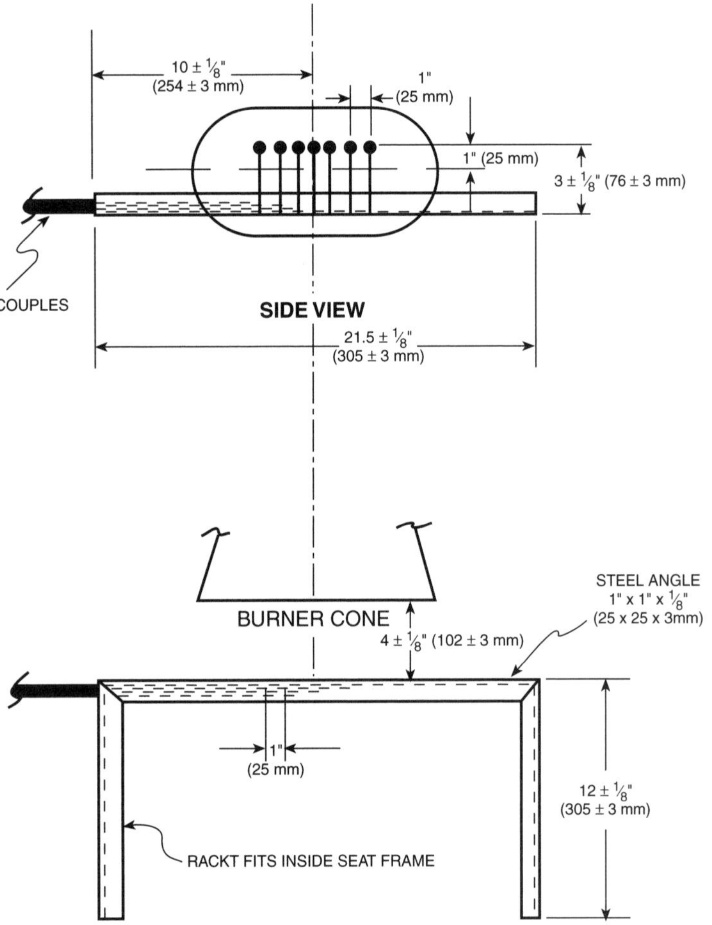


**TOP VIEW**
THERMOCOUPLE RAKE BRACKET

PART III — TEST METHOD TO DETERMINE FLAME PENETRATION RESISTANCE OF CARGO COMPARTMENT LINERS.

**(a)** *Criteria for Acceptance.*

(1) At least three specimens of cargo compartment sidewall or ceiling liner panels must be tested.

(2) Each specimen tested must simulate the cargo compartment sidewall or ceiling liner panel, including any design features, such as joints, lamp assemblies, etc., the failure of which would affect the capability of the liner to safely contain a fire.

(3) There must be no flame penetration of any specimen within 5 minutes after application of the flame source, and the peak temperature measured at 4 inches above the upper surface of the horizontal test sample must not exceed 400°F.

**(b)** *Summary of Method.* This method provides a laboratory test procedure for measuring the capability of cargo compartment lining materials to resist flame penetration with a 2 gallon per hour (GPH) #2 Grade kerosene or equivalent burner fire source. Ceiling and sidewall liner panels may be tested individually provided a baffle is used to simulate the missing panel. Any specimen that passes the test as a ceiling liner panel may be used as a sidewall liner panel.

**(c)** *Test Specimens.*

(1) The specimen to be tested must measure 16±1⁄8 inches (406±3 mm) by 24+1⁄8 inches (610±3 mm).

(2) The specimens must be conditioned at 70°F ±5°F (21°C ±2°C) and 55% ±5% humidity for at least 24 hours before testing.

**(d)** *Test Apparatus.* The arrangement of the test apparatus, which is shown in Figure 3 of Part II and Figures 1 through 3 of this part of Appendix F, must include the components described in this section. Minor details of the apparatus may vary, depending on the model of the burner used.

(1) *Specimen Mounting Stand.* The mounting stand for the test specimens consists of steel angles as shown in Figure 1.

(2) *Test Burner.* The burner to be used in testing must—

(i) Be a modified gun type.

(ii) Use a suitable nozzle and maintain fuel pressure to yield a 2 GPH fuel flow. For example: an 80 degree nozzle nominally rated at 2.25 GPH and operated at 85 pounds per square inch (PSI) gage to deliver 2.03 GPH.

(iii) Have a 12 inch (305 mm) burner extension installed at the end of the draft tube with an opening 6 inches (152 mm) high and 11 inches (280 mm) wide as shown in Figure 3 of Part II of this appendix.

(iv) Have a burner fuel pressure regulator that is adjusted to deliver a nominal 2.0 GPH of #2 Grade kerosene or equivalent.

Burner models which have been used successfully in testing are the Lenox Model OB-32, Carlin Model 200 CRD and Park Model DPL. The basic burner is described in FAA Powerplant Engineering Report No. 3A, Standard Fire Test Apparatus and Procedure for Flexible Hose Assemblies, dated March 1978; however, the test settings specified in this appendix differ in some instances from those specified in the report.

(3) *Calorimeter.*

(i) The calorimeter to be used in testing must be a total heat flux Foil Type Gardon Gage of an appropriate range (approximately 0 to 15.0 British thermal unit (BTU) per ft.$^2$ sec., 0-17.0 watts/cm$^2$). The calorimeter must be mounted in a 6 inch by 12 inch (152 by 305 mm) by 3⁄4 inch (19 mm) thick insulating block which is attached to a steel angle bracket for placement in the test stand during burner calibration as shown in Figure 2 of this part of this appendix.

(ii) The insulating block must be monitored for deterioration and the mounting shimmed as necessary to ensure that the calorimeter face is parallel to the exit plane of the test burner cone.

(4) *Thermocouples.* The seven thermocouples to be used for testing must be 1⁄16 inch ceramic sheathed, type K, grounded thermocouples with a nominal 30 American wire gage (AWG) size conductor. The seven thermocouples must be attached to a steel angle bracket to form a thermocouple rake for placement in the test stand during burner calibration as shown in Figure 3 of this part of this appendix.

(5) *Apparatus Arrangement.* The test burner must be mounted on a suitable stand to position the exit of the burner cone a distance of 8 inches from the ceiling liner panel and 2 inches from the sidewall liner panel. The burner stand should have the capability of allowing the burner to be swung away from the test specimen during warm-up periods.

(6) *Instrumentation.* A recording potentiometer or other suitable instrument with an appropriate range must be used to measure and record the outputs of the calorimeter and the thermocouples.

(7) *Timing Device.* A stopwatch or other device must be used to measure the time of flame application and the time of flame penetration, if it occurs.

**(e)** *Preparation of Apparatus.* Before calibration, all equipment must be turned on and allowed to stabilize, and the burner fuel flow must be adjusted as specified in paragraph (d)(2).

**(f)** *Calibration.* To ensure the proper thermal output of the burner the following test must be made:

(1) Remove the burner extension from the end of the draft tube. Turn on the blower portion of the

burner without turning the fuel or igniters on. Measure the air velocity using a hot wire anemometer in the center of the draft tube across the face of the opening. Adjust the damper such that the air velocity is in the range of 1550 to 1800 ft./min. If tabs are being used at the exit of the draft tube, they must be removed prior to this measurement. Reinstall the draft tube extension cone.

(2) Place the calorimeter on the test stand as shown in Figure 2 at a distance of 8 inches (203 mm) from the exit of the burner cone to simulate the position of the horizontal test specimen.

(3) Turn on the burner, allow it to run for 2 minutes for warm-up, and adjust the damper to produce a calorimeter reading of 8.0 ±0.5 BTU per ft.$^2$ sec. (9.1 ± 0.6 Watts/cm$^2$).

(4) Replace the calorimeter with the thermocouple rake (see Figure 3).

(5) Turn on the burner and ensure that each of the seven thermocouples reads 1700°F ±100°F (927°C ±38°C.) to ensure steady state conditions have been achieved. If the temperature is out of this range, repeat steps 2 through 5 until proper readings are obtained.

(6) Turn off the burner and remove the thermocouple rake.

(7) Repeat (1) to ensure that the burner is in the correct range.

**(g)** *Test Procedure.*

(1) Mount a thermocouple of the same type as that used for calibration at a distance of 4 inches (102 mm) above the horizontal (ceiling) test specimen. The thermocouple should be centered over the burner cone.

(2) Mount the test specimen on the test stand shown in Figure 1 in either the horizontal or vertical position. Mount the insulating material in the other position.

(3) Position the burner so that flames will not impinge on the specimen, turn the burner on, and allow it to run for 2 minutes. Rotate the burner to apply the flame to the specimen and simultaneously start the timing device.

(4) Expose the test specimen to the flame for 5 minutes and then turn off the burner. The test may be terminated earlier if flame penetration is observed.

(5) When testing ceiling liner panels, record the peak temperature measured 4 inches above the sample.

(6) Record the time at which flame penetration occurs if applicable.

**(h)** *Test Report.* The test report must include the following:

(1) A complete description of the materials tested including type, manufacturer, thickness, and other appropriate data.

(2) Observations of the behavior of the test specimens during flame exposure such as delamination, resin ignition, smoke, etc., including the time of such occurrence.

(3) The time at which flame penetration occurs, if applicable, for each of the three specimens tested.

(4) Panel orientation (ceiling or sidewall).

*[SEE FIGURES BEGINNING ON THE NEXT PAGE]*

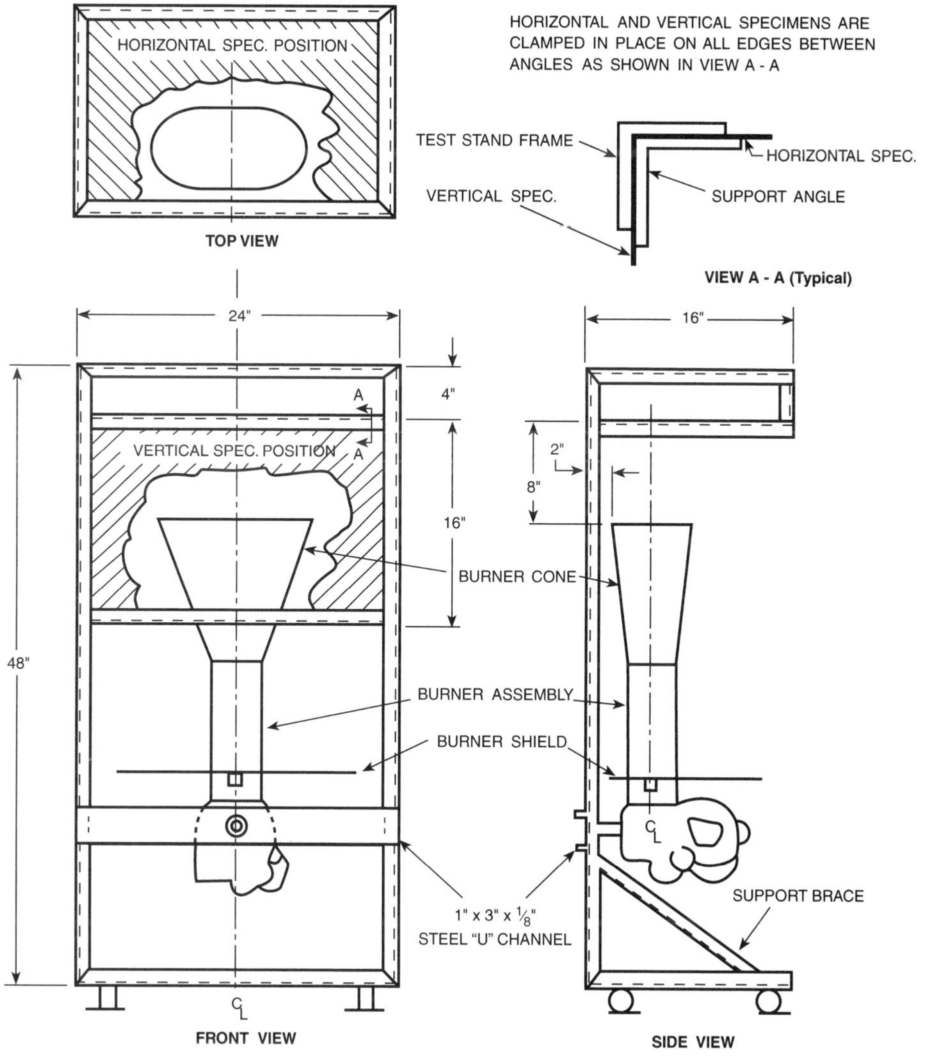


TEST STAND IS CONSTRUCTED WITH 1" x 1" x 1/8" STEEL ANGLES, ALL JOINTS WELDED
SUPPORT ANGLES ARE 1" x 1" x 1/8" CUT TO FIT

FIGURE 1. Test apparatus for horizontal and vertical mounting.

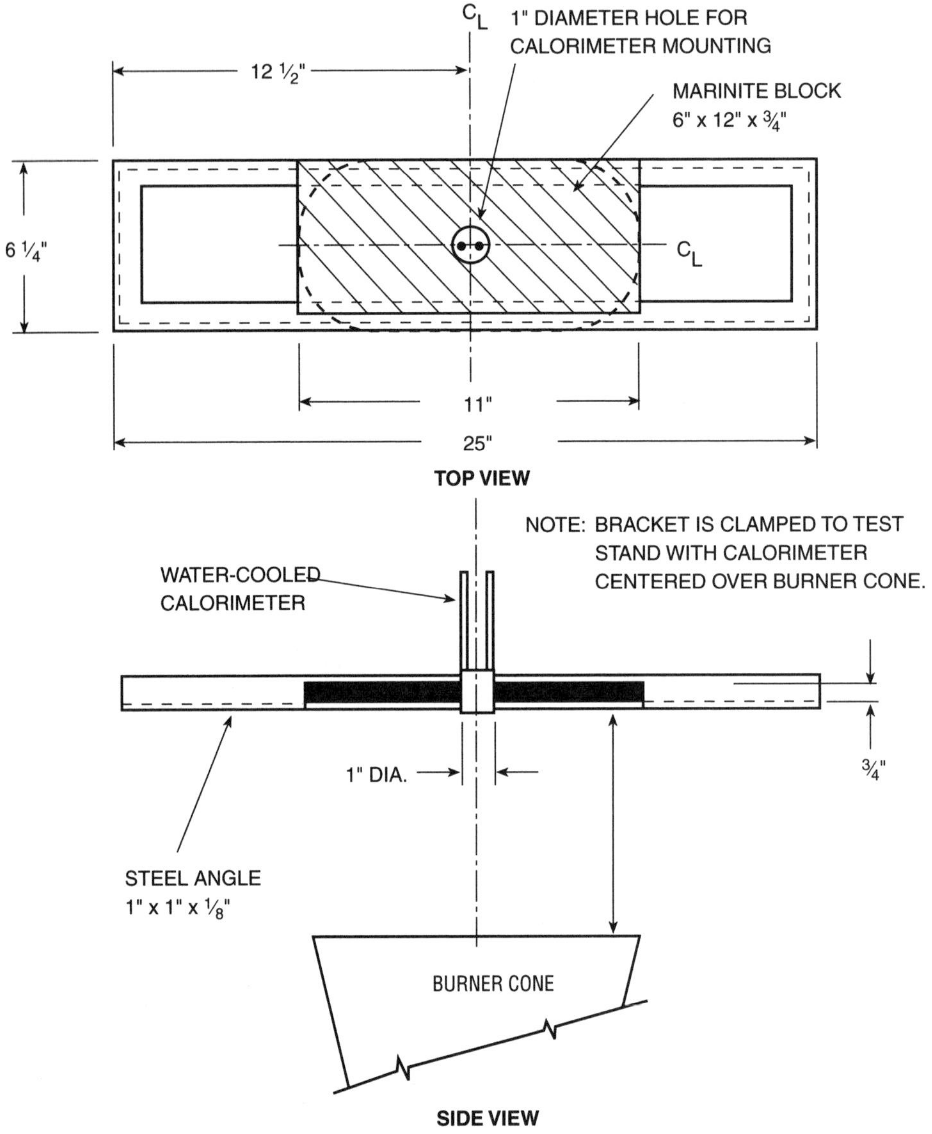


FIGURE 2. Calorimeter bracket

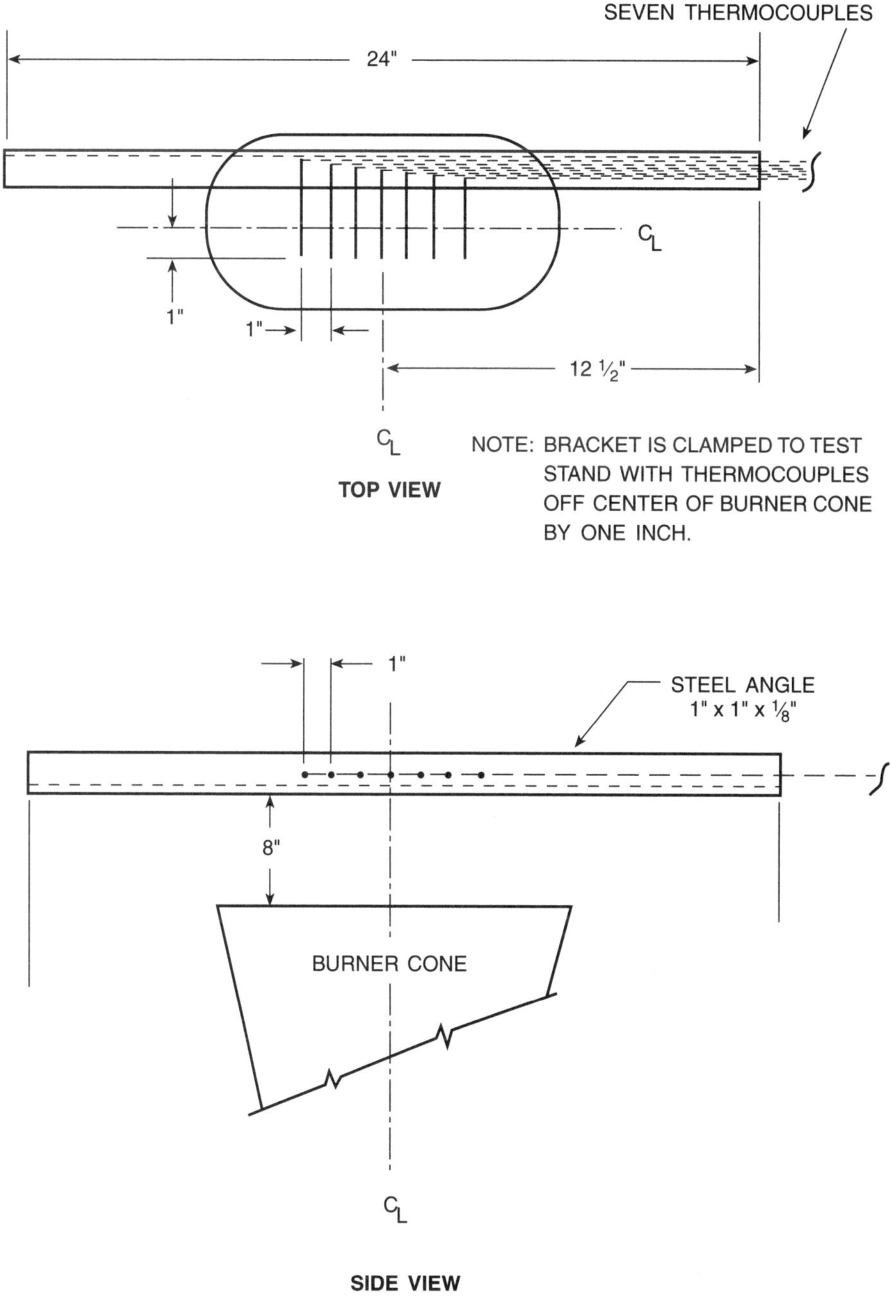


FIGURE 3. Thermocouple rake bracket

*PART IV—TEST METHOD TO DETERMINE THE HEAT RELEASE RATE FROM CABIN MATERIALS EXPOSED TO RADIANT HEAT.*

**(a)** *Summary of Method.* Three or more specimens representing the completed aircraft component are tested. Each test specimen is injected into an environmental chamber through which a constant flow of air passes. The specimen's exposure is determined by a radiant heat source adjusted to produce, on the specimen, the desired total heat flux of 3.5 W/cm$^2$. The specimen is tested with the exposed surface vertical. Combustion is initiated by piloted ignition. The combustion products leaving the chamber are monitored in order to calculate the release rate of heat.

**(b)** *Apparatus.* The Ohio State University (OSU) rate of heat release apparatus, as described below, is used. This is a modified version of the rate of heat release apparatus standardized by the American Society of Testing and Materials (ASTM), ASTM E-906.

(1) This apparatus is shown in Figures 1A and 1B of this part IV. All exterior surfaces of the apparatus, except the holding chamber, must be insulated with 1 inch (25 mm) thick, low density, high temperature, fiberglass board insulation. A gasketed door, through which the sample injection rod slides, must be used to form an airtight closure on the specimen hold chamber.

(2) *Thermopile.* The temperature difference between the air entering the environmental chamber and that leaving must be monitored by a thermopile having five hot, and five cold, 24-gauge Chromel-Alumel junctions. The hot junctions must be spaced across the top of the exhaust stack, .38 inches (10 mm) below the top of the chimney. The thermocouples must have a .050 ±.010 inch (1.3 ±.3 mm) diameter, ball-type, welded tip. One thermocouple must be located in the geometric center, with the other four located 1.18 inch (30 mm) from the center along the diagonal toward each of the corners (Figure 5 of this part IV). The cold junctions must be located in the pan below the lower air distribution plate (see paragraph (b)(4) of this part IV). Thermopile hot junctions must be cleared of soot deposits as needed to maintain the calibrated sensitivity.

(3) *Radiation Source.* A radiant heat source incorporating four Type LL silicon carbide elements, 20 inches (508 mm) long by .63 inch (16 mm) O.D., must be used, as shown in Figures 2A and 2B of this part IV. The heat source must have a nominal resistance of 1.4 ohms and be capable of generating a flux up to 100 kW/m$^2$. The silicone carbide elements must be mounted in the stainless steel panel box by inserting them through .63 inch (16 mm) holes in .03 inch (1 mm) thick ceramic fiber or calcium-silicate millboard. Locations of the holes in the pads and stainless steel cover plates are shown in Figure 2B of this part IV. The truncated diamond-shaped mask of .042 ±.002 inch (1.07 ±.05 mm) stainless steel must be added to provide uniform heat flux density over the area occupied by the vertical sample.

(4) *Air Distribution System.* The air entering the environmental chamber must be distributed by a .25 inch (6.3 mm) thick aluminum plate having eight No. 4 drill-holes, located 2 inches (51 mm) from sides on 4 inch (102 mm) centers, mounted at the base of the environmental chamber. A second plate of 18 gauge stainless steel having 120, evenly spaced, No. 28 drill holes must be mounted 6 inches (152 mm) above the aluminum plate. A well-regulated air supply is required. The air-supply manifold at the base of the pyramidal section must have 48, evenly spaced, No. 26 drill holes located .38 inch (10 mm) from the inner edge of the manifold, resulting in an airflow split of approximately three to one within the apparatus.

(5) *Exhaust Stack.* An exhaust stack, 5.25 x 2.75 inches (133 x 70 mm) in cross section, and 10 inches (254 mm) long, fabricated from 28 gauge stainless steel must be mounted on the outlet of the pyramidal section. A 1.0 x 3.0 inch (25 x 76 mm) baffle plate of 0.18 ±.002 inch (.50 ±.05 mm) stainless steel must be centered inside the stack, perpendicular to the air flow, 3 inches (76 mm) above the base of the stack.

(6) *Specimen Holders.* (i) The specimen must be tested in a vertical orientation. The specimen holder (Figure 3 of this part IV) must incorporate a frame that touches the specimen (which is wrapped with aluminum foil as required by paragraph (d)(3) of this Part) along only the .25 inch (6 mm) perimeter. A "V" shaped spring is used to hold the assembly together. A detachable .50 x .50 x 5.91 inch (12 x 12 x 150 mm) drip pan and two .020 inch (.5 mm) stainless steel wires (as shown in Figure 3 of this part IV) must be used for testing materials prone to melting and dripping. The positioning of the spring and frame may be changed to accommodate different specimen thicknesses by inserting the retaining rod in different holes on the specimen holder.

(ii) Since the radiation shield described in ASTM E-906 is not used, a guide pin must be added to the injection mechanism. This fits into a slotted metal plate on the injection mechanism outside of the holding chamber. It can be used to provide accurate positioning of the specimen face after injection. The front surface of the specimen must be 3.9 inches (100 mm) from the closed radiation doors after injection.

(iii) The specimen holder clips onto the mounted bracket (Figure 3 of this part IV). The mounting bracket must be attached to the injection rod by three screws that pass through a wide-area washer welded onto a ½-inch (13 mm) nut. The end of the injection rod must be threaded to screw into the nut, and a .020 inch (5.1 mm) thick wide area washer must be held between two ½-inch (13 mm) nuts that are adjusted to tightly cover the hole in the radiation doors through which the injection rod or calibration calorimeter pass.

(7) *Calorimeter.* A total-flux type calorimeter must be mounted in the center of a ½-inch Kaowool "M" board inserted in the sample holder to measure the total heat flux. The calorimeter must have a view angle of 180 degrees and be calibrated for incident flux. The calorimeter calibration must be acceptable to the Administrator.

(8) *Pilot-Flame Positions.* Pilot ignition of the specimen must be accomplished by simultaneously exposing the specimen to a lower pilot burner and an upper pilot burner, as described in paragraph (b)(8)(i) and (b)(8)(ii) or (b)(8)(iii) of this part IV, respectively. Since intermittent pilot flame extinguishment for more than 3 seconds would invalidate the test results, a spark ignitor may be installed to ensure that the lower pilot burner remains lighted.

(i) *Lower Pilot Burner.* The pilot-flame tubing must be .25 inch (6.3 mm) O.D., .03 inch (0.8 mm) wall, stainless steel tubing. A mixture of 120 $cm^3$/min. of methane and 850 $cm^3$/min. of air must be fed to the lower pilot flame burner. The normal position of the end of the pilot burner tubing is .40 inch (10 mm) from and perpendicular to the exposed vertical surface of the specimen. The centerline at the outlet of the burner tubing must intersect the vertical centerline of the sample at a point .20 inch (5 mm) above the lower exposed edge of the specimen.

(ii) *Standard Three-Hole Upper Burner.* The pilot burner must be a straight length of .25 inch (6.3 mm) O.D., .03 inch (0.8 mm) wall, stainless steel tubing that is 14 inches (360 mm) long. One end of the tubing must be closed, and three No. 40 drill holes must be drilled into the tubing, 2.38 inch (60 mm) apart, for gas ports, all radiating in the same direction. The first hole must be .19 inch (5 mm) from the closed end of the tubing. The tube must be positioned .75 inch (19 mm) above and .75 inch (19 mm) behind the exposed upper edge of the specimen. The middle hole must be in the vertical plane perpendicular to the exposed surface of the specimen which passes through its vertical centerline and must be pointed toward the radiation source. The gas supplied to the burner must be methane and must be adjusted to produce flame lengths of 1 inch (25 mm).

(iii) *Optional Fourteen-Hole Upper Pilot Burner.* This burner may be used in lieu of the standard three-hole burner described in paragraph (b)(8)(ii) of this part IV. The pilot burner must be a straight length of .25 inch (6.3 mm) O.D., .03 inch (0.8 mm) wall, stainless steel tubing that is 15.75 inches (400 mm) long. One end of the tubing must be closed, and 14 No. 59 drill holes must be drilled into the tubing, .50 inch (13 mm) apart, for gas ports, all radiating in the same direction. The first hole must be .50 inch (13 mm) from the closed end of the tubing. The tube must be positioned above the specimen holder so that the holes are placed above the specimen as shown in Figure 1B of this part IV. The fuel supplied to the burner must be methane mixed with air in a ratio of approximately 50/50 by volume. The total gas flow must be adjusted to produce flame lengths of 1 inch (25 mm). When the gas/air ratio and the flow rate are properly adjusted, approximately .25 inch (6 mm) of the flame length appears yellow in color.

**(c)** *Calibration of Equipment.*

(1) *Heat Release Rate.* A calibration burner, as shown in Figure 4, must be placed over the end of the lower pilot flame tubing using a gas tight connection. The flow of gas to the pilot flame must be at least 99 percent methane and must be accurately metered. Prior to usage, the wet test meter must be properly leveled and filled with distilled water to the tip of the internal pointer while no gas is flowing. Ambient temperature and pressure of the water are based on the internal wet test meter temperature. A baseline flow rate of approximately 1 liter/min. must be set and increased to higher preset flows of 4, 6, 8, 6 and 4 liters/min. Immediately prior to recording methane flow rates, a flow rate of 8 liters/min. must be used for 2 minutes to precondition the chamber. This is not recorded as part of calibration. The rate must be determined by using a stopwatch to time a complete revolution of the wet test meter for both the baseline and higher flow, with the flow returned to baseline before changing to the next higher flow. The thermopile baseline voltage must be measured. The gas flow to the burner must be increased to the higher preset flow and allowed to burn for 2.0 minutes, and the thermopile voltage must be measured. The sequence must be repeated until all five values have been determined. The average of the five values must be used as the calibration factor. The procedure must be repeated if the percent relative standard deviation is greater than 5 percent. Calculations are shown in paragraph (f) of this part IV.

(2) *Flux Uniformity.* Uniformity of flux over the specimen must be checked periodically and after each heating element change to determine if it is within acceptable limits of plus or minus 5 percent.

(3) As noted in paragraph (b)(2) of this part IV, thermopile hot junctions must be cleared of soot deposits as needed to maintain the calibrated sensitivity.

**(d)** *Preparation of Test Specimens.*

(1) The test specimens must be representative of the aircraft component in regard to materials and construction methods. The standard size for

the test specimens is 5.91 ±.03 x 5.91 ±.03 inches (149 ±1 x 149 ±1 mm). The thickness of the specimen must be the same as that of the aircraft component it represents up to a maximum thickness of 1.75 inches (45 mm). Test specimens representing thicker components must be 1.75 inches (45 mm).

(2) *Conditioning.* Specimens must be conditioned as described in Part 1 of this appendix.

(3) *Mounting.* Each test specimen must be wrapped tightly on all sides of the specimen, except for the one surface that is exposed with a single layer of .001 inch (.025 mm) aluminum foil.

**(e)** *Procedure.*

(1) The power supply to the radiant panel must be set to produce a radiant flux of 3.5 ±.05 W/cm$^2$, as measured at the point the center of the specimen surface will occupy when positioned for the test. The radiant flux must be measured after the air flow through the equipment is adjusted to the desired rate.

(2) After the pilot flames are lighted, their position must be checked as described in paragraph (b)(8) of this part IV.

(3) Air flow through the apparatus must be controlled by a circular plate orifice located in a 1.5 inch (38.1 mm) I.D. pipe with two pressure measuring points, located 1.5 inches (38 mm) upstream and .75 inches (19 mm) downstream of the orifice plate. The pipe must be connected to a manometer set at a pressure differential of 7.87 inches (200 mm) of Hg. (See Figure 1B of this part IV.) The total air flow to the equipment is approximately .04 m$^3$/seconds. The stop on the vertical specimen holder rod must be adjusted so that the exposed surface of the specimen is positioned 3.9 inches (100 mm) from the entrance when injected into the environmental chamber.

(4) The specimen must be placed in the hold chamber with the radiation doors closed. The airtight outer door must be secured, and the recording devices must be started. The specimen must be retained in the hold chamber for 60 seconds, plus or minus 10 seconds, before injection. The thermopile "zero" value must be determined during the last 20 seconds of the hold period. The sample must not be injected before completion of the "Zero" value determination.

(5) When the specimen is to be injected, the radiation doors must be opened. After the specimen is injected into the environmental chamber, the radiation doors must be closed behind the specimen.

(6) [Reserved]

(7) Injection of the specimen and closure of the inner door marks time zero. A record of the thermopile output with at least one data point per second must be made during the time the specimen is in the environmental chamber.

(8) The test duration time is five minutes. The lower pilot burner and the upper pilot burner must remain lighted for the entire duration of the test, except that there may be intermittent flame extinguishment for periods that do not exceed 3 seconds. Furthermore, if the optional three-hole upper burner is used, at least two flamelets must remain lighted for the entire duration of the test, except that there may be intermittent flame extinguishment of all three flamelets for periods that do not exceed 3 seconds.

(9) A minimum of three specimens must be tested.

**(f)** *Calculations.*

(1) The calibration factor is calculated as follows:

$$K_h = \frac{(F_1 - F_0)}{(V_1 - V_0)} \times \frac{(210.8 - 22)k_{cal}}{\text{mole}} \times \frac{273}{T_a} \times \frac{P - P_V}{760} \times \frac{\text{mole CH4STP}}{22.41} \times \frac{\text{WATT min}}{.01433\ k_{cal}} \times \frac{kw}{1000w}$$

$F_0$ = flow of methane at baseline (1pm)
$F_1$ = higher preset flow of methane (1pm)
$V_0$ = thermopile voltage at baseline (mv)
$V_1$ = thermopile voltage at higher flow (mv)
$T_a$ = Ambient temperature (K)
P = Ambient pressure (mm Hg)
$P_v$ = Water vapor pressure (mm Hg)

(2) Heat release rates may be calculated from the reading of the thermopile output voltage at any instant of time as:

$$HRR = \frac{(V_m - V_b)Kn}{.02323\text{m}^2}$$

HRR = Heat release rate (kw/m$^2$)
$V_b$ = baseline voltage (mv)
$V_m$ = measured thermopile voltage (mv)
$K_h$ = calibration factor (kw/mv)

(3) The integral of the heat release rate is the total heat release as a function of time and is calculated by multiplying the rate by the data sampling frequency in minutes and summing the time from zero to two minutes.

**(g)** *Criteria.* The total positive heat release over the first two minutes of exposure for each of the three or more samples tested must be averaged, and the peak heat release rate for each of the samples must be averaged. The average total heat release must not exceed 65 kilowatt-minutes per square meter, and the average peak heat release rate must not exceed 65 kilowatts per square meter.

**(h)** *Report.* The test report must include the following for each specimen tested:

(1) Description of the specimen.

(2) Radiant heat flux to the specimen, expressed in $W/cm^2$.

(3) Data giving release rates of heat (in $kW/m^2$) as a function of time, either graphically or tabulated at intervals no greater than 10 seconds. The calibration factor ($k_n$) must be recorded.

(4) If melting, sagging, delaminating, or other behavior that affects the exposed surface area or the mode of burning occurs, these behaviors must be reported, together with the time at which such behaviors were observed.

(5) The peak heat release and the 2-minute integrated heat release rate must be reported.

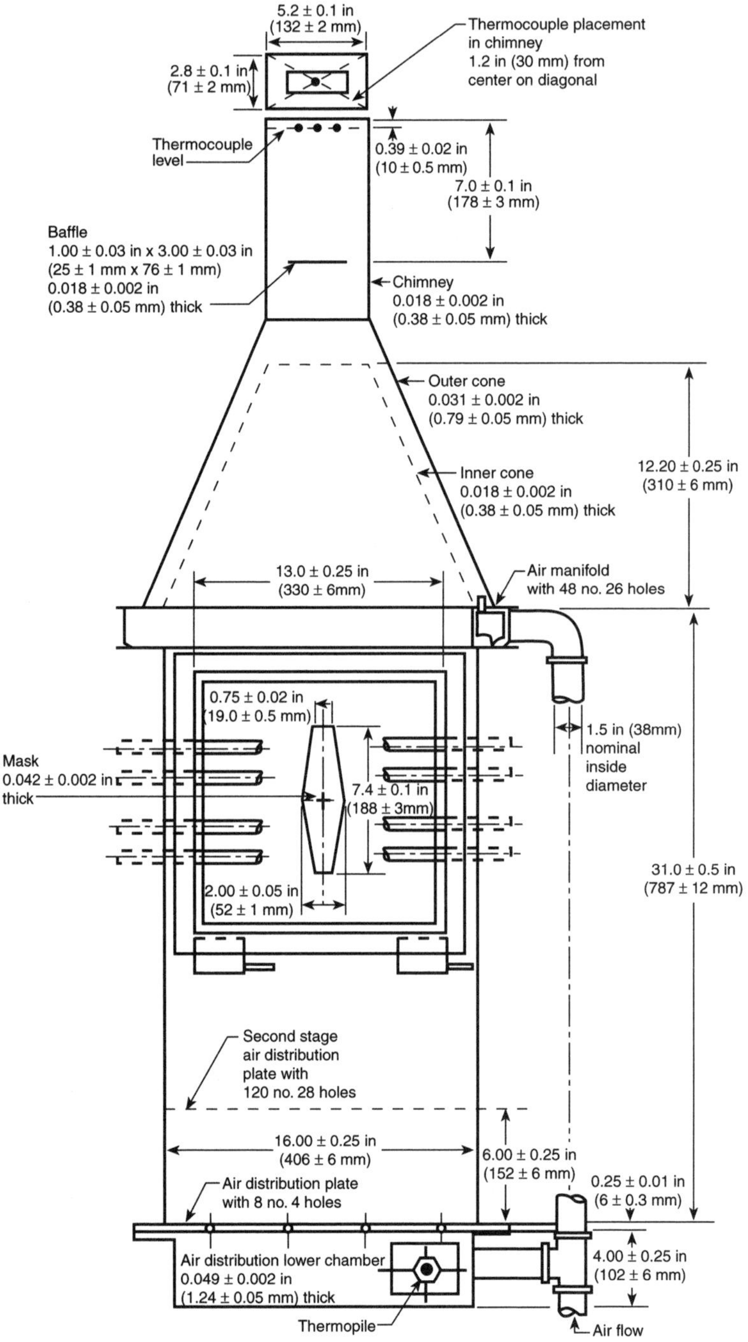


Figure 1A Rate of Heat Release Apparatus

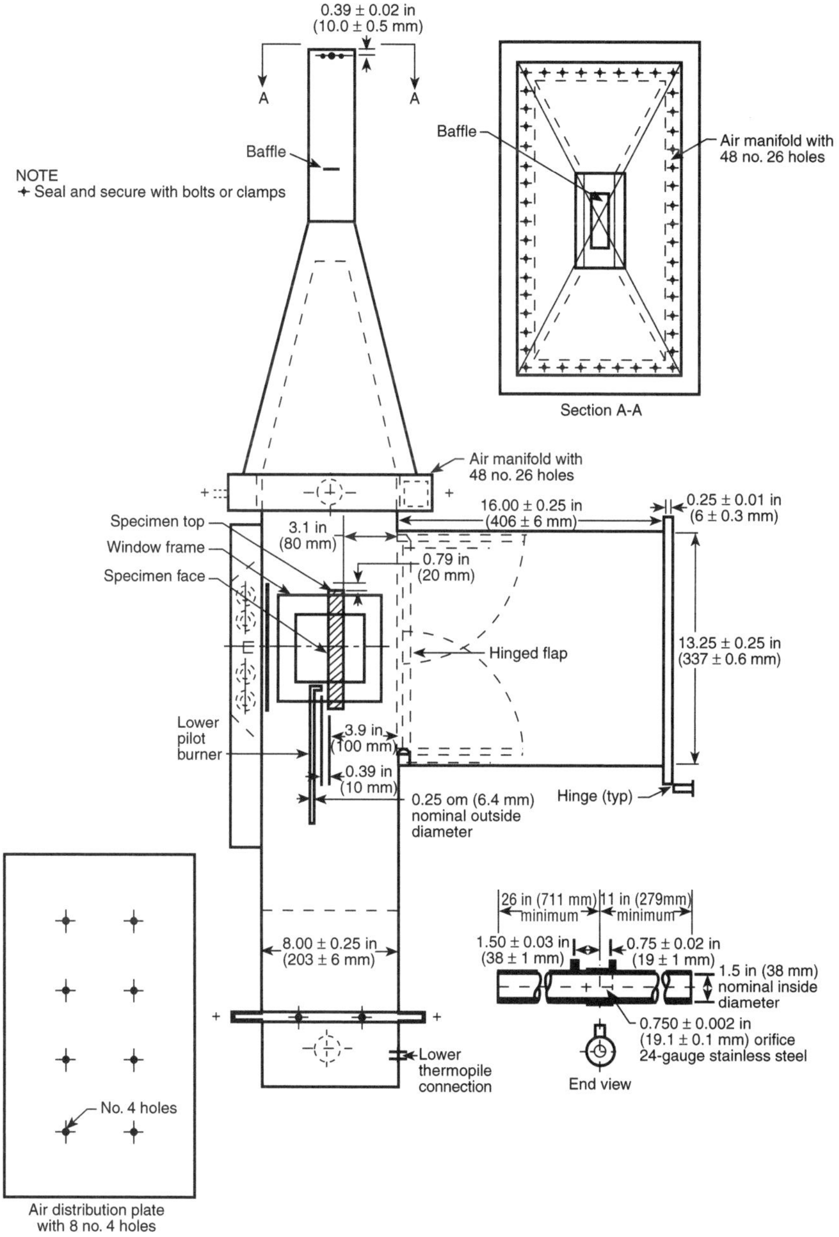


Figure 1B Rate of Heat Release Apparatus

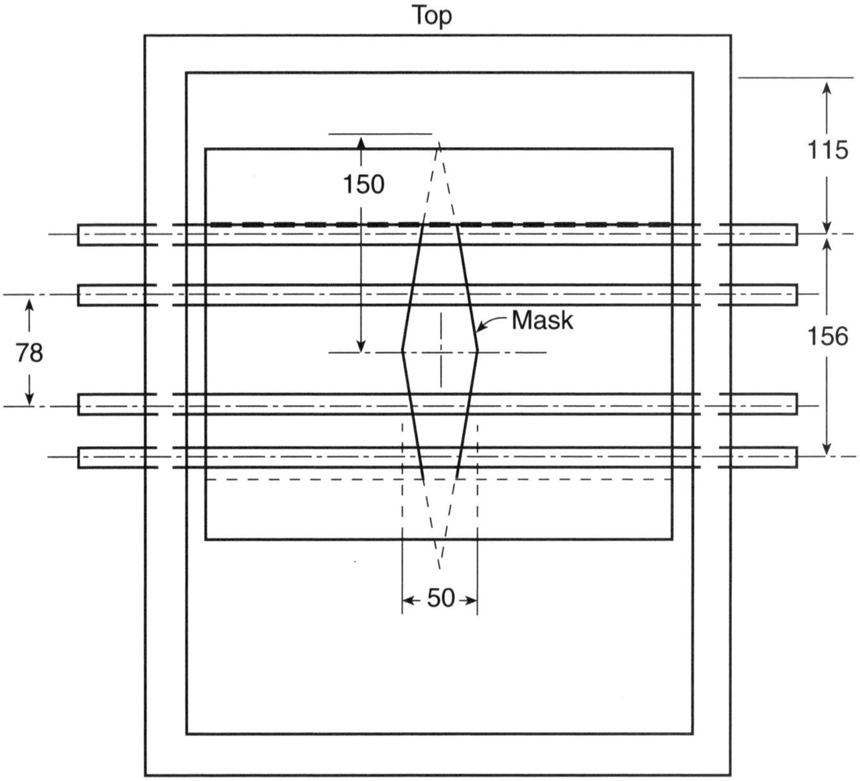


(Unless denoted otherwise all dimensions are in millimeters.)
Figure 2A. "Globar" Radiant Panel

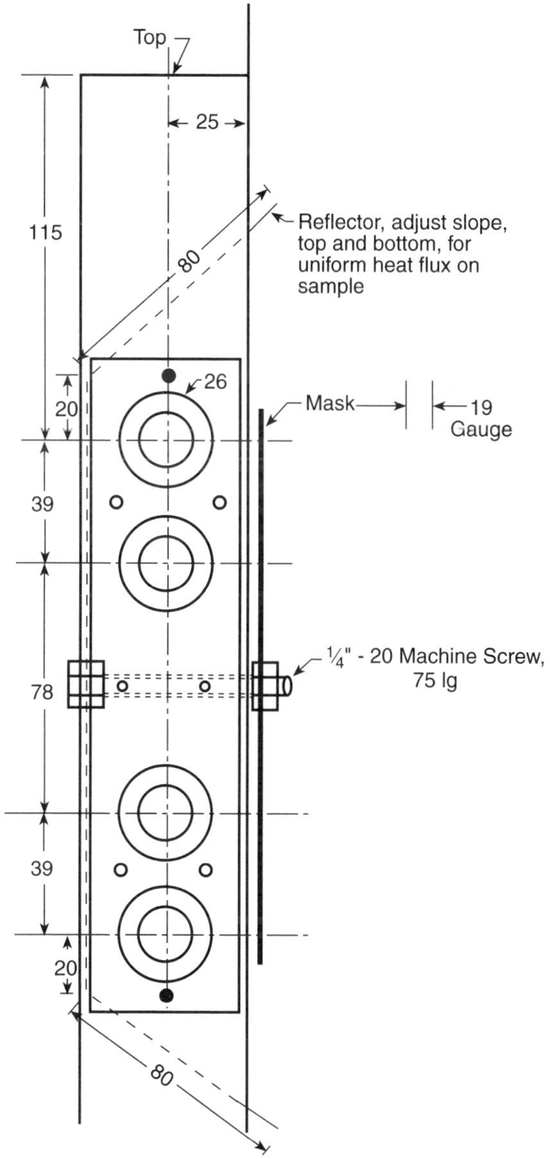


(Unless denoted otherwise all dimensions are in millimeters.)
Figure 2B. "Globar" Radiant Panel

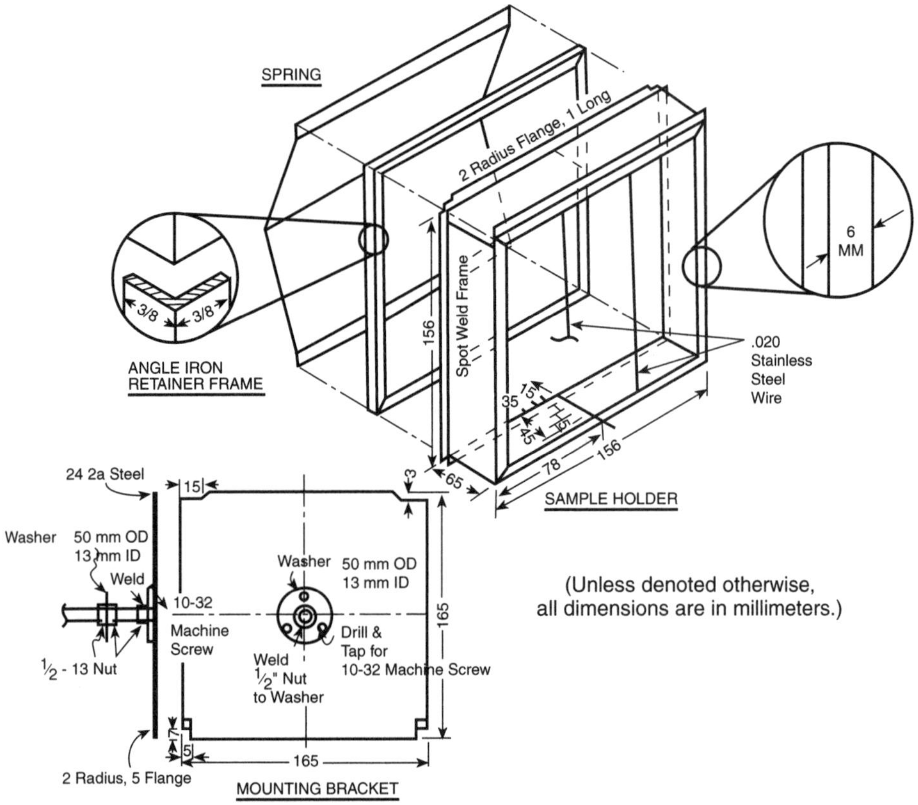


(Unless denoted otherwise, all dimensions are in millimeters.)

Figure 3.

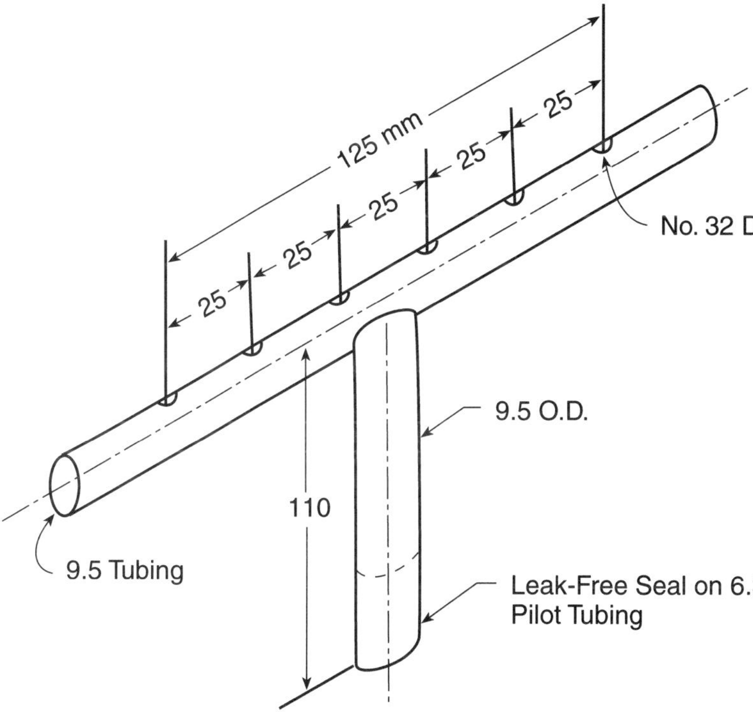


(Unless denoted otherwise, all dimensions are in millimeters.)

Figure 4.

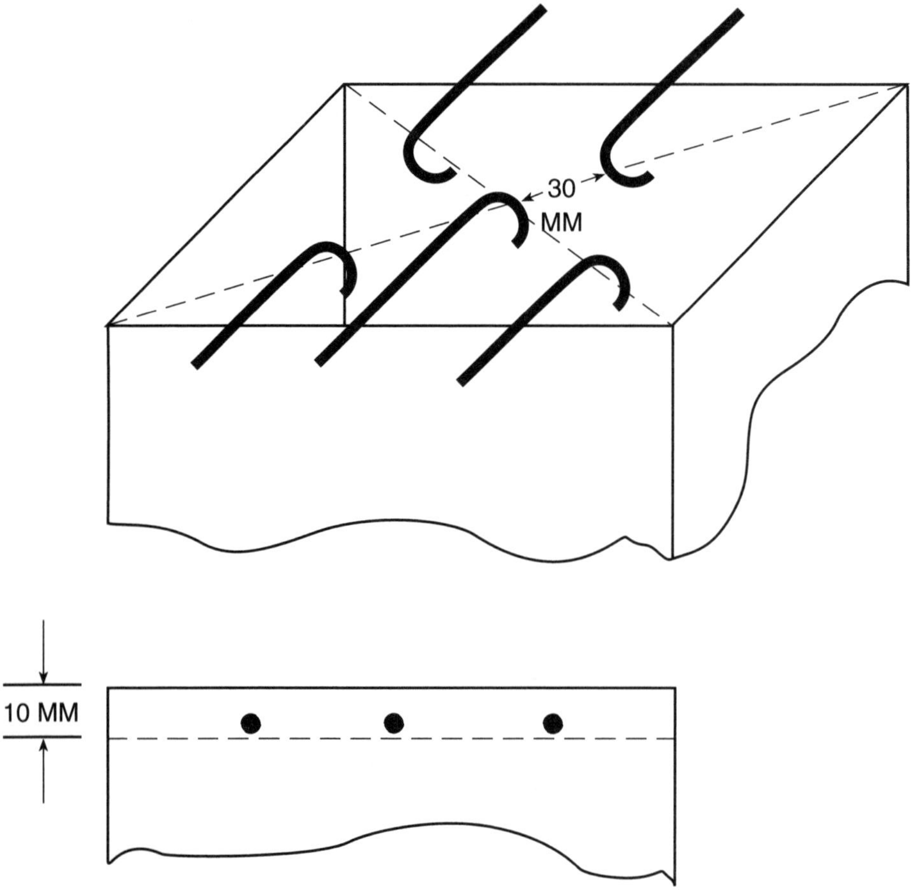


Figure 5. Thermocouple Position

### PART V — TEST METHOD TO DETERMINE THE SMOKE EMISSION CHARACTERISTICS OF CABIN MATERIALS

**(a)** *Summary of Method.* The specimens must be constructed, conditioned, and tested in the flaming mode in accordance with American Society of Testing and Materials (ASTM) Standard Test Method ASTM F814-83.

**(b)** *Acceptance Criteria.* The specific optical smoke density ($D_S$), which is obtained by averaging the reading obtained after 4 minutes with each of the three specimens, shall not exceed 200.

[Docket No. 5066, 29 FR 18291, Dec. 24, 1964; as amended by Amdt. 25–32, 37 FR 3972, Feb. 24, 1972; 37 FR 5284, March 14, 1972; Amdt. 25–55, 47 FR 13315, March 29, 1982; Amdt. 25–59, 49 FR 43193, Oct. 26, 1984; Amdt. 25–60, 51 FR 18243, May 16, 1986; Amdt. 25–61, 51 FR 26214, July 21, 1986; 51 FR 28322, Aug. 7, 1986; Amdt. 25–66, 53 FR 32573, Aug. 25, 1988; 53 FR 37542, 37671, Sept. 27, 1988; Amdt. 25–72, 55 FR 29787, July 20, 1990; Amdt. 25–83, 60 FR 6623, Feb. 2, 1995; Amdt. 25–83, 60 FR 11194, March 1, 1995; Amdt. 25–94, 63 FR 8848, Feb. 23, 1998]

## APPENDIX G TO PART 25

### CONTINUOUS GUST DESIGN CRITERIA

The continuous gust design criteria in this appendix must be used in establishing the dynamic response of the airplane to vertical and lateral continuous turbulence unless a more rational criteria is used. The following gust load requirements apply to mission analysis and design envelope analysis:

**(a)** The limit gust loads utilizing the continuous turbulence concept must be determined in accordance with the provisions of either paragraph (b) or paragraphs (c) and (d) of this appendix.

**(b)** *Design envelope analysis.* The limit loads must be determined in accordance with the following:

(1) All critical altitudes, weights, and weight distributions, as specified in §25.321(b), and all critical speeds within the ranges indicated in paragraph (b)(3) of this appendix must be considered.

(2) Values of $\bar{A}$8 (ratio of root-mean-square incremental load root-mean-square gust velocity) must be determined by dynamic analysis. The power spectral density of the atmospheric turbulence must be as given by the equation—

$$\phi(\Omega) = \sigma^2 L/\pi \quad \frac{1 + 8/3\,(1.339\,L\Omega)^2}{[1 + (1.339L\Omega)^2]^{11/6}}$$

where:

$\phi$ = power-spectral density (ft./sec.)$^2$/rad./ft.

$\sigma$ = root-mean-square gust velocity, ft./sec.

$\Omega$ = reduced frequency, radians per foot.

L = 2,500 ft.

(3) The limit loads must be obtained by multiplying the $\bar{A}$ values determined by the dynamic analysis by the following values of the gust velocity $U\sigma$:

(i) At speed $V_C$: $U\sigma$ = 85 fps true gust velocity in the interval 0 to 30,000 ft. altitude and is linearly decreased to 30 fps true gust velocity at 80,000 ft. altitude. Where the Administrator finds that a design is comparable to a similar design with extensive satisfactory service experience, it will be acceptable to select $U\sigma$ at $V_C$, less than 85 fps, but not less than 75 fps, with linear decrease from that value at 20,000 feet to 30 fps at 80,000 feet. The following factors will be taken into account when assessing comparability to a similar design:

(1) The transfer function of the new design should exhibit no unusual characteristics as compared to the similar design which will significantly affect response to turbulence; e.g., coalescence of modal response in the frequency regime which can result in a significant increase of loads.

(2) The typical mission of the new airplane is substantially equivalent to that of the similar design.

(3) The similar design should demonstrate the adequacy of the $U\sigma$ selected.

(ii) At speed $V_B$: $U\sigma$ is equal to 1.32 times the values obtained under paragraph (b)(3)(i) of this appendix.

(iii) At speed $V_D$: $U\sigma$ is equal to ½ the values obtained under paragraph (b)(3)(i) of this appendix.

(iv) At speeds between $V_B$ and $V_C$ and between $V_C$ and $V_D$: $U\sigma$ is equal to a value obtained by linear interpolation.

(4) When a stability augmentation system is included in the analysis, the effect of system nonlinearities on loads at the limit load level must be realistically or conservatively accounted for.

**(c)** *Mission analysis.* Limit loads must be determined in accordance with the following:

(1) The expected utilization of the airplane must be represented by one or more night profiles in which the load distribution and the variation with time of speed, altitude, gross weight, and center of gravity position are defined. These profiles must be divided into mission segments or blocks, for analysis, and average or effective values of the pertinent parameters defined for each segment.

(2) For each of the mission segments defined under paragraph (c)(1) of this appendix, values of $\bar{A}$ and $N_O$ must be determined by analysis. $\bar{A}$ is defined as the ratio of root-mean-square incremental load to root-mean-square gust velocity and $N_O$ is the radius of gyration of the load power spectral density function about zero frequency. The power spectral density of the atmospheric turbulence must be given by the equation set forth in paragraph (b)(2) of this appendix.

(3) For each of the load and stress quantities selected, the frequency of exceedance must be determined as a function of load level by means of the equation—

$$N_{(y)} = \Sigma t N_0 \left[ P_1 \exp\left( -\frac{|y - y_{one-g}|}{b_1 \bar{A}} \right) + P_2 \exp\left( \frac{|y - y_{one-g}|}{b_2 \bar{A}} \right) \right]$$

where:

t = selected time interval.

y = net value of the load or stress.

$Y_{one-g}$ = value of the load or stress in one-g level flight.

N(y) = average number of exceedances of the indicated value of the load or stress in unit time.

Σ = symbol denoting summation over all mission segments.

$N_0$, $\bar{A}$ = parameters determined by dynamic analysis as defined in paragraph (c)(2) of this appendix.

$P_1$, $P_2$, $b_1$, $b_2$ = parameters defining the probability distributions of root-mean-square gust velocity, to be read from Figures 1 and 2 of this appendix.

The limit gust loads must be read form the frequency of exceedance curves at a frequency of exceedance of 2 x $10^{-5}$ exceedances per hour. Both positive and negative load directions must be considered in determining the limit loads.

(4) If a stability augmentation system is utilized to reduce the gust loads, consideration must be given to the fraction of flight time that the system may be inoperative. The flight profiles of paragraph (c)(1) of this appendix must include flight with the system inoperative for this fraction of the flight time. When a stability augmentation system is included in the analysis, the effect of system nonlinearities on loads at the limit load level must be conservatively accounted for.

**(d)** *Supplementary design envelope analysis.* In addition to the limit loads defined by paragraph (c) of this appendix, limit loads must also be determined in accordance with paragraph (b) of this appendix, except that—

(1) In paragraph (b)(3)(i) of this appendix, the value of Uσ = 85 fps true gust velocity is replaced by Uσ = 60 fps true gust velocity on the interval 0 to 30,000 ft. altitude, and is linearly decreased to 25 fps true gust velocity at 80,000 ft. altitude; and

(2) In paragraph (b) of this appendix, the reference to paragraphs (b)(3)(i) through (b)(3)(iii) of this appendix is to be understood as referring to the paragraph as modified by paragraph (d)(1).

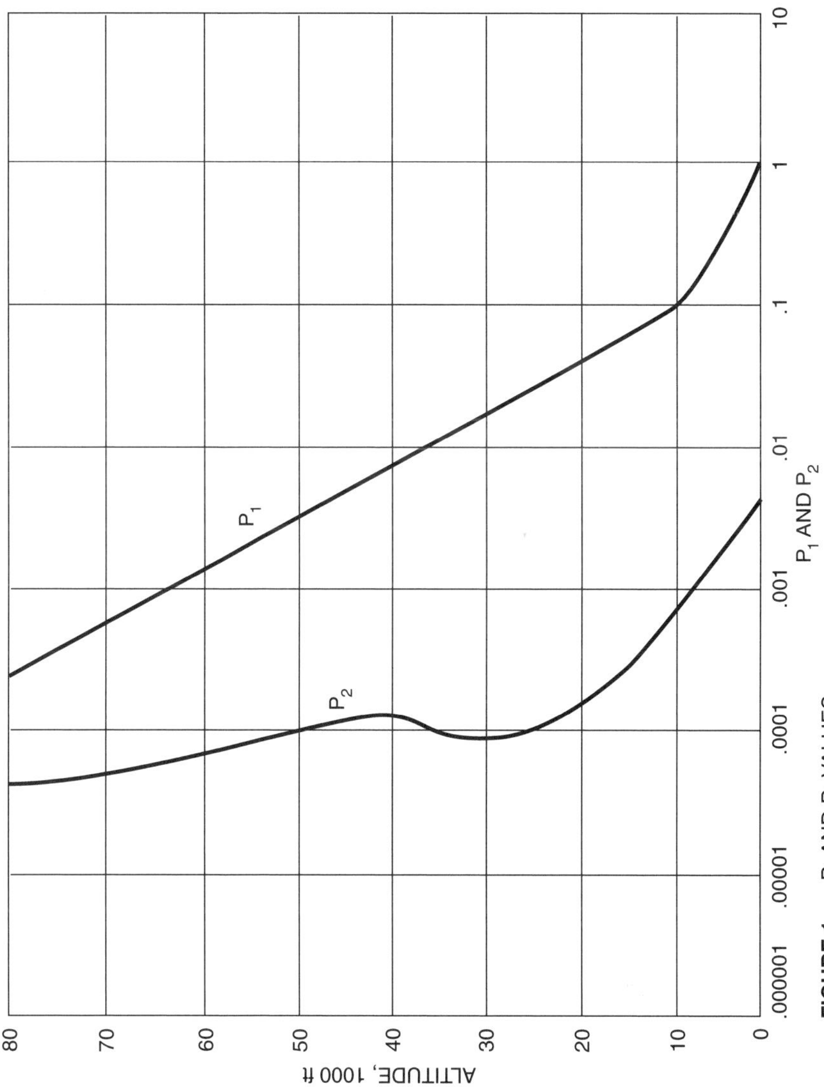


**FIGURE 1** $P_1$ AND $P_2$ VALUES

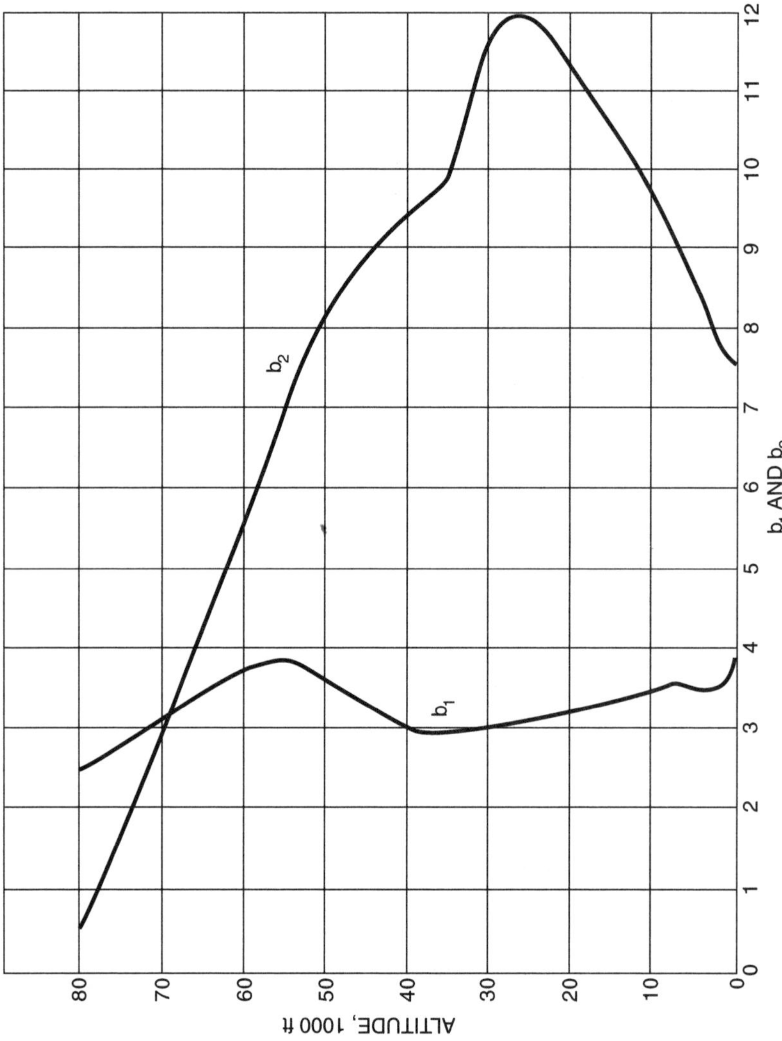


**FIGURE 2** $b_1$ AND $b_2$ VALUES

## APPENDIX H TO PART 25

### INSTRUCTIONS FOR CONTINUED AIRWORTHINESS

#### *H25.1 GENERAL.*

**(a)** This appendix specifies requirements for the preparation of Instructions for Continued Airworthiness as required by §25.1529.

**(b)** The Instructions for Continued Airworthiness for each airplane must include the Instructions for Continued Airworthiness for each engine and propeller (hereinafter designated "products"), for each appliance required by this chapter, and any required information relating to the interface of those appliances and products with the airplane. If Instructions for Continued Airworthiness are not supplied by the manufacturer of an appliance or product installed in the airplane, the Instructions for Continued Airworthiness for the airplane must include the information essential to the continued airworthiness of the airplane.

**(c)** The applicant must submit to the FAA a program to show how changes to the Instructions for Continued Airworthiness made by the applicant or by the manufacturers or products and appliances installed in the airplane will be distributed.

#### *H25.2 FORMAT.*

**(a)** The Instructions for Continued Airworthiness must be in the form of a manual or manuals as appropriate for the quantity of data to be provided.

**(b)** The format of the manual or manuals must provide for a practical arrangement.

#### *H25.3 CONTENT.*

The contents of the manual or manuals must be prepared in the English language. The Instructions for Continued Airworthiness must contain the following manuals or sections, as appropriate, and information:

**(a)** *Airplane maintenance manual or section.*

(1) Introduction information that includes an explanation of the airplane's features and data to the extent necessary for maintenance or preventive maintenance.

(2) A description of the airplane and its systems and installations including its engines, propellers, and appliances.

(3) Basic control and operation information describing how the airplane components and systems are controlled and how they operate, including any special procedures and limitations that apply.

(4) Servicing information that covers details regarding servicing points, capacities of tanks, reservoirs, types of fluids to be used, pressures applicable to the various systems, location of access panels for inspection and servicing, locations of lubrication points, lubricants to be used, equipment required for servicing, tow instructions and limitations, mooring, jacking, and leveling information.

**(b)** *Maintenance instructions.*

(1) Scheduling information for each part of the airplane and its engines, auxiliary power units, propellers, accessories, instruments, and equipment that provides the recommended periods at which they should be cleaned, inspected, adjusted, tested, and lubricated, and the degree of inspection, the applicable wear tolerances, and work recommended at these periods. However, the applicant may refer to an accessory, instrument, or equipment manufacturer as the source of this information if the applicant shows that the item has an exceptionally high degree of complexity requiring specialized maintenance techniques, test equipment, or expertise. The recommended overhaul periods and necessary cross references to the Airworthiness Limitations section of the manual must also be included. In addition, the applicant must include an inspection program that includes the frequency and extent of the inspections necessary to provide for the continued airworthiness of the airplane.

(2) Troubleshooting information describing probable malfunctions, how to recognize those malfunctions, and the remedial action for those malfunctions.

(3) Information describing the order and method of removing and replacing products and parts with any necessary precautions to be taken.

(4) Other general procedural instructions including procedures for system testing during ground running, symmetry checks, weighing and determining the center of gravity, lifting and shoring, and storage limitations.

**(c)** Diagrams of structural access plates and information needed to gain access for inspections when access plates are not provided.

**(d)** Details for the application of special inspection techniques including radiographic and ultrasonic testing where such processes are specified.

**(e)** Information needed to apply protective treatments to the structure after inspection.

**(f)** All data relative to structural fasteners such as identification, discard recommendations, and torque values.

**(g)** A list of special tools needed.

#### *H25.4 AIRWORTHINESS LIMITATIONS SECTION.*

**(a)** The Instructions for Continued Airworthiness must contain a section titled Airworthiness Limitations that is segregated and clearly distinguishable from the rest of the document. This section must set forth—

(1) Each mandatory replacement time, structural inspection interval, and related structural in-

spection procedures approved under §25.571; and

(2) Each mandatory replacement time, inspection interval, related inspection procedure, and all critical design configuration control limitations approved under §25.981 for the fuel tank system.

**(b)** If the Instructions for Continued Airworthiness consist of multiple documents, the section required by this paragraph must be included in the principal manual. This section must contain a legible statement in a prominent location that reads: "The Airworthiness Limitations section is FAA-approved and specifies maintenance required under §§43.16 and 91.403 of the Federal Aviation Regulations, unless an alternative program has been FAA approved."

[Docket No. 5066, 29 FR 18291, Dec. 24, 1964; as amended by Amdt. 25–54, 45 FR 60177, Sept. 11, 1980; Amdt. 25–68, 54 FR 34329, Aug. 18, 1989; Amdt. 25–102, 66 FR 23130, May 7, 2001]

## APPENDIX I TO PART 25

### INSTALLATION OF AN AUTOMATIC TAKEOFF THRUST CONTROL SYSTEM (ATTCS)

### *I25.1 GENERAL.*

**(a)** This appendix specifies additional requirements for installation of an engine power control system that automatically resets thrust or power on operating engine(s) in the event of any one engine failure during takeoff.

**(b)** With the ATTCS and associated systems functioning normally as designed, all applicable requirements of Part 25, except as provided in this appendix, must be met without requiring any action by the crew to increase thrust or power.

### *I25.2 DEFINITIONS.*

**(a)** *Automatic Takeoff Thrust Control System (ATTCS).* An ATTCS is defined as the entire automatic system used on takeoff, including all devices, both mechanical and electrical, that sense engine failure, transmit signals, actuate fuel controls or power levers or increase engine power by other means on operating engines to achieve scheduled thrust or power increases, and furnish cockpit information on system operation.

**(b)** *Critical Time Interval.* When conducting an ATTCS takeoff, the critical time interval is between $V_1$ minus 1 second and a point on the minimum performance, all-engine flight path where, assuming a simultaneous occurrence of an engine and ATTCS failure, the resulting minimum flight path thereafter intersects the Part 25 required actual flight path at no less than 400 feet above the takeoff surface. This time interval is shown in the following illustration:

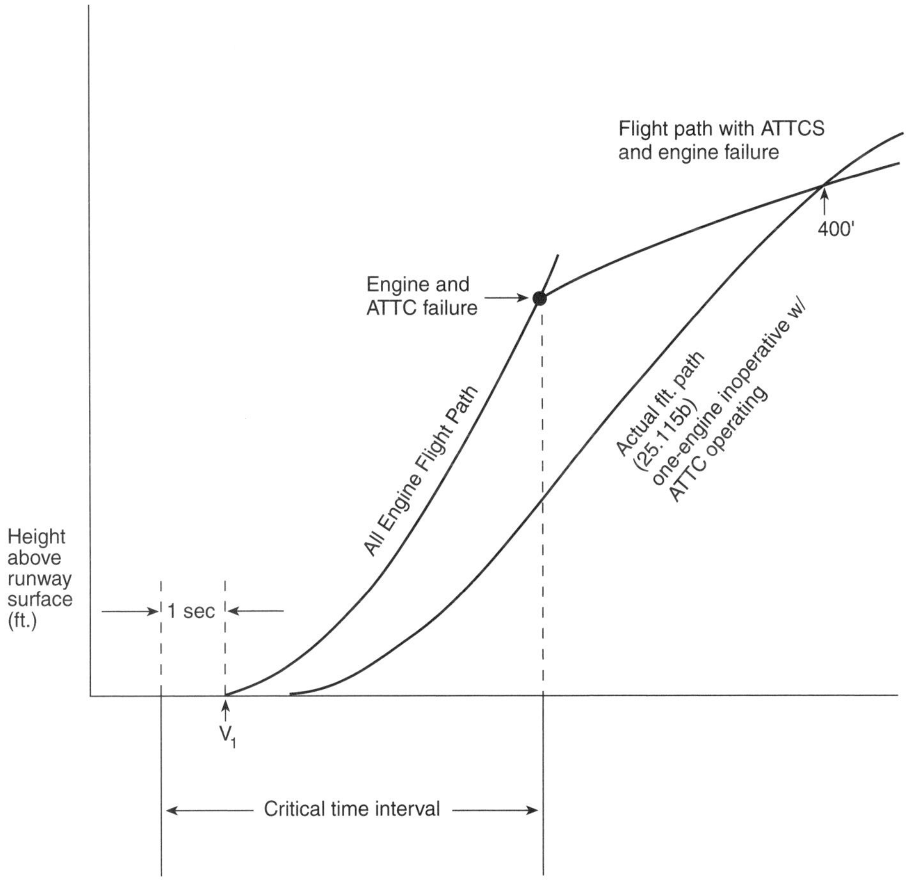
Flight path with ATTCS
and engine failure
400'
Engine and
ATTC failure
All Engine Flight Path
Actual flt. path
(25.115b)
one-engine inoperative w/
ATTC operating
Height
above
runway
surface
(ft.)
1 sec
V1
Critical time interval

*I25.3 Performance and System Reliability Requirements.*

The applicant must comply with the performance and ATTCS reliability requirements as follows:

**(a)** An ATTCS failure or a combination of failures in the ATTCS during the critical time interval:

(1) Shall not prevent the insertion of the *maximum approved takeoff* thrust or power, or must be shown to be an improbable event.

(2) Shall not result in a significant loss or reduction in thrust or power, or must be shown to be an extremely improbable event.

**(b)** The concurrent existence of an ATTCS failure and an engine failure during the critical time interval must be shown to be extremely improbable.

**(c)** All applicable performance requirements of Part 25 must be met with an engine failure occurring at the most critical point during takeoff with the ATTCS system functioning.

*I25.4 Thrust Setting.*

The initial takeoff thrust or power setting on each engine at the beginning of the takeoff roll may not be less than any of the following:

**(a)** Ninety (90) percent of the thrust or power set by the ATTCS (the maximum takeoff thrust or power approved for the airplane under existing ambient conditions);

**(b)** That required to permit normal operation of all safety-related systems and equipment dependent upon engine thrust or power lever position; or

**(c)** That shown to be free of hazardous engine response characteristics when thrust or power is advanced from the initial takeoff thrust or power to the maximum approved takeoff thrust or power.

*I25.5 Powerplant Controls.*

**(a)** In addition to the requirements of §25.1141, no single failure or malfunction, or probable combination thereof, of the ATTCS, including associated systems, may cause the failure of any powerplant function necessary for safety.

**(b)** The ATTCS must be designed to:

(1) Apply thrust or power on the operating engine(s), following any one engine failure during takeoff, to achieve the maximum approved takeoff thrust or power without exceeding engine operating limits;

(2) Permit manual decrease or increase in thrust or power up to the maximum takeoff thrust or power approved for the airplane under existing conditions through the use of the power lever. For airplanes equipped with limiters that automatically prevent engine operating limits from being exceeded under existing ambient conditions, other means may be used to increase the thrust or power in the event of an ATTCS failure provided the means is located on or forward of the power levers; is easily identified and operated under all operating conditions by a single action of either pilot with the hand that is normally used to actuate the power levers; and meets the requirements of §25.777 (a), (b), and (c);

(3) Provide a means to verify to the flightcrew before takeoff that the ATTCS is in a condition to operate; and

(4) Provide a means for the flightcrew to deactivate the automatic function. This means must be designed to prevent inadvertent deactivation.

*I25.6 Powerplant Instruments.*

In addition to the requirements of §25.1305:

**(a)** A means must be provided to indicate when the ATTCS is in the armed or ready condition; and

**(b)** If the inherent flight characteristics of the airplane do not provide adequate warning that an engine has failed, a warning system that is independent of the ATTCS must be provided to give the pilot a clear warning of any engine failure during takeoff.

[Docket No. 5066, 29 FR 18291, Dec. 24, 1964; as amended by Amdt. 25–62, 52 FR 43156, Nov. 9, 1987]

## Appendix J to Part 25

Emergency Demonstration

The following test criteria and procedures must be used for showing compliance with §25.803:

**(a)** The emergency evacuation must be conducted either during the dark of the night or during daylight with the dark of night simulated. If the demonstration is conducted indoors during daylight hours, it must be conducted with each window covered and each door closed to minimize the daylight effect. Illumination on the floor or ground may be used, but it must be kept low and shielded against shining into the airplane's windows or doors.

**(b)** The airplane must be in a normal attitude with landing gear extended.

**(c)** Unless the airplane is equipped with an off-wing descent means, stands or ramps may be used for descent from the wing to the ground. Safety equipment such as mats or inverted life rafts may be placed on the floor or ground to protect participants. No other equipment that is not part of the emergency evacuation equipment of the airplane may be used to aid the participants in reaching the ground.

**(d)** Except as provided in paragraph (a) of this Appendix, only the airplane's emergency lighting system may provide illumination.

**(e)** All emergency equipment required for the planned operation of the airplane must be installed.

**(f)** Each external door and exit, and each internal door or curtain, must be in the takeoff configuration.

**(g)** Each crewmember must be seated in the normally assigned seat for takeoff and must remain in the seat until receiving the signal for commencement of the demonstration. Each crewmember must be a person having knowledge of the operation of exits and emergency equipment and, if compliance with §121.291 is also being demonstrated, each flight attendant must be a member of a regularly scheduled line crew.

**(h)** A representative passenger load of persons in normal health must be used as follows:

(1) At least 40 percent of the passenger load must be female.

(2) At least 35 percent of the passenger load must be over 50 years of age.

(3) At least 15 percent of the passenger load must be female and over 50 years of age.

(4) Three life-size dolls, not included as part of the total passenger load, must be carried by passengers to simulate live infants 2 years old or younger.

(5) Crewmembers, mechanics, and training personnel, who maintain or operate the airplane in the normal course of their duties, may not be used as passengers.

**(i)** No passenger may be assigned a specific seat except as the Administrator may require. Except as required by subparagraph (g) of this paragraph, no employee of the applicant may be seated next to an emergency exit.

**(j)** Seat belts and shoulder harnesses (as required) must be fastened.

**(k)** Before the start of the demonstration, approximately one-half of the total average amount of carry-on baggage, blankets, pillows, and other similar articles must be distributed at several locations in aisles and emergency exit access ways to create minor obstructions.

**(l)** No prior indication may be given to any crewmember or passenger of the particular exits to be used in the demonstration.

**(m)** The applicant may not practice, rehearse, or describe the demonstration for the participants nor may any participant have taken part in this type of demonstration within the preceding 6 months.

**(n)** The pretakeoff passenger briefing required by §121.571 may be given. The passengers may also be advised to follow directions of crewmembers but not be instructed on the procedures to be followed in the demonstration.

**(o)** If safety equipment as allowed by paragraph (c) of this appendix is provided, either all passenger and cockpit windows must be blacked out or all of the emergency exits must have safety equipment in order to prevent disclosure of the available emergency exits.

**(p)** Not more than 50 percent of the emergency exits in the sides of the fuselage of an airplane that meets all of the requirements applicable to the required emergency exits for that airplane may be used for the demonstration. Exits that are not to be used in the demonstration must have the exit handle deactivated or must be indicated by red lights, red tape, or other acceptable means placed outside the exits to indicate fire or other reason why they are unusable. The exits to be used must be representative of all of the emergency exits on the airplane and must be designated by the applicant, subject to approval by the Administrator. At least one floor level exit must be used.

**(q)** Except as provided in paragraph (c) of this section, all evacuees must leave the airplane by a means provided as part of the airplane's equipment.

**(r)** The applicant's approved procedures must be fully utilized, except the flightcrew must take no active role in assisting others inside the cabin during the demonstration.

**(s)** The evacuation time period is completed when the last occupant has evacuated the airplane and is on the ground. Provided that the acceptance rate of the stand or ramp is no greater than the acceptance rate of the means available on the airplane for descent from the wing during an actual crash situation, evacuees using stands or ramps allowed by paragraph (c) of this Appendix are considered to be on the ground when they are on the stand or ramp.

[Docket No. 5066, 29 FR 18291, Dec. 24, 1964; as amended by Amdt. 25–72, 55 FR 29788, July 20, 1990; Amdt. 25–79, Aug. 26, 1993]

# PART 63
## CERTIFICATION: FLIGHT CREWMEMBERS OTHER THAN PILOTS

SPECIAL FEDERAL AVIATION REGULATIONS



**Authority:** 49 U.S.C. 106(g), 40113, 44701–44703, 44707, 44709–44711, 45102–45103, 45301–45302.

SPECIAL FEDERAL AVIATION REGULATIONS

## SFAR NO. 58 TO PART 63

### ADVANCED QUALIFICATION PROGRAM

*EDITORIAL NOTE: FOR THE TEXT OF SFAR NO. 58, SEE PART 121 OF THIS CHAPTER.*

## SFAR NO. 93 TO PART 63

### TEMPORARY EXTENSION OF TIME TO ALLOW FOR CERTAIN TRAINING AND TESTING

*EDITORIAL NOTE: FOR THE TEXT OF SFAR NO. 93, SEE PART 121 OF THIS CHAPTER.*

## SFAR NO. 96 TO PART 63

### RELIEF FOR PARTICIPANTS IN OPERATION ENDURING FREEDOM

**1.** *Applicability.* Flight Standards District Offices are authorized to accept from an eligible person, as described in paragraph 2 of this SFAR, the following:

(a) An expired flight instructor certificate to show eligibility for renewal of a flight instructor certificate under §61.197, or an expired written test report to show eligibility under part 61 to take a practical test;

(b) An expired written test report to show eligibility under §§63.33 and 63.57 to take a practical test; and

(c) An expired written test report to show eligibility to take a practical test required under part 65 or an expired authorization to show eligibility for renewal under §65.93.

**2.** *Eligibility.* A person is eligible for the relief described in paragraph 1 of this SFAR if:

(a) The person served in a civilian or military capacity in support of Operation Enduring Freedom outside the United States during some period of time from September, 11, 2001, to May 6, 2004;

(b) The person's flight instructor certificate, airman written test report, or inspection authorization expired some time between September 11, 2001, and 6 calendar months after returning to the United States, or May 6, 2004, whichever is earlier; and

(c) The person complies with §61.197 or §65.93 of this chapter, as appropriate, or completes the appropriate practical test within 6 calendar months after returning to the United States, or May 6, 2004, whichever is earlier.

**3.** *Required documents.* The person must present in person the Airman Certificate and/or Rating Application (FAA Form 8710-1) to the appropriate Flight Standards District Office. The person must include with the application one of the following documents, which must show the date of assignment outside the United States and the date of return to the United States.

(a) An official U.S. Government notification of personnel action, or equivalent document, showing the person was a civilian on official duty for the U.S. Government in support of Operation Enduring Freedom outside the United States some time between September 11, 2001, and May 6, 2004;

(6) Military orders showing the person was assigned to duty outside the United States in support of Operation Enduring Freedom some time between September 11, 2001, and May 6, 2004;

(c) A letter from the person's unit commander or supervisor providing the dates during which the person served outside the United States in support of Operation Enduring Freedom some time between September 11, 2001, and May 6, 2004.

**4.** *Expiration date.* This Special Federal Aviation Regulation No. 96 expires May 6, 2004, unless sooner superseded or rescinded.

[Docket No. FAA–2002–12199, 67 FR 30527, May 6, 2002]

# Subpart A—General

**Source:** Docket No. 1179, 27 FR 7969, Aug. 10, 1962, unless otherwise noted.

## §63.1 Applicability.

This part prescribes the requirements for issuing flight engineer and flight navigator certificates and the general operating rules for holders of those certificates.

## §63.2 Certification of foreign flight crewmembers other than pilots.

A person who is neither a United States citizen nor a resident alien is issued a certificate under this part (other than under §63.23 or §63.42) outside the United States only when the Administrator finds that the certificate is needed for the operation of a U.S.-registered civil aircraft.

(Secs. 313, 601, 602, Federal Aviation Act of 1958; as amended (49 U.S.C. 1354, 1421, and 1422); sec. 6(c), Department of Transportation Act (49 U.S.C. 1655(c)); Title V, Independent Offices Appropriations Act of 1952 (31 U.S.C. 483(a)); sec. 28, International Air Transportation Competition Act of 1979 (49 U.S.C. 1159(b)))

[Docket No. 22052, 47 FR 35693, Aug. 18, 1982]

## §63.3 Certificates and ratings required.

**(a)** No person may act as a flight engineer of a civil aircraft of U.S. registry unless he has in his personal possession a current flight engineer certificate with appropriate ratings issued to him under this part and a second-class (or higher) medical certificate issued to him under part 67 of this chapter within the preceding 12 months. However, when the aircraft is operated within a foreign country, a current flight engineer certificate issued by the country in which the aircraft is operated, with evidence of current medical qualification for that certificate, may be used. Also, in the case of a flight engineer certificate issued under §63.42, evidence of current medical qualification accepted for the issue of that certificate is used in place of a medical certificate.

**(b)** No person may act as a flight navigator of a civil aircraft of U.S. registry unless he has in his personal possession a current flight navigator certificate issued to him under this part and a second-class (or higher) medical certificate issued to him under part 67 of this chapter within the preceding 12 months. However, when the aircraft is operated within a foreign country, a current flight navigator certificate issued by the country in which the aircraft is operated, with evidence of current medical qualification for that certificate, may be used.

**(c)** Each person who holds a flight engineer or flight navigator certificate, or medical certificate, shall present either or both for inspection upon the request of the Administrator or an authorized representative of the National Transportation Safety Board, or of any Federal, State, or local law enforcement officer.

(Secs. 3, 6, 9, 80 Stat. 931, 49 U.S.C. 1652, 1655, 1657)

[Docket No. 1179, 27 FR 7969, Aug. 10, 1962; as amended by Amdt. 63–1, 27 FR 10410, Oct. 25, 1962; Amdt. 63–3, 30 FR 14559, Nov. 23, 1965; Amdt. 63–7, 31 FR 13523, Oct. 20, 1966; Docket No. 8084, 32 FR 5769, April 11, 1967; Amdt. 63–9, 33 FR 18613, Dec. 17, 1968; Amdt. 63–11, 35 FR 5320, March 31, 1970]

## §63.11 Application and issue.

**(a)** An application for a certificate and appropriate class rating, or for an additional rating, under this part must be made on a form and in a manner prescribed by the Administrator. Each person who is neither a United States citizen nor a resident alien and applies for a written or practical test to be administered outside the United States for any certificate or rating issued under this part must show evidence that the fee prescribed in Appendix A of part 187 of this chapter has been paid.

**(b)** An applicant who meets the requirements of this part is entitled to an appropriate certificate and appropriate class ratings.

**(c)** Unless authorized by the Administrator, a person whose flight engineer certificate is suspended may not apply for any rating to be added to that certificate during the period of suspension.

**(d)** Unless the order of revocation provides otherwise, a person whose flight engineer or flight navigator certificate is revoked may not apply for the same kind of certificate for 1 year after the date of revocation.

(Secs. 313, 601, 602, Federal Aviation Act of 1958; as amended (49 U.S.C. 1354, 1421, and 1422); sec. 6(c), Department of Transportation Act (49 U.S.C. 1655(c)); Title V, Independent Offices Appropriations Act of 1952 (31 U.S.C. 483(a)); sec. 28, International Air Transportation Competition Act of 1979 (49 U.S.C. 1159(b)))

[Docket No. 1179, 27 FR 7969, Aug. 10, 1962; as amended by Amdt. 63–3, 30 FR 14559, Nov. 23, 1965; Amdt. 63–7, 31 FR 13523, Oct. 20, 1966; Amdt. 63–22, 47 FR 35693, Aug. 16, 1982]

## §63.12 Offenses involving alcohol or drugs.

**(a)** A conviction for the violation of any Federal or state statute relating to the growing, processing, manufacture, sale, disposition, possession, transportation, or importation of narcotic drugs, marihuana, or depressant or stimulant drugs or substances is grounds for—

(1) Denial of an application for any certificate or rating issued under this part for a period of up to 1 year after the date of final conviction; or

(2) Suspension or revocation of any certificate or rating issued under this part.

**(b)** The commission of an act prohibited by §91.17(a) or §91.19(a) of this chapter is grounds for—

(1) Denial of an application for a certificate or rating issued under this part for a period of up to 1 year after the date of that act; or

(2) Suspension or revocation of any certificate or rating issued under this part.

[Docket No. 21956, 50 FR 15379, April 17, 1985; as amended by Amdt. 63–27, 54 FR 34330, Aug. 18, 1989]

## §63.12a Refusal to submit to an alcohol test or to furnish test results.

A refusal to submit to a test to indicate the percentage by weight of alcohol in the blood, when requested by a law enforcement officer in accordance with §91.11(c) of this chapter, or a refusal to furnish or authorize the release of the test results when requested by the Administrator in accordance with §91.17 (c) or (d) of this chapter, is grounds for—

**(a)** Denial of an application for any certificate or rating issued under this part for a period of up to 1 year after the date of that refusal; or

**(b)** Suspension or revocation of any certificate or rating issued under this part.

[Docket No. 21956, 51 FR 1229, Jan. 9, 1986; as amended by Amdt. 63–27, 54 FR 34330, Aug. 18, 1989]

## §63.12b Refusal to submit to a drug or alcohol test.

**(a)** This section applies to an employee who performs a function listed in Appendix I or Appendix J to part 121 of this chapter directly or by contract for a part 121 certificate holder, a part 135 certificate holder, or an operator as defined in §135.1(c) of this chapter.

**(b)** Refusal by the holder of a certificate issued under this part to take a drug test required under the provisions of Appendix I to part 121 or an alcohol test required under the provisions of Appendix J to part 121 is grounds for—

(1) Denial of an application for any certificate or rating issued under this part for a period of up to 1 year after the date of such a refusal; and

(2) Suspension or revocation of any certificate or rating issued under this part.

[Docket No. 1179, 27 FR 7969, Aug. 10, 1962; as amended by Amdt. 63–29, 59 FR 7389, Feb. 15, 1994]

## §63.13 Temporary certificate.

A certificate effective for a period of not more than 120 days may be issued to a qualified applicant, pending review of his application and supplementary documents and the issue of the certificate for which he applied.

[Docket No. 1179, 27 FR 7969, Aug. 10, 1962; as amended by Amdt. 63–19, 43 FR 22639, May 25, 1978]

## §63.15 Duration of certificates.

**(a)** Except as provided in §63.23 and paragraph (b) of this section, a certificate or rating issued under this part is effective until it is surrendered, suspended, or revoked.

**(b)** A flight engineer certificate (with any amendment thereto) issued under §63.42 expires at the end of the 24th month after the month in which the certificate was issued or renewed. However, the holder may exercise the privileges of that certificate only while the foreign flight engineer license on which that certificate is based is effective.

**(c)** Any certificate issued under this part ceases to be effective if it is surrendered, suspended, or revoked. The holder of any certificate issued under this part that is suspended or revoked shall, upon the Administrator's request, return it to the Administrator.

(Sec. 6, 80 Stat. 937, 49 U.S.C. 1655; secs. 313, 601, 602, Federal Aviation Act of 1958; as amended (49 U.S.C. 1354, 1421, and 1422); sec. 6(c), Department of Transportation Act (49 U.S.C. 1655(c)); Title V, Independent Offices Appropriations Act of 1952 (31 U.S.C. 483(a)); sec. 28, International Air Transportation Competition Act of 1979 (49 U.S.C. 1159(b)))

[Docket No. 8846, 33 FR 18613, Dec. 17, 1968; as amended by Amdt. 63–22, 47 FR 35693, Aug. 16, 1982]

## §63.15a [Reserved]

## §63.16 Change of name; replacement of lost or destroyed certificate.

**(a)** An application for a change of name on a certificate issued under this part must be accompanied by the applicant's current certificate and the marriage license, court order, or other document verifying the change. The documents are returned to the applicant after inspection.

**(b)** An application for a replacement of a lost or destroyed certificate is made by letter to the Department of Transportation, Federal Aviation Administration, Airman Certification Branch, Post Office Box 25082, Oklahoma City, OK 73125. The letter must—

(1) Contain the name in which the certificate was issued, the permanent mailing address (including zip code), social security number (if any), and date and place of birth of the certificate holder, and any available information regarding the grade, number, and date of issue of the certificate, and the ratings on it; and

(2) Be accompanied by a check or money order for $2, payable to the Federal Aviation Administration.

**(c)** An application for a replacement of a lost or destroyed medical certificate is made by letter to the Department of Transportation, Federal Aviation Administration, Civil Aeromedical Institute, Aeromedical Certification Branch, Post Office Box 25082, Oklahoma City, OK 73125, accompanied by a check or money order for $2.00.

**(d)** A person whose certificate issued under this part or medical certificate, or both, has been lost may obtain a telegram from the Federal Aviation Administration confirming that it was issued. The telegram may be carried as a certificate for a period not to exceed 60 days pending his receiving a duplicate under paragraph (b) or (c) of this section, unless he has been notified that the certificate has been suspended or revoked. The request for such a telegram may be made by prepaid telegram, stating the date upon which a duplicate certificate was requested, or including the request for a duplicate and a money order for the necessary amount. The request for a telegraphic certificate should be sent to the office prescribed in paragraph (b) or (c) of this section, as appropriate. However, a request for both at the same time should be sent to the office prescribed in paragraph (b) of this section.

[Docket No. 7258, 31 FR 13523, Oct. 20, 1966; as amended by Docket No. 8084, 32 FR 5769, Apr. 11, 1967; Amdt. 63–12, 35 FR 14075, Sept. 4, 1970; Amdt. 63–13, 36 FR 28654, Feb. 11, 1971]

## §63.17 Tests: General procedure.

**(a)** Tests prescribed by or under this part are given at times and places, and by persons, designated by the Administrator.

**(b)** The minimum passing grade for each test is 70 percent.

## §63.18 Written tests: Cheating or other unauthorized conduct.

**(a)** Except as authorized by the Administrator, no person may—

(1) Copy, or intentionally remove, a written test under this part;

(2) Give to another, or receive from another, any part or copy of that test;

(3) Give help on that test to, or receive help on that test from, any person during the period that test is being given.

(4) Take any part of that test in behalf of another person;

(5) Use any material or aid during the period that test is being given; or

(6) Intentionally cause, assist, or participate in any act prohibited by this paragraph.

**(b)** No person who commits an act prohibited by paragraph (a) of this section is eligible for any airman or ground instructor certificate or rating under this chapter for a period of 1 year after the date of that act. In addition, the commission of that act is a basis for suspending or revoking any airman or ground instructor certificate or rating held by that person.

[Docket No. 4086, 30 FR 2196, Feb. 18, 1965]

## §63.19 Operations during physical deficiency.

No person may serve as a flight engineer or flight navigator during a period of known physical deficiency, or increase in physical deficiency, that would make him unable to meet the physical requirements for his current medical certificate.

## §63.20 Applications, certificates, logbooks, reports, and records; falsification, reproduction, or alteration.

**(a)** No person may make or cause to be made—

(1) Any fraudulent or intentionally false statement on any application for a certificate or rating under this part;

(2) Any fraudulent or intentionally false entry in any logbook, record, or report that is required to be kept, made, or used, to show compliance with any requirement for any certificate or rating under this part;

(3) Any reproduction, for fraudulent purpose, of any certificate or rating under this part; or

(4) Any alteration of any certificate or rating under this part.

**(b)** The commission by any person of an act prohibited under paragraph (a) of this section is a basis for suspending or revoking any airman or ground instructor certificate or rating held by that person.

[Docket No. 4086, 30 FR 2196, Feb. 18, 1965]

## §63.21 Change of address.

Within 30 days after any change in his permanent mailing address, the holder of a certificate issued under this part shall notify the Department of Transportation, Federal Aviation Administration, Airman Certification Branch, Post Office Box 25082, Oklahoma City, OK 73125, in writing, of his new address.

[Docket No. 10536, 35 FR 14075, Sept. 4, 1970]

## §63.23 Special purpose flight engineer and flight navigator certificates: Operation of U.S.-registered civil airplanes leased by a person not a U.S. citizen.

**(a)** *General.* The holder of a current foreign flight engineer or flight navigator certificate, license, or authorization issued by a foreign contracting State to the Convention on International Civil Aviation, who meets the requirements of this section, may hold a special purpose flight engineer or flight navigator certificate, as appropriate, authorizing the holder to perform flight engineer or flight navigator duties on a civil airplane of U.S. registry, leased to a person not a citizen of the United States, carrying persons or property for compensation or hire. Special purpose flight engineer and flight navigator certificates are issued under this section only for airplane types that can have a maximum passenger seating configuration, excluding any flight crewmember seat, of more than 30 seats or a maximum payload capacity (as defined in §135.2(e) of this chapter) of more than 7,500 pounds.

**(b)** *Eligibility.* To be eligible for the issuance, or renewal, of a certificate under this section, an applicant must present the following to the Administrator:

(1) A current foreign flight engineer or flight navigator certificate, license, or authorization issued by the aeronautical authority of a foreign contracting State to the Convention on International Civil Aviation or a facsimile acceptable to the Administrator. The certificate or license must authorize the applicant to perform the flight engineer or flight navigator duties to be authorized by a certificate issued under this section on the same airplane type as the leased airplane.

(2) A current certification by the lessee of the airplane—

(i) Stating that the applicant is employed by the lessee;

(ii) Specifying the airplane type on which the applicant will perform flight engineer or flight navigator duties; and

(iii) Stating that the applicant has received ground and flight instruction which qualifies the applicant to perform the duties to be assigned on the airplane.

(3) Documentation showing that the applicant currently meets the medical standards for the foreign flight engineer or flight navigator certificate, license, or authorization required by paragraph (b)(1) of this section, except that a U.S. medical certificate issued under part 67 of this chapter is not evidence that the applicant meets those standards unless the State which issued the applicant's foreign flight engineer or flight navigator certificate, license, or authorization accepts a U.S. medical certificate as evidence of medical fitness for a flight engineer or flight navigator certificate, license, or authorization.

**(c)** *Privileges.* The holder of a special purpose flight engineer or flight navigator certificate issued under this section may exercise the same privileges as those shown on the certificate, license, or authorization specified in paragraph (b)(1) of this section, subject to the limitations specified in this section.

**(d)** *Limitations.* Each certificate issued under this section is subject to the following limitations:

(1) It is valid only—

(i) For flights between foreign countries and for flights in foreign air commerce;

(ii) While it and the certificate, license, or authorization required by paragraph (b)(1) of this section are in the certificate holder's personal possession and are current;

(iii) While the certificate holder is employed by the person to whom the airplane described in the certification required by paragraph (b)(2) of this section is leased;

(iv) While the certificate holder is performing flight engineer or flight navigator duties on the U.S.-registered civil airplane described in the certification required by paragraph (b)(2) of this section; and

(v) While the medical documentation required by paragraph (b)(3) of this section is in the certificate holder's personal possession and is currently valid.

(2) Each certificate issued under this section contains the following:

(i) The name of the person to whom the U.S.-registered civil airplane is leased.

(ii) The type of airplane.

(iii) The limitation: "Issued under, and subject to, §63.23 of the Federal Aviation Regulations."

(iv) The limitation: "Subject to the privileges and limitations shown on the holder's foreign flight (engineer or navigator) certificate, license, or authorization."

(3) Any additional limitations placed on the certificate which the Administrator considers necessary.

**(e)** *Termination.* Each special purpose flight engineer or flight navigator certificate issued under this section terminates—

(1) When the lease agreement for the airplane described in the certification required by paragraph (b)(2) of this section terminates;

(2) When the foreign flight engineer or flight navigator certificate, license, or authorization, or the medical documentation required by paragraph (b) of this section is suspended, revoked, or no longer valid; or

(3) After 24 months after the month in which the special purpose flight engineer or flight navigator certificate was issued.

**(f)** *Surrender of certificate.* The certificate holder shall surrender the special purpose flight engineer or flight navigator certificate to the Administrator within 7 days after the date it terminates.

**(g)** *Renewal.* The certificate holder may have the certificate renewed by complying with the requirements of paragraph (b) of this section at the time of application for renewal.

(Secs. 313(a), 601, and 602, Federal Aviation Act of 1958; as amended (49 U.S.C. 1354(a), 1421, and 1422); sec. 6(c), Department of Transportation Act (49 U.S.C. 1655(c)))

[Docket No. 19300, 45 FR 5672, Jan. 24, 1980]

# Subpart B— Flight Engineers

**Authority:** Secs. 313(a), 601, and 602, Federal Aviation Act of 1958; 49 U.S.C. 1354, 1421, 1422.

**Source:** Docket No. 6458, 30 FR 14559, Nov. 23, 1965, unless otherwise noted.

## §63.31 Eligibility requirements; general.

To be eligible for a flight engineer certificate, a person must—

**(a)** Be at least 21 years of age;

**(b)** Be able to read, speak, and understand the English language, or have an appropriate limitation placed on his flight engineer certificate;

**(c)** Hold at least a second-class medical certificate issued under part 67 of this chapter within the 12 months before the date he applies, or other evidence of medical qualification accepted for the issue of a flight engineer certificate under §63.42; and

**(d)** Comply with the requirements of this subpart that apply to the rating he seeks.

(Sec. 6, 80 Stat. 937, 49 U.S.C. 1655)

[Docket No. 6458, 30 FR 14559, Nov. 23, 1965; as amended by Amdt. 63–9, 33 FR 18614, Dec. 17, 1968]

## §63.33 Aircraft ratings.

**(a)** The aircraft class ratings to be placed on flight engineer certificates are—

(1) Reciprocating engine powered;

(2) Turbopropeller powered; and

(3) Turbojet powered.

**(b)** To be eligible for an additional aircraft class rating after his flight engineer certificate with a class rating is issued to him, an applicant must pass the written test that is appropriate to the class of airplane for which an additional rating is sought, and—

(1) Pass the flight test for that class of aircraft; or

(2) Satisfactorily complete an approved flight engineer training program that is appropriate to the additional class rating sought.

## §63.35 Knowledge requirements.

**(a)** An applicant for a flight engineer certificate must pass a written test on the following:

(1) The regulations of this chapter that apply to the duties of a flight engineer.

(2) The theory of flight and aerodynamics.

(3) Basic meteorology with respect to engine operations.

(4) Center of gravity computations.

**(b)** An applicant for the original or additional issue of a flight engineer class rating must pass a written test for that airplane class on the following:

(1) Preflight.

(2) Airplane equipment.
(3) Airplane systems.
(4) Airplane loading.
(5) Airplane procedures and engine operations with respect to limitations.
(6) Normal operating procedures.
(7) Emergency procedures.
(8) Mathematical computation of engine operations and fuel consumption.

**(c)** Before taking the written tests prescribed in paragraphs (a) and (b) of this section, an applicant for a flight engineer certificate must present satisfactory evidence of having completed one of the experience requirements of §63.37. However, he may take the written tests before acquiring the flight training required by §63.37.

**(d)** An applicant for a flight engineer certificate or rating must have passed the written tests required by paragraphs (a) and (b) of this section since the beginning of the 24th calendar month before the month in which the flight is taken. However, this limitation does not apply to an applicant for a flight engineer certificate or rating if—

(1) The applicant—

(i) Within the period ending 24 calendar months after the month in which the applicant passed the written test, is employed as a flight crewmember or mechanic by a U.S. air carrier or commercial operator operating either under part 121 or as a commuter air carrier under part 135 (as defined in part 298 of this title) and is employed by such a certificate holder at the time of the flight test;

(ii) If employed as a flight crewmember, has completed initial training, and, if appropriate, transition or upgrade training; and

(iii) Meets the recurrent training requirements of the applicable part or, for mechanics, meets the recency of experience requirements of part 65; or

(2) Within the period ending 24 calendar months after the month in which the applicant passed the written test, the applicant participated in a flight engineer or maintenance training program of a U.S. scheduled military air transportation service and is currently participating in that program.

**(e)** An air carrier or commercial operator with an approved training program under part 121 of this chapter may, when authorized by the Administrator, provide as part of that program a written test that it may administer to satisfy the test required for an additional rating under paragraph (b) of this section.

(Sec. 6, 80 Stat. 937, 49 U.S.C. 1655; Secs. 313(a), 601 through 605 of the Federal Aviation Act of 1958 (49 U.S.C. 1354(a), 1421 through 1425); sec. 6(c), Department of Transportation Act (49 U.S.C. 1655(c)); and 14 CFR 11.49)

[Docket No. 1179, 27 FR 7969, Aug. 10, 1962; as amended by Amdt. 63–17, 40 FR 32830, Aug. 5, 1975; Amdt. 63–21, 47 FR 13316, March 29, 1982]

## §63.37 Aeronautical experience requirements.

**(a)** Except as otherwise specified therein, the flight time used to satisfy the aeronautical experience requirements of paragraph (b) of this section must have been obtained on an airplane—

(1) On which a flight engineer is required by this chapter; or

(2) That has at least three engines that are rated at least 800 horsepower each or the equivalent in turbine-powered engines.

**(b)** An applicant for a flight engineer certificate with a class rating must present, for the class rating sought, satisfactory evidence of one of the following:

(1) At least 3 years of diversified practical experience in aircraft and aircraft engine maintenance (of which at least 1 year was in maintaining multiengine aircraft with engines rated at least 800 horsepower each, or the equivalent in turbine engine powered aircraft), and at least 5 hours of flight training in the duties of a flight engineer.

(2) Graduation from at least a 2-year specialized aeronautical training course in maintaining aircraft and aircraft engines (of which at least 6 calendar months were in maintaining multiengine aircraft with engines rated at least 800 horsepower each or the equivalent in turbine engine powered aircraft), and at least 5 hours of flight training in the duties of a flight engineer.

(3) A degree in aeronautical, electrical, or mechanical engineering from a recognized college, university, or engineering school; at least 6 calendar months of practical experience in maintaining multiengine aircraft with engines rated at least 800 horsepower each, or the equivalent in turbine engine powered aircraft; and at least 5 hours of flight training in the duties of a flight engineer.

(4) At least a commercial pilot certificate with an instrument rating and at least 5 hours of flight training in the duties of a flight engineer.

(5) At least 200 hours of flight time in a transport category airplane (or in a military airplane with at least two engines and at least equivalent weight and horsepower) as pilot in command or second in command performing the functions of a pilot in command under the supervision of a pilot in command.

(6) At least 100 hours of flight time as a flight engineer.

(7) Within the 90-day period before he applies, successful completion of an approved flight engineer ground and flight course of instruction as provided in Appendix C of this part.

(Sec. 6, 80 Stat. 937, 49 U.S.C. 1655)

63

[Docket No. 6458, 30 FR 14559, Nov. 23, 1965; as amended by Amdt. 63–5, 31 FR 9047, July 1, 1966; Amdt. 63–17, 40 FR 32830, Aug. 5, 1975]

## §63.39 Skill requirements.

**(a)** An applicant for a flight engineer certificate with a class rating must pass a practical test on the duties of a flight engineer in the class of airplane for which a rating is sought. The test may only be given on an airplane specified in §63.37(a).

**(b)** The applicant must—

(1) Show that he can satisfactorily perform preflight inspection, servicing, starting, pretakeoff, and postlanding procedures;

(2) In flight, show that he can satisfactorily perform the normal duties and procedures relating to the airplane, airplane engines, propellers (if appropriate), systems, and appliances; and

(3) In flight, in an airplane simulator, or in an approved flight engineer training device, show that he can satisfactorily perform emergency duties and procedures and recognize and take appropriate action for malfunctions of the airplane, engines, propellers (if appropriate), systems and appliances.

## §63.41 Retesting after failure.

An applicant for a flight engineer certificate who fails a written test or practical test for that certificate may apply for retesting—

**(a)** After 30 days after the date he failed that test; or

**(b)** After he has received additional practice or instruction (flight, synthetic trainer, or ground training, or any combination thereof) that is necessary, in the opinion of the Administrator or the applicant's instructor (if the Administrator has authorized him to determine the additional instruction necessary) to prepare the applicant for retesting.

## §63.42 Flight engineer certificate issued on basis of a foreign flight engineer license.

**(a)** *Certificates issued.* The holder of a current foreign flight engineer license issued by a contracting State to the Convention on International Civil Aviation, who meets the requirements of this section, may have a flight engineer certificate issued to him for the operation of civil aircraft of U.S. registry. Each flight engineer certificate issued under this section specifies the number and State of issuance of the foreign flight engineer license on which it is based. If the holder of the certificate cannot read, speak, or understand the English language, the Administrator may place any limitation on the certificate that he considers necessary for safety.

**(b)** *Medical standards and certification.* An applicant must submit evidence that he currently meets the medical standards for the foreign flight engineer license on which the application for a certificate under this section is based. A current medical certificate issued under part 67 of this chapter will be excepted as evidence that the applicant meets those standards. However, a medical certificate issued under part 67 of this chapter is not evidence that the applicant meets those standards outside the United States unless the State that issued the applicant's foreign flight engineer license also accepts that medical certificate as evidence of the applicant's physical fitness for his foreign flight engineer license.

**(c)** *Ratings issued.* Aircraft class ratings listed on the applicant's foreign flight engineer license, in addition to any issued to him after testing under the provisions of this part, are placed on the applicant's flight engineer certificate. An applicant without an aircraft class rating on his foreign flight engineer license may be issued a class rating if he shows that he currently meets the requirements for exercising the privileges of his foreign flight engineer license on that class of aircraft.

**(d)** *Privileges and limitations.* The holder of a flight engineer certificate issued under this section may act as a flight engineer of a civil aircraft of U.S. registry subject to the limitations of this part and any additional limitations placed on his certificate by the Administrator. He is subject to these limitations while he is acting as a flight engineer of the aircraft within or outside the United States. However, he may not act as flight engineer or in any other capacity as a required flight crewmember, of a civil aircraft of U.S. registry that is carrying persons or property for compensation or hire.

**(e)** *Renewal of certificate and ratings.* The holder of a certificate issued under this section may have that certificate and the ratings placed thereon renewed if, at the time of application for renewal, the foreign flight engineer license on which that certificate is based is in effect. Application for the renewal of the certificate and ratings thereon must be made before the expiration of the certificate.

(Sec. 6, 80 Stat. 937, 49 U.S.C. 1655)

[Docket No. 8846, 33 FR 18614, Dec. 17, 1968; as amended by Amdt. 63–20, 45 FR 5673, Jan. 24, 1980]

## §63.43 Flight engineer courses.

An applicant for approval of a flight engineer course must submit a letter to the Administrator requesting approval, and must also submit three copies of each course outline, a description of the facilities and equipment, and a list of the instructors and their qualifications. An air carrier or commercial operator with an approved flight engineer training course under part 121 of this chapter may apply for approval of a training course under this part by letter without submitting the additional information required by this paragraph. Minimum requirements for obtaining approval of a flight engineer course are set forth in Appendix C of this part.

# Subpart C— Flight Navigators

**Authority:** Secs. 313(a), 314, 601, and 607; 49 U.S.C. 1354(a), 1355, 1421, and 1427.

**Source:** Docket No. 1179, 27 FR 7970, Aug. 10, 1962, unless otherwise noted.

## §63.51 Eligibility requirements; general.

To be eligible for a flight navigator certificate, a person must—

**(a)** Be at least 21 years of age;

**(b)** Be able to read, write, speak, and understand the English language;

**(c)** Hold at least a second-class medical certificate issued under part 67 of this chapter within the 12 months before the date he applies; and

**(d)** Comply with §§63.53, 63.55, and 63.57.

## §63.53 Knowledge requirements.

**(a)** An applicant for a flight navigator certificate must pass a written test on—

(1) The regulations of this chapter that apply to the duties of a flight navigator;

(2) The fundamentals of flight navigation, including flight planning and cruise control;

(3) Practical meteorology, including analysis of weather maps, weather reports, and weather forecasts; and weather sequence abbreviations, symbols, and nomenclature;

(4) The types of air navigation facilities and procedures in general use;

(5) Calibrating and using air navigation instruments;

(6) Navigation by dead reckoning;

(7) Navigation by celestial means;

(8) Navigation by radio aids;

(9) Pilotage and map reading; and

(10) Interpretation of navigation aid identification signals.

**(b)** A report of the test is mailed to the applicant. A passing grade is evidence, for a period of 24 months after the test, that the applicant has complied with this section.

[Docket No. 1179, 27 FR 7970, Aug. 10 1962; as amended by Amdt. 63–19, 43 FR 22639, May 25, 1978]

## §63.55 Experience requirements.

**(a)** An applicant for a flight navigator certificate must be a graduate of a flight navigator course approved by the Administrator or present satisfactory documentary evidence of—

(1) Satisfactory determination of his position in flight at least 25 times by night by celestial observations and at least 25 times by day by celestial observations in conjunction with other aids; and

(2) At least 200 hours of satisfactory flight navigation including celestial and radio navigation and dead reckoning.

A pilot who has logged 500 hours of cross-country flight time, of which at least 100 hours were at night, may be credited with not more than 100 hours for the purposes of paragraph (a)(2) of this section.

**(b)** Flight time used exclusively for practicing long-range navigation methods, with emphasis on celestial navigation and dead reckoning, is considered to be satisfactory navigation experience for the purposes of paragraph (a) of this section. It must be substantiated by a logbook, by records of an armed force or a certificated air carrier, or by a letter signed by a certificated flight navigator and attached to the application.

## §63.57 Skill requirements.

**(a)** An applicant for a flight navigator certificate must pass a practical test in navigating aircraft by—

(1) Dead reckoning;

(2) Celestial means; and

(3) Radio aids to navigation.

**(b)** An applicant must pass the written test prescribed by §63.53 before taking the test under this section. However, if a delay in taking the test under this section would inconvenience the applicant or an air carrier, he may take it before he receives the result of the written test, or after he has failed the written test.

**(c)** The test requirements for this section are set forth in Appendix A of this part.

[Docket No. 1179, 27 FR 7970, Aug. 10, 1962; as amended by Amdt. 63–19, 43 FR 22639, May 25, 1978]

## §63.59 Retesting after failure.

**(a)** An applicant for a flight navigator certificate who fails a written or practical test for that certificate may apply for retesting—

(1) After 30 days after the date he failed that test; or

(2) Before the 30 days have expired if the applicant presents a signed statement from a certificated flight navigator, certificated ground instructor, or any other qualified person approved by the Administrator, certifying that person has given the

63

applicant additional instruction in each of the subjects failed and that person considers the applicant ready for retesting.

**(b)** A statement from a certificated flight navigator, or from an operations official of an approved navigator course, is acceptable, for the purposes of paragraph (a)(2) of this section, for the written test and for the flight test. A statement from a person approved by the Administrator is acceptable for the written tests. A statement from a supervising or check navigator with the United States Armed Forces is acceptable for the written test and for the practical test.

**(c)** If the applicant failed the flight test, the additional instruction must have been administered in flight.

[Docket No. 1179, 27 FR 7970, Aug. 10, 1962; as amended by Amdt. 63–19, 43 FR 22640, May 25, 1978]

## §63.61 Flight navigator courses.

An applicant for approval of a flight navigator course must submit a letter to the Administrator requesting approval, and must also submit three copies of the course outline, a description of his facilities and equipment, and a list of the instructors and their qualifications. Requirements for the course are set forth in Appendix B to this part.

## Appendix A to Part 63

### Test Requirements for Flight Navigator Certificate

**(a)** *Demonstration of skill.* An applicant will be required to pass practical tests on the prescribed subjects. These tests may be given by FAA inspectors and designated flight navigator examiners.

**(b)** *The examination.* The practical examination consists of a ground test and a flight test as itemized on the examination check sheet. Each item must be completed satisfactorily in order for the applicant to obtain a passing grade. Items 5, 6, 7 of the ground test may be completed orally, and items 17, 22, 23, 34, 36, 37, 38, and 39 of the flight test may be completed by an oral examination when a lack of ground facilities or navigation equipment makes such procedure necessary. In these cases a notation to that effect shall be made in the "Remarks" space on the check sheet.

**(c)** *Examination procedure.*

(1) An applicant will provide an aircraft in which celestial observations can be taken in all directions. Minimum equipment shall include a table for plotting, a drift meter or absolute altimeter, an instrument for taking visual bearings, and a radio direction finder.

(2) More than one flight may be used to complete the flight test and any type of flight pattern may be used. The test will be conducted chiefly over water whenever practicable, and without regard to radio range legs or radials. If the test is conducted chiefly over land, a chart should be used which shows very little or no topographical and aeronautical data. The total flight time will cover a period of at least four hours. Only one applicant may be examined at one time, and no applicant may perform other than navigator duties during the examination.

(3) When the test is conducted with an aircraft belonging to an air carrier, the navigation procedures should conform with those set forth in the carrier's operations manual. Items of the flight test which are not performed during the routine navigation of the flight will be completed by oral examination after the flight or at times during flight which the applicant indicates may be used for tests on those items. Since in-flight weather conditions, the reliability of the weather forecast, and the stability of the aircraft will have considerable effect on an applicant's performance, good judgment must be used by the agent or examiner in evaluating the tests.

**(d)** *Ground test.* For the ground test, in the order of the numbered items on the examination check sheet, an applicant will be required to:

(1) Identify without a star identifier, at least six navigational stars and all planets available for navigation at the time of the examination and explain the method of identification.

(2) Identify two additional stars with a star identifier or sky diagrams and explain identification procedure.

(3) Precompute a time-altitude curve for a period of about 20 minutes and take 10 single observations of a celestial body which is rising or setting rapidly. The intervals between observations should be at least one minute. Mark each observation on the graph to show accuracy. All observations, after corrections, shall plot within 8 minutes of arc from the time-altitude curve, and the average error shall not exceed 5 minutes of arc.

(4) Take and plot one 3-star fix and 3 LOP's of the sun. Plotted fix or an average of LOP's must fall within 5 miles of the actual position of the observer.

(5) Demonstrate or explain the compensation and swinging of a liquid-type magnetic compass.

(6) Demonstrate or explain a method of aligning one type of drift meter.

(7) Demonstrate or explain a method of aligning an astro-compass or periscopic sextant.

**(e)** *Flight test.* For the flight test, in the order of the numbered items on the examination check sheet, an applicant will be required to:

(1) Demonstrate his ability to read weather symbols and interpret synoptic surface and upper air weather maps with particular emphasis being placed on winds.

(2) Prepare a flight plan by zones from the forecast winds or pressure data of an upper air chart and the operator's data.

(3) Compute from the operator's data the predicted fuel consumption for each zone of the flight, including the alternate.

(4) Determine the point-of-no-return for the flight with all engines running and the equitime point with one engine inoperative. Graphical methods which are part of the company's operations manual may be used for these computations.

(5) Prepare a cruise control (howgozit) chart from the operator's data.

(6) Enter actual fuel consumed on the cruise control chart and interpret the variations of the actual curve from the predicted curve.

(7) Check the presence on board and operating condition of all navigation equipment. Normally a check list will be used. This check will include a time tick or chronometer comparison. Any lack of thoroughness during this check will justify this item being graded unsatisfactory.

(8) Locate emergency equipment, such as, the nearest fire extinguisher, life preserver, life rafts, exits, axe, first aid kits, etc.

(9) Recite the navigator's duties and stations during emergencies for the type of aircraft used for the test.

(10) Demonstrate the proper use of a flux gate compass or gyrosyn compass (when available), with special emphasis on the caging methods and the location of switches, circuit breakers, and fuses. If these compasses are not part of the aircraft's equipment, an oral examination will be given.

(11) Be accurate and use good judgment when setting and altering headings. Erroneous application of variation, deviation, or drift correction, or incorrect measurement of course on the chart will be graded as unsatisfactory.

(12) Demonstrate or explain the use of characteristics of various chart projections used in long-range air navigation, including the plotting of courses and bearings, and the measuring of distances.

(13) Demonstrate ability to identify designated landmarks by the use of a sectional or WAC chart.

(14) Use a computer with facility and accuracy for the computation of winds, drift correction and drift angles, ground speeds, ETA's, fuel loads, etc.

(15) Determine track, ground speed, and wind by the double drift method. When a drift meter is not part of the aircraft's equipment, an oral examination on the use of the drift meter and a double drift problem shall be completed.

(16) Determine ground speed and wind by the timing method with a drift meter. When a drift meter is not part of the aircraft's equipment, an oral examination on the procedure and a problem shall be completed.

(17) Demonstrate the use of air plot for determining wind between fixes and for plotting pressure lines of position when using pressure and absolute altimeter comparisons.

(18) Give ETA's to well defined check points at least once each hour after the second hour of flight. The average error shall not be more than 5 percent of the intervening time intervals, and the maximum error of any one ETA shall not be more than 10 percent.

(19) Demonstrate knowledge and use of D/F equipment and radio facility information. Grading on this item will be based largely on the applicant's selection of those radio aids which will be of most value to his navigation, the manner with which he uses equipment, including filter box controls, and the precision with which he reads bearings. The aircraft's compass heading and all compass corrections must be considered for each bearing.

(20) Use care in tuning to radio stations to insure maximum reception of signal and check for interference signals. Receiver will be checked to ascertain that antenna and BFO (Voice-CW) switches are in correct positions.

(21) Identify at least three radio stations using International Morse code only for identification. The agent or examiner will tune in these stations so that the applicant will have no knowledge of the direction, distance, or frequency of the stations.

(22) Take at least one radio bearing by manual use of the loop. The agent or examiner will check the applicant's bearing by taking a manual bearing on the same station immediately after the applicant.

(23) Show the use of good judgment in evaluating radio bearings, and explain why certain bearings may be of doubtful value.

(24) Determine and apply correctly the correction required to be made to radio bearings before plotting them on a Mercator chart, and demonstrate the ability to plot bearings accurately on charts of the Mercator and Lambert conformal projections.

(25) Compute the compass heading, ETA, and fuel remaining if it is assumed that the flight would be diverted to an alternate airport at a time specified by the agent or examiner.

(26) Check the counter scales of a Loran receiver for accuracy, and explain the basic (face) adjustments which affect tuning and counter alignment. A guide sheet may be used for this test.

(27) Demonstrate a knowledge of the basic principle of Loran and the ability to tune a Loran receiver, to match signals, to read time differences, to plot Loran LOP's, and to identify and use sky waves.

(28) Take and plot bearings from a consol station and explain the precautions which must be

taken when tuning a radio receiver for consol signals. Also, discuss those conditions which affect the reliability of consol bearings.

(29) Demonstrate the ability to properly operate and read an absolute altimeter.

(30) Determine the "D" factors for a series of compared readings of an absolute altimeter and a pressure altimeter.

(31) Determine drift angle or lateral displacement from the true headingline by application of Bellamy's formula or a variation thereof.

(32) Interpret the altimeter comparison data with respect to the pressure system found at flight level. From this data evaluate the accuracy of the prognostic weather map used for flight planning and apply this analysis to the navigation of the flight.

(33) Interpret single LOP's for most probable position, and show how a series of single LOP's of the same body may be used to indicate the probable track and ground speed. Also, show how a series of single LOP's (celestial or radio) from the same celestial body or radio station may be used to determine position when the change of azimuth or bearing is 30° or more between observations.

(34) Select one of the celestial LOP's used during the flight and explain how to make a single line of position approach to a point selected by the agent or examiner, giving headings, times, and ETA's.

(35) Demonstrate the proper use of an astrocompass or periscopic sextant for taking bearings.

(36) Determine compass deviation as soon as possible after reaching cruising altitude and whenever there is a change of compass heading of 15° or more.

(37) Take celestial fixes at hourly intervals when conditions permit. The accuracy of these fixes shall be checked by means of a Loran, radio, or visual fix whenever practicable. After allowing for the probable error of a Loran, radio, or visual fix, a celestial fix under favorable conditions should plot within 10 miles of the actual position.

(38) Select celestial bodies for observation, when possible, whose azimuths will differ by approximately 120° for a 3-body fix and will differ by approximately 90° for a 2-body fix. The altitudes of the selected bodies should be between 25° and 75° whenever practicable.

(39) Have POMAR and any other required reports ready for transmission at time of schedule, and be able to inform the pilot in command promptly with regard to the aircraft's position and progress in comparison with the flight plan.

(40) Keep a log with sufficient legible entries to provide a record from which the flight could be retraced.

(41) Note significant weather changes which might influence the drift or ground speed of the aircraft, such as, temperature, "D" factors, frontal conditions, turbulence, etc.

(42) Determine the wind between fixes as a regular practice.

(43) Estimate the time required and average ground speed during a letdown, under conditions specified by the pilot in command.

(44) Work with sufficient speed to determine the aircraft's position hourly by celestial means and also make all other observations and records pertinent to the navigation. The applicant should be able to take the observation, compute, and plot a celestial LOP within a time limit of 8 minutes; take and plot a Loran LOP within a time limit of 3 minutes for ground waves and 4 minutes for sky waves; observe the absolute and pressure altimeters and compute the drift or lateral displacement within a time limit of 3 minutes.

(45) Be accurate in reading instruments and making computations. Errors which are made and corrected without affecting the navigation will be disregarded unless they cause considerable loss of time.

An uncorrected error in computation (including reading instruments and books) which will affect the reported position more than 25 miles, the heading more than 3°, or any ETA more than 15 minutes will cause this item to be graded unsatisfactory.

(46) Be alert to changing weather or other conditions during flight which might affect the navigation. An applicant should not fail to take celestial observations just prior to encountering a broken or overcast sky condition; and he should not fail to take a bearing on a radio station, which operates at scheduled intervals and which would be a valuable aid to the navigation.

(47) Show a logical choice and sequence in using the various navigation methods according to time and accuracy, and check the positions determined by one method against positions determined by other methods.

(48) Use a logical sequence in performing the various duties of a navigator and plan work according to a schedule. The more important duties should not be neglected for others of less importance.

## Appendix B to Part 63

### Flight Navigator Training Course Requirements

**(a)** *Training course outline*—

(1) *Format.* The ground course outline and the flight course outline shall be combined in one looseleaf binder and shall include a table of contents, divided into two parts—ground course and flight course. Each part of the table of contents

must contain a list of the major subjects, together with hours allotted to each subject and the total classroom and flight hours.

(2) *Ground course outline.*

(i) It is not mandatory that a course outline have the subject headings arranged exactly as listed in this paragraph. Any arrangement of general headings and subheadings will be satisfactory provided all the subject material listed here is included and the acceptable minimum number of hours is assigned to each subject. Each general subject shall be broken down into detail showing items to be covered.

(ii) If any agency desires to include additional subjects in the ground training curriculum, such as international law, flight hygiene, or others which are not required, the hours allotted these additional subjects may not be included in the minimum classroom hours.

(iii) The following subjects with classroom hours are considered the minimum coverage for a ground training course for flight navigators:

| Subject | Classroom hours |
|---|---|
| **Federal Aviation Regulations**<br>To include Parts 63, 91, and 121 of this chapter. | 5 |
| **Meteorology**<br>To include:<br>Basic weather principles<br>Temperature<br>Pressure<br>Winds<br>Moisture in the atmosphere<br>Stability<br>Clouds<br>Hazards<br>Air masses<br>Front weather<br>Fog<br>Thunderstorms<br>Icing<br>World weather and climate<br>Weather maps and weather reports<br>Forecasting | 40 |
| **International Morse code:**<br>Ability to receive code groups of letters and numerals at a speed of eight words per minute | |
| **Navigation instruments**<br>(exclusive of radio and radar)<br>To include:<br>Compasses<br>Pressure altimeters<br>Airspeed indicators<br>Driftmeters<br>Bearing indicators<br>Aircraft octants<br>Instrument calibration and alignment | 20 |
| **Charts and pilotage**<br>To include:<br>Chart projections<br>Chart symbols<br>Principles of pilotage | 15 |
| **Dead Reckoning**<br>To include:<br>Air plot<br>Ground plot<br>Calculation of ETA<br>Vector analysis<br>Use of computer<br>Search | 30 |
| **Absolute altimeter with: Applications**<br>To include:<br>Principles of construction<br>Operating instructions<br>Use of Bellamy's formula<br>Flight planning with single drift correction | 15 |
| **Radio and long-range navigational aids**<br>To include:<br>Principles of radio transmission and reception<br>Radio aids to navigation<br>Government publications<br>Airborne D/F equipment<br>Errors of radio bearings<br>Quadrantal correction<br>Plotting radio bearings<br>ICAO Q code for direction finding<br>Loran<br>Consol | 35 |
| **Celestial navigation**<br>To include:<br>The solar system<br>The celestial sphere<br>The astronomical triangle<br>Theory of lines of position<br>Use of the Air Almanac<br>Time and its applications<br>Navigation tables<br>Precomputation<br>Celestial line of position approach<br>Star identification<br>Corrections to celestial observations | 150 |
| **Flight planning and cruise control**<br>To include:<br>The flight plan<br>Fuel consumption charts<br>Methods of cruise control<br>Flight progress chart<br>Point-of-no-return<br>Equitime point | 25 |
| **Long-range flight problems** | 15 |
| **Total (exclusive of final examinations)** | 350 |

(3) *Flight course outline.*

(i) A minimum of 150 hours of supervised flight training shall be given, of which at least 50 hours of flight training must be given at night, and celestial navigation must be used during flights which total at least 125 hours.

63

(ii) A maximum of 50 hours of the required flight training may be obtained in acceptable types of synthetic flight navigator training devices.

(iii) Flights should be at least four hours in length and should be conducted off civil airways. Some training on long-range flights is desirable, but is not required. There is no limit to the number of students that may be trained on one flight, but at least one astrodome or one periscopic sextant mounting must be provided for each group of four students.

(iv) Training must be given in dead reckoning, pilotage, radio navigation, celestial navigation, and the use of the absolute altimeter.

**(b)** *Equipment.*

(1) Classroom equipment shall include one table at least 24" x 32" in dimensions for each student.

(2) Aircraft suitable for the flight training must be available to the approved course operator to insure that the flight training may be completed without undue delay.

The approved course operator may contract or obtain written agreements with aircraft operators for the use of suitable aircraft. A copy of the contract or written agreement with an aircraft operator shall be attached to each of the three copies of the course outline submitted for approval. In all cases, the approved course operator is responsible for the nature and quality of instruction given during flight.

**(c)** *Instructors.*

(1) Sufficient classroom instructors must be available to prevent an excessive ratio of students to instructors. Any ratio in excess of 20 to 1 will be considered unsatisfactory.

(2) At least one ground instructor must hold a valid flight navigator certificate, and be utilized to coordinate instruction of ground school subjects.

(3) Each instructor who conducts flight training must hold a valid flight navigator certificate.

**(d)** *Revision of training course.*

(1) Requests for revisions to course outlines, facilities, and equipment shall follow procedures for original approval of the course. Revisions should be submitted in such form that an entire page or pages of the approved outline can be removed and replaced by the revisions.

(2) The list of instructors may be revised at any time without request for approval, provided the minimum requirement of paragraph (e) of this section is maintained.

**(e)** *Credit for previous training and experience.*

(1) Credit may be granted by an operator to students for previous training and experience which is provable and comparable to portions of the approved curriculum. When granting such credit, the approved course operator should be fully cognizant of the fact that he is responsible for the proficiency of his graduates in accordance with subdivision (i) of paragraph (3) of this section.

(2) Where advanced credit is allowed, the operator shall evaluate the student's previous training and experience in accordance with the normal practices of accredited technical schools. Before credit is given for any ground school subject or portion thereof, the student must pass an appropriate examination given by the operator. The results of the examination, the basis for credit allowance, and the hours credited shall be incorporated as a part of the student's records.

(3) Credit up to a maximum of 50 hours toward the flight training requirement may be given to pilots who have logged at least 500 hours while a member of a flight crew which required a certificated flight navigator or the Armed Forces equivalent. A similar credit may also be given to a licensed deck officer of the Maritime Service who has served as such for at least one year on ocean-going vessels. One-half of the flight time credited under the terms of this paragraph may be applied toward the 50 hours of flight training required at night.

**(f)** *Students records and reports.* Approval of a course shall not be continued in effect unless the course operator keeps an accurate record of each student, including a chronological log of all instruction, subjects covered and course examinations and grades, and unless he prepares and transmits to the local Flight Standards District Office not later than January 31 of each year, a report containing the following information for the previous calendar year:

(1) The names of all students graduated, together with their school grades for ground and flight subjects.

(2) The names of all students failed or dropped, together with their school grades and reasons for dropping.

**(g)** *Quality of instruction.* Approval of a course shall not be continued in effect unless at least 80 percent of the students who apply within 90 days after graduation are able to qualify on the first attempt for certification as flight navigators.

**(h)** *Statement of graduation.* Each student who successfully completes an approved flight navigator course shall be given a statement of graduation.

**(i)** *Inspections.* Approved course operations will be inspected by authorized representatives of the Administrator as often as deemed necessary to insure that instruction is maintained at the required standards, but the period between inspections shall not exceed 12 months.

**(j)** *Change of ownership, name, or location—*

(1) *Change of ownership.* Approval of a flight navigator course shall not be continued in effect after the course has changed ownership. The new

owner must obtain a new approval by following the procedure prescribed for original approval.

(2) *Change in name.* An approved course changed in name but not changed in ownership shall remain valid if the change is reported by the approved course operator to the local Flight Standards District Office. A letter of approval under the new name will be issued by the regional office.

(3) *Change in location.* An approved course shall remain in effect even though the approved course operator changes location if the change is reported without delay by the operator to the local Flight Standards District Office, which will inspect the facilities to be used. If they are found to be adequate, a letter of approval showing the new location will be issued by the regional office.

**(k)** *Cancellation of approval.*

(1) Failure to meet or maintain any of the requirements set forth in this section for the approval or operation of an approved flight navigator course shall be considered sufficient reason for cancellation of the approval.

(2) If an operator should desire voluntary cancellation of his approved course, he should submit the effective letter of approval and a written request for cancellation to the Administrator through the local Flight Standards District Office.

**(l)** *Duration.* The authority to operate an approved flight navigator course shall expire 24 months after the last day of the month of issuance.

**(m)** *Renewal.* Application for renewal of authority to operate an approved flight navigator course may be made by letter to the local Flight Standards District Office at any time within 60 days before to the expiration date. Renewal of approval will depend upon the course operator meeting the current conditions for approval and having a satisfactory record as an operator.

[Docket No. 1179, 27 FR 7970, Aug. 10, 1962; as amended by Amdt. 63–6, 31 FR 9211, July 6, 1966; Amdt. 63–28, 54 FR 39291, Sept. 25, 1989]

## Appendix C to Part 63

### Flight Engineer Training Course Requirements

**(a)** *Training course outline—*

(1) *Format.* The ground course outline and the flight course outline are independent. Each must be contained in a looseleaf binder to include a table of contents. If an applicant desires approval of both a ground school course and a flight school course, they must be combined in one looseleaf binder that includes a separate table of contents for each course. Separate course outlines are required for each type of airplane.

(2) *Ground course outline.*

(i) It is not mandatory that the subject headings be arranged exactly as listed in this paragraph. Any arrangement of subjects is satisfactory if all the subject material listed here is included and at least the minimum programmed hours are assigned to each subject. Each general subject must be broken down into detail showing the items to be covered.

(ii) If any course operator desires to include additional subjects in the ground course curriculum, such as international law, flight hygiene, or others that are not required, the hours allotted these additional subjects may not be included in the minimum programmed classroom hours.

(iii) The following subjects and classroom hours are the minimum programmed coverage for the initial approval of a ground training course for flight engineers. Subsequent to initial approval of a ground training course an applicant may apply to the Administrator for a reduction in the programmed hours. Approval of a reduction in the approved programmed hours is based on improved training effectiveness due to improvements in methods, training aids, quality of instruction, or any combination thereof.

| Subject | Classroom hours |
|---|---|
| **Federal Aviation Regulations**<br>To include the regulations of this chapter that apply to flight engineers | 10 |
| **Theory of Flight and Aerodynamics** | 10 |
| **Airplane Familiarization**<br>To include as appropriate:<br>Specifications<br>Construction features<br>Flight controls<br>Hydraulic systems<br>Pneumatic systems<br>Electrical systems<br>Anti-icing and de-icing systems<br>Pressurization and air-conditioning systems<br>Pilot static systems<br>Instrument systems<br>Fuel and oil systems<br>Emergency equipment | 90 |
| **Engine Familiarization**<br>To include as appropriate:<br>Specifications<br>Construction features<br>Lubrication<br>Ignition<br>Carburetor and induction, supercharging and fuel control systems<br>Accessories<br>Propellers<br>Instrumentation<br>Emergency equipment | 45 |

| Subject | Classroom hours |
|---|---|
| **Normal Operations (Ground and Flight)**<br>To include as appropriate:<br>Servicing methods and procedures<br>Operation of all the airplane systems<br>Operation of all the engine systems<br>Loading and center of gravity computations<br>Cruise control (normal, long range, maximum endurance)<br>Power and fuel computation<br>Meteorology as applicable to engine operation | **50** |
| **Emergency Operations**<br>To include as appropriate:<br>Landing gear, brakes, flaps, speed brakes, and leading edge devices<br>Pressurization and air-conditioning<br>Portable fire extinguishers<br>Fuselage fire and smoke control<br>Loss of electrical power<br>Engine fire control<br>Engine shut-down and restart<br>Oxygen | **80** |
| **Total (exclusive of final examinations)** | **235** |

The above subjects, except Theory of Flight and Aerodynamics, and Regulations must apply to the same type of airplane in which the student flight engineer is to receive flight training.

(3) *Flight Course Outline.*

(i) The flight training curriculum must include at least 10 hours of flight instruction in an airplane specified in §63.37(a). The flight time required for the practical test may not be credited as part of the required flight instruction.

(ii) All of the flight training must be given in the same type airplane.

(iii) As appropriate to the airplane type, the following subjects must be taught in the flight training course:

SUBJECT

*NORMAL DUTIES, PROCEDURES AND OPERATIONS*

To include as appropriate:
- Airplane preflight.
- Engine starting, power checks, pretakeoff, postlanding and shut-down procedures.
- Power control.
  - Temperature control.
  - Engine operation analysis.
  - Operation of all systems.
  - Fuel management.
  - Logbook entries.
  - Pressurization and air conditioning.

*RECOGNITION AND CORRECTION OF IN-FLIGHT MALFUNCTIONS*

To include:
- Analysis of abnormal engine operation.
- Analysis of abnormal operation of all systems.
- Corrective action.

*EMERGENCY OPERATIONS IN FLIGHT*

To include as appropriate:
- Engine fire control.
- Fuselage fire control.
- Smoke control.
- Loss of power or pressure in each system.
- Engine overspeed.
- Fuel dumping.
- Landing gear, spoilers, speed brakes, and flap extension and retraction.
- Engine shut-down and restart.
- Use of oxygen.

(iv) If the Administrator finds a simulator or flight engineer training device to accurately reproduce the design, function, and control characteristics, as pertaining to the duties and responsibilities of a flight engineer on the type of airplane to be flown, the flight training time may be reduced by a ratio of 1 hour of flight time to 2 hours of airplane simulator time, or 3 hours of flight engineer training device time, as the case may be, subject to the following limitations:

*(a)* Except as provided in subdivision (b) of this paragraph, the required flight instruction time in an airplane may not be less than 5 hours.

*(b)* As to a flight engineer student holding at least a commercial pilot certificate with an instrument rating, airplane simulator or a combination of airplane simulator and flight engineer training device time may be submitted for up to all 10 hours of the required flight instruction time in an airplane. However, not more than 15 hours of flight engineer training device time may be substituted for flight instruction time.

(v) To obtain credit for flight training time, airplane simulator time, or flight engineer training device time, the student must occupy the flight engineer station and operate the controls.

**(b)** *Classroom equipment.* Classroom equipment should consist of systems and procedural training devices, satisfactory to the Administrator, that duplicate the operation of the systems of the airplane in which the student is to receive his flight training.

**(c)** *Contracts or agreements.*

(1) An approved flight engineer course operator may contract with other persons to obtain suitable airplanes, airplane simulators, or other training devices or equipment.

(2) An operator who is approved to conduct both the flight engineer ground course and the flight engineer flight course may contract with others to conduct one course or the other in its entirety but may not contract with others to conduct both courses for the same airplane type.

(3) An operator who has approval to conduct a flight engineer ground course or flight course for a type of airplane, but not both courses, may not contract with another person to conduct that course in whole or in part.

(4) An operator who contracts with another to conduct a flight engineer course may not authorize or permit the course to be conducted in whole or in part by a third person.

(5) In all cases, the course operator who is approved to operate the course is responsible for the nature and quality of the instruction given.

(6) A copy of each contract authorized under this paragraph must be attached to each of the 3 copies of the course outline submitted for approval.

**(d)** *Instructors.*

(1) Only certificated flight engineers may give the flight instruction required by this appendix in an airplane, simulator, or flight engineer training device.

(2) There must be a sufficient number of qualified instructors available to prevent an excess ratio of students to instructors.

**(e)** *Revisions.*

(1) Requests for revisions of the course outlines, facilities or equipment must follow the procedures for original approval of the course. Revisions must be submitted in such form that an entire page or pages of the approved outline can be removed and replaced by the revisions.

(2) The list of instructors may be revised at any time without request for approval, if the requirements of paragraph (d) of this appendix are maintained.

**(f)** *Ground school credits.*

(1) Credit may be granted a student in the ground school course by the course operator for comparable previous training or experience that the student can show by written evidence: however, the course operator must still meet the quality of instruction as described in paragraph (h) of this appendix.

(2) Before credit for previous training or experience may be given, the student must pass a test given by the course operator on the subject for which the credit is to be given. The course operator shall incorporate results of the test, the basis for credit allowance, and the hours credited as part of the student's records.

**(g)** *Records and reports.*

(1) The course operator must maintain, for at least two years after a student graduates, fails, or drops from a course, a record of the student's training, including a chronological log of the subject course, attendance examinations, and grades.

(2) Except as provided in paragraph (3) of this section, the course operator must submit to the Administrator, not later than January 31 of each year, a report for the previous calendar year's training, to include:

(i) Name, enrollment and graduation date of each student;

(ii) Ground school hours and grades of each student;

(iii) Flight, airplane simulator, flight engineer training device hours, and grades of each student; and

(iv) Names of students failed or dropped, together with their school grades and reasons for dropping.

(3) Upon request, the Administrator may waive the reporting requirements of paragraph (2) of this section for an approved flight engineer course that is part of an approved training course under subpart N of part 121 of this chapter.

**(h)** *Quality of instruction.*

(1) Approval of a ground course is discontinued whenever less than 80 percent of the students pass the FAA written test on the first attempt.

(2) Approval of a flight course is discontinued whenever less than 80 percent of the students pass the FAA practical test on the first attempt.

(3) Notwithstanding paragraphs (1) and (2) of this section, approval of a ground or flight course may be continued when the Administrator finds—

(i) That the failure rate was based on less than a representative number of students; or

(ii) That the course operator has taken satisfactory means to improve the effectiveness of the training.

**(i)** *Time limitation.* Each student must apply for the written test and the flight test within 90 days after completing the ground school course.

**(j)** *Statement of course completion.*

(1) The course operator shall give to each student who successfully completes an approved flight engineer ground school training course, and passes the FAA written test, a statement of successful completion of the course that indicates the date of training, the type of airplane on which the ground course training was based, and the number of hours received in the ground school course.

(2) The course operator shall give each student who successfully completes an approved flight engineer flight course, and passed the FAA practical test, a statement of successful completion of the flight course that indicates the dates of the training, the type of airplane used in the flight course, and the number of hours received in the flight course.

(3) A course operator who is approved to conduct both the ground course and the flight course may include both courses in a single statement of course completion if the provisions of paragraphs (1) and (2) of this section are included.

(4) The requirements of this paragraph do not apply to an air carrier or commercial operator with an approved training course under part 121 of this

chapter providing the student receives a flight engineer certificate upon completion of that course.

**(k)** *Inspections.* Each course operator shall allow the Administrator at any time or place, to make any inspection necessary to ensure that the quality and effectiveness of the instruction are maintained at the required standards.

**(l)** *Change of ownership, name, or location.*

(1) Approval of a flight engineer ground course or flight course is discontinued if the ownership of the course changes. The new owner must obtain a new approval by following the procedure prescribed for original approval.

(2) Approval of a flight engineer ground course or flight course does not terminate upon a change in the name of the course that is reported to the Administrator within 30 days. The Administrator issues a new letter of approval, using the new name, upon receipt of notice within that time.

(3) Approval of a flight engineer ground course or flight course does not terminate upon a change in location of the course that is reported to the Administrator within 30 days. The Administrator issues a new letter of approval, showing the new location, upon receipt of notice within that time, if he finds the new facilities to be adequate.

**(m)** *Cancellation of approval.*

(1) Failure to meet or maintain any of the requirements of this appendix for the approval of a flight engineer ground course or flight course is reason for cancellation of the approval.

(2) If a course operator desires to voluntarily terminate the course, he should notify the Administrator in writing and return the last letter of approval.

**(n)** *Duration.* Except for a course operated as part of an approved training course under subpart N of part 121 of this chapter, the approval to operate a flight engineer ground course or flight course terminates 24 months after the last day of the month of issue.

**(o)** *Renewal.*

(1) Renewal of approval to operate a flight engineer ground course or flight course is conditioned upon the course operator's meeting the requirements of this appendix.

(2) Application for renewal may be made to the Administrator at any time after 60 days before the termination date.

**(p)** *Course operator approvals.* An applicant for approval of a flight engineer ground course, or flight course, or both, must meet all of the requirements of this appendix concerning application, approval, and continuing approval of that course or courses.

**(q)** *Practical test eligibility.* An applicant for a flight engineer certificate and class rating under the provisions of §63.37(b)(6) is not eligible to take the practical test unless he has successfully completed an approved flight engineer ground school course in the same type of airplane for which he has completed an approved flight engineer flight course.

[Docket No. 6458, 30 FR 14560, Nov. 23, 1965; as amended by Amdt. 63–15, 37 FR 9758, May 17, 1972]

# PART 65

# CERTIFICATION: AIRMEN OTHER THAN FLIGHT CREWMEMBERS

Appendix A to Part 65—
Aircraft Dispatcher Courses

**Authority:** 49 U.S.C. 106(g), 40113, 44701–44703, 44707, 44709–44711, 45102–45103, 45301–45302.

**Source:** Docket No. 1179, 27 FR 7973, Aug. 10, 1962, unless otherwise noted.

SPECIAL FEDERAL AVIATION REGULATIONS

## SFAR No. 58 TO PART 65

ADVANCED QUALIFICATION PROGRAM

*EDITORIAL NOTE: FOR THE TEXT OF SFAR NO. 58, SEE PART 121 OF THIS CHAPTER.*

## SFAR No. 96 TO PART 65

RELIEF FOR PARTICIPANTS IN OPERATION ENDURING FREEDOM

**1.** *Applicability.* Flight Standards District Offices are authorized to accept from an eligible person, as described in paragraph 2 of this SFAR, the following:

(a) An expired flight instructor certificate to show eligibility for renewal of a flight instructor certificate under §61.197, or an expired written test report to show eligibility under part 61 to take a practical test;

(b) An expired written test report to show eligibility under §§63.33 and 63.57 to take a practical test; and

(c) An expired written test report to show eligibility to take a practical test required under part 65 or an expired authorization to show eligibility for renewal under §65.93.

**2.** *Eligibility.* A person is eligible for the relief described in paragraph 1 of this SFAR if:

(a) The person served in a civilian or military capacity in support of Operation Enduring Freedom outside the United States during some period of time from September, 11, 2001, to May 6, 2004;

(b) The person's flight instructor certificate, airman written test report, or inspection authorization expired some time between September 11, 2001, and 6 calendar months after returning to the United States, or May 6, 2004, whichever is earlier; and

(c) The person complies with §61.197 or §65.93 of this chapter, as appropriate, or completes the appropriate practical test within 6 calendar months after returning to the United States, or May 6, 2004, whichever is earlier.

**3.** *Required documents.* The person must present in person the Airman Certificate and/or Rating Application (FAA Form 8710-1) to the appropriate Flight Standards District Office. The person must include with the application one of the following documents, which must show the date of assignment outside the United States and the date of return to the United States.

(a) An official U.S. Government notification of personnel action, or equivalent document, showing the person was a civilian on official duty for the U.S. Government in support of Operation Enduring Freedom outside the United States some time between September 11, 2001, and May 6, 2004;

(6) Military orders showing the person was assigned to duty outside the United States in support of Operation Enduring Freedom some time between September 11, 2001, and May 6, 2004;

(c) A letter from the person's unit commander or supervisor providing the dates during which the person served outside the United States in support of Operation Enduring Freedom some time between September 11, 2001, and May 6, 2004.

**4.** *Expiration date.* This Special Federal Aviation Regulation No. 96 expires May 6, 2004, unless sooner superseded or rescinded.

[Docket No. FAA–2002–12199, 67 FR 30527, May 6, 2002]

# Subpart A—General

### §65.1 Applicability.

This part prescribes the requirements for issuing the following certificates and associated ratings and the general operating rules for the holders of those certificates and ratings:

**(a)** Air-traffic control-tower operators.
**(b)** Aircraft dispatchers.
**(c)** Mechanics.
**(d)** Repairmen.
**(e)** Parachute riggers.

### §65.3 Certification of foreign airmen other than flight crewmembers.

A person who is neither a U.S. citizen nor a resident alien is issued a certificate under subpart D of this part, outside the United States, only when the Administrator finds that the certificate is needed for the operation or continued airworthiness of a U.S.-registered civil aircraft.

[Docket 65–28, FR 35693, Aug. 16, 1982]

### §65.11 Application and issue.

**(a)** Application for a certificate and appropriate class rating, or for an additional rating, under this part must be made on a form and in a manner prescribed by the Administrator. Each person who is neither a U.S. citizen nor a resident alien and who applies for a written or practical test to be administered outside the United States or for any certificate or rating issued under this part must show evidence that the fee prescribed in Appendix A of part 187 of this chapter has been paid.

**(b)** An applicant who meets the requirements of this part is entitled to an appropriate certificate and rating.

**(c)** Unless authorized by the Administrator, a person whose air traffic control tower operator, mechanic, or parachute rigger certificate is suspended may not apply for any rating to be added to that certificate during the period of suspension.

**(d)** Unless the order of revocation provides otherwise—

(1) A person whose air traffic control tower operator, aircraft dispatcher, or parachute rigger certificate is revoked may not apply for the same kind of certificate for 1 year after the date of revocation; and

(2) A person whose mechanic or repairman certificate is revoked may not apply for either of those kinds of certificates for 1 year after the date of revocation.

[Docket No. 1179, 27 FR 7973, Aug. 10, 1962; as amended by Amdt. 65–9, 31 FR 13524, Oct. 20, 1966; Amdt. 65–28, 47 FR 35693, Aug. 16, 1982]

## §65.12 Offenses involving alcohol or drugs.

**(a)** A conviction for the violation of any Federal or state statute relating to the growing, processing, manufacture, sale, disposition, possession, transportation, or importation of narcotic drugs, marihuana, or depressant or stimulant drugs or substances is grounds for—

(1) Denial of an application for any certificate or rating issued under this part for a period of up to 1 year after the date of final conviction; or

(2) Suspension or revocation of any certificate or rating issued under this part.

**(b)** The commission of an act prohibited by §91.19(a) of this chapter is grounds for—

(1) Denial of an application for a certificate or rating issued under this part for a period of up to 1 year after the date of that act; or

(2) Suspension or revocation of any certificate or rating issued under this part.

[Docket No. 21956, 50 FR 15379, Apr. 17, 1985; as amended by Amdt. 65–34, 54 FR 34330, Aug. 18, 1989]

## §65.13 Temporary certificate.

A certificate and ratings effective for a period of not more than 120 days may be issued to a qualified applicant, pending review of his application and supplementary documents and the issue of the certificate and ratings for which he applied.

[Docket No. 1179, 27 FR 7973, Aug. 10, 1962; as amended by Amdt. 65–23, 43 FR 22640, May 25, 1978]

## §65.15 Duration of certificates.

**(a)** Except for repairman certificates, a certificate or rating issued under this part is effective until it is surrendered, suspended, or revoked.

**(b)** Unless it is sooner surrendered, suspended, or revoked, a repairman certificate is effective until the holder is relieved from the duties for which the holder was employed and certificated.

**(c)** The holder of a certificate issued under this part that is suspended, revoked, or no longer effective shall return it to the Administrator.

[Docket No. 22052, 47 FR 35693, Aug. 16, 1982]

## §65.16 Change of name: Replacement of lost or destroyed certificate.

**(a)** An application for a change of name on a certificate issued under this part must be accompanied by the applicant's current certificate and the marriage license, court order, or other document verifying the change. The documents are returned to the applicant after inspection.

**(b)** An application for a replacement of a lost or destroyed certificate is made by letter to the Department of Transportation, Federal Aviation Administration, Airman Certification Branch, Post Office Box 25082, Oklahoma City, OK 73125. The letter must—

(1) Contain the name in which the certificate was issued, the permanent mailing address (including zip code), social security number (if any), and date and place of birth of the certificate holder, and any available information regarding the grade, number, and date of issue of the certificate, and the ratings on it; and

(2) Be accompanied by a check or money order for $2, payable to the Federal Aviation Administration.

**(c)** An application for a replacement of a lost or destroyed medical certificate is made by letter to the Department of Transportation, Federal Aviation Administration, Civil Aeromedical Institute, Aeromedical Certification Branch, Post Office Box 25082, Oklahoma City, OK 73125, accompanied by a check or money order for $2.00.

**(d)** A person whose certificate issued under this part or medical certificate, or both, has been lost may obtain a telegram from the FAA confirming that it was issued. The telegram may be carried as a certificate for a period not to exceed 60 days pending his receiving a duplicate certificate under paragraph (b) or (c) of this section, unless he has been notified that the certificate has been suspended or revoked. The request for such a telegram may be made by prepaid telegram, stating the date upon which a duplicate certificate was requested, or including the request for a duplicate and a money order for the necessary amount. The request for a telegraphic certificate

65

should be sent to the office prescribed in paragraph (b) or (c) of this section, as appropriate. However, a request for both at the same time should be sent to the office prescribed in paragraph (b) of this section.

[Docket No. 7258, 31 FR 13524, Oct. 20, 1966; as amended by Docket No. 8084, 32 FR 5769, Apr. 11, 1967; Amdt. 65–16, 35 FR 14075, Sept. 4, 1970; Amdt. 65–17, 36 FR 2865, Feb. 11, 1971]

## §65.17 Tests: General procedure.

**(a)** Tests prescribed by or under this part are given at times and places, and by persons, designated by the Administrator.

**(b)** The minimum passing grade for each test is 70 percent.

## §65.18 Written tests: Cheating or other unauthorized conduct.

**(a)** Except as authorized by the Administrator, no person may—

(1) Copy, or intentionally remove, a written test under this part;

(2) Give to another, or receive from another, any part or copy of that test;

(3) Give help on that test to, or receive help on that test from, any person during the period that test is being given;

(4) Take any part of that test in behalf of another person;

(5) Use any material or aid during the period that test is being given; or

(6) Intentionally cause, assist, or participate in any act prohibited by this paragraph.

**(b)** No person who commits an act prohibited by paragraph (a) of this section is eligible for any airman or ground instructor certificate or rating under this chapter for a period of 1 year after the date of that act. In addition, the commission of that act is a basis for suspending or revoking any airman or ground instructor certificate or rating held by that person.

[Docket No. 4086, 30 FR 2196, Feb. 18, 1965]

## §65.19 Retesting after failure.

An applicant for a written, oral, or practical test for a certificate and rating, or for an additional rating under this part, may apply for retesting—

**(a)** After 30 days after the date the applicant failed the test; or

**(b)** Before the 30 days have expired if the applicant presents a signed statement from an airman holding the certificate and rating sought by the applicant, certifying that the airman has given the applicant additional instruction in each of the subjects failed and that the airman considers the applicant ready for retesting.

[Docket No. 16383, 43 FR 22640, May 25, 1978]

## §65.20 Applications, certificates, logbooks, reports, and records: Falsification, reproduction, or alteration.

**(a)** No person may make or cause to be made—

(1) Any fraudulent or intentionally false statement on any application for a certificate or rating under this part;

(2) Any fraudulent or intentionally false entry in any logbook, record, or report that is required to be kept, made, or used, to show compliance with any requirement for any certificate or rating under this part;

(3) Any reproduction, for fraudulent purpose, of any certificate or rating under this part; or

(4) Any alteration of any certificate or rating under this part.

**(b)** The commission by any person of an act prohibited under paragraph (a) of this section is a basis for suspending or revoking any airman or ground instructor certificate or rating held by that person.

[Docket No. 4086, 30 FR 2196, Feb. 18, 1965]

## §65.21 Change of address.

Within 30 days after any change in his permanent mailing address, the holder of a certificate issued under this part shall notify the Department of Transportation, Federal Aviation Administration, Airman Certification Branch, Post Office Box 25082, Oklahoma City, OK 73125, in writing, of his new address.

[Docket No. 10536, 35 FR 14075, Sept. 4, 1970]

## §65.23 Refusal to submit to a drug or alcohol test.

**(a)** *General.* This section applies to an employee who performs a function listed in Appendix I or Appendix J to part 121 of this chapter directly or by contract for a part 121 certificate holder, a part 135 certificate holder, an operator as defined in §135.1(c) of this chapter, or an air traffic control facility not operated by the FAA or the U.S. military.

**(b)** Refusal by the holder of a certificate issued under this part to take a drug test required under the provisions of Appendix I to part 121 or an alcohol test required under the provisions of Appendix J to part 121 is grounds for—

(1) Denial of an application for any certificate or rating issued under this part for a period of up to 1 year after the date of such refusal; and

(2) Suspension or revocation of any certificate or rating issued under this part.

[Docket No. 25148, 53 FR 47056, Nov. 21, 1988; 54 FR 1289, Jan. 12, 1989; as amended by Amdt. 65–33, 54 FR 15152, April 14, 1989; Amdt. 65–37, 59 FR 7389, Feb. 15, 1994]

# Subpart B—Air Traffic Control Tower Operators

**Source:** Docket No. 10193, 35 FR 12326, Aug. 1, 1970, unless otherwise noted.

## §65.31 Required certificates, and rating or qualification.

No person may act as an air traffic control tower operator at an air traffic control tower in connection with civil aircraft unless he—

**(a)** Holds an air traffic control tower operator certificate issued to him under this subpart;

**(b)** Holds a facility rating for that control tower issued to him under this subpart, or has qualified for the operating position at which he acts and is under the supervision of the holder of a facility rating for that control tower; and

For the purpose of this subpart, *operating position* means an air traffic control function performed within or directly associated with the control tower;

**(c)** Except for a person employed by the FAA or employed by, or on active duty with, the Department of the Air Force, Army, or Navy or the Coast Guard, holds at least a second-class medical certificate issued under part 67 of this chapter.

[Docket No. 10193, 35 FR 12326, Aug. 1, 1970; as amended by Amdt. 65–25, 45 FR 18911, March 24, 1980; Amdt. 65–31, 52 FR 17518, May 8, 1987]

## §65.33 Eligibility requirements: General.

To be eligible for an air traffic control tower operator certificate a person must—

**(a)** Be at least 18 years of age;

**(b)** Be of good moral character;

**(c)** Be able to read, write, and understand the English language and speak it without accent or impediment of speech that would interfere with two-way radio conversation;

**(d)** Except for a person employed by the FAA or employed by, or on active duty with, the Department of the Air Force, Army, or Navy or the Coast Guard, holds at least a second-class medical certificate issued under part 67 of this chapter within the 12 months before the date application is made; and

**(e)** Comply with §65.35.

[Docket No. 10193, 35 FR 12326, Aug. 1, 1970; as amended by Amdt. 65–25, 45 FR 18911, March 24, 1980; Amdt. 65–31, 52 FR 17518, May 8, 1987]

## §65.35 Knowledge requirements.

Each applicant for an air traffic control tower operator certificate must pass a written test on—

**(a)** The flight rules in part 91 of this chapter:

**(b)** Airport traffic control procedures, and this subpart:

**(c)** En route traffic control procedures;

**(d)** Communications operating procedures;

**(e)** Flight assistance service;

**(f)** Air navigation, and aids to air navigation; and

**(g)** Aviation weather.

## §65.37 Skill requirements: Operating positions.

No person may act as an air traffic control tower operator at any operating position unless he has passed a practical test on—

**(a)** Control tower equipment and its use;

**(b)** Weather reporting procedures and use of reports;

**(c)** Notices to Airmen, and use of the Airman's Information Manual;

**(d)** Use of operational forms;

**(e)** Performance of noncontrol operational duties; and

**(f)** Each of the following procedures that is applicable to that operating position and is required by the person performing the examination:

(1) The airport, including rules, equipment, runways, taxiways, and obstructions.

(2) The terrain features, visual check-points, and obstructions within the lateral boundaries of the surface areas of Class B, Class C, Class D, or Class E airspace designated for the airport.

(3) Traffic patterns and associated procedures for use of preferential runways and noise abatement.

(4) Operational agreements.

(5) The center, alternate airports, and those airways, routes, reporting points, and air navigation aids used for terminal air traffic control.

(6) Search and rescue procedures.

(7) Terminal air traffic control procedures and phraseology.

(8) Holding procedures, prescribed instrument approach, and departure procedures.

(9) Radar alignment and technical operation.

(10) The application of the prescribed radar and nonradar separation standard, as appropriate.

[Docket No. 10193, 35 FR 12326, Aug. 1, 1991; as amended by Amdt. 65–36, 56 FR 65653, Dec. 17, 1991]

### §65.39 Practical experience requirements: Facility rating.

Each applicant for a facility rating at any air traffic control tower must have satisfactorily served—

**(a)** As an air traffic control tower operator at that control tower without a facility rating for at least 6 months; or

**(b)** As an air traffic control tower operator with a facility rating at a different control tower for at least 6 months before the date he applies for the rating.

However, an applicant who is a member of an Armed Force of the United States meets the requirements of this section if he has satisfactorily served as an air traffic control tower operator for at least 6 months.

[Docket No. 1179, 27 FR 7973, Aug. 10, 1962; as amended by Amdt. 65–19, 36 FR 21280, Nov. 5, 1971]

### §65.41 Skill requirements: Facility ratings.

Each applicant for a facility rating at an air traffic control tower must have passed a practical test on each item listed in §65.37 of this part that is applicable to each operating position at the control tower at which the rating is sought.

### §65.43 Rating privileges and exchange.

**(a)** The holder of a senior rating on August 31, 1970, may at any time after that date exchange his rating for a facility rating at the same air traffic control tower. However, if he does not do so before August 31, 1971, he may not thereafter exercise the privileges of his senior rating at the control tower concerned until he makes the exchange.

**(b)** The holder of a junior rating on August 31, 1970, may not control air traffic, at any operating position at the control tower concerned, until he has met the applicable requirements of §65.37 of this part. However, before meeting those requirements he may control air traffic under the supervision, where required, of an operator with a senior rating (or facility rating) in accordance with §65.41 of this part in effect before August 31, 1970.

### §65.45 Performance of duties.

**(a)** An air traffic control tower operator shall perform his duties in accordance with the limitations on his certificate and the procedures and practices prescribed in air traffic control manuals of the FAA, to provide for the safe, orderly, and expeditious flow of air traffic.

**(b)** An operator with a facility rating may control traffic at any operating position at the control tower at which he holds a facility rating. However, he may not issue an air traffic clearance for IFR flight without authorization from the appropriate facility exercising IFR control at that location.

**(c)** An operator who does not hold a facility rating for a particular control tower may act at each operating position for which he has qualified, under the supervision of an operator holding a facility rating for that control tower.

[Docket No. 10193, 35 FR 12326, Aug. 1, 1970; as amended by Amdt. 65–16, 35 FR 14075, Sept. 4, 1970]

### §65.46 Use of prohibited drugs.

**(a)** The following definitions apply for the purposes of this section:

(1) An *employee* is a person who performs an air traffic control function for an employer. For the purpose of this section, a person who performs such a function pursuant to a contract with an employer is considered to be performing that function for the employer.

(2) An "employer" means an air traffic control facility not operated by the FAA or by or under contract to the U.S. military that employs a person to perform an air traffic control function.

**(b)** Each employer shall provide each employee performing a function listed in Appendix I to part 121 of this chapter and his or her supervisor with the training specified in that appendix. No employer may use any contractor to perform an air traffic control function unless that contractor provides each of its employees and his or her supervisor with the training specified in that appendix.

**(c)** No employer may knowingly use any person to perform, nor may any person perform for an employer, either directly or by contract, any air traffic control function while that person has a prohibited drug, as defined in Appendix I to part 121 of this chapter, in his or her system.

**(d)** No employer shall knowingly use any person to perform, nor may any person perform for an employer, either directly or by contract, any air traffic control function if the person has a verified positive drug test result on or has refused to submit to a drug test required by Appendix I to part 121 of this chapter and the person has not met the requirements of Appendix I to part 121 of this chapter for returning to the performance of safety-sensitive duties.

**(e)** Each employer shall test each of its employees who perform any air traffic control function in accordance with Appendix I to part 121 of this chapter. No employer may use any contractor to perform any air traffic control function unless that contractor tests each employee performing such a function for the employer in accordance with that appendix.

[Docket No. 25148, 53 FR 47056, Nov. 21, 1988; as amended by Amdt. 65–38, 59 FR 42977, Aug. 19, 1994]

## §65.46a Misuse of alcohol.

**(a)** This section applies to employees who perform air traffic control duties directly or by contract for an employer that is an air traffic control facility not operated by the FAA or the U.S. military *(covered employees).*

**(b)** *Alcohol concentration.* No covered employee shall report for duty or remain on duty requiring the performance of safety-sensitive functions while having an alcohol concentration of 0.04 or greater. No employer having actual knowledge that an employee has an alcohol concentration of 0.04 or greater shall permit the employee to perform or continue to perform safety-sensitive functions.

**(c)** *On-duty use.* No covered employee shall use alcohol while performing safety-sensitive functions. No employer having actual knowledge that a covered employee is using alcohol while performing safety-sensitive functions shall permit the employee to perform or continue to perform safety-sensitive functions.

**(d)** *Pre-duty use.* No covered employees shall perform air traffic control duties within 8 hours after using alcohol. No employer having actual knowledge that such an employee has used alcohol within 8 hours shall permit the employee to perform or continue to perform air traffic control duties.

**(e)** *Use following an accident.* No covered employee who has actual knowledge of an accident involving an aircraft for which he or she performed a safety-sensitive function at or near the time of the accident shall use alcohol for 8 hours following the accident, unless he or she has been given a post-accident test under Appendix J to part 121 of this chapter, or the employer has determined that the employee's performance could not have contributed to the accident.

**(f)** *Refusal to submit to a required alcohol test.* No covered employee shall refuse to submit to a post-accident, random, reasonable suspicion, or follow-up alcohol test required under Appendix J to part 121 of this chapter. No employer shall permit an employee who refuses to submit to such a test to perform or continue to perform safety-sensitive functions.

[Docket No. 27065, 59 FR 7389, Feb. 15, 1994]

## §65.46b Testing for alcohol.

**(a)** Each air traffic control facility not operated by the FAA or the U.S. military (hereinafter *employer*) must establish an alcohol misuse prevention program in accordance with the provisions of Appendix J to part 121 of this chapter.

**(b)** No employer shall use any person who meets the definition of *covered employee* in Appendix J to part 121 to perform a safety-sensitive function listed in that appendix unless such person is subject to testing for alcohol misuse in accordance with the provisions of Appendix J.

[Docket No. 27065, 59 FR 7389, Feb. 15, 1994]

## §65.47 Maximum hours.

Except in an emergency, a certificated air traffic control tower operator must be relieved of all duties for at least 24 consecutive hours at least once during each 7 consecutive days. Such an operator may not serve or be required to serve—

**(a)** For more than 10 consecutive hours; or

**(b)** For more than 10 hours during a period of 24 consecutive hours, unless he has had a rest period of at least 8 hours at or before the end of the 10 hours of duty.

## §65.49 General operating rules.

**(a)** Except for a person employed by the FAA or employed by, or on active duty with, the Department of the Air Force, Army, or Navy, or the Coast Guard, no person may act as an air traffic control tower operator under a certificate issued to him or her under this part unless he or she has in his or her personal possession an appropriate current medical certificate issued under part 67 of this chapter.

**(b)** Each person holding an air traffic control tower operator certificate shall keep it readily available when performing duties in an air traffic control tower, and shall present that certificate or his medical certificate or both for inspection upon the request of the Administrator or an authorized representative of the National Transportation Safety Board, or of any Federal, State, or local law enforcement officer.

**(c)** A certificated air traffic control tower operator who does not hold a facility rating for a particular control tower may not act at any operating position at the control tower concerned unless there is maintained at that control tower, readily available to persons named in paragraph (b) of this section, a current record of the operating positions at which he has qualified.

**(d)** An air traffic control tower operator may not perform duties under his certificate during any period of known physical deficiency that would make him unable to meet the physical requirements for his current medical certificate. However, if the deficiency is temporary, he may perform duties that are not affected by it whenever another certificated and qualified operator is present and on duty.

**(e)** A certificated air traffic control tower operator may not control air traffic with equipment that the Administrator has found to be inadequate.

**(f)** The holder of an air traffic control tower operator certificate, or an applicant for one, shall, upon the reasonable request of the Administrator, cooperate fully in any test that is made of him.

[Docket No. 1179, 27 FR 7973, Aug. 10, 1962; as amended by Amdt. 65–31, 52 FR 17519, May 8, 1987]

### §65.50 Currency requirements.

The holder of an air traffic control tower operator certificate may not perform any duties under that certificate unless—

**(a)** He has served for at least three of the preceding 6 months as an air traffic control tower operator at the control tower to which his facility rating applies, or at the operating positions for which he has qualified; or

**(b)** He has shown that he meets the requirements for his certificate and facility rating at the control tower concerned, or for operating at positions for which he has previously qualified.

## Subpart C— Aircraft Dispatchers

**Source:** Docket No. FAA–1998–4553, 64 FR 68923, Dec. 8, 1999, unless otherwise noted.

### §65.51 Certificate required.

**(a)** No person may act as an aircraft dispatcher (exercising responsibility with the pilot in command in the operational control of a flight) in connection with any civil aircraft in air commerce unless that person has in his or her personal possession an aircraft dispatcher certificate issued under this subpart.

**(b)** Each person who holds an aircraft dispatcher certificate must present it for inspection upon the request of the Administrator or an authorized representative of the National Transportation Safety Board, or of any Federal, State, or local law enforcement officer.

### §65.53 Eligibility requirements: General.

**(a)** To be eligible to take the aircraft dispatcher knowledge test, a person must be at least 21 years of age.

**(b)** To be eligible for an aircraft dispatcher certificate, a person must—

(1) Be at least 23 years of age;

(2) Be able to read, speak, write, and understand the English language;

(3) Pass the required knowledge test prescribed by §65.55 of this part;

(4) Pass the required practical test prescribed by §65.59 of this part; and

(5) Comply with the requirements of §65.57 of this part.

### §65.55 Knowledge requirements.

**(a)** A person who applies for an aircraft dispatcher certificate must pass a knowledge test on the following aeronautical knowledge areas:

(1) Applicable Federal Aviation Regulations of this chapter that relate to airline transport pilot privileges, limitations, and flight operations;

(2) Meteorology, including knowledge of and effects of fronts, frontal characteristics, cloud formations, icing, and upper-air data;

(3) General system of weather and NOTAM collection, dissemination, interpretation, and use;

(4) Interpretation and use of weather charts, maps, forecasts, sequence reports, abbreviations, and symbols;

(5) National Weather Service functions as they pertain to operations in the National Airspace System;

(6) Windshear and microburst awareness, identification, and avoidance;

(7) Principles of air navigation under instrument meteorological conditions in the National Airspace System;

(8) Air traffic control procedures and pilot responsibilities as they relate to enroute operations, terminal area and radar operations, and instrument departure and approach procedures;

(9) Aircraft loading, weight and balance, use of charts, graphs, tables, formulas, and computations, and their effect on aircraft performance;

(10) Aerodynamics relating to an aircraft's flight characteristics and performance in normal and abnormal flight regimes;

(11) Human factors;

(12) Aeronautical decision making and judgment; and

(13) Crew resource management, including crew communication and coordination.

**(b)** The applicant must present documentary evidence satisfactory to the administrator of having passed an aircraft dispatcher knowledge test within the preceding 24 calendar months.

### §65.57 Experience or training requirements.

An applicant for an aircraft dispatcher certificate must present documentary evidence satisfactory to the Administrator that he or she has the experience prescribed in paragraph (a) of this section or has accomplished the training described in paragraph (b) of this section as follows:

**(a)** A total of at least 2 years experience in the 3 years before the date of application, in any one or in any combination of the following areas:

(1) In military aircraft operations as a—

(i) Pilot;
(ii) Flight navigator; or
(iii) Meteorologist.
(2) In aircraft operations conducted under part 121 of this chapter as—
(i) An assistant in dispatching air carrier aircraft, under the direct supervision of a dispatcher certificated under this subpart;
(ii) A pilot;
(iii) A flight engineer; or
(iv) A meteorologist.
(3) In aircraft operations as—
(i) An Air Traffic Controller; or
(ii) A Flight Service Specialist.
(4) In aircraft operations, performing other duties that the Administrator finds provide equivalent experience.
**(b)** A statement of graduation issued or revalidated in accordance with §65.70(b) of this part, showing that the person has successfully completed an approved aircraft dispatcher course.

### §65.59 Skill requirements.

An applicant for an aircraft dispatcher certificate must pass a practical test given by the Administrator, with respect to any one type of large aircraft used in air carrier operations. The practical test must be based on the aircraft dispatcher practical test standards, as published by the FAA, on the items outlined in appendix A of this part.

### §65.61 Aircraft dispatcher certification courses: Content and minimum hours.

**(a)** An approved aircraft dispatcher certification course must:
(1) Provide instruction in the areas of knowledge and topics listed in appendix A of this part;
(2) Include a minimum of 200 hours of instruction.
**(b)** An applicant for approval of an aircraft dispatcher course must submit an outline that describes the major topics and subtopics to be covered and the number of hours proposed for each.
**(c)** Additional subject headings for an aircraft dispatcher certification course may also be included, however the hours proposed for any subjects not listed in appendix A of this part must be in addition to the minimum 200 course hours required in paragraph (a) of this section.
**(d)** For the purpose of completing an approved course, a student may substitute previous experience or training for a portion of the minimum 200 hours of training. The course operator determines the number of hours of credit based on an evaluation of the experience or training to determine if it is comparable to portions of the approved course curriculum. The credit allowed, including the total hours and the basis for it, must be placed in the student's record required by §65.70(a) of this part.

### §65.63 Aircraft dispatcher certification courses: Application, duration, and other general requirements.

**(a)** Application. Application for original approval of an aircraft dispatcher certification course or the renewal of approval of an aircraft dispatcher certification course under this part must be:
(1) Made in writing to the Administrator;
(2) Accompanied by two copies of the course outline required under §65.61(b) of this part, for which approval is sought;
(3) Accompanied by a description of the equipment and facilities to be used; and
(4) Accompanied by a list of the instructors and their qualifications.
**(b)** Duration. Unless withdrawn or canceled, an approval of an aircraft dispatcher certification course of study expires:
(1) On the last day of the 24th month from the month the approval was issued; or
(2) Except as provided in paragraph (f) of this section, on the date that any change in ownership of the school occurs.
**(c)** Renewal. Application for renewal of an approved aircraft dispatcher certification course must be made within 30 days preceding the month the approval expires, provided the course operator meets the following requirements:
(1) At least 80 percent of the graduates from that aircraft dispatcher certification course, who applied for the practical test required by §65.59 of this part, passed the practical test on their first attempt; and
(2) The aircraft dispatcher certification course continues to meet the requirements of this subpart for course approval.
**(d)** Course revisions. Requests for approval of a revision of the course outline, facilities, or equipment must be in accordance with paragraph (a) of this section. Proposed revisions of the course outline or the description of facilities and equipment must be submitted in a format that will allow an entire page or pages of the approved outline or description to be removed and replaced by any approved revision. The list of instructors may be revised at any time without request for approval, provided the minimum requirements of §65.67 of this part are maintained and the Administrator is notified in writing.
**(e)** Withdrawal or cancellation of approval. Failure to continue to meet the requirements of this subpart for the approval or operation of an approved aircraft dispatcher certification course is grounds for withdrawal of approval of the course. A course operator may request cancellation of course approval by a letter to the Administrator.

The operator must forward any records to the FAA as requested by the Administrator.

**(f)** Change in ownership. A change in ownership of a part 65, appendix A-approved course does not terminate that aircraft dispatcher certification course approval if, within 10 days after the date that any change in ownership of the school occurs:

(1) Application is made for an appropriate amendment to the approval; and

(2) No change in the facilities, personnel, or approved aircraft dispatcher certification course is involved.

**(g)** Change in name or location. A change in name or location of an approved aircraft dispatcher certification course does not invalidate the approval if, within 10 days after the date that any change in name or location occurs, the course operator of the part 65, appendix A-approved course notifies the Administrator, in writing, of the change.

### §65.65 Aircraft dispatcher certification courses: Training facilities.

An applicant for approval of authority to operate an aircraft dispatcher course of study must have facilities, equipment, and materials adequate to provide each student the theoretical and practical aspects of aircraft dispatching. Each room, training booth, or other space used for instructional purposes must be temperature controlled, lighted, and ventilated to conform to local building, sanitation, and health codes. In addition, the training facility must be so located that the students in that facility are not distracted by the instruction conducted in other rooms.

### §65.67 Aircraft dispatcher certification courses: Personnel.

**(a)** Each applicant for an aircraft dispatcher certification course must meet the following personnel requirements:

(1) Each applicant must have adequate personnel, including one instructor who holds an aircraft dispatcher certificate and is available to coordinate all training course instruction.

(2) Each applicant must not exceed a ratio of 25 students for one instructor.

**(b)** The instructor who teaches the practical dispatch applications area of the appendix A course must hold an aircraft dispatchers certificate.

### §65.70 Aircraft dispatcher certification courses: Records.

**(a)** The operator of an aircraft dispatcher course must maintain a record for each student, including a chronological log of all instructors, subjects covered, and course examinations and results. The record must be retained for at least 3 years after graduation. The course operator also must prepare, for its records, and transmit to the Administrator not later than January 31 of each year, a report containing the following information for the previous year:

(1) The names of all students who graduated, together with the results of their aircraft dispatcher certification courses.

(2) The names of all the students who failed or withdrew, together with the results of their aircraft dispatcher certification courses or the reasons for their withdrawal.

**(b)** Each student who successfully completes the approved aircraft dispatcher certification course must be given a written statement of graduation, which is valid for 90 days. After 90 days, the course operator may revalidate the graduation certificate for an additional 90 days if the course operator determines that the student remains proficient in the subject areas listed in appendix A of this part.

## Subpart D—Mechanics

### §65.71 Eligibility requirements: General.

**(a)** To be eligible for a mechanic certificate and associated ratings, a person must—

(1) Be at least 18 years of age;

(2) Be able to read, write, speak, and understand the English language, or in the case of an applicant who does not meet this requirement and who is employed outside of the United States by a U.S. air carrier, have his certificate endorsed "Valid only outside the United States";

(3) Have passed all of the prescribed tests within a period of 24 months; and

(4) Comply with the sections of this subpart that apply to the rating he seeks.

**(b)** A certificated mechanic who applies for an additional rating must meet the requirements of §65.77 and, within a period of 24 months, pass the tests prescribed by §§65.75 and 65.79 for the additional rating sought.

[Docket No. 1179, 27 FR 7973, Aug. 10, 1962; as amended by Amdt. 65–6, 31 FR 5950, April 19, 1966]

### §65.73 Ratings.

**(a)** The following ratings are issued under this subpart:

(1) Airframe.

(2) Powerplant.

**(b)** A mechanic certificate with an aircraft or aircraft engine rating, or both, that was issued before, and was valid on, June 15, 1952, is equal to a mechanic certificate with an airframe or powerplant rating, or both, as the case may be, and may be exchanged for such a corresponding certificate and rating or ratings.

## §65.75 Knowledge requirements.

**(a)** Each applicant for a mechanic certificate or rating must, after meeting the applicable experience requirements of §65.77, pass a written test covering the construction and maintenance of aircraft appropriate to the rating he seeks, the regulations in this subpart, and the applicable provisions of parts 43 and 91 of this chapter. The basic principles covering the installation and maintenance of propellers are included in the powerplant test.

**(b)** The applicant must pass each section of the test before applying for the oral and practical tests prescribed by §65.79. A report of the written test is sent to the applicant.

[Docket No. 1179, 27 FR 7973, Aug. 10, 1962; as amended by Amdt. 65–1, 27 FR 10410, Oct. 25, 1962; Amdt. 65–6, 31 FR 5950, April 19, 1966]

## §65.77 Experience requirements.

Each applicant for a mechanic certificate or rating must present either an appropriate graduation certificate or certificate of completion from a certificated aviation maintenance technician school or documentary evidence, satisfactory to the Administrator, of—

**(a)** At least 18 months of practical experience with the procedures, practices, materials, tools, machine tools, and equipment generally used in constructing, maintaining, or altering airframes, or powerplants appropriate to the rating sought; or

**(b)** At least 30 months of practical experience concurrently performing the duties appropriate to both the airframe and powerplant ratings.

[Docket No. 1179, 27 FR, 7973, Aug. 10, 1962; as amended by Amdt. 65–14, 35 FR, 5533, April 3, 1970]

## §65.79 Skill requirements.

Each applicant for a mechanic certificate or rating must pass an oral and a practical test on the rating he seeks. The tests cover the applicant's basic skill in performing practical projects on the subjects covered by the written test for that rating. An applicant for a powerplant rating must show his ability to make satisfactory minor repairs to, and minor alterations of, propellers.

## §65.80 Certificated aviation maintenance technician school students.

Whenever an aviation maintenance technician school certificated under part 147 of this chapter shows to an FAA inspector that any of its students has made satisfactory progress at the school and is prepared to take the oral and practical tests prescribed by §65.79, that student may take those tests during the final subjects of his training in the approved curriculum, before he meets the applicable experience requirements of §65.77 and before he passes each section of the written test prescribed by §65.75.

[Docket No. 1179, 27 FR, 7973, Aug. 10, 1962; as amended by Amdt. 65–14, 35 FR, 5533, April 3, 1970]

## §65.81 General privileges and limitations.

**(a)** A certificated mechanic may perform or supervise the maintenance, preventive maintenance or alteration of an aircraft or appliance, or a part thereof, for which he is rated (but excluding major repairs to, and major alterations of, propellers, and any repair to, or alteration of, instruments), and may perform additional duties in accordance with §§65.85, 65.87, and 65.95. However, he may not supervise the maintenance, preventive maintenance, or alteration of, or approve and return to service, any aircraft or appliance, or part thereof, for which he is rated unless he has satisfactorily performed the work concerned at an earlier date. If he has not so performed that work at an earlier date, he may show his ability to do it by performing it to the satisfaction of the Administrator or under the direct supervision of a certificated and appropriately rated mechanic, or a certificated repairman, who has had previous experience in the specific operation concerned.

**(b)** A certificated mechanic may not exercise the privileges of his certificate and rating unless he understands the current instructions of the manufacturer, and the maintenance manuals, for the specific operation concerned.

[Docket No. 1179, 27 FR 7973, Aug. 10, 1962; as amended by Amdt. 65–2, 29 FR 5451, April 23, 1964; Amdt. 65–26, 45 FR 46737, July 10, 1980]

## §65.83 Recent experience requirements.

A certificated mechanic may not exercise the privileges of his certificate and rating unless, within the preceding 24 months—

**(a)** The Administrator has found that he is able to do that work; or

**(b)** He has, for at least 6 months—

(1) Served as a mechanic under his certificate and rating;

(2) Technically supervised other mechanics;

(3) Supervised, in an executive capacity, the maintenance or alteration of aircraft; or

(4) Been engaged in any combination of paragraph (b) (1), (2), or (3) of this section.

## §65.85 Airframe rating; additional privileges.

A certificated mechanic with an airframe rating may approve and return to service an airframe, or any related part or appliance, after he has performed, supervised, or inspected its maintenance or alteration (excluding major repairs and major alterations). In addition, he may perform the 100-hour inspection required by part 91 of this chapter on an airframe, or any related part or appliance, and approve and return it to service.

[Docket No. 1179, 27 FR 7973, Aug. 10, 1962; as amended by Amdt. 65–10, 32 FR 5770, April 11, 1967]

## §65.87 Powerplant rating; additional privileges.

A certificated mechanic with a powerplant rating may approve and return to service a powerplant or propeller or any related part or appliance, after he has performed, supervised, or inspected its maintenance or alteration (excluding major repairs and major alterations). In addition, he may perform the 100-hour inspection required by part 91 of this chapter on a powerplant or propeller, or any part thereof, and approve and return it to service.

[Docket No. 1179, 27 FR 7973, Aug. 10, 1962; as amended by Amdt. 65–10, 32 FR 5770, April 11, 1967]

## §65.89 Display of certificate.

Each person who holds a mechanic certificate shall keep it within the immediate area where he normally exercises the privileges of the certificate and shall present it for inspection upon the request of the Administrator or an authorized representative of the National Transportation Safety Board, or of any Federal, State, or local law enforcement officer.

[Docket No. 7258, 31 FR 13524, Oct. 20, 1966; as amended by Docket No. 8084, 32 FR 5769, April 11, 1967]

## §65.91 Inspection authorization.

**(a)** An application for an inspection authorization is made on a form and in a manner prescribed by the Administrator.

**(b)** An applicant who meets the requirements of this section is entitled to an inspection authorization.

**(c)** To be eligible for an inspection authorization, an applicant must—

(1) Hold a currently effective mechanic certificate with both an airframe rating and a powerplant rating, each of which is currently effective and has been in effect for a total of at least 3 years;

(2) Have been actively engaged, for at least the 2-year period before the date he applies, in maintaining aircraft certificated and maintained in accordance with this chapter;

(3) Have a fixed base of operations at which he may be located in person or by telephone during a normal working week but it need not be the place where he will exercise his inspection authority;

(4) Have available to him the equipment, facilities, and inspection data necessary to properly inspect airframes, powerplants, propellers, or any related part or appliance; and

(5) Pass a written test on his ability to inspect according to safety standards for returning aircraft to service after major repairs and major alterations and annual and progressive inspections performed under part 43 of this chapter.

An applicant who fails the test prescribed in paragraph (c)(5) of this section may not apply for retesting until at least 90 days after the date he failed the test.

[Docket No. 1179, 27 FR 7973, Aug. 10, 1962; as amended by Amdt. 65–5, 31 FR 3337, March 3, 1966; Amdt. 65–22, 42 FR 46279, Sept. 15, 1977; Amdt. 65–30, 50 FR 15700, April 19, 1985]

## §65.92 Inspection authorization: Duration.

**(a)** Each inspection authorization expires on March 31 of each year. However, the holder may exercise the privileges of that authorization only while he holds a currently effective mechanic certificate with both a currently effective airframe rating and a currently effective powerplant rating.

**(b)** An inspection authorization ceases to be effective whenever any of the following occurs:

(1) The authorization is surrendered, suspended, or revoked.

(2) The holder no longer has a fixed base of operation.

(3) The holder no longer has the equipment, facilities, and inspection data required by §65.91(c) (3) and (4) for issuance of his authorization.

**(c)** The holder of an inspection authorization that is suspended or revoked shall, upon the Administrator's request, return it to the Administrator.

[Docket No. 12537, 42 FR 46279, Sept. 15, 1977]

## §65.93 Inspection authorization: Renewal.

**(a)** To be eligible for renewal of an inspection authorization for a 1-year period an applicant must present evidence annually, during the month of March, at an FAA Flight Standards District Office or an International Field Office that the applicant still meets the requirements of §65.91(c) (1) through (4) and must show that, during the current period that the applicant held the inspection authorization, the applicant—

(1) Has performed at least one annual inspection for each 90 days that the applicant held the current authority; or

(2) Has performed inspections of at least two major repairs or major alterations for each 90 days that the applicant held the current authority; or

(3) Has performed or supervised and approved at least one progressive inspection in accordance with standards prescribed by the Administrator; or

(4) Has attended and successfully completed a refresher course, acceptable to the Administrator, of not less than 8 hours of instruction during the 12-month period preceding the application for renewal; or

(5) Has passed on oral test by an FAA inspector to determine that the applicant's knowledge of applicable regulations and standards is current.

**(b)** The holder of an inspection authorization that has been in effect for less than 90 days before the expiration date need not comply with paragraphs (a) (1) through (5) of this section.

[Docket No. 18241, 45 FR 46738, July 10, 1980; as amended by Amdt. 65–35, 54 FR 39292, Sept. 25, 1989]

### §65.95 Inspection authorization: Privileges and limitations.

**(a)** The holder of an inspection authorization may—

(1) Inspect and approve for return to service any aircraft or related part or appliance (except any aircraft maintained in accordance with a continuous airworthiness program under part 121 of this chapter) after a major repair or major alteration to it in accordance with part 43 [New] of this chapter, if the work was done in accordance with technical data approved by the Administrator; and

(2) Perform an annual, or perform or supervise a progressive inspection according to §§43.13 and 43.15 of this chapter.

**(b)** When he exercises the privileges of an inspection authorization the holder shall keep it available for inspection by the aircraft owner, the mechanic submitting the aircraft, repair, or alteration for approval (if any), and shall present it upon the request of the Administrator or an authorized representative of the National Transportation Safety Board, or of any Federal, State, or local law enforcement officer.

**(c)** If the holder of an inspection authorization changes his fixed base of operation, he may not exercise the privileges of the authorization until he has notified the FAA Flight Standards District Office or International Field Office for the area in which the new base is located, in writing, of the change.

[Docket No. 1179, 27 FR 7973, Aug. 10, 1962; as amended by Amdt. 65–2, 29 FR 5451, April 23, 1964; Amdt. 65–4, 30 FR 3638, March 14, 1965; Amdt. 65–5, 31 FR 3337, March 3, 1966; Amdt. 65–9, 31 FR 13524, Oct. 20, 1966; 32 FR 5769, April 11, 1967; Amdt. 65–35, 54 FR 39292, Sept. 25, 1989; Amdt. 65–41, 66 FR 21066, April 27, 2001]

## Subpart E—Repairmen

### §65.101 Eligibility requirements: General.

**(a)** To be eligible for a repairman certificate a person must—

(1) Be at least 18 years of age;

(2) Be specially qualified to perform maintenance on aircraft or components thereof, appropriate to the job for which he is employed;

(3) Be employed for a specific job requiring those special qualifications by a certificated repair station, or by a certificated commercial operator or certificated air carrier, that is required by its operating certificate or approved operations specifications to provide a continuous airworthiness maintenance program according to its maintenance manuals;

(4) Be recommended for certification by his employer, to the satisfaction of the Administrator, as able to satisfactorily maintain aircraft or components, appropriate to the job for which he is employed;

(5) Have either—

(i) At least 18 months of practical experience in the procedures, practices, inspection methods, materials, tools, machine tools, and equipment generally used in the maintenance duties of the specific job for which the person is to be employed and certificated; or

(ii) Completed formal training that is acceptable to the Administrator and is specifically designed to qualify the applicant for the job on which the applicant is to be employed; and

(6) Be able to read, write, speak, and understand the English language, or, in the case of an applicant who does not meet this requirement and who is employed outside the United States by a certificated repair station, a certificated U.S. commercial operator, or a certificated U.S. air carrier, described in paragraph (c) of this section, have his certificate endorsed "Valid only outside the United States."

**(b)** This section does not apply to the issuance of repairman certificates (experimental aircraft builder) under §65.104.

[Docket No. 1179, 27 FR 7973, Aug. 10, 1962; as amended by Amdt. 65–11, 32 FR 13506, Sept. 27, 1967; Amdt. 65–24, 44 FR 46781, Aug. 9, 1979; Amdt. 65–27, 47 FR 13316, March 29, 1982]

### §65.103 Repairman certificate: Privileges and limitations.

**(a)** A certificated repairman may perform or supervise the maintenance, preventive maintenance, or alteration of aircraft or aircraft components appropriate to the job for which the repairman was employed and certificated, but only in connection with duties for the certificate holder by whom the repairman was employed and recommended.

**(b)** A certificated repairman may not perform or supervise duties under the repairman certificate unless the repairman understands the current instructions of the certificate holder by whom the repairman is employed and the manufacturer's instructions for continued airworthiness relating to the specific operations concerned.

[Docket No. 18241, 45 FR 46738, July 10, 1980]

### §65.104 Repairman certificate—experimental aircraft builder—Eligibility, privileges and limitations.

**(a)** To be eligible for a repairman certificate (experimental aircraft builder), an individual must—

(1) Be at least 18 years of age;

(2) Be the primary builder of the aircraft to which the privileges of the certificate are applicable;

(3) Show to the satisfaction of the Administrator that the individual has the requisite skill to determine whether the aircraft is in a condition for safe operations; and

(4) Be a citizen of the United States or an individual citizen of a foreign country who has lawfully been admitted for permanent residence in the United States.

**(b)** The holder of a repairman certificate (experimental aircraft builder) may perform condition inspections on the aircraft constructed by the holder in accordance with the operating limitations of that aircraft.

**(c)** Section 65.103 does not apply to the holder of a repairman certificate (experimental aircraft builder) while performing under that certificate.

[Docket No. 18739, 44 FR 46781, Aug. 9, 1979]

### §65.105 Display of certificate.

Each person who holds a repairman certificate shall keep it within the immediate area where he normally exercises the privileges of the certificate and shall present it for inspection upon the request of the Administrator or an authorized representative of the National Transportation Safety Board, or of any Federal, State, or local law enforcement officer.

[Docket No. 7258, 31 FR 13524, Oct. 20, 1966; as amended by Docket No. 8084, 32 FR 5769, April 11, 1967]

## Subpart F— Parachute Riggers

### §65.111 Certificate required.

**(a)** No person may pack, maintain, or alter any personnel-carrying parachute intended for emergency use in connection with civil aircraft of the United States (including the reserve parachute of a dual parachute system to be used for intentional parachute jumping) unless that person holds an appropriate current certificate and type rating issued under this subpart and complies with §§65.127 through 65.133.

**(b)** No person may pack, maintain, or alter any main parachute of a dual-parachute system to be used for intentional parachute jumping in connection with civil aircraft of the United States unless that person—

(1) Has an appropriate current certificate issued under this subpart;

(2) Is under the supervision of a current certificated parachute rigger;

(3) Is the person making the next parachute jump with that parachute in accordance with §105.43(a) of this chapter; or

(4) Is the parachutist in command making the next parachute jump with that parachute in a tandem parachute operation conducted under §105.45(b)(1) of this chapter.

**(c)** Each person who holds a parachute rigger certificate shall present it for inspection upon the request of the Administrator or an authorized representative of the National Transportation Safety Board, or of any Federal, State, or local law enforcement officer.

**(d)** The following parachute rigger certificates are issued under this part:

(1) Senior parachute rigger.

(2) Master parachute rigger.

**(e)** Sections 65.127 through 65.133 do not apply to parachutes packed, maintained, or altered for the use of the armed forces.

[Docket No. 1179, 27 FR 7973, Aug. 10, 1962; as amended by Amdt. 65–9, 31 FR 13524, Oct. 20, 1966; 32 FR 5769, April 11, 1967; Amdt. 65–42, 66 FR 23553, May 9, 2001]

## §65.113 Eligibility requirements: General.

**(a)** To be eligible for a parachute rigger certificate, a person must—

(1) Be at least 18 years of age;

(2) Be able to read, write, speak, and understand the English language, or, in the case of a citizen of Puerto Rico, or a person who is employed outside of the United States by a U.S. air carrier, and who does not meet this requirement, be issued a certificate that is valid only in Puerto Rico or while he is employed outside of the United States by that air carrier, as the case may be; and

(3) Comply with the sections of this subpart that apply to the certificate and type rating he seeks.

**(b)** Except for a master parachute rigger certificate, a parachute rigger certificate that was issued before, and was valid on, October 31, 1962, is equal to a senior parachute rigger certificate, and may be exchanged for such a corresponding certificate.

## §65.115 Senior parachute rigger certificate: Experience, knowledge, and skill requirements.

Except as provided in §65.117, an applicant for a senior parachute rigger certificate must—

**(a)** Present evidence satisfactory to the Administrator that he has packed at least 20 parachutes of each type for which he seeks a rating, in accordance with the manufacturer's instructions and under the supervision of a certificated parachute rigger holding a rating for that type or a person holding an appropriate military rating;

**(b)** Pass a written test, with respect to parachutes in common use, on—

(1) Their construction, packing, and maintenance;

(2) The manufacturer's instructions;

(3) The regulations of this subpart; and

**(c)** Pass an oral and practical test showing his ability to pack and maintain at least one type of parachute in common use, appropriate to the type rating he seeks.

[Docket No. 10468, 37 FR 13251, July 6, 1972]

## §65.117 Military riggers or former military riggers: Special certification rule.

In place of the procedure in §65.115, an applicant for a senior parachute rigger certificate is entitled to it if he passes a written test on the regulations of this subpart and presents satisfactory documentary evidence that he—

**(a)** Is a member or civilian employee of an Armed Force of the United States, is a civilian employee of a regular armed force of a foreign country, or has, within the 12 months before he applies, been honorably discharged or released from any status covered by this paragraph;

**(b)** Is serving, or has served within the 12 months before he applies, as a parachute rigger for such an Armed Force; and

**(c)** Has the experience required by §65.115(a).

## §65.119 Master parachute rigger certificate: Experience, knowledge, and skill requirements.

An applicant for a master parachute rigger certificate must meet the following requirements:

**(a)** Present evidence satisfactory to the Administrator that he has had at least 3 years of experience as a parachute rigger and has satisfactorily packed at least 100 parachutes of each of two types in common use, in accordance with the manufacturer's instructions—

(1) While a certificated and appropriately rated senior parachute rigger; or

(2) While under the supervision of a certificated and appropriately rated parachute rigger or a person holding appropriate military ratings.

An applicant may combine experience specified in paragraphs (a) (1) and (2) of this section to meet the requirements of this paragraph.

**(b)** If the applicant is not the holder of a senior parachute rigger certificate, pass a written test, with respect to parachutes in common use, on—

(1) Their construction, packing, and maintenance;

(2) The manufacturer's instructions; and

(3) The regulations of this subpart.

**(c)** Pass an oral and practical test showing his ability to pack and maintain two types of parachutes in common use, appropriate to the type ratings he seeks.

[Docket No. 10468, 37 FR 13252, July 6, 1972]

## §65.121 Type ratings.

**(a)** The following type ratings are issued under this subpart:

(1) Seat.

(2) Back.

(3) Chest.

(4) Lap.

**(b)** The holder of a senior parachute rigger certificate who qualifies for a master parachute rigger certificate is entitled to have placed on his master parachute rigger certificate the ratings that were on his senior parachute rigger certificate.

## §65.123 Additional type ratings: Requirements.

A certificated parachute rigger who applies for an additional type rating must—

**(a)** Present evidence satisfactory to the Administrator that he has packed at least 20 parachutes of the type for which he seeks a rating, in accordance with the manufacturer's instructions and under the supervision of a certificated parachute rigger holding a rating for that type or a person holding an appropriate military rating; and

**(b)** Pass a practical test, to the satisfaction of the Administrator, showing his ability to pack and maintain the type of parachute for which he seeks a rating.

[Docket No. 1179, 27 FR 7973, Aug. 10, 1962; as amended by Amdt. 65–20, 37 FR 13251, July 6, 1972]

## §65.125 Certificates: Privileges.

**(a)** A certificated senior parachute rigger may—

(1) Pack or maintain (except for major repair) any type of parachute for which he is rated; and

(2) Supervise other persons in packing any type of parachute for which that person is rated in accordance with §105.43(a) or §105.45(b)(1) of this chapter.

**(b)** A certificated master parachute rigger may—

(1) Pack, maintain, or alter any type of parachute for which he is rated; and

(2) Supervise other persons in packing, maintaining, or altering any type of parachute for which the certificated parachute rigger is rated in accordance with §105.43(a) or §105.45(b)(1) of this chapter.

**(c)** A certificated parachute rigger need not comply with §§65.127 through 65.133 (relating to facilities, equipment, performance standards, records, recent experience, and seal) in packing, maintaining, or altering (if authorized) the main parachute of a dual parachute pack to be used for intentional jumping.

[Docket No. 1179, 27 FR 7973, Aug. 10, 1962; as amended by Amdt. 65–20, 37 FR 13252, July 6, 1972; Amdt. 65–42, 66 FR 23553, May 9, 2001]

## §65.127 Facilities and equipment.

No certificated parachute rigger may exercise the privileges of his certificate unless he has at least the following facilities and equipment available to him:

**(a)** A smooth top table at least three feet wide by 40 feet long.

**(b)** Suitable housing that is adequately heated, lighted, and ventilated for drying and airing parachutes.

**(c)** Enough packing tools and other equipment to pack and maintain the types of parachutes that he services.

**(d)** Adequate housing facilities to perform his duties and to protect his tools and equipment.

[Docket No. 1179, 27 FR 7973, Aug. 10, 1962; as amended by Amdt. 65–27, 47 FR 13316, March 29, 1982]

## §65.129 Performance standards.

No certificated parachute rigger may—

**(a)** Pack, maintain, or alter any parachute unless he is rated for that type;

**(b)** Pack a parachute that is not safe for emergency use;

**(c)** Pack a parachute that has not been thoroughly dried and aired;

**(d)** Alter a parachute in a manner that is not specifically authorized by the Administrator or the manufacturer;

**(e)** Pack, maintain, or alter a parachute in any manner that deviates from procedures approved by the Administrator or the manufacturer of the parachute; or

**(f)** Exercise the privileges of his certificate and type rating unless he understands the current manufacturer's instructions for the operation involved and has—

(1) Performed duties under his certificate for at least 90 days within the preceding 12 months; or

(2) Shown the Administrator that he is able to perform those duties.

## §65.131 Records.

**(a)** Each certificated parachute rigger shall keep a record of the packing, maintenance, and alteration of parachutes performed or supervised by him. He shall keep in that record, with respect to each parachute worked on, a statement of—

(1) Its type and make;

(2) Its serial number;

(3) The name and address of its owner;

(4) The kind and extent of the work performed;

(5) The date when and place where the work was performed; and

(6) The results of any drop tests made with it.

**(b)** Each person who makes a record under paragraph (a) of this section shall keep it for at least 2 years after the date it is made.

**(c)** Each certificated parachute rigger who packs a parachute shall write, on the parachute packing record attached to the parachute, the date and place of the packing and a notation of any defects he finds on inspection. He shall sign that record with his name and the number of his certificate.

## §65.133 Seal.

Each certificated parachute rigger must have a seal with an identifying mark prescribed by the Administrator, and a seal press. After packing a parachute he shall seal the pack with his seal in accordance with the manufacturer's recommendation for that type of parachute.

## Appendix A to Part 65

### Aircraft Dispatcher Courses

**Source:** Docket No. FAA–1998–4553, 64 FR 68925, Dec. 8, 1999, unless otherwise noted.

*Overview*

This appendix sets forth the areas of knowledge necessary to perform dispatcher functions. The items listed below indicate the minimum set of topics that must be covered in a training course for aircraft dispatcher certification. The order of coverage is at the discretion of the approved school. For the latest technological advancements refer to the Practical Test Standards as published by the FAA.

***I. Regulations***

A. Subpart C of this part;
B. Parts 1, 25, 61, 71, 91, 121, 139, and 175, of this chapter;
C. 49 CFR part 830;
D. General Operating Manual.

***II. Meteorology***

A. Basic Weather Studies
(1) The earth's motion and its effects on weather.
(2) Analysis of the following regional weather types, characteristics, and structures, or combinations thereof:
(a) Maritime.
(b) Continental.
(c) Polar.
(d) Tropical.
(3) Analysis of the following local weather types, characteristics, and structures or combinations thereof:
(a) Coastal.
(b) Mountainous.
(c) Island.
(d) Plains.
(4) The following characteristics of the atmosphere:
(a) Layers.
(b) Composition.
(c) Global Wind Patterns.
(d) Ozone.
(5) Pressure:
(a) Units of Measure.
(b) Weather Systems Characteristics.
(c) Temperature Effects on Pressure.
(d) Altimeters.
(e) Pressure Gradient Force.
(f) Pressure Pattern Flying Weather.
(6) Wind:
(a) Major Wind Systems and Coriolis Force.
(b) Jetstreams and their Characteristics.
(c) Local Wind and Related Terms.
(7) States of Matter:
(a) Solids, Liquid, and Gases.
(b) Causes of change of state.
(8) Clouds:
(a) Composition, Formation, and Dissipation.
(b) Types and Associated Precipitation.
(c) Use of Cloud Knowledge in Forecasting.
(9) Fog:
(a) Causes, Formation, and Dissipation.
(b) Types.
(10) Ice:
(a) Causes, Formation, and Dissipation.
(b) Types.
(11) Stability/Instability:
(a) Temperature Lapse Rate, Convection.
(b) Adiabatic Processes.
(c) Lifting Processes.
(d) Divergence.
(e) Convergence.
(12) Turbulence:
(a) Jetstream Associated.
(b) Pressure Pattern Recognition.
(c) Low Level Windshear.
(d) Mountain Waves.
(e) Thunderstorms.
(f) Clear Air Turbulence.
(13) Airmasses:
(a) Classification and Characteristics.
(b) Source Regions.
(c) Use of Airmass Knowledge in Forecasting.
(14) Fronts:
(a) Structure and Characteristics, Both Vertical and Horizontal.
(b) Frontal Types.
(c) Frontal Weather Flying.
(15) Theory of Storm Systems:
(a) Thunderstorms.
(b) Tornadoes.
(c) Hurricanes and Typhoons.
(d) Microbursts.
(e) Causes, Formation, and Dissipation.
B. Weather, Analysis, and Forecasts
(1) Observations:
(a) Surface Observations.
(i) Observations made by certified weather observer.
(ii) Automated Weather Observations.
(b) Terminal Forecasts.
(c) Significant En route Reports and Forecasts.
(i) Pilot Reports.
(ii) Area Forecasts.

(iii) Sigmets, Airmets.
(iv) Center Weather Advisories.
(d) Weather Imagery.
(i) Surface Analysis.
(ii) Weather Depiction.
(iii) Significant Weather Prognosis.
(iv) Winds and Temperature Aloft.
(v) Tropopause Chart.
(vi) Composite Moisture Stability Chart.
(vii) Surface Weather Prognostic Chart.
(viii) Radar Meteorology.
(ix) Satellite Meteorology.
(x) Other charts as applicable.
(e) Meteorological Information Data Collection Systems.
(2) Data Collection, Analysis, and Forecast Facilities.
(3) Service Outlets Providing Aviation Weather Products.
C. Weather Related Aircraft Hazards
(1) Crosswinds and Gusts.
(2) Contaminated Runways.
(3) Restrictions to Surface Visibility.
(4) Turbulence and Windshear.
(5) Icing.
(6) Thunderstorms and Microburst.
(7) Volcanic Ash.

***III. Navigation***

A. Study of the Earth
(1) Time reference and location (0 Longitude, UTC).
(2) Definitions.
(3) Projections.
(4) Charts.
B. Chart Reading, Application, and Use.
C. National Airspace Plan.
D. Navigation Systems.
E. Airborne Navigation Instruments.
F. Instrument Approach Procedures.
(1) Transition Procedures.
(2) Precision Approach Procedures.
(3) Non-precision Approach Procedures.
(4) Minimums and the relationship to weather.
G. Special Navigation and Operations.
(1) North Atlantic.
(2) Pacific.
(3) Global Differences.

***IV. Aircraft***

A. Aircraft Flight Manual.
B. Systems Overview.
(1) Flight controls.
(2) Hydraulics.
(3) Electrical.
(4) Air Conditioning and Pressurization.
(5) Ice and Rain protection.
(6) Avionics, Communication, and Navigation.
(7) Powerplants and Auxiliary Power Units.
(8) Emergency and Abnormal Procedures.
(9) Fuel Systems and Sources.
C. Minimum Equipment List/Configuration Deviation List (MEL/CDL) and Applications.
D. Performance.
(1) Aircraft in general.
(2) Principles of flight:
(a) Group one aircraft.
(b) Group two aircraft.
(3) Aircraft Limitations.
(4) Weight and Balance.
(5) Flight instrument errors.
(6) Aircraft performance:
(a) Take-off performance.
(b) En route performance.
(c) Landing performance.

***V. Communications***

A. Regulatory requirements.
B. Communication Protocol.
C. Voice and Data Communications.
D. Notice to Airmen (NOTAMs).
E. Aeronautical Publications.
F. Abnormal Procedures.

***VI. Air Traffic Control***

A. Responsibilities.
B. Facilities and Equipment.
C. Airspace classification and route structure.
D. Flight Plans.
(1) Domestic.
(2) International.
E. Separation Minimums.
F. Priority Handling.
G. Holding Procedures.
H. Traffic Management.

***VII. Emergency and Abnormal Procedures***

A. Security measures on the ground.
B. Security measures in the air.
C. FAA responsibility and services.
D. Collection and dissemination of information on overdue or missing aircraft.
E. Means of declaring an emergency.
F. Responsibility for declaring an emergency.
G. Required reporting of an emergency.
H. NTSB reporting requirements.

***VIII. Practical Dispatch Applications***

A. Human Factors.
(1) Decisionmaking:
(a) Situation Assessment.
(b) Generation and Evaluation of Alternatives.
(i) Tradeoffs and Prioritization.
(ii) Contingency Planning.
(c) Support Tools and Technologies.
(2) Human Error:
(a) Causes.
(i) Individual and Organizational Factors.
(ii) Technology-Induced Error.
(b) Prevention.

(c) Detection and Recovery.
(3) Teamwork:
(a) Communication and Information Exchange.
(b) Cooperative and Distributed Problem-Solving.
(c) Resource Management.
(i) Air Traffic Control (ATC) activities and workload.
(ii) Flightcrew activities and workload.
(iii) Maintenance activities and workload.
(iv) Operations Control Staff activities and workload.
B. Applied Dispatching.
(1) Briefing techniques, Dispatcher, Pilot.
(2) Preflight:
(a) Safety.
(b) Weather Analysis.
(i) Satellite imagery.
(ii) Upper and lower altitude charts.
(iii) Significant en route reports and forecasts.
(iv) Surface charts.
(v) Surface observations.
(vi) Terminal forecasts and orientation to Enhanced Weather Information System (EWINS).
(c) NOTAMs and airport conditions.
(d) Crew.
(i) Qualifications.
(ii) Limitations.
(e) Aircraft.
(i) Systems.
(ii) Navigation instruments and avionics systems.
(iii) Flight instruments.
(iv) Operations manuals and MEL/CDL.
(v) Performance and limitations.
(f) Flight Planning.
(i) Route of flight.
1. Standard Instrument Departures and Standard Terminal Arrival Routes.
2. En route charts.
3. Operational altitude.
4. Departure and arrival charts.
(ii) Minimum departure fuel.
1. Climb.
2. Cruise.
3. Descent.
(g) Weight and balance.
(h) Economics of flight overview (Performance, Fuel Tankering).
(i) Decision to operate the flight.
(j) ATC flight plan filing.
(k) Flight documentation.
(i) Flight plan.
(ii) Dispatch release.
(3) Authorize flight departure with concurrence of pilot in command.
(4) In-flight operational control:
(a) Current situational awareness.
(b) Information exchange.
(c) Amend original flight release as required.
(5) Post-Flight:
(a) Arrival verification.
(b) Weather debrief.
(c) Flight irregularity reports as required.

# SUBCHAPTER G AIR CARRIERS AND OPERATORS FOR COMPENSATION OR HIRE: CERTIFICATION AND OPERATIONS

# PART 121

## OPERATING REQUIREMENTS: DOMESTIC, FLAG, AND SUPPLEMENTAL OPERATIONS

SPECIAL FEDERAL AVIATION REGULATIONS



### Subpart A—General

Sec.


### Subpart B—[Reserved]

### Subpart C—[Reserved]

### Subpart D—[Reserved]

### Subpart E—Approval of Routes: Domestic and Flag Operations



### Subpart F—Approval of Areas and Routes for Supplemental Operations



### Subpart G—Manual Requirements



### Subpart H—Aircraft Requirements



### Subpart I— Airplane Performance Operating Limitations

### Subpart J— Special Airworthiness Requirements



### Subpart K— Instrument and Equipment Requirements

**Subpart L—Maintenance, Preventive Maintenance, and Alterations**



**Subpart M—
Airman and Crewmember Requirements**



**Subpart N—Training Program**

### Subpart O—Crewmember Qualifications



### Subpart P—Aircraft Dispatcher Qualifications and Duty Time



### Subpart Q—Flight Time Limitations and Rest Requirements: Domestic Operations



### Subpart R—Flight Time Limitations: Flag Operations



### Subpart S—Flight Time Limitations: Supplemental Operations

**Subpart V—Records and Reports**



**Subpart W—**
**Crewmember Certificate: International**



**Subpart X—**
**Emergency Medical Equipment and Training**



APPENDICES TO PART 121



**Authority:** 49 U.S.C. 106(g), 40113, 40119, 41706, 44101, 44701–44702, 44705, 44709–44711, 44713, 44716–44717, 44722, 46105.

SPECIAL FEDERAL AVIATION REGULATIONS

## SFAR No. 14 to Part 121

Contrary performance provisions of the Civil Air Regulations notwithstanding, the Administrator may grant performance credit for the use of standby power on transport category airplanes. Such credit shall be applicable only to the maximum certificated take-off and landing weights, and the take-off distance, and the take-off paths, and shall not exceed that found by the Administrator to result in an over-all level of safety in the take-off, approach, and landing regimes of flight equivalent to that prescribed in the regulations under which the airplane was originally certificated without standby power. (Note: Standby power is power and/or thrust obtained from rocket engines for a relatively short period and actuated only in cases of emergency.) The following provisions shall apply:

**(1)** *Take-off; general.* The take-off data prescribed in sections (2) and (3) shall be determined at all weights and altitudes, and at ambient temperatures if applicable, at which performance credit is to be applied.

**(2)** *Take-off path.*

(a) The one-engine-inoperative take-off path with standby power in use shall be determined in accordance with the performance requirements of the applicable airworthiness regulations.

(b) The one-engine-inoperative take-off path (excluding that portion where the airplane is on or just above the take-off surface-determined in accordance with paragraph (a) of this section shall lie above the one/engine-inoperative take-off path without standby power at the maximum take/off weight at which all of the applicable airworthiness requirements are met. For the purpose of this comparison, the flight path shall be considered to extend to at least a height of 400 feet above the take-off surface.

(c) The take-off path with all engines operating, but without the use of standby power, shall reflect a conservatively greater over-all level of performance than the one-engine-inoperative take-off path established in accordance with paragraph (a) of this section. The aforementioned margin shall be established by the Administrator to insure safe day-to-day operations, but in no case shall it be less than 15 percent. The all-engines-operating take-off path shall be determined by a procedure consistent with that established in complying with paragraph (a) of this section.

(d) For reciprocating-engine-powered airplanes, the take-off path to be scheduled in the Airplane Flight Manual shall represent the one-engine-inoperative take-off path determined in accordance with paragraph (a) of this section and modified to reflect the procedure (see section (6)) established by the applicant for flap retraction and attainment of the en route speed. The scheduled take-off path shall have a positive slope at all points of the airborne portion and at no point shall it lie above the take-off path specified in paragraph (a) of this section.

**(3)** *Take-off distance.* The take-off distance shall be the horizontal distance along the one/engine-inoperative take-off to the point where the airplane attains a height of 50 feet above the take-off surface for reciprocating-engine-powered airplanes and a height of 35 feet above the take-off surface for turbine-powered airplanes.

**(4)** *Maximum certificated take-off weights.* The maximum certificated take-off weights shall be determined at all altitudes, and at ambient temperatures if applicable, at which performance credit is to be applied and shall not exceed the weights established in compliance with paragraphs (a) and (b) of this section.

(a) The conditions of section (2)(b) through (d) shall be met at the maximum certificated take-off weight.

(b) Without the use of standby power, the airplane shall meet all of the en route requirements of the applicable airworthiness regulations under which the airplane was originally certificated. In addition, turbine-powered airplanes without the use of standby power shall meet the final take-off climb requirements prescribed in the applicable airworthiness regulations.

**(5)** *Maximum certificated landing weights.*

(a) The maximum certificated landing weights (one-engine/inoperative approach and all-engine/operating landing climb) shall be determined at all altitudes, and at ambient temperatures if applicable, at which performance credit is to be applied and shall not exceed that established in compliance with the provisions of paragraph (b) of this section.

(b) The flight path, with the engines operating at the power and/or thrust appropriate to the airplane configuration and with standby power in use, shall lie above the flight path without standby power in use at the maximum weight at which all of the applicable airworthiness requirements are met. In addition, the flight paths shall comply with the provisions of paragraphs (i) and (ii) of this paragraph (b).

(i) The flight paths shall be established without changing the appropriate airplane configuration.

(ii) The flight paths shall be carried out for a minimum height of 400 feet above the point where standby power is actuated.

**(6)** *Airplane configuration, speed, and power and/or thrust; general.* Any change in the airplane's configuration, speed, and power and/or thrust shall be made in accordance with the procedures established by the applicant for the operation of the airplane in service and shall comply with the provisions of paragraphs (a) through (c) of this section. In addition, procedures shall be established for the execution of balked landings and missed approaches.

(a) The Administrator shall find that the procedure can be consistently executed in service by crews of average skill.

(b) The procedure shall not involve methods or the use of devices which have not been proven to be safe and reliable.

(c) Allowances shall be made for such time delays in the execution of the procedures as may be reasonably expected to occur during service.

**(7)** *Installation and operation; standby power.* The standby power unit and its installation shall comply with the provisions of paragraphs (a) and (b) of this section.

(a) The standby power unit and its installation shall not adversely affect the safety of the airplane.

(b) The operation of the standby power unit and its control shall have proven to be safe and reliable.

[23 FR 7454, Sept. 25, 1958. Redesignated at 29 FR 19099, Dec. 30, 1964]

## SFAR No. 36 to Part 121

**1.** *Definitions.* For purposes of this Special Federal Aviation Regulation—

(a) A product is an aircraft, airframe, aircraft engine, propeller, or appliance;

(b) An article is an airframe, powerplant, propeller, instrument, radio, or accessory; and

(c) A component is a part of a product or article.

**2.** *General.*

(a) Contrary provisions of §121.379(b) and §135.437(b) of this chapter notwithstanding, the holder of an air carrier certificate or operating certificate, that operates large aircraft, and that has been issued operations specifications for operations required to be conducted in accordance with 14 CFR part 121 or 135, may perform a major repair on a product as described in §121.379(b) or §135.437(a), using technical data that have not been approved by the Administrator, and approve that product for return to service, if authorized in accordance with this Special Federal Aviation Regulation.

(b) Reserved.

(c) Contrary provisions of §145.201(c)(2) notwithstanding, the holder of a repair station certificate under 14 CFR part 145 that is located in the United States may perform a major repair on an article for which it is rated using technical data not approved by the FAA and approve that article for return to service, if authorized in accordance with this Special Federal Aviation Regulation. If the certificate holder holds a rating limited to a component of a product or article, the holder may not, by virtue of this Special Federal Aviation Regulation, approve that product or article for return to service.

**3.** *Major Repair Data and Return to Service.*

(a) As referenced in section 2 of this Special Federal Aviation Regulation, a certificate holder may perform a major repair on a product or article using technical data that have not been approved by the Administrator, and approve that product or article for return to service, if the certificate holder—

(1) Has been issued an authorization under, and a procedures manual that complies with, Special Federal Aviation Regulation No. 36–7, effective on January 23, 1999;

(2) Has developed the technical data in accordance with the procedures manual;

(3) Has developed the technical data specifically for the product or article being repaired; and

(4) Has accomplished the repair in accordance with the procedures manual and the procedures approved by the Administrator for the certificate.

(b) For purposes of this section, an authorization holder may develop technical data to perform a major repair on a product or article and use that data to repair a subsequent product or article of the same type as long as the holder—

(1) Evaluates each subsequent repair and the technical data to determine that performing the subsequent repair with the same data will return the product or article to its original or properly altered condition, and that the repaired product or article conforms with applicable airworthiness requirements; and

(2) Records each evaluation in the records referenced in paragraph (a) of section 13 of this Special Federal Aviation Regulation.

**4.** *Application.* The applicant for an authorization under this Special Federal Aviation Regulation must submit an application, in writing and signed by an officer of the applicant, to the FAA Flight Standards District Office charged with the overall inspection of the applicant's operations under its certificate. The application must contain—

(a) If the applicant is

(1) The holder of an air carrier operating or commercial operating certificate, or the holder of an air taxi operating certificate that operates large aircraft, the—

(i) The applicant's certificate number; and

(ii) The specific product(s) the applicant is authorized to maintain under its certificate, operations specifications, and maintenance manual; or

(2) The holder of a domestic repair station certificate—

(i) The applicant's certificate number;

(ii) A copy of the applicant's operations specifications; and

(iii) The specific article(s) for which the applicant is rated;

(b) The name, signature, and title of each person for whom authorization to approve, on behalf of the authorization holder, the use of technical data for major repairs is requested; and

(c) The qualifications of the applicant's staff that show compliance with section 5 of this Special Federal Aviation Regulation.

**5.** *Eligibility.*

(a) To be eligible for an authorization under this Special Federal Aviation Regulation, the applicant, in addition to having the authority to repair products or articles must—

(1) Hold an air carrier certificate or operating certificate, operate large aircraft, and have been issued operations specifications for operations required to be conducted in accordance with 14 CFR part 121 or 135, or hold a domestic repair station certificate under 14 CFR part 145;

(2) Have an adequate number of sufficiently trained personnel in the United States to develop data and repair the products that the applicant is authorized to maintain under its operating certificate or the articles for which it is rated under its domestic repair station certificate;

(3) Employ, or have available, a staff of engineering personnel that can determine compliance with the applicable airworthiness requirements of the Federal Aviation Regulations.

(b) At least one member of the staff required by paragraph (a)(3) of this section must—

(1) Have a thorough working knowledge of the applicable requirements of the Federal Aviation Regulations;

(2) Occupy a position on the applicant's staff that has the authority to establish a repair program that ensures that each repaired product or article meets the applicable requirements of the Federal Aviation Regulations;

(3) Have at least one year of satisfactory experience in processing engineering work, in direct contact with the FAA, for type certification or major repair projects; and

(4) Have at least eight years of aeronautical engineering experience (which may include the one year of experience in processing engineering work for type certification or major repair projects).

(c) The holder of an authorization issued under this Special Federal Aviation Regulation shall notify the Administrator within 48 hours of any change (including a change of personnel) that could affect the ability of the holder to meet the requirements of this Special Federal Aviation Regulation.

**6.** *Procedures Manual.*

(a) A certificate holder may not approve a product or article for return to service under section 2 of this Special Federal Aviation Regulation unless the holder—

(1) Has a procedures manual that has been approved by the Administrator as complying with paragraph (b) of this section; and

(2) Complies with the procedures contained in this procedures manual.

(b) The approved procedures manual must contain—

(1) The procedures for developing and determining the adequacy of technical data for major repairs;

(2) The identification (names, signatures, and responsibilities) of officials and of each staff member described in section 5 of this Special Federal Aviation Regulation who—

(i) Has the authority to make changes in procedures that require a revision to the procedures manual; and

(ii) Prepares or determines the adequacy of technical data, plans or conducts tests, and approves, on behalf of the authorization holder, test results; and

(3) A "log of revisions" page that identifies each revised item, page, and date of revision, and contains the signature of the person approving the change for the Administrator.

(c) The holder of an authorization issued under this Special Federal Aviation Regulation may not approve a product or article for return to service after a change in staff necessary to meet the requirements of section 5 of this regulation or a change in procedures from those approved under paragraph (a) of this section, unless that change has been approved by the FAA and entered in the procedures manual.

**7.** *Duration of Authorization.* Each authorization issued under this Special Federal Aviation Regulation is effective from the date of issuance until January 23, 2004, unless it is earlier surrendered, suspended, revoked, or otherwise terminated. Upon termination of such authorization, the terminated authorization holder must:

(a) Surrender to the FAA all data developed pursuant to Special Federal Aviation Regulation No. 36; or

(b) Maintain indefinitely all data developed pursuant to Special Federal Aviation Regulation No. 36, and make that data available to the FAA for inspection upon request.

**8.** *Transferability.* An authorization issued under this Special Federal Aviation Regulation is not transferable.

**9.** *Inspections.* Each holder of an authorization issued under this Special Federal Aviation Regulation and each applicant for an authorization must allow the Administrator to inspect its personnel, facilities, products and articles, and records upon request.

**10.** *Limits of Applicability.* An authorization issued under this Special Federal Aviation Regulation applies only to—

(a) A product that the air carrier, commercial, or air taxi operating certificate holder is authorized to maintain pursuant to its continuous airworthiness maintenance program or maintenance manual; or

(b) An article for which the domestic repair station certificate holder is rated. If the certificate holder is rated for a component of an article, the holder may not, in accordance with this Special Federal Aviation Regulation, approve that article for return to service.

**11.** *Additional Authorization Limitations.* Each holder of an authorization issued under this Special Federal Aviation Regulation must comply with any additional limitations prescribed by the Administrator and made a part of the authorization.

**12.** *Data Review and Service Experience.* If the Administrator finds that a product or article has been approved for return to service after a major repair has been performed under this Special Federal Aviation Regulation, that the product or article may not conform to the applicable airworthiness requirements or that an unsafe feature or characteristic of the product or article may exist, and that the nonconformance or unsafe feature or characteristic may be attributed to the repair performed, the holder of the authorization, upon notification by the Administrator, shall—

(a) Investigate the matter;

(b) Report to the Administrator the results of the investigation and any action proposed or taken; and

(c) If notified that an unsafe condition exists, provide within the time period stated by the Administrator, the information necessary for the FAA to issue an airworthiness directive under part 39 of the Federal Aviation Regulations.

*13. Current Records.* Each holder of an authorization issued under this Special Federal Aviation Regulation shall maintain, at its facility, current records containing—

(a) For each product or article for which it has developed and used major repair data, a technical data file that includes all data and amendments thereto (including drawings, photographs, specifications, instructions, and reports) necessary to accomplish the major repair;

(b) A list of products or articles by make, model, manufacturer's serial number (including specific part numbers and serial numbers of components) and, if applicable, FAA Technical Standard Order (TSO) or Parts Manufacturer Approval (PMA) identification, that have been repaired under the authorization; and

(c) A file of information from all available sources on difficulties experienced with products and articles repaired under the authorization.

This Special Federal Aviation Regulation terminates January 23, 2004.

[Docket No. 17551, 43 FR 3085, Jan. 23, 1978; as amended by SFAR 36–1, 45 FR 5677, Jan. 24, 1980; SFAR 36–2, 46 FR 58652, Dec. 3, 1981; SFAR 36–3, 49 FR 4354, Feb. 3, 1984; SFAR 36–4, 53 FR 49523, Dec. 7, 1988; SFAR 36–5, 54 FR 39293, Sept. 25, 1989; SFAR 36–6, 59 FR 3940, Jan. 27, 1994; SFAR 36–7, 64 FR 960, Jan. 6, 1999; SFAR 36–7, 66 FR 41116, Aug. 6, 2001]

## SFAR No. 50–2 to Part 121

### Special Flight Rules in the Vicinity of the Grand Canyon National Park, AZ

**Section 1.** *Applicability.* This rule prescribes special operating rules for all persons operating aircraft in the following airspace, designated as the Grand Canyon National Park Special Flight Rules Area:

That airspace extending upward from the surface up to but not including 14,500 feet MSL within an area bounded by a line beginning at lat. 36°09'30" N., long. 114°03'00" W.; northeast to lat. 36°14'00" N., long. 113°09'50" W.; thence northeast along the boundary of the Grand Canyon National Park to lat. 36°24'47" N., long. 112°52'00" W.; to lat. 36°30'30" N., long. 112°36'15" W. to lat. 36°21'30" N., long. 112°00'00" W. to lat. 36°35'30" N., long. 111°53'10" W., to lat. 36°53'00" N., long. 111°36'45" W. to lat. 36°53'00" N., long. 111°33'00" W.; to lat. 36°19'00" N., long. 111°50'50" W.; to lat. 36°17'00" N., long. 111°42'00" W.; to lat. 35°59'30" N., long. 111°42'00" W.; to lat. 35°57'30" N., long. 112°03'55" W.; thence counterclockwise via the 5 statute mile radius of the Grand Canyon Airport airport reference point (lat. 35°57'09" N., long. 112°08'47" W.) to lat. 35°57'30" N., long. 112°14'00" W.; to lat. 35°57'30" N., long. 113°11'00" W.; to lat. 35°42'30" N., long. 113°11'00" W.; to 35°38'30" N.; long. 113°27'30" W.; thence counterclockwise via the 5 statute mile radius of the Peach Springs VORTAC to lat. 35°41'20" N., long. 113°36'00" W.; to lat. 35°55'25" N., long. 113°49'10" W.; to lat. 35°57'45" N., 113°45'20" W.; thence northwest along the park boundary to lat. 36°02'20" N., long. 113°50'15" W.; to 36°00'10" N., long. 113°53'45" W.; thence to the point of beginning.

**Section 3.** *Aircraft operations: general.* Except in an emergency, no person may operate an aircraft in the Special Flight Rules, Area under VFR on or after September 22, 1988, or under IFR on or after April 6, 1989, unless the operation—

(a) Is conducted in accordance with the following procedures:

**Note:** The following procedures do not relieve the pilot from see-and-avoid responsibility or compliance with FAR 91.119.

(1) Unless necessary to maintain a safe distance from other aircraft or terrain—

(i) Remain clear of the areas described in Section 4; and

(ii) Remain at or above the following altitudes in each sector of the canyon:

Eastern section from Lees Ferry to North Canyon and North Canyon to Boundary Ridge: as prescribed in Section 5.

Boundary Ridge to Supai Point (Yumtheska Point): 10,000 feet MSL.

Western section from Diamond Creek to the Grant Wash Cliffs: 8,000 feet MSL.

(2) Proceed through the four flight corridors describe in Section 4 at the following altitudes unless otherwise authorized in writing by the Flight Standards District Office:

*Northbound*
11,500 or
13,500 feet MSL

*Southbound*
>10,500 or
>12,500 feet MSL

(b) Is authorized in writing by the Flight Standards District Office and is conducted in compliance with the conditions contained in that authorization. Normally authorization will be granted for operation in the areas described in Section 4 or below the altitudes listed in Section 5 only for operations of aircraft necessary for law enforcement, firefighting, emergency medical treatment/evacuation of persons in the vicinity of the Park; for support of Park maintenance or activities; or for aerial access to and maintenance of other property located within the Special Flight Rules Area. Authorization may be issued on a continuing basis.

(c)(1) Prior to November 1, 1988, is conducted in accordance with a specific authorization to operate in that airspace incorporated in the operator's part 135 operations specifications in accordance with the provisions of SFAR 50–1, notwithstanding the provisions of Sections 4 and 5; and

(2) On or after November 1, 1988, is conducted in accordance with a specific authorization to operate in that airspace incorporated in the operated in the operator's operations specifications and approved by the Flight Standards District Office in accordance with the provisions of SFAR 50–2.

(d) Is a search and rescue mission directed by the U.S. Air Force Rescue Coordination Center.

(e) Is conducted within 3 nautical miles of Whitmore Airstrip, Pearce Ferry Airstrip, North Rim Airstrip, Cliff Dwellers Airstrip, or Marble Canyon Airstrip at an altitudes less than 3,000 feet above airport elevation, for the purpose of landing at or taking off from that facility. Or

(f) Is conducted under an IFR clearance and the pilot is acting in accordance with ATC instructions. An IFR flight plan may not be filed on a route or at an altitude that would require operation in an area described in Section 4.

**Section 4.** *Flight-free zones.* Except in an emergency or if otherwise necessary for safety of flight, or unless otherwise authorized by the Flight Standards District Office for a purpose listed in Section 3(b), no person may operate an aircraft in the Special Flight Rules Area within the following areas:

(a) Desert View Flight-Free Zone. Within an area bounded by a line beginning at Lat. 35°59'30" N., Long. 111°46'20" W. to 35°59'30" N., Long. 111°52'45" W.; to Lat. 36°04'50" N., Long. 111°52'00" W.; to Lat. 36°06'00" N., Long. 111°46'20" W.; to the point of origin; but not including the airspace at and above 10,500 feet MSL within 1 mile of the western boundary of the zone. The area between the Desert View and Bright Angel Flight-Free Zones is designated the "Zuni Point Corridor."

(b) Bright Angel Flight-Free Zone. Within an area bounded by a line beginning at Lat. 35°59'30" N., Long. 111°55'30" W.; to Lat. 35°59'30" N., Long. 112°04'00" W.; thence counterclockwise via the 5 statute mile radius of the Grand Canyon Airport point (Lat. 35°57'09" N., Long. 112°08'47" W.) to Lat. 36°01'30" N., Long. 112°11'00" W.; to Lat. 36°06'15" N., Long. 112°12'50" W.; to Lat. 36°14'40" N., Long. 112°08'50" W.; to Lat. 36°14'40" N., Long. 111°57'30" W.; to Lat. 36°12'30" N., Long. 111°53'50" W.; to the point of origin; but not including the airspace at and above 10,500 feet MSL within 1 mile of the eastern boundary between the southern boundary and Lat. 36°04'50" N. or the airspace at and above 10,500 feet MSL within 2 miles of the northwest boundary. The area bounded by the Bright Angel and Shinumo Flight-Free Zones is designated the "Dragon Corridor."

(c) Shinumo Flight-Free Zone. Within an area bounded by a line beginning at Lat. 36°04'00" N., Long. 112°16'40" W.; northwest along the park boundary to a point at Lat. 36°12'47" N., Long. 112°30'53" W.; to Lat. 36°21'15" N., Long. 112°20'20" W.; east along the park boundary to Lat. 36°21'15" N., Long. 112°13'55" W.; to Lat. 36°14'40" N., Long. 112°11'25" W.; to the point of origin. The area between the Thunder River/Toroweap and Shinumo Flight Free Zones is designated the "Fossil Canyon Corridor."

(d) Toroweap/Thunder River Flight-Free Zone. Within an area bounded by a line beginning at Lat. 36°22'45" N., Long. 112°20'35" W.; thence northwest along the boundary of the Grand Canyon National Park to Lat. 36°17'48" N., Long. 113°03'15" W.; to Lat. 36°15'00" N., Long. 113°07'10" W.; to Lat. 36°10'30" N., Long. 113°07'10" W.; thence east along the Colorado

121

River to the confluence of Havasu Canyon (Lat. 36°18'40" N., Long. 112°45'45" W.;) including that area within a 1.5 nautical mile radius of Toroweap Overlook (Lat. 36°12'45" N., Long. 113°03'30" W.); to the point of origin; but not including the following airspace designated as the "Tuckup Corridor": at or above 10,500 feet MSL within 2 nautical miles either side of a line extending between Lat. 36°24'47" N., Long. 112°48'50" W. and Lat. 36°17'10" N., Long. 112°48'50" W.; to the point of origin.

**Section 5.** *Minimum flight altitudes.* Except in an emergency or if otherwise necessary for safety of flight, or unless otherwise authorized by the Flight Standards District Office for a purpose listed in Section 3(b), no person may operate an aircraft in the Special Flight Rules Area at an altitude lower than the following:

(a) Eastern section from Lees Ferry to North Canyon: 5,000 feet MSL.

(b) Eastern section from North Canyon to Boundary Ridge: 6,000 feet MSL.

(c) Boundary Ridge to Supai (Yumtheska) Point: 7,500 feet MSL.

(d) Supai Point to Diamond Creek: 6,500 feet MSL.

(e) Western section from Diamond Creek to the Grand Wash Cliffs: 5,000 feet MSL.

**Section 9.** *Termination date.* Section 1. Applicability, Section 4, Flight-free zones, and Section 5. Minimum flight altitudes, expire on April 19, 2001.

**Note:** An informational map of the special flight rules areas defined by SFAR 50–2 is available on the Office of Rulemaking's website at http://www.faa.gov/avr/armhome.htm. A paper copy is available from the Office of Rulemaking by calling Linda Williams at (202) 267-9685.

[Docket No. FAA–1999–5926, 66 FR 1003, Jan. 4, 2001; as amended by FAA–2001–9218, 66 FR 16584, March 26, 2001]

## SFAR No. 58 to Part 121

### Advanced Qualification Program

Section

1. Purpose and eligibility
2. Definitions
3. Required Curriculums
4. Indoctrination Curriculums
5. Qualification Curriculums
6. Continuing Qualification Curriculums
7. Other Requirements
8. Certification
9. Training Devices and Simulators
10. Approval of Advanced Qualification Program
11. Approval of Training, Qualification, or Evaluation by a Person Who Provides Training by Arrangement
12. Recordkeeping requirements
13. Expiration

Contrary provisions of parts 61, 63, 65, 121, and 135 of the Federal Aviation Regulations notwithstanding—

**1.** *Purpose and Eligibility.*

(a) This Special Federal Aviation Regulation provides for approval of an alternate method (known as "Advanced Qualification Program" or "AQP") for qualifying, training, certifying, and otherwise ensuring competency of crewmembers, aircraft dispatchers, other operations personnel, instructors, and evaluators who are required to be trained or qualified under parts 121 and 135 of the FAR or under this SFAR.

(b) A certificate holder is eligible under this Special Federal Aviation Regulation if the certificate holder is required to have an approved training program under §121.401 or §135.341 of the FAR, or elects to have an approved training program under §135.341.

(c) A certificate holder obtains approval of each proposed curriculum under this AQP as specified in section 10 of this SFAR.

(d) A curriculum approved under the AQP may include elements of present Part 121 and Part 135 training programs. Each curriculum must specify the make, model, and series aircraft (or variant) and each crewmember position or other positions to be covered by that curriculum. Positions to be covered by the AQP must include all flight crewmember positions, instructors, and evaluators and may include other positions, such as flight attendants, aircraft dispatchers, and other operations personnel.

(e) Each certificate holder that obtains approval of an AQP under this SFAR shall comply with all of the requirements of that program.

**2.** *Definitions.* As used in this SFAR:

*Curriculum* means a portion of an Advanced Qualification Program that covers one of three program areas: (1) indoctrination, (2) qualification, or (3) continuing qualification. A qualification or continuing qualification curriculum addresses the required training and qualification activities for a specific make, model, and series aircraft (or variant) and for a specific duty position.

*Evaluator* means a person who has satisfactorily completed training and evaluation that qualifies that person to evaluate the performance of crewmembers, instructors, other evaluators, aircraft dispatchers, and other operations personnel.

*Facility* means the physical environment required for training and qualification (e.g., buildings, classrooms).

*Training center* means an organization certificated under part 142 of this chapter or an organization approved by the Administrator to operate under the terms of this SFAR to provide training as described in section 1(b) of SFAR 58.

*Variant* means a specifically configured aircraft for which the FAA has identified training and qualification requirements that are significantly different from those applicable to other aircraft of the same make, model and series.

**3.** *Required Curriculums.* Each AQP must have separate curriculums for indoctrination, qualification, and continuing qualification as specified in sections 4, 5, and 6 of this SFAR.

**4.** *Indoctrination Curriculums.* Each indoctrination curriculum must include the following:

(a) For newly hired persons being trained under an AQP: Company policies and operating practices and general operational knowledge.

(b) For newly hired flight crewmembers and air dispatchers: General aeronautical knowledge.

(c) For instructors: The fundamental principles of the teaching and learning process; methods and theories of instruction; and the knowledge necessary to use aircraft flight training devices, flight simulators, and other training equipment in advanced qualification curriculums.

(d) For evaluators: Evaluation requirements specified in each approved curriculum; methods of evaluating crewmembers and aircraft dispatchers and other operations personnel; and policies and practices used to conduct the kinds of evaluations particular to an advanced qualification curriculum.(e.g., proficiency and online).

**5.** *Qualification Curriculums.* Each qualification curriculum must include the following:

(a) The certificate holder's planned hours of training, evaluation, and supervised operating experience.

(b) A list of and text describing the training, qualification and certification activities, as applicable for specific positions subject to the AQP, as follows:

(1) *Crewmembers, aircraft dispatchers, and other operations personnel.* Training, evaluation and certification activities which are aircraft- and equipment-specific to qualify a person for a particular duty position on, or duties related to the operation of a specific make, model, and series aircraft (or variant); a list of and text describing the knowledge requirements, subject materials, job skills, and each maneuver procedure to be trained and evaluated; the practical test requirements in addition to or in place of the requirements of parts 61, 63, and 65; and a list of and text describing supervised operating experience.

(2) *Instructors.* Training and evaluation to qualify a person to impart instruction on how to operate, or on how to ensure the safe operation of a particular make, model and series aircraft (or variant).

(3) *Evaluators.* Training, evaluation, and certification activities that are aircraft and equipment specific to qualify a person to evaluate the performance of persons who operate or who ensure the safe operation of, a particular make, model, and series aircraft (or variant).

**6.** *Continuing Qualification Curriculums.* Continuing qualification curriculums must comply with the following requirements:

(a) *General.* A continuing qualification curriculum must be based on—

(1) A continuing qualification cycle that ensures that during each cycle each person qualified under an AQP, including instructors and evaluators, will receive a balanced mix of training and evaluation on all events and subjects necessary to ensure that each person maintains the minimum proficiency levels of knowledge, skills and attitudes required for original qualification: and

(2) If applicable, flight crewmember or aircraft dispatcher recency of experience requirements.

(b) *Continuing Qualification Cycle Content.* Each continuing qualification cycle must include at least the following:

(1) *Evaluation period.* An evaluation period during which each person qualified under an AQP must receive at least one training session and a proficiency evaluation at a training facility. The number and frequency of training sessions must be approved by the Administrator. A training session, including any proficiency evaluation completed at that session, that occurs any time during the two calendar months before the last date for completion of an evaluation period can be considered by the certificate holder to be completed in the last calendar month.

(2) *Training.* Continuing qualification must include training in all events and major subjects required for original qualification, as follows:

(i) For pilots in command, seconds in command, flight engineers, and instructors and evaluators: Ground training including a general review of knowledge and skills covered in qualification training, updated information on newly developed procedures, and safety information.

(ii) For crewmembers, aircraft dispatchers, instructors, evaluators, and other operation personnel who conduct their duties in flight; Proficiency training in an aircraft, flight training device, or flight simulator on normal, abnormal, and emergency flight procedures and maneuvers.

(iii) For instructors and evaluators who are limited to conducting their duties in flight simulators and flight training devices: Proficiency training in a flight training device and/or flight simulator regarding operation of this training equipment and in operational flight procedures and maneuvers (normal, abnormal, and emergency).

(3) *Evaluations.* Continuing qualification must include evaluation in all events and major subjects

required for original qualification, and online evaluations for pilots in command and other eligible flight crewmembers. Each person qualified under an AQP must successfully complete a proficiency evaluation and, if applicable, an online evaluation during each evaluation period. An individual's proficiency evaluation may be accomplished over several training sessions if a certificate holder provides more than one training session in an evaluation period. The following evaluation requirements apply:

(i) Proficiency evaluations as follows:

(A) For pilots in command, seconds in command, and flight engineers: A proficiency evaluation, portions of which may be conducted in an aircraft flight simulator, or flight training device as approved in the certificate holder's curriculum which must be completed during each evaluation period.

(B) For any other persons covered by an AQP a means to evaluate their proficiency in the performance of their duties in their assigned tasks in an operational setting.

(ii) Online evaluations as follows:

(A) For pilots in command: An online evaluation conducted in an aircraft during actual flight operations under part 121 or part 135 or during operationally (line) oriented flights or proving flights. An online evaluation in an aircraft must be completed in the calendar month that includes the midpoint of the evaluation period. An online evaluation that is satisfactorily completed in the calendar month after the calendar month in which it becomes due is considered to have been completed during the calendar month it became due. However, in no case is an evaluation under this paragraph required more often than once during an evaluation period.

(B) During the online evaluations required under paragraph (b)(3)(ii)(A) of this section, each person performing duties as a pilot in command, second in command, or flight engineer for that flight, must be individually evaluated to determine whether he or she—

(1) Remains adequately trained and currently proficient with respect to the particular aircraft, crew position, and type of operation in which he or she serves; and

(2) Has sufficient knowledge and skills to operate effectively as part of the crew.

(4) *Recency of experience.* For pilots in command and seconds in command and if the certificate holder elects, flight engineers and flight dispatchers, approved recency of experience requirements.

(c) *Duration periods.* Initially the continuing qualification cycle approved for an AQP may not exceed 26 calendar months and the evaluation period may not exceed 13 calendar months. Thereafter, upon demonstration by a certificate holder that an extension is warranted, the Administrator may approve extensions of the continuing qualification cycle and the evaluation period in increments not exceeding 3 calendar months. However, a continuing qualification cycle may not exceed 39 calendar months and an evaluation period may not exceed 26 calendar months.

(d) *Requalification.* Each continuing qualification curriculum must include a curriculum segment that covers the requirements for requalifying a crewmember, aircraft dispatcher, or other operations personnel who has not maintained continuing qualification.

**7.** *Other Requirements.* In addition to the requirements of sections 4, 5, and 6, each AQP qualification and continuing qualification curriculum must include the following requirements:

(a) Approved Cockpit Resource Management (CRM) Training applicable to each position for which training is provided under an AQP.

(b) Approved training on an evaluation of skills and proficiency of each person being trained under an AQP to use their cockpit resource management skills and technical (piloting or other) skills in an actual or simulated operations scenario. For flight crew members this training and evaluation must be conducted in an approved flight training device or flight simulator.

(c) Data collection procedures that will ensure that the certificate holder provides information from its crew members, instructors, and evaluators that will enable the FAA to determine whether the training and evaluations are working to accomplish the overall objectives of the curriculum.

**8.** *Certification.* A person enrolled in an AQP is eligible to receive a commercial or airline transport pilot, flight engineer, aircraft dispatcher certificate or appropriate rating based on the successful completion of training and evaluation events accomplished under that program if the following requirements are met:

(a) Training and evaluation of required knowledge and skills under the AQP must meet minimum certification and rating criteria established by the Administrator in parts 61, 63, or 65. The Administrator may accept substitutes for the practical test requirements of parts 61, 63, or 65, as applicable.

(b) The applicant satisfactorily completes the appropriate qualification curriculum.

(c) The applicant shows competence in required technical knowledge and skills (e.g., piloting) and cockpit resource management knowledge and skills in scenarios that test both types of knowledge and skills together.

(d) the applicant is otherwise eligible under the applicable requirements of part 61, 63, or 65.

**9.** *Training Devices and Simulators.*

(a) *Qualification and approval of flight training devices and flight simulators.*

(1) Any training device or simulator that will be used in an AQP for one of the following purposes must be evaluated by the Administrator for assignment of a flight training device or flight simulator qualification level:

(i) Required evaluation of individual or crew proficiency.

(ii) Training activities that determine if an individual or crew is ready for a proficiency evaluation.

(iii) Activities used to meet recency of experience requirements.

(iv) Line Operations Simulations (LOS).

(2) To be eligible to request evaluation for a qualification level of a flight training device or flight simulator an applicant must—

(i) Hold an operating certificate; or

(ii) Be a training center that has applied for authorization to the administrator or has been authorized by the Administrator to conduct training or qualification under an AQP.

(3) Each flight training device or flight simulator to be used by a certificate holder or training center for any of the purposes set forth in paragraph (a)(1) of this section must—

(i) Be, or have been, evaluated against a set of criteria established by the Administrator for a particular qualification level of simulation;

(ii) Be approved for its intended use in a specified AQP, and

(iii) Be part of a flight simulator or flight training device continuing qualification program approved by the Administrator.

(b) *Approval of other Training Equipment.*

(1) Any training device that is intended to be used in an AQP for purposes other than those set forth in paragraph (a)(1) of this section must be approved by the Administrator for its intended use.

(2) An applicant for approval of a training device under this paragraph must identify the device by its nomenclature and describe its intended use.

(3) Each training device approved for use in an AQP must be part of a continuing program to provide for its serviceability and fitness to perform its intended function as approved by the Administrator.

**10.** *Approval of Advanced Qualification Program.*

(a) *Approval Process.* Each applicant for approval of an AQP curriculum under this SFAR shall apply for approval of that curriculum. Application for approval is made to the certificate holder's FAA Flight Standards District Office.

(b) *Approval Criteria.* An application for approval of an AQP curriculum will be approved if the program meets the following requirements:

(1) It must be submitted in a form and manner acceptable to the Administrator.

(2) It must meet all of the requirements of this SFAR.

(3) It must indicate specifically the requirements of parts 61, 63, 65, 121 or 135, as applicable, that would be replaced by an AQP curriculum. If a requirement of parts 61, 63, 65, 121, or 135 is replaced by an AQP curriculum, the certificate holder must show how the AQP curriculum provides an equivalent level of safety for each requirement that is replaced. Each applicable requirement of parts 61, 63, 65, 121 or 135 that is not specifically addressed in an AQP curriculum continues to apply to the certificate holder.

(c) *Application and Transition.* Each certificate holder that applies for one or more advanced qualification curriculums or for a revision to a previously approved curriculum must comply with §121.405 or §135.325, as applicable, and must include as part of its application a proposed transition plan (containing a calendar of events) for moving from its present approved training to the advanced qualification training.

(d) *Advanced Qualification Program Revisions or Rescissions of Approval.* If after a certificate holder begins operations under an AQP, the Administrator finds that the certificate holder is not meeting the provisions of its approved AQP, the Administrator may require the certificate holder to make revisions in accordance with §121.405 or §135.325, as applicable, or to submit and obtain approval for a plan (containing a schedule of events) that the certificate holder must comply with and use to transition to an approved part 121 or part 135 training program, as appropriate.

**11.** *Approval of Training, Qualification, or Evaluation by a Person who Provides Training by Arrangement.*

(a) A certificate holder under part 121 or part 135 may arrange to have AQP required training, qualification, or evaluation functions performed by another person (a *training center*) if the following requirements are met:

(1) The training center's training and qualification curriculums, curriculum segments, or portions of curriculum segments must be provisionally approved by the Administrator. A training center may apply for provisional approval independently or in conjunction with a certificate holder's application for AQP approval. Application for provisional approval must be made to the FAA's Flight Standards District Office that has responsibility for the training center.

(2) The specific use of provisionally approved curriculums, curriculum segments, or portions of curriculum segments in a certificate holder's AQP must be approved by the Administrator as set forth in section 10 of this SFAR.

(b) An applicant for provisional approval of a curriculum, curriculum segment, or portion of a curriculum segment under this paragraph must show that the following requirements are met:

(1) The applicant must have a curriculum for the qualification and continuing qualification of

each instructor or evaluator employed by the applicant.

(2) The applicant's facilities must be found by the Administrator to be adequate for any planned training, qualification, or evaluation for a part 121 or part 135 certificate holder.

(3) Except for indoctrination curriculums, the curriculum, curriculum segment, or portion of a curriculum segment must identify the specific make, model, and series aircraft (or variant) and crewmember or other positions for which it is designed.

(c) A certificate holder who wants approval to use a training center's provisionally approved curriculum, curriculum segment, or portion of a curriculum segment in its AQP, must show that the following requirements are met:

(1) Each instructor or evaluator used by the training center must meet all of the qualification and continuing qualification requirements that apply to employees of the certificate holder that has arranged for the training, including knowledge of the certificate holder's operations.

(2) Each provisionally approved curriculum, curriculum segment, or portion of a curriculum segment must be approved by the Administrator for use in the certificate holder's AQP. The Administrator will either provide approval or require modifications to ensure that each curriculum, curriculum segment, or portion of a curriculum segment is applicable to the certificate holder's AQP.

(d) Approval for the training, qualification, or evaluation by a person who provides training by arrangement authorized by this section expires on August 3, 1998 unless that person meets the eligibility requirements specified under §121.402 or §135.324 of this chapter. After August 2, 1998 approval for the training, qualification, or evaluation, by a person who provides training by arrangement authorized by this section, shall be granted only to persons who meet the eligibility requirements specified under §121.402 or §135.234 of this chapter.

**12.** *Recordkeeping Requirements.* Each certificate holder and each training center holding AQP provisional approval shall show that it will establish and maintain records in sufficient detail to establish the training, qualification, and certification of each person qualified under an AQP in accordance with the training, qualification, and certification requirements of this SFAR.

**13.** *Expiration.* This Special Federal Aviation Regulation terminates on October 2, 2005, unless sooner terminated.

[SFAR 58, 55 FR 40275, Oct. 2, 1990; as amended by 60 FR 51851, Oct. 3, 1995; SFAR 58–2, 61 FR 34560, July 2, 1996; Amdt. 121–280, 65 FR 60336, Oct. 10, 2000]

## SFAR No. 78 to Part 121

### Special Operating Rules for Commercial Air Tour Operators in the Vicinity of the Rocky Mountain National Park

**Section 1.** *Applicability.* This Special Federal Aviation Regulation prescribes operating rules for commercial air tour flight operations within the lateral boundaries of the Rocky Mountain National Park, CO.

**Section 2.** *Definition.* For the purpose of this SFAR: "commercial air tour" means: the operation of an aircraft carrying passengers for compensation or hire for aerial sightseeing.

**Section 3.** *Restriction.* No person may conduct a commercial air tour operation in the airspace over Rocky Mountain National Park, CO.

*Expiration:* This SFAR will expire on the adoption date of a final rule in Docket No. 27643.

[Docket No. 28577, 62 FR 1205, Jan. 8, 1997]

## SFAR No. 80 to Part 121

### Alternative Communications and Dispatching Procedures

**1.** *Applicability.* This Special Federal Aviation Regulation applies to each holder of an air carrier or operating certificate (hereafter, certificate holder) that meets one of the following eligibility requirements:

a. The certificate holder conducts scheduled operations with airplanes having a passenger-seat configuration of 30 seats or fewer, excluding each crewmember seat, and a payload capacity of 7,500 pounds or less under part 121 of this chapter.

b. The certificate holder conducts domestic operations in Alaska under part 121 of this chapter.

**2.** *Alternative requirements.*

a. If an operator described in paragraph 1.a. of this SFAR is conducting a flight with an airplane described in 1.a. and if communications cannot be maintained over the entire route (which would be contrary to the requirements of §121.99 of this chapter), such an operator may continue to operate over such a route subject to approval by the Administrator. In granting such approval the Administrator considers the following:

i. The operator has an established dispatch communication system.

ii. Gaps in communication are not over the entire route, but only over portions of the route.

iii. When communication gaps occur, they occur due to one or more of the following:

A. Lack of infrastructure.

B. Geographical considerations.

C. Assigned operating altitude.

iv. Procedures are established for the prompt re-establishment of communications.

v. The operator has presented a plan or schedule for coming into compliance with the requirements in §121.99 of this chapter.

b. A certificate holder who conducts domestic operations in Alaska may, notwithstanding the requirements of §121.99 of this chapter, use a communications system operated by the United States for those operations.

c. An operator described in paragraph 1.a. of this SFAR who conducts operations in Alaska may share the aircraft dispatcher required by §121.395 with another operator described in paragraph 1.a. of this SFAR who conducts operations in Alaska if authorized to do so by the Administrator. Before granting such an authorization, the Administrator considers:

i. The operators' joint plans for complying with the aircraft dispatcher training rules in Subpart N of part 121 of this chapter and the aircraft dispatcher qualification and duty time limitation rules in Subpart P of part 121 of this chapter.

ii. The number of flights for which the aircraft dispatcher would be responsible.

iii. Whether the responsibilities of the dispatcher would be beyond the capability of a single dispatcher.

**3.** *Expiration.* This Special Federal Aviation Regulation terminates on March 12, 2001, unless sooner terminated.

[SFAR 80, 62 FR 13255, March 19, 1997; as amended by 62 FR 15570, April 1, 1997]

## SFAR No. 89 to Part 121

### Suspension of Certain Flight Data Recorder Requirements

**1.** *Applicability.* This Special Federal Aviation Regulation provides relief to operators of the airplanes listed in paragraph 2 of this regulation. Relief under this regulation is limited to suspension of the resolution requirements only as listed in appendix M to part 121, appendix E to part 125, or appendix F to part 135, for the flight data recorder parameters noted for individual airplane models.

**2.** *Airplanes Affected.*

(a) Boeing model 717 airplanes—resolution requirement of appendix M to Part 121 or appendix E to part 125 for parameter number 5. Normal Acceleration (Vertical);

(b) Boeing model 757 airplanes—resolution requirements of appendix M to Part 121 or appendix E to part 125 for parameter number 12a. Pitch Control(s) position (non-fly-by-wire systems); number 14a. Yaw Control position(s) (non-fly-by-wire); number 19. Pitch Trim Surface Position; and number 23. Ground Spoiler Position or Speed Brake Selection.

(c) Boeing Model 767 airplanes—resolution requirements of appendix M to Part 121 or appendix E to part 125 for parameter number 12a. Pitch Control(s) position (non-fly-by-wire systems); number 14a. Yaw Control position(s) (non-fly-by-wire); number 16. Lateral Control Surface(s) Position (for inboard aileron(s) only); number 19. Pitch Trim Surface Position; and number 23. Ground Spoiler Position or Speed Brake Selection.

(d) Dassault Model Falcon 900 EX and Model Mystere-Falcon 900 (with modification M1975 or M2695 installed) airplanes—resolution requirements of appendix M to Part 121, appendix E to part 125 or appendix F to part 135 for parameter number 5. Normal Acceleration (Vertical); and number 26. Radio Altitude.

(e) Other airplanes for which notification under paragraph 3(b) of this regulation is made to the FAA regarding flight data recorder resolution requirement noncompliance.

**3.** *Requirements for use.*

(a) An operator of an airplane described in paragraphs 2(a) through 2(d) of this regulation may make immediate use of the relief granted by this SFAR.

(b) An operator seeking relief for another airplane model under paragraph 2(e) of this SFAR must notify the FAA immediately in writing as to the nature and extent of the resolution problem found, and must comply with all other requirements of this SFAR, including the report required in paragraph 3(d) of this SFAR. Operators may make immediate use of this relief, but relief may be withdrawn by the FAA after a review of the information filed. Additional information may be required.

(c) An operator of an affected airplane must continue to record all affected parameters to the maximum resolution possible using the installed equipment; that equipment must be maintained in proper working order.

(d) An operator of an affected airplane must, within 30 days of using the relief granted by this regulation, report the following information:

(1) The operator's name and address, and the name and phone number of a contact person for the information reported;

(2) The model and registration number of each affected airplane;

(3) For each affected airplane, the parameter(s) for which resolution relief is being used, and the actual resolution being recorded;

(4) Any additional information requested by the FAA.

(e) Reports must be filed with the FAA Flight Standards Service, Denise Cashmere, Administrative Officer, AFS-200, 800 Independence Ave. SW, Washington, DC 20591. Additionally, each operator must file a copy of the report with its

Principal Avionics Inspector or Principal Operations Inspector, as appropriate.

**4.** *Expiration.* This Special Federal Aviation Regulation expires on August 18, 2003.

[Docket No. FAA–2001–10428, SFAR 89, 66 FR 44273, Aug. 22, 2001]

## SFAR No. 92-4 to Part 121

### Flightcrew Compartment Access and Door Designs

**1.** *Applicability.* This Special Federal Aviation Regulation (SFAR) applies to all operators that hold an air carrier certificate or operating certificate issued under 14 CFR part 119 and that conduct operations under this part 121 and to operators of U.S. registered transport category aircraft operated under 14 CFR part 129, except paragraph 5 of this SFAR does not apply to cargo operations and 14 CFR part 129 operations. It applies to the operators specified in this SFAR that modify airplanes to improve the flightcrew compartment door installations to restrict the unwanted entry of persons into the flightcrew compartment. This SFAR also applies to production certificate holders applicants for airworthiness certificates for airplanes to be operated by operators specified in this SFAR, and producers of parts to be used in such modification.

**2.** *Regulatory Relief.* Contrary provisions of this part 21, and §§121.313(h), 121.153(a)(2), 121.153(c), 121.379(b), 121.583(b)(1) and (2) and 14 CFR 129.13 notwithstanding:

(a) An operator may operate airplanes modified to improve the flightcrew compartment door installations to restrict the unauthorized entry of persons into the flightcrew compartment without regard to the applicable airworthiness requirements and may modify those airplanes for that purpose, using technical data for previously approved by the Administrator, subject to the following conditions:

(i) Not later than February 15, 2002, submit to the Director, Aircraft Certification Service, a detailed description of the changes to the airplanes that have been accomplished before that date to enhance the intrusion resistance of the flightcrew compartment including identification of what major alterations have been done without previously approved data.

(ii) If, upon reviewing the data submitted in paragraph 2(a)(i) of this SFAR, the Administrator determines that a door modification presents an unacceptable safety risk, the FAA may issue an order requiring changes to such modifications.

(b) An applicant for an airworthiness certificate may obtain such a certificate for modified airplanes to be operated by operators described in this SFAR.

(c) A holder of a production certificate may submit for airworthiness certification or approval, modified airplanes to be operated by operators described in this SFAR.

(d) A person may produce parts for installation on airplanes in connection with modifications described in this SFAR, without FAA parts manufacturer approval (PMA).

**3.** *Report of Modifications.* Not later than April 22, 2002, all operators who are required to install flightdeck door modifications in accordance with §121.313(j) must submit a report to the Director, Aircraft Certification Service. The report must describe the modifications to be made and provide a schedule for the changes necessary to restore compliance with all applicable airworthiness requirements and to meet the requirements of §121.313(j). The schedule may not extend beyond the termination data of this SFAR.

**4.** *Return to Service Documentation.* Where operators have modified airplanes as authorized in this SFAR, the affected airplane must be returned to service with a note that it was done under the provisions of this SFAR.

**5.** *Provision for Flightdeck Door Compartment Key.* Contrary to provisions of §121.313(g), the following provision applies: A key for each door that separates a passenger compartment from an emergency exist must be identified to passengers in the briefing required by §121.571(a)(1)(ii). The key required for access to the emergency exit must be readily available for each crewmember. No key to the flightcrew compartment shall be available to any crewmember during flight, except for flight crewmembers, unless an internal flightdeck locking device such as a deadbolt or bar is installed, operative, and in use.

**6.** *Door Modification Requirement.* After March 1, 2002, for each airplane required under §121.313(f) to have a door between the passenger and pilot compartments, and for transport category all-cargo airplanes that have a door installed between the pilot compartment and any other occupied compartment on or after January 15, 2002, such door must be equipped with an internal locking device installed, operative, and in use. Such internal locking device has to be designed so that it can only be unlocked from inside the flightdeck.

**7.** *Termination.* This SFAR terminates on April 9, 2003.

[Docket No. FAA–2001–10770, 67 FR 12823, March 19, 2002]

## SFAR No. 93 to Part 121

### Temporary Extension of Time To Allow for Certain Training and Testing

**1.** *Applicability.* This SFAR applies to all part 121 and 135 check airmen (simulator) and flight instructors (simulator), part 121 aircraft dispatchers, and part 142 training center instructors who were required to complete qualification requirements, an inflight line observation program, or operating familiarization in September 2001 to become qualified, or remain qualified, to perform their assigned duties. It also applies to persons who have satisfactorily accomplished the part 61 aeronautical knowledge test or the part 63 written test, either one of which has an expiration date of September 2001 for pilot, flight instructor, or flight engineer certification.

**2.** *Special Qualification Requirements.* The sections of 14 CFR that prescribes these requirements are sections 61.39(a)(1); 63.35(d); 121.411(f); 121.412(f); 121.463(a)(2); 121.463(c); 135.337(f); 135.338(f); 142.53(b)(2) and (b)(3).

**3.** *Extension of Time to Fulfill Certain Qualification Requirements.* Persons identified in paragraph 1 of this SFAR who had until the end of September 2001 to complete the specified qualification requirements in September 2001 will be deemed to have completed those requirements in September 2001 provided they satisfactorily complete those requirements by November 30, 2001. For those persons identified in paragraph 1, who are qualifying for the first time to be a check airmen (simulator), flight instructor (simulator), aircraft dispatcher, or training center instructor, they must fulfill the applicable qualification requirements before they may serve as a check airmen (simulator), flight instructor (simulator), aircraft dispatcher, or training center instructor, as appropriate. This extension does not change the 12-calendar-month requirement for aircraft dispatchers or the anniversary month for check airmen, flight instructors and training center instructors. Therefore, if you were due for qualification in September 2001 you will be due for qualification September 2002, regardless of this extension for 2001.

**4.** *Termination Date.* This Special Federal Aviation Regulation expires November 30, 2001.

[Docket No. FAA–2001–10797; 66 FR 52279, Oct. 12, 2001]

# Subpart A—General

## §121.1 Applicability.

This part prescribes rules governing—

**(a)** The domestic, flag, and supplemental operations of each person who holds or is required to hold an Air Carrier Certificate or Operating Certificate under part 119 of this chapter.

**(b)** Each person employed or used by a certificate holder conducting operations under this part including maintenance, preventive maintenance, and alteration of aircraft.

**(c)** Each person who applies for provisional approval of an Advanced Qualification Program curriculum, curriculum segment, or portion of a curriculum segment under SFAR No. 58 of 14 CFR part 121, and each person employed or used by an air carrier or commercial operator under this part to perform training, qualification, or evaluation functions under an Advanced Qualification Program under SFAR No. 58 of 14 CFR part 121.

**(d)** Nonstop sightseeing flights conducted with airplanes having a passenger-seat configuration of 30 seats or fewer and a maximum payload capacity of 7,500 pounds or less that begin and end at the same airport, and are conducted within a 25 statute mile radius of that airport; however, except for operations subject to SFAR 50–2 of 14 CFR part 121, these operations, when conducted for compensation or hire, must comply only with §§121.455 and 121.457, except that an operator who does not hold an air carrier certificate or an operating certificate is permitted to use a person who is otherwise authorized to perform aircraft maintenance or preventive maintenance duties and who is not subject to FAA-approved anti-drug and alcohol misuse prevention programs to perform—

(1) Aircraft maintenance or preventive maintenance on the operator's aircraft if the operator would otherwise be required to transport the aircraft more than 50 nautical miles further than the repair point closest to the operator's principal base of operations to obtain these services; or

(2) Emergency repairs on the operator's aircraft if the aircraft cannot be safely operated to a location where an employee subject to FAA-approved programs can perform the repairs.

**(e)** Each person who is on board an aircraft being operated under this part.

**(f)** Each person who is an applicant for an Air Carrier Certificate or an Operating Certificate under part 119 of this chapter, when conducting proving tests.

[Docket No. 28154, 60 FR 65925, Dec. 20, 1995]

## §121.2 Compliance schedule for operators that transition to part 121; certain new entrant operators.

**(a)** *Applicability:* This section applies to the following:

(1) Each certificate holder that was issued an air carrier or operating certificate and operations specifications under the requirements of part 135 of this chapter or under SFAR No. 38–2 of 14

CFR part 121 before January 19, 1996, and that conducts scheduled passenger-carrying operations with:

(i) Nontransport category turbopropeller powered airplanes type certificated after December 31, 1964, that have a passenger seat configuration of 10–19 seats;

(ii) Transport category turbopropeller powered airplanes that have a passenger seat configuration of 20–30 seats; or

(iii) Turbojet engine powered airplanes having a passenger seat configuration of 1–30 seats.

(2) Each person who, after January 19, 1996, applies for or obtains an initial air carrier or operating certificate and operations specifications to conduct scheduled passenger-carrying operations in the kinds of airplanes described in paragraphs (a)(1)(i), (a)(1)(ii), or paragraph (a)(1)(iii) of this section.

**(b)** *Obtaining operations specifications.* A certificate holder described in paragraph (a)(1) of this section may not, after March 20, 1997, operate an airplane described in paragraphs (a)(1)(i), (a)(1)(ii), or (a)(1)(iii) of this section in scheduled passenger-carrying operations, unless it obtains operations specifications to conduct its scheduled operations under this part on or before March 20, 1997.

**(c)** *Regular or accelerated compliance.* Except as provided in paragraphs (d), (e), and (i) of this section, each certificate holder described in paragraphs (a)(1) of this section shall comply with each applicable requirement of this part on and after March 20, 1997, or on and after the date on which the certificate holder is issued operations specifications under this part, whichever occurs first. Except as provided in paragraphs (d) and (e) of this section, each person described in paragraph (a)(2) of this section shall comply with each applicable requirement of this part on and after the date on which that person is issued a certificate and operations specifications under this part.

**(d)** *Delayed compliance dates.* Unless paragraph (e) of this section specifies an earlier compliance date, no certificate holder that is covered by paragraph (a) of this section may operate an airplane in 14 CFR part 121 operations on or after a date listed in this paragraph unless that airplane meets the applicable requirement of this paragraph (d):

(1) *Nontransport category turbopropeller powered airplanes type certificated after December 31, 1964, that have a passenger seat configuration of 10–19 seats.* No certificate holder may operate under this part an airplane that is described in paragraph (a)(10)(i) of this section on or after a date listed in paragraph (d)(1) of this section unless that airplane meets the applicable requirement listed in this paragraph (d)(1) of this section:

(i) December 20, 1997:

(A) Section 121.289, Landing gear aural warning.

(B) Section 121.308, Lavatory fire protection.

(C) Section 121.310(e), Emergency exit handle illumination.

(D) Section 121.337(b)(8), Protective breathing equipment.

(E) Section 121.340, Emergency flotation means.

(ii) December 20, 1999: Section 121.342, Pitot heat indication system.

(iii) December 20, 2010:

(A) For airplanes described in §121.157(f), the Airplane Performance Operating Limitations in §§121.189 through 121.197.

(B) Section 121.161(b), Ditching approval.

(C) Section 121.305(j), Third attitude indicator.

(D) Section 121.312(c), Passenger seat cushion flammability.

(iv) March 12, 1999: Section 121.310(b)(1), Interior emergency exit locating sign.

(2) *Transport category turbopropeller powered airplanes that have a passenger seat configuration of 20–30 seats.* No certificate holder may operate under this part an airplane that is described in paragraph (a)(1)(ii) of this section on or after a date listed in paragraph (d)(2) of this section unless that airplane meets the applicable requirement listed in paragraph (d)(2) of this section:

(i) December 20, 1997:

(A) Section 121.308, Lavatory fire protection.

(B) Section 121.337(b)(8) and (9), Protective breathing equipment.

(C) Section 121.340, Emergency flotation means.

(ii) December 20, 2010: §121.305(j), third attitude indicator.

**(e)** *Newly manufactured airplanes.* No certificate holder that is described in paragraph (a) of this section may operate under this part an airplane manufactured on or after a date listed in this paragraph unless that airplane meets the applicable requirement listed in this paragraph (e).

(1) For nontransport category turbopropeller powered airplanes type certificated after December 31, 1964, that have a passenger seat configuration of 10–19 seats:

(i) Manufactured on or after March 20, 1997:

(A) Section 121.305(j), Third attitude indicator.

(B) Section 121.311(f), Safety belts and shoulder harnesses.

(ii) Manufactured on or after December 20, 1997: Section 121.317(a), Fasten seat belt light.

(iii) Manufactured on or after December 20, 1999: Section 121.293, Takeoff warning system.

(iv) Manufactured on or after March 12, 1999: Section 121.310(b)(1), Interior emergency exit locating sign.

(2) For transport category turbopropeller powered airplanes that have a passenger seat config-

uration of 20–30 seats manufactured on or after March 20, 1997: Section 121.305(j), Third attitude indicator.

**(f)** *New type certification requirements.* No person may operate an airplane for which the application for a type certificate was filed after March 29, 1995, in 14 CFR part 121 operations unless that airplane is type certificated under part 25 of this chapter.

**(g)** *Transition plan.* Before March 19, 1996, each certificate holder described in paragraph (a)(1) of this section must submit to the FAA a transition plan (containing a calendar of events) for moving from conducting its scheduled operations under the commuter requirements of part 135 of this chapter to the requirements for domestic or flag operations under this part. Each transition plan must contain details on the following:

(1) Plans for obtaining new operations specifications authorizing domestic or flag operations;

(2) Plans for being in compliance with the applicable requirements of this part on or before March 20, 1997; and

(3) Plans for complying with the compliance date schedules contained in paragraphs (d) and (e) of this section.

**(h)** *Continuing requirements.* A certificate holder described in paragraph (a) of this section shall comply with the applicable airplane operating and equipment requirements of part 135 of this chapter for the airplanes described in paragraph (a)(1) of this section, until the airplane meets the specific compliance dates in paragraphs (d) and (e) of this section.

**(i)** *Delayed pilot age limitation:*

(1) Notwithstanding §121.383(c), and except as provided in paragraph (i)(2) of this section, a certificate holder may use the services of a person as a pilot in operations covered by paragraph (a)(1) of this section after that person has reached his or her 60th birthday, until December 20, 1999. Notwithstanding §121.383(c), and except as provided in paragraph (i)(2) of this section, a person may serve as a pilot in operations covered by paragraph (a)(1) of this section after that person has reached his or her 60th birthday, until December 20, 1999.

(2) This paragraph applies only to persons who were employed as pilots by a certificate holder in operations covered by paragraph (a)(1) of this section on March 20, 1997.

**(j)** Any training or qualification obtained by a crewmember under part 135 of this chapter before March 20, 1997, is entitled to credit under this part for the purpose of meeting the requirements of this part, as determined by the Administrator. Records kept by a certificate holder under part 135 of this chapter before March 20, 1997, can be annotated with the approval of the Administrator, to reflect crewmember training and qualification credited toward part 121 requirements.

[Docket No. 28154, 60 FR 65925, Dec. 20, 1995; as amended by Amdt. 121–253, 61 FR 2609, Jan. 26, 1996; Amdt. 121–256, 61 FR 30434, June 14, 1996; Amdt. 121–262, 62 FR 13256, March 19, 1997]

### §121.4 Applicability of rules to unauthorized operators.

The rules in this part which refer to a person certificated under Part 119 of this chapter apply also to any person who engages in an operation governed by this part without the appropriate certificate and operations specifications required by Part 119 of this chapter.

[Docket No. 11675, 37 FR 20937, Oct. 5, 1972; as amended by Amdt. 121–251, 60 FR 65926, Dec 20, 1995]

### §121.11 Rules applicable to operations in a foreign country.

Each certificate holder shall, while operating an airplane within a foreign country, comply with the air traffic rules of the country concerned and the local airport rules, except where any rule of this part is more restrictive and may be followed without violating the rules of that country.

[Docket No. 16383, 43 FR 22641, May 25, 1978]

### §121.15 Carriage of narcotic drugs, marihuana, and depressant or stimulant drugs or substances.

If a certificate holder operating under this part permits any aircraft owned or leased by that holder to be engaged in any operation that the certificate holder knows to be in violation of §91.19(a) of this chapter, that operation is a basis for suspending or revoking the certificate.

[Docket No. 28154, 60 FR 65926, Dec. 20, 1995]

## Subpart B—[Reserved]

## Subpart C—[Reserved]

## Subpart D—[Reserved]

# Subpart E— Approval of Routes: Domestic and Flag Operations

**Source:** Docket No. 6258, 29 FR 19194, Dec. 31, 1964, unless otherwise noted.

## §121.91 Applicability.

This subpart prescribes rules for obtaining approval of routes by certificate holders conducting domestic or flag operations.

[Docket No. 28154, 61 FR 2610, Jan. 26, 1996]

## §121.93 Route requirements: General.

**(a)** Each certificate holders conducting flag operations seeking a route approval must show—

(1) That it is able to conduct satisfactorily scheduled operations between each regular, provisional, and refueling airport over that route or route segment; and

(2) That the facilities and services required by §§121.97 through 121.107 are available and adequate for the proposed operation.

The Administrator approves a route outside of controlled airspace if he determines that traffic density is such that an adequate level of safety can be assured.

**(b)** Paragraph (a) of this section does not require actual flight over a route or route segment if the certificate holder shows that the flight is not essential to safety, considering the availability and adequacy of airports, lighting, maintenance, communication, navigation, fueling, ground, and airplane radio facilities, and the ability of the personnel to be used in the proposed operation.

[Docket No. 6258, 29 FR 19194, Dec. 31, 1964; as amended by Amdt. 121–3, 30 FR 3638, March 19, 1965; Amdt. 121–253, 61 FR 2610, Jan. 26, 1996]

## §121.95 Route width.

**(a)** Approved routes and route segments over U.S. Federal airways or foreign airways (and advisory routes in the case of certificate holders conducting flag operations) have a width equal to the designated width of those airways or routes. Whenever the Administrator finds it necessary to determine the width of other approved routes, he considers the following:

(1) Terrain clearance.

(2) Minimum en route altitudes.

(3) Ground and airborne navigation aids.

(4) Air traffic density.

(5) ATC procedures.

**(b)** Any route widths of other approved routes determined by the Administrator are specified in the certificate holder's operations specifications.

[Docket No. 6258, 29 FR 19194, Dec. 31, 1964; as amended by Amdt. 121–253, 61 FR 2610, Jan. 26, 1996]

## §121.97 Airports: Required data.

**(a)** Each certificate holder conducting domestic or flag operations must show that each route it submits for approval has enough airports that are properly equipped and adequate for the proposed operation, considering such items as size, surface, obstructions, facilities, public protection, lighting, navigational and communications aids, and ATC.

**(b)** Each certificate holder conducting domestic or flag operations must show that it has an approved system for obtaining, maintaining, and distributing to appropriate personnel current aeronautical data for each airport it uses to ensure a safe operation at that airport. The aeronautical data must include the following:

(1) Airports.

(i) Facilities.

(ii) Public protection.

(iii) Navigational and communications aids.

(iv) Construction affecting takeoff, landing, or ground operations.

(v) Air traffic facilities.

(2) Runways, clearways and stopways.

(i) Dimensions.

(ii) Surface.

(iii) Marking and lighting systems.

(iv) Elevation and gradient.

(3) Displaced thresholds.

(i) Location.

(ii) Dimensions.

(iii) Takeoff or landing or both.

(4) Obstacles.

(i) Those affecting takeoff and landing performance computations in accordance with Subpart I of this part.

(ii) Controlling obstacles.

(5) Instrument flight procedures.

(i) Departure procedure.

(ii) Approach procedure.

(iii) Missed approach procedure.

(6) Special information.

(i) Runway visual range measurement equipment.

(ii) Prevailing winds under low visibility conditions.

**(c)** If the certificate-holding district office charged with the overall inspection of the certificate holder's operations finds that revisions are necessary for the continued adequacy of the certificate holder's system for collection, dissemination, and usage of aeronautical data that has been granted approval, the certificate holder shall, after notification by the certificate-holding district office, make those revisions in the system. Within 30 days after the certificate holder receives such notice, the certificate holder may file a petition to reconsider the notice with the Director, Flight Standards Service. This filing of a petition to reconsider stays the notice pending a decision by the Director, Flight Standards Service. However, if the certificate-holding district office finds that there is an emergency that requires immediate action in the interest of safety in air transportation, the Director, Flight Standards Service may, upon statement of the reasons, require a change effective without stay.

[Docket No. 6258, 29 FR 19194, Dec. 31, 1964; as amended by Amdt. 121–162, 45 FR 46738, July 10, 1980; Amdt. 121–207, 54 FR 39293, Sept. 25, 1989; Amdt. 121–253, 61 FR 2610, Jan. 26, 1996]

## §121.99 Communication facilities.

**(a)** Each certificate holder conducting domestic or flag operations must show that a two-way radio communication system or other means of communication approved by the Administrator is available at points that will ensure reliable and rapid communications, under normal operating conditions over the entire route (either direct or via approved point-to-point circuits) between each airplane and the appropriate dispatch office, and between each airplane and the appropriate air traffic control unit, except as specified as §121.351(c).

**(b)** For the following types of operations, the communications systems between each airplane and the dispatch office must be independent of any system operated by the United States:

(1) All domestic operations;

(2) Flag operations in the 48 contiguous States and the District of Columbia; and

(3) After March 12, 2001, flag operations outside the 48 contiguous States and the District of Columbia.

[Docket No. 28154, 62 FR 13256, March 19, 1997]

## §121.101 Weather reporting facilities.

**(a)** Each certificate holder conducting domestic or flag operations must show that enough weather reporting services are available along each route to ensure weather reports and forecasts necessary for the operation.

**(b)** Except as provided in paragraph (d) of this section, no certificate holder conducting domestic or flag operations may use any weather report to control flight unless—

(1) For operations within the 48 contiguous States and the District of Columbia, it was prepared by the U.S. National Weather Service or a source approved by the U.S. National Weather Service; or

(2) For operations conducted outside the 48 contiguous States and the District of Columbia, it was prepared by a source approved by the Administrator.

**(c)** Each certificate holder conducting domestic or flag operations that uses forecasts to control flight movements shall use forecasts prepared from weather reports specified in paragraph (b) of this section and from any source approved under its system adopted pursuant to paragraph (d) of this section.

**(d)** Certificate holder conducting domestic or flag operations shall adopt and put into use an approved system for obtaining forecasts and reports of adverse weather phenomena, such as clear air turbulence, thunderstorms, and low altitude wind shear, that may affect safety of flight on each route to be flown and at each airport to be used.

[Docket No. 6258, 29 FR 19194, Dec. 31, 1964; as amended by Amdt. 121–27, 36 FR 13911, July 28, 1971; Amdt. 121–134, 42 FR 27573, May 31, 1977; Amdt. 121–253, 61 FR 2610, Jan. 26, 1996]

## §121.103 En route navigational facilities.

**(a)** Except as provided in paragraph (b) of this section, each certificate holder conducting domestic or flag operations must show, for each proposed route, that nonvisual ground aids are—

(1) Available over the route for navigating aircraft within the degree of accuracy required for ATC; and

(2) Located to allow navigation to any regular, provisional, refueling, or alternate airport, within the degree of accuracy necessary for the operation involved.

Except for those aids required for routes to alternate airports, nonvisual ground aids required for approval of routes outside of controlled airspace are listed in the certificate holder's operations specifications.

**(b)** Nonvisual ground aids are not required for—

(1) Day VFR operations that the certificate holder shows can be conducted safely by pilotage because of the characteristics of the terrain;

(2) Night VFR operations on routes that the certificate holder shows have reliably lighted landmarks adequate for safe operation; and

(3) Operations on route segments where the use of celestial or other specialized means of navigation is approved by the Administrator.

[Docket No. 6258, 29 FR 19194, Dec. 31, 1964; as amended by Amdt. 121–253, 61 FR 2610, Jan. 26, 1996]

## §121.105 Servicing and maintenance facilities.

Each certificate holder conducting domestic or flag operations must show that competent personnel and adequate facilities and equipment (including spare parts, supplies, and materials) are available at such points along the certificate holder's route as are necessary for the proper servicing, maintenance, and preventive maintenance of airplanes and auxiliary equipment.

[Docket No. 28154, 61 FR 2610, Jan. 26, 1996]

## §121.107 Dispatch centers.

Each certificate holder conducting domestic or flag operations must show that it has enough dispatch centers, adequate for the operations to be conducted, that are located at points necessary to ensure proper operational control of each flight.

[Docket No. 28154, 61 FR 2610, Jan. 26, 1996]

# Subpart F— Approval of Areas and Routes for Supplemental Operations

**Source:** Docket No. 6258, 29 FR 19195, Dec. 31, 1964, unless otherwise noted.

## §121.111 Applicability.

This subpart prescribes rules for obtaining approval of areas and routes by certificate holders conducting supplemental operations.

[Docket No. 28154, 61 FR 2610, Jan. 26, 1996]

## §121.113 Area and route requirements: General.

**(a)** Each certificate holder conducting supplemental operations seeking route and area approval must show—

(1) That it is able to conduct operations within the United States in accordance with paragraphs (a)(3) and (4) of this section;

(2) That it is able to conduct operations in accordance with the applicable requirements for each area outside the United States for which authorization is requested;

(3) That it is equipped and able to conduct operations over, and use the navigational facilities associated with, the Federal airways, foreign airways, or advisory routes (ADR's) to be used; and

(4) That it will conduct all IFR and night VFR operations over Federal airways, foreign airways, controlled airspace, or advisory routes (ADR's).

**(b)** Notwithstanding paragraph (a)(4) of this section, the Administrator may approve a route outside of controlled airspace if the certificate holder conducting supplemental operations shows the route is safe for operations and the Administrator finds that traffic density is such that an adequate level of safety can be assured. The certificate holder may not use such a route unless it is approved by the Administrator and is listed in the certificate holder's operations specifications.

[Docket No. 6258, 29 FR 19195, Dec. 31, 1964; as amended by Amdt. 121–253, 61 FR 2610, Jan. 26, 1996]

## §121.115 Route width.

**(a)** Routes and route segments over Federal airways, foreign airways, or advisory routes have a width equal to the designated width of those airways or advisory routes. Whenever the Administrator finds it necessary to determine the width of other routes, he considers the following:

(1) Terrain clearance.

(2) Minimum en route altitudes.

(3) Ground and airborne navigation aids.

(4) Air traffic density.

(5) ATC procedures.

**(b)** Any route widths of other routes determined by the Administrator are specified in the certificate holder's operations specifications.

[Docket No. 6258, 29 FR 19195, Dec. 31, 1964; as amended by Amdt. 121–253, 61 FR 2610, Jan. 26, 1996]

## §121.117 Airports: Required data.

**(a)** No certificate holder conducting supplemental operations may use any airport unless it is properly equipped and adequate for the proposed operation, considering such items as size, surface, obstructions, facilities, public protection, lighting, navigational and communications aids, and ATC.

**(b)** Each supplemental certificate holder conducting supplemental operations must show that it has an approved system for obtaining, maintaining, and distributing to appropriate personnel current aeronautical data for each airport it uses to ensure a safe operation at that airport. The aeronautical data must include the following:

(1) Airports.

(i) Facilities.

(ii) Public protection.

(iii) Navigational and communications aids.

(iv) Construction affecting takeoff, landing, or ground operations.

(v) Air traffic facilities.

(2) Runways, clearways, and stopways.

(i) Dimensions.

(ii) Surface.

(iii) Marking and lighting systems.

(iv) Elevation and gradient.

(3) Displaced thresholds.
(i) Location.
(ii) Dimensions.
(iii) Takeoff or landing or both.
(4) Obstacles.
(i) Those affecting takeoff and landing performance computations in accordance with Subpart I of this part.
(ii) Controlling obstacles.
(5) Instrument flight procedures.
(i) Departure procedure.
(ii) Approach procedure.
(iii) Missed approach procedure.
(6) Special information.
(i) Runway visual range measurement equipment.
(ii) Prevailing winds under low visibility conditions.

**(c)** If the certificate-holding district office charged with the overall inspection of the certificate holder's operations finds that revisions are necessary for the continued adequacy of the certificate holder's system for collection, dissemination, and usage of aeronautical data that has been granted approval, the certificate holder shall, after notification by the certificate-holding district office, make those revisions in the system. Within 30 days after the certificate holder receives such notice, the certificate holder may file a petition to reconsider the notice with the Director, Flight Standards Service. This filing of a petition to reconsider stays the notice pending a decision by the Director, Flight Standards Service. However, if the certificate-holding district office finds that there is an emergency that requires immediate action in the interest of safety in air transportation, the Director, Flight Standards Service may, upon a statement of the reasons, require a change effective without stay.

[Docket No. 6258, 29 FR 19195, Dec. 31, 1964; as amended by Amdt. 121–162, 45 FR 46738, July 10, 1980; Amdt. 121–207, 54 FR 39293, Sept. 25, 1989; Amdt. 121–253, 61 FR 2610, Jan. 26, 1996]

## §121.119 Weather reporting facilities.

**(a)** No certificate holder conducting supplemental operations may use any weather report to control flight unless it was prepared and released by the U.S. National Weather Service or a source approved by the Weather Bureau. For operations outside the U.S., or at U.S. Military airports, where those reports are not available, the certificate holder must show that its weather reports are prepared by a source found satisfactory by the Administrator.

**(b)** Each certificate holder conducting supplemental operations that uses forecasts to control flight movements shall use forecasts prepared from weather reports specified in paragraph (a) of this section.

[Docket No. 6258, 29 FR 19195, Dec. 31, 1964; as amended by Amdt. 121–76, 36 FR 13911, July 28, 1971; Amdt. 121–253, 61 FR 2611, Jan. 26, 1996]

## §121.121 En route navigational facilities.

**(a)** Except as provided in paragraph (b) of this section, no certificate holder conducting supplemental operations may conduct any operation over a route unless nonvisual ground aids are—
(1) Available over the route for navigating airplanes within the degree of accuracy required for ATC; and
(2) Located to allow navigation to any airport of destination, or alternate airport, within the degree of accuracy necessary for the operation involved.

**(b)** Nonvisual ground aids are not required for—
(1) Day VFR operations that can be conducted safely by pilotage because of the characteristics of the terrain;
(2) Night VFR operations on lighted airways or on routes that the Administrator determines have reliable landmarks adequate for safe operation; or
(3) Operations on route segments where the use of celestial or other specialized means of navigation is approved.

**(c)** Except for those aids required for routes to alternate airports, the nonvisual ground navigational aids that are required for approved of routes outside of controlled airspace are specified in the certificate holders operations specifications.

[Docket No. 6258, 29 FR 19195, Dec. 31, 1964; as amended by Amdt. 121–253, 61 FR 2611, Jan. 26, 1996]

## §121.123 Servicing maintenance facilities.

Each certificate holder conducting supplemental operations must show that competent personnel and adequate facilities and equipment (including spare parts, supplies, and materials) are available for the proper servicing, maintenance, and preventive maintenance of aircraft and auxiliary equipment.

[Docket No. 28154, 61 FR 2611, Jan. 26, 1996]

## §121.125 Flight following system.

**(a)** Each certificate holder conducting supplemental operations must show that it has—
(1) An approved flight following system established in accordance with Subpart U of this part and adequate for the proper monitoring of each flight, considering the operations to be conducted; and
(2) Flight following centers located at those points necessary—

121

(i) To ensure the proper monitoring of the progress of each flight with respect to its departure at the point of origin and arrival at its destination, including intermediate stops and diversions therefrom, and maintenance or mechanical delays encountered at those points or stops; and

(ii) To ensure that the pilot in command is provided with all information necessary for the safety of the flight.

**(b)** A certificate holder conducting supplemental operations may arrange to have flight following facilities provided by persons other than its employees, but in such a case the certificate holder continues to be primarily responsible for operational control of each flight.

**(c)** A flight following system need not provide for in-flight monitoring by a flight following center.

**(d)** The certificate holder's operations specifications specify the flight following system it is authorized to use and the location of the centers.

[Docket No. 6258, 29 FR 19195, Dec. 31, 1964; as amended by Amdt. 121–253, 61 FR 2611, Jan. 26, 1996]

### §121.127 Flight following system; requirements.

**(a)** Each certificate holder conducting supplemental operations using a flight following system must show that—

(1) The system has adequate facilities and personnel to provide the information necessary for the initiation and safe conduct of each flight to—

(i) The flight crew of each aircraft; and

(ii) The persons designated by the certificate holder to perform the function of operational control of the aircraft; and

(2) The system has a means of communication by private or available public facilities (such as telephone, telegraph, or radio) to monitor the progress of each flight with respect to its departure at the point of origin and arrival at its destination, including intermediate stops and diversions therefrom, and maintenance or mechanical delays encountered at those points or stops.

**(b)** The certificate holder conducting supplemental operations must show that the personnel specified in paragraph (a) of this section, and those it designates to perform the function of operational control of the aircraft, are able to perform their required duties.

[Docket No. 6258, 29 FR 19195, Dec. 31, 1964; as amended by Amdt. 121–253, 61 FR 2611, Jan. 26, 1996]

## Subpart G— Manual Requirements

### §121.131 Applicability.

This subpart prescribes requirements for preparing and maintaining manuals by all certificate holders.

[Docket No. 6258, 29 FR 19196, Dec. 31, 1964]

### §121.133 Preparation.

**(a)** Each certificate holder shall prepare and keep current a manual for the use and guidance of flight, ground operations, and management personnel in conducting its operations.

**(b)** For the purpose of this subpart, the certificate holder may prepare that part of the manual containing maintenance information and instructions, in whole or in part, in printed form or other form acceptable to the Administrator.

[Docket No. 28154, 60 FR 65926, Dec. 20, 1995]

### §121.135 Contents.

**(a)** Each manual required by §121.133 must—

(1) Include instructions and information necessary to allow the personnel concerned to perform their duties and responsibilities with a high degree of safety;

(2) Be in a form that is easy to revise;

(3) Have the date of last revision on each page concerned; and

(4) Not be contrary to any applicable Federal regulation and, in the case of a flag or supplemental operation, any applicable foreign regulation, or the certificate holder's operations specifications or operating certificate.

**(b)** The manual may be in two or more separate parts, containing together all of the following information, but each part must contain that part of the information that is appropriate for each group of personnel:

(1) General policies.

(2) Duties and responsibilities of each crewmember, appropriate members of the ground organization, and management personnel.

(3) Reference to appropriate Federal Aviation Regulations.

(4) Flight dispatching and operational control, including procedures for coordinated dispatch or flight control or flight following procedures, as applicable.

(5) En route flight, navigation, and communication procedures, including procedures for the dispatch or release or continuance of flight if any item of equipment required for the particular type of operation becomes inoperative or unserviceable en route.

(6) For domestic or flag operations, appropriate information from the en route operations specifications, including for each approved route the types of airplanes authorized, the type of operation such as VFR, IFR, day, night, etc., and any other pertinent information.

(7) For supplemental operations, appropriate information from the operations specifications, including the area of operations authorized, the types of airplanes authorized, the type of operation such as VFR, IFR, day, night, etc., and any other pertinent information.

(8) Appropriate information from the airport operations specifications, including for each airport—

(i) Its location (domestic and flag operations only);

(ii) Its designation (regular, alternate, provisional, etc.) (domestic and flag operations only);

(iii) The types of airplanes authorized (domestic and flag operations only);

(iv) Instrument approach procedures;

(v) Landing and takeoff minimums; and

(vi) Any other pertinent information.

(9) Takeoff, en route, and landing weight limitations.

(10) Procedures for familiarizing passengers with the use of emergency equipment, during flight.

(11) Emergency equipment and procedures.

(12) The method of designating succession of command of flight crewmembers.

(13) Procedures for determining the usability of landing and takeoff areas, and for disseminating pertinent information thereon to operations personnel.

(14) Procedures for operating in periods of ice, hail, thunderstorms, turbulence, or any potentially hazardous meteorological condition.

(15) Each training program curriculum required by §121.403.

(16) Instructions and procedures for maintenance, preventive maintenance, and servicing.

(17) Time limitations, or standards for determining time limitations, for overhauls, inspections, and checks of airframes, engines, propellers, appliances and emergency equipment.

(18) Procedures for refueling aircraft, eliminating fuel contamination, protection from fire (including electrostatic protection), and supervising and protecting passengers during refueling.

(19) Airworthiness inspections, including instructions covering procedures, standards, responsibilities, and authority of inspection personnel.

(20) Methods and procedures for maintaining the aircraft weight and center of gravity within approved limits.

(21) Where applicable, pilot and dispatcher route and airport qualification procedures.

(22) Accident notification procedures.

(23) Procedures and information to assist personnel to identify packages marked or labeled as containing hazardous materials and, if these materials are to be carried, stored, or handled, procedures and instructions relating to the carriage, storage, or handling of hazardous materials, including the following:

(i) Procedures for determining the proper shipper certification required by 49 CFR Subchapter C, proper packaging, marking, labeling, shipping documents, compatibility of materials, and instructions on the loading, storage, and handling.

(ii) Notification procedures for reporting hazardous material incidents as required by 49 CFR Subchapter C.

(iii) Instructions and procedures for the notification of the pilot in command when there are hazardous materials aboard, as required by 49 CFR Subchapter C.

(24) Other information or instructions relating to safety.

**(c)** Each certificate holder shall maintain at least one complete copy of the manual at its principal base of operations.

[Docket No. 6258, 29 FR 19196, Dec. 31, 1964; as amended by Amdt. 121–104, 38 FR 14915, June 7, 1973; Amdt. 121–106, 38 FR 22377, Aug. 20, 1973; Amdt. 121–143, 43 FR 22641, May 25, 1978; Amdt. 121–162, 45 FR 46739, July 10, 1980; Amdt. 121–251, 60 FR 65926, Dec. 20, 1995; Amdt. 121–250, 60 FR 65948, Dec. 20, 1995]

## §121.137 Distribution and availability.

**(a)** Each certificate holder shall furnish copies of the manual required by §121.133 (and the changes and additions thereto) or appropriate parts of the manual to—

(1) Its appropriate ground operations and maintenance personnel;

(2) Crewmembers; and

(3) Representatives of the Administrator assigned to it.

**(b)** Each person to whom a manual or appropriate parts of it are furnished under paragraph (a) of this section shall keep it up-to-date with the changes and additions furnished to that person and shall have the manual or appropriate parts of it accessible when performing assigned duties.

**(c)** For the purpose of complying with paragraph (a) of this section, a certificate holder may furnish the persons listed therein the maintenance part of the manual in printed form or other form, acceptable to the Administrator, that is retrievable in the English language.

[Docket No. 6258, 29 FR 19196, Dec. 31, 1964; as amended by Amdt. 121–71, 35 FR 17176, Nov. 7, 1970; Amdt. 121–162, 45 FR 46739, July 10, 1980; Amdt. 121–262, 62 FR 13256, March 19, 1997]

### §121.139 Requirement for manual aboard aircraft: Supplemental operations.

**(a)** Except as provided in paragraph (b) of this section, each certificate holder conducting supplemental operations shall carry appropriate parts of the manual on each airplane when away from the principal base of operations. The appropriate parts must be available for use by ground or flight personnel. If the certificate holder carries aboard an airplane all or any portion of the maintenance part of its manual in other than printed form, it must carry a compatible reading device that produces a legible image of the maintenance information and instructions or a system that is able to retrieve the maintenance information and instructions in the English language.

**(b)** If a certificate holder conducting supplemental operations is able to perform all scheduled maintenance at specified stations where it keeps maintenance parts of the manual, it does not have to carry those parts of the manual aboard the aircraft en route to those stations.

[Docket No. 6258, 29 FR 19196, Dec. 31, 1964; as amended by Amdt. 121–71, 35 FR 17176, Nov. 7, 1970; Amdt. 121–253, 61 FR 2611, Jan. 26, 1996; Amdt. 121–262, 62 FR 13256, March 19, 1997; Amdt. 121–262, 62 FR 15570, April 1, 1997]

### §121.141 Airplane flight manual.

**(a)** Each certificate holder shall keep a current approved airplane flight manual for each type of airplane that it operates except for nontransport category airplanes certificated before January 1, 1965.

**(b)** In each airplane required to have an airplane flight manual in paragraph (a) of this section, the certificate holder shall carry either the manual required by §121.133, if it contains the information required for the applicable flight manual and this information is clearly identified as flight manual requirements, or an approved Airplane Manual. If the certificate holder elects to carry the manual required by §121.133, the certificate holder may revise the operating procedures sections and modify the presentation of performance data from the applicable flight manual if the revised operating procedures and modified performance date presentation are—

(1) Approved by the Administrator; and

(2) Clearly identified as airplane flight manual requirements.

[Docket No. 28154, 60 FR 65927, Dec. 20, 1995]

## Subpart H— Aircraft Requirements

**Source:** Docket No. 6258, 29 FR 19197, Dec. 31, 1964, unless otherwise noted.

### §121.151 Applicability.

This subpart prescribes aircraft requirements for all certificate holders.

### §121.153 Aircraft requirements: General.

**(a)** Except as provided in paragraph (c) of this section, no certificate holder may operate an aircraft unless that aircraft—

(1) Is registered as a civil aircraft of the United States and carries an appropriate current airworthiness certificate issued under this chapter; and

(2) Is in an airworthy condition and meets the applicable airworthiness requirements of this chapter, including those relating to identification and equipment.

**(b)** A certificate holder may use an approved weight and balance control system based on average, assumed, or estimated weight to comply with applicable airworthiness requirements and operating limitations.

**(c)** A certificate holder may operate in common carriage, and for the carriage of mail, a civil aircraft which is leased or chartered to it without crew and is registered in a country which is a party to the Convention on International Civil Aviation if—

(1) The aircraft carries an appropriate airworthiness certificate issued by the country of registration and meets the registration and identification requirements of that country;

(2) The aircraft is of a type design which is approved under a U.S. type certificate and complies with all of the requirements of this chapter (14 CFR Chapter 1) that would be applicable to that aircraft were it registered in the United States, including the requirements which must be met for issuance of a U.S. standard airworthiness certificate (including type design conformity, condition for safe operation, and the noise, fuel venting, and engine emission requirements of this chapter), except that a U.S. registration certificate and a U.S. standard airworthiness certificate will not be issued for the aircraft;

(3) The aircraft is operated by U.S.-certificated airmen employed by the certificate holder; and

(4) The certificate holder files a copy of the aircraft lease or charter agreement with the FAA Aircraft Registry, Department of Transportation, 6400 South MacArthur Boulevard, Oklahoma City, OK (Mailing address: P.O. Box 25504, Oklahoma City, OK 73125).

[Docket No. 6258, 29 FR 19197, Dec. 31, 1964; as amended by Amdt. 121–165, 45 FR 68649, Oct. 16, 1980]

## §121.155 [Reserved]

## §121.157 Aircraft certification and equipment requirements.

**(a)** *Airplanes certificated before July 1, 1942.* No certificate holder may operate an airplane that was type certificated before July 1, 1942, unless—

(1) That airplane meets the requirements of §121.173(c), or

(2) That airplane and all other airplanes of the same or related type operated by that certificate holder meet the performance requirements of sections 4a.737-T through 4a.750-T of the Civil Air Regulations as in effect on January 31, 1965; or §§25.45 through 25.75 and §121.173 (a), (b), (d), and (e) of this title.

**(b)** *Airplanes certificated after June 30, 1942.* Except as provided in paragraphs (c), (d), (e), and (f) of this section, no certificate holder may operate an airplane that was type certificated after June 30, 1942, unless it is certificated as a transport category airplane and meets the requirements of §121.173(a), (b), (d), and (e).

**(c)** *C–46 type airplanes: passenger-carrying operations.* No certificate holder may operate a C–46 airplane in passenger-carrying operations unless that airplane is operated in accordance with the operating limitations for transport category airplanes and meets the requirements of paragraph (b) of this section or meets the requirements of part 4b, as in effect July 20, 1950, and the requirements of §121.173(a), (b), (d) and (e), except that—

(1) The requirements of sections 4b.0 through 4b.19 as in effect May 18, 1954, must be complied with;

(2) The birdproof windshield requirements of section 4b.352 need not be complied with;

(3) The provisions of sections 4b.480 through 4b.490 (except sections 4b.484(a)(1) and 4b.487(e)), as in effect May 16, 1953, must be complied with; and

(4) The provisions of paragraph 4b.484(a)(1), as in effect July 20, 1950, must be complied with.

In determining the takeoff path in accordance with section 4b.116 and the one-engine inoperative climb in accordance with section 4b.120 (a) and (b), the propeller of the inoperative engine may be assumed to be feathered if the airplane is equipped with either an approved means for automatically indicating when the particular engine has failed or an approved means for automatically feathering the propeller of the inoperative engine. The Administrator may authorize deviations from compliance with the requirements of sections 4b.130 through 4b.190 and Subparts C, D, E, and F of part 4b (as designated in this paragraph) if he finds that (considering the effect of design changes) compliance is extremely difficult to accomplish and that service experience with the C–46 airplane justifies the deviation.

**(d)** *C–46 type airplanes: cargo operations.* No certificate holder may use a nontransport category C-46 type airplane in cargo operations unless—

(1) It is certificated at a maximum gross weight that is not greater than 48,000 pounds;

(2) It meets the requirements of §§121.199 through 121.205 using the performance data in Appendix C to this part;

(3) Before each flight, each engine contains at least 25 gallons of oil; and

(4) After December 31, 1964—

(i) It is powered by a type and model engine as set forth in Appendix C of this part, when certificated at a maximum gross takeoff weight greater than 45,000 pounds; and

(ii) It complies with the special airworthiness requirement set forth in §§121.213 through 121.287 of this part or in Appendix C of this part.

**(e)** *Commuter category airplanes.* Except as provided in paragraph (f) of this section, no certificate holder may operate under this part a nontransport category airplane type certificated after December 31, 1964, and before March 30, 1995, unless it meets the applicable requirements of §121.173(a), (b), (d), and (e) and was type certificated in the commuter category.

**(f)** *Other nontransport category airplanes.* No certificate holder may operate under this part a nontransport category airplane type certificated after December 31, 1964, unless it meets the applicable requirements of §121.173(a), (b), (d), and (e), was manufactured before March 20, 1997, and meets one of the following:

(1) Until December 20, 2010:

(i) The airplane was type certificated in the normal category before July 1, 1970, and meets special conditions issued by the Administrator for airplanes intended for use in operations under part 135 of this chapter.

(ii) The airplane was type certificated in the normal category before July 19, 1970, and meets the additional airworthiness standards in SFAR No. 23, 14 CFR part 23.

(iii) The airplane was type certificated in the normal category and meets the additional airworthiness standards in Appendix A of part 135 of this chapter.

(iv) The airplane was type certificated in the normal category and complies with either section 1.(a) or 1.(b) of SFAR No. 41 of 14 CFR part 21.

(2) The airplane was type certificated in the normal category, meets the additional requirements described in paragraphs (f)(1)(i) through (f)(1)(iv) of this section, and meets the performance requirements in Appendix K of this part.

**(g)** *Certain newly manufactured airplanes.* No certificate holder may operate an airplane under this part that was type certificated as described in paragraphs (f)(1)(i) through (f)(1)(iv) of this section and that was manufactured after March 20, 1997, unless it meets the performance requirements in Appendix K of this part.

**(h)** *Newly type certificated airplanes.* No person may operate under this part an airplane for which the application for a type certificate is submitted after March 29, 1995, unless the airplane is type certificated under part 25 of this chapter.

[Docket No. 6258, 29 FR 19197, Dec. 31, 1964; as amended by Amdt. 121–251, 60 FR 65927, Dec. 20, 1995; Amdt. 121–256, 61 FR 30434, June 14, 1996]

## §121.159 Single-engine airplanes prohibited.

No certificate holder may operate a single-engine airplane under this part.

[Docket No. 28154, 60 FR 65927, Dec. 20, 1995]

## §121.161 Airplane limitations: Type of route.

**(a)** Unless authorized by the Administrator, based on the character of the terrain, the kind of operation, or the performance of the airplane to be used, no certificate holder may operate two-engine or three-engine airplanes (except a three-engine turbine powered airplane) over a route that contains a point farther than 1 hour flying time (in still air at normal cruising speed with one engine inoperative) from an adequate airport.

**(b)** Except as provided in paragraph (c) of this section, no certificate holder may operate a land airplane (other than a DC-3, C-46, CV-240, CV-340, CV-440, CV-580, CV-600, CV-640, or Martin 404) in an extended overwater operation unless it is certificated or approved as adequate for ditching under the ditching provisions of part 25 of this chapter.

**(c)** Until December 20, 2010, a certificate holder may operate, in an extended overwater operation, a nontransport category land airplane type certificated after December 31, 1964, that was not certificated or approved as adequate for ditching under the ditching provisions of part 25 of this chapter.

[Docket No. 7329, 31 FR 13078, Oct. 8, 1966; as amended by Amdt. 121–162, 45 FR 46739, July 10, 1980; Amdt. 121–251, 60 FR 65927, Dec. 20, 1995]

## §121.163 Aircraft proving tests.

**(a)** *Initial airplane proving tests.* No person may operate an airplane not before proven for use in a kind of operation under this part or part 135 of this chapter unless an airplane of that type has had, in addition to the airplane certification tests, at least 100 hours of proving tests acceptable to the Administrator, including a representative number of flights into en route airports. The requirement for at least 100 hours of proving tests may be reduced by the Administrator if the Administrator determines that a satisfactory level of proficiency has been demonstrated to justify the reduction. At least 10 hours of proving flights must be flown at night; these tests are irreducible.

**(b)** *Proving tests for kinds of operations.* Unless otherwise authorized by the Administrator, for each type of airplane, a certificate holder must conduct at least 50 hours of proving tests acceptable to the Administrator for each kind of operation it intends to conduct, including a representative number of flights into en route airports.

**(c)** *Proving tests for materially altered airplanes.* Unless otherwise authorized by the Administrator, for each type of airplane that is materially altered in design, a certificate holder must conduct at least 50 hours of proving tests acceptable to the Administrator for each kind of operation it intends to conduct with that airplane, including a representative number of flights into en route airports.

**(d)** *Definition of materially altered.* For the purposes of paragraph (c) of this section, a type of airplane is considered to be materially altered in design if the alteration includes—

(1) The installation of powerplants other than those of a type similar to those with which it is certificated; or

(2) Alterations to the aircraft or its components that materially affect flight characteristics.

**(e)** No certificate holder may carry passengers in an aircraft during proving tests, except for those needed to make the test and those designated by the Administrator. However, it may carry mail, express, or other cargo, when approved.

[Docket No. 6258, 29 FR 19197, Dec. 31, 1964; as amended by Amdt. 121–42, 33 FR 10330, July 19, 1968; 34 FR 13468, Aug. 21, 1969; Amdt. 121–162, 45 FR 46739, July 10, 1980; Amdt. 121–251, 60 FR 65927, Dec. 20, 1995]

# Subpart I— Airplane Performance Operating Limitations

SOURCE: Docket No. 6258, 29 FR 19198, Dec. 31, 1964; 30 FR 130, Jan. 7, 1965, unless otherwise noted.

## §121.171 Applicability.

**(a)** This subpart prescribes airplane performance operating limitations for all certificate holders.

**(b)** For purposes of this part, *effective length of the runway for landing* means the distance from the point at which the obstruction clearance plane associated with the approach end of the runway intersects the centerline of the runway to the far end thereof.

**(c)** For the purposes of this subpart, *obstruction clearance plane* means a plane sloping upward from the runway at a slope of 1:20 to the horizontal, and tangent to or clearing all obstructions within a specified area surrounding the runway as shown in a profile view of that area. In the plan view, the centerline of the specified area coincides with the centerline of the runway, beginning at the point where the obstruction clearance plane intersects the centerline of the runway and proceeding to a point at least 1,500 feet from the beginning point. Thereafter the centerline coincides with the takeoff path over the ground for the runway (in the case of takeoffs) or with the instrument approach counterpart (for landings), or, where the applicable one of these paths has not been established, it proceeds consistent with turns of at least 4,000 foot radius until a point is reached beyond which the obstruction clearance plane clears all obstructions. This area extends laterally 200 feet on each side of the centerline at the point where the obstruction clearance plane intersects the runway and continues at this width to the end of the runway; then it increases uniformly to 500 feet on each side of the centerline at a point 1,500 feet from the intersection of the obstruction clearance plane with the runway; thereafter it extends laterally 500 feet on each side of the centerline.

[Docket No. 6258, 29 FR 19198, Dec. 31, 1964; as amended by Amdt. 121–132, 41 FR 55475, Dec. 20, 1976]

## §121.173 General.

**(a)** Except as provided in paragraph (c) of this section, each certificate holder operating a reciprocating-engine-powered airplane shall comply with §§121.175 through 121.187.

**(b)** Except as provided in paragraph (c) of this section, each certificate holder operating a turbine-engine-powered airplane shall comply with the applicable provisions of §§121.189 through 121.197, except that when it operates—

(1) A turbo-propeller-powered airplane type certificated after August 29, 1959, but previously type certificated with the same number of reciprocating engines, the certificate holder may comply with §§121.175 through 121.187; or

(2) Until December 20, 2010, a turbo-propeller-powered airplane described in §121.157(f), the certificate holder may comply with the applicable performance requirements of Appendix K of this part.

**(c)** Each certificate holder operating a large nontransport category airplane type certificated before January 1, 1965, shall comply with §§121.199 through 121.205 and any determination of compliance must be based only on approved performance data.

**(d)** The performance data in the Airplane Flight Manual applies in determining compliance with §§121.175 through 121.197. Where conditions are different from those on which the performance data is based, compliance is determined by interpolation or by computing the effects of changes in the specific variables if the results of the interpolation or computations are substantially as accurate as the results of direct tests.

**(e)** Except as provided in paragraph (c) of this section, no person may take off a reciprocating-engine-powered airplane at a weight that is more than the allowable weight for the runway being used (determined under the runway takeoff limitations of the transport category operating rules of 14 CFR part 121, Subpart I) after taking into account the temperature operating correction factors in the applicable Airplane Flight Manual.

**(f)** The Administrator may authorize in the operations specifications deviations from the requirements in the subpart if special circumstances make a literal observance of a requirement unnecessary for safety.

**(g)** The ten-mile width specified in §§121.179 through 121.183 may be reduced to five miles, for not more than 20 miles, when operating VFR or where navigation facilities furnish reliable and accurate identification of high ground and obstructions located outside of five miles, but within ten miles, on each side of the intended track.

[Docket No. 6258, 29 FR 19198, Dec. 31, 1964; as amended by Amdt. 121–251, 60 FR 65928, Dec. 20, 1995]

## §121.175 Airplanes: Reciprocating engine-powered: Weight limitations.

**(a)** No person may take off a reciprocating engine powered airplane from an airport located at an elevation outside of the range for which maximum takeoff weights have been determined for that airplane.

121

**(b)** No person may take off a reciprocating engine powered airplane for an airport of intended destination that is located at an elevation outside of the range for which maximum landing weights have been determined for that airplane.

**(c)** No person may specify, or have specified, an alternate airport that is located at an elevation outside of the range for which maximum landing weights have been determined for the reciprocating engine powered airplane concerned.

**(d)** No person may take off a reciprocating engine powered airplane at a weight more than the maximum authorized takeoff weight for the elevation of the airport.

**(e)** No person may take off a reciprocating engine powered airplane if its weight on arrival at the airport of destination will be more than the maximum authorized landing weight for the elevation of that airport, allowing for normal consumption of fuel and oil en route.

**(f)** This section does not apply to large nontransport category airplanes operated under §121.173(c).

[Docket No. 6258, 29 FR 19195, Dec. 31, 1964; as amended by Amdt. 121–251, 60 FR 65928, Dec. 20, 1995]

## §121.177 Airplanes: Reciprocating engine-powered: Takeoff limitations.

**(a)** No person operating a reciprocating engine powered airplane may takeoff that airplane unless it is possible—

(1) To stop the airplane safely on the runway, as shown by the accelerate stop distance data, at any time during takeoff until reaching critical-engine failure speed;

(2) If the critical engine fails at any time after the airplane reaches critical-engine failure speed $V_1$, to continue the takeoff and reach a height of 50 feet, as indicated by the takeoff path data, before passing over the end of the runway; and

(3) To clear all obstacles either by at least 50 feet vertically (as shown by the takeoff path data) or 200 feet horizontally within the airport boundaries and 300 feet horizontally beyond the boundaries, without banking before reaching a height of 50 feet (as shown by the takeoff path data) and thereafter without banking more than 15 degrees.

**(b)** In applying this section, corrections must be made for the effective runway gradient. To allow for wind effect, takeoff data based on still air may be corrected by taking into account not more than 50 percent of any reported headwind component and not less than 150 percent of any reported tailwind component.

**(c)** This section does not apply to large nontransport category airplanes operated under §121.173(c).

[Docket No. 6258, 29 FR 19198, Dec. 31, 1964; as amended by Amdt. 121–159, 45 FR 41593, June 19, 1980; Amdt. 121–251, 60 FR 65928, Dec. 20, 1995]

## §121.179 Airplanes: Reciprocating engine-powered: En route limitations: All engines operating.

**(a)** No person operating a reciprocating engine powered airplane may take off that airplane at a weight, allowing for normal consumption of fuel and oil, that does not allow a rate of climb (in feet per minute), with all engines operating, of at least 6.90 $V_{SO}$ (that is, the number of feet per minute is obtained by multiplying the number of knots by 6.90) at an altitude of at least 1,000 feet above the highest ground or obstruction within ten miles of each side of the intended track.

**(b)** This section does not apply to airplanes certificated under part 4a of the Civil Air Regulations.

**(c)** This section does not apply to large nontransport category airplanes operated under §121.173(c).

[Docket No. 6258, 29 FR 19195, Dec. 31, 1964; as amended by Amdt. 121–251, 60 FR 65928, Dec. 20, 1995]

## §121.181 Airplanes: Reciprocating engine-powered: En route limitations: One engine inoperative.

**(a)** Except as provided in paragraph (b) of this section, no person operating a reciprocating engine powered airplane may take off that airplane at a weight, allowing for normal consumption of fuel and oil, that does not allow a rate of climb (in feet per minute), with one engine inoperative, of at least

$$(0.079 - 0.106/N)\ V_{SO}^2$$

(where N is the number of engines installed and $V_{SO}$ is expressed in knots) at an altitude of at least 1,000 feet above the highest ground or obstruction within 10 miles of each side of the intended track. However, for the purposes of this paragraph the rate of climb for airplanes certificated under part 4a of the Civil Air Regulations is 0.026 $V_{SO}^2$.

**(b)** In place of the requirements of paragraph (a) of this section, a person may, under an approved procedure, operate a reciprocating engine powered airplane, at an all-engines-operating altitude that allows the airplane to continue, after an engine failure, to an alternate airport where a landing can be made in accordance with §121.187, allowing for normal consumption of fuel and oil. After the assumed failure, the flight path must clear the ground and any obstruction within five miles on each side of the intended track by at least 2,000 feet.

**(c)** If an approved procedure under paragraph (b) of this section is used, the certificate holder shall comply with the following:

(1) The rate of climb (as prescribed in the Airplane Flight Manual for the appropriate weight and altitude) used in calculating the airplane's flight path shall be diminished by an amount, in feet per minute, equal to

$$(0.079-0.106/N)\ V_{SO}^2$$

(when N is the number of engines installed and $V_{SO}$ is expressed in knots) for airplanes certificated under part 25 of this chapter and by 0.026 $V_{SO}^2$ for airplanes certificated under part 4a of the Civil Air Regulations.

(2) The all-engines-operating altitude shall be sufficient so that in the event the critical engine becomes inoperative at any point along the route, the flight will be able to proceed to a predetermined alternate airport by use of this procedure. In determining the takeoff weight, the airplane is assumed to pass over the critical obstruction following engine failure at a point no closer to the critical obstruction than the nearest approved radio navigational fix, unless the Administrator approves a procedure established on a different basis upon finding that adequate operational safeguards exist.

(3) The airplane must meet the provisions of paragraph (a) of this section at 1,000 feet above the airport used as an alternate in this procedure.

(4) The procedure must include an approved method of accounting for winds and temperatures that would otherwise adversely affect the flight path.

(5) In complying with this procedure fuel jettisoning is allowed if the certificate holder shows that it has an adequate training program, that proper instructions are given to the flight crew, and all other precautions are taken to insure a safe procedure.

(6) The certificate holder shall specify in the dispatch or flight release an alternate airport that meets the requirements of §121.625.

**(d)** This section does not apply to large nontransport category airplanes operated under §121.173(c).

[Docket No. 6258, 29 FR 19198, Dec. 31, 1964; as amended by Amdt. 121–251, 60 FR 65928, Dec. 20, 1995]

## §121.183 Part 25 airplanes with four or more engines: Reciprocating engine-powered: En route limitations: Two engines inoperative.

**(a)** No person may operate an airplane certificated under part 25 and having four or more engines unless—

(1) There is no place along the intended track that is more than 90 minutes (with all engines operating at cruising power) from an airport that meets the requirements of §121.187; or

(2) It is operated at a weight allowing the airplane, with the two critical engines inoperative, to climb at 0.013 $V_{SO}^2$ feet per minute (that is, the number of feet per minute is obtained by multiplying the number of knots squared by 0.013) at an altitude of 1,000 feet above the highest ground or obstruction within 10 miles on each side of the intended track, or at an altitude of 5,000 feet, whichever is higher.

**(b)** For the purposes of paragraph (a)(2) of this section, it is assumed that—

(1) The two engines fail at the point that is most critical with respect to the takeoff weight:

(2) Consumption of fuel and oil is normal with all engines operating up to the point where the two engines fail and with two engines operating beyond that point;

(3) Where the engines are assumed to fail at an altitude above the prescribed minimum altitude, compliance with the prescribed rate of climb at the prescribed minimum altitude need not be shown during the descent from the cruising altitude to the prescribed minimum altitude, if those requirements can be met once the prescribed minimum altitude is reached, and assuming descent to be along a net flight path and the rate of descent to be 0.013 $V_{SO}^2$ greater than the rate in the approved performance data; and

(4) If fuel jettisoning is provided, the airplane's weight at the point where the two engines fail is considered to be not less than that which would include enough fuel to proceed to an airport meeting the requirements of §121.187 and to arrive at an altitude of at least 1,000 feet directly over that airport.

[Docket No. 6258, 29 FR 19198, Dec. 31, 1964; as amended by Amdt. 121–251, 60 FR 65928, Dec. 20, 1995]

## §121.185 Airplanes: Reciprocating engine-powered: Landing limitations: Destination airport.

**(a)** Except as provided in paragraph (b) of this section no person operating a reciprocating engine powered airplane may take off that airplane, unless its weight on arrival, allowing for normal consumption of fuel and oil in flight, would allow a full stop landing at the intended destination within 60 percent of the effective length of each runway described below from a point 50 feet directly above the intersection of the obstruction clearance plane and the runway. For the purposes of determining the allowable landing weight at the destination airport the following is assumed:

(1) The airplane is landed on the most favorable runway and in the most favorable direction in still air.

(2) The airplane is landed on the most suitable runway considering the probable wind velocity and direction (forecast for the expected time of arrival), the ground handling characteristics of the type of airplane, and other conditions such as landing aids and terrain, and allowing for the effect of the landing path and roll of not more than 50 percent of the headwind component or not less than 150 percent of the tailwind component.

**(b)** An airplane that would be prohibited from being taken off because it could not meet the requirements of paragraph (a)(2) of this section may be taken off if an alternate airport is specified that meets all of the requirements of this section except that the airplane can accomplish a full stop landing within 70 percent of the effective length of the runway.

**(c)** This section does not apply to large nontransport category airplanes operated under §121.173(c).

[Docket No. 6258, 29 FR 19198, Dec. 31, 1964; as amended by Amdt. 121–251, 60 FR 65928, Dec. 20, 1995]

## §121.187 Airplanes: Reciprocating-engine-powered: Landing limitations: Alternate airport.

**(a)** No person may list an airport as an alternate airport in a dispatch or flight release unless the airplane (at the weight anticipated at the time of arrival at the airport), based on the assumptions in §121.185, can be brought to a full stop landing, within 70 percent of the effective length of the runway.

**(b)** This section does not apply to large nontransport category airplanes operated under §121.173(c).

[Docket No. 6258, 29 FR 19198, Dec. 31, 1964; as amended by Amdt. 121–251, 60 FR 65928, Dec. 20, 1995]

## §121.189 Airplanes: Turbine engine-powered: Takeoff limitations.

**(a)** No person operating a turbine engine powered airplane may take off that airplane at a weight greater than that listed in the Airplane Flight Manual for the elevation of the airport and for the ambient temperature existing at takeoff.

**(b)** No person operating a turbine engine powered airplane certificated after August 26, 1957, but before August 30, 1959 (SR422, 422A), may take off that airplane at a weight greater than that listed in the Airplane Flight Manual for the minimum distances required for takeoff. In the case of an airplane certificated after September 30, 1958 (SR422A, 422B), the takeoff distance may include a clearway distance but the clearway distance included may not be greater than ½ of the takeoff run.

**(c)** No person operating a turbine engine powered airplane certificated after August 29, 1959 (SR422B), may take off that airplane at a weight greater than that listed in the Airplane Flight Manual at which compliance with the following may be shown:

(1) The accelerate-stop distance must not exceed the length of the runway plus the length of any stopway.

(2) The takeoff distance must not exceed the length of the runway plus the length of any clearway except that the length of any clearway included must not be greater than one-half the length of the runway.

(3) The takeoff run must not be greater than the length of the runway.

**(d)** No person operating a turbine engine powered airplane may take off that airplane at a weight greater than that listed in the Airplane Flight Manual—

(1) In the case of an airplane certificated after August 26, 1957, but before October 1, 1958 (SR422), that allows a takeoff path that clears all obstacles either by at least (35+0.01D) feet vertically (D is the distance along the intended flight path from the end of the runway in feet), or by at least 200 feet horizontally within the airport boundaries and by at least 300 feet horizontally after passing the boundaries; or

(2) In the case of an airplane certificated after September 30, 1958 (SR 422A, 422B), that allows a net takeoff flight path that clears all obstacles either by a height of at least 35 feet vertically, or by at least 200 feet horizontally within the airport boundaries and by at least 300 feet horizontally after passing the boundaries.

**(e)** In determining maximum weights, minimum distances, and flight paths under paragraphs (a) through (d) of this section, correction must be made for the runway to be used, the elevation of the airport, the effective runway gradient, the ambient temperature and wind component at the time of takeoff, and, if operating limitations exist for the minimum distances required for takeoff from wet runways, the runway surface condition (dry or wet). Wet runway distances associated with grooved or porous friction course runways, if provided in the Airplane Flight Manual, may be used only for runways that are grooved or treated with a porous friction course (PFC) overlay, and that the operator determines are designed, constructed, and maintained in a manner acceptable to the Administrator.

**(f)** For the purposes of this section, it is assumed that the airplane is not banked before reaching a height of 50 feet, as shown by the takeoff path or net takeoff flight path data (as appropriate) in the Airplane Flight Manual, and thereafter that the maximum bank is not more than 15 degrees.

**(g)** For the purposes of this section the terms, *takeoff distance, takeoff run, net takeoff flight path* and *takeoff path* have the same meanings as set forth in the rules under which the airplane was certificated.

[Docket No. 6258, 29 FR 19197, Dec. 31, 1964; as amended by Amdt. 121–268, 63 FR 8321, Feb. 18, 1998]

## §121.191 Airplanes: Turbine engine-powered: En route limitations: One engine inoperative.

**(a)** No person operating a turbine engine powered airplane may take off that airplane at a weight, allowing for normal consumption of fuel and oil, that is greater than that which (under the approved, one engine inoperative, en route net flight path data in the Airplane Flight Manual for that airplane) will allow compliance with paragraph (a)(1) or (2) of this section, based on the ambient temperatures expected en route:

(1) There is a positive slope at an altitude of at least 1,000 feet above all terrain and obstructions within five statute miles on each side of the intended track, and, in addition, if that airplane was certificated after August 29, 1959 (SR 422B) there is a positive slope at 1,500 feet above the airport where the airplane is assumed to land after an engine fails.

(2) The net flight path allows the airplane to continue flight from the cruising altitude to an airport where a landing can be made under §121.197, clearing all terrain and obstructions within five statute miles of the intended track by at least 2,000 feet vertically and with a positive slope at 1,000 feet above the airport where the airplane lands after an engine fails, or, if that airplane was certificated after September 30, 1958 (SR 422A, 422B), with a positive slope at 1,500 feet above the airport where the airplane lands after an engine fails.

**(b)** For the purposes of paragraph (a)(2) of this section, it is assumed that—

(1) The engine fails at the most critical point en route;

(2) The airplane passes over the critical obstruction, after engine failure at a point that is no closer to the obstruction than the nearest approved radio navigation fix, unless the Administrator authorizes a different procedure based on adequate operational safeguards;

(3) An approved method is used to allow for adverse winds:

(4) Fuel jettisoning will be allowed if the certificate holder shows that the crew is properly instructed, that the training program is adequate, and that all other precautions are taken to insure a safe procedure;

(5) The alternate airport is specified in the dispatch or flight release and meets the prescribed weather minimums; and

(6) The consumption of fuel and oil after engine failure is the same as the consumption that is allowed for in the approved net flight path data in the Airplane Flight Manual.

[Docket No. 6258, 29 FR 19198, Dec. 31, 1964; 30 FR 130, Jan. 7, 1965; as amended by Amdt. 121–143, 43 FR 22641, May 25, 1978]

## §121.193 Airplanes: Turbine engine-powered: En route limitations: Two engines inoperative.

**(a)** *Airplanes certificated after August 26, 1957, but before October 1, 1958* (SR 422). No person may operate a turbine engine powered airplane along an intended route unless he complies with either of the following:

(1) There is no place along the intended track that is more than 90 minutes (with all engines operating at cruising power) from an airport that meets the requirements of §121.197.

(2) Its weight, according to the two-engine-inoperative, en route, net flight path data in the Airplane Flight Manual, allows the airplane to fly from the point where the two engines are assumed to fail simultaneously to an airport that meets the requirements of §121.197, with a net flight path (considering the ambient temperature anticipated along the track) having a positive slope at an altitude of at least 1,000 feet above all terrain and obstructions within five miles on each side of the intended track, or at an altitude of 5,000 feet, whichever is higher.

For the purposes of paragraph (a)(2) of this section, it is assumed that the two engines fail at the most critical point en route, that if fuel jettisoning is provided, the airplane's weight at the point where the engines fail includes enough fuel to continue to the airport and to arrive at an altitude of at least 1,000 feet directly over the airport, and that the fuel and oil consumption after engine failure is the same as the consumption allowed for in the net flight path data in the Airplane Flight Manual.

**(b)** *Aircraft certificated after September 30, 1958, but before August 30, 1959* (SR 422A). No person may operate a turbine engine powered airplane along an intended route unless he complies with either of the following:

(1) There is no place along the intended track that is more than 90 minutes (with all engines operating at cruising power) from an airport that meets the requirements of §121.197.

(2) Its weight, according to the two-engine-inoperative, en route, net flight path data in the Air-

121

plane Flight Manual, allows the airplane to fly from the point where the two engines are assumed to fail simultaneously to an airport that meets the requirements of §121.197, with a net flight path (considering the ambient temperatures anticipated along the track) having a positive slope at an altitude of at least 1,000 feet above all terrain and obstructions within 5 miles on each side of the intended track, or at an altitude of 2,000 feet, whichever is higher.

For the purposes of paragraph (b)(2) of this section, it is assumed that the two engines fail at the most critical point en route, that the airplane's weight at the point where the engines fail includes enough fuel to continue to the airport, to arrive at an altitude of at least 1,500 feet directly over the airport, and thereafter to fly for 15 minutes at cruise power or thrust, or both, and that the consumption of fuel and oil after engine failure is the same as the consumption allowed for in the net flight path data in the Airplane Flight Manual.

**(c)** *Aircraft certificated after August 29, 1959* (SR 422B). No person may operate a turbine engine powered airplane along an intended route unless he complies with either of the following:

(1) There is no place along the intended track that is more than 90 minutes (with all engines operating at cruising power) from an airport that meets the requirements of §121.197.

(2) Its weight, according to the two-engine inoperative, en route, net flight path data in the Airplane Flight Manual, allows the airplane to fly from the point where the two engines are assumed to fail simultaneously to an airport that meets the requirements of §121.197, with the net flight path (considering the ambient temperatures anticipated along the track) clearing vertically by at least 2,000 feet all terrain and obstructions within five statute miles (4.34 nautical miles) on each side of the intended track. For the purposes of this subparagraph, it is assumed that—

(i) The two engines fail at the most critical point en route;

(ii) The net flight path has a positive slope at 1,500 feet above the airport where the landing is assumed to be made after the engines fail;

(iii) Fuel jettisoning will be approved if the certificate holder shows that the crew is properly instructed, that the training program is adequate, and that all other precautions are taken to ensure a safe procedure;

(iv) The airplane's weight at the point where the two engines are assumed to fail provides enough fuel to continue to the airport, to arrive at an altitude of at least 1,500 feet directly over the airport, and thereafter to fly for 15 minutes at cruise power or thrust, or both; and

(v) The consumption of fuel and oil after the engine failure is the same as the consumption that is allowed for in the net flight path data in the Airplane Flight Manual.

## §121.195 Airplanes: Turbine engine-powered: Landing limitations: Destination airports.

**(a)** No person operating a turbine engine powered airplane may take off that airplane at such a weight that (allowing for normal consumption of fuel and oil in flight to the destination or alternate airport) the weight of the airplane on arrival would exceed the landing weight set forth in the Airplane Flight Manual for the elevation of the destination or alternate airport and the ambient temperature anticipated at the time of landing.

**(b)** Except as provided in paragraph (c), (d), or (e) of this section, no person operating a turbine engine powered airplane may take off that airplane unless its weight on arrival, allowing for normal consumption of fuel and oil in flight (in accordance with the landing distance set forth in the Airplane Flight Manual for the elevation of the destination airport and the wind conditions anticipated there at the time of landing), would allow a full stop landing at the intended destination airport within 60 percent of the effective length of each runway described below from a point 50 feet above the intersection of the obstruction clearance plane and the runway. For the purpose of determining the allowable landing weight at the destination airport the following is assumed:

(1) The airplane is landed on the most favorable runway and in the most favorable direction, in still air.

(2) The airplane is landed on the most suitable runway considering the probable wind velocity and direction and the ground handling characteristics of the airplane, and considering other conditions such as landing aids and terrain.

**(c)** A turbopropeller powered airplane that would be prohibited from being taken off because it could not meet the requirements of paragraph (b)(2) of this section, may be taken off if an alternate airport is specified that meets all the requirements of this section except that the airplane can accomplish a full stop landing within 70 percent of the effective length of the runway.

**(d)** Unless, based on a showing of actual operating landing techniques on wet runways, a shorter landing distance (but never less than that required by paragraph (b) of this section) has been approved for a specific type and model airplane and included in the Airplane Flight Manual, no person may takeoff a turbojet powered airplane when the appropriate weather reports and forecasts, or a combination thereof, indicate that the runways at the destination airport may be wet or slippery at the estimated time of arrival unless the effective runway length at the destination airport is at least 115 percent of the runway length required under paragraph (b) of this section.

**(e)** A turbojet powered airplane that would be prohibited from being taken off because it could not meet the requirements of paragraph (b)(2) of this section may be taken off if an alternate airport is specified that meets all the requirements of paragraph (b) of this section.

[Docket No. 6258, 29 FR 19198, Dec. 31, 1964; as amended by Amdt. 121–9, 30 FR 8572, July 7, 1965]

## §121.197 Airplanes: Turbine engine-powered: Landing limitations: Alternate airports.

No person may list an airport as an alternate airport in a dispatch or flight release for a turbine engine powered airplane unless (based on the assumptions in §121.195(b)) that airplane at the weight anticipated at the time of arrival can be brought to a full stop landing within 70 percent of the effective length of the runway for turbopropeller powered airplanes and 60 percent of the effective length of the runway for turbojet powered airplanes, from a point 50 feet above the intersection of the obstruction clearance plane and the runway. In the case of an alternate airport for departure, as provided in §121.617, allowance may be made for fuel jettisoning in addition to normal consumption of fuel and oil when determining the weight anticipated at the time of arrival.

[Docket No. 6258, 29 FR 19198, Dec. 31, 1964; as amended by Amdt. 121–9, 30 FR 8572, July 7, 1965; Amdt. 121–179, 47 FR 33390, Aug. 2, 1982]

## §121.198 Cargo service airplanes: Increased zero fuel and landing weights.

**(a)** Notwithstanding the applicable structural provisions of the airworthiness regulations but subject to paragraphs (b) through (g) of this section, a certificate holder may operate (for cargo service only) any of the following airplanes (certificated under part 4b of the Civil Air Regulations effective before March 13, 1956) at increased zero fuel and landing weights—

(1) DC–6A, DC–6B, DC–7B, and DC–7C; and

(2) L1049B, C, D, E, F, G, and H, and the L1649A when modified in accordance with supplemental type certificate SA 4–1402.

**(b)** The zero fuel weight (maximum weight of the airplane with no disposable fuel and oil) and the structural landing weight may be increased beyond the maximum approved in full compliance with applicable regulations only if the Administrator finds that—

(1) The increase is not likely to reduce seriously the structural strength;

(2) The probability of sudden fatigue failure is not noticeably increased;

(3) The flutter, deformation, and vibration characteristics do not fall below those required by applicable regulations; and

(4) All other applicable weight limitations will be met.

**(c)** No zero fuel weight may be increased by more than five percent, and the increase in the structural landing weight may not exceed the amount, in pounds, of the increase in zero fuel weight.

**(d)** Each airplane must be inspected in accordance with the approved special inspection procedures, for operations at increased weights, established and issued by the manufacturer of the type of airplane.

**(e)** Each airplane operated under this section must be operated in accordance with the passenger-carrying performance operating limitations prescribed in this part.

**(f)** The Airplane Flight Manual for each airplane operated under this section must be appropriately revised to include the operating limitations and information needed for operation at the increased weights.

**(g)** Except as provided for the carrying of persons under §121.583 each airplane operated at an increased weight under this section must, before it is used in passenger service, be inspected under the special inspection procedures for return to passenger service established and issued by the manufacturer and approved by the Administrator.

## §121.199 Nontransport category airplanes: Takeoff limitations.

**(a)** No person operating a nontransport category airplane may take off that airplane at a weight greater than the weight that would allow the airplane to be brought to a safe stop within the effective length of the runway, from any point during the takeoff before reaching 105 percent of minimum control speed (the minimum speed at which an airplane can be safely controlled in flight after an engine becomes inoperative) or 115 percent of the power off stalling speed in the takeoff configuration, whichever is greater.

**(b)** For the purposes of this section—

(1) It may be assumed that takeoff power is used on all engines during the acceleration;

(2) Not more than 50 percent of the reported headwind component, or not less than 150 percent of the reported tailwind component, may be taken into account;

(3) The average runway gradient (the difference between the elevations of the endpoints of the runway divided by the total length) must be considered if it is more than one-half of 1 percent;

(4) It is assumed that the airplane is operating in standard atmosphere; and

(5) The *effective length of the runway* for takeoff means the distance from the end of the runway at which the takeoff is started to a point at which the obstruction clearance plane associated with the other end of the runway intersects the runway centerline.

[Docket No. 6258, 29 FR 19198, Dec. 31, 1964; as amended by Amdt. 121–132, 41 FR 55475, Dec. 20, 1976]

### §121.201 Nontransport category airplanes: En route limitations: One engine inoperative.

**(a)** Except as provided in paragraph (b) of this section, no person operating a nontransport category airplane may take off that airplane at a weight that does not allow a rate of climb of at least 50 feet a minute, with the critical engine inoperative, at an altitude of at least 1,000 feet above the highest obstruction within five miles on each side of the intended track, or 5,000 feet, whichever is higher.

**(b)** Notwithstanding paragraph (a) of this section, if the Administrator finds that safe operations are not impaired, a person may operate the airplane at an altitude that allows the airplane, in case of engine failure, to clear all obstructions within 5 miles on each side of the intended track by 1,000 feet. If this procedure is used, the rate of descent for the appropriate weight and altitude is assumed to be 50 feet a minute greater than the rate in the approved performance data. Before approving such a procedure, the Administrator considers the following for the route, route segment, or area concerned:

(1) The reliability of wind and weather forecasting.

(2) The location and kinds of navigation aids.

(3) The prevailing weather conditions, particularly the frequency and amount of turbulence normally encountered.

(4) Terrain features.

(5) Air traffic control problems.

(6) Any other operational factors that affect the operation.

**(c)** For the purposes of this section, it is assumed that—

(1) The critical engine is inoperative;

(2) The propeller of the inoperative engine is in the minimum drag position;

(3) The wing flaps and landing gear are in the most favorable position;

(4) The operating engines are operating at the maximum continuous power available;

(5) The airplane is operating in standard atmosphere; and

(6) The weight of the airplane is progressively reduced by the anticipated consumption of fuel and oil.

### §121.203 Nontransport category airplanes: Landing limitations: Destination airport.

**(a)** No person operating a nontransport category airplane may take off that airplane at a weight that—

(1) Allowing for anticipated consumption of fuel and oil, is greater than the weight that would allow a full stop landing within 60 percent of the effective length of the most suitable runway at the destination airport; and

(2) Is greater than the weight allowable if the landing is to be made on the runway—

(i) With the greatest effective length in still air; and

(ii) Required by the probable wind, taking into account not more than 50 percent of the headwind component or not less than 150 percent of the tailwind component.

**(b)** For the purposes of this section, it is assumed that—

(1) The airplane passes directly over the intersection of the obstruction clearance plane and the runway at a height of 50 feet in a steady gliding approach at a true indicated airspeed of at least 1.3 $V_{SO}$;

(2) The landing does not require exceptional pilot skill; and

(3) The airplane is operating in standard atmosphere.

### §121.205 Nontransport category airplanes: Landing limitations: Alternate airport.

No person may list an airport as an alternate airport in a dispatch or flight release for a nontransport category airplane unless that airplane (at the weight anticipated at the time of arrival) based on the assumptions contained in §121.203, can be brought to a full stop landing within 70 percent of the effective length of the runway.

### §121.207 Provisionally certificated airplanes: Operating limitations.

In addition to the limitations in §91.317 of this chapter, the following limitations apply to the operation of provisionally certificated airplanes by certificate holders:

**(a)** In addition to crewmembers, each certificate holder may carry on such an airplane only those persons who are listed in §121.547(c) or who are specifically authorized by both the certificate holder and the Administrator.

**(b)** Each certificate holder shall keep a log of each flight conducted under this section and shall keep accurate and complete records of each inspection made and all maintenance performed on the airplane. The certificate holder shall make the log and records made under this section available to the manufacturer and the Administrator.

[Docket No. 28154, 61 FR 2611, Jan. 26, 1996]

# Subpart J— Special Airworthiness Requirements

SOURCE: Docket No. 6258, 29 FR 19202, Dec. 31, 1964, unless otherwise noted.

## §121.211 Applicability.

**(a)** This subpart prescribes special airworthiness requirements applicable to certificate holders as stated in paragraphs (b) through (e) of this section.

**(b)** Except as provided in paragraph (d) of this section, each airplane type certificated under Aero Bulletin 7A or part 04 of the Civil Air Regulations in effect before November 1, 1946 must meet the special airworthiness requirements in §§121.215 through 121.283.

**(c)** Each certificate holder must comply with the requirements of §§121.285 through 121.291.

**(d)** If the Administrator determines that, for a particular model of airplane used in cargo service, literal compliance with any requirement under paragraph (b) of this section would be extremely difficult and that compliance would not contribute materially to the objective sought, he may require compliance only with those requirements that are necessary to accomplish the basic objectives of this part.

**(e)** No person may operate under this part a nontransport category airplane type certificated after December 31, 1964, unless the airplane meets the special airworthiness requirements in §121.293.

[Docket No. 28154, 60 FR 65928, Dec. 20, 1995]

## §121.213 [Reserved]

## §121.215 Cabin interiors.

**(a)** Except as provided in §121.312, each compartment used by the crew or passengers must meet the requirements of this section.

**(b)** Materials must be at least flash resistant.

**(c)** The wall and ceiling linings and the covering of upholstering, floors, and furnishings must be flame resistant.

**(d)** Each compartment where smoking is to be allowed must be equipped with self-contained ash trays that are completely removable and other compartments must be placarded against smoking.

**(e)** Each receptacle for used towels, papers, and wastes must be of fire-resistant material and must have a cover or other means of containing possible fires started in the receptacles.

[Docket No. 6258, 29 FR 19202, Dec. 31, 1964; as amended by Amdt. 121–84, 37 FR 3974, Feb. 24, 1972]

## §121.217 Internal doors.

In any case where internal doors are equipped with louvres or other ventilating means, there must be a means convenient to the crew for closing the flow of air through the door when necessary.

## §121.219 Ventilation.

Each passenger or crew compartment must be suitably ventilated. Carbon monoxide concentration may not be more than one part in 20,000 parts of air, and fuel fumes may not be present. In any case where partitions between compartments have louvres or other means allowing air to flow between compartments, there must be a means convenient to the crew for closing the flow of air through the partitions, when necessary.

## §121.221 Fire precautions.

**(a)** Each compartment must be designed so that, when used for storing cargo or baggage, it meets the following requirements:

(1) No compartment may include controls, wiring, lines, equipment, or accessories that would upon damage or failure, affect the safe operation of the airplane unless the item is adequately shielded, isolated, or otherwise protected so that it cannot be damaged by movement of cargo in the compartment and so that damage to or failure of the item would not create a fire hazard in the compartment.

(2) Cargo or baggage may not interfere with the functioning of the fire-protective features of the compartment.

(3) Materials used in the construction of the compartments, including tie-down equipment, must be at least flame resistant.

(4) Each compartment must include provisions for safeguarding against fires according to the classifications set forth in paragraphs (b) through (f) of this section.

**(b)** *Class A.* Cargo and baggage compartments are classified in the "A" category if—

(1) A fire therein would be readily discernible to a member of the crew while at his station; and

(2) All parts of the compartment are easily accessible in flight.

There must be a hand fire extinguisher available for each Class A compartment.

**(c)** *Class B.* Cargo and baggage compartments are classified in the "B" category if enough access is provided while in flight to enable a member of the crew to effectively reach all of the compartment and its contents with a hand fire extinguisher and the compartment is so designed that, when the access provisions are being used, no hazardous amount of smoke, flames, or extinguishing agent enters any compartment occupied by the crew or passengers. Each Class B compartment must comply with the following:

(1) It must have a separate approved smoke or fire detector system to give warning at the pilot or flight engineer station.

(2) There must be a hand fire extinguisher available for the compartment.

(3) It must be lined with fire-resistant material, except that additional service lining of flame-resistant material may be used.

**(d)** *Class C.* Cargo and baggage compartments are classified in the "C" category if they do not conform with the requirements for the "A," "B," "D," or "E" categories. Each Class C compartment must comply with the following:

(1) It must have a separate approved smoke or fire detector system to give warning at the pilot or flight engineer station.

(2) It must have an approved built-in fire-extinguishing system controlled from the pilot or flight engineer station.

(3) It must be designed to exclude hazardous quantities of smoke, flames, or extinguishing agents from entering into any compartment occupied by the crew or passengers.

(4) It must have ventilation and draft controlled so that the extinguishing agent provided can control any fire that may start in the compartment.

(5) It must be lined with fire-resistant material, except that additional service lining of flame-resistant material may be used.

**(e)** *Class D.* Cargo and baggage compartments are classified in the "D" category if they are so designed and constructed that a fire occurring therein will be completely confined without endangering the safety of the airplane or the occupants. Each Class D compartment must comply with the following:

(1) It must have a means to exclude hazardous quantities of smoke, flames, or noxious gases from entering any compartment occupied by the crew or passengers.

(2) Ventilation and drafts must be controlled within each compartment so that any fire likely to occur in the compartment will not progress beyond safe limits.

(3) It must be completely lined with fire-resistant material.

(4) Consideration must be given to the effect of heat within the compartment on adjacent critical parts of the airplane.

**(f)** *Class E.* On airplanes used for the carriage of cargo only, the cabin area may be classified as a Class "E" compartment. Each Class E compartment must comply with the following:

(1) It must be completely lined with fire-resistant material.

(2) It must have a separate system of an approved type smoke or fire detector to give warning at the pilot or flight engineer station.

(3) It must have a means to shut off the ventilating air flow to or within the compartment and the controls for that means must be accessible to the flight crew in the crew compartment.

(4) It must have a means to exclude hazardous quantities of smoke, flames, or noxious gases from entering the flight crew compartment.

(5) Required crew emergency exits must be accessible under all cargo loading conditions.

## §121.223 Proof of compliance with §121.221.

Compliance with those provisions of §121.221 that refer to compartment accessibility, the entry of hazardous quantities of smoke or extinguishing agent into compartments occupied by the crew or passengers, and the dissipation of the extinguishing agent in Class "C" compartments must be shown by tests in flight. During these tests it must be shown that no inadvertent operation of smoke or fire detectors in other compartments within the airplane would occur as a result of fire contained in any one compartment, either during the time it is being extinguished, or thereafter, unless the extinguishing system floods those compartments simultaneously.

## §121.225 Propeller deicing fluid.

If combustible fluid is used for propeller deicing, the certificate holder must comply with §121.255.

## §121.227 Pressure cross-feed arrangements.

**(a)** Pressure cross-feed lines may not pass through parts of the airplane used for carrying persons or cargo unless—

(1) There is a means to allow crewmembers to shut off the supply of fuel to these lines; or

(2) The lines are enclosed in a fuel and fume-proof enclosure that is ventilated and drained to the exterior of the airplane.

However, such an enclosure need not be used if those lines incorporate no fittings on or within the personnel or cargo areas and are suitably routed or protected to prevent accidental damage.

**(b)** Lines that can be isolated from the rest of the fuel system by valves at each end must incorporate provisions for relieving excessive pressures that may result from exposure of the isolated line to high temperatures.

## §121.229 Location of fuel tanks.

**(a)** Fuel tanks must be located in accordance with §121.255.

**(b)** No part of the engine nacelle skin that lies immediately behind a major air outlet from the engine compartment may be used as the wall of an integral tank.

**(c)** Fuel tanks must be isolated from personnel compartments by means of fume- and fuel-proof enclosures.

## §121.231 Fuel system lines and fittings.

**(a)** Fuel lines must be installed and supported so as to prevent excessive vibration and so as to be adequate to withstand loads due to fuel pressure and accelerated flight conditions.

**(b)** Lines connected to components of the airplanes between which there may be relative motion must incorporate provisions for flexibility.

**(c)** Flexible connections in lines that may be under pressure and subject to axial loading must use flexible hose assemblies rather than hose clamp connections.

**(d)** Flexible hose must be of an acceptable type or proven suitable for the particular application.

## §121.233 Fuel lines and fittings in designated fire zones.

Fuel lines and fittings in each designated fire zone must comply with §121.259.

## §121.235 Fuel valves.

Each fuel valve must—

**(a)** Comply with §121.257;

**(b)** Have positive stops or suitable index provisions in the "on" and "off" positions; and

**(c)** Be supported so that loads resulting from its operation or from accelerated flight conditions are not transmitted to the lines connected to the valve.

## §121.237 Oil lines and fittings in designated fire zones.

Oil line and fittings in each designated fire zone must comply with §121.259.

## §121.239 Oil valves.

**(a)** Each oil valve must—

(1) Comply with §121.257;

(2) Have positive stops or suitable index provisions in the "on" and "off" positions; and

(3) Be supported so that loads resulting from its operation or from accelerated flight conditions are not transmitted to the lines attached to the valve.

**(b)** The closing of an oil shutoff means must not prevent feathering the propeller, unless equivalent safety provisions are incorporated.

## §121.241 Oil system drains.

Accessible drains incorporating either a manual or automatic means for positive locking in the closed position, must be provided to allow safe drainage of the entire oil system.

## §121.243 Engine breather lines.

**(a)** Engine breather lines must be so arranged that condensed water vapor that may freeze and obstruct the line cannot accumulate at any point.

**(b)** Engine breathers must discharge in a location that does not constitute a fire hazard in case foaming occurs and so that oil emitted from the line does not impinge upon the pilots' windshield.

**(c)** Engine breathers may not discharge into the engine air induction system.

## §121.245 Fire walls.

Each engine, auxiliary power unit, fuel-burning heater, or other item of combustion equipment that is intended for operation in flight must be isolated from the rest of the airplane by means of firewalls or shrouds, or by other equivalent means.

## §121.247 Fire-wall construction.

Each fire wall and shroud must—

**(a)** Be so made that no hazardous quantity of air, fluids, or flame can pass from the engine compartment to other parts of the airplane;

**(b)** Have all openings in the fire wall or shroud sealed with close-fitting fire-proof grommets, bushings, or firewall fittings;

**(c)** Be made of fireproof material; and

**(d)** Be protected against corrosion.

## §121.249 Cowling.

**(a)** Cowling must be made and supported so as to resist the vibration inertia, and air loads to which it may be normally subjected.

**(b)** Provisions must be made to allow rapid and complete drainage of the cowling in normal ground and flight attitudes. Drains must not discharge in locations constituting a fire hazard. Parts of the cowling that are subjected to high temperatures because they are near exhaust system parts or because of exhaust gas impingement must be made of fireproof material. Unless otherwise specified in these regulations all other parts of the cowling must be made of material that is at least fire resistant.

## §121.251 Engine accessory section diaphragm.

Unless equivalent protection can be shown by other means, a diaphragm that complies with §121.247 must be provided on air-cooled engines to isolate the engine power section and all parts of the exhaust system from the engine accessory compartment.

## §121.253 Powerplant fire protection.

**(a)** Designated fire zones must be protected from fire by compliance with §§121.255 through 121.261.

**(b)** Designated fire zones are—

(1) Engine accessory sections;

(2) Installations where no isolation is provided between the engine and accessory compartment; and

(3) Areas that contain auxiliary power units, fuel-burning heaters, and other combustion equipment.

## §121.255 Flammable fluids.

**(a)** No tanks or reservoirs that are a part of a system containing flammable fluids or gases may be located in designated fire zones, except where the fluid contained, the design of the system, the materials used in the tank, the shutoff means, and the connections, lines, and controls provide equivalent safety.

**(b)** At least one-half inch of clear airspace must be provided between any tank or reservoir and a firewall or shroud isolating a designated fire zone.

## §121.257 Shutoff means.

**(a)** Each engine must have a means for shutting off or otherwise preventing hazardous amounts of fuel, oil, deicer, and other flammable fluids from flowing into, within, or through any designated fire zone. However, means need not be provided to shut off flow in lines that are an integral part of an engine.

**(b)** The shutoff means must allow an emergency operating sequence that is compatible with the emergency operation of other equipment, such as feathering the propeller, to facilitate rapid and effective control of fires.

**(c)** Shutoff means must be located outside of designated fire zones, unless equivalent safety is provided, and it must be shown that no hazardous amount of flammable fluid will drain into any designated fire zone after a shut off.

**(d)** Adequate provisions must be made to guard against inadvertent operation of the shutoff means and to make it possible for the crew to reopen the shutoff means after it has been closed.

## §121.259 Lines and fittings.

**(a)** Each line, and its fittings, that is located in a designated fire zone, if it carries flammable fluids or gases under pressure, or is attached directly to the engine, or is subject to relative motion between components (except lines and fittings forming an integral part of the engine), must be flexible and fire-resistant with fire-resistant, factory-fixed, detachable, or other approved fire-resistant ends.

**(b)** Lines and fittings that are not subject to pressure or to relative motion between components must be of fire-resistant materials.

## §121.261 Vent and drain lines.

All vent and drain lines and their fittings, that are located in a designated fire zone must, if they carry flammable fluids or gases, comply with §121.259, if the Administrator finds that the rupture or breakage of any vent or drain line may result in a fire hazard.

## §121.263 Fire-extinguishing systems.

**(a)** Unless the certificate holder shows that equivalent protection against destruction of the airplane in case of fire is provided by the use of fireproof materials in the nacelle and other components that would be subjected to flame, fire-extinguishing systems must be provided to serve all designated fire zones.

**(b)** Materials in the fire-extinguishing system must not react chemically with the extinguishing agent so as to be a hazard.

## §121.265 Fire-extinguishing agents.

Only methyl bromide, carbon dioxide, or another agent that has been shown to provide equivalent extinguishing action may be used as a fire-extinguishing agent. If methyl bromide or any other toxic extinguishing agent is used, provisions must be made to prevent harmful concentrations of fluid or fluid vapors from entering any personnel compartment either because of leakage during normal operation of the airplane or because of discharging the fire extinguisher on the ground or in flight when there is a defect in the extinguishing system. If a methyl bromide system is used, the containers must be charged with dry agent and sealed by the fire-extinguisher manufacturer or some other person using satisfactory recharging equipment. If carbon dioxide is used, it must not be possible to discharge enough gas into the personnel compartments to create a danger of suffocating the occupants.

## §121.267 Extinguishing agent container pressure relief.

Extinguishing agent containers must be provided with a pressure relief to prevent bursting of the container because of excessive internal pressures. The discharge line from the relief connection must terminate outside the airplane in a place convenient for inspection on the ground. An indicator must be provided at the discharge end of the line to provide a visual indication when the container has discharged.

## §121.269 Extinguishing agent container compartment temperature.

Precautions must be taken to insure that the extinguishing agent containers are installed in places where reasonable temperatures can be maintained for effective use of the extinguishing system.

## §121.271 Fire-extinguishing system materials.

**(a)** Except as provided in paragraph (b) of this section, each component of a fire-extinguishing system that is in a designated fire zone must be made of fireproof materials.

**(b)** Connections that are subject to relative motion between components of the airplane must be made of flexible materials that are at least fire-resistant and be located so as to minimize the probability of failure.

## §121.273 Fire-detector systems.

Enough quick-acting fire detectors must be provided in each designated fire zone to assure the detection of any fire that may occur in that zone.

## §121.275 Fire detectors.

Fire detectors must be made and installed in a manner that assures their ability to resist, without failure, all vibration, inertia, and other loads to which they may be normally subjected. Fire detectors must be unaffected by exposure to fumes, oil, water, or other fluids that may be present.

## §121.277 Protection of other airplane components against fire.

**(a)** Except as provided in paragraph (b) of this section, all airplane surfaces aft of the nacelles in the area of one nacelle diameter on both sides of the nacelle centerline must be made of material that is at least fire resistant.

**(b)** Paragraph (a) of this section does not apply to tail surfaces lying behind nacelles unless the dimensional configuration of the airplane is such that the tail surfaces could be affected readily by heat, flames, or sparks emanating from a designated fire zone or from the engine compartment of any nacelle.

## §121.279 Control of engine rotation.

**(a)** Except as provided in paragraph (b) of this section, each airplane must have a means of individually stopping and restarting the rotation of any engine in flight.

**(b)** In the case of turbine engine installations, a means of stopping the rotation need be provided only if the Administrator finds that rotation could jeopardize the safety of the airplane.

## §121.281 Fuel system independence.

**(a)** Each airplane fuel system must be arranged so that the failure of any one component does not result in the irrecoverable loss of power of more than one engine.

**(b)** A separate fuel tank need not be provided for each engine if the certificate holder shows that the fuel system incorporates features that provide equivalent safety.

## §121.283 Induction system ice prevention.

A means for preventing the malfunctioning of each engine due to ice accumulation in the engine air induction system must be provided for each airplane.

## §121.285 Carriage of cargo in passenger compartments.

**(a)** Except as provided in paragraph (b), (c), or (d) or this section, no certificate holder may carry cargo in the passenger compartment of an airplane.

**(b)** Cargo may be carried anywhere in the passenger compartment if it is carried in an approved cargo bin that meets the following requirements:

(1) The bin must withstand the load factors and emergency landing conditions applicable to the passenger seats of the airplane in which the bin is installed, multiplied by a factor of 1.15, using the combined weight of the bin and the maximum weight of cargo that may be carried in the bin.

(2) The maximum weight of cargo that the bin is approved to carry and any instructions necessary to insure proper weight distribution within the bin must be conspicuously marked on the bin.

(3) The bin may not impose any load on the floor or other structure of the airplane that exceeds the load limitations of that structure.

(4) The bin must be attached to the seat tracks or to the floor structure of the airplane, and its attachment must withstand the load factors and emergency landing conditions applicable to the

passenger seats of the airplane in which the bin is installed, multiplied by either the factor 1.15 or the seat attachment factor specified for the airplane, whichever is greater, using the combined weight of the bin and the maximum weight of cargo that may be carried in the bin.

(5) The bin may not be installed in a position that restricts access to or use of any required emergency exit, or of the aisle in the passenger compartment.

(6) The bin must be fully enclosed and made of material that is at least flame resistant.

(7) Suitable safeguards must be provided within the bin to prevent the cargo from shifting under emergency landing conditions.

(8) The bin may not be installed in a position that obscures any passenger's view of the "seat belt" sign "no smoking" sign, or any required exit sign, unless an auxiliary sign or other approved means for proper notification of the passenger is provided.

**(c)** Cargo may be carried aft of a bulkhead or divider in any passenger compartment provided the cargo is restrained to the load factors in §25.561(b)(3) and is loaded as follows:

(1) It is properly secured by a safety belt or other tiedown having enough strength to eliminate the possibility of shifting under all normally anticipated flight and ground conditions.

(2) It is packaged or covered in a manner to avoid possible injury to passengers and passenger compartment occupants.

(3) It does not impose any load on seats or the floor structure that exceeds the load limitation for those components.

(4) Its location does not restrict access to or use of any required emergency or regular exit, or of the aisle in the passenger compartment.

(5) Its location does not obscure any passenger's view of the "seat belt" sign, "no smoking" sign, or required exit sign, unless an auxiliary sign or other approved means for proper notification of the passenger is provided.

**(d)** Cargo, including carry-on baggage, may be carried anywhere in the passenger compartment of a nontransport category airplane type certificated after December 31, 1964, if it is carried in an approved cargo rack, bin, or compartment installed in or on the airplane, if it is secured by an approved means, or if it is carried in accordance with each of the following:

(1) For cargo, it is properly secured by a safety belt or other tie-down having enough strength to eliminate the possibility of shifting under all normally anticipated flight and ground conditions, or for carry-on baggage, it is restrained so as to prevent its movement during air turbulence.

(2) It is packaged or covered to avoid possible injury to occupants.

(3) It does not impose any load on seats or in the floor structure that exceeds the load limitation for those components.

(4) It is not located in a position that obstructs the access to, or use of, any required emergency or regular exit, or the use of the aisle between the crew and the passenger compartment, or is located in a position that obscures any passenger's view of the "seat belt" sign, "no smoking" sign or placard, or any required exit sign, unless an auxiliary sign or other approved means for proper notification of the passengers is provided.

(5) It is not carried directly above seated occupants.

(6) It is stowed in compliance with this section for takeoff and landing.

(7) For cargo-only operations, paragraph (d)(4) of this section does not apply if the cargo is loaded so that at least one emergency or regular exit is available to provide all occupants of the airplane a means of unobstructed exit from the airplane if an emergency occurs.

[Docket No. 6258, 29 FR 19202, Dec. 31, 1964; as amended by Amdt. 121–179, 47 FR 33390, Aug. 2, 1982; Amdt. 121–251, 60 FR 65928, Dec. 20, 1995]

## §121.287 Carriage of cargo in cargo compartments.

When cargo is carried in cargo compartments that are designed to require the physical entry of a crewmember to extinguish any fire that may occur during flight, the cargo must be loaded so as to allow a crewmember to effectively reach all parts of the compartment with the contents of a hand fire extinguisher.

## §121.289 Landing gear: Aural warning device.

**(a)** Except for airplanes that comply with the requirements of §25.729 of this chapter on or after January 6, 1992, each airplane must have a landing gear aural warning device that functions continuously under the following conditions:

(1) For airplanes with an established approach wing-flap position, whenever the wing flaps are extended beyond the maximum certificated approach climb configuration position in the Airplane Flight Manual and the landing gear is not fully extended and locked.

(2) For airplanes without an established approach climb wing-flap position, whenever the wing flaps are extended beyond the position at which landing gear extension is normally performed and the landing gear is not fully extended and locked.

**(b)** The warning system required by paragraph (a) of this section—

(1) May not have a manual shutoff;

(2) Must be in addition to the throttle-actuated device installed under the type certification airworthiness requirements; and

(3) May utilize any part of the throttle-actuated system including the aural warning device.

**(c)** The flap position sensing unit may be installed at any suitable place in the airplane.

[Docket No. 6258, 29 FR 19202, Dec. 31, 1964; as amended by Amdt. 121–3, 30 FR 3638, March 19, 1965; Amdt. 121–130, 41 FR 47229, Oct. 28, 1976; Amdt. 121–227, 56 FR 63762, Dec. 5, 1991; Amdt. 121–251, 60 FR 65929, Dec. 20, 1995]

## §121.291 Demonstration of emergency evacuation procedures.

**(a)** Except as provided in paragraph (a)(1) of this section, each certificate holder must conduct an actual demonstration of emergency evacuation procedures in accordance with paragraph (a) of Appendix D to this part to show that each type and model of airplane with a seating capacity of more than 44 passengers to be used in its passenger-carrying operations allows the evacuation of the full capacity, including crewmembers, in 90 seconds or less.

(1) An actual demonstration need not be conducted if that airplane type and model has been shown to be in compliance with this paragraph in effect on or after October 24, 1967, or, if during type certification, with §25.803 of this chapter in effect on or after December 1, 1978.

(2) Any actual demonstration conducted after September 27, 1993, must be in accordance with paragraph (a) of Appendix D to this part in effect on or after that date or with §25.803 in effect on or after that date.

**(b)** Each certificate holder conducting operations with airplanes with a seating capacity of more than 44 passengers must conduct a partial demonstration of emergency evacuation procedures in accordance with paragraph (c) of this section upon:

(1) Initial introduction of a type and model of airplane into passenger/carrying operation, if the certificate holder has not conducted an actual demonstration under paragraph (a) of this section;

(2) Changing the number, location, or emergency evacuation duties or procedures of flight attendants who are required by §121.391; or

(3) Changing the number, location, type of emergency exits, or type of opening mechanism on emergency exits available for evacuation.

**(c)** In conducting the partial demonstration required by paragraph (b) of this section, each certificate holder must:

(1) Demonstrate the effectiveness of its crewmember emergency training and evacuation procedures by conducting a demonstration, not requiring passengers and observed by the Administrator, in which the flight attendants for that type and model of airplane, using that operator's line operating procedures, open 50 percent of the required floor-level emergency exits and 50 percent of the required non-floor-level emergency exits whose opening by a flight attendant is defined as an emergency evacuation duty under §121.397, and deploy 50 percent of the exit slides. The exits and slides will be selected by the administrator and must be ready for use within 15 seconds;

(2) Apply for and obtain approval from the certificate-holding district office before conducting the demonstration;

(3) Use flight attendants in this demonstration who have been selected at random by the Administrator, have completed the certificate holder's FAA-approved training program for the type and model of airplane, and have passed a written or practical examination on the emergency equipment and procedures; and

(4) Apply for and obtain approval from the certificate-holding district office before commencing operations with this type and model airplane.

**(d)** Each certificate holder operating or proposing to operate one or more landplanes in extended overwater operations, or otherwise required to have certain equipment under §121.339, must show, by simulated ditching conducted in accordance with paragraph (b) of Appendix D to this part, that it has the ability to efficiently carry out its ditching procedures. For certificate holders subject to §121.2(a)(1), this paragraph applies only when a new type or model airplane is introduced into the certificate holder's operations after January 19, 1996.

**(e)** For a type and model airplane for which the simulated ditching specified in paragraph (d) has been conducted by a part 121 certificate holder, the requirements of paragraphs (b)(2), (b)(4), and (b)(5) of Appendix D to this part are complied with if each life raft is removed from stowage, one life raft is launched and inflated (or one slide life raft is inflated) and crewmembers assigned to the inflated life raft display and describe the use of each item of required emergency equipment. The life raft or slide life raft to be inflated will be selected by the Administrator.

[Docket No. 21269, 46 FR 61453, Dec. 17, 1981; as amended by Amdt. 121–233, 58 FR 45230, Aug. 26, 1993; Amdt. 121–251, 60 FR 65929, Dec. 20, 1995]

## §121.293 Special airworthiness requirements for nontransport category airplanes type certificated after December 31, 1964.

No certificate holder may operate a nontransport category airplane manufactured after December 20, 1999, unless the airplane contains a takeoff warning system that meets the requirements of 14 CFR 25.703. However, the takeoff warning system does not have to cover any device for which it has been demonstrated that takeoff with that device in the most adverse position would not create a hazardous condition.

[Docket No. 28154, 60 FR 65929, Dec. 20, 1995]

# Subpart K—Instrument and Equipment Requirements

**Source:** Docket No. 6258, 29 FR 19205, Dec. 31, 1964, unless otherwise noted.

## §121.301 Applicability.

This subpart prescribes instrument and equipment requirements for all certificate holders.

## §121.303 Airplane instruments and equipment.

**(a)** Unless otherwise specified, the instrument and equipment requirements of this subpart apply to all operations under this part.

**(b)** Instruments and equipment required by §§121.305 through 121.359 and 121.803 must be approved and installed in accordance with the airworthiness requirements applicable to them.

**(c)** Each airspeed indicator must be calibrated in knots, and each airspeed limitation and item of related information in the Airplane Flight Manual and pertinent placards must be expressed in knots.

**(d)** Except as provided in §§121.627(b) and 121.628, no person may take off any airplane unless the following instruments and equipment are in operable condition:

(1) Instruments and equipment required to comply with airworthiness requirements under which the airplane is type certificated and as required by §§121.213 through 121.283 and 121.289.

(2) Instruments and equipment specified in §§121.305 through 121.321, 121.359, 121.360, and 121.803 for all operations, and the instruments and equipment specified in §§121.323 through 121.351 for the kind of operation indicated, wherever these items are not already required by paragraph (d)(1) of this section.

[Docket No. 6258, 29 FR 19202, Dec. 31, 1964; as amended by Amdt. 121–44, 33 FR 14406, Sept. 25, 1968; Amdt. 121–65, 35 FR 12709, Aug. 11, 1970; Amdt. 121–114, 39 FR 44440, Dec. 24, 1974; Amdt. 121–126, 40 FR 55314, Nov. 28, 1975; Amdt. 121–222, 56 FR 12310, March 22, 1991; Amdt. 121–253, 61 FR, Jan. 26, 1996; Amdt. 121–281, 66 FR 19043, April 12, 2001]

## §121.305 Flight and navigational equipment.

No person may operate an airplane unless it is equipped with the following flight and navigational instruments and equipment:

**(a)** An airspeed indicating system with heated pitot tube or equivalent means for preventing malfunctioning due to icing.

**(b)** A sensitive altimeter.

**(c)** A sweep-second hand clock (or approved equivalent).

**(d)** A free-air temperature indicator.

**(e)** A gyroscopic bank and pitch indicator (artificial horizon).

**(f)** A gyroscopic rate-of-turn indicator combined with an integral slip-skid indicator (turn-and-bank indicator) except that only a slip-skid indicator is required when a third attitude instrument system usable through flight attitudes of 360° of pitch and roll is installed in accordance with paragraph (k) of this section.

**(g)** A gyroscopic direction indicator (directional gyro or equivalent).

**(h)** A magnetic compass.

**(i)** A vertical speed indicator (rate-of-climb indicator).

**(j)** On the airplane described in this paragraph, in addition to two gyroscopic bank and pitch indicators (artificial horizons) for use at the pilot stations, a third such instrument is installed in accordance with paragraph (k) of this section:

(1) On each turbojet powered airplane.

(2) On each turbopropeller powered airplane having a passenger-seat configuration of more than 30 seats, excluding each crewmember seat, or a payload capacity of more than 7,500 pounds.

(3) On each turbopropeller powered airplane having a passenger-seat configuration of 30 seats or fewer, excluding each crewmember seat, and a payload capacity of 7,500 pounds or less that is manufactured on or after March 20, 1997.

(4) After December 20, 2010, on each turbopropeller powered airplane having a passenger seat configuration of 10–30 seats and a payload capacity of 7,500 pounds or less that was manufactured before March 20, 1997.

**(k)** When required by paragraph (j) of this section, a third gyroscopic bank-and-pitch indicator (artificial horizon) that:

(1) Is powered from a source independent of the electrical generating system;

(2) Continues reliable operation for a minimum of 30 minutes after total failure of the electrical generating system;

(3) Operates independently of any other attitude indicating system;

(4) Is operative without selection after total failure of the electrical generating system;

(5) Is located on the instrument panel in a position acceptable to the Administrator that will make it plainly visible to and usable by each pilot at his or her station; and

(6) Is appropriately lighted during all phases of operation.

[Docket No. 6258, 29 FR 19205, Dec. 31, 1964; as amended by Amdt. 121–57, 35 FR 304, Jan. 8, 1970; Amdt. 121–60, 35 FR 7108, May 6, 1970; Amdt. 121–81, 36 FR 23050, Dec. 3, 1971; Amdt. 121–130, 41 FR 47229, Oct. 28, 1976; Amdt. 121–230, 58 FR 12158, March 3, 1993; Amdt. 121–251, 60 FR 65929, Dec. 20, 1995; Amdt. 121–262, 62 FR 13256, March 19, 1997]

## §121.306 Portable electronic devices.

**(a)** Except as provided in paragraph (b) of this section, no person may operate, nor may any operator or pilot in command of an aircraft allow the operation of, any portable electronic device on any U.S.-registered civil aircraft operating under this part.

**(b)** Paragraph (a) of this section does not apply to—

(1) Portable voice recorders;

(2) Hearing aids;

(3) Heart pacemakers;

(4) Electric shavers; or

(5) Any other portable electronic device that the part 119 certificate holder has determined will not cause interference with the navigation or communication system of the aircraft on which it is to be used.

**(c)** The determination required by paragraph (b)(5) of this section shall be made by that part 119 certificate holder operating the particular device to be used.

[Docket No. 6258, 29 FR 19205, Dec. 31, 1964; as amended by Amdt. 121–270, 64 FR 1080, Jan. 7, 1999]

## §121.307 Engine instruments.

Unless the Administrator allows or requires different instrumentation for turbine engine powered airplanes to provide equivalent safety, no person may conduct any operation under this part without the following engine instruments:

**(a)** A carburetor air temperature indicator for each engine.

**(b)** A cylinder head temperature indicator for each air-cooled engine.

**(c)** A fuel pressure indicator for each engine.

**(d)** A fuel flowmeter or fuel mixture indicator for each engine not equipped with an automatic altitude mixture control.

**(e)** A means for indicating fuel quantity in each fuel tank to be used.

**(f)** A manifold pressure indicator for each engine.

**(g)** An oil pressure indicator for each engine.

**(h)** An oil quantity indicator for each oil tank when a transfer or separate oil reserve supply is used.

**(i)** An oil-in temperature indicator for each engine.

**(j)** A tachometer for each engine.

**(k)** An independent fuel pressure warning device for each engine or a master warning device for all engines with a means for isolating the individual warning circuits from the master warning device.

**(l)** A device for each reversible propeller, to indicate to the pilot when the propeller is in reverse pitch, that complies with the following:

(1) The device may be actuated at any point in the reversing cycle between the normal low pitch stop position and full reverse pitch, but it may not give an indication at or above the normal low pitch stop position.

(2) The source of indication must be actuated by the propeller blade angle or be directly responsive to it.

## §121.308 Lavatory fire protection.

**(a)** Except as provided in paragraphs (c) and (d) of this section, no person may operate a passenger-carrying airplane unless each lavatory in the airplane is equipped with a smoke detector system or equivalent that provides a warning light in the cockpit or provides a warning light or audio warning in the passenger cabin which would be readily detected by a flight attendant, taking into consideration the positioning of flight attendants throughout the passenger compartment during various phases of flight.

**(b)** Except as provided in paragraph (c) of this section, no person may operate a passenger-carrying airplane unless each lavatory in the airplane is equipped with a built-in fire extinguisher for each disposal receptacle for towels, paper, or waste located within the lavatory. The built-in fire extinguisher must be designed to discharge automatically into each disposal receptacle upon occurrence of a fire in the receptacle.

**(c)** Until December 22, 1997, a certificate holder described in §121.2(a)(1) or (2) may operate an airplane with a passenger seat configuration of 30 or fewer seats that does not comply with the smoke detector system requirements described in paragraph (a) of this section and the fire extinguisher requirements described in paragraph (b) of this section.

**(d)** After December 22, 1997, no person may operate a nontransport category airplane type certificated after December 31, 1964, with a passenger seat configuration of 10–19 seats unless that airplane complies with the smoke detector system requirements described in paragraph (a) of this section, except that the smoke detector system or equivalent must provide a warning light in the cockpit or an audio warning that would be readily detected by the flightcrew.

[Docket No. 28154, 60 FR 65929, Dec. 20, 1995]

## §121.309 Emergency equipment.

**(a)** *General:* No person may operate an airplane unless it is equipped with the emergency equipment listed in this section and in §121.310.

**(b)** Each item of emergency and flotation equipment listed in this section and in §§121.310, 121.339, and 121.340—

(1) Must be inspected regularly in accordance with inspection periods established in the operations specifications to ensure its condition for continued serviceability and immediate readiness to perform its intended emergency purposes;

(2) Must be readily accessible to the crew and, with regard to equipment located in the passenger compartment, to passengers;

(3) Must be clearly identified and clearly marked to indicate its method of operation; and

(4) When carried in a compartment or container, must be carried in a compartment or container marked as to contents and the compartment or container, or the item itself, must be marked as to date of last inspection.

**(c)** *Hand fire extinguishers for crew, passenger, and cargo and galley compartments.* Hand fire extinguishers of an approved type must be provided for use in crew, passenger, cargo and galley compartments in accordance with the following:

(1) The type and quantity of extinguishing agent must be suitable for the kinds of fires likely to occur in the compartment where the extinguisher is intended to be used and, for passenger compartments, must be designed to minimize the hazard of toxic gas concentrations.

(2) *Cargo compartments.* At least one hand fire extinguisher must be conveniently located for use in each Class E cargo compartment that is accessible to crew members during flight.

(3) *Galley compartments.* At least one hand fire extinguisher must be conveniently located for use in each galley located in a compartment other than a passenger, cargo or crew compartment.

(4) *Flightcrew compartment.* At least one hand extinguisher must be conveniently located on the flight deck for use by the flightcrew.

(5) *Passenger compartment*s. hand fire extinguishers for use in passenger compartments must be conveniently located and, when two or more are required, uniformly distributed throughout each compartment. Hand fire extinguishers shall be provided in passenger compartment as follows:

(i) For airplanes having passenger seats accommodating more than 6 but fewer than 31 passengers, at least one.

(ii) For airplanes having passenger seats accommodating more than 30 but fewer than 61 passengers, at least two.

(iii) For airplanes having passenger seats accommodating more than 60 passengers, there must be at least the following number of hand fire extinguishers:

**Minimum Number of Hand Fire Extinguishers**

| Passenger seating accommodations: | |
|---|---|
| 61 through 200 | 3 |
| 201 through 300 | 4 |
| 301 through 400 | 5 |
| 401 through 500 | 6 |
| 501 through 600 | 7 |
| 601 or more | 8 |

(6) Notwithstanding the requirement for uniform distribution of hand fire extinguishers as prescribed in paragraph (c)(5) of this section, for those cases where a galley is located in a passenger compartment, at least one hand fire extinguisher must be conveniently located and easily accessible for use in the galley.

(7) At least two of the required hand fire extinguisher installed in passenger-carrying airplanes must contain Halon 1211 (bromochlorofluoromethane) or equivalent as the extinguishing agent. At least one hand fire extinguisher in the passenger compartment must contain Halon 1211 or equivalent.

**(d)** Reserved

**(e)** *Crash ax.* Except for nontransport category airplanes type certificated after December 31, 1964, each airplane must be equipped with a crash ax.

**(f)** *Megaphones.* Each passenger-carrying airplane must have a portable battery-powered megaphone or megaphones readily accessible to the crewmembers assigned to direct emergency evacuation, installed as follows:

(1) One megaphone on each airplane with a seating capacity of more than 60 and less than 100 passengers, at the most rearward location in the passenger cabin where it would be readily accessible to a normal flight attendant seat. However, the Administrator may grant a deviation from the requirements of this subparagraph if he finds that a different location would be more useful for evacuation of persons during an emergency.

(2) Two megaphones in the passenger cabin on each airplane with a seating capacity of more than 99 passengers, one installed at the forward end and the other at the most rearward location where it would be readily accessible to a normal flight attendant seat.

[Docket No. 6258, 29 FR 19205, Dec. 31, 1964; as amended by Amdt. 121–30, 32 FR 13267, Sept. 20, 1967; Amdt. 121–48, 34 FR 11489, July 11, 1969; Amdt. 121–106, 38 FR 22377, Aug. 20, 1973; Amdt. 121–185, 50 FR 12733, March 29, 1985; 50 FR 14373, April 12, 1985; Amdt. 121–188, 51 FR 1223, Jan. 9, 1986; Amdt. 121–230, 57 FR 42672, Sept. 15, 1992; Amdt. 121–242, 59 FR 52642, Oct. 18, 1994; Amdt. 121–242, 59 FR 55208, Nov. 4, 1994; Amdt. 121–251, 60 FR 65930, Dec. 20, 1995; Amdt. 121–281, 66 FR 19043, April 12, 2001]

## §121.310 Additional emergency equipment.

**(a)** *Means for emergency evacuation.* Each passenger-carrying landplane emergency exit (other than over-the-wing) that is more than 6 feet from the ground with the airplane on the ground and the landing gear extended, must have an approved means to assist the occupants in descending to the ground. The assisting means for a floor-level emergency exit must meet the requirements of §25.809(f)(1) of this chapter in effect on April 30, 1972, except that, for any airplane for which the application for the type certificate was filed after that date, it must meet the requirements under which the airplane was type certificated. An assisting means that deploys automatically must be armed during taxiing, takeoffs, and landings. However, if the Administrator finds that the design of the exit makes compliance impractical, he may grant a deviation from the requirement of automatic deployment if the assisting means automatically erects upon deployment and, with respect to required emergency exits, if an emergency evacuation demonstration is conducted in accordance with §121.291(a). This paragraph does not apply to the rear window emergency exit of DC-3 airplanes operated with less than 36 occupants, including crewmembers and less than five exits authorized for passenger use.

**(b)** *Interior emergency exit marking.* The following must be complied with for each passenger-carrying airplane:

(1) Each passenger emergency exit, its means of access, and its means of opening must be conspicuously marked. The identity and location of each passenger emergency exit must be recognizable from a distance equal to the width of the cabin. The location of each passenger emergency exit must be indicated by a sign visible to occupants approaching along the main passenger aisle. There must be a locating sign—

(i) Above the aisle near each over-the-wing passenger emergency exit, or at another ceiling location if it is more practical because of low headroom;

(ii) Next to each floor level passenger emergency exit, except that one sign may serve two such exits if they both can be seen readily from that sign; and

(iii) On each bulkhead or divider that prevents fore and aft vision along the passenger cabin, to indicate emergency exits beyond and obscured by it, except that if this is not possible the sign may be placed at another appropriate location.

(2) Each passenger emergency exit marking and each locating sign must meet the following:

(i) Except as provided in paragraph (b)(2)(iii) of this section, for an airplane for which the application for the type certificate was filed prior to May 1, 1972, each passenger emergency exit marking and each locating sign must be manufactured to meet the requirements of §25.812(b) of this chapter in effect on April 30, 1972. On these airplanes, no sign may continue to be used if its luminescence (brightness) decreases to below 100 microlamberts. The colors may be reversed if it increases the emergency illumination of the passenger compartment. However, the Administrator may authorize deviation from the 2-inch background requirements if he finds that special circumstances exist that make compliance impractical and that the proposed deviation provides an equivalent level of safety.

(ii) For a transport category airplane for which the application for the type certificate was filed on or after May 1, 1972, each passenger emergency exit marking and each locating sign must be manufactured to meet the interior emergency exit marking requirements under which the airplane was type certificated. On these airplanes, no sign may continue to be used if its luminescence (brightness) decreases to below 250 microlamberts.

(iii) For a nontransport category turbopropeller powered airplane type certificated after December 31, 1964, each passenger emergency exit marking and each locating sign must be manufactured to meet the requirements of §23.811(b) of this chapter. On these airplanes, no sign may continue to be used if its luminescence (brightness) decreases to below 100 microlamberts.

**(c)** *Lighting for interior emergency exit markings.* Except for nontransport category airplanes type certificated after December 31, 1964, each passenger-carrying airplane must have an emergency lighting system, independent of the main lighting system. However, sources of general cabin illumination may be common to both the emergency and the main lighting systems if the power supply to the emergency lighting system is independent of the power supply to the main lighting system. The emergency lighting system must—

(1) Illuminate each passenger exit marking and locating sign;

(2) Provide enough general lighting in the passenger cabin so that the average illumination when measured at 40-inch intervals at seat arm-

rest height, on the centerline of the main passenger aisle, is at least 0.05 foot-candles; and

(3) For airplanes type certificated after January 1, 1958, after November 26, 1986, include floor proximity emergency escape path marking which meets the requirements of §25.812(e) of this chapter in effect on November 26, 1984.

**(d)** *Emergency light operation.* Except for lights forming part of emergency lighting subsystems provided in compliance with §25.812(h) of this chapter (as prescribed in paragraph (h) of this section) that serve no more than one assist means, are independent of the airplane's main emergency lighting systems, and are automatically activated when the assist means is deployed, each light required by paragraphs (c) and (h) of this section must comply with the following:

(1) Each light must—

(i) Be operable manually both from the flightcrew station and, for airplanes on which a flight attendant is required, from a point in the passenger compartment that is readily accessible to a normal flight attendant seat;

(ii) Have a means to prevent inadvertent operation of the manual controls; and

(iii) When armed or turned on at either station, remain lighted or become lighted upon interruption of the airplane's normal electric power.

(2) Each light must be armed or turned on during taxiing, takeoff, and landing. In showing compliance with this paragraph a transverse vertical separation of the fuselage need not be considered.

(3) Each light must provide the required level of illumination for at least 10 minutes at the critical ambient conditions after emergency landing.

(4) Each light must have a cockpit control device that has an "on," "off," and "armed" position

**(e)** *Emergency exit operating handles.*

(1) For a passenger-carrying airplane for which the application for the type certificate was filed prior to May 1, 1972, the location of each passenger emergency exit operating handle, and instructions for opening the exit, must be shown by a marking on or near the exit that is readable from a distance of 30 inches. In addition, for each Type I and Type II emergency exit with a locking mechanism released by rotary motion of the handle, the instructions for opening must be shown by—

(i) A red arrow with a shaft at least three-fourths inch wide and a head twice the width of the shaft, extending along at least 70° of arc at a radius approximately equal to three-fourths of the handle length; and

(ii) The word "open" in red letters 1 inch high placed horizontally near the head of the arrow.

(2) For a passenger-carrying airplane for which the application for the type certificate was filed on or after May 1, 1972, the location of each passenger emergency exit operating handle and instructions for opening the exit must be shown in accordance with the requirements under which the airplane was type certificated. On these airplanes, no operating handle or operating handle cover may continue to be used if its luminescence (brightness) decreases to below 100 microlamberts.

**(f)** *Emergency exit access.* Access to emergency exits must be provided as follows for each passenger-carrying transport category airplane:

(1) Each passage way between individual passenger areas, or leading to a Type I or Type II emergency exit, must be unobstructed and at least 20 inches wide.

(2) There must be enough space next to each Type I or Type II emergency exit to allow a crewmember to assist in the evacuation of passengers without reducing the unobstructed width of the passageway below that required in paragraph (f)(1) of this section. However the Administrator may authorize deviation from this requirement for an airplane certificated under the provisions of part 4b of the Civil Air Regulations in effect before December 20, 1951, if he finds that special circumstances exist that provide an equivalent level of safety.

(3) There must be access from the main aisle to each Type III and Type IV exit. The access from the aisle to these exits must not be obstructed by seats, berths, or other protrusions in a manner that would reduce the effectiveness of the exit. In addition—

(i) For an airplane for which the application for the type certificate was filed prior to May 1, 1972, the access must meet the requirements of §25.813(c) of this chapter in effect on April 30, 1972; and

(ii) For an airplane for which the application for the type certificate was filed on or after May 1, 1972, the access must meet the emergency exit access requirements under which the airplane was type certificated; except that,

(iii) After December 3, 1992, the access for an airplane type certificated after January 1, 1958, must meet the requirements of §25.813(c) of this chapter, effective June 3, 1992.

(iv) Contrary provisions of this section notwithstanding, the Manager of the Transport Airplane Directorate, Aircraft Certification Service, Federal Aviation Administration, may authorize deviation from the requirements of paragraph (f)(3)(iii) of this section if it is determined that special circumstances make compliance impractical. Such special circumstances include, but are not limited to, the following conditions when they preclude achieving compliance with §25.813(c)(1)(i) or (ii) without a reduction in the total number of passenger seats; emergency exits located in close proximity to each other; fixed installations such as lavatories, galleys, etc.; permanently mounted bulkheads; an insufficient number of rows ahead of or behind the exit to enable compliance without a reduction in the seat row pitch of more than one inch; or an insufficient number of rows to enable compliance without a reduction in the seat row pitch to less than 30 inches. A request for such grant of deviation must include credible reasons as to why literal compliance with §25.813(c)(1)(i) or (ii) is impractical and a description of the steps taken to achieve a level of safety as close to that intended by §25.813(c)(1)(i) or (ii) as is practical.

(v) The Manager of the Transport Airplane Directorate, Aircraft Certification Service, Federal Aviation Administration, may also authorize a compliance date later than December 3, 1992, if it is determined that special circumstances make compliance by that date impractical. A request for such grant of deviation must outline the airplanes for which compliance will be achieved by December 3, 1992, and include a proposed schedule for incremental compliance of the remaining airplanes in the operator's fleet. In addition, the request must include credible reasons why compliance cannot be achieved earlier.

(4) If it is necessary to pass through a passageway between passenger compartments to reach any required emergency exit from any seat in the passenger cabin, the passageway must not be obstructed. However, curtains may be used if they allow free entry through the passageway.

(5) No door may be installed in any partition between passenger compartments.

(6) If it is necessary to pass through a doorway separating the passenger cabin from other areas to reach required emergency exit from any passenger seat, the door must have a means to latch it in open position, and the door must be latched open during each takeoff and landing. The latching means must be able to withstand the loads imposed upon it when the door is subjected to the ultimate inertia forces, relative to the surrounding structure, listed in §25.561(b) of this chapter.

**(g)** *Exterior exit markings.* Each passenger emergency exit and the means of opening that exit from the outside must be marked on the outside of the airplane. There must be a 2-inch colored band outlining each passenger emergency exit on the side of the fuselage. Each outside marking, including the band, must be readily distinguishable from the surrounding fuselage area by contrast in color. The markings must comply with the following:

(1) If the reflectance of the darker color is 15 percent or less, the reflectance of the lighter color must be at least 45 percent.

(2) If the reflectance of the darker color is greater than 15 percent, at least a 30 percent difference between its reflectance and the reflectance of the lighter color must be provided.

(3) Exits that are not in the side of the fuselage must have the external means of opening and applicable instructions marked conspicuously in red or, if red is inconspicuous against the background color, in bright chrome yellow and, when the opening means for such an exit is located on only one side of the fuselage, a conspicuous marking to that effect must be provided on the other side. *Reflectance* is the ratio of the luminous flux reflected by a body to the luminous flux it receives.

**(h)** *Exterior emergency lighting and escape route.*

(1) Except for nontransport category airplanes certificated after December 31, 1964, each passenger-carrying airplane must be equipped with exterior lighting that meets the following requirements:

(i) For an airplane for which the application for the type certificate was filed prior to May 1, 1972, the requirements of §25.812(f) and (g) of this chapter in effect on April 30, 1972.

(ii) For an airplane for which the application for the type certificate was filed on or after May 1, 1972, the exterior emergency lighting requirements under which the airplane was type certificated.

(2) Each passenger-carrying airplane must be equipped with a slip-resistant escape route that meets the following requirements:

(i) For an airplane for which the application for the type certificate was filed prior to May 1, 1972, the requirements of §25.803(e) of this chapter in effect on April 30, 1972.

(ii) For an airplane for which the application for the type certificate was filed on or after May 1, 1972, the slip-resistant escape route requirements under which the airplane was type certificated.

**(i)** *Floor level exits.* Each floor level door or exit in the side of the fuselage (other than those leading into a cargo or baggage compartment that is not accessible from the passenger cabin) that is 44 or more inches high and 20 or more inches wide, but not wider than 46 inches, each passenger ventral exit (except the ventral exits on M-404 and CV-240 airplanes), and each tail cone exit, must meet the requirements of this section for floor level emergency exits. However, the Administrator may grant a deviation from this paragraph if he finds that circumstances make full compliance impractical and that an acceptable level of safety has been achieved.

**(j)** *Additional emergency exits.* Approved emergency exits in the passenger compartments that are in excess of the minimum number of required emergency exits must meet all of the applicable provisions of this section except paragraphs (f)(1), (2), and (3) of this section and must be readily accessible.

**(k)** On each large passenger-carrying turbojet-powered airplane, each ventral exit and tailcone exit must be—

(1) Designed and constructed so that it cannot be opened during flight; and

(2) Marked with a placard readable from a distance of 30 inches and installed at a conspicuous location near the means of opening the exit, stating that the exit has been designed and constructed so that it cannot be opened during flight.

**(l)** *Portable lights.* No person may operate a passenger-carrying airplane unless it is equipped with flashlight stowage provisions accessible from each flight attendant seat.

**(m)** Except for an airplane used in operations under this part on October 16, 1987, and having an emergency exit configuration installed and authorized for operation prior to October 16, 1987, for an airplane that is required to have more than one passenger emergency exit for each side of the fuselage, no passenger emergency exit shall be more than 60 feet from any adjacent passenger emergency exit on the same side of the same deck of the fuselage, as measured parallel to the airplane's longitudinal axis between the nearest exit edges.

[Docket No. 2033, 30 FR 3205, March 9, 1965; as amended by Amdt. 121–251, 60 FR 65930, Dec. 20, 1995; Amdt. 121–262, 62 FR 13256, March 19, 1997; Amdt. 121–283, 66 FR 20740, April 25, 2001]

## §121.311 Seats, safety belts, and shoulder harnesses.

**(a)** No person may operate an airplane unless there are available during the takeoff, en route flight, and landing—

(1) An approved seat or berth for each person on board the airplane who has reached his second birthday; and

(2) An approved safety belt for separate use by each person on board the airplane who has reached his second birthday, except that two persons occupying a berth may share one approved safety belt and two persons occupying a multiple lounge or divan seat may share one approved safety belt during en route flight only.

**(b)** Except as provided in this paragraph, each person on board an airplane operated under this part shall occupy an approved seat or berth with a separate safety belt properly secured about him or her during movement on the surface, takeoff, and landing. A safety belt provided for the occupant of a seat may not be used by more than one person who has reached his or her second birthday. Notwithstanding the preceding requirements, a child may:

(1) Be held by an adult who is occupying an approved seat or berth, provided the child has not reached his or her second birthday and the child does not occupy or use any restraining device; or

(2) Notwithstanding any other requirement of this chapter, occupy an approved child restraint system furnished by the certificate holder or one of the persons described in paragraph (b)(2)(i) of this section, provided:

(i) The child is accompanied by a parent, guardian, or attendant designated by the child's parent or guardian to attend to the safety of the child during the flight;

(ii) Except as provided in paragraph (b)(2)(ii)(D) of this section, the approved child restraint system bears one or more labels as follows:

(A) Seats manufactured to U.S. standards between January 1, 1981 and February 25, 1985, must bear the label: "This child restraint system conforms to all applicable Federal motor vehicle safety standards.";

(B) Seats manufactured to U.S. standards on or after February 26, 1985, must bear two labels:

*(1)* "This child restraint system conforms to all applicable Federal motor vehicle safety standards;" and

*(2)* "THIS RESTRAINT IS CERTIFIED FOR USE IN MOTOR VEHICLES AND AIRCRAFT" in red lettering;

(C) Seats that do not qualify under paragraphs (b)(2)(ii)(A) and b)(2)(ii)(B) of this section must bear either a label showing approval of a foreign government or a label showing that the seat was manufactured under the standards of the United Nations;

(D) Notwithstanding any other provisions of this section, booster-type child restraint systems (as defined in Federal Motor Vehicle Standard No. 213 (49 CFR 571.213)), vest- and harness-type child restraint systems, and lap held child restraints are not approved for use in aircraft; and

(iii) The certificate holder complies with the following requirements:

(A) The restraint system must be properly secured to an approved forward-facing seat or berth;

(B) The child must be properly secured in the restraint system and must not exceed the specified weight limit for the restraint system; and

(C) The restraint system must bear the appropriate label(s).

**(c)** Except as provided in paragraph (c)(3) of this section, the following prohibitions apply to certificate holders:

(1) No certificate holder may permit a child, in an aircraft, to occupy a booster-type child restraint system, a vest-type child restraint system, a harness-type child restraint system, or a lap held child restraint system during take off, landing, and movement on the surface.

(2) Except as required in paragraph (c)(1) of this section, no certificate holder may prohibit a

child, if requested by the child's parent, guardian, or designated attendant, from occupying a child restraint system furnished by the child's parent, guardian, or designated attendant provided—

(i) The child holds a ticket for an approved seat or berth or such seat or berth is otherwise made available by the certificate holder for the child's use;

(ii) The requirements of paragraph (b)(2)(i) of this section are met;

(iii) The requirements of paragraph (b)(2)(iii) of this section are met; and

(iv) The child restraint system has one or more of the labels described in paragraphs (b)(2)(ii)(A) through (b)(2)(ii)(C) of this section.

(3) This section does not prohibit the certificate holder from providing child restraint systems authorized by this section or, consistent with safe operating practices, determining the most appropriate passenger seat location for the child restraint system.

**(d)** Each sideward facing seat must comply with the applicable requirements of §25.785(c) of this chapter.

**(e)** Except as provided in paragraphs (e)(1) through (e)(3) of this section, no certificate holder may take off or land an airplane unless each passenger seat back is in the upright position. Each passenger shall comply with instructions given by a crewmember in compliance with this paragraph.

(1) This paragraph does not apply to seat backs placed in other than the upright position in compliance with §121.310(f)(3).

(2) This paragraph does not apply to seats on which cargo or persons who are unable to sit erect for a medical reason are carried in accordance with procedures in the certificate holder's manual if the seat back does not obstruct any passenger's access to the aisle or to any emergency exit.

(3) On airplanes with no flight attendant, the certificate holder may take off or land as long as the flightcrew instructs each passenger to place his or her seat back in the upright position for takeoff and landing.

**(f)** No person may operate a transport category airplane that was type certificated after January 1, 1958, or a nontransport category airplane manufactured after March 20, 1997, unless it is equipped at each flight deck station with a combined safety belt and shoulder harness that meets the applicable requirements specified in §25.785 of this chapter, effective March 6, 1980, except that—

(1) Shoulder harnesses and combined safety belt and shoulder harnesses that were approved and installed before March 6, 1980, may continue to be used; and

(2) Safety belt and shoulder harness restraint systems may be designed to the inertia load factors established under the certification basis of the airplane.

**(g)** Each flight attendant must have a seat for takeoff and landing in the passenger compartment that meets the requirements of §25.785 of this chapter, effective March 6, 1980, except that—

(1) Combined safety belt and shoulder harnesses that were approved and installed before March, 6, 1980, may continue to be used; and

(2) Safety belt and shoulder harness restraint systems may be designed to the inertia load factors established under the certification basis of the airplane.

(3) The requirements of §25.785(h) do not apply to passenger seats occupied by flight attendants not required by §121.391.

**(h)** Each occupant of a seat equipped with a shoulder harness or with a combined safety belt and shoulder harness must have the shoulder harness or combined safety belt and shoulder harness properly secured about that occupant during takeoff and landing, except that a shoulder harness that is not combined with a safety belt may be unfastened if the occupant cannot perform the required duties with the shoulder harness fastened.

**(i)** At each unoccupied seat, the safety belt and shoulder harness, if installed, must be secured so as not to interfere with crewmembers in the performance of their duties or with the rapid egress of occupants in an emergency.

[Docket No. 7522, 32 FR 13267, Sept. 20, 1967; as amended by Amdt. 121–41, 33 FR 9067, June 20, 1968; Amdt. 121–75, 36 FR 12512, July 1, 1971; Amdt. 121–133, 42 FR 18394, April 7, 1977; Amdt. 121–155, 45 FR 7756, Feb. 4, 1980; Amdt. 121–170, 46 FR 15482, March 5, 1981; Amdt. 121–177, 47 FR 10516, March 11, 1982; Amdt. 121–230, 57 FR 42673, Sept. 15, 1992; Amdt. 121–251, 60 FR 65930, Dec. 20, 1995; Amdt. 121–255, 61 FR 28421, June 4, 1996]

## §121.312 Materials for compartment interiors.

**(a)** All interior materials; transport category airplanes and nontransport category airplanes type certificated before January 1, 1965. Except for the materials covered by paragraph (b) of this section, all materials in each compartment of a transport category airplane, or a nontransport category airplane type certificated before January 1, 1965, used by the crewmembers and passengers, must meet the requirements of §25.853 of this chapter in effect as follows, or later amendment thereto:

(1) Airplane with passenger seating capacity of 20 or more.

(i) Manufactured after August 19, 1988, but prior to August 20, 1990. Except as provided in paragraph (a)(3)(ii) of this section, each airplane with a passenger capacity of 20 or more and manufactured after August 19, 1988, but prior to August 20, 1990, must comply with the heat release rate testing provisions of §25.853(d) in effect March 6, 1995 (formerly §25.853(a-1) in effect on August 20, 1986) (see Appendix L of this part), except that the total heat release over the first 2 minutes of sample exposure must not exceed 100 kilowatt minutes per square meter and the peak heat release rate must not exceed 100 kilowatts per square meter.

(ii) Manufactured after August 19, 1990. Each airplane with a passenger capacity of 20 or more and manufactured after August 19, 1990, must comply with the heat release rate and smoke testing provisions of §25.853(d) in effect March 6, 1995 (formerly §25.853(a-1) (see Appendix L of this part) in effect on September 26, 1988).

(2) Substantially complete replacement of the cabin interior on or after May 1, 1972.

(i) Airplane for which the application for type certificate was filed prior to May 1, 1972. Except as provided in paragraph (a)(3)(i) or (a)(3)(ii) of this section, each airplane for which the application for type certificate was filed prior to May 1, 1972, must comply with the provisions of §25.853 in effect on April 30, 1972, regardless of passenger capacity, if there is a substantially complete replacement of the cabin interior after April 30, 1972.

(ii) Airplane for which the application for type certificate was filed on or after May 1, 1972. Except as provided in paragraph (a)(3)(i) or (a)(3)(ii) of this section, each airplane for which the application for type certificate was filed on or after May 1, 1972, must comply with the material requirements under which the airplane was type certificated, regardless of passenger capacity, if there is a substantially complete replacement of the cabin interior on or after that date.

(3) Airplane type certificated after January 1, 1958, with passenger capacity of 20 or more.

(i) Substantially complete replacement of the cabin interior on or after March 6, 1995. Except as provided in paragraph (a)(3)(ii) of this section, each airplane that was type certificated after January 1, 1958, and has a passenger capacity of 20 or more, must comply with the heat release rate testing provisions of §25.853(d) in effect March 6, 1995 (formerly §25.853(a-1) in effect on August 20, 1986) (see Appendix L of this part), if there is a substantially complete replacement of the cabin interior components identified in §25.853(d), on or after that date, except that the total heat release over the first 2 minutes of sample exposure shall not exceed 100 kilowatt-minutes per square meter and the peak heat release rate must not exceed 100 kilowatts per square meter.

(ii) Substantially complete replacement of the cabin interior on or after August 20, 1990. Each airplane that was type certificated after January 1, 1958, and has a passenger capacity of 20 or more, must comply with the heat release rate and smoke testing provisions of §25.853(d) in effect March 6, 1995 (formerly §25.853(a-1) in effect on September 26, 1988) (see Appendix L of this part), if there is a substantially complete replacement of the cabin interior components identified in §25.853(d), on or after August 20, 1990.

(4) Contrary provisions of this section notwithstanding, the Manager of the Transport Airplane Directorate, Aircraft Certification Service, Federal Aviation Administration, may authorize deviation from the requirements of paragraph (a)(1)(i), (a)(1)(ii), (a)(3)(i), or (a)(3)(ii) of this section for specific components of the cabin interior that do not meet applicable flammability and smoke emission requirements, if the determination is made that special circumstances exist that make compliance impractical. Such grants of deviation will be limited to those airplanes manufactured within 1 year after the applicable date specified in this section and those airplanes in which the interior is replaced within 1 year of that date. A request for such grant of deviation must include a thorough and accurate analysis of each component subject to §25.853(a-1), the steps being taken to achieve compliance, and, for the few components for which timely compliance will not be achieved, credible reasons for such noncompliance.

(5) Contrary provisions of this section notwithstanding, galley carts and galley standard containers that do not meet the flammability and smoke emission requirements of §25.853(d) in effect March 6, 1995 (formerly §25.853(a-1)) (see Appendix L of this part) may be used in airplanes that must meet the requirements of paragraphs (a)(1)(i), (a)(1)(ii), (a)(3)(i), or (a)(3)(ii) of this section, provided the galley carts or standard containers were manufactured prior to March 6, 1995.

**(b)** *Seat cushions.* Seat cushions, except those on flight crewmember seats, in each compartment occupied by crew or passengers, must comply with the requirements pertaining to seat cushions in §25.853(c) effective on November 26, 1984, on each airplane as follows:

(1) Each transport category airplane type certificated after January 1, 1958; and

(2) On or after December 20, 2010, each non-transport category airplane type certificated after December 31, 1964.

**(c)** All interior materials; airplanes type certificated in accordance with SFAR No. 41 of 14 CFR part 21. No person may operate an airplane that conforms to an amended or supplemental type certificate issued in accordance with SFAR No. 41 of 14 CFR part 21 for a maximum certificated takeoff weight in excess of 12,500 pounds unless the airplane meets the compartment interior requirements set forth in §25.853(a) in effect March

6, 1995 (formerly §25.853(a), (b), (b-1), (b-2), and (b-3) of this chapter in effect on September 26, 1978)(see Appendix L of this part).

**(d)** All interior materials; other airplanes. For each material or seat cushion to which a requirement in paragraphs (a), (b), or (c) of this section does not apply, the material and seat cushion in each compartment used by the crewmembers and passengers must meet the applicable requirement under which the airplane was type certificated.

[Docket No. 28154, 60 FR 65930, Dec. 20, 1995]

## §121.313 Miscellaneous equipment.

No person may conduct any operation unless the following equipment is installed in the airplane:

**(a)** If protective fuses are installed on an airplane, the number of spare fuses approved for that airplane and appropriately described in the certificate holder's manual.

**(b)** A windshield wiper or equivalent for each pilot station.

**(c)** A power supply and distribution system that meets the requirements of §§25.1309, 25.1331, 25.1351(a) and (b)(1) through (4), 25.1353, 25.1355, and 25.1431(b) or that is able to produce and distribute the load for the required instruments and equipment, with use of an external power supply if any one power source or component of the power distribution system fails. The use of common elements in the system may be approved if the Administrator finds that they are designed to be reasonably protected against malfunctioning. Engine-driven sources of energy, when used, must be on separate engines.

**(d)** A means for indicating the adequacy of the power being supplied to required flight instruments.

**(e)** Two independent static pressure systems, vented to the outside atmospheric pressure so that they will be least affected by air flow variation or moisture or other foreign matter, and installed so as to be airtight except for the vent. When a means is provided for transferring an instrument from its primary operating system to an alternate system, the means must include a positive positioning control and must be marked to indicate clearly which system is being used.

**(f)** A door between the passenger and pilot compartments (i.e., flightdeck door), with a locking means to prevent passengers from opening it without the pilot's permission, except that non-transport category airplanes certificated after December 31, 1964, are not required to comply with this paragraph. For airplanes equipped with a crew rest area having separate entries from the flightdeck and the passenger compartment, a door with such a locking means must be provided between the crew rest area and the passenger compartment.

**(g)** A key for each door that separates a passenger compartment from another compartment that has emergency exit provisions. Except for flightdeck doors, a key must be readily available for each crewmember. Except as provided below, no person other than a person who is assigned to perform duty on the flightdeck may have a key to the flightdeck door. Before April 22, 2003, any crewmember may have a key to the flightdeck door but only if the flightdeck door has an internal flightdeck locking device installed, operative, and in use. Such "internal flightdeck locking device" has to be designed so that it can only be unlocked from inside the flightdeck.

**(h)** A placard on each door that is the means of access to a required passenger emergency exit, to indicate that it must be open during takeoff and landing.

**(i)** A means for the crew, in an emergency to unlock each door that leads to a compartment that is normally accessible to passengers and that can be locked by passengers.

**(j)** After April 9, 2003, for airplanes required by paragraph (f) of this section to have a door between the passenger and pilot or crew rest compartments, and for transport category, all-cargo airplanes that have a door installed between the pilot compartment and any other occupied compartment on January 15, 2002;

(1) Each such door must meet the requirements of §§25.795 (a)(1) and (2) in effect on January 15, 2002; and

(2) Each operator must establish methods to enable a flight attendant to enter the pilot compartment in the event that a flightcrew member becomes incapacitated. Any associated signal or confirmation system must be operable by each flightcrew member from that flightcrew member's duty station.

[Docket No. 6258, 29 FR 19205, Dec. 31, 1964; as amended by Amdt. 121–5, 30 FR 6113, April 30, 1965; Amdt. 121–251, 60 FR 65931, Dec. 20, 1995; Amdt. 121–288, 67 FR 2127, Jan. 15, 2002]

## §121.314 Cargo and baggage compartments.

For each transport category airplane type certificated after January 1, 1958:

**(a)** Each Class C or Class D compartment, as defined in §25.857 of this Chapter in effect on June 16, 1986 (see Appendix L to this part), that is greater than 200 cubic feet in volume must have ceiling and sidewall liner panels which are constructed of:

(1) Glass fiber reinforced resin;

(2) Materials which meet the test requirements of part 25, Appendix F, part III of this chapter; or

(3) In the case of liner installations approved prior to March 20, 1989, aluminum.

**(b)** For compliance with paragraph (a) of this section, the term "liner" includes any design feature, such as a joint or fastener, which would affect the capability of the liner to safely contain a fire.

**(c)** After March 19, 2001, each Class D compartment, regardless of volume, must meet the standards of §§25.857(c) and 25.858 of this Chapter for a Class C compartment unless the operation is an all-cargo operation in which case each Class D compartment may meet the standards in §25.857(e) for a Class E compartment.

**(d)** Reports of conversions and retrofits.

(1) Until such time as all Class D compartments in aircraft operated under this part by the certificate have been converted or retrofitted with appropriate detection and suppression systems, each certificate holder must submit written progress reports to the FAA that contain the information specified below.

(i) The serial number of each airplane listed in the operations specifications issued to the certificate holder for operation under this part in which all Class D compartments have been converted to Class C or Class E compartments;

(ii) The serial number of each airplane listed in the operations specification issued to the certificate holder for operation under this part, in which all Class D compartments have been retrofitted to meet the fire detection and suppression requirements for Class C or the fire detection requirements for Class E; and

(iii) The serial number of each airplane listed in the operations specifications issued to the certificate holder for operation under this part that has at least one Class D compartment that has not been converted or retrofitted.

(2) The written report must be submitted to the Certificate Holding District Office by July 1, 1998, and at each three-month interval thereafter.

[Docket No. 28937, 63 FR 8049, Feb. 17, 1998]

## §121.315 Cockpit check procedure.

**(a)** Each certificate holder shall provide an approved cockpit check procedure for each type of aircraft.

**(b)** The approved procedures must include each item necessary for flight crewmembers to check for safety before starting engines, taking off, or landing, and in engine and systems emergencies. The procedures must be designed so that a flight crewmember will not need to rely upon his memory for items to be checked.

**(c)** The approved procedures must be readily usable in the cockpit of each aircraft and the flight crew shall follow them when operating the aircraft.

## §121.316 Fuel tanks.

Each turbine powered transport category airplane operated after October 30, 1991, must meet the requirements of §25.963(e) of this chapter in effect on October 30, 1989.

[Docket No. 25614, 54 FR 40354, Sept. 29, 1989]

## §121.317 Passenger information requirements, smoking prohibitions, and additional seat belt requirements.

**(a)** Except as provided in paragraph (l) of this section, no person may operate an airplane unless it is equipped with passenger information signs that meet the requirements of §25.791 of this chapter. Except as provided in paragraph (l) of this section, the signs must be constructed so that the crewmembers can turn them on and off.

**(b)** Except as provided in paragraph (l) of this section, the "Fasten Seat Belt" sign shall be turned on during any movement on the surface, for each takeoff, for each landing, and at any other time considered necessary by the pilot in command.

**(c)** No person may operate an airplane on a flight on which smoking is prohibited by part 252 of this title unless either the "No Smoking" passenger information signs are lighted during the entire flight, or one or more "No Smoking" placards meeting the requirements of §25.1541 of this chapter are posted during the entire flight segment. If both the lighted signs and the placards are used, the signs must remain lighted during the entire flight segment.

**(d)** No person may operate a passenger-carrying airplane under this part unless at least one legible sign or placard that reads "Fasten Seat Belt While Seated" is visible from each passenger seat. These signs or placards need not meet the requirements of paragraph (a) of this section.

**(e)** No person may operate an airplane unless there is installed in each lavatory a sign or placard that reads: "Federal law provides for a penalty of up to $2,000 for tampering with the smoke detector installed in this lavatory." These signs or placards need not meet the requirements of paragraph (a) of this section.

**(f)** Each passenger required by §121.311(b) to occupy a seat or berth shall fasten his or her safety belt about him or her and keep it fastened while the "Fasten Seat Belt" sign is lighted.

**(g)** No person may smoke while a "No Smoking" sign is lighted or while "No Smoking" placards are posted, except as follows:

(1) *Supplemental operations.* The pilot in command of an airplane engaged in a supplemental operation may authorize smoking on the flight deck (if it is physically separated from any passenger compartment), but not in any of the following situations:

(i) During airplane movement on the surface or during takeoff or landing;

(ii) During scheduled passenger-carrying public charter operations conducted under part 380 of this title; or

(iii) During any operation where smoking is prohibited by part 252 of this title or by international agreement.

(2) *Certain intrastate domestic operations.* Except during airplane movement on the surface or during takeoff or landing, a pilot in command of an airplane engaged in a domestic operation may authorize smoking on the flight deck (if it is physically separated from the passenger compartment) if—

(i) Smoking on the flight deck is not otherwise prohibited by part 252 of this title;

(ii) The flight is conducted entirely within the same State of the United States (a flight from one place in Hawaii to another place in Hawaii through the airspace over a place outside of Hawaii is not entirely within the same State); and

(iii) The airplane is either not turbojet-powered or the airplane is not capable of carrying at least 30 passengers.

**(h)** No person may smoke in any airplane lavatory.

**(i)** No person may tamper with, disable, or destroy any smoke detector installed in any airplane lavatory.

**(j)** On flight segments other than those described in paragraph (c) of this section, the "No Smoking" sign must be turned on during any movement on the surface, for each takeoff, for each landing, and at any other time considered necessary by the pilot in command.

**(k)** Each passenger shall comply with instructions given him or her by a crewmember regarding compliance with paragraphs (f), (g), (h), and (l) of this section.

**(l)** A certificate holder may operate a nontransport category airplane type certificated after December 31, 1964, that is manufactured before December 20, 1997, if it is equipped with at least one placard that is legible to each person seated in the cabin that states "Fasten Seat Belt," and if, during any movement on the surface, for each takeoff, for each landing, and at any other time considered necessary by the pilot in command, a crewmember orally instructs the passengers to fasten their seat belts.

[Docket No. 25590, 53 FR 12361, April 13, 1988; as amended by Amdt. 121–196, 53 FR 44182, Nov 2, 1988; Amdt. 121–213, 55 FR 8367, March 7, 1990; Amdt. 121–230, 57 FR 42673, Sept. 15, 1992; Amdt. 121–251, 60 FR 65931, Dec. 20, 1995; Amdt. 121–256, 61 FR 30434, June 14, 1996; Amdt. 121–277, 65 FR 36779, June 9, 2000]

## §121.318 Public address system.

No person may operate an airplane with a seating capacity of more than 19 passengers unless it is equipped with a public address system which—

**(a)** Is capable of operation independent of the crewmember interphone system required by §121.319, except for handsets, headsets, microphones, selector switches, and signaling devices;

**(b)** Is approved in accordance with §21.305 of this chapter;

**(c)** Is accessible for immediate use from each of two flight crewmember stations in the pilot compartment;

**(d)** For each required floor-level passenger emergency exit which has an adjacent flight attendant seat, has a microphone which is readily accessible to the seated flight attendant, except that one microphone may serve more than one exit, provided the proximity of the exits allows unassisted verbal communication between seated flight attendants;

**(e)** Is capable of operation within 10 seconds by a flight attendant at each of those stations in the passenger compartment from which its use is accessible;

**(f)** Is audible at all passenger seats, lavatories, and flight attendant seats and work stations; and

**(g)** For transport category airplanes manufactured on or after November 27, 1990, meets the requirements of §25.1423 of this chapter.

[Docket No. 24995, 54 FR 43926, Oct. 27, 1989]

## §121.319 Crewmember interphone system.

**(a)** No person may operate an airplane with a seating capacity of more than 19 passengers unless the airplane is equipped with a crewmember interphone system that:

(1) [Reserved]

(2) Is capable of operation independent of the public address system required by §121.318(a) except for handsets, headsets, microphones, selector switches, and signaling devices; and

(3) Meets the requirements of paragraph (b) of this section.

**(b)** The crewmember interphone system required by paragraph (a) of this section must be approved in accordance with §21.305 of this chapter and meet the following requirements:

(1) It must provide a means of two-way communication between the pilot compartment and—

(i) Each passenger compartment; and

(ii) Each galley located on other than the main passenger deck level.

(2) It must be accessible for immediate use from each of two flight crewmember stations in the pilot compartment;

(3) It must be accessible for use from at least one normal flight attendant station in each passenger compartment;

(4) It must be capable of operation within 10 seconds by a flight attendant at those stations in each passenger compartment from which its use is accessible; and

(5) For large turbojet-powered airplanes:

(i) It must be accessible for use at enough flight attendant stations so that all floor-level emergency exits (or entryways to those exits in the case of exits located within galleys) in each passenger compartment are observable from one or more of those stations so equipped;

(ii) It must have an alerting system incorporating aural or visual signals for use by flight crewmembers to alert flight attendants and for use by flight attendants to alert flight crewmembers;

(iii) The alerting system required by paragraph (b)(5)(ii) of this section must have a means for the recipient of a call to determine whether it is a normal call or an emergency call; and

(iv) When the airplane is on the ground, it must provide a means of two-way communication between ground personnel and either of at least two flight crewmembers in the pilot compartment. The interphone system station for use by ground personnel must be so located that personnel using the system may avoid visible detection from within the airplane.

[Docket No. 10865, 38 FR 21494, Aug. 9, 1973; as amended by Amdt. 121–121, 40 FR 42186, Sept. 11, 1975; Amdt. 121–149, 43 FR 50602, Oct. 30, 1978; Amdt. 121–178, 47 FR 13316, March 29, 1982; Amdt. 121–253, 61 FR 611, Jan. 26, 1996]

## §121.321 [Reserved]

## §121.323 Instruments and equipment for operations at night.

No person may operate an airplane at night under this part unless it is equipped with the following instruments and equipment in addition to those required by §§121.305 through 121.321 and 121.803:

**(a)** Position lights.

**(b)** An anti-collision light.

**(c)** Two landing lights, except that only one landing light is required for nontransport category airplanes type certificated after December 31, 1964.

**(d)** Instrument lights providing enough light to make each required instrument, switch, or similar instrument, easily readable and installed so that the direct rays are shielded from the flight crewmembers' eyes and that no objectionable reflections are visible to them. There must be a means of controlling the intensity of illumination unless it is shown that nondimming instrument lights are satisfactory.

**(e)** An airspeed-indicating system with heated pitot tube or equivalent means for preventing malfunctioning due to icing.

**(f)** A sensitive altimeter.

[Docket No. 6258, 29 FR 19205, Dec. 31, 1964; as amended by Amdt. 121–251, 60 FR 65932, Dec. 20, 1995; Amdt. 121–281, 66 FR 19043, April 12, 2001]

## §121.325 Instruments and equipment for operations under IFR or over-the-top.

No person may operate an airplane under IFR or over-the-top conditions under this part unless it is equipped with the following instruments and equipment, in addition to those required by §§121.305 through 121.321 and 121.803:

**(a)** An airspeed indicating system with heated pitot tube or equivalent means for preventing malfunctioning due to icing.

**(b)** A sensitive altimeter.

**(c)** Instrument lights providing enough light to make each required instrument, switch, or similar instrument, easily readable and so installed that the direct rays are shielded from the flight crewmembers' eyes and that no objectionable reflections are visible to them, and a means of controlling the intensity of illumination unless it is shown that nondimming instrument lights are satisfactory.

[Docket No. 6258, 29 FR 19205, Dec. 31, 1964; as amended by Amdt. 121–281, 66 FR 19043, April 12, 2001]

## §121.327 Supplemental oxygen: Reciprocating engine powered airplanes.

**(a)** *General.* Except where supplemental oxygen is provided in accordance with §121.331, no person may operate an airplane unless supplemental oxygen is furnished and used as set forth in paragraphs (b) and (c) of this section. The amount of supplemental oxygen required for a particular operation is determined on the basis of flight altitudes and flight duration, consistent with the operation procedures established for each operation and route.

**(b)** *Crewmembers.*

(1) At cabin pressure altitudes above 10,000 feet up to and including 12,000 feet, oxygen must be provided for, and used by, each member of the flight crew on flight deck duty, and must be provided for other crewmembers, for that part of the flight at those altitudes that is of more than 30 minutes duration.

(2) At cabin pressure altitudes above 12,000 feet, oxygen must be provided for, and used by, each member of the flight crew on flight deck duty,

and must be provided for other crewmembers, during the entire flight time at those altitudes.

(3) When a flight crewmember is required to use oxygen, he must use it continuously, except when necessary to remove the oxygen mask or other dispenser in connection with his regular duties. Standby crewmembers who are on call or are definitely going to have flight deck duty before completing the flight must be provided with an amount of supplemental oxygen equal to that provided for crewmembers on duty other than on flight deck duty. If a standby crewmember is not on call and will not be on flight deck duty during the remainder of the flight, he is considered to be a passenger for the purposes of supplemental oxygen requirements.

**(c)** *Passengers.* Each certificate holder shall provide a supply of oxygen, approved for passenger safety, in accordance with the following:

(1) For flights of more than 30 minutes duration at cabin pressure altitudes above 8,000 feet up to and including 14,000 feet, enough oxygen for 30 minutes for 10 percent of the passengers.

(2) For flights at cabin pressure altitudes above 14,000 feet up to and including 15,000 feet, enough oxygen for that part of the flight at those altitudes for 30 percent of the passengers.

(3) For flights at cabin pressure altitudes above 15,000 feet, enough oxygen for each passenger carried during the entire flight at those altitudes.

**(d)** For the purposes of this subpart *cabin pressure altitude* means the pressure altitude corresponding with the pressure in the cabin of the airplane, and *flight altitude* means the altitude above sea level at which the airplane is operated. For airplanes without pressurized cabins, "cabin pressure altitude" and "flight altitude" mean the same thing.

## §121.329 Supplemental oxygen for sustenance: Turbine engine powered airplanes.

**(a)** *General.* When operating a turbine engine powered airplane, each certificate holder shall equip the airplane with sustaining oxygen and dispensing equipment for use as set forth in this section:

(1) The amount of oxygen provided must be at least the quantity necessary to comply with paragraphs (b) and (c) of this section.

(2) The amount of sustaining and first-aid oxygen required for a particular operation to comply with the rules in this part is determined on the basis of cabin pressure altitudes and flight duration, consistent with the operating procedures established for each operation and route.

(3) The requirements for airplanes with pressurized cabins are determined on the basis of cabin pressure altitude and the assumption that a cabin pressurization failure will occur at the altitude or point of flight that is most critical from the standpoint of oxygen need, and that after the failure the airplane will descend in accordance with the emergency procedures specified in the Airplane Flight Manual, without exceeding its operating limitations, to a flight altitude that will allow successful termination of the flight.

(4) Following the failure, the cabin pressure altitude is considered to be the same as the flight altitude unless it is shown that no probable failure of the cabin or pressurization equipment will result in a cabin pressure altitude equal to the flight altitude. Under those circumstances, the maximum cabin pressure altitude attained may be used as a basis for certification or determination of oxygen supply, or both.

**(b)** *Crewmembers.* Each certificate holder shall provide a supply of oxygen for crewmembers in accordance with the following:

(1) At cabin pressure altitudes above 10,000 feet, up to and including 12,000 feet, oxygen must be provided for and used by each member of the flight crew on flight deck duty and must be provided for other crewmembers for that part of the flight at those altitudes that is of more than 30 minutes duration.

(2) At cabin pressure altitudes above 12,000 feet, oxygen must be provided for, and used by, each member of the flight crew on flight deck duty, and must be provided for other crewmembers during the entire flight at those altitudes.

(3) When a flight crewmember is required to use oxygen, he must use it continuously except when necessary to remove the oxygen mask or other dispenser in connection with his regular duties. Standby crewmembers who are on call or are definitely going to have flight deck duty before completing the flight must be provided with an amount of supplemental oxygen equal to that provided for crewmembers on duty other than on flight duty. If a standby crewmember is not on call and will not be on flight deck duty during the remainder of the flight, he is considered to be a passenger for the purposes of supplemental oxygen requirements.

**(c)** *Passengers.* Each certificate holder shall provide a supply of oxygen for passengers in accordance with the following:

(1) For flights at cabin pressure altitudes above 10,000 feet, up to and including 14,000 feet, enough oxygen for that part of the flight at those altitudes that is of more than 30 minutes duration, for 10 percent of the passengers.

(2) For flights at cabin pressure altitudes above 14,000 feet, up to and including 15,000 feet, enough oxygen for that part of the flight at those altitudes for 30 percent of the passengers.

(3) For flights at cabin pressure altitudes above 15,000 feet, enough oxygen for each passenger carried during the entire flight at those altitudes.

## §121.331 Supplemental oxygen requirements for pressurized cabin airplanes: Reciprocating engine powered airplanes.

**(a)** When operating a reciprocating engine powered airplane pressurized cabin, each certificate holder shall equip the airplane to comply with paragraphs (b) through (d) of this section in the event of cabin pressurization failure.

**(b)** *For crewmembers.* When operating at flight altitudes above 10,000 feet, the certificate holder shall provide enough oxygen for each crewmember for the entire flight at those altitudes and not less than a two-hour supply for each flight crewmember on flight deck duty. The required two hours supply is that quantity of oxygen necessary for a constant rate of descent from the airplane's maximum certificated operating altitude to 10,000 feet in ten minutes and followed by 110 minutes at 10,000 feet. The oxygen required by §121.337 may be considered in determining the supplemental breathing supply required for flight crewmembers on flight deck duty in the event of cabin pressurization failure.

**(c)** *For passengers.* When operating at flight altitudes above 8,000 feet, the certificate holder shall provide oxygen as follows:

(1) When an airplane is not flown at a flight altitude above flight level 250, enough oxygen for 30 minutes for 10 percent of the passengers, if at any point along the route to be flown the airplane can safely descend to a flight altitude of 14,000 feet or less within four minutes.

(2) If the airplane cannot descend to a flight altitude of 14,000 feet or less within four minutes, the following supply of oxygen must be provided:

(i) For that part of the flight that is more than four minutes duration at flight altitudes above 15,000 feet, the supply required by §121.327(c)(3).

(ii) For that part of the flight at flight altitudes above 14,000 feet, up to and including 15,000 feet, the supply required by §121.327(c)(2).

(iii) For flight at flight altitudes above 8,000 feet up to and including 14,000 feet, enough oxygen for 30 minutes for 10 percent of the passengers.

(3) When an airplane is flown at a flight altitude above flight level 250, enough oxygen for 30 minutes for 10 percent of the passengers for the entire flight (including emergency descent) above 8,000 feet, up to and including 14,000 feet, and to comply with §121.327(c)(2) and (3) for flight above 14,000 feet.

**(d)** For the purposes of this section it is assumed that the cabin pressurization failure occurs at a time during flight that is critical from the standpoint of oxygen need and that after the failure the airplane will descend, without exceeding its normal operating limitations, to flight altitudes allowing safe flight with respect to terrain clearance.

[Docket No. 6258, 29 FR 19205, Dec. 31, 1964; as amended by Amdt. 121–132, 41 FR 55475, Dec. 20, 1976]

## §121.333 Supplemental oxygen for emergency descent and for first aid; turbine engine powered airplanes with pressurized cabins.

**(a)** *General.* When operating a turbine engine powered airplane with a pressurized cabin, the certificate holder shall furnish oxygen and dispensing equipment to comply with paragraphs (b) through (e) of this section in the event of cabin pressurization failure.

**(b)** *Crewmembers.* When operating at flight altitudes above 10,000 feet, the certificate holder shall supply enough oxygen to comply with §121.329, but not less than a two-hour supply for each flight crewmember on flight deck duty. The required two hours supply is that quantity of oxygen necessary for a constant rate of descent from the airplane's maximum certificated operating altitude to 10,000 feet in ten minutes and followed by 110 minutes at 10,000 feet. The oxygen required in the event of cabin pressurization failure by §121.337 may be included in determining the supply required for flight crewmembers on flight deck duty.

**(c)** *Use of oxygen masks by flight crewmembers.*

(1) When operating at flight altitudes above flight level 250, each flight crewmember on flight deck duty must be provided with an oxygen mask so designed that it can be rapidly placed on his face from its ready position, properly secured, sealed, and supplying oxygen upon demand; and so designed that after being placed on the face it does not prevent immediate communication between the flight crewmember and other crewmembers over the airplane intercommunication system. When it is not being used at flight altitudes above flight level 250, the oxygen mask must be kept in condition for ready use and located so as to be within the immediate reach of the flight crewmember while at his duty station.

(2) When operating at flight altitudes above flight level 250, one pilot at the controls of the airplane shall at all times wear and use an oxygen mask secured, sealed, and supplying oxygen, in accordance with the following:

(i) The one pilot need not wear and use an oxygen mask at or below the following flight levels if each flight crewmember on flight deck duty has a quick-donning type of oxygen mask that the certificate holder has shown can be placed on the face from its ready position, properly secured, sealed, and supplying oxygen upon demand, with one hand and within five seconds:

(A) For airplanes having a passenger seat configuration of more than 30 seats, excluding any required crewmember seat, or a payload capacity of more than 7,500 pounds, at or below flight level 410.

(B) For airplanes having a passenger seat configuration of less than 31 seats, excluding any required crewmember seat, and a payload capacity of 7,500 pounds or less, at or below flight level 350.

(ii) Whenever a quick-donning type of oxygen mask is to be used under this section, the certificate holder shall also show that the mask can be put on without disturbing eye glasses and without delaying the flight crewmember from proceeding with his assigned emergency duties. The oxygen mask after being put on must not prevent immediate communication between the flight crewmember and other crewmembers over the airplane intercommunication system.

(3) Notwithstanding paragraph (c)(2) of this section, if for any reason at any time it is necessary for one pilot to leave his station at the controls of the airplane when operating at flight altitudes above flight level 250, the remaining pilot at the controls shall put on and use his oxygen mask until the other pilot has returned to his duty station.

(4) Before the takeoff of a flight, each flight crewmember shall personally preflight his oxygen equipment to insure that the oxygen mask is functioning, fitted properly, and connected to appropriate supply terminals, and that the oxygen supply and pressure are adequate for use.

**(d)** *Use of portable oxygen equipment by cabin attendants.* Each attendant shall, during flight above flight level 250 flight altitude, carry portable oxygen equipment with at least a 15-minute supply of oxygen unless it is shown that enough portable oxygen units with masks or spare outlets and masks are distributed throughout the cabin to insure immediate availability of oxygen to each cabin attendant, regardless of his location at the time of cabin depressurization.

**(e)** *Passenger cabin occupants.* When the airplane is operating at flight altitudes above 10,000 feet, the following supply of oxygen must be provided for the use of passenger cabin occupants:

(1) When an airplane certificated to operate at flight altitudes up to and including flight level 250, can at any point along the route to be flown, descend safely to a flight altitude of 14,000 feet or less within four minutes, oxygen must be available at the rate prescribed by this part for a 30-minute period for at least 10 percent of the passenger cabin occupants.

(2) When an airplane is operated at flight altitudes up to and including flight level 250 and cannot descend safely to a flight altitude of 14,000 feet within four minutes, or when an airplane is operated at flight altitudes above flight level 250, oxygen must be available at the rate prescribed by this part for not less than 10 percent of the passenger cabin occupants for the entire flight after cabin depressurization, at cabin pressure altitudes above 10,000 feet up to and including 14,000 feet and, as applicable, to allow compliance with §121.329(c)(2) and (3), except that there must be not less than a 10-minute supply for the passenger cabin occupants.

(3) For first-aid treatment of occupants who for physiological reasons might require undiluted oxygen following descent from cabin pressure altitudes above flight level 250, a supply of oxygen in accordance with the requirements of §25.1443(d) must be provided for two percent of the occupants for the entire flight after cabin depressurization at cabin pressure altitudes above 8,000 feet, but in no case to less than one person. An appropriate number of acceptable dispensing units, but in no case less than two, must be provided, with a means for the cabin attendants to use this supply.

**(f)** *Passenger briefing.* Before flight is conducted above flight level 250, a crewmember shall instruct the passengers on the necessity of using oxygen in the event of cabin depressurization and shall point out to them the location and demonstrate the use of the oxygen-dispensing equipment.

[Docket No. 6258, 29 FR 19205, Dec. 31, 1964; as amended by Amdt. 121–11, 30 FR 12466, Sept. 30, 1965; Amdt. 121–132, 41 FR 55475, Dec. 20, 1976; Amdt. 121–262, 62 FR 13256, March 19, 1997; Amdt. 121–262, 62 FR 15570, April 1, 1997]

## §121.335 Equipment standards.

**(a)** *Reciprocating engine powered airplanes.* The oxygen apparatus, the minimum rates of oxygen flow, and the supply of oxygen necessary to comply with §121.327 must meet the standards established in section 4b.651 of the Civil Air Regulations as in effect on July 20, 1950, except that if the certificate holder shows full compliance with those standards to be impracticable, the Administrator may authorize any change in those standards that he finds will provide an equivalent level of safety.

**(b)** *Turbine engine powered airplanes.* The oxygen apparatus, the minimum rate of oxygen flow, and the supply of oxygen necessary to comply with §§121.329 and 121.333 must meet the standards established in section 4b.651 of the Civil Air Regulations as in effect on September 1, 1958, except that if the certificate holder shows full compliance with those standards to be impracticable, the Administrator may authorize any changes in those standards that he finds will provide an equivalent level of safety.

## §121.337 Protective breathing equipment.

**(a)** The certificate holder shall furnish approved protective breathing equipment (PBE) meeting the equipment, breathing gas, and communication requirements contained in paragraph (b) of this section.

**(b)** *Pressurized and nonpressurized cabin airplanes.* Except as provided in paragraph (f) of this section, no person may operate an airplane unless protective breathing equipment meeting the requirements of this section is provided as follows:

(1) *General.* The equipment must protect the flightcrew from the effects of smoke, carbon dioxide or other harmful gases or an oxygen deficient environment caused by other than an airplane depressurization while on flight deck duty and must protect crewmembers from the above effects while combating fires on board the airplane.

(2) The equipment must be inspected regularly in accordance with inspection guidelines and the inspection periods established by the equipment manufacturer to ensure its condition for continued serviceability and immediate readiness to perform its intended emergency purposes. The inspection periods may be changed upon a showing by the certificate holder that the changes would provide an equivalent level of safety.

(3) That part of the equipment protecting the eyes must not impair the wearer's vision to the extent that a crewmember's duties cannot be accomplished and must allow corrective glasses to be worn without impairment of vision or loss of the protection required by paragraph (b)(1) of this section.

(4) The equipment, while in use, must allow the flightcrew to communicate using the airplane radio equipment and to communicate by interphone with each other while at their assigned duty stations. The equipment, while in use, must also allow crewmember interphone communications between each of two flight crewmember stations in the pilot compartment and at least one normal flight attendant station in each passenger compartment.

(5) The equipment, while in use, must allow any crewmember to use the airplane interphone system at any of the flight attendant stations referred to in paragraph (b)(4) of this section.

(6) The equipment may also be used to meet the supplemental oxygen requirements of this part provided it meets the oxygen equipment standards of §121.335 of this part.

(7) Protective breathing gas duration and supply system equipment requirements are as follows:

(i) The equipment must supply breathing gas for 15 minutes at a pressure altitude of 8,000 feet for the following:

(A) Flight crewmembers while performing flight deck duties; and

(B) Crewmembers while combating an in-flight fire.

(ii) The breathing gas system must be free from hazards in itself, in its method of operation, and in its effect upon other components.

(iii) For breathing gas systems other than chemical oxygen generators, there must be a means to allow the crew to readily determine, during the equipment preflight described in paragraph (c) of this section, that the gas supply is fully charged.

(iv) For each chemical oxygen generator, the supply system equipment must meet the requirements of §25.1450(b) and (c) of this chapter.

(8) *Smoke and fume protection.* Protective breathing equipment with a fixed or portable breathing gas supply meeting the requirements of this section must be conveniently located on the flight deck and be easily accessible for immediate use by each required flight crewmember at his or her assigned duty station.

(9) *Fire combatting.* Except for nontransport category airplanes type certificated after December 31, 1964, protective breathing equipment with a portable breathing gas supply meeting the requirements of this section must be easily accessible and conveniently located for immediate use by crewmembers in combatting fires as follows:

(i) One PBE is required for each hand fire extinguisher located for use in a galley other than a galley located in a passenger, cargo, or crew compartment.

(ii) One on the flight deck, except that the Administrator may authorize another location for this PBE if special circumstances exist that make compliance impractical and the proposed deviation would provide an equivalent level of safety.

(iii) In each passenger compartment, one for each hand fire extinguisher required by §121.309 of this part, to be located within 3 feet of each required hand fire extinguisher, except that the Administrator may authorize a deviation allowing locations of PBE more than 3 feet from required hand fire extinguisher locations if special circumstances exist that make compliance impractical and if the proposed deviation provides an equivalent level of safety.

**(c)** *Equipment preflight.*

(1) Before each flight, each item of PBE at flight crewmember duty stations must be checked by the flight crewmember who will use the equipment to ensure that the equipment—

(i) For other than chemical oxygen generator systems, is functioning, is serviceable, fits properly (unless a universal-fit type), and is connected to supply terminals and that the breathing gas supply and pressure are adequate for use; and

(ii) For chemical oxygen generator systems, is serviceable and fits properly (unless a universal-fit type).

(2) Each item of PBE located at other than a flight crewmember duty station must be checked by a designated crewmember to ensure that each is properly stowed and serviceable, and, for other than chemical oxygen generator systems, the breathing gas supply is fully charged. Each certificate holder, in its operations manual, must designate at least one crewmember to perform those checks before he or she takes off in that airplane for his or her first flight of the day.

[Docket No. 24792, 52 FR 20957, June 3, 1987; as amended by Amdt. 121–204, 54 FR 22271, May 22, 1989; Amdt. 121–212, 55 FR 5551, Feb. 15, 1990; Amdt. 121–218, 55 FR 31565, Aug. 2, 1990; Amdt. 121–230, 57 FR 42674, Sept. 15, 1992; Amdt. 121–251, 60 FR 65932, Dec. 20, 1995; Amdt. 121–261, 61 FR 43921, Aug. 26, 1996; Amdt. 121–261, 61 FR 57585, Nov. 7, 1996]

## §121.339 Emergency equipment for extended over-water operations.

**(a)** Except where the Administrator, by amending the operations specifications of the certificate holder, requires the carriage of all or any specific items of the equipment listed below for any overwater operation, or upon application of the certificate holder, the Administrator allows deviation for a particular extended overwater operation, no person may operate an airplane in extended overwater operations without having on the airplane the following equipment:

(1) A life preserver equipped with an approved survivor locator light, for each occupant of the airplane.

(2) Enough life rafts (each equipped with an approved survivor locator light) of a rated capacity and buoyancy to accommodate the occupants of the airplane. Unless excess rafts of enough capacity are provided, the buoyancy and seating capacity beyond the rated capacity of the rafts must accommodate all occupants of the airplane in the event of a loss of one raft of the largest rated capacity.

(3) At least one pyrotechnic signaling device for each life raft.

(4) An approved survival type emergency locator transmitter. Batteries used in this transmitter must be replaced (or recharged, if the battery is rechargeable) when the transmitter has been in use for more than 1 cumulative hour, or when 50 percent of their useful life (or for rechargeable batteries, 50 percent of their useful life of charge) has expired, as established by the transmitter manufacturer under its approval. The new expiration date for replacing (or recharging) the battery must be legibly marked on the outside of the transmitter. The battery useful life (or useful life of charge) requirements of this paragraph do not apply to batteries (such as water-activated batteries) that are essentially unaffected during probable storage intervals.

**(b)** The required life rafts, life preservers, and survival type emergency locator transmitter must be easily accessible in the event of a ditching without appreciable time for preparatory procedures. This equipment must be installed in conspicuously marked, approved locations.

**(c)** A survival kit, appropriately equipped for the route to be flown, must be attached to each required life raft.

[Docket No. 6258, 29 FR 19205, Dec. 31, 1964; as amended by Amdt. 121–53, 34 FR 15244, Sept. 30, 1969; Amdt. 121–79, 36 FR 18724, Sept. 21, 1971; Amdt. 121–93, 37 FR 14294, June 19, 1972; Amdt. 121–106, 38 FR 22378, Aug. 20, 1973; Amdt. 121–149, 43 FR 50603, Oct. 30, 1978; Amdt. 121–158, 45 FR 38348, June 9, 1980; Amdt. 121–239, 59 FR 32057, June 21, 1994]

## §121.340 Emergency flotation means.

**(a)** Except as provided in paragraph (b) of this section, no person may operate an airplane in any overwater operation unless it is equipped with life preservers in accordance with §121.339(a)(1) or with an approved flotation means for each occupant. This means must be within easy reach of each seated occupant and must be readily removable from the airplane.

**(b)** Upon application by the air carrier or commercial operator, the Administrator may approve the operation of an airplane over water without the life preservers or flotation means required by paragraph (a) of this section, if the air carrier or commercial operator shows that the water over which the airplane is to be operated is not of such size and depth that life preservers or flotation means would be required for the survival of its occupants in the event the flight terminates in that water.

[Docket No. 6713, 31 FR 1147, Jan. 28, 1966; as amended by Amdt. 121–25, 32 FR 3223, Feb. 24, 1967; Amdt. 121–251, 60 FR 65932, Dec. 20, 1995]

## §121.341 Equipment for operations in icing conditions.

**(a)** Except as permitted in paragraph (c)(2) of this section, unless an airplane is type certificated under the transport category airworthiness requirements relating to ice protection, or unless an airplane is a non-transport category airplane type certificated after December 31, 1964, that has the ice protection provisions that meet section 34 of Appendix A of part 135 of this chapter, no person may operate an airplane in icing conditions unless it is equipped with means for the prevention or removal of ice on windshields, wings, empennage, propellers, and other parts of the airplane where ice formation will adversely affect the safety of the airplane.

**(b)** No person may operate an airplane in icing conditions at night unless means are provided for illuminating or otherwise determining the formation of ice on the parts of the wings that are critical from the standpoint of ice accumulation. Any illuminating that is used must be of a type that will not cause glare or reflection that would handicap crewmembers in the performance of their duties.

**(c)** Non-transport category airplanes type certificated after December 31, 1964. Except for an airplane that has ice protection provisions that meet section 34 of Appendix A of part 135 of this chapter, or those for transport category airplane type certification, no person may operate—

(1) Under IFR into known or forecast light or moderate icing conditions;

(2) Under VFR into known light or moderate icing conditions; unless the airplane has functioning deicing anti-icing equipment protecting each propeller, windshield, wing, stabilizing or control surface, and each airspeed, altimeter, rate of climb, or flight attitude instrument system; or

(3) Into known or forecast severe icing conditions.

**(d)** If current weather reports and briefing information relied upon by the pilot in command indicate that the forecast icing condition that would otherwise prohibit the flight will not be encountered during the flight because of changed weather conditions since the forecast, the restrictions in paragraph (c) of this section based on forecast conditions do not apply.

[Docket No. 6258, 29 FR 18205, Dec. 31, 1964; as amended by Amdt. 121–251, 60 FR 65932, Dec. 20, 1995]

## §121.342 Pitot heat indication systems.

No person may operate a transport category airplane or, after December 20, 1999, a nontransport category airplane type certificated after December 31, 1964, that is equipped with a flight instrument pitot heating system unless the airplane is also equipped with an operable pitot heat indication system that complies §25.1326 of this chapter in effect on April 12, 1978.

[Docket No. 28154, 60 FR 65932, Dec. 20, 1995]

## §121.343 Flight recorders.

**(a)** Except as provided in paragraphs (b), (c), (d), (e), and (f) of this section, no person may operate a large airplane that is certificated for operations above 25,000 feet altitude or is turbine-engine powered unless it is equipped with one or more approved flight recorders that record data from which the following may be determined within the ranges, accuracies, and recording intervals specified in Appendix B of this part:

(1) Time;

(2) Altitude;

(3) Airspeed;

(4) Vertical acceleration;

(5) Heading; and

(6) Time of each radio transmission either to or from air traffic control.

**(b)** No person may operate a large airplane type certificated up to and including September 30, 1969, for operations above 25,000 feet altitude, or a turbine-engine powered airplane certificated before the same date, unless it is equipped before May 26, 1989 with one or more approved flight recorders that utilize a digital method of recording and storing data and a method of readily retrieving that data from the storage medium. The following information must be able to be determined within the ranges, accuracies, and recording intervals specified in Appendix B of this part:

(1) Time;

(2) Altitude;

(3) Airspeed;

(4) Vertical acceleration;

(5) Heading; and

(6) Time of each radio transmission either to or from air traffic control.

**(c)** Except as provided in paragraph (l) of this section, no person may operate an airplane specified in paragraph (b) of this section unless it is equipped, before May 26, 1994, with one or more approved flight recorders that utilize a digital method of recording and storing data and a method of readily retrieving that data from the storage medium. The following information must be able to be determined within the ranges, accuracies and recording intervals specified in Appendix B of this part:

(1) Time;

(2) Altitude;

(3) Airspeed;

(4) Vertical acceleration;

(5) Heading;

(6) Time of each radio transmission either to or from air traffic control;

(7) Pitch attitude;

(8) Roll attitude;

(9) Longitudinal acceleration;

(10) Control column or pitch control surface position; and

(11) Thrust of each engine.

**(d)** No person may operate an airplane specified in paragraph (b) of this section that is manufactured after May 26, 1989, as well as airplanes specified in paragraph (a) of this section that have been type certificated after September 30, 1969, unless it is equipped with one or more approved flight recorders that utilize a digital method of recording and storing data and a method of readily retrieving that data from the storage medium. The following information must be able to be determined within the ranges, accuracies, and recording intervals specified in Appendix B of this part:

(1) Time;

(2) Altitude;

(3) Airspeed;

(4) Vertical acceleration;

(5) Heading;

(6) Time of each radio transmission either to or from air traffic control;

(7) Pitch attitude;

(8) Roll attitude;

(9) Longitudinal acceleration;

(10) Pitch trim position;

(11) Control column or pitch control surface position;

(12) Control wheel or lateral control surface position;

(13) Rudder pedal or yaw control surface position;

(14) Thrust of each engine;

(15) Position of each thrust reverser;

(16) Trailing edge flap or cockpit flap control position; and

(17) Leading edge flap or cockpit flap control position.

For the purpose of this section, *manufactured* means the point in time at which the airplane inspection acceptance records reflect that the airplane is complete and meets the FAA-approved type design data.

**(e)** After October 11, 1991, no person may operate a large airplane equipped with a digital data bus and ARINC 717 digital flight data acquisition unit (DFDAU) or equivalent unless it is equipped with one or more approved flight recorders that utilize a digital method of recording and storing data and a method of readily retrieving that data from the storage medium. Any parameters specified in Appendix B of this part that are available on the digital data bus must be recorded within the ranges, accuracies, resolutions, and sampling intervals specified.

**(f)** After October 11, 1991, no person may operate an airplane specified in paragraph (b) of this section that is manufactured after October 11, 1991, nor an airplane specified in paragraph (a) of this section that has been type certificated after September 30, 1969, and manufactured after October 11, 1991, unless it is equipped with one or more flight recorders that utilize a digital method of recording and storing data and a method of readily retrieving that data from the storage medium. The parameters specified in Appendix B of this part must be recorded within the ranges, accuracies, resolutions, and sampling intervals specified.

**(g)** Whenever a flight recorder required by this section is installed, it must be operated continuously from the instant the airplane begins the takeoff roll until it has completed the landing roll at an airport.

**(h)** Except as provided in paragraph (i) of this section, and except for recorded data erased as authorized in this paragraph, each certificate holder shall keep the recorded data prescribed in paragraph (a), (b), (c), or (d) of this section, as appropriate, until the airplane has been operated for at least 25 hours of the operating time specified in §121.359(a). A total of 1 hour of recorded data may be erased for the purpose of testing the flight recorder or the flight recorder system. Any erasure made in accordance with this paragraph must be of the oldest recorded data accumulated at the time of testing. Except as provided in paragraph (i) of this section, no record need be kept more than 60 days.

**(i)** In the event of an accident or occurrence that requires immediate notification of the National Transportation Safety Board under part 830 of its regulations and that results in termination of the flight, the certificate holder shall remove the recording media from the airplane and keep the recorded data required by paragraph (a), (b), (c), or (d) of this section, as appropriate, for at least 60 days or for a longer period upon the request of the Board or the Administrator.

**(j)** Each flight recorder required by this section must be installed in accordance with the requirements of §25.1459 of this chapter in effect on August 31, 1977. The correlation required by §25.1459(c) of this chapter need be established only on one airplane of any group of airplanes—

(1) That are of the same type;

(2) On which the model flight recorder and its installation are the same; and

(3) On which there is no difference in the type design with respect to the installation of those first pilot's instruments associated with the flight recorder. The most recent instrument calibration, including the recording medium from which this calibration is derived, and the recorder correlation must be retained by the certificate holder.

**(k)** Each flight recorder required by this section that records the data specified in paragraph (a), (b), (c), or (d) of this section, as appropriate, must have an approved device to assist in locating that recorder under water.

**(l)** No person may operate an airplane specified in paragraph (b) of this section that meets the Stage 2 noise levels of part 36 of this chapter and is subject to §91.801(c) of this chapter unless it is equipped with one or more approved flight data recorders that utilize a digital method of recording and storing data and a method of readily retrieving that data from the storage medium. The information specified in paragraphs (c)(1) through (c)(11) of this section must be able to be determined within the ranges, accuracies and recording intervals specified in Appendix B of this part. In addition—

(1) This flight data recorder must be installed at the next heavy maintenance check after May 26, 1994, but no later than May 26, 1995. A heavy maintenance check is considered to be any time an aircraft is scheduled to be out of service for 4 or more days.

(2) By June 23, 1994, each carrier must submit to the FAA Flight Standards Service, Air Transportation Division (AFS-200), documentation listing those airplanes covered under this paragraph and evidence that it has ordered a sufficient number of flight data recorders to meet the May 26, 1995, compliance date for all aircraft on that list.

(3) After May 26, 1994, any aircraft that is modified to meet Stage 3 noise levels must have the flight data recorder described in paragraph (c) of this section installed before operating under this part.

[Docket No. 24418, 52 FR 9636, March 25, 1987; as amended by Amdt. 121–197, 53 FR 26147, July 11, 1988; Amdt. 121–238, 59 FR 26900, May 24, 1994]

## §121.344 Digital flight data recorders for transport category airplanes.

**(a)** Except as provided in paragraph (l) of this section, no person may operate under this part a turbine-engine-powered transport category airplane unless it is equipped with one or more approved flight recorders that use a digital method of recording and storing data and a method of readily retrieving that data from the storage medium. The operational parameters required to be recorded by digital flight data recorders required by this section are as follows: the phrase "when an information source is installed" following a parameter indicates that recording of that parameter is not intended to require a change in installed equipment:

(1) Time;
(2) Pressure altitude;
(3) Indicated airspeed;
(4) Heading—primary flight crew reference (if selectable, record discrete, true or magnetic);
(5) Normal acceleration (Vertical);
(6) Pitch attitude;
(7) Roll attitude;
(8) Manual radio transmitter keying, or CVR/DFDR synchronization reference;
(9) Thrust/power of each engine—primary flight crew reference;
(10) Autopilot engagement status;
(11) Longitudinal acceleration;
(12) Pitch control input;
(13) Lateral control input;
(14) Rudder pedal input;
(15) Primary pitch control surface position;
(16) Primary lateral control surface position;
(17) Primary yaw control surface position;
(18) Lateral acceleration;
(19) Pitch trim surface position or parameters of paragraph (a)(82) of this section if currently recorded;
(20) Trailing edge flap or cockpit flap control selection (except when parameters of paragraph (a)(85) of this section apply);
(21) Leading edge flap or cockpit flap control selection (except when parameters of paragraph (a)(86) of this section apply);
(22) Each Thrust reverser position (or equivalent for propeller airplane);
(23) Ground spoiler position or speed brake selection (except when parameters of paragraph (a)(87) of this section apply);
(24) Outside or total air temperature;
(25) Automatic Flight Control System (AFCS) modes and engagement status, including autothrottle;
(26) Radio altitude (when an information source is installed);
(27) Localizer deviation, MLS Azimuth;
(28) Glideslope deviation, MLS Elevation;
(29) Marker beacon passage;
(30) Master warning;
(31) Air/ground sensor (primary airplane system reference nose or main gear);
(32) Angle of attack (when information source is installed);
(33) Hydraulic pressure low (each system);
(34) Ground speed (when an information source is installed);
(35) Ground proximity warning system;
(36) Landing gear position or landing gear cockpit control selection;
(37) Drift angle (when an information source is installed);
(38) Wind speed and direction (when an information source is installed);
(39) Latitude and longitude (when an information source is installed);
(40) Stick shaker/pusher (when an information source is installed);
(41) Windshear (when an information source is installed);
(42) Throttle/power lever position;
(43) Additional engine parameters (as designated in Appendix M of this part);
(44) Traffic alert and collision avoidance system;
(45) DME 1 and 2 distances;
(46) Nav 1 and 2 selected frequency;
(47) Selected barometric setting (when an information source is installed);

(48) Selected altitude (when an information source is installed);
(49) Selected speed (when an information source is installed);
(50) Selected mach (when an information source is installed);
(51) Selected vertical speed (when an information source is installed);
(52) Selected heading (when an information source is installed);
(53) Selected flight path (when an information source is installed);
(54) Selected decision height (when an information source is installed);
(55) EFIS display format;
(56) Multi-function/engine/alerts display format;
(57) Thrust command (when an information source is installed);
(58) Thrust target (when an information source is installed);
(59) Fuel quantity in CG trim tank (when an information source is installed);
(60) Primary Navigation System Reference;
(61) Icing (when an information source is installed);
(62) Engine warning each engine vibration (when an information source is installed);
(63) Engine warning each engine over temp. (when an information source is installed);
(64) Engine warning each engine oil pressure low (when an information source is installed);
(65) Engine warning each engine over speed (when an information source is installed);
(66) Yaw trim surface position;
(67) Roll trim surface position;
(68) Brake pressure (selected system);
(69) Brake pedal application (left and right);
(70) Yaw or sideslip angle (when an information source is installed);
(71) Engine bleed valve position (when an information source is installed);
(72) De-icing or anti-icing system selection (when an information source is installed);
(73) Computed center of gravity (when an information source is installed);
(74) AC electrical bus status;
(75) DC electrical bus status;
(76) APU bleed valve position (when an information source is installed);
(77) Hydraulic pressure (each system);
(78) Loss of cabin pressure;
(79) Computer failure;
(80) Heads-up display (when an information source is installed);
(81) Para-visual display (when an information source is installed);
(82) Cockpit trim control input position—pitch;
(83) Cockpit trim control input position—roll;
(84) Cockpit trim control input position—yaw;
(85) Trailing edge flap and cockpit flap control position;
(86) Leading edge flap and cockpit flap control position;
(87) Ground spoiler position and speed brake selection; and
(88) All cockpit flight control input forces (control wheel, control column, rudder pedal).

**(b)** For all turbine-engine powered transport category airplanes manufactured on or before October 11, 1991, by August 20, 2001.

(1) For airplanes not equipped as of July 16, 1996, with a flight data acquisition unit (FDAU), the parameters listed in paragraphs (a)(1) through (a)(18) of this section must be recorded within the ranges and accuracies specified in Appendix B of this part, and—

(i) For airplanes with more than two engines, the parameter described in paragraph (a)(18) is not required unless sufficient capacity is available on the existing recorder to record that parameter;

(ii) Parameters listed in paragraphs (a)(12) through (a)(17) each may be recorded from a single source.

(2) For airplanes that were equipped as of July 16, 1996, with a flight data acquisition unit (FDAU), the parameters listed in paragraphs (a)(1) through (a)(22) of this section must be recorded within the ranges, accuracies, and recording intervals specified in Appendix M of this part. Parameters listed in paragraphs (a)(12) through (a)(17) each may be recorded from a single source.

(3) The approved flight recorder required by this section must be installed at the earliest time practicable, but no later than the next heavy maintenance check after August 18, 1999 and no later than August 20, 2001. A heavy maintenance check is considered to be any time an airplane is scheduled to be out of service for 4 or more days and is scheduled to include access to major structural components.

**(c)** For all turbine-engine powered transport category airplanes manufactured on or before October 11, 1991—

(1) That were equipped as of July 16, 1996, with one or more digital data bus(es) and an ARINC 717 digital flight data acquisition unit (DFDAU) or equivalent, the parameters specified in paragraphs (a)(1) through (a)(22) of this section must be recorded within the ranges, accuracies, resolutions, and sampling intervals specified in Appendix M of this part by August 20, 2001. Parameters listed in paragraphs (a)(12) through (a)(14) each may be recorded from a single source.

(2) Commensurate with the capacity of the recording system (DFDAU or equivalent and the DFDR), all additional parameters for which information sources are installed and which are con-

nected to the recording system must be recorded within the ranges, accuracies, resolutions, and sampling intervals specified in Appendix M of this part by August 20, 2001.

(3) That were subject to §121.343(e) of this part, all conditions of §121.343(e) must continue to be met until compliance with paragraph (c)(1) of this section is accomplished.

**(d)** For all turbine-engine-powered transport category airplanes that were manufactured after October 11, 1991—

(1) The parameters listed in paragraph (a)(1) through (a)(34) of this section must be recorded within the ranges, accuracies, resolutions, and recording intervals specified in Appendix M of this part by August 20, 2001. Parameters listed in paragraphs (a)(12) through (a)(14) each may be recorded from a single source.

(2) Commensurate with the capacity of the recording system, all additional parameters for which information sources are installed and which are connected to the recording system must be recorded within the ranges, accuracies, resolutions, and sampling intervals specified in Appendix M of this part by August 20, 2001.

**(e)** For all turbine-engine-powered transport category airplanes that are manufactured after August 18, 2000—

(1) The parameters listed in paragraph (a)(1) through (57) of this section must be recorded within the ranges, accuracies, resolutions, and recording intervals specified in Appendix M of this part.

(2) Commensurate with the capacity of the recording system, all additional parameters for which information sources are installed and which are connected to the recording system, must be recorded within the ranges, accuracies, resolutions, and sampling intervals specified in Appendix M of this part.

**(f)** For all turbine-engine-powered transport category airplanes that are manufactured after August 19, 2002 the parameters listed in paragraph (a)(1) through (a)(88) of this section must be recorded within the ranges, accuracies, resolutions, and recording intervals specified in Appendix M of this part.

**(g)** Whenever a flight data recorder required by this section is installed, it must be operated continuously from the instant the airplane begins its takeoff roll until it has completed its landing roll.

**(h)** Except as provided in paragraph (i) of this section, and except for recorded data erased as authorized in this paragraph, each certificate holder shall keep the recorded data prescribed by this section, as appropriate, until the airplane has been operated for at least 25 hours of the operating time specified in §121.359(a) of this part. A total of 1 hour of recorded data may be erased for the purpose of testing the flight recorder or the flight recorder system. Any erasure made in accordance with this paragraph must be of the oldest recorded data accumulated at the time of testing. Except as provided in paragraph (i) of this section, no record need be kept more than 60 days.

**(i)** In the event of an accident or occurrence that requires immediate notification of the National Transportation Safety Board under 49 CFR 830 of its regulations and that results in termination of the flight, the certificate holder shall remove the recorder from the airplane and keep the recorder data prescribed by this section, as appropriate, for at least 60 days or for a longer period upon the request of the Board or the Administrator.

**(j)** Each flight data recorder system required by this section must be installed in accordance with the requirements of §25.1459 (a), (b), (d), and (e) of this chapter. A correlation must be established between the values recorded by the flight data recorder and the corresponding values being measured. The correlation must contain a sufficient number of correlation points to accurately establish the conversion from the recorded values to engineering units or discrete state over the full operating range of the parameter. Except for airplanes having separate altitude and airspeed sensors that are an integral part of the flight data recorder system, a single correlation may be established for any group of airplanes—

(1) That are of the same type;

(2) On which the flight recorder system and its installation are the same; and

(3) On which there is no difference in the type design with respect to the installation of those sensors associated with the flight data recorder system. Documentation sufficient to convert recorded data into the engineering units and discrete values specified in the applicable appendix must be maintained by the certificate holder.

**(k)** Each flight data recorder required by this section must have an approved device to assist in locating that recorder under water.

**(l)** The following airplanes that were manufactured before August 18, 1997 need not comply with this section, but must continue to comply with applicable paragraphs of §121.343 of this chapter, as appropriate:

(1) Airplanes that meet the State 2 noise levels of part 36 of this chapter and are subject to §91.801(c) of this chapter, until January 1, 2000. On and after January 1, 2000, any Stage 2 airplane otherwise allowed to be operated under Part 91 of this chapter must comply with the applicable flight data recorder requirements of this section for that airplane.

(2) General Dynamics Convair 580, General Dynamics Convair 600, General Dynamics Convair 640, deHavilland Aircraft Company Ltd. DHC-7, Fairchild Industries FH 227, Fokker F-27 (except Mark 50), F-28 Mark 1000 and Mark 4000, Gulfstream Aerospace G-159, Lockheed Aircraft Cor-

poration Electra 10-A, Lockheed Aircraft Corporation Electra 10-B, Lockheed Aircraft Corporation Electra 10-E, Lockheed Aircraft Corporation Electra L-188, Maryland Air Industries, Inc. F27, Mitsubishi Heavy Industries, Ltd. YS-11, Short Bros. Limited SD3-30, Short Bros. Limited SD3-60.

[Docket No. 28109, 62 FR 38378, July 17, 1997; as amended by Amdt. 121–266, 62 FR 48135, Sept. 12, 1997]

## §121.344a Digital flight data recorders for 10–19 seat airplanes.

**(a)** Except as provided in paragraph (f) of this section, no person may operate under this part a turbine-engine-powered airplane having a passenger seating configuration, excluding any required crewmember seat, of 10 to 19 seats, that was brought onto the U.S. register after, or was registered outside the United States and added to the operator's U.S. operations specifications after, October 11, 1991, unless it is equipped with one or more approved flight recorders that use a digital method of recording and storing data and a method of readily retrieving that data from the storage medium. On or before August 20, 2001, airplanes brought onto the U.S. register after October 11, 1991, must comply with either the requirements in this section or the applicable paragraphs in §135.152 of this chapter. In addition, by August 20, 2001:

(1) The parameters listed in §§121.344(a)(1) through 121.344(a)(18) of this part must be recorded with the ranges, accuracies, and resolutions specified in Appendix B of part 135 of this chapter, except that—

(i) Either the parameter listed in §121.344 (a)(12) or (a)(15) of this part must be recorded; either the parameters listed in §121.344(a)(13) or (a)(16) of this part must be recorded; and either the parameter listed in §121.344(a)(14) or (a)(17) of this part must be recorded.

(ii) For airplanes with more than two engines, the parameter described in §121.344(a)(18) of this part must also be recorded if sufficient capacity is available on the existing recorder to record that parameter;

(iii) Parameters listed in §§121.344(a)(12) through 121.344(a)(17) of this part each may be recorded from a single source;

(iv) Any parameter for which no value is contained in Appendix B of part 135 of this chapter must be recorded within the ranges, accuracies, and resolutions specified in Appendix M of this part.

(2) Commensurate with the capacity of the recording system (FDAU or equivalent and the DFDR), the parameters listed in §§121.344(a)(19) through 121.344(a)(22) of this part also must be recorded within the ranges, accuracies, resolutions, and recording intervals specified in Appendix B of part 135 of this chapter.

(3) The approved flight recorder required by this section must be installed as soon as practicable, but no later than the next heavy maintenance check or equivalent after August 18, 1999. A heavy maintenance check is considered to be any time an airplane is scheduled to be out of service for 4 more days and is scheduled to include access to major structural components.

**(b)** For a turbine-engine-powered airplanes having a passenger seating configuration, excluding any required crewmember seat, of 10 to 19 seats, that are manufactured after August 18, 2000.

(1) The parameters listed in §§121.344(a)(1) through 121.344(a)(57) of this part, must be recorded within the ranges, accuracies, resolutions, and recording intervals specified in Appendix M of this part.

(2) Commensurate with the capacity of the recording system, all additional parameters listed in §121.344(a) of this part for which information sources are installed and which are connected to the recording system, must be recorded within the ranges, accuracies, resolutions, and sampling intervals specified in Appendix M of this part by August 20, 2001.

**(c)** For all turbine-engine-powered airplanes having a passenger seating configuration, excluding any required crewmember seats, of 10 to 19 seats, that are manufactured after August 19, 2002, the parameters listed in §121.344(a)(1) through (a)(88) of this part must be recorded within the ranges, accuracies, resolutions, and recording intervals specified in Appendix M of this part.

**(d)** Each flight data recorder system required by this section must be installed in accordance with the requirements of §23.1459(a), (b), (d), and (e) of this chapter. A correlation must be established between the values recorded by the flight data recorder and the corresponding values being measured. The correlation must contain a sufficient number of correlation points to accurately establish the conversion from the recorded values to engineering units or discrete state over the full operating range of the parameter. A single correlation may be established for any group of airplanes—

(1) That are of the same type;

(2) On which the flight recorder system and its installation are the same; and

(3) On which there is no difference in the type design with respect to the installation of those sensors associated with the flight data recorder system. Correlation documentation must be maintained by the certificate holder.

**(e)** All airplanes subject to this section are also subject to the requirements and exceptions stated in §§121.344(g) through 121.344(k) of this part.

**(f)** For airplanes that were manufactured before August 18, 1997, the following airplane types need not comply with this section, but must continue to comply with applicable paragraphs of §135.152 of this chapter, as appropriate: Beech Aircraft-99 Series, Beech Aircraft 1300, Beech Aircraft 1900C, Construcciones Aeronauticas, S.A. (CASA) C-212, deHavilland DHC-6, Dornier 228, HS-748, Embraer EMB 110, Jetstream 3101, Jetstream 3201, Fairchild Aircraft SA-226.

[Docket No. 28109, 62 FR 38380, July 17, 1997; as amended by Amdt. 121–266, 62 FR 48135, Sept. 12, 1997; Amdt. 121–266, 62 FR 65202, Dec. 11, 1997]

## §121.345 Radio equipment.

**(a)** No person may operate an airplane unless it is equipped with radio equipment required for the kind of operation being conducted.

**(b)** Where two independent (separate and complete) radio systems are required by §§121.347 and 121.349, each system must have an independent antenna installation except that, where rigidly supported nonwire antennas or other antenna installations of equivalent reliability are used, only one antenna is required.

**(c)** ATC transponder equipment installed within the time periods indicated below must meet the performance and environmental requirements of the following TSO's:

(1) *Through January 1, 1992:*

(i) Any class of TSO–C74b or any class of TSO–C74c as appropriate, provided that the equipment was manufactured before January 1, 1990; or

(ii) The appropriate class of TSO–C112 (Mode S).

(2) *After January 1, 1992*: The appropriate class of TSO–C112 (Mode S). For purposes of paragraph (c)(2) of this section, "installation" does not include—

(i) Temporary installation of TSO–C74b or TSO–C74c substitute equipment, as appropriate, during maintenance of the permanent equipment;

(ii) Reinstallation of equipment after temporary removal for maintenance; or

(iii) For fleet operations, installation of equipment in a fleet aircraft after removal of the equipment for maintenance from another aircraft in the same operator's fleet.

[Docket No. 6258, 29 FR 19205, Dec. 31, 1964; as amended by Amdt. 121–101, 37 FR 28499, Dec. 27, 1972; Amdt. 121–190, 52 FR 3391, Feb. 3, 1987]

## §121.347 Radio equipment for operations under VFR over routes navigated by pilotage.

**(a)** No person may operate an airplane under VFR over routes that can be navigated by pilotage, unless it is equipped with the radio equipment necessary under normal operating conditions to fulfill the following:

(1) Communicate with at least one appropriate ground station from any point on the route.

(2) Communicate with appropriate traffic control facilities from any point within the lateral boundaries of the surface areas of Class B, Class C, Class D, or Class E airspace designated for an airport in which flights are intended.

(3) Receive meteorological information from any point en route by either of two independent systems. One of the means provided to comply with this subparagraph may be used to comply with paragraphs (a)(1) and (2) of this section.

**(b)** No person may operate an airplane at night under VFR over routes than can be navigated by pilotage unless that airplane is equipped with the radio equipment necessary under normal operating conditions to fulfill the functions specified in paragraph (a) of this section and to receive radio navigational signals applicable to the route flown, except that a marker beacon receiver or ILS receiver is not required.

[Docket No. 6258, 29 FR 19205, Dec. 30, 1964; as amended by Amdt. 121–226, 56 FR 65663, Dec. 17, 1991]

## §121.349 Radio equipment for operations under VFR over routes not navigated by pilotage or for operations under IFR or over-the-top.

**(a)** No person may operate an airplane under VFR over routes that cannot be navigated by pilotage or for operations conducted under IFR or over-the-top, unless the airplane is equipped with that radio equipment necessary under normal operating conditions to fulfill the functions specified in §121.347(a) and to receive satisfactorily by either of two independent systems radio navigational signals from all primary en route and approach navigational facilities intended to be used. However, only one marker beacon receiver providing visual and aural signals and one ILS receiver need be provided. Equipment provided to receive signals en route may be used to receive signals on approach, if it is capable of receiving both signals.

**(b)** In the case of operation over routes on which navigation is based on low frequency radio range or automatic direction finding, only one low frequency radio range or ADF receiver need be installed if the airplane is equipped with two VOR receivers, and VOR navigational aids are so located and the airplane is so fueled that, in the case of failure of the low frequency radio range receiver or ADF receiver, the flight may proceed safely to a suitable airport, by means of VOR aids, and complete an instrument approach by use of the remaining airplane radio system.

**(c)** Whenever VOR navigational receivers are required by paragraph (a) or (b) of this section, at least one approved distance measuring equipment unit (DME) capable of receiving and indicating distance information from VORTAC facilities must be installed on each airplane when operated in the 50 states and the District of Columbia.

**(d)** If the distance measuring equipment (DME) becomes inoperative en route, the pilot shall notify ATC of that failure as soon as it occurs.

**(e)** No person may operate an airplane having a passenger seat configuration of 10 to 30 seats, excluding each crewmember seat, and a payload of 7,500 pounds or less under IFR or in extended overwater operations unless it has, in addition to any other required radio communications and navigational equipment appropriate to the facilities to be used which are capable of transmitting to, and receiving from, at any place on the route to be flown, at least one ground facility, two microphones, and two headsets or one headset and one speaker.

[Docket No. 6258, 29 FR 19205, Dec. 31, 1964; as amended by Amdt. 121–19, 31 FR 6265, April 23, 1966; Amdt. 121–130, 41 FR 47229, Oct. 28, 1976; Amdt. 121–251, 60 FR 65932, Dec. 20, 1995]

## §121.351 Radio and navigation equipment for extended overwater operations and for certain other operations.

**(a)** Except as provided in paragraph (c) of this section, no person may conduct an extended overwater operation unless the airplane is equipped with the radio communication equipment necessary to comply with Sec. 121.349, an independent system that complies with Sec. 121.347 (a)(1), and two long-range navigation systems when VOR or ADF radio navigation equipment is unusable along a portion of the route.

**(b)** No certificate holder conducting a flag or supplemental operation or a domestic operation within the State of Alaska may conduct an operation without the equipment specified in paragraph (a) of this section, if the Administrator finds that equipment to be necessary for search and rescue operations because of the nature of the terrain to be flown over.

**(c)** Notwithstanding the requirements of paragraph (a) of this section, installation and use of a single LRNS and a single LRCS may be authorized by the Administrator and approved in the certificate holder's operations specifications for operations and routes in certain geographic areas. The following are among the operational factors the Administrator may consider in granting an authorization:

(1) The ability of the flightcrew to reliably fix the position of the airplane within the degree of accuracy required by ATC,

(2) The length of the route being flown, and

(3) The duration of the very high frequency communications gap.

[Docket No. 6258, 29 FR 19205, Dec. 31, 1964; as amended by Amdt. 121–253, 61 FR, Jan. 26, 1996; Amdt. 121–254, 61 FR, Feb. 26, 1996]

## §121.353 Emergency equipment for operations over uninhabited terrain areas: Flag, supplemental and certain domestic operations.

Unless the airplane has the following equipment, no person may conduct a flag or supplemental operation or a domestic operation within the States of Alaska or Hawaii over an uninhabited area or any other area that (in its operations specifications) the Administrator specifies required equipment for search and rescue in case of an emergency:

**(a)** Suitable pyrotechnic signaling devices.

**(b)** An approved survival type emergency locator transmitter. Batteries used in this transmitter must be replaced (or recharged, if the battery is rechargeable) when the transmitter has been in use for more than 1 cumulative hour, or when 50 percent of their useful life (or for rechargeable batteries, 50 percent of their useful life of charge) has expired, as established by the transmitter manufacturer under its approval. The new expiration date for replacing (or recharging) the battery must be legibly marked on the outside of the transmitter. The battery useful life (or useful life of charge) requirements of this paragraph do not apply to batteries (such as water-activated batteries) that are essentially unaffected during probable storage intervals.

**(c)** Enough survival kits, appropriately equipped for the route to be flown for the number of occupants of the airplane.

[Docket No. 6258, 29 FR 19205, Dec. 31, 1964; as amended by Amdt. 121–79, 36 FR 18724, Sept. 21, 1971; Amdt. 121–106, 38 FR 22378 Aug. 20, 1973; Amdt. 121–158, 45 FR 38348, June 9, 1980; Amdt. 121–239, 59 FR 32057, June 21, 1994; Amdt. 121–251, 60 FR 65932, Dec. 20, 1995]

## §121.354 Terrain awareness and warning system.

**(a)** *Airplanes manufactured after March 29, 2002.* No person may operate a turbine-powered airplane unless that airplane is equipped with an approved terrain awareness and warning system that meets the requirements for Class A equipment in Technical Standard Order (TSO)–C151.

The airplane must also include an approved terrain situational awareness display.

**(b)** *Airplanes manufactured on or before March 29, 2002.* No person may operate a turbine-powered airplane after March 29, 2005, unless that airplane is equipped with an approved terrain awareness and warning system that meets the requirements for Class A equipment in Technical Standard Order (TSO)–C151. The airplane must also include an approved terrain situational awareness display.

(Approved by the Office of Management and Budget under control number 2120-0631)

**(c)** *Airplane Flight Manual.* The Airplane Flight Manual shall contain appropriate procedures for—

(1) The use of the terrain awareness and warning system; and

(2) Proper flight crew reaction in response to the terrain awareness and warning system audio and visual warnings.

[Docket No. 29312, 65 FR 16755, March 29, 2000]

## §121.355 Equipment for operations on which specialized means of navigation are used.

**(a)** No certificate holder may conduct an operation—

(1) Using Doppler Radar or an Inertial Navigation System outside the 48 contiguous States and the District of Columbia, unless such systems have been approved in accordance with Appendix G to this part; or

(2) Using Doppler Radar or an Inertial Navigation System within the 48 contiguous States and the District of Columbia, or any other specialized means of navigation, unless it shows that an adequate airborne system is provided for the specialized navigation authorized for the particular operation.

**(b)** Notwithstanding paragraph (a) of this section, Doppler Radar and Inertial Navigation Systems, and the training programs, maintenance programs, relevant operations manual material, and minimum equipment lists prepared in accordance therewith, approved before April 29, 1972, are not required to be approved in accordance with that paragraph.

[Docket No. 10204, 37 FR 6464, March 30, 1972]

## §121.356 Traffic Alert and Collision Avoidance System.

**(a)** Unless otherwise authorized by the Administrator, each certificate holder operating a large airplane that has a passenger seating configuration, excluding any pilot seat, of more than 30 seats, shall equip its airplanes with an approved TCAS II traffic alert and collision avoidance system and the appropriate class of Mode S transponder according to the following schedule:

| Date | Required equipage |
|---|---|
| December 30, 1990 | At least 20% of all covered airplanes, if the certificate holder operates more than 30 such airplanes. |
| December 30, 1991 | 50% of all covered airplanes. |
| December 30, 1993 | 100% of all covered airplanes. |

**(b)** Unless otherwise authorized by the Administrator, after December 31, 1995, no person may operate a passenger or combination cargo/passenger (combi) airplane that has a passenger seat configuration, excluding any pilot seat, of 10 to 30 seats unless it is equipped with an approved traffic alert and collision avoidance system. If a TCAS II system is installed, it must be capable of coordinating with TCAS units that meet TSO C-119.

**(c)** The appropriate manuals required by §121.131 shall contain the following information on the TCAS II System or TCAS I System, as appropriate, as required by this section:

(1) Appropriate procedures for—

(i) The operation of the equipment; and

(ii) Proper flightcrew action with respect to the equipment.

(2) An outline of all input sources that must be operative for the TCAS to function properly.

[Docket No. 25355, 54 FR 951, Jan. 10, 1989; as amended by Amdt. 121–217, 55 FR 13247, April 9, 1990; Amdt. 121–246, 59 FR 67586, Dec. 29, 1994; Amdt. 121–247, 60 FR 3303, Jan. 13, 1995; Amdt. 121–251, 60 FR 65932, Dec. 20, 1995]

## §121.357 Airborne weather radar equipment requirements.

**(a)** No person may operate any transport category airplane (except C-46 type airplanes) or a nontransport category airplane certificated after December 31, 1964, unless approved airborne weather radar equipment has been installed in the airplane.

**(b)** [Reserved]

**(c)** Each person operating an airplane required to have approved airborne weather radar equipment installed shall, when using it under this part, operate it in accordance with the following:

(1) *Dispatch.* No person may dispatch an airplane (or begin the flight of an airplane in the case of an air carrier or commercial operator that does not use a dispatch system) under IFR or night VFR conditions when current weather reports indicate that thunderstorms, or other potentially hazardous weather conditions that can be detected with airborne weather radar, may reasonably be expected along the route to be flown, un-

less the airborne weather radar equipment is in satisfactory operating condition.

(2) If the airborne weather radar becomes inoperative en route, the airplane must be operated in accordance with the approved instructions and procedures specified in the operations manual for such an event.

**(d)** This section does not apply to airplanes used solely within the State of Hawaii or within the State of Alaska and that part of Canada west of longitude 130 degrees W, between latitude 70 degrees N, and latitude 53 degrees N, or during any training, test, or ferry flight.

**(e)** Notwithstanding any other provision of this chapter, an alternate electrical power supply is not required for airborne weather radar equipment.

[Docket No. 6258, 29 FR 19205, Dec. 31, 1964; as amended by Amdt. 121–18, 31 FR 5825, April 15, 1966; Amdt. 121–130, 41 FR 47229, Oct. 28, 1976; Amdt. 121–251, 60 FR 65932, Dec. 20, 1995]

## §121.358 Low-altitude windshear system equipment requirements.

**(a)** *Airplanes manufactured after January 2, 1991.* No person may operate a turbine-powered airplane manufactured after January 2, 1991, unless it is equipped with either an approved airborne windshear warning and flight guidance system, an approved airborne detection and avoidance system, or an approved combination of these systems.

**(b)** *Airplanes manufactured before January 3, 1991.* Except as provided in paragraph (c) of this section, after January 2, 1991, no person may operate a turbine-powered airplane manufactured before January 3, 1991 unless it meets one of the following requirements as applicable.

(1) The makes/models/series listed below must be equipped with either an approved airborne windshear warning and flight guidance system, an approved airborne detection and avoidance system, or an approved combination of these systems:

(i) A–300—600;

(ii) A–310—all series;

(iii) A–320—all series;

(iv) B–737—300, 400, and 500 series;

(v) B–747—400;

(vi) B–757—all series;

(vii) B–767—all series;

(viii) F–100—all series;

(ix) MD–11—all series; and

(x) MD–80 series equipped with an EFIS and Honeywell–970 digital flight guidance computer.

(2) All other turbine-powered airplanes not listed above must be equipped with as a minimum requirement, an approved airborne windshear warning system. These airplanes may be equipped with an approved airborne windshear detection and avoidance system, or an approved combination of these systems.

**(c)** *Extension of the compliance date.* A certificate holder may obtain an extension of the compliance date in paragraph (b) of this section if it obtains FAA approval of a retrofit schedule. To obtain approval of a retrofit schedule and show continued compliance with that schedule, a certificate holder must do the following:

(1) Submit a request for approval of a retrofit schedule by June 1, 1990, to the Flight Standards Division Manager in the region of the certificate holding district office.

(2) Show that all of the certificate holder's airplanes required to be equipped in accordance with this section will be equipped by the final compliance date established for TCAS II retrofit.

(3) Comply with its retrofit schedule and submit status reports containing information acceptable to the Administrator. The initial report must be submitted by January 2, 1991, and subsequent reports must be submitted every six months thereafter until completion of the schedule. The reports must be submitted to the certificate holder's assigned Principal Avionics Inspector.

**(d)** *Definitions.* For the purposes of this section the following definitions apply—

(1) *Turbine-powered airplane* includes, e.g., turbofan-, turbojet-, propfan-, and ultra-high bypass fan-powered airplanes. The definition specifically excludes turbopropeller-powered airplanes.

(2) An airplane is considered manufactured on the date the inspection acceptance records reflect that the airplane is complete and meets the FAA Approved Type Design data.

[Docket No. 25954, 55 FR 13242, April 9, 1990]

## §121.359 Cockpit voice recorders.

**(a)** No certificate holder may operate a large turbine engine powered airplane or a large pressurized airplane with four reciprocating engines unless an approved cockpit voice recorder is installed in that airplane and is operated continuously from the start of the use of the checklist (before starting engines for the purpose of flight), to completion of the final checklist at the termination of the flight.

**(b)** [Reserved]

**(c)** The cockpit voice recorder required by paragraph (a) of this section must meet the following application standards:

(1) The requirements of part 25 of this chapter in affect on August 31, 1977.

(2) After September 1, 1980, each recorder container must—

(i) Be either bright orange or bright yellow;

(ii) Have reflective tape affixed to the external surface to facilitate its location under water; and

(iii) Have an approved underwater locating device on or adjacent to the container which is secured in such a manner that they are not likely to be separated during crash impact, unless the cockpit voice recorder, and the flight recorder required by §121.343, are installed adjacent to each other in such a manner that they are not likely to be separated during crash impact.

**(d)** No person may operate a multiengine, turbine-powered airplane having a passenger seat configuration of 10–19 seats unless it is equipped with an approved cockpit voice recorder that:

(1) Is installed in compliance with §23.1457 (a)(1) and (2), (b), (c), (d), (e), (f), and (g); §25.1457 (a)(1) and (2), (b), (c), (d), (e), (f), and (g) of this chapter, as applicable; and

(2) Is operated continuously from the use of the checklist before the flight to completion of the final checklist at the end of the flight.

**(e)** No person may operate a multiengine, turbine-powered airplane having a passenger seat configuration of 20 to 30 seats unless it is equipped with an approved cockpit voice recorder that—

(1) Is installed in compliance with §23.1457 or §25.1457 of this chapter, as applicable; and

(2) Is operated continuously from the use of the checklist before the flight to completion of the final checklist at the end of the flight.

**(f)** In complying with this section, an approved cockpit voice recorder having an erasure feature may be used, so that at any time during the operation of the recorder, information recorded more than 30 minutes earlier may be erased or otherwise obliterated.

**(g)** For those aircraft equipped to record the uninterrupted audio signals received by a boom or a mask microphone, the flight crewmembers are required to use the boom microphone below 18,000 feet mean sea level. No person may operate a large turbine engine powered airplane or a large pressurized airplane with four reciprocating engines manufactured after October 11, 1991, unless it is equipped to record the uninterrupted audio signal received by a boom or mask microphone in accordance with §25.1457(c)(5) of this chapter.

**(h)** In the event of an accident or occurrence requiring immediate notification of the National Transportation Safety Board under part 830 of its regulations, which results in the termination of the flight, the certificate holder shall keep the recorded information for at least 60 days or, if requested by the Administrator or the Board, for a longer period. Information obtained from the record is used to assist in determining the cause of accidents or occurrences in connection with investigations under part 830. The Administrator does not use the record in any civil penalty or certificate action.

[Docket No. 6258, 29 FR 19205, Dec. 31, 1964; as amended by Amdt. 121–20, 31 FR 8912, June 28, 1966; Amdt. 121–23, 31 FR 15192, Dec. 3, 1966; Amdt. 121–32, 32 FR 13914, Oct. 6, 1967; Amdt. 121–130, 41 FR 47229, Oct. 28, 1976; Amdt. 121–135, 42 FR 36973, July 18, 1977; Amdt. 121–143, 43 FR 22642, May 25, 1978; Amdt. 121–197, 53 FR 26147, July 11, 1988; Amdt. 121–251, 60 FR 65933, Dec. 20, 1995]

## §121.360 Ground proximity warning-glide slope deviation alerting system.

**(a)** No person may operate a turbine-powered airplane unless it is equipped with a ground proximity warning system that meets the performance and environmental standards of TSO-C92 (available from the FAA, 800 Independence Avenue SW, Washington, DC 20591) or incorporates TSO-approved ground proximity warning equipment:

**(b)** For the ground proximity warning system required by this section, the Airplane Flight Manual shall contain—

(1) Appropriate procedures for—

(i) The use of the equipment;

(ii) Proper flightcrew action with respect to the equipment;

(iii) Deactivation for planned abnormal and emergency conditions;

(iv) Inhibition of Mode 4 warnings based on flaps being in other than the landing configuration if the system incorporates a Mode 4 flap warning inhibition control; and

(2) An outline of all input sources that must be operating.

**(c)** No person may deactivate a ground proximity warning system required by this section except in accordance with the procedures contained in the Airplane Flight Manual.

**(d)** Whenever a ground proximity warning system required by this section is deactivated, an entry shall be made in the airplane maintenance record that includes the date and time of deactivation.

**(e)** No person may operate a turbine-powered airplane unless it is equipped with a ground proximity warning/glide slope deviation alerting system that meets the performance and environmental standards contained in TSO-C92a or TSO-C92b or incorporates TSO-approved ground proximity warning-glide slope deviation alerting equipment.

**(f)** No person may operate a turbojet powered airplane equipped with a system required by paragraph (e) of this section, that incorporates equipment that meets the performance and environmental standards of TSO-C92b or is approved under that TSO, using other than Warning Envelopes 1 or 3 for Warning Modes 1 and 4.

**(g)** This section expires on March 29, 2005.

[Docket No. 28154, 60 FR 65933, Dec. 20, 1995; as amended by Amdt. 121–273, 65 FR 16755, March 29, 2000]

# Subpart L— Maintenance, Preventive Maintenance, and Alterations

**Source:** Docket No. 6258, 29 FR 19210, Dec. 31, 1964, unless otherwise noted.

## §121.361 Applicability.

**(a)** Except as provided by paragraph (b) of this section, this subpart prescribes requirements for maintenance, preventive maintenance, and alternations for all certificate holders.

**(b)** The Administrator may amend a certificate holder's operations specifications to permit deviation from those provisions of this subpart that would prevent the return to service and use of airframe components, powerplants, appliances, and spare parts thereof because those items have been maintained, altered, or inspected by persons employed outside the United States who do not hold U.S. airman certificates. Each certificate holder who uses parts under this deviation must provide for surveillance of facilities and practices to assure that all work performed on these parts is accomplished in accordance with the certificate holder's manual.

[Docket No. 8754, 33 FR 14406, Sept. 25, 1968]

## §121.363 Responsibility for airworthiness.

**(a)** Each certificate holder is primarily responsible for—

(1) The airworthiness of its aircraft, including airframes, aircraft engines, propellers, appliances, and parts thereof; and

(2) The performance of the maintenance, preventive maintenance, and alteration of its aircraft, including airframes, aircraft engines, propellers, appliances, emergency equipment, and parts thereof, in accordance with its manual and the regulations of this chapter.

**(b)** A certificate holder may make arrangements with another person for the performance of any maintenance, preventive maintenance, or alterations. However, this does not relieve the certificate holder of the responsibility specified in paragraph (a) of this section.

[Docket No. 6258, 29 FR 19210, Dec. 31, 1964; as amended by Amdt. 121–106, 38 FR 22378, Aug. 20, 1973]

## §121.365 Maintenance, preventive maintenance, and alteration organization.

**(a)** Each certificate holder that performs any of its maintenance (other than required inspections), preventive maintenance, or alterations, and each person with whom it arranges for the performance of that work must have an organization adequate to perform the work.

**(b)** Each certificate holder that performs any inspections required by its manual in accordance with §121.369(b)(2) or (3) (in this subpart referred to as *required inspections*) and each person with whom it arranges for the performance of that work must have an organization adequate to perform that work.

**(c)** Each person performing required inspections in addition to other maintenance, preventive maintenance, or alterations, shall organize the performance of those functions so as to separate the required inspection functions from the other maintenance, preventive maintenance, and alteration functions. The separation shall be below the level of administrative control at which overall responsibility for the required inspection functions and other maintenance, preventive maintenance, and alteration functions are exercised.

[Docket No. 6258, 29 FR 19210, Dec. 31, 1964; as amended by Amdt. 121–3, 30 FR 3639, March 19, 1965]

## §121.367 Maintenance, preventive maintenance, and alterations programs.

Each certificate holder shall have an inspection program and a program covering other maintenance, preventive maintenance, and alterations that ensures that—

**(a)** Maintenance, preventive maintenance, and alterations performed by it, or by other persons, are performed in accordance with the certificate holder's manual;

**(b)** Competent personnel and adequate facilities and equipment are provided for the proper performance of maintenance, preventive maintenance, and alterations; and

**(c)** Each aircraft released to service is airworthy and has been properly maintained for operation under this part.

[Docket No. 6258, 29 FR 19210, Dec. 31, 1964; as amended by Amdt. 121–100, 37 FR 28053, Dec. 20, 1972]

## §121.369 Manual requirements.

**(a)** The certificate holder shall put in its manual a chart or description of the certificate holder's organization required by §121.365 and a list of persons with whom it has arranged for the performance of any of its required inspections, other maintenance, preventive maintenance, or alterations, including a general description of that work.

**(b)** The certificate holder's manual must contain the programs required by §121.367 that must be followed in performing maintenance, preventive maintenance, and alterations of that certificate holder's airplanes, including airframes, aircraft engines, propellers, appliances, emergency equipment, and parts thereof, and must include at least the following:

(1) The method of performing routine and nonroutine maintenance (other than required inspections), preventive maintenance, and alterations.

(2) A designation of the items of maintenance and alteration that must be inspected (required inspections), including at least those that could result in a failure, malfunction, or defect endangering the safe operation of the aircraft, if not performed properly or if improper parts or materials are used.

(3) The method of performing required inspections and a designation by occupational title of personnel authorized to perform each required inspection.

(4) Procedures for the reinspection of work performed pursuant to previous required inspection findings (*buy-back procedures*).

(5) Procedures, standards, and limits necessary for required inspections and acceptance or rejection of the items required to be inspected and for periodic inspection and calibration of precision tools, measuring devices, and test equipment.

(6) Procedures to ensure that all required inspections are performed.

(7) Instructions to prevent any person who performs any item of work from performing any required inspection of that work.

(8) Instructions and procedures to prevent any decision of an inspector, regarding any required inspection from being countermanded by persons other than supervisory personnel of the inspection unit, or a person at that level of administrative control that has overall responsibility for the management of both the required inspection functions and the other maintenance, preventive maintenance, and alterations functions.

(9) Procedures to ensure that required inspections, other maintenance, preventive maintenance, and alterations that are not completed as a result of shift changes or similar work interruptions are properly completed before the aircraft is released to service.

**(c)** The certificate holder must set forth in its manual a suitable system (which may include a coded system) that provides for preservation and retrieval of information in a manner acceptable to the Administrator and that provides—

(1) A description (or reference to data acceptable to the Administrator) of the work performed;

(2) The name of the person performing the work if the work is performed by a person outside the organization of the certificate holder; and

(3) The name or other positive identification of the individual approving the work.

[Docket No. 6258, 29 FR 19210, Dec. 31, 1964; as amended by Amdt. 121–94, 37 FR 15983, Aug. 9, 1972; Amdt. 121–106, 38 FR 22378, Aug. 20, 1973]

## §121.370 Special maintenance program requirements.

**(a)** No certificate holder may operate an Airbus Model A300 (excluding the –600 series), British Aerospace Model BAC 1–11, Boeing Model 707, 720, 727, 737, or 747, McDonnell Douglas Model DC–8, DC–9/MD–80 or DC–10, Fokker Model F28, or Lockheed Model L–1011 airplane beyond the applicable flight cycle implementation time specified below, or May 25, 2001, whichever occurs later, unless operations specifications have been issued to reference repair assessment guidelines applicable to the fuselage pressure boundary (fuselage skin, door skin, and bulkhead webs), and those guidelines are incorporated in its maintenance program. The repair assessment guidelines must be approved by the FAA Aircraft Certification Office (ACO), or office of the Transport Airplane Directorate, having cognizance over the type certificate for the affected airplane.

(1) For the Airbus Model A300 (excluding the –600 series), the flight cycle implementation time is:

(i) Model B2: 36,000 flights.

(ii) Model B4–100 (including Model B4–2C): 30,000 flights above the window line, and 36,000 flights below the window line.

(iii) Model B4–200: 25,500 flights above the window line, and 34,000 flights below the window line.

(2) For all models of the British Aerospace BAC 1–11, the flight cycle implementation time is 60,000 flights.

(3) For all models of the Boeing 707, the flight cycle implementation time is 15,000 flights.

(4) For all models of the Boeing 720, the flight cycle implementation time is 23,000 flights.

(5) For all models of the Boeing 727, the flight cycle implementation time is 45,000 flights.

(6) For all models of the Boeing 737, the flight cycle implementation time is 60,000 flights.

(7) For all models of the Boeing 747, the flight cycle implementation time is 15,000 flights.

(8) For all models of the McDonnell Douglas DC–8, the flight cycle implementation time is 30,000 flights.

(9) For all models of the McDonnell Douglas DC–9/MD–80, the flight cycle implementation time is 60,000 flights.

(10) For all models of the McDonnell Douglas DC–10, the flight cycle implementation time is 30,000 flights.

(11) For all models of the Lockheed L–1011, the flight cycle implementation time is 27,000 flights.

(12) For the Fokker F–28 Mark 1000, 2000, 3000, and 4000, the flight cycle implementation time is 60,000 flights.

**(b)** After June 7, 2004, no certificate holder may operate a turbine-powered transport category airplane with a type certificate issued after January 1, 1958, and either a maximum type certificated passenger capacity of 30 or more, or a maximum type certificated payload capacity of 7,500 pounds or more, unless instructions for maintenance and inspection of the fuel tank system are incorporated in its maintenance program. These instructions must address the actual configuration of the fuel tank systems of each affected airplane and must be approved by the FAA Aircraft Certification Office (ACO), or office of the Transport Airplane Directorate, having cognizance over the type certificate for the affected airplane. Operators must submit their request through an appropriate FAA Principal Maintenance Inspector, who may add comments and then send it to the manager of the appropriate office. Thereafter, the approved instructions can be revised only with the approval of the FAA Aircraft Certification Office (ACO), or office of the Transport Airplane Directorate, having cognizance over the type certificate for the affected airplane. Operators must submit their requests for revisions through an appropriate FAA Principal Maintenance Inspector, who may add comments and then send it to the manager of the appropriate office.

[Docket No. 29104, 65 FR 24125, April 25, 2000; as amended by Amdt. 121–275, 65 FR 50744, Aug. 21, 2000; Amdt. 121–282, 66 FR 23130, May 7, 2001]

## §121.371 Required inspection personnel.

**(a)** No person may use any person to perform required inspections unless the person performing the inspection is appropriately certificated, properly trained, qualified, and authorized to do so.

**(b)** No person may allow any person to perform a required inspection unless, at that time, the person performing that inspection is under the supervision and control of an inspection unit.

**(c)** No person may perform a required inspection if he performed the item of work required to be inspected.

**(d)** Each certificated holder shall maintain, or shall determine that each person with whom it arranges to perform its required inspections maintains, a current listing of persons who have been trained, qualified, and authorized to conduct required inspections. The persons must be identified by name, occupational title, and the inspections that they are authorized to perform. The certificated holder (or person with whom it arranges to perform its required inspections) shall give written information to each person so authorized describing the extent of his responsibilities authorities, and inspectional limitations. The list shall be made available for inspection by the Administrator upon request.

## §121.373 Continuing analysis and surveillance.

**(a)** Each certificate holder shall establish and maintain a system for the continuing analysis and surveillance of the performance and effectiveness of its inspection program and the program covering other maintenance, preventive maintenance, and alterations and for the correction of any deficiency in those programs, regardless of whether those programs are carried out by the certificate holder or by another person.

**(b)** Whenever the Administrator finds that either or both of the programs described in paragraph (a) of this section does not contain adequate procedures and standards to meet the requirements of this part, the certificate holder shall, after notification by the Administrator, make any changes in those programs that are necessary to meet those requirements.

**(c)** A certificate holder may petition the Administrator to reconsider the notice to make a change in a program. The petition must be filed with the FAA certificate-holding district office charged with the overall inspection of the certificate holder's operations within 30 days after the certificate holder receives the notice. Except in the case of an emergency requiring immediate action in the interest of safety, the filing of the petition stays the notice pending a decision by the Administrator.

[Docket No. 6258, 29 FR 19210, Dec. 31, 1964; as amended by Amdt. 121–207, 54 FR 39293, Sept. 25, 1989; Amdt. 121–253, 61 FR, Jan. 26, 1996]

## §121.375 Maintenance and preventive maintenance training program.

Each certificate holder or person performing maintenance or preventive maintenance functions for it shall have a training program to ensure that each person (including inspection personnel) who determines the adequacy of work done is fully informed about procedures and techniques and new equipment in use and is competent to perform his duties.

## §121.377 Maintenance and preventive maintenance personnel duty time limitations.

Within the United States, each certificate holder (or person performing maintenance or preventive maintenance functions for it) shall relieve each person performing maintenance or preventive maintenance from duty for a period of at least 24 consecutive hours during any seven consecutive days, or the equivalent thereof within any one calendar month.

## §121.378 Certificate requirements.

**(a)** Except for maintenance, preventive maintenance, alterations, and required inspections performed by a certificated repair station that is located outside the United States, each person who is directly in charge of maintenance, preventive maintenance, or alterations, and each person performing required inspections must hold an appropriate airman certificate.

**(b)** For the purposes of this section, a person *directly in charge* is each person assigned to a position in which he is responsible for the work of a shop or station that performs maintenance, preventive maintenance, alterations, or other functions affecting aircraft airworthiness. A person who is *directly in charge* need not physically observe and direct each worker constantly but must be available for consultation and decision on matters requiring instruction or decision from higher authority than that of the persons performing the work.

[Docket No. 6258, 29 FR 19210, Dec. 31, 1964; as amended by Amdt. 121–21, 31 FR 10618, Aug. 9, 1966; Amdt. 121–286, 66 FR 41116, Aug. 6, 2001]

## §121.379 Authority to perform and approve maintenance, preventive maintenance, and alterations.

**(a)** A certificate holder may perform, or it may make arrangements with other persons to perform, maintenance, preventive maintenance, and alterations as provided in its continuous airworthiness maintenance program and its maintenance manual. In addition, a certificate holder may perform these functions for another certificate holder as provided in the continuous airworthiness maintenance program and maintenance manual of the other certificate holder.

**(b)** A certificate holder may approve any aircraft, airframe, aircraft engine, propeller, or appliance for return to service after maintenance, preventive maintenance, or alterations that are performed under paragraph (a) of this section. However, in the case of a major repair or major alteration, the work must have been done in accordance with technical data approved by the Administrator.

[Docket No. 10289, 35 FR 16793, Oct. 30, 1970]

## §121.380 Maintenance recording requirements.

**(a)** Each certificate holder shall keep (using the system specified in the manual required in §121.369) the following records for the periods specified in paragraph (c) of this section:

(1) All the records necessary to show that all requirements for the issuance of an airworthiness release under §121.709 have been met.

(2) Records containing the following information:

(i) The total time in service of the airframe.

(ii) Except as provided in paragraph (b) of this section, the total time in service of each engine and propeller.

(iii) The current status of life-limited parts of each airframe, engine, propeller, and appliance.

(iv) The time since last overhaul of all items installed on the aircraft which are required to be overhauled on a specified time basis.

(v) The identification of the current inspection status of the aircraft, including the times since the last inspections required by the inspection program under which the aircraft and its appliances are maintained.

(vi) The current status of applicable airworthiness directives, including the date and methods of compliance, and, if the airworthiness directive involves recurring action, the time and date when the next action is required.

(vii) A list of current major alterations to each airframe, engine, propeller, and appliance.

**(b)** A certificate holder need not record the total time in service of an engine or propeller on a transport category airplane that has a passenger seat configuration of more than 30 seats or a nontransport category airplane type certificated before January 1, 1958, until the following, whichever occurs first:

(1) March 20, 1997; or

(2) The date of the first overhaul of the engine or propeller, as applicable, after January 19, 1996

**(c)** Each certificate holder shall retain the records required to be kept by this section for the following periods:

(1) Except for the records of the last complete overhaul of each airframe, engine, propeller, and appliance, the records specified in paragraph (a)(1) of this section shall be retained until the

work is repeated or superseded by other work or for one year after the work is performed.

(2) The records of the last complete overhaul of each airframe, engine, propeller, and appliance shall be retained until the work is superseded by work of equivalent scope and detail.

(3) The records specified in paragraph (a)(2) of this section shall be retained and transferred with the aircraft at the time the aircraft is sold.

**(d)** The certificate holder shall make all maintenance records required to be kept by this section available for inspection by the Administrator or any authorized representative of the National Transportation Safety Board (NTSB).

[Docket No. 10658, 37 FR 15983, Aug. 9, 1972; as amended by Amdt. 121–251, 60 FR 65933, Dec. 20, 1995]

### §121.380a Transfer of maintenance records.

Each certificate holder who sells a U.S. registered aircraft shall transfer to the purchaser, at the time of sale, the following records of that aircraft, in plain language form or in coded form at the election of the purchaser, if the coded form provides for the preservation and retrieval of information in a manner acceptable to the Administrator:

**(a)** The record specified in §121.380(a)(2).

**(b)** The records specified in §121.380(a)(1) which are not included in the records covered by paragraph (a) of this section, except that the purchaser may permit the seller to keep physical custody of such records. However, custody of records in the seller does not relieve the purchaser of his responsibility under §121.380(c) to make the records available for inspection by the Administrator or any authorized representative of the National Transportation Safety Board (NTSB).

[Docket No. 10658, 37 FR 15984, Aug. 9, 1972]

## Subpart M— Airman and Crewmember Requirements

**Source:** Docket No. 6258, 29 FR 19212, Dec. 31, 1964, unless otherwise noted.

### §121.381 Applicability.

This subpart prescribes airman and crewmember requirements for all certificate holders.

### §121.383 Airman: Limitations on use of services.

**(a)** No certificate holder may use any person as an airman nor may any person serve as an airman unless that person—

(1) Holds an appropriate current airman certificate issued by the FAA;

(2) Has any required appropriate current airman and medical certificates in his possession while engaged in operations under this part; and

(3) Is otherwise qualified for the operation for which he is to be used.

**(b)** Each airman covered by paragraph (a)(2) of this section shall present either or both certificates for inspection upon the request of the Administrator.

**(c)** No certificate holder may use the services of any person as a pilot on an airplane engaged in operations under this part if that person has reached his 60th birthday. No person may serve as a pilot on an airplane engaged in operations under this part if that person has reached his 60th birthday.

[Docket No. 6258, 29 FR 19212, Dec. 31, 1964; as amended by Amdt. 121–144, 43 FR 22646, May 25, 1978]

### §121.385 Composition of flight crew.

**(a)** No certificate holder may operate an airplane with less than the minimum flight crew in the airworthiness certificate or the airplane Flight Manual approved for that type airplane and required by this part for the kind of operation being conducted.

**(b)** In any case in which this part requires the performance of two or more functions for which an airman certificate is necessary, that requirement is not satisfied by the performance of multiple functions at the same time by one airman.

**(c)** The minimum pilot crew is two pilots and the certificate holder shall designate one pilot as pilot in command and the other second in command.

**(d)** On each flight requiring a flight engineer at least one flight crewmember, other than the flight engineer, must be qualified to provide emergency performance of the flight engineer's functions for the safe completion of the flight if the flight engineer becomes ill or is otherwise incapacitated. A pilot need not hold a flight engineer's certificate to perform the flight engineer's functions in such a situation.

[Docket No. 6258, 29 FR 19212, Dec. 31, 1964; as amended by Amdt. 121–178, 47 FR 13316, March 29, 1982; Amdt. 121–253, 61 FR, Jan. 26, 1996; Amdt. 121–256, 61 FR 30434, June 14, 1996]

## §121.387 Flight engineer.

No certificate holder may operate an airplane for which a type certificate was issued before January 2, 1964, having a maximum certificated takeoff weight of more than 80,000 pounds without a flight crewmember holding a current flight engineer certificate. For each airplane type certificated after January 1, 1964, the requirement for a flight engineer is determined under the type certification requirements of §25.1523.

[Docket No. 5025, 30 FR 6067, April 29, 1965]

## §121.389 Flight navigator and specialized navigation equipment.

**(a)** No certificate holder may operate an airplane outside the 48 contiguous States and the District of Columbia, when its position cannot be reliably fixed for a period of more than 1 hour, without—

(1) A flight crewmember who holds a current flight navigator certificate; or

(2) Specialized means of navigation approved in accordance with §121.355 which enables a reliable determination to be made of the position of the airplane by each pilot seated at his duty station.

**(b)** Notwithstanding paragraph (a) of this section, the Administrator may also require a flight navigator or special navigation equipment, or both, when specialized means of navigation are necessary for 1 hour or less. In making this determination, the Administrator considers—

(1) The speed of the airplane;

(2) Normal weather conditions en route;

(3) Extent of air traffic control;

(4) Traffic congestion;

(5) Area of navigational radio coverage at destination;

(6) Fuel requirements;

(7) Fuel available for return to point of departure or alternates;

(8) Predication of flight upon operation beyond the point of no return; and

(9) Any other factors he determines are relevant in the interest of safety.

**(c)** Operations where a flight navigator or special navigation equipment, or both, are required are specified in the operations specifications of the air carrier or commercial operator.

[Docket No. 10204, 37 FR 6464, March 30, 1972; as amended by Amdt. 121–178, 47 FR 13316, March 29, 1982]

## §121.391 Flight attendants.

**(a)** Each certificate holder shall provide at least the following flight attendants on each passenger-carrying airplane used:

(1) For airplanes having a maximum payload capacity of more than 7,500 pounds and having a seating capacity of more than 9 but less than 51 passengers—one flight attendant.

(2) For airplanes having a maximum payload capacity of 7,500 pounds or less and having a seating capacity of more than 19 but less than 51 passengers—one flight attendant.

(3) For airplanes having a seating capacity of more than 50 but less than 101 passengers—two flight attendants.

(4) For airplanes having a seating capacity of more than 100 passengers—two flight attendants plus one additional flight attendant for each unit (or part of a unit) of 50 passenger seats above a seating capacity of 100 passengers.

**(b)** If, in conducting the emergency evacuation demonstration required under §121.291(a) or (b), the certificate holder used more flight attendants than is required under paragraph (a) of this section for the maximum seating capacity of the airplane used in the demonstration, he may not, thereafter, take off that airplane—

(1) In its maximum seating capacity configuration with fewer flight attendants than the number used during the emergency evacuation demonstration; or

(2) In any reduced seating capacity configuration with fewer flight attendants than the number required by paragraph (a) of this section for that seating capacity plus the number of flight attendants used during the emergency evacuation demonstration that were in excess of those required under paragraph (a) of this section.

**(c)** The number of flight attendants approved under paragraphs (a) and (b) of this section are set forth in the certificate holder's operations specifications.

**(d)** During takeoff and landing, flight attendants required by this section shall be located as near as practicable to required floor level exists and shall be uniformly distributed throughout the airplane in order to provide the most effective egress of passengers in event of an emergency evacuation. During taxi, flight attendants required by this section must remain at their duty stations with safety belts and shoulder harnesses fastened except to perform duties related to the safety of the airplane and its occupants.

[Docket No. 2033, 30 FR 3206, March 9, 1965; as amended by Amdt. 121–30, 32 FR 13268, Sept. 20, 1967; Amdt. 121–46, 34 FR 5545, March 22, 1969; Amdt. 121–84, 37 FR 3975, Feb. 24, 1972; Amdt. 121–88, 37 FR 5606, March 17, 1972; Amdt. 121–159, 45 FR 41593, June 19, 1980; Amdt. 121–176, 46 FR 61454, Dec. 17, 1981; Amdt. 121–180, 47 FR 56463, Dec. 16, 1982; Amdt. 121–251, 60 FR 65933, Dec. 20, 1995]

## §121.393 Crewmember requirements at stops where passengers remain on board.

At stops where passengers remain on board, the certificate holder must meet the following requirements:

**(a)** On each airplane for which a flight attendant is not required by §121.391(a), the certificate holder must ensure that a person who is qualified in the emergency evacuation procedures for the airplane, as required in §121.417, and who is identified to the passengers, remains:

(1) On board the airplane; or

(2) Nearby the airplane, in a position to adequately monitor passenger safety, and:

(i) The airplane engines are shut down; and

(ii) At least one floor level exit remains open to provide for the deplaning of passengers.

**(b)** On each airplane for which flight attendants are required by §121.391(a), but the number of flight attendants remaining on board is fewer than required by §121.391(a), the certificate holder must meet the following requirements:

(1) The certificate holder shall ensure that:

(i) The airplane engines are shut down;

(ii) At least one floor level exit remains open to provide for the deplaning of passengers; and

(iii) The number of flight attendants on board is at least half the number required by §121.391(a), rounded down to the next lower number in the case of fractions, but never fewer than one.

(2) The certificate holder may substitute for the required flight attendants other persons qualified in the emergency evacuation procedures for that aircraft as required in §121.417, if these persons are identified to the passengers.

(3) If only one flight attendant or other qualified person is on board during a stop, that flight attendant or other qualified person shall be located in accordance with the certificate holder's FAA-approved operating procedures. If more than one flight attendant or other qualified person is on board, the flight attendants or other qualified persons shall be spaced throughout the cabin to provide the most effective assistance for the evacuation in case of an emergency.

[Docket No. 28154, 60 FR 65934, Dec. 20, 1995]

## §121.395 Aircraft dispatcher: Domestic and flag operations.

Each certificate holder conducting domestic or flag operations shall provide enough qualified aircraft dispatchers at each dispatch center to ensure proper operational control of each flight.

[Docket No. 28154, 61 FR 2611, Jan. 26, 1996]

## §121.397 Emergency and emergency evacuation duties.

**(a)** Each certificate holder shall, for each type and model of airplane, assigned to each category of required crewmember, as appropriate, the necessary functions to be performed in an emergency or a situation requiring emergency evacuation. The certificate holder shall show those functions are realistic, can be practically accomplished, and will meet any reasonably anticipated emergency including the possible incapacitation of individual crewmembers or their inability to reach the passenger cabin because of shifting cargo in combination cargo-passenger airplanes.

**(b)** The certificate holder shall describe in its manual the functions of each category of required crewmembers under paragraph (a) of this section.

[Docket No. 2033, 30 FR 3206, March 9, 1965; as amended by Amdt. 121–7, 30 FR 6727, May 18, 1965]

# Subpart N— Training Program

SOURCE: Docket No. 9509, 35 FR 90, Jan. 3, 1970, unless otherwise noted.

## §121.400 Applicability and terms used.

**(a)** This subpart prescribes the requirements applicable to each certificate holder for establishing and maintaining a training program for crewmembers, aircraft dispatchers, and other operations personnel, and for the approval and use of training devices in the conduct of the program.

**(b)** For the purpose of this subpart, airplane groups are as follows:

(1) *Group I.* Propeller driven, including—

(i) Reciprocating powered; and

(ii) Turbopropeller powered.

(2) *Group II.* Turbojet powered.

**(c)** For the purpose of this subpart, the following terms and definitions apply:

(1) *Initial training.* The training required for crewmembers and dispatchers who have not qualified and served in the same capacity on another airplane of the same group.

(2) *Transition training.* The training required for crewmembers and dispatchers who have qualified and served in the same capacity on another airplane of the same group.

(3) *Upgrade training.* The training required for crewmembers who have qualified and served as second in command or flight engineer on a particular airplane type, before they serve as pilot in command or second in command, respectively, on that airplane.

(4) *Differences training.* The training required for crewmembers and dispatchers who have qual-

ified and served on a particular type airplane, when the Administrator finds differences training is necessary before a crewmember serves in the same capacity on a particular variation of that airplane.

(5) *Programmed hours.* The hours of training prescribed in this subpart which may be reduced by the Administrator upon a showing by the certificate holder that circumstances justify a lesser amount.

(6) *Inflight.* Refers to maneuvers, procedures, or functions that must be conducted in the airplane.

(7) *Training center.* An organization governed by the applicable requirements of part 142 of this chapter that provides training, testing, and checking under contract or other arrangement to certificate holders subject to the requirements of this part.

(8) *Requalification training.* The training required for crewmembers previously trained and qualified, but who have become unqualified due to not having met within the required period the recurrent training requirements of §121.427 or the proficiency check requirements of §121.441.

[Docket No. 9509, 35 FR 90, Jan. 3, 1970; 35 FR 2819, Feb. 11, 1970; as amended by Amdt. 121–104, 38 FR 14915, June 7, 1973; Amdt. 121–259, 61 FR 34560, July 2, 1996]

## §121.401 Training program: General.

**(a)** Each certificate holder shall:

(1) Establish, obtain the appropriate initial and final approval of, and provide, a training program that meets the requirements of this subpart and appendixes E and F and that insures that each crewmember, aircraft dispatcher, flight instructor, and check airman, and each person assigned duties for the carriage and handling of dangerous articles and magnetized materials, is adequately trained to perform his assigned duties.

(2) Provide adequate ground and flight training facilities and properly qualified ground instructors for the training required by this subpart;

(3) Provide and keep current with respect to each airplane type and, if applicable, the particular variations within that airplane type, appropriate training material, examinations, forms, instructions, and procedures for use in conducting the training and checks required by this part; and

(4) Provide enough flight instructors, simulator instructors, and approved check airmen to conduct required flight training and flight checks, and simulator training courses permitted under this part.

**(b)** Whenever a crewmember or aircraft dispatcher who is required to take recurrent training, a flight check, or a competence check, takes the check or completes the training in the calendar month before or after the calendar month in which that training or check is required, he is considered to have taken or completed it in the calendar month in which it was required.

**(c)** Each instructor, supervisor, or check airman who is responsible for a particular ground training subject, segment of flight training, course of training, flight check, or competence check under this part shall certify as to the proficiency and knowledge of the crewmember, aircraft dispatcher, flight instructor, or check airman concerned upon completion of that training or check. That certification shall be made a part of the crewmember's or dispatcher's record. When the certification required by this paragraph is made by an entry in a computerized recordkeeping system, the certifying instructor, supervisor, or check airman must be identified with that entry. However, the signature of the certifying instructor, supervisor, or check airman is not required for computerized entries.

**(d)** Training subjects that are applicable to more than one airplane or crewmember position and that have been satisfactorily completed in connection with prior training for another airplane or another crewmember position, need not be repeated during subsequent training other than recurrent training.

**(e)** A person who progresses successfully through flight training, is recommended by his instructor or a check airman, and successfully completes the appropriate flight check for a check airman or the Administrator, need not complete the programmed hours of flight training for the particular airplane. However, whenever the Administrator finds that 20 percent of the flight checks given at a particular training base during the previous 6 months under this paragraph are unsuccessful, this paragraph may not be used by the certificate holder at that base until the Administrator finds that the effectiveness of the flight training there has improved.

In the case of a certificate holder using a course of training permitted in §121.409(c), the Administrator may require the programmed hours of inflight training in whole or in part, until he finds the effectiveness of the flight training has improved as provided in paragraph (e) of this section.

[Docket No. 9509, 35 FR 90, Jan. 3, 1970; as amended by Amdt. 121–104, 38 FR 14915, June 7, 1973; Amdt. 121–108, 38 FR 35446, Dec. 28, 1973; Amdt. 121–143, 43 FR 22642, May 25, 1978]

## §121.402 Training program: Special rules.

**(a)** Other than the certificate holder, only another certificate holder certificated under this part or a flight training center certificated under part 142 of this chapter is eligible under this subpart to provide flight training, testing, and checking under contract or other arrangement to those persons subject to the requirements of this subpart.

**(b)** A certificate holder may contract with, or otherwise arrange to use the services of, a training center certificated under part 142 of this chapter to provide training, testing, and checking required by this part only if the training center—

(1) Holds applicable training specifications issued under part 142 of this chapter;

(2) Has facilities, training equipment, and courseware meeting the applicable requirements of part 142 of this chapter;

(3) Has approved curriculums, curriculum segments, and portions of curriculum segments applicable for use in training courses required by this subpart; and

(4) Has sufficient instructor and check airmen qualified under the applicable requirements of §§121.411 or 121.413 to provide training, testing, and checking to persons subject to the requirements of this subpart.

[Docket No. 26933, 61 FR 34560, July 2, 1996; as amended by Amdt. 121–263, 62 FR 13791, March 21, 1997]

## §121.403 Training program: Curriculum.

**(a)** Each certificate holder must prepare and keep current a written training program curriculum for each type of airplane with respect to dispatchers and each crewmember required for that type airplane. The curriculum must include ground and flight training required by this subpart.

**(b)** Each training program curriculum must include:

(1) A list of principal ground training subjects, including emergency training subjects, that are provided.

(2) A list of all the training devices mockups, systems trainers, procedures trainers, or other training aids that the certificate holder will use.

(3) Detailed descriptions or pictorial displays of the approved normal, abnormal, and emergency maneuvers, procedures and functions that will be performed during each flight training phase or flight check, indicating those maneuvers, procedures and functions that are to be performed during the inflight portions of flight training and flight checks.

(4) A list of airplane simulators or other training devices approved under §121.407, including approvals for particular maneuvers, procedures, or functions.

(5) The programmed hours of training that will be applied to each phase of training.

(6) A copy of each statement issued by the Administrator under §121.405(d) for reduction of programmed hours of training.

## §121.404 Compliance dates: Crew and dispatcher resource management training.

After March 19, 1998, no certificate holder may use a person as a flight crewmember, and after March 19, 1999, no certificate holder may use a person as a flight attendant or aircraft dispatcher unless that person has completed approved crew resource management (CRM) or dispatcher resource management (DRM) initial training, as applicable, with that certificate holder or with another certificate holder.

[Docket No. 28154, 61 FR 30435, June 14, 1996]

## §121.405 Training program and revision: Initial and final approval.

**(a)** To obtain initial and final approval of a training program, or a revision to an approved training program, each certificate holder must submit to the Administrator—

(1) An outline of the proposed program or revision, including an outline of the proposed or revised curriculum, that provides enough information for a preliminary evaluation of the proposed training program or revised training program; and

(2) Additional relevant information as may be requested by the Administrator.

**(b)** If the proposed training program or revision complies with this subpart the Administrator grants initial approval in writing after which the certificate holder may conduct the training in accordance with that program. The Administrator then evaluates the effectiveness of the training program and advises the certificate holder of deficiencies, if any, that must be corrected.

**(c)** The Administrator grants final approval of the training program or revision if the certificate holder shows that the training conducted under the initial approval set forth in paragraph (b) of this section ensures that each person that successfully completes the training is adequately trained to perform his assigned duties.

**(d)** In granting initial and final approval of training programs or revisions, including reductions in programmed hours specified in this subpart, the Administrator considers the training aids, devices, methods, and procedures listed in the certificate holder's curriculum as set forth in §121.403 that increase the quality and effectiveness of the teaching-learning process.

If approval of reduced programmed hours of training is granted, the Administrator provides the certificate holder with a statement of the basis for the approval.

**(e)** Whenever the Administrator finds that revisions are necessary for the continued adequacy of a training program that has been granted final approval, the certificate holder shall, after notification by the Administrator, make any changes in the program that are found necessary by the Administrator. Within 30 days after the certificate holder receives such notice, it may file a petition to reconsider the notice with the certificate-holding district office. The filing of a petition to reconsider stays the notice pending a decision by the Administrator. However, if the Administrator finds that there is an emergency that requires immediate action in the interest of safety in air transportation, he may, upon a statement of the reasons, require a change effective without stay.

**(f)** Each certificate holder described in §135.3(b) and (c) of this chapter must include the material required by §121.403 in the manual required by §135.21 of this chapter.

**(g)** The Administrator may grant a deviation to certificate holders described in §135.3(b) and (c) of this chapter to allow reduced programmed hours of ground training required by §121.419 if it is found that a reduction is warranted based on the certificate holder's operations and the complexity of the make, model, and series of the aircraft used.

[Docket No. 9509, 35 FR 90, Jan. 3, 1970; as amended by Amdt. 121–207, 54 FR 39293, Sept. 25, 1989; Amdt. 121–250, 60 FR 65948, Dec. 20, 1995; Amdt. 121–253, 61 FR 2612, Jan. 26, 1996]

## §121.406 Credit for previous CRM/DRM training.

**(a)** For flightcrew members, the Administrator may credit CRM training received before March 19, 1998, toward all or part of the initial ground CRM training required by §121.419.

**(b)** For flight attendants, the Administrator may credit CRM training received before March 19, 1999, toward all or part of the initial ground CRM training required by §121.421.

**(c)** For aircraft dispatchers, the Administrator may credit CRM training received before March 19, 1999, toward all or part of the initial ground CRM training required by §121.422.

**(d)** In granting credit for initial ground CRM or DRM training, the Administrator considers training aids, devices, methods, and procedures used by the certificate holder in a voluntary CRM or DRM program or in an AQP program that effectively meets the quality of an approved CRM or DRM initial ground training program under sections 121.419, 121.421, or 121.422 as appropriate.

[Docket No. 27993, 60 FR 65949, Dec. 20, 1995; as amended by Amdt. 121–256, 61 FR 30435, June 14, 1996]

## §121.407 Training program: Approval of airplane simulators and other training devices.

**(a)** Each airplane simulator and other training device that is used in a training course permitted under §121.409, in checks required under Subpart O of this part or as permitted in Appendices E and F to this part must:

(1) Be specifically approved for—

(i) The certificate holder;

(ii) The type airplane and, if applicable, the particular variation within type, for which the training or check is being conducted; and

(iii) The particular maneuver, procedure, or crewmember function involved.

(2) Maintain the performance, functional, and other characteristics that are required for approval.

(3) Be modified to conform with any modification to the airplane being simulated that results in changes to performance, functional, or other characteristics required for approval.

(4) Be given a daily functional preflight check before being used.

(5) Have a daily discrepancy log kept with each discrepancy entered in that log by the appropriate instructor or check airman at the end of each training or check flight.

**(b)** A particular airplane simulator or other training device may be approved for use by more than one certificate holder.

**(c)** An airplane simulator may be used instead of the airplane to satisfy the in-flight requirements of §§121.439 and 121.441 and Appendices E and F of this part, if the simulator—

(1) Is approved under this section and meets the appropriate simulator requirements of Appendix H of this part; and

(2) Is used as part of an approved program that meets the training requirements of §121.424(a) and (c) and Appendix H of this part.

**(d)** An airplane simulator approved under this section must be used instead of the airplane to satisfy the pilot flight training requirements prescribed in the certificate holder's approved low-altitude windshear flight training program set forth in §121.409(d) of this part.

[Docket No. 9509, 35 FR 90, Jan. 3, 1970; as amended by Amdt. 121–161, 45 FR 44183, June 30, 1980; Amdt. 121–199, 53 FR 37696, Sept. 27, 1988]

## §121.409 Training courses using airplane simulators and other training devices.

**(a)** Training courses utilizing airplane simulators and other training devices may be included in the certificate holder's approved training program for use as provided in this section.

**(b)** A course of training in an airplane simulator may be included for use as provided in §121.441 if that course—

(1) Provides at least 4 hours of training at the pilot controls of an airplane simulator as well as a proper briefing before and after the training;

(2) Provides training in at least the procedures and maneuvers set forth in Appendix F to this part; or

(3) Provides line-oriented training that—

(i) Utilizes a complete flight crew;

(ii) Includes at least the maneuvers and procedures (abnormal and emergency) that may be expected in line operations;

(iii) Is representative of the flight segment appropriate to the operations being conducted by the certificate holder; and

(4) Is given by an instructor who meets the applicable requirements of §121.412.

The satisfactory completion of the course of training must be certified by either the Administrator or a qualified check airman.

**(c)** The programmed hours of flight training set forth in this subpart do not apply if the training program for the airplane type includes—

(1) A course of pilot training in an airplane simulator as provided in §121.424(d); or

(2) A course of flight engineer training in an airplane simulator or other training device as provided in §121.425(c).

**(d)** Each certificate holder required to comply with §121.358 of this part must use an approved simulator for each airplane type in each of its pilot training courses that provides training in at least the procedures and maneuvers set forth in the certificate holder's approved low-altitude windshear flight training program. The approved low-altitude windshear flight training, if applicable, must be included in each of the pilot flight training courses prescribed in §§121.409(b), 121.418, 121.424, and 121.427 of this part.

[Docket No. 9509, 35 FR 90, Jan. 3, 1970; as amended by Amdt. 121–130, 41 FR 47229, Oct. 28, 1976; Amdt. 121–144, 43 FR 22646, May 25, 1978; Amdt. 121–199, 53 FR 37696, Sept. 27, 1988; Amdt. 121–264, 62 FR 23120, April 28, 1997]

## §121.411 Qualifications: Check airmen (airplane) and check airmen (simulator).

**(a)** For the purposes of this section and §121.413:

(1) A check airman (airplane) is a person who is qualified, and permitted, to conduct flight checks or instruction in an airplane, in a flight simulator, or in a flight training device for a particular type airplane.

(2) A check airman (simulator) is a person who is qualified to conduct flight checks or instruction, but only in a flight simulator or in a flight training device for a particular type airplane.

(3) Check airmen (airplane) and check airmen (simulator) are those check airmen who perform the functions described in §121.401(a)(4).

**(b)** No certificate holder may use a person, nor may any person serve as a check airman (airplane) in a training program established under this subpart unless, with respect to the airplane type involved, that person—

(1) Holds the airman certificates and ratings required to serve as a pilot in command, a flight engineer, or a flight navigator, as applicable, in operations under this part;

(2) Has satisfactorily completed the appropriate training phases for the airplane, including recurrent training, that are required to serve as a pilot in command, flight engineer, or flight navigator, as applicable, in operations under this part;

(3) Has satisfactorily completed the appropriate proficiency or competency checks that are required to serve as a pilot in command, flight engineer, or flight navigator, as applicable, in operations under this part;

(4) Has satisfactorily completed the applicable training requirements of §121.413 including in-flight training and practice for initial and transition training;

(5) Holds at least a Class III medical certificate unless serving as a required crewmember, in which case holds a Class I or Class II medical certificate as appropriate;

(6) Has satisfied the recency of experience requirements of §121.439; and

(7) Has been approved by the Administrator for the check airman duties involved.

**(c)** No certificate holder may use a person nor may any person serve as a check airman (simulator) in a training program established under this subpart unless, with respect to the airplane type involved, that person meets the provisions of paragraph (b) of this section, or—

(1) Holds the airman certificates and ratings, except medical certificate, required to serve as a pilot in command, a flight engineer, or a flight navigator, as applicable, in operations under this part;

(2) Has satisfactorily completed the appropriate training phases for the airplane, including recurrent training, that are required to serve as a pilot in command, flight engineer, or flight navigator in operations under this part;

(3) Has satisfactorily completed the appropriate proficiency or competency checks that are required to serve as a pilot in command, flight engineer, or flight navigator in operations under this part;

(4) Has satisfactorily completed the applicable training requirements of §121.413; and

(5) Has been approved by the Administrator for the check airman (simulator) duties involved.

**(d)** Completion of the requirements in paragraphs (b)(2), (3), and (4) or (c)(2), (3), and (4) of this section, as applicable, shall be entered in the individual's training record maintained by the certificate holder.

**(e)** Check airmen who have reached their 60th birthday or who do not hold an appropriate medical certificate may function as check airmen, but may not serve as pilot flightcrew members in operations under this part.

**(f)** A check airman (simulator) must accomplish the following—

(1) Fly at least two flight segments as a required crewmember for the type airplane involved within the 12-month period preceding the performance of any check airman duty in a flight simulator; or

(2) Satisfactorily complete an approved line-observation program within the period prescribed by that program and that must precede the performance of any check airman duty in a flight simulator.

**(g)** The flight segments or line-observation program required in paragraph (f) of this section are considered to be completed in the month required if completed in the calendar month before or in the calendar month after the month in which it is due.

[Docket No. 28471, 61 FR 30741, June 17, 1996]

## §121.412 Qualifications: Flight instructors (airplane) and flight instructors (simulator).

**(a)** For the purposes of this section and §121.414:

(1) A flight instructor (airplane) is a person who is qualified to instruct in an airplane, in a flight simulator, or in a flight training device for a particular type airplane.

(2) A flight instructor (simulator) is a person who is qualified to instruct, but only in a flight simulator, in a flight training device, or both, for a particular type airplane.

(3) Flight instructors (airplane) and flight instructors (simulator) are those instructors who perform the functions described in §121.401(a)(4).

**(b)** No certificate holder may use a person nor may any person serve as a flight instructor (airplane) in a training program established under this subpart unless, with respect to the airplane type involved, that person—

(1) Holds the airman certificates and rating required to serve as a pilot in command, a flight engineer, or a flight navigator, as applicable, in operations under this part;

(2) Has satisfactorily completed the appropriate training phases for the airplane, including recurrent training, that are required to serve as a pilot in command, flight engineer, or flight navigator, as applicable, in operations under this part;

(3) Has satisfactorily completed the appropriate proficiency or competency checks that are required to serve as a pilot in command, flight engineer, or flight navigator, as applicable, in operations under this part;

(4) Has satisfactorily completed the applicable training requirements of §121.414, including in-flight training and practice for initial and transition training;

(5) Holds at least a Class III medical certificate unless serving as a required crewmember, in which case holds a Class I or a Class II medical certificate as appropriate.

(6) Has satisfied the recency of experience requirements of §121.439.

**(c)** No certificate holder may use a person, nor may any person serve as a flight instructor (simulator) in a training program established under this subpart, unless, with respect to the airplane type involved, that person meets the provisions of paragraph (b) of this section, or—

(1) Holds the airman certificates and ratings, except medical certificate, required to serve as a pilot in command, a flight engineer, or a flight navigator, as applicable, in operations under this part except before March 19, 1997 that person need not hold a type rating for the airplane type involved provided that he or she only provides the instruction described in §§121.409(b) and 121.441;

(2) Has satisfactorily completed the appropriate training phases for the airplane, including recurrent training, that are required to serve as a pilot in command, flight engineer, or flight navigator, as applicable, in operations under this part;

(3) Has satisfactorily completed the appropriate proficiency or competency checks that are required to serve as a pilot in command, flight engineer, or flight navigator, as applicable, in operations under this part; and

(4) Has satisfactorily completed the applicable training requirements of §121.414.

**(d)** Completion of the requirements in paragraphs (b)(2), (3), and (4) or (c)(2), (3), and (4) of this section as applicable shall be entered in the individual's training record maintained by the certificate holder.

**(e)** Flight instructors who have reached their 60th birthday, or who do not hold an appropriate medical certificate, may function as flight instructors, but may not serve as pilot flight crewmembers in operations under this part.

**(f)** A flight instructor (simulator) must accomplish the following—

(1) Fly at least two flight segments as a required crewmember for the type of airplane within the 12-month period preceding the performance of any flight instructor duty in a flight simulator

(and must hold a Class I or Class II medical certificate as appropriate); or

(2) Satisfactorily complete an approved line-observation program within the period prescribed by that program and that must precede the performance of any check airman duty in a flight simulator.

**(g)** The flight segments or line-observation program required in paragraph (f) of this section is considered completed in the month required if completed in the calendar month before, or the calendar month after the month in which it is due.

[Docket No. 28471, 61 FR 30742, June 17, 1996; as amended by Amdt. 121–257, 62 FR 3739, Jan. 24, 1997; Amdt. 121–264, 62 FR 23120, April 28, 1997]

## §121.413 Initial and transition training and checking requirements: Check airmen (airplane), check airmen (simulator).

**(a)** No certificate holder may use a person nor may any person serve as a check airman unless—

(1) That person has satisfactorily completed initial or transition check airman training; and

(2) Within the preceding 24 calendar months that person satisfactorily conducts a proficiency or competency check under the observation of an FAA inspector or an aircrew designated examiner employed by the operator. The observation check may be accomplished in part or in full in an airplane, in a flight simulator, or in a flight training device. This paragraph applies after March 19, 1997.

**(b)** The observation check required by paragraph (a)(2) of this section is considered to have been completed in the month required if completed in the calendar month before, or the calendar month after, the month in which it is due.

**(c)** The initial ground training for check airmen must include the following:

(1) Check airman duties, functions, and responsibilities.

(2) The applicable Code of Federal Regulations and the certificate holder's policies and procedures.

(3) The appropriate methods, procedures, and techniques for conducting the required checks.

(4) Proper evaluation of student performance including the detection of—

(i) Improper and insufficient training; and

(ii) Personal characteristics of an applicant that could adversely affect safety.

(5) The appropriate corrective action in the case of unsatisfactory checks.

(6) The approved methods, procedures, and limitations for performing the required normal, abnormal, and emergency procedures in the airplane.

**(d)** The transition ground training for check airmen must include approved methods, procedures, and limitations for performing the required normal, abnormal, and emergency procedures applicable to the airplane to which the check airman is in transition.

**(e)** The initial and transition flight training for pilot check airmen (airplane), flight engineer check airmen (airplane), and flight navigator check airmen (airplane) must include the following:

(1) The safety measures for emergency situations that are likely to develop during a check.

(2) The potential results of improper, untimely, or non-execution of safety measures during a check.

(3) For pilot check airman (airplane)—

(i) Training and practice in conducting flight checks from the left and right pilot seats in the required normal, abnormal, and emergency procedures to ensure competence to conduct the pilot flight checks required by this part; and

(ii) The safety measures to be taken from either pilot seat for emergency situations that are likely to develop during a check.

(4) For flight engineer check airmen (airplane) and flight navigator check airmen (airplane), training to ensure competence to perform assigned duties.

**(f)** The requirements of paragraph (e) of this section may be accomplished in full or in part in flight, in a flight simulator, or in a flight training device, as appropriate.

**(g)** The initial and transition flight training for check airmen (simulator) must include the following:

(1) Training and practice in conducting flight checks in the required normal, abnormal, and emergency procedures to ensure competence to conduct the flight checks required by this part. This training and practice must be accomplished in a flight simulator or in a flight training device.

(2) Training in the operation of flight simulators or flight training devices, or both, to ensure competence to conduct the flight checks required by this part.

[Docket No. 28471, 61 FR 30743, June 17, 1996; as amended by Amdt. 121–257, 62 FR 3739, Jan. 24, 1997; Amdt. 121–264, 62 FR 23120, April 28, 1997]

## §121.414 Initial and transition training and checking requirements: flight instructors (airplane), flight instructors (simulator).

**(a)** No certificate holder may use a person nor may any person serve as a flight instructor unless—

(1) That person has satisfactorily completed initial or transition flight instructor training; and

(2) Within the preceding 24 calendar months, that person satisfactorily conducts instruction under the observation of an FAA inspector, an operator check airman, or an aircrew designated examiner employed by the operator. The observation check may be accomplished in part or in full in an airplane, in a flight simulator, or in a flight training device. This paragraph applies after March 19, 1997.

**(b)** The observation check required by paragraph (a)(2) of this section is considered to have been completed in the month required if completed in the calendar month before, or the calendar month after, the month in which it is due.

**(c)** The initial ground training for flight instructors must include the following:

(1) Flight instructor duties, functions, and responsibilities.

(2) The applicable Code of Federal Regulations and the certificate holder's policies and procedures.

(3) The appropriate methods, procedures, and techniques for conducting flight instruction.

(4) Proper evaluation of student performance including the detection of—

(i) Improper and insufficient training; and

(ii) Personal characteristics of an applicant that could adversely affect safety.

(5) The corrective action in the case of unsatisfactory training progress.

(6) The approved methods, procedures, and limitations for performing the required normal, abnormal, and emergency procedures in the airplane.

(7) Except for holders of a flight instructor certificate—

(i) The fundamental principles of the teaching-learning process;

(ii) Teaching methods and procedures; and

(iii) The instructor-student relationship.

**(d)** The transition ground training for flight instructors must include the approved methods, procedures, and limitations for performing the required normal, abnormal, and emergency procedures applicable to the airplane to which the flight instructor is in transition.

**(e)** The initial and transition flight training for flight instructors (airplane), flight engineer instructors (airplane), and flight navigator instructors (airplane) must include the following:

(1) The safety measures for emergency situations that are likely to develop during instruction.

(2) The potential results of improper, untimely, or non-execution of safety measures during instruction.

(3) For pilot flight instructor (airplane)—

(i) In-flight training and practice in conducting flight instruction from the left and right pilot seats in the required normal, abnormal, and emergency procedures to ensure competence as an instructor; and

(ii) The safety measures to be taken from either pilot seat for emergency situations that are likely to develop during instruction.

(4) For flight engineer instructors (airplane) and flight navigator instructors (airplane), in-flight training to ensure competence to perform assigned duties.

**(f)** The requirements of paragraph (e) of this section may be accomplished in full or in part in flight, in a flight simulator, or in a flight training device, as appropriate.

**(g)** The initial and transition flight training for flight instructors (simulator) must include the following:

(1) Training and practice in the required normal, abnormal, and emergency procedures to ensure competence to conduct the flight instruction required by this part. This training and practice must be accomplished in full or in part in a flight simulator or in a flight training device.

(2) Training in the operation of flight simulators or flight training devices, or both, to ensure competence to conduct the flight instruction required by this part.

[Docket No. 28471, 61 FR 30743, June 17, 1996; as amended by Amdt. 121–257, 62 FR 3739, Jan. 24, 1997]

## §121.415 Crewmember and dispatcher training requirements.

**(a)** Each training program must provide the following ground training as appropriate to the particular assignment of the crewmember or dispatcher:

(1) Basic indoctrination ground training for newly hired crewmembers or dispatchers including 40 programmed hours of instruction, unless reduced under §121.405 or as specified in §121.401(d), in at least the following—

(i) Duties and responsibilities of crewmembers or dispatchers, as applicable;

(ii) Appropriate provisions of the Federal Aviation Regulations;

(iii) Contents of the certificate holder's operating certificate and operations specifications (not required for flight attendants); and

(iv) Appropriate portions of the certificate holder's operating manual.

(2) The initial and transition ground training specified in §§121.419 through 121.422, as applicable.

(3) For crewmembers, emergency training as specified in §121.417 and 121.805.

**(b)** Each training program must provide the flight training specified in §§121.424 through 121.426, as applicable.

**(c)** Each training program must provide recurrent ground and flight training as provided in §121.427.

**(d)** Each training program must provide the differences training specified in §121.418 if the Administrator finds that, due to differences between airplanes of the same type operated by the certificate holder, additional training is necessary to insure that each crewmember and dispatcher is adequately trained to perform his assigned duties.

**(e)** Upgrade training as specified in §§121.419 and 121.424 for a particular type airplane may be included in the training program for crewmembers who have qualified and served as second in command pilot or flight engineer on that airplane.

**(f)** Particular subjects, maneuvers, procedures, or parts thereof specified in §§121.419 through 121.425 for transition or upgrade training, as applicable, may be omitted, or the programmed hours of ground instruction or inflight training may be reduced, as provided in §121.405.

**(g)** In addition to initial, transition, upgrade, recurrent and differences training, each training program must also provide ground and flight training, instruction, and practice as necessary to insure that each crewmember and dispatcher—

(1) Remains adequately trained and currently proficient with respect to each airplane, crewmember position, and type of operation in which he serves; and

(2) Qualifies in new equipment, facilities, procedures, and techniques, including modifications to airplanes.

[Docket No. 9509, 35 FR 90, Jan. 3, 1970; as amended by Amdt. 121–130, 41 FR 47229, Oct. 28, 1976; Amdt. 121–281, 66 FR 19043, April 12, 2001]

## §121.417 Crewmember emergency training.

**(a)** Each training program must provide the emergency training set forth in this section with respect to each airplane type, model, and configuration, each required crewmember, and each kind of operation conducted, insofar as appropriate for each crewmember and the certificate holder.

**(b)** Emergency training must provide the following:

(1) Instruction in emergency assignments and procedures, including coordination among crewmembers.

(2) Individual instruction in the location, function, and operation of emergency equipment including—

(i) Equipment used in ditching and evacuation;

(ii) [Reserved]

(iii) Portable fire extinguishers, with emphasis on type of extinguisher to be used on different classes of fires; and

(iv) Emergency exits in the emergency mode with the evacuation slide/raft pack attached (if applicable), with training emphasis on the operation of the exits under adverse conditions.

(3) Instruction in the handling of emergency situations including—

(i) Rapid decompression;

(ii) Fire in flight or on the surface, and smoke control procedures with emphasis on electrical equipment and related circuit breakers found in cabin areas including all galleys, service centers, lifts, lavatories and movie screens;

(iii) Ditching and other evacuation, including the evacuation of persons and their attendants, if any, who may need the assistance of another person to move expeditiously to an exit in the event of an emergency.

(iv) [Reserved]

(v) Hijacking and other unusual situations.

(4) Review and discussion of previous aircraft accidents and incidents pertaining to actual emergency situations.

**(c)** Each crewmember must accomplish the following emergency training during the specified training periods, using those items of installed emergency equipment for each type of airplane in which he or she is to serve (Alternate recurrent training required by §121.433(c) of this part may be accomplished by approved pictorial presentation or demonstration):

(1) One-time emergency drill requirements to be accomplished during initial training. Each crewmember must perform—

(i) At least one approved protective breathing equipment (PBE) drill in which the crewmember combats an actual or simulated fire using at least one type of installed hand fire extinguisher or approved fire extinguisher that is appropriate for the type of actual fire or simulated fire to be fought while using the type of installed PBE required by Part 121.337 or approved PBE simulation device as defined by paragraph (d) of this section for combating fires aboard airplanes.

(ii) At least one approved firefighting drill in which the crewmember combats an actual fire using at least one type of installed hand fire extinguisher or approved fire extinguisher that is appropriate for the type of fire to be fought. This firefighting drill is not required if the crewmember performs the PBE drill of paragraph (c)(1)(i) by combating an actual fire; and

(iii) An emergency evacuation drill with each person egressing the airplane or approved training device using at least one type of installed emergency evacuation slide. The crewmember may either observe the airplane exits being opened in the emergency mode and the associated exit slide/raft pack being deployed and inflated, or perform the tasks resulting in the accomplishment of these actions.

121

(2) Additional emergency drill requirements to be accomplished during initial training and once each 24 calendar months during recurrent training. Each crewmember must—

(i) Perform the following emergency drills and operate the following equipment:

(A) Each type of emergency exit in the normal and emergency modes, including the actions and forces required in the deployment of the emergency evacuation slides;

(B) Each type of installed hand fire extinguisher;

(C) Each type of emergency oxygen system to include protective breathing equipment;

(D) Donning, use, and inflation of individual flotation means, if applicable; and

(E) Ditching, if applicable, including but not limited to, as appropriate:

(1) Cockpit preparation and procedures;

(2) Crew coordination;

(3) Passenger briefing and cabin preparation;

(4) Donning and inflation of life preservers;

(5) Use of life-lines; and

(6) Boarding of passengers and crew into raft or a slide/raft pack.

(ii) Observe the following drills:

(A) Removal from the airplane (or training device) and inflation of each type of life raft, if applicable;

(B) Transfer of each type of slide/raft pack from one door to another;

(C) Deployment, inflation, and detachment from the airplane (or training device) of each type of slide/raft pack; and

(D) Emergency evacuation including the use of a slide.

**(d)** After September 1, 1993, no crewmember may serve in operations under this part unless that crewmember has performed the PBE drill and the firefighting drill described by paragraphs (c)(1)(i) and (c)(1)(ii) of this section, as part of a one-time training requirement of paragraphs (c)(1) or (c)(2) of this section as appropriate. Any crewmember who performs the PBE drill and the firefighting drill prescribed in paragraphs (c)(1)(i) and (c)(1)(ii) of this section after May 26, 1987, is deemed to be in compliance with this regulation upon presentation of information or documentation, in a form and manner acceptable to the Director, Flight Standards Service, showing that the appropriate drills have been accomplished.

**(e)** Crewmembers who serve in operations above 25,000 feet must receive instruction in the following:

(1) Respiration.

(2) Hypoxia.

(3) Duration of consciousness without supplemental oxygen at altitude.

(4) Gas expansion.

(5) Gas bubble formation.

(6) Physical phenomena and incidents of decompression.

**(f)** For the purpose of this section the following definitions apply:

(1) *Actual fire* means an ignited combustible material, in controlled conditions, of sufficient magnitude and duration to accomplish the training objectives outlined in paragraphs (c)(1)(i) and (c)(1)(ii) of this section.

(2) *Approved fire extinguisher* means a training device that has been approved by the Administrator for use in meeting the training requirements of Part 121.417(c).

(3) *Approved PBE simulation device* means a training device that has been approved by the Administrator for use in meeting the training requirements of Part 121.417(c).

(4) *Combats*, in this context, means to properly fight an actual or simulated fire using an appropriate type of fire extinguisher until that fire is extinguished.

(5) *Observe* means to watch without participating actively in the drill.

(6) *PBE drill* means an emergency drill in which a crewmember demonstrates the proper use of protective breathing equipment while fighting an actual or simulated fire.

(7) *Perform* means to satisfactorily accomplish a prescribed emergency drill using established procedures that stress the skill of the persons involved in the drill.

(8) *Simulated fire* means an artificial duplication of smoke or flame used to create various aircraft firefighting scenarios, such as lavatory, galley oven, and aircraft seat fires.

[Docket No. 9509, 35 FR 90, Jan. 3, 1970; as amended by Amdt. 121–281, 66 FR 19043, April 12, 2001]

## §121.418 Differences training: Crewmembers and dispatchers.

**(a)** Differences training for crewmembers and dispatchers must consist of at least the following as applicable to their assigned duties and responsibilities:

(1) Instruction in each appropriate subject or part thereof required for initial ground training in the airplane unless the Administrator finds that particular subjects are not necessary.

(2) Flight training in each appropriate maneuver or procedure required for initial flight training in the airplane unless the Administrator finds that particular maneuvers or procedures are not necessary.

(3) The number of programmed hours of ground and flight training determined by the Administrator to be necessary for the airplane, the operation, and the crewmember or aircraft dispatcher involved.

Differences training for all variations of a particular type airplane may be included in initial, transition, upgrade, and recurrent training for the airplane.

## §121.419 Pilots and flight engineers: Initial, transition, and upgrade ground training.

**(a)** Initial, transition, and upgrade ground training for pilots and flight engineers must include instruction in at least the following as applicable to their assigned duties:

(1) General subjects—

(i) The certificate holder's dispatch or flight release procedures;

(ii) Principles and methods for determining weight and balance, and runway limitations for takeoff and landing;

(iii) Enough meteorology to insure a practical knowledge of weather phenomena, including the principles of frontal systems, icing, fog, thunderstorms, and high altitude weather situations;

(iv) Air traffic control systems, procedures, and phraseology;

(v) Navigation and the use of navigation aids, including instrument approach procedures;

(vi) Normal and emergency communication procedures;

(vii) Visual cues prior to and during descent below DH or MDA;

(viii) Approved crew resource management initial training; and

(ix) Other instructions as necessary to ensure his competence.

(2) For each airplane type—

(i) A general description;

(ii) Performance characteristics;

(iii) Engines and propellers;

(iv) Major components;

(v) Major airplane systems (i.e., flight controls, electrical, hydraulic); other systems as appropriate; principles of normal, abnormal, and emergency operations; appropriate procedures and limitations;

(vi) Procedures for—

(A) Recognizing and avoiding severe weather situations;

(B) Escaping from severe weather situations, in case of inadvertent encounters, including low-altitude windshear, and

(C) Operating in or near thunderstorms (including best penetrating altitudes), turbulent air (including clear air turbulence), icing, hail, and other potentially hazardous meteorological conditions;

(vii) Operating limitations;

(viii) Fuel consumption and cruise control;

(ix) Flight planning;

(x) Each normal and emergency procedure; and

(xi) The approved Airplane Flight Manual.

**(b)** Initial ground training for pilots and flight engineers must consist of at least the following programmed hours of instruction in the required subjects specified in paragraph (a) of this section and in §121.415(a) unless reduced under §121.405:

(1) Group I airplanes—

(i) Reciprocating powered, 64 hours; and

(ii) Turbopropeller powered, 80 hours.

(2) Group II airplanes, 120 hours.

[Docket No. 9509, 35 FR 90, Jan. 3, 1970; as amended by Amdt. 121–199, 53 FR 37696, Sept. 27, 1988; Amdt. 121–250, 60 FR 65949, Dec. 20, 1995; Amdt. 121–250, 61 FR, Jan. 29, 1996]

## §121.420 Flight navigators: Initial and transition ground training.

**(a)** Initial and transition ground training for flight navigators must include instruction in the subjects specified in §121.419(a) as appropriate to his assigned duties and responsibilities and in the following with respect to the particular type airplane:

(1) Limitations on climb, cruise, and descent speeds.

(2) Each item of navigational equipment installed including appropriate radio, radar, and other electronic equipment.

(3) Airplane performance.

(4) Airspeed, temperature, and pressure indicating instruments or systems.

(5) Compass limitations and methods of compensation.

(6) Cruise control charts and data, including fuel consumption rates.

(7) Any other instruction as necessary to ensure his competence.

**(b)** Initial ground training for flight navigators must consist of at least the following programmed hours of instruction in the subjects specified in paragraph (a) of this section and in §121.415(a) unless reduced under §121.405:

(1) Group I airplanes—

(i) Reciprocating powered, 16 hours; and

(ii) Turbopropeller powered; 32 hours.

(2) Group II airplanes, 32 hours.

## §121.421 Flight attendants: Initial and transition ground training.

**(a)** Initial and transition ground training for flight attendants must include instruction in at least the following:

(1) General subjects—

(i) The authority of the pilot in command;

(ii) Passenger handling, including the procedures to be followed in the case of deranged persons or other persons whose conduct might jeopardize safety; and

(iii) Approved crew resource management initial training.

(2) For each airplane type—

(i) A general description of the airplane emphasizing physical characteristics that may have a bearing on ditching, evacuation, and inflight emergency procedures and on other related duties;

(ii) The use of both the public address system and the means of communicating with other flight crewmembers, including emergency means in the case of attempted hijacking or other unusual situations; and

(iii) Proper use of electrical galley equipment and the controls for cabin heat and ventilation.

**(b)** Initial and transition ground training for flight attendants must include a competence check to determine ability to perform assigned duties and responsibilities.

**(c)** Initial ground training for flight attendants must consist of at least the following programmed hours of instruction in the subjects specified in paragraph (a) of this section and in §121.415(a) unless reduced under §121.405.

(1) Group I airplanes—

(i) Reciprocating powered, 8 hours; and

(ii) Turbopropeller powered, 8 hours.

(2) Group II airplanes, 16 hours.

[Docket No. 9509, 35 FR 90, Jan. 3, 1970; as amended by Amdt. 121–250, 60 FR 65949, Dec. 20, 1995]

## §121.422 Aircraft dispatchers: Initial and transition ground training.

**(a)** Initial and transition ground training for aircraft dispatchers must include instruction in at least the following:

(1) General subjects—

(i) Use of communications systems including the characteristics of those systems and the appropriate normal and emergency procedures;

(ii) Meteorology, including various types of meteorological information and forecasts, interpretation of weather data (including forecasting of en route and terminal temperatures and other weather conditions), frontal systems, wind conditions, and use of actual and prognostic weather charts for various altitudes;

(iii) The NOTAM system;

(iv) Navigational aids and publications;

(v) Joint dispatcher-pilot responsibilities;

(vi) Characteristics of appropriate airports;

(vii) Prevailing weather phenomena and the available sources of weather information;

(viii) Air traffic control and instrument approach procedures; and

(ix) Approved dispatcher resource management (DRM) initial training.

(2) For each airplane—

(i) A general description of the airplane emphasizing operating and performance characteristics, navigation equipment, instrument approach and communication equipment, emergency equipment and procedures, and other subjects having a bearing on dispatcher duties and responsibilities;

(ii) Flight operation procedures including procedures specified in §121.419(a)(2)(vi);

(iii) Weight and balance computations;

(iv) Basic airplane performance dispatch requirements and procedures;

(v) Flight planning including track selection, flight time analysis, and fuel requirements; and

(vi) Emergency procedures.

(3) Emergency procedures must be emphasized, including the alerting of proper governmental, company, and private agencies during emergencies to give maximum help to an airplane in distress.

**(b)** Initial and transition ground training for aircraft dispatchers must include a competence check given by an appropriate supervisor or ground instructor that demonstrates knowledge and ability with the subjects set forth in paragraph (a) of this section.

**(c)** Initial ground training for aircraft dispatchers must consist of at least the following programmed hours of instruction in the subjects specified in paragraph (a) of this section and in §121.415(a) unless reduced under §121.405:

(1) Group I airplanes—

(i) Reciprocating powered, 30 hours; and

(ii) Turbopropeller powered, 40 hours.

(2) Group II airplanes, 40 hours.

[Docket No. 9509, 35 FR 90, Jan. 3, 1970; as amended by Amdt. 121–250, 60 FR 65949, Dec. 20, 1995]

## §121.424 Pilots: Initial, transition, and upgrade flight training.

**(a)** Initial, transition, and upgrade training for pilots must include flight training and practice in the maneuvers and procedures set forth in the certificate holder's approved low-altitude windshear flight training program and in Appendix E to this part, as applicable.

**(b)** The maneuvers and procedures required by paragraph (a) of this section must be performed inflight except—

(1) That windshear maneuvers and procedures must be performed in a simulator in which the maneuvers and procedures are specifically authorized to be accomplished; and

(2) To the extent that certain other maneuvers and procedures may be performed in an airplane simulator, an appropriate training device, or a static airplane as permitted in Appendix E to this part.

**(c)** Except as permitted in paragraph (d) of this section, the initial flight training required by paragraph (a) of this section must include at least the following programmed hours of inflight training and practice unless reduced under §121.405;

(1) Group I airplanes—

(i) *Reciprocating powered.* Pilot in command, 10 hours; second in command, 6 hours; and

(ii) *Turbopropeller powered.* Pilot in command, 15 hours; second in command, 7 hours.

(2) *Group II airplanes.* Pilot in command, 20 hours; second in command, 10 hours.

**(d)** If the certificate holder's approved training program includes a course of training utilizing an airplane simulator under §121.409(c) and (d) of this part, each pilot must successfully complete—

(1) With respect to §121.409(c) of this part—

(i) Training and practice in the simulator in at least all of the maneuvers and procedures set forth in Appendix E to this part for initial flight training that are capable of being performed in an airplane simulator without a visual system; and

(ii) A flight check in the simulator or the airplane to the level of proficiency of a pilot in command or second in command, as applicable, in at least the maneuvers and procedures set forth in Appendix F to this part that are capable of being performed in an airplane simulator without a visual system.

(2) With respect to §121.409(d) of this part, training and practice in at least the maneuvers and procedures set forth in the certificate holder's approved low-altitude windshear flight training program that are capable of being performed in an airplane simulator in which the maneuvers and procedures are specifically authorized.

[Docket No. 9509, 35 FR 90, Jan. 3, 1970; as amended by Amdt. 121–199, 53 FR 37697, Sept. 27, 1988]

## §121.425 Flight engineers: Initial and transition flight training.

**(a)** Initial and transition flight training for flight engineers must include at least the following:

(1) Training and practice in procedures related to the carrying out of flight engineer duties and functions. This training and practice may be accomplished either inflight, in an airplane simulator, or in a training device.

(2) A flight check that includes—

(i) Preflight inspection;

(ii) Inflight performance of assigned duties accomplished from the flight engineer station during taxi, runup, takeoff, climb, cruise, descent, approach, and landing;

(iii) Accomplishment of other functions, such as fuel management and preparation of fuel consumption records, and normal and emergency or alternate operation of all airplane flight systems, performed either inflight, in an airplane simulator, or in a training device.

Flight engineers possessing a commercial pilot certificate with an instrument, category and class rating, or pilots already qualified as second in command and reverting to flight engineer, may complete the entire flight check in an approved airplane simulator.

**(b)** Except as permitted in paragraph (c) of this section, the initial flight training required by paragraph (a) of this section must include at least the same number of programmed hours of flight training and practice that are specified for a second in command pilot under §121.424(c) unless reduced under §121.405.

**(c)** If the certificate holder's approved training program includes a course of training utilizing an airplane simulator or other training device under §121.409(c), each flight engineer must successfully complete in the simulator or other training device—

(1) Training and practice in at least all of the assigned duties, procedures, and functions required by paragraph (a) of this section; and

(2) A flight check to a flight engineer level of proficiency in the assigned duties, procedures, and functions.

[Docket No. 9509, Amdt. 121–55, 35 FR 90, Jan. 3, 1970; as amended by Amdt. 121–144, 43 FR 22647, May 25, 1978]

## §121.426 Flight navigators: Initial and transition flight training.

**(a)** Initial and transition flight training for flight navigators must include flight training and a flight check that are adequate to insure his proficiency in the performance of his assigned duties.

**(b)** The flight training and checks specified in paragraph (a) of this section must be performed—

(1) Inflight or in an appropriate training device; or

(2) In operations under this part if performed under supervision of a qualified flight navigator.

## §121.427 Recurrent training.

**(a)** Recurrent training must ensure that each crew member or dispatcher is adequately trained and currently proficient with respect to the type airplane (including differences training, if applicable) and crewmember position involved.

**(b)** Recurrent ground training for crewmembers and dispatchers must include at least the following:

(1) A quiz or other review to determine the state of the crewmember's or dispatcher's knowledge with respect to the airplane and position involved.

(2) Instruction as necessary in the subjects required for initial ground training by §§121.415(a) and 121.805, as appropriate, including emergency training (not required for aircraft dispatchers).

(3) For flight attendants and dispatchers, a competence check as required by §§121.421(b) and 121.422(b), respectively.

(4) Approved recurrent CRM training. For flight crewmembers, this training or portions thereof may be accomplished during an approved simulator line operational flight training (LOFT) session. The recurrent CRM training requirement does not apply until a person has completed the applicable initial CRM training required by §§121.419, 121.421, or 121.422.

**(c)** Recurrent ground training for crewmembers and dispatchers must consist of at least the following programmed hours unless reduced under §121.405:

(1) For pilots and flight engineers—

(i) Group I, reciprocating powered airplanes, 16 hours;

(ii) Group I turbopropeller powered airplanes, 20 hours; and

(iii) Group II airplanes, 25 hours.

(2) For flight navigators—

(i) Group I reciprocating powered airplanes, 12 hours;

(ii) Group I turbopropeller powered airplanes, 16 hours; and

(iii) Group II airplanes, 16 hours.

(3) For flight attendants—

(i) Group I reciprocating powered airplanes, 4 hours;

(ii) Group I turbopropeller powered airplanes, 5 hours; and

(iii) Group II airplanes, 12 hours.

(4) For aircraft dispatchers—

(i) Group I reciprocating powered airplanes, 8 hours;

(ii) Group I turbopropeller powered airplanes, 10 hours; and

(iii) Group II airplanes, 20 hours.

**(d)** Recurrent flight training for flight crewmembers must include at least the following:

(1) For pilots, flight training in an approved simulator in maneuvers and procedures set forth in the certificate holder's approved low-altitude windshear flight training program and flight training in maneuvers and procedures set forth in Appendix F to this part, or in a flight training program approved by the Administrator, except as follows—

(i) The number of programmed inflight hours is not specified; and

(ii) Satisfactory completion of a proficiency check may be substituted for recurrent flight training as permitted in §121.433(c).

(2) For flight engineers, flight training as provided by §121.425(a) except as follows—

(i) The specified number of inflight hours is not required; and

(ii) The flight check, other than the preflight inspection, may be conducted in an airplane simulator or other training device. The preflight inspection may be conducted in an airplane, or by using an approved pictorial means that realistically portrays the location and detail or preflight inspection items and provides for the portrayal of abnormal conditions. Satisfactory completion of an approved line-oriented simulator training program may be substituted for the flight check.

(3) For flight navigators, enough inflight training and an inflight check to insure competency with respect to operating procedures and navigation equipment to be used and familiarity with essential navigation information pertaining to the certificate holder's routes that require a flight navigator.

[Docket No. 9509, 35 FR 90, Jan. 30, 1970; as amended by Amdt. 121–80, 36 FR 19362, Oct. 5, 1971; Amdt. 121–144, 43 FR 22647, May 25, 1978; Amdt. 121–199, 53 FR 37697, Sept. 27, 1988; Amdt. 121–250, 60 FR 65949, Dec. 20, 1995; Amdt. 121–281, 66 FR 19043, April 12, 2001]

### §121.429 Prohibited drugs.

**(a)** Each certificate holder shall provide each employee performing a function listed in Appendix I to this part and his or her supervisor with the training specified in that appendix.

**(b)** No certificate holder may use any contractor to perform a function listed in Appendix I to this part unless that contractor provides each of its employees performing that function for the certificate holder and his or her supervisor with the training specified in that appendix.

[Docket No. 25148, 53 FR 47057, Nov. 21, 1988]

## Subpart O—Crewmember Qualifications

### §121.431 Applicability.

**(a)** This subpart:

(1) Prescribes crewmember qualifications for all certificate holders except where otherwise specified. The qualification requirements of this subpart also apply to each certificate holder that conducts commuter operations under part 135 of this chapter with airplanes for which two pilots are required by the aircraft type certification rules of this chapter. The Administrator may authorize any other certificate holder that conducts operations under part 135 of this chapter to comply with the training and qualification requirements of this subpart instead of Subparts E, G, and H of part 135 of this chapter, except that these certificate holders may choose to comply with the operating experience requirements of §135.344 of this chapter, instead of the requirements of §121.434; and

(2) Permits training center personnel authorized under part 142 of this chapter who meet the

requirements of §§121.411 through 121.414 to provide training, testing, and checking under contract or other arrangement to those persons subject to the requirements of this subpart.

**(b)** For the purpose of this subpart, the airplane groups and terms and definitions prescribed in §121.400 and the following definitions apply:

*Consolidation* is the process by which a person through practice and practical experience increases proficiency in newly acquired knowledge and skills.

*Line operating flight time* is flight time performed in operations under this part.

*Operating cycle* is a complete flight segment consisting of a takeoff, climb, enroute portion, descent, and a landing.

[Docket No. 10171, 36 FR 12284, June 30, 1971; as amended by Amdt. 121–248, 60 FR 20869, April 27, 1995; Amdt. 121–250, 60 FR 65949, Dec. 20, 1995; Amdt. 121–256, 61 FR 30435, June 14, 1996; Amdt 121–259, 61 FR 34561, July 2, 1996; Amdt. 121–263, 62 FR 13791, March 21, 1997]

## §121.432 General.

**(a)** Except in the case of operating experience under §121.434, a pilot who serves as second in command of an operation that requires three or more pilots must be fully qualified to act as pilot in command of that operation.

**(b)** No certificate holder may conduct a check or any training in operations under this part, except for the following checks and training required by this part or the certificate holder:

(1) Line checks for pilots.

(2) Flight navigator training conducted under the supervision of a flight navigator flight instructor.

(3) Flight navigator flight checks.

(4) Flight engineer checks (except for emergency procedures), if the person being checked is qualified and current in accordance with §121.453(a).

(5) Flight attendant training and competence checks.

Except for pilot line checks and flight engineer flight checks, the person being trained or checked may not be used as a required crewmember.

**(c)** For the purposes of this subpart the airplane groups prescribed in §121.400 apply.

**(d)** For the purposes of this subpart the terms and definitions in §121.400 apply.

[Docket No. 9509, 35 FR 95, Jan. 3, 1970; as amended by Amdt. 121–130, 41 FR 47229, Oct. 28, 1976]

## §121.433 Training required.

**(a)** *Initial training.* No certificate holder may use any person nor may any person serve as a required crewmember on an airplane unless that person has satisfactorily completed, in a training program approved under Subpart N of this part, initial ground and flight training for that type airplane and for the particular crewmember position, except as follows:

(1) Crewmembers who have qualified and served as a crewmember on another type airplane of the same group may serve in the same crewmember capacity upon completion of transition training as provided in §121.415.

(2) Crewmembers who have qualified and served as second in command or flight engineer on a particular type airplane may serve as pilot in command or second in command, respectively, upon completion of upgrade training for that airplane as provided in §121.415.

**(b)** *Differences training.* No certificate holder may use any person nor may any person serve as a required crewmember on an airplane of a type for which differences training is included in the certificate holder's approved training program unless that person has satisfactorily completed, with respect to both the crewmember position and the particular variation of the airplane in which he serves, either initial or transition ground and flight training, or differences training, as provided in §121.415.

**(c)** *Recurrent training.*

(1) No certificate holder may use any person nor may any person serve as a required crewmember on an airplane unless, within the preceding 12 calendar months—

(i) For flight crewmembers, he has satisfactorily completed recurrent ground and flight training for that airplane and crewmember position and a flight check as applicable;

(ii) For flight attendants and dispatchers, he has satisfactorily completed recurrent ground training and a competence check; and

(iii) In addition, for pilots in command he has satisfactorily completed, within the preceding 6 calendar months, recurrent flight training in addition to the recurrent flight training required in paragraph (c)(1)(i) of this section, in an airplane in which he serves as pilot in command in operations under this part.

(2) For pilots, a proficiency check as provided in §121.441 of this part may be substituted for the recurrent flight training required by this paragraph and the approved simulator course of training under §121.409(b) of this part may be substituted for alternate periods of recurrent flight training required in that airplane, except as provided in paragraphs (d) and (e) of this section.

**(d)** For each airplane in which a pilot serves as pilot in command, he must satisfactorily complete either recurrent flight training or a proficiency check within the preceding 12 calendar months.

**(e)** Notwithstanding paragraphs (c)(2) and (d) of this section, a proficiency check as provided in §121.441 of this part may not be substituted for training in those maneuvers and procedures set forth in a certificate holder's approved low-altitude windshear flight training program when that program is included in a recurrent flight training course as required by §121.409(d) of this part.

[Docket No. 9509, 35 FR 95, Jan. 3, 1970; as amended by Amdt. 121–91, 37 FR 10729, May 27, 1972; Amdt. 121–199, 53 FR 37697, Sept. 27, 1988]

## §121.433a Training requirements: Handling and carriage of dangerous articles and magnetized materials.

**(a)** No certificate holder may use any person to perform and no person may perform, any assigned duties and responsibilities for the handling or carriage of dangerous articles and magnetized materials governed by Title 49 CFR, unless within the preceding 12 calendar months that person has satisfactorily completed training in a program established and approved under this subpart which includes instructions regarding the proper packaging, marking, labeling, and documentation of dangerous articles and magnetized materials, as required by Title 49 CFR and instructions regarding their compatibility, loading, storage, and handling characteristics. A person who satisfactorily completes training in the calendar month before, or the calendar month after, the month in which it becomes due, is considered to have taken that training during the month it became due.

**(b)** Each certificate holder shall maintain a record of the satisfactory completion of the initial and recurrent training given to crewmembers and ground personnel who perform assigned duties and responsibilities for the handling and carriage of dangerous articles and magnetized materials.

**(c)** A certificate holder operating in a foreign country where the loading and unloading of aircraft must be performed by personnel of the foreign country, may use personnel not meeting the requirements of paragraphs (a) and (b) of this section if they are supervised by a person qualified under paragraphs (a) and (b) of this section to supervise the loading, offloading and handling of hazardous materials.

[Docket No. 12124, 38 FR 14915, June 7, 1973; as amended by Amdt. 121–144, 43 FR 22647, May 25, 1978]

## §121.434 Operating experience, operating cycles, and consolidation of knowledge and skills.

**(a)** No certificate holder may use a person nor may any person serve as a required crewmember of an airplane unless the person has satisfactorily completed, on that type airplane and in that crewmember position, the operating experience, operating cycles, and the line operating flight time for consolidation of knowledge and skills, required by this section, except as follows:

(1) Crewmembers other than pilots in command may serve as provided herein for the purpose of meeting the requirements of this section.

(2) Pilots who are meeting the pilot in command requirements may serve as second in command.

(3) Separate operating experience, operating cycles, and line operating flight time for consolidation of knowledge and skills are not required for variations within the same type airplane.

**(b)** In acquiring the operating experience, operating cycles, and line operating flight time for consolidation of knowledge and skills, crewmembers must comply with the following:

(1) In the case of a flight crewmember, he must hold the appropriate certificates and ratings for the crewmember position and the airplane, except that a pilot who is meeting the pilot in command requirements must hold the appropriate certificates and ratings for a pilot in command in the airplane.

(2) The operating experience, operating cycles, and line operating flight time for consolidation of knowledge and skills must be acquired after satisfactory completion of the appropriate ground and flight training for the particular airplane type and crewmember position.

(3) The experience must be acquired in flight during operations under this part. However, in the case of an aircraft not previously used by the certificate holder in operations under this part, operating experience acquired in the aircraft during proving flights or ferry flights may be used to meet this requirement.

**(c)** Pilot crewmembers must acquire operating experience and operating cycles as follows:

(1) A pilot in command must—

(i) Perform the duties of a pilot in command under the supervision of a check pilot; and

(ii) In addition, if a qualifying pilot in command is completing initial or upgrade training specified in §121.424, be observed in the performance of prescribed duties by an FAA inspector during at least one flight leg which includes a takeoff and landing. During the time that a qualifying pilot in command is acquiring the operating experience in paragraphs (c)(1) (i) and (ii) of this section, a check pilot who is also serving as the pilot in command must occupy a pilot station. However, in the case of a transitioning pilot in command the check pilot serving as pilot in command may occupy the observer's seat, if the transitioning pilot has made at least two takeoffs and landings in the type airplane used, and has satisfactorily demonstrated to the check pilot that he is qualified to perform the duties of a pilot in command of that type of airplane.

(2) A second in command pilot must perform the duties of a second in command under the supervision of an appropriately qualified check pilot.

(3) The hours of operating experience and operating cycles for all pilots are as follows:

(i) For initial training, 15 hours in Group I reciprocating powered airplanes, 20 hours in Group I turbopropeller powered airplanes, and 25 hours in Group II airplanes. Operating experience in both airplane groups must include at least 4 operating cycles (at least 2 as the pilot flying the airplane).

(ii) For transition training, except as provided in paragraph (c)(3)(iii) of this section, 10 hours in Group I reciprocating powered airplanes, 12 hours in Group I turbopropeller powered airplanes, 25 hours for pilots in command in Group II airplanes, and 15 hours for second in command pilots in Group II airplanes. Operating experience in both airplane groups must include at least 4 operating cycles (at least 2 as the pilot flying the airplane).

(iii) In the case of transition training where the certificate holder's approved training program includes a course of training in an airplane simulator under §121.409(c), each pilot in command must comply with the requirements prescribed in paragraph (c)(3)(i) of this section for initial training.

**(d)** A flight engineer must perform the duties of a flight engineer under the supervision of a check airman or a qualified flight engineer for at least the following number of hours:

(1) Group I reciprocating powered airplanes, 8 hours.

(2) Group I turbopropeller powered airplanes, 10 hours.

(3) Group II airplanes, 12 hours.

**(e)** A flight attendant must, for at least 5 hours, perform the assigned duties of a flight attendant under the supervision of a flight attendant supervisor qualified under this part who personally observes the performance of these duties. However, operating experience is not required for a flight attendant who has previously acquired such experience on any large passenger carrying airplane of the same group, if the certificate holder shows that the flight attendant has received sufficient ground training for the airplane in which the flight attendant is to serve. Flight attendants receiving operating experience may not be assigned as a required crewmember. Flight attendants who have satisfactorily completed training time acquired in an approved training program conducted in a full-scale (except for length) cabin training device of the type airplane in which they are to serve may substitute this time for 50 percent of the hours required by this paragraph.

**(f)** Flight crewmembers may substitute one additional takeoff and landing for each hour of flight to meet the operating experience requirements of this section, up to a maximum reduction of 50% of flight hours, except those in Group II initial training, and second in command pilots in Group II transition training. Notwithstanding the reductions in programmed hours permitted under §§121.405 and 121.409, the hours of operating experience for flight crewmembers are not subject to reduction other than as provided in this paragraph and paragraph (e) of this section.

**(g)** Except as provided in paragraph (h) of this section, pilot in command and second in command crewmembers must each acquire at least 100 hours of line operating flight time for consolidation of knowledge and skills (including operating experience required under paragraph (c) of this section) within 120 days after the satisfactory completion of:

(1) Any part of the flight maneuvers and procedures portion of either an airline transport pilot certificate with type rating practical test or an additional type rating practical test, or

(2) A §121.441 proficiency check.

**(h)** The following exceptions apply to the consolidation requirement of paragraph (g) of this section:

(1) Pilots who have qualified and served as pilot in command or second in command on a particular type airplane in operations under this part before August 25, 1995 are not required to complete line operating flight time for consolidation of knowledge and skills.

(2) Pilots who have completed the line operating flight time requirement for consolidation of knowledge and skills while serving as second in command on a particular type airplane in operations under this part after August 25, 1995 are not required to repeat the line operating flight time before serving as pilot in command on the same type airplane.

(3) If, before completing the required 100 hours of line operating flight time, a pilot serves as pilot in another airplane type operated by the certificate holder, the pilot may not serve as a pilot in the airplane for which the pilot has newly qualified unless the pilot satisfactorily completes refresher training as provided in the certificate holder's approved training program and that training is conducted by an appropriately qualified instructor or check pilot.

(4) If the required 100 hours of line operating flight time are not completed within 120 days, the certificate holder may extend the 120-day period to no more than 150 days if—

(i) The pilot continues to meet all other applicable requirements of Subpart O of this part; and

(ii) On or before the 120th day the pilot satisfactorily completes refresher training conducted by an appropriately qualified instructor or check pilot as provided in the certificate holder's approved training program, or a check pilot determines that the pilot has retained an adequate level of proficiency after observing that pilot in a supervised line operating flight.

(5) The Administrator, upon application by the certificate holder, may authorize deviations from the requirements of paragraph (g) of this section, by an appropriate amendment to the operations specifications, to the extent warranted by any of the following circumstances:

(i) A newly certificated certificate holder does not employ any pilots who meet the minimum requirements of paragraph (g) of this section.

(ii) An existing certificate holder adds to its fleet an airplane type not before proven for use in its operations.

(iii) A certificate holder establishes a new domicile to which it assigns pilots who will be required to become qualified on the airplanes operated from that domicile.

**(i)** Notwithstanding the reductions in programmed hours permitted under §§121.405 and 121.409 of Subpart N of this part, the hours of operating experience for flight crewmembers are not subject to reduction other than as provided in paragraphs (e) and (f) of this section.

[Docket No. 9509, 35 FR 95, Jan. 3, 1970; as amended by Amdt. 121–74, 36 FR 12284, June 30, 1971; Amdt. 121–91, 37 FR 10729, May 27, 1972; Amdt. 121–140, 43 FR 9599, March 9, 1978; Amdt. 121–144, 43 FR 22647, May 25, 1978; Amdt. 121–159, 45 FR 41593, June 19, 1980; Amdt. 121–248, 60 FR 20870, April 27, 1995]

## §121.437 Pilot qualification: Certificates required.

**(a)** No pilot may act as pilot in command of an aircraft (or as second in command of an aircraft in a flag or supplemental operation that requires three or more pilots) unless he holds an airline transport pilot certificate and an appropriate type rating for that aircraft.

**(b)** No certificate holder may use nor may any pilot act as a pilot in a capacity other than those specified in paragraph (a) of this section unless the pilot holds at least a commercial pilot certificate with appropriate category and class ratings for the aircraft concerned, and an instrument rating. Notwithstanding the requirements of §61.63 (b) and (c) of this chapter, a pilot who is currently employed by a certificate holder and meets applicable training requirements of Subpart N of this part, and the proficiency check requirements of §121.441, may be issued the appropriate category and class ratings by presenting proof of compliance with those requirements to a Flight Standards District Office.

[Docket No. 6258, 29 FR 19215, Dec. 31, 1964; as amended by Amdt. 121–148, 43 FR 46235, Oct. 5, 1978; 44 FR 25202, April 30, 1979; Amdt. 121–207, 54 FR 39293, Sept. 25, 1989; Amdt. 121–253, 61 FR 2612, Jan. 26, 1996; Amdt. 121–262, 62 FR 13257, March 19, 1997]

## §121.438 Pilot operating limitations and pairing requirements.

**(a)** If the second in command has fewer than 100 hours of flight time as second in command in operations under this part in the type airplane being flown, and the pilot in command is not an appropriately qualified check pilot, the pilot in command must make all takeoffs and landings in the following situations:

(1) At special airports designated by the Administrator or at special airports designated by the certificate holder; and

(2) In any of the following conditions:

(i) The prevailing visibility value in the latest weather report for the airport is at or below $3/4$ mile.

(ii) The runway visual range for the runway to be used is at or below 4,000 feet.

(iii) The runway to be used has water, snow, slush or similar conditions that may adversely affect airplane performance.

(iv) The braking action on the runway to be used is reported to be less than "good."

(v) The crosswind component for the runway to be used is in excess of 15 knots.

(vi) Windshear is reported in the vicinity of the airport.

(vii) Any other condition in which the PIC determines it to be prudent to exercise the PIC's prerogative.

**(b)** No person may conduct operations under this part unless, for that type airplane, either the pilot in command or the second in command has at least 75 hours of line operating flight time, either as pilot in command or second in command. The Administrator may, upon application by the certificate holder, authorize deviations from the requirements of this paragraph (b) by an appropriate amendment to the operations specifications in any of the following circumstances:

(1) A newly certificated certificate holder does not employ any pilots who meet the minimum requirements of this paragraph.

(2) An existing certificate holder adds to its fleet a type airplane not before proven for use in its operations.

(3) An existing certificate holder establishes a new domicile to which it assigns pilots who will be required to become qualified on the airplanes operated from that domicile.

[Docket No. 27210, 60 FR 20870, April 27, 1995]

## §121.439 Pilot qualification: Recent experience.

**(a)** No certificate holder may use any person nor may any person serve as a required pilot flight crewmember, unless within the preceding 90 days, that person has made at least three takeoffs and landings in the type airplane in which that person is to serve. The takeoffs and landings required by this paragraph may be performed in a visual simulator approved under §121.407 to include takeoff and landing maneuvers. In addition, any person who fails to make the three required takeoffs and landings within any consecutive 90-day period must reestablish recency of experience as provided in paragraph (b) of this section.

**(b)** In addition to meeting all applicable training and checking requirements of this part, a required pilot flight crewmember who has not met the requirements of paragraph (a) of this section must reestablish recency of experience as follows:

(1) Under the supervision of a check airman, make at least three takeoffs and landings in the type airplane in which that person is to serve or in an advanced simulator or visual simulator. When a visual simulator is used, the requirements of paragraph (c) of this section must be met.

(2) The takeoffs and landings required in paragraph (b)(1) of this section must include—

(i) At least one takeoff with a simulated failure of the most critical powerplant;

(ii) At least one landing from an ILS approach to the lowest ILS minimum authorized for the certificate holder; and

(iii) At least one landing to a full stop.

**(c)** A required pilot flight crewmember who performs the maneuvers prescribed in paragraph (b) of this section in a visual simulator must—

(1) Have previously logged 100 hours of flight time in the same type airplane in which he is to serve;

(2) Be observed on the first two landings made in operations under this part by an approved check airman who acts as pilot in command and occupies a pilot seat. The landings must be made in weather minimums that are not less than those contained in the certificate holder's operations specifications for Category I Operations, and must be made within 45 days following completion of simulator training.

**(d)** When using a simulator to accomplish any of the requirements of paragraph (a) or (b) of this section, each required flight crewmember position must be occupied by an appropriately qualified person and the simulator must be operated as if in a normal in-flight environment without use of the repositioning features of the simulator.

**(e)** A check airman who observes the takeoffs and landings prescribed in paragraphs (b)(1) and (c) of this section shall certify that the person being observed is proficient and qualified to perform flight duty in operations under this part and may require any additional maneuvers that are determined necessary to make this certifying statement.

[Docket No. 16383, 43 FR 22648, May 25, 1978; as amended by Amdt. 121–148, 43 FR 46235, Oct. 5, 1978; Amdt. 121–179, 47 FR 33390, Aug. 2, 1982]

## §121.440 Line checks.

**(a)** No certificate holder may use any person nor may any person serve as pilot in command of an airplane unless, within the preceding 12 calendar months, that person has passed a line check in which he satisfactorily performs the duties and responsibilities of a pilot in command in one of the types of airplanes he is to fly.

**(b)** A pilot in command line check for domestic and flag operations must—

(1) Be given by a pilot check airman who is currently qualified on both the route and the airplane; and

(2) Consist of at least one flight over a typical part of the certificate holder's route, or over a foreign or Federal airway, or over a direct route.

**(c)** A pilot in command line check for supplemental operations must—

(1) Be given by a pilot check airman who is currently qualified on the airplane; and

(2) Consist of at least one flight over a part of a Federal airway, foreign airway, or advisory route over which the pilot may be assigned.

[Docket No. 9509, 35 FR 96, Jan. 3, 1970; as amended by Amdt. 121–143, 43 FR 22642, May 25, 1978; Amdt. 121–253, 61 FR 2612, Jan. 26, 1996]

## §121.441 Proficiency checks.

**(a)** No certificate holder may use any person nor may any person serve as a required pilot flight crewmember unless that person has satisfactorily completed either a proficiency check, or an approved simulator course of training under §121.409, as follows:

(1) For a pilot in command, a proficiency check within the preceding 12 calendar months and, in addition, within the preceding 6 calendar months, either a proficiency check or the simulator training.

(2) For all other pilots—

(i) Within the preceding 24 calendar months either a proficiency check or the line-oriented simulator training course under §121.409; and

(ii) Within the preceding 12 calendar months, either a proficiency check or any simulator training course under §121.409.

**(b)** Except as provided in paragraphs (c) and (d) of this section, a proficiency check must meet the following requirements:

(1) It must include at least the procedures and maneuvers set forth in Appendix F to this part unless otherwise specifically provided in that appendix.

(2) It must be given by the Administrator or a pilot check airman.

**(c)** An approved airplane simulator or other appropriate training device may be used in the conduct of a proficiency check as provided in Appendix F to this part.

**(d)** A person giving a proficiency check may, in his discretion, waive any of the maneuvers or procedures for which a specific waiver authority is set forth in Appendix F to this part if—

(1) The Administrator has not specifically required the particular maneuver or procedure to be performed;

(2) The pilot being checked is, at the time of the check, employed by a certificate holder as a pilot; and

(3) The pilot being checked is currently qualified for operations under this part in the particular type airplane and flight crewmember position or has, within the preceding six calendar months, satisfactorily completed an approved training program for the particular type airplane.

**(e)** If the pilot being checked fails any of the required maneuvers, the person giving the proficiency check may give additional training to the pilot during the course of the proficiency check. In addition to repeating the maneuvers failed, the person giving the proficiency check may require the pilot being checked to repeat any other maneuvers he finds are necessary to determine the pilot's proficiency. If the pilot being checked is unable to demonstrate satisfactory performance to the person conducting the check, the certificate holder may not use him nor may he serve in operations under this part until he has satisfactorily completed a proficiency check.

However, the entire proficiency check (other than the initial second-in-command proficiency check) required by this section may be conducted in an approved visual simulator if the pilot being checked accomplishes at least two landings in the appropriate airplane during a line check or other check conducted by a pilot check airman (a pilot-in-command may observe and certify the satisfactory accomplishment of these landings by a second-in-command). If a pilot proficiency check is conducted in accordance with this paragraph, the next required proficiency check for that pilot must be conducted in the same manner, or in accordance with Appendix F of this part, or a course of training in an airplane visual simulator under §121.409 may be substituted therefor.

[Docket No. 9509, 35 FR 96, Jan. 3, 1970; as amended by Amdt. 121–103, 38 FR 12203, May 10, 1973, Amdt. 121–108, 38 FR 35446, Dec. 28, 1973; Amdt. 121–144, 43 FR 22648, May 25, 1978; Amdt. 121–263, 62 FR 13791, March 21, 1997]

## §121.443 Pilot in command qualification: Route and airports.

**(a)** Each certificate holder shall provide a system acceptable to the Administrator for disseminating the information required by paragraph (b) of this section to the pilot in command and appropriate flight operation personnel. The system must also provide an acceptable means for showing compliance with §121.445.

**(b)** No certificate holder may use any person, nor may any person serve, as pilot in command unless the certificate holder has provided that person current information concerning the following subjects pertinent to the areas over which that person is to serve, and to each airport and terminal area into which that person is to operate, and ensures that person has adequate knowledge of, and the ability to use, the information:

(1) Weather characteristics appropriate to the season.

(2) Navigation facilities.

(3) Communication procedures, including airport visual aids.

(4) Kinds of terrain and obstructions.

(5) Minimum safe flight levels.

(6) En route and terminal area arrival and departure procedures, holding procedures and authorized instrument approach procedures for the airports involved.

(7) Congested areas and physical layout of each airport in the terminal area in which the pilot will operate.

(8) Notices to Airmen.

[Docket No. 17897, 45 FR 41594, June 19, 1980; as amended by Amdt. 121–159, 45 FR 43154, June 26, 1980]

## §121.445 Pilot in command airport qualification: Special areas and airports.

**(a)** The Administrator may determine that certain airports (due to items such as surrounding terrain, obstructions, or complex approach or departure procedures) are special airports requiring special airport qualifications and that certain areas or routes, or both, require a special type of navigation qualification.

**(b)** Except as provided in paragraph (c) of this section, no certificate holder may use any person, nor may any person serve, as pilot in command to or from an airport determined to require special airport qualifications unless, within the preceding 12 calendar months:

(1) The pilot in command or second in command has made an entry to that airport (including a takeoff and landing) while serving as a pilot flight crewmember; or

(2) The pilot in command has qualified by using pictorial means acceptable to the Administrator for that airport.

**(c)** Paragraph (b) of this section does not apply when an entry to that airport (including a takeoff or a landing) is being made if the ceiling at that airport is at least 1,000 feet above the lowest MEA or MOCA, or initial approach altitude prescribed for the instrument approach procedure for that airport, and the visibility at that airport is at least 3 miles.

**(d)** No certificate holder may use any person, nor may any person serve, as pilot in command between terminals over a route or area that requires a special type of navigation qualification unless, within the preceding 12 calendar months, that person has demonstrated qualification on the applicable navigation system in a manner acceptable to the Administrator, by one of the following methods:

(1) By flying over a route or area as pilot in command using the applicable special type of navigation system.

(2) By flying over a route or area as pilot in command under the supervision of a check airman using the special type of navigation system.

(3) By completing the training program requirements of Appendix G of this part.

[Docket No. 17897, 45 FR 41594, June 19, 1980]

## §121.447 [Reserved]

## §121.453 Flight engineer qualifications.

**(a)** No certificate holder may use any person nor may any person serve as a flight engineer on an airplane unless, within the preceding 6 calendar months, he has had at least 50 hours of flight time as a flight engineer on that type airplane or the certificate holder or the Administrator has checked him on that type airplane and determined that he is familiar and competent with all essential current information and operating procedures.

**(b)** A flight check given in accordance with §121.425(a)(2) satisfies the requirements of paragraph (a) of this section.

[Docket No. 9509, 35 FR 96, Jan. 3, 1970]

## §121.455 Use of prohibited drugs.

**(a)** This section applies to persons who perform a function listed in Appendix I to this part for the certificate holder or operator. For the purpose of this section, a person who performs such a function pursuant to a contract with the certificate holder or operator is considered to be performing that function for the certificate holder or operator.

**(b)** No certificate holder or operator may knowingly use any person to perform, nor may any person perform for a certificate holder or operator, either directly or by contract, any function listed in Appendix I to this part while that person has a prohibited drug, as defined in that appendix, in his or her system.

**(c)** No certificate holder or operator shall knowingly use any person to perform, nor shall any person perform for a certificate holder or operator, either directly or by contract, any safety-sensitive function if the person has a verified positive drug test result on or has refused to submit to a drug test required by Appendix I to part 121 of this chapter and the person has not met the requirements of Appendix I for returning to the performance of safety-sensitive duties.

[Docket No. 25148, 53 FR 47057, Nov. 21, 1988; as amended by Amdt. 121–240, 59 FR 42928, Aug. 19, 1994; Amdt. 121–251, 60 FR 65934, Dec. 20, 1995]

## §121.457 Testing for prohibited drugs.

**(a)** Each certificate holder or operator shall test each of its employees who performs a function listed in Appendix I to this part in accordance with that appendix.

**(b)** No certificate holder or operator may use any contractor to perform a function listed in Appendix I to this part unless that contractor tests each employee performing such a function for the certificate holder or operator in accordance with that appendix.

[Docket No. 25148, 53 FR 47057, Nov. 21, 1988; as amended by Amdt. 121–251, 60 FR 65934, Dec. 20, 1995]

121

## §121.458 Misuse of alcohol.

**(a)** *General.* This section applies to employees who perform a function listed in Appendix J to this part for a certificate holder *(covered employees).* For the purpose of this section, a person who meets the definition of covered employee in Appendix J is considered to be performing the function for the certificate holder.

**(b)** *Alcohol concentration.* No covered employee shall report for duty or remain on duty requiring the performance of safety-sensitive functions while having an alcohol concentration of 0.04 or greater. No certificate holder having actual knowledge that an employee has an alcohol concentration of 0.04 or greater shall permit the employee to perform or continue to perform safety-sensitive functions.

**(c)** *On-duty use.* No covered employee shall use alcohol while performing safety-sensitive functions. No certificate holder having actual knowledge that a covered employee is using alcohol while performing safety-sensitive functions shall permit the employee to perform or continue to perform safety-sensitive functions.

**(d)** *Pre-duty use.*

(1) No covered employee shall perform flight crewmember or flight attendant duties within 8 hours after using alcohol. No certificate holder having actual knowledge that such an employee has used alcohol within 8 hours shall permit the employee to perform or continue to perform the specified duties.

(2) No covered employee shall perform safety-sensitive duties other than those specified in paragraph (d)(1) of this section within 4 hours after using alcohol. No certificate holder having actual knowledge that such an employee has used alcohol within 4 hours shall permit the employee to perform or continue to perform safety-sensitive functions.

**(e)** *Use following an accident.* No covered employee who has actual knowledge of an accident involving an aircraft for which he or she performed a safety-sensitive function at or near the time of the accident shall use alcohol for 8 hours following the accident, unless he or she has been given a post-accident test under Appendix J of this part, or the employer has determined that the employee's performance could not have contributed to the accident.

**(f)** *Refusal to submit to a required alcohol test.* No covered employee shall refuse to submit to a post-accident, random, reasonable suspicion, or follow-up alcohol test required under Appendix J to this part. No certificate holder shall permit an employee who refuses to submit to such a test to perform or continue to perform safety-sensitive functions.

[Docket No. 27065, 59 FR 7389, Feb. 15, 1994]

## §121.459 Testing for alcohol.

**(a)** Each certificate holder must establish an alcohol misuse prevention program in accordance with the provisions of Appendix J to this part.

**(b)** No certificate holder shall use any person who meets the definition of *covered employee* in Appendix J to this part to perform a safety-sensitive function listed in that appendix unless such person is subject to testing for alcohol misuse in accordance with the provisions of Appendix J.

[Docket No. 27065, 59 FR 7390, Feb. 15, 1994]

# Subpart P — Aircraft Dispatcher Qualifications and Duty Time

***Limitations:*** **Domestic and Flag Operations; Flight Attendant Duty Period Limitations and Rest Requirements: Domestic, Flag, and Supplemental Operations**

## §121.461 Applicability.

This subpart prescribes—

**(a)** Qualifications and duty time limitations for aircraft dispatchers for certificate holders conducting domestic flag operations; and

**(b)** Duty period limitations and rest requirements for flight attendants used by certificate holders conducting domestic, flag, or supplemental operations.

[Docket No. 28154, 61 FR 2612, Jan. 26, 1996]

## §121.463 Aircraft dispatcher qualifications.

**(a)** No certificate holder conducting domestic or flag operations may use any person, nor may any person serve, as an aircraft dispatcher for a particular airplane group unless that person has, with respect to an airplane of that group, satisfactorily completed the following:

(1) Initial dispatcher training, except that a person who has satisfactorily completed such training for another type airplane of the same group need only complete the appropriate transition training.

(2) Operating familiarization consisting of at least 5 hours observing operations under this part from the flight deck or, for airplanes without an observer seat on the flight deck, from a forward passenger seat with headset or speaker. This requirement may be reduced to a minimum of $2\frac{1}{2}$ hours by the substitution of one additional takeoff and landing for an hour of flight. A person may serve as an aircraft dispatcher without meeting the requirement of this paragraph (a) for 90 days after initial introduction of the airplane into operations under this part.

**(b)** No certificate holder conducting domestic or flag operations may use any person, nor may any person serve, as an aircraft dispatcher for a particular type airplane unless that person has, with respect to that airplane, satisfactorily completed differences training, if applicable.

**(c)** No certificate holder conducting domestic or flag operations may use any person, nor may any person serve, as an aircraft dispatcher unless within the preceding 12 calendar months the aircraft dispatcher has satisfactorily completed operating familiarization consisting of at least 5 hours observing operations under this part, in one of the types of airplanes in each group to be dispatched. This observation shall be made from the flight deck or, for airplanes without an observer seat on the flight deck, from a forward passenger seat with headset or speaker. The requirement of paragraph (a) of this section may be reduced to a minimum of $2\frac{1}{2}$ hours by the substitution of one additional takeoff and landing for an hour of flight. The requirement of this paragraph may be satisfied by observation of 5 hours of simulator training for each airplane group in one of the simulators approved under §121.407 for the group. However, if the requirement of paragraph (a) is met by the use of a simulator, no reduction in hours is permitted.

**(d)** No certificate holder conducting domestic or flag operations may use any person, nor may any person serve as an aircraft dispatcher to dispatch airplanes in operations under this part unless the certificate holder has determined that he is familiar with all essential operating procedures for that segment of the operation over which he exercises dispatch jurisdiction. However, a dispatcher who is qualified to dispatch airplanes through one segment of an operation may dispatch airplanes through other segments of the operation after coordinating with dispatchers who are qualified to dispatch airplanes through those other segments.

**(e)** For the purposes of this section, the airplane groups, terms, and definitions in §121.400 apply.

[Docket No. 7325, 37 FR 5607, March 17, 1972; as amended by Amdt. 121–251, 60 FR 65934, Dec. 20, 1995]

## §121.465 Aircraft dispatcher duty time limitations: Domestic and flag operations.

**(a)** Each certificate holder conducting domestic or flag operations shall establish the daily duty period for a dispatcher so that it begins at a time that allows him or her to become thoroughly familiar with existing and anticipated weather conditions along the route before he or she dispatches any airplane. He or she shall remain on duty until each airplane dispatched by him or her has completed its flight, or has gone beyond his or her jurisdiction, or until he or she is relieved by another qualified dispatcher.

**(b)** Except in cases where circumstances or emergency conditions beyond the control of the certificate holder require otherwise—

(1) No certificate holder conducting domestic or flag operations may schedule a dispatcher for more than 10 consecutive hours of duty;

(2) If a dispatcher is scheduled for more than 10 hours of duty in 24 consecutive hours, the certificate holder shall provide him or her a rest period of at least eight hours at or before the end of 10 hours of duty.

(3) Each dispatcher must be relieved of all duty with the certificate holder for at least 24 consecutive hours during any seven consecutive days or the equivalent thereof within any calendar month.

**(c)** Notwithstanding paragraphs (a) and (b) of this section, a certificate holder conducting flag operations may, if authorized by the Administrator, schedule an aircraft dispatcher at a duty station outside of the 48 contiguous States and the District of Columbia, for more than 10 consecutive hours of duty in a 24-hour period if that aircraft dispatcher is relieved of all duty with the certificate holder for at least eight hours during each 24-hour period.

[Docket No. 28154, 61 FR 2612, Jan. 26, 1996]

## §121.467 Flight attendant duty period limitations and rest requirements: Domestic, flag, and supplemental operations.

**(a)** For purposes of this section—

*Calendar day* means the period of elapsed time, using Coordinated Universal Time or local time, that begins at midnight and ends 24 hours later at the next midnight.

*Duty period* means the period of elapsed time between reporting for an assignment involving flight time and release from that assignment by the certificate holder conducting domestic, flag, or supplemental operations. The time is calculated using either Coordinated Universal Time or local time to reflect the total elapsed time.

*Flight attendant* means an individual, other than a flight crewmember, who is assigned by a certificate holder conducting domestic, flag, or supplemental operations, in accordance with the required minimum crew complement under the certificate holder's operations specification or in addition to that minimum complement, to duty in an aircraft during flight time and whose duties include but are not necessarily limited to cabin-safety-related responsibilities.

*Rest period* means the period free of all restraint or duty for a certificate holder conducting domestic, flag, or supplemental operations and free of all responsibility for work or duty should the occasion arise.

**(b)** Except as provided in paragraph (c) of this section, a certificate holder conducting domestic, flag, or supplemental operations may assign a duty period to a flight attendant only when the applicable duty period limitations and rest requirements of this paragraph are met.

(1) Except as provided in paragraphs (b)(4), (b)(5), and (b)(6) of this section, no certificate holder conducting domestic, flag, or supplemental

operations may assign a flight attendant to a scheduled duty period of more than 14 hours.

(2) Except as provided in paragraph (b)(3) of this section, a flight attendant scheduled to a duty period of 14 hours or less as provided under paragraph (b)(1) of this section must be given a scheduled rest period of at least 9 consecutive hours. This rest period must occur between the completion of the scheduled duty period and the commencement of the subsequent duty period.

(3) The rest period required under paragraph (b)(2) of this section may be scheduled or reduced to 8 consecutive hours if the flight attendant is provided a subsequent rest period of at least 10 consecutive hours; this subsequent rest period must be scheduled to begin no later than 24 hours after the beginning of the reduced rest period and must occur between the completion of the scheduled duty period and the commencement of the subsequent duty period.

(4) A certificate holder conducting domestic, flag, or supplemental operations may assign a flight attendant to a scheduled duty period of more than 14 hours, but no more than 16 hours, if the certificate holder has assigned to the flight or flights in that duty period at least one flight attendant in addition to the minimum flight attendant complement required for the flight or flights in that duty period under the certificate holder's operations specifications.

(5) A certificate holder conducting domestic, flag, or supplemental operations may assign a flight attendant to a scheduled duty period of more than 16 hours, but no more than 18 hours, if the certificate holder has assigned to the flight or flights in that duty period at least two flight attendants in addition to the minimum flight attendant complement required for the flight or flights in that duty period under the certificate holder's operations specifications.

(6) A certificate holder conducting domestic, flag, or supplemental operations may assign a flight attendant to a scheduled duty period of more than 18 hours, but no more than 20 hours, if the scheduled duty period includes one or more flights that land or take off outside the 48 contiguous states and the District of Columbia, and if the certificate holder has assigned to the flight or flights in that duty period at least three flight attendants in addition to the minimum flight attendant complement required for the flight or flights in that duty period under the domestic certificate holder's operations specifications.

(7) Except as provided in paragraph (b)(8) of this section, a flight attendant scheduled to a duty period of more than 14 hours but no more than 20 hours, as provided in paragraphs (b)(4), (b)(5), and (b)(6) of this section, must be given a scheduled rest period of at least 12 consecutive hours. This rest period must occur between the completion of the scheduled duty period and the commencement of the subsequent duty period.

(8) The rest period required under paragraph (b)(7) of this section may be scheduled or reduced to 10 consecutive hours if the flight attendant is provided a subsequent rest period of at least 14 consecutive hours; this subsequent rest period must be scheduled to begin no later than 24 hours after the beginning of the reduced rest period and must occur between the completion of the scheduled duty period and the commencement of the subsequent duty period.

(9) Notwithstanding paragraphs (b)(4), (b)(5), and (b)(6) of this section, if a certificate holder conducting domestic, flag, or supplemental operations elects to reduce the rest period to 10 hours as authorized by paragraph (b)(8) of this section, the certificate holder may not schedule a flight attendant for a duty period of more than 14 hours during the 24-hour period commencing after the beginning of the reduced rest period.

(10) No certificate holder conducting domestic, flag, or supplemental operations may assign a flight attendant any duty period with the certificate holder unless the flight attendant has had at least the minimum rest required under this section.

(11) No certificate holder conducting domestic, flag, or supplemental operations may assign a flight attendant to perform any duty with the certificate holder during any required rest period.

(12) Time spent in transportation, not local in character, that a certificate holder conducting domestic, flag, or supplemental operations requires of a flight attendant and provides to transport the flight attendant to an airport at which that flight attendant is to serve on a flight as a crewmember, or from an airport at which the flight attendant was relieved from duty to return to the flight attendant's home station, is not considered part of a rest period.

(13) Each certificate holder conducting domestic, flag, or supplemental operations must relieve each flight attendant engaged in air transportation and each commercial operator must relieve each flight attendant engaged in air commerce from all further duty for at least 24 consecutive hours during any 7 consecutive calendar days.

(14) A flight attendant is not considered to be scheduled for duty in excess of duty period limitations if the flights to which the flight attendant is assigned are scheduled and normally terminate within the limitations but due to circumstances beyond the control of the certificate holder conducting domestic, flag, or supplemental operations (such as adverse weather conditions) are not at the time of departure expected to reach their destination within the scheduled time.

**(c)** Notwithstanding paragraph (b) of this section, a certificate holder conducting domestic, flag, or supplemental operations may apply the flight crewmember flight time and duty limitations and rest requirements of this part to flight attendants for all operations conducted under this part provided that—

(1) The certificate holder established written procedures than—

(i) Apply to all flight attendants used in the certificate holder's operation;

(ii) Include the flight crewmember requirements contained in Subparts Q, R, or S of this part, as appropriate to the operation being conducted, except that rest facilities on board the aircraft are not required;

(iii) Include provisions to add one flight attendant to the minimum flight attendant complement for each flight crewmember who is in excess of the minimum number required in the aircraft type certificate data sheet and who is assigned to the aircraft under the provisions of Subparts Q, R, and S, as applicable, of this part;

(iv) Are approved by the Administrator and are described or referenced in the certificate holder's operations specifications; and

(2) Whenever the Administrator finds that revisions are necessary for the continued adequacy of the written procedures that are required by paragraph (c)(1) of this section and that had been granted final approval, the certificate holder must, after notification by the Administrator, make any changes in the procedures that are found necessary by the Administrator. Within 30 days after the certificate holder receives such notice, it may file a petition to reconsider the notice with the certificate-holding district office. The filing of a petition to reconsider stays the notice, pending decision by the Administrator. However, if the Administrator finds that an emergency requires immediate action in the interest of safety, the Administrator may, upon a statement of the reasons, require a change effective without stay.

[Docket No. 27229, 59 FR 42991, August 19, 1994; as amended by Amdt. 121–253, 61 FR 2612, Jan. 26, 1996]

# Subpart Q— Flight Time Limitations and Rest Requirements: Domestic Operations

**Source:** Docket No. 23634, 50 FR 29319, July 18, 1985, unless otherwise noted.

## §121.470 Applicability.

This subpart prescribes flight time limitations and rest requirements for domestic operations, except that:

**(a)** Certificate holders conducting operations with airplanes having a passenger seat configuration of 30 seats or fewer, excluding each crewmember seat, and a payload capacity of 7,500 pounds or less, may comply with the applicable requirements of §§135.261 through 135.273 of this chapter.

**(b)** Certificate holders conducting scheduled operations entirely within the States of Alaska or Hawaii with airplanes having a passenger seat configuration of more than 30 seats, excluding each crewmember seat, or a payload capacity of more than 7,500 pounds, may comply with the requirements of Subpart R of this part for those operations.

[Docket No. 28154, 60 FR 65934, Dec. 20, 1995]

## §121.471 Flight time limitations and rest requirements: All flight crewmembers.

**(a)** No certificate holder conducting domestic operations may schedule any flight crewmember and no flight crewmember may accept an assignment for flight time in scheduled air transportation or in other commercial flying if that crewmember's total flight time in all commercial flying will exceed—

(1) 1,000 hours in any calendar year;

(2) 100 hours in any calendar month;

(3) 30 hours in any 7 consecutive days;

(4) 8 hours between required rest periods.

**(b)** Except as provided in paragraph (c) of this section, no certificate holder conducting domestic operations may schedule a flight crewmember and no flight crewmember may accept an assignment for flight time during the 24 consecutive hours preceding the scheduled completion of any flight segment without a scheduled rest period during that 24 hours of at least the following:

(1) 9 consecutive hours of rest for less than 8 hours of scheduled flight time.

(2) 10 consecutive hours of rest for 8 or more but less than 9 hours of scheduled flight time.

(3) 11 consecutive hours of rest for 9 or more hours of scheduled flight time.

**(c)** An certificate holder may schedule a flight crewmember for less than the rest required in paragraph (b) of this section or may reduce a scheduled rest under the following conditions:

(1) A rest required under paragraph (b)(1) of this section may be scheduled for or reduced to a minimum of 8 hours if the flight crewmember is given a rest period of at least 10 hours that must begin no later than 24 hours after the commencement of the reduced rest period.

(2) A rest required under paragraph (b)(2) of this section may be scheduled for or reduced to a minimum of 8 hours if the flight crewmember is given a rest period of at least 11 hours that must

begin no later than 24 hours after the commencement of the reduced rest period.

(3) A rest required under paragraph (b)(3) of this section may be scheduled for or reduced to a minimum of 9 hours if the flight crewmember is given a rest period of at least 12 hours that must begin no later than 24 hours after the commencement of the reduced rest period.

(4) No certificate holder may assign, nor may any flight crewmember perform any flight time with the certificate holder unless the flight crewmember has had at least the minimum rest required under this paragraph.

**(d)** Each certificate holder conducting domestic operations shall relieve each flight crewmember engaged in scheduled air transportation from all further duty for at least 24 consecutive hours during any 7 consecutive days.

**(e)** No certificate holder conducting domestic operations may assign any flight crewmember and no flight crewmember may accept assignment to any duty with the air carrier during any required rest period.

**(f)** Time spent in transportation, not local in character, that an certificate holder requires of a flight crewmember and provides to transport the crewmember to an airport at which he is to serve on a flight as a crewmember, or from an airport at which he was relieved from duty to return to his home station, is not considered part of a rest period.

**(g)** A flight crewmember is not considered to be scheduled for flight time in excess of flight time limitations if the flights to which he is assigned are scheduled and normally terminate within the limitations, but due to circumstances beyond the control of the certificate holder (such as adverse weather conditions), are not at the time of departure expected to reach their destination within the scheduled time.

[Docket No. 23634, 50 FR 29319, July 18, 1985; as amended by Amdt. 121–253, 61 FR 2612, Jan. 26, 1996]

# Subpart R— Flight Time Limitations: Flag Operations

SOURCE: Docket No. 6258, 29 FR 19217, Dec. 31, 1964; 30 FR 3639, March 19, 1965, unless otherwise noted.

## §121.480 Applicability.

This subpart prescribes flight time limitations and rest requirements for flag operations, except that certificate holders conducting operations with airplanes having a passenger seat configuration of 30 seats or fewer, excluding each crewmember seat, and a payload capacity of 7,500 pounds or less, may comply with the applicable requirements of §§135.261 through 135.273 of this chapter.

[Docket No. 28154, 60 FR 65934, Dec. 20, 1995]

## §121.481 Flight time limitations: One or two pilot crews.

**(a)** A certificate holder conducting flag operations may schedule a pilot to fly in an airplane that has a crew of one or two pilots for eight hours or less during any 24 consecutive hours without a rest period during these eight hours.

**(b)** If a certificate holder conducting flag operations schedules a pilot to fly more than eight hours during any 24 consecutive hours, it shall give him an intervening rest period, at or before the end of eight scheduled hours of flight duty. This rest period must be at least twice the number of hours flown since the preceding rest period, but not less than eight hours. The certificate holder shall relieve that pilot of all duty with it during that rest period.

**(c)** Each pilot who has flown more than eight hours during 24 consecutive hours must be given at least 18 hours of rest before being assigned to any duty with the certificate holder.

**(d)** No pilot may fly more than 32 hours during any seven consecutive days, and each pilot must be relieved from all duty for at least 24 consecutive hours at least once during any seven consecutive days.

**(e)** No pilot may fly as a member of a crew more than 100 hours during any one calendar month.

**(f)** No pilot may fly as a member of a crew more than 1,000 hours during any 12-calendar-month period.

[Docket No. 6258, 29 FR 19217, Dec. 31, 1964; as amended by Amdt. 121–253, 61 FR 2612, Jan. 26, 1996]

## §121.483 Flight time limitations: Two pilots and one additional flight crewmember.

**(a)** No certificate holder conducting flag operations may schedule a pilot to fly, in an airplane that has a crew of two pilots and at least one additional flight crewmember, for a total of more than 12 hours during any 24 consecutive hours.

**(b)** If a pilot has flown 20 or more hours during any 48 consecutive hours or 24 or more hours during any 72 consecutive hours, he must be given at least 18 hours of rest before being assigned to any duty with the air carrier. In any case, he must be given at least 24 consecutive hours of rest during any seven consecutive days.

**(c)** No pilot may fly as a flight crewmember more than—

(1) 120 hours during any 30 consecutive days;

(2) 300 hours during any 90 consecutive days; or

(3) 1,000 hours during any 12-calendar-month period.

[Docket No. 6258, 29 FR 19217, Dec. 31, 1964; as amended by Amdt. 121–253, 61 FR 2612, Jan. 26, 1996]

## §121.485 Flight time limitations: Three or more pilots and an additional flight crewmember.

**(a)** Each certificate holder conducting flag operations shall schedule its flight hours to provide adequate rest periods on the ground for each pilot who is away from his base and who is a pilot on an airplane that has a crew of three or more pilots and an additional flight crewmember. It shall also provide adequate sleeping quarters on the airplane whenever a pilot is scheduled to fly more than 12 hours during any 24 consecutive hours.

**(b)** The certificate holder conducting flag operations shall give each pilot, upon return to his base from any flight or series of flights, a rest period that is at least twice the total number of hours he flew since the last rest period at his base. During the rest period required by this paragraph, the air carrier may not require him to perform any duty for it. If the required rest period is more than seven days, that part of the rest period in excess of seven days may be given at any time before the pilot is again scheduled for flight duty on any route.

**(c)** No pilot may fly as a flight crewmember more than—

(1) 350 hours during any 90 consecutive days; or

(2) 1,000 hours during any 12-calendar-month period.

[Docket No. 6258, 29 FR 19217, Dec. 31, 1964; as amended by Amdt. 121–253, 61 FR 2612, Jan. 26, 1996]

## §121.487 Flight time limitations: Pilots not regularly assigned.

**(a)** Except as provided in paragraphs (b) through (e) of this section, a pilot who is not regularly assigned as a flight crewmember for an entire calendar month under §121.483 or 121.485 may not fly more than 100 hours in any 30 consecutive days.

**(b)** The monthly flight time limitations for a pilot who is scheduled for duty aloft for more than 20 hours in two-pilot crews in any calendar month, or whose assignment in such a crew is interrupted more than once in that calendar month by assignment to a crew consisting of two or more pilots and an additional flight crewmember, are those set forth in §121.481.

**(c)** Except for a pilot covered by paragraph (b) of this section, the monthly and quarterly flight time limitations for a pilot who is scheduled for duty aloft for more than 20 hours in two-pilot and additional flight crewmember crews in any calendar month, or whose assignment in such a crew is interrupted more than once in that calendar month by assignment to a crew consisting of three pilots and additional flight crewmember, are those set forth in §121.483.

**(d)** The quarterly flight time limitations for a pilot to whom paragraphs (b) and (c) of this section do not apply and who is scheduled for duty aloft for a total of not more than 20 hours within any calendar month in two-pilot crews (with or without additional flight crewmembers) are those set forth in §121.485.

**(e)** The monthly and quarterly flight time limitations for a pilot assigned to each of two-pilot, two-pilot and additional flight crewmember, and three-pilot and additional flight crewmember crews in a given calendar month, and who is not subject to paragraph (b), (c), or (d) of this section, are those set forth in §121.483.

[Docket No. 6258, 29 FR 19217, Dec. 31, 1964; as amended by Amdt. 121–3, 30 FR 3639, March 19, 1965; Amdt. 121–137, 42 FR 43973, Sept. 1, 1977]

## §121.489 Flight time limitations: Other commercial flying.

No pilot that is employed as a pilot by a certificate holder conducting flag operations may do any other commercial flying if that commercial flying plus his flying in air transportation will exceed any flight time limitation in this part.

[Docket No. 28154, 61 FR 2612, Jan. 26, 1996]

## §121.491 Flight time limitations: Deadhead transportation.

Time spent in deadhead transportation to or from duty assignment is not considered to be a part of a rest period.

## §121.493 Flight time limitations: Flight engineers and flight navigators.

**(a)** In any operation in which one flight engineer or flight navigator is required, the flight time limitations in §121.483 apply to that flight engineer or flight navigator.

**(b)** In any operation in which more than one flight engineer or flight navigator is required, the flight time limitations in §121.485 apply to those flight engineers or flight navigators.

# Subpart S— Flight Time Limitations: Supplemental Operations

SOURCE: Docket No. 6258, 29 FR 19218, Dec. 31, 1964; 30 FR 3639, March 19, 1965, unless otherwise noted.

## §121.500 Applicability.

This subpart prescribes flight time limitations and rest requirements for supplemental operations, except that certificate holders conducting operations with airplanes having a passenger seat configuration of 30 seats or fewer, excluding each crewmember seat, and a payload capacity of 7,500 pounds or less, may comply with the applicable requirements of §§135.261 through 135.273 of this chapter.

[Docket No. 28154, 60 FR 65934, Dec. 20, 1995]

## §121.503 Flight time limitations: Pilots: airplanes.

**(a)** A certificate holder conducting supplemental operations may schedule a pilot to fly in an airplane for eight hours or less during any 24 consecutive hours without a rest period during those eight hours.

**(b)** Each pilot who has flown more than eight hours during any 24 consecutive hours must be given at least 16 hours of rest before being assigned to any duty with the certificate holder.

**(c)** Each certificate holder conducting supplemental operations shall relieve each pilot from all duty for at least 24 consecutive hours at least once during any seven consecutive days.

**(d)** No pilot may fly as a crewmember in air transportation more than 100 hours during any 30 consecutive days.

**(e)** No pilot may fly as a crewmember in air transportation more than 1,000 hours during any calendar year.

**(f)** Notwithstanding paragraph (a) of this section, the certificate holder may, in conducting a transcontinental nonstop flight, schedule a flight crewmember for more than eight but not more than 10 hours of continuous duty aloft without an intervening rest period, if—

(1) The flight is in an airplane with a pressurization system that is operative at the beginning of the flight;

(2) The flight crew consists of at least two pilots and a flight engineer; and

(3) The certificate holder uses, in conducting the operation, an air/ground communication service that is independent of systems operated by the United States, and a dispatch organization, both of which are approved by the Administrator as adequate to serve the terminal points concerned.

[Docket No. 6258, 29 FR 19218, Dec. 31, 1964; as amended by Amdt. 121–253, 61 FR 2613, Jan. 26, 1996]

## §121.505 Flight time limitations: Two pilot crews: airplanes.

**(a)** If a certificate holder conducting supplemental operations schedules a pilot to fly more than eight hours during any 24 consecutive hours, it shall give him an intervening rest period at or before the end of eight scheduled hours of flight duty. This rest period must be at least twice the number of hours flown since the preceding rest period, but not less than eight hours. The certificate holder conducting supplemental operations shall relieve that pilot of all duty with it during that rest period.

**(b)** No pilot of an airplane that has a crew of two pilots may be on duty for more than 16 hours during any 24 consecutive hours.

[Docket No. 6258, 29 FR 19218, Dec. 31, 1964; as amended by Amdt. 121–253, 61 FR 2613, Jan. 26, 1996]

## §121.507 Flight time limitations: Three pilot crews: Airplanes.

**(a)** No certificate holder conducting supplemental operations may schedule a pilot—

(1) For flight deck duty in an airplane that has a crew of three pilots for more than eight hours in any 24 consecutive hours; or

(2) To be aloft in an airplane that has a crew of three pilot for more than 12 hours in any 24 consecutive hours.

**(b)** No pilot of an airplane that has a crew of three pilots may be on duty for more than 18 hours in any 24 consecutive hours.

[Docket No. 6258, 29 FR 19218, Dec. 31, 1964; as amended by Amdt. 121–253, 61 FR 2613, Jan. 26, 1996]

## §121.509 Flight time limitations: Four pilot crews: airplanes.

**(a)** No certificate holder conducting supplemental operations may schedule a pilot—

(1) For flight deck duty in an airplane that has a crew of four pilots for more than eight hours in any 24 consecutive hours; or

(2) To be aloft in an airplane that has a crew of four pilots for more than 16 hours in any 24 consecutive hours.

**(b)** No pilot of an airplane that has a crew of four pilots may be on duty for more than 20 hours in any 24 consecutive hours.

[Docket No. 6258, 29 FR 19218, Dec. 31, 1964; as amended by Amdt. 121–253, 61 FR 2613, Jan. 26, 1996]

### §121.511 Flight time limitations: Flight engineers: airplanes.

**(a)** In any operation in which one flight engineer is serving the flight time limitations in §§121.503 and 121.505 apply to that flight engineer.

**(b)** In any operation in which more than one flight engineer is serving and the flight crew contains more than two pilots the flight time limitations in §121.509 apply in place of those in §121.505.

### §121.513 Flight time limitations: Overseas and international operations: airplanes.

In place of the flight time limitations in §§121.503 through 121.511, a certificate holder conducting supplemental operations may elect to comply with the flight time limitations of §§121.515 and 121.521 through 121.525 for operations conducted—

**(a)** Between a place in the 48 contiguous States and the District of Columbia, or Alaska, and any place outside thereof;

**(b)** Between any two places outside the 48 contiguous States, the District of Columbia, and Alaska; or

**(c)** Between two places within the State of Alaska or the State of Hawaii.

[Docket No. 6258, 29 FR 19218, Dec. 31, 1964; as amended by Amdt. 121–253, 61 FR 2613, Jan. 26, 1996]

### §121.515 Flight time limitations: All airmen: airplanes.

No airman may be aloft as a flight crewmember more than 1,000 hours in any 12-calendar-month period.

### §121.517 Flight time limitations: Other commercial flying: airplanes.

No airman who is employed by a certificate holder conducting supplemental operations may do any other commercial flying, if that commercial flying plus his flying in operations under this part will exceed any flight time limitation in this part.

[Docket No. 28154, 61 FR 2613, Jan. 26, 1996]

### §121.519 Flight time limitations: Deadhead transportation: Airplanes.

Time spent by an airman in deadhead transportation to or from a duty assignment is not considered to be part of any rest period.

### §121.521 Flight time limitations: Crew of two pilots and one additional airman as required.

**(a)** No certificate holder conducting supplemental operations may schedule an airman to be aloft as a member of the flight crew in an airplane that has a crew of two pilots and at least one additional flight crewmember for more than 12 hours during any 24 consecutive hours.

**(b)** If an airman has been aloft as a member of a flight crew for 20 or more hours during any 48 consecutive hours or 24 or more hours during any 72 consecutive hours, he must be given at least 18 hours of rest before being assigned to any duty with the certificate holder. In any case, he must be relieved of all duty for at least 24 consecutive hours during any seven consecutive days.

**(c)** No airman may be aloft as a flight crewmember more than—

(1) 120 hours during any 30 consecutive days; or

(2) 300 hours during any 90 consecutive days.

[Docket No. 6258, 29 FR 19218, Dec. 31, 1964; as amended by Amdt. 121–17, 31 FR 1147, Jan. 28, 1966; Amdt. 121–253, 61 FR 2613, Jan. 26, 1996]

### §121.523 Flight time limitations: Crew of three or more pilots and additional airmen as required.

**(a)** No certificate holder conducting supplemental operations may schedule an airman for flight deck duty as a flight engineer, or navigator in a crew of three or more pilots and additional airmen for a total of more than 12 hours during any 24 consecutive hours.

**(b)** Each certificate holder conducting supplemental operations shall schedule its flight hours to provide adequate rest periods on the ground for each airman who is away from his principal operations base. It shall also provide adequate sleeping quarters on the airplane whenever an airman is scheduled to be aloft as a flight crewmember for more than 12 hours during any 24 consecutive hours.

**(c)** No certificate holder conducting supplemental operations may schedule any flight crewmember to be on continuous duty for more than 30 hours. Such a crewmember is considered to be on continuous duty from the time he reports for duty until the time he is released from duty for a rest period of at least 10 hours on the ground. If a flight crewmember is on continuous duty for more than 24 hours (whether scheduled or not) duty any scheduled duty period, he must be given at least 16 hours for rest on the ground after completing the last flight scheduled for that scheduled duty period before being assigned any further flight duty.

121

**(d)** If a flight crewmember is required to engage in deadhead transportation for more than four hours before beginning flight duty, one half of the time spent in deadhead transportation must be treated as duty time for the purpose of complying with duty time limitations, unless he is given at least 10 hours of rest on the ground before being assigned to flight duty.

**(e)** Each certificate holder conducting supplemental operations shall give each airman, upon return to his operations base from any flight or series of flights, a rest period that is at least twice the total number of hours he was aloft as a flight crewmember since the last rest period at his base, before assigning him to any further duty. If the required rest period is more than seven days, that part of the rest period that is more than seven days may be given at any time before the pilot is again scheduled for flight duty.

**(f)** No airman may be aloft as a flight crewmember for more than 350 hours in any 90 consecutive days.

[Docket No. 6258, 29 FR 19218, Dec. 31, 1964; as amended by Amdt. 121–253, 61 FR 2613, Jan. 26, 1996]

### §121.525 Flight time limitations: Pilots serving in more than one kind of flight crew.

**(a)** This section applies to each pilot assigned during any 30 consecutive days to more than one type of flight crew.

**(b)** The flight time limitations for a pilot who is scheduled for duty aloft for more than 20 hours in two-pilot crews in 30 consecutive days, or whose assignment in such a crew is interrupted more than once in any 30 consecutive days by assignment to a crew of two or more pilots and an additional flight crewmember, are those listed in §§121.503 through 121.509, as appropriate.

**(c)** Except for a pilot covered by paragraph (b) of this section, the flight time limitations for a pilot scheduled for duty aloft for more than 20 hours in two-pilot and additional flight crewmember crews in 30 consecutive days or whose assignment in such a crew is interrupted more than once in any 30 consecutive days by assignment to a crew consisting of three pilots and an additional flight crewmember, are those set forth in §121.521.

**(d)** The flight time limitations for a pilot to whom paragraphs (b) and (c) of this section do not apply, and who is scheduled for duty aloft for a total of not more than 20 hours within 30 consecutive days in two-pilot crews (with or without additional flight crewmembers) are those set forth in §121.523.

**(e)** The flight time limitations for a pilot assigned to each of two-pilot, two-pilot and additional flight crewmember, and three-pilot and additional flight crewmember crews in 30 consecutive days, and who is not subject to paragraph (b), (c), or (d) of this section, are those listed in §121.523.

## Subpart T— Flight Operations

**Source:** Docket No. 6258, 29 FR 19219, Dec. 31, 1964, unless otherwise noted.

### §121.531 Applicability.

This subpart prescribes requirements for flight operations applicable to all certificate holders, except where otherwise specified.

### §121.533 Responsibility for operational control: Domestic operations.

**(a)** Each certificate holder conducting domestic operations is responsible for operational control.

**(b)** The pilot in command and the aircraft dispatcher are jointly responsible for the preflight planning, delay, and dispatch release of a flight in compliance with this chapter and operations specifications.

**(c)** The aircraft dispatcher is responsible for—

(1) Monitoring the progress of each flight;

(2) Issuing necessary information for the safety of the flight; and

(3) Canceling or redispatching a flight if, in his opinion or the opinion of the pilot in command, the flight cannot operate or continue to operate safely as planned or released.

**(d)** Each pilot in command of an aircraft is, during flight time, in command of the aircraft and crew and is responsible for the safety of the passengers, crewmembers, cargo, and airplane.

**(e)** Each pilot in command has full control and authority in the operation of the aircraft, without limitation, over other crewmembers and their duties during flight time, whether or not he holds valid certificates authorizing him to perform the duties of those crewmembers.

[Docket No. 6258, 29 FR 19219, Dec. 31, 1964; as amended by Amdt. 121–253, 61 FR 2613, Jan. 26, 1996]

### §121.535 Responsibility for operational control: Flag operations.

**(a)** Each certificate holder conducting flag operations is responsible for operational control.

**(b)** The pilot in command and the aircraft dispatcher are jointly responsible for the preflight planning, delay, and dispatch release of a flight in compliance with this chapter and operations specifications.

**(c)** The aircraft dispatcher is responsible for—

(1) Monitoring the progress of each flight;

(2) Issuing necessary instructions and information for the safety of the flight; and

(3) Canceling or redispatching a flight if, in his opinion or the opinion of the pilot in command, the flight cannot operate or continue to operate safely as planned or released.

**(d)** Each pilot in command of an aircraft is, during flight time, in command of the aircraft and crew and is responsible for the safety of the passengers, crewmembers, cargo, and airplane.

**(e)** Each pilot in command has full control and authority in the operation of the aircraft, without limitation, over other crewmembers and their duties during flight time, whether or not he holds valid certificates authorizing him to perform the duties of those crewmembers.

**(f)** No pilot may operate an aircraft in a careless or reckless manner so as to endanger life or property.

[Docket No. 6258, 29 FR 19219, Dec. 31, 1964; as amended by Amdt. 121–253, 61 FR 2613, Jan. 26, 1996]

## §121.537 Responsibility for operational control: Supplemental operations.

**(a)** Each certificate holder conducting supplemental operations—

(1) Is responsible for operational control; and

(2) Shall list each person authorized by it to exercise operational control in its operator's manual.

**(b)** The pilot in command and the director of operations are jointly responsible for the initiation, continuation, diversion, and termination of a flight in compliance with this chapter and the operations specifications. The director of operations may delegate the functions for the initiation, continuation, diversion, and termination of a flight but he may not delegate the responsibility for those functions.

**(c)** The director of operations is responsible for canceling, diverting, or delaying a flight if in his opinion or the opinion of the pilot in command the flight cannot operate or continue to operate safely as planned or released. The director of operations is responsible for assuring that each flight is monitored with respect to at least the following:

(1) Departure of the flight from the place of origin and arrival at the place of destination, including intermediate stops and any diversions therefrom.

(2) Maintenance and mechanical delays encountered at places of origin and destination and intermediate stops.

(3) Any known conditions that may adversely affect the safety of flight.

**(d)** Each pilot in command of an aircraft is, during flight time, in command of the aircraft and crew and is responsible for the safety of the passengers, crewmembers, cargo, and aircraft. The pilot in command has full control and authority in the operation of the aircraft, without limitation, over other crewmembers and their duties during flight time, whether or not he holds valid certificates authorizing him to perform the duties of those crewmembers.

**(e)** Each pilot in command of an aircraft is responsible for the preflight planning and the operation of the flight in compliance with this chapter and the operations specifications.

**(f)** No pilot may operate an aircraft, in a careless or reckless manner, so as to endanger life or property.

[Docket No. 6258, 29 FR 19219, Dec. 31, 1964; as amended by Amdt. 121–253, 61 FR 2613, Jan. 26, 1996]

## §121.538 Airplane security.

Certificate holders conducting operations under this part must comply with the applicable security requirements in 49 CFR chapter XII.

[Docket No. TSA–2002–11602, 67 FR 8350, Feb. 22, 2002]

## §121.539 Operations notices.

Each certificate holder shall notify its appropriate operations personnel of each change in equipment and operating procedures, including each known change in the use of navigation aids, airports, air traffic control procedures and regulations, local airport traffic control rules, and known hazards to flight, including icing and other potentially hazardous meteorological conditions and irregularities in ground and navigation facilities.

## §121.541 Operations schedules: Domestic and flag operations.

In establishing flight operations schedules, each certificate holder conducting domestic or flag operations shall allow enough time for the proper servicing of aircraft at intermediate stops, and shall consider the prevailing winds en route and the cruising speed of the type of aircraft used. This cruising speed may not be more than that resulting from the specified cruising output of the engines.

[Docket No. 28154, 61 FR 2613, Jan. 26, 1996]

## §121.542 Flight crewmember duties.

**(a)** No certificate holder shall require, nor may any flight crewmember perform, any duties during a critical phase of flight except those duties required for the safe operation of the aircraft. Duties such as company required calls made for such nonsafety related purposes as ordering galley supplies and confirming passenger connections, announcements made to passengers promoting the air carrier or pointing out sights of interest, and filling out company payroll and related records are not required for the safe operation of the aircraft.

**(b)** No flight crewmember may engage in, nor may any pilot in command permit, any activity during a critical phase of flight which could distract any flight crewmember from the performance of his or her duties or which could interfere in any way with the proper conduct of those duties. Activities such as eating meals, engaging in nonessential conversations within the cockpit and nonessential communications between the cabin and cockpit crews, and reading publications not related to the proper conduct of the flight are not required for the safe operation of the aircraft.

**(c)** For the purposes of this section, critical phases of flight includes all ground operations involving taxi, takeoff and landing, and all other flight operations conducted below 10,000 feet, except cruise flight.

NOTE: Taxi is defined as "movement of an airplane under its own power on the surface of an airport."

[Docket No. 20661, 46 FR 5502, Jan. 19, 1981]

## §121.543 Flight crewmembers at controls.

**(a)** Except as provided in paragraph (b) of this section, each required flight crewmember on flight deck duty must remain at the assigned duty station with seat belt fastened while the aircraft is taking off or landing, and while it is en route.

**(b)** A required flight crewmember may leave the assigned duty station—

(1) If the crewmember's absence is necessary for the performance of duties in connection with the operation of the aircraft;

(2) If the crewmember's absence is in connection with physiological needs; or

(3) If the crewmember is taking a rest period, and relief is provided—

(i) In the case of the assigned pilot in command during the en route cruise portion of the flight, by a pilot who holds an airline transport pilot certificate and an appropriate type rating, is currently qualified as pilot in command or second in command, and is qualified as pilot in command of that aircraft during the en route cruise portion of the flight. A second in command qualified to act as a pilot in command en route need not have completed the following pilot in command requirements: The 6-month recurrent flight training required by §121.433(c)(1)(iii); the operating experience required by §121.434; the takeoffs and landings required by §121.439; the line check required by §121.440; and the 6-month proficiency check or simulator training required by §121.441(a)(1); and

(ii) In the case of the assigned second in command, by a pilot qualified to act as second in command of that aircraft during en route operations. However, the relief pilot need not meet the recent experience requirements of §121.439(b).

[Docket No. 16383, 43 FR 22648, May 25, 1978; as amended by Amdt. 121–179, 47 FR 33390, Aug. 2, 1982]

## §121.545 Manipulation of controls.

No pilot in command may allow any person to manipulate the controls of an aircraft during flight nor may any person manipulate the controls during flight unless that person is—

**(a)** A qualified pilot of the certificate holder operating that aircraft.

**(b)** An authorized pilot safety representative of the Administrator or of the National Transportation Safety Board who has the permission of the pilot in command, is qualified in the aircraft, and is checking flight operations; or

**(c)** A pilot of another certificate holder who has the permission of the pilot in command, is qualified in the aircraft, and is authorized by the certificate holder operating the aircraft.

[Docket No. 6258, 29 FR 19220, Dec. 31, 1964; as amended by Docket No. 8084, 32 FR 5769, April 11, 1967; Amdt. 121–144, 43 FR 22648, May 25, 1978]

## §121.547 Admission to flight deck.

**(a)** No person may admit any person to the flight deck of an aircraft unless the person being admitted is—

(1) A crewmember;

(2) An FAA air carrier inspector, or an authorized representative of the National Transportation Safety Board, who is performing official duties;

(3) Any person who—

(i) Has permission of the pilot in command, an appropriate management official of the part 119 certificate holder, and the Administrator; and

(ii) Is an employee of—

(A) The United States, or

(B) A part 119 certificate holder and whose duties are such that admission to the flightdeck is necessary or advantageous for safe operation; or

(C) An aeronautical enterprise certificated by the Administrator and whose duties are such that admission to the flightdeck is necessary or advantageous for safe operation.

(4) Any person who has the permission of the pilot in command, an appropriate management official of the part 119 certificate holder and the Administrator. Paragraph (a)(2) of this section does not limit the emergency authority of the pilot in command to exclude any person from the flightdeck in the interests of safety.

**(b)** For the purposes of paragraph (a)(3) of this section, employees of the United States who deal responsibly with matters relating to safety and employees of the certificate holder whose efficiency would be increased by familiarity with flight conditions, may be admitted by the certificate holder. However, the certificate holder may not admit employees of traffic, sales, or other departments that are not directly related to flight operations, unless they are eligible under paragraph (a)(4) of this section.

**(c)** No person may admit any person to the flight deck unless there is a seat available for his use in the passenger compartment, except—

(1) An FAA air carrier inspector or an authorized representative of the Administrator or National Transportation Safety Board who is checking or observing flight operations;

(2) An air traffic controller who is authorized by the Administrator to observe ATC procedures;

(3) A certificated airman employed by the certificate holder whose duties require an airman certificate;

(4) A certificated airman employed by another part 119 certificate holder whose duties with that part 119 certificate holder require an airman certificate and who is authorized by the part 119 certificate holder operating the aircraft to make specific trips over a route;

(5) An employee of the part 119 certificate holder operating the aircraft whose duty is directly related to the conduct or planning of flight operations or the in-flight monitoring of aircraft equipment or operating procedures, if his presence on the flightdeck is necessary to perform his duties and he has been authorized in writing by a responsible supervisor, listed in the Operations Manual as having that authority; and

(6) A technical representative of the manufacturer of the aircraft or its components whose duties are directly related to the in-flight monitoring of aircraft equipment or operating procedures, if his presence on the flightdeck is necessary to perform his duties and he has been authorized in writing by the Administrator and by a responsible supervisor of the operations department of the part 119 certificate holder, listed in the Operations Manual as having that authority.

[Docket No. 6258, 29 FR 19220, Dec. 31, 1964; as amended by Docket No. 8084, 32 FR 5769, April 11, 1967; Amdt. 121–253, 61 FR 2613, Jan. 26, 1996; Amdt. 121–288, 67 FR 2127, Jan. 15, 2002]

## §121.548 Aviation safety inspector's credentials: Admission to pilot's compartment.

Whenever, in performing the duties of conducting an inspection, an inspector of the Federal Aviation Administration presents form FAA 110A, "Aviation Safety Inspector's Credential," to the pilot in command of an aircraft operated by a certificate holder, the inspector must be given free and uninterrupted access to the pilot's compartment of that aircraft.

[Docket No. 28154, 61 FR 3613, Jan. 26, 1996]

## §121.549 Flying equipment.

**(a)** The pilot in command shall ensure that appropriate aeronautical charts containing adequate information concerning navigation aids and instrument approach procedures are aboard the aircraft for each flight.

**(b)** Each crewmember shall, on each flight, have readily available for his use a flashlight that is in good working order.

## §121.550 Secret Service Agents: Admission to flight deck.

Whenever an Agent of the Secret Service who is assigned the duty of protecting a person aboard an aircraft operated by a certificate holder considers it necessary in the performance of his duty to ride on the flight deck of the aircraft, he must, upon request and presentation of his Secret Service credentials to the pilot in command of the aircraft, be admitted to the flight deck and permitted to occupy an observer seat thereon.

[Docket No. 9031, 35 FR 12061, July 28, 1970; as amended by Amdt. 121–253, 61 FR 2613, Jan. 26, 1996]

## §121.551 Restriction or suspension of operation: Domestic and flag operations.

When a certificate holder conducting domestic or flag operations knows of conditions, including airport and runway conditions, that are a hazard to safe operations, it shall restrict or suspend operations until those conditions are corrected.

[Docket No. 28154, 61 FR 2613, Jan. 26, 1996]

## §121.553 Restriction or suspension of operation: Supplemental operations.

When a certificate holder conducting supplemental operations or pilot in command knows of conditions, including airport and runway conditions, that are a hazard to safe operations, the certificate holder or pilot in command, as the case may be, shall restrict or suspend operations until those conditions are corrected.

[Docket No. 28154, 61 FR 2613, Jan. 26, 1996]

## §121.555 Compliance with approved routes and limitations: Domestic and flag operations.

No pilot may operate an airplane in scheduled air transportation—

**(a)** Over any route or route segment unless it is specified in the certificate holder's operations specifications; or

**(b)** Other than in accordance with the limitations in the operations specifications.

[Docket No. 6258, 29 FR 19219, Dec. 31, 1964; as amended by Amdt. 121–253, 61 FR 2614, Jan. 26, 1996]

## §121.557 Emergencies: Domestic and flag operations.

**(a)** In an emergency situation that requires immediate decision and action the pilot in command may take any action that he considers necessary under the circumstances. In such a case he may deviate from prescribed operations procedures and methods, weather minimums, and this chapter, to the extent required in the interests of safety.

**(b)** In an emergency situation arising during flight that requires immediate decision and action by an aircraft dispatcher, and that is known to him, the aircraft dispatcher shall advise the pilot in command of the emergency, shall ascertain the decision of the pilot in command, and shall have the decision recorded. If the aircraft dispatcher cannot communicate with the pilot, he shall declare an emergency and take any action that he considers necessary under the circumstances.

**(c)** Whenever a pilot in command or dispatcher exercises emergency authority, he shall keep the appropriate ATC facility and dispatch centers fully informed of the progress of the flight. The person declaring the emergency shall send a written report of any deviation through the certificate holder's operations manager, to the Administrator. A dispatcher shall send his report within 10 days after the date of the emergency, and a pilot in command shall send his report within 10 days after returning to his home base.

[Docket No. 6258, 29 FR 19219, Dec. 31, 1964; as amended by Amdt. 121–253, 61 FR 2614, Jan. 26, 1996]

## §121.559 Emergencies: Supplemental operations.

**(a)** In an emergency situation that requires immediate decision and action, the pilot in command may take any action that he considers necessary under the circumstances. In such a case, he may deviate from prescribed operations, procedures and methods, weather minimums, and this chapter, to the extent required in the interests of safety.

**(b)** In an emergency situation arising during flight that requires immediate decision and action by appropriate management personnel in the case of operations conducted with a flight following service and which is known to them, those personnel shall advise the pilot in command of the emergency, shall ascertain the decision of the pilot in command, and shall have the decision recorded. If they cannot communicate with the pilot, they shall declare an emergency and take any action that they consider necessary under the circumstances.

**(c)** Whenever emergency authority is exercised, the pilot in command or the appropriate management personnel shall keep the appropriate ground radio station fully informed of the progress of the flight. The person declaring the emergency shall send a written report of any deviation, through the certificate holder's director of operations, to the Administrator within 10 days after the flight is completed or, in the case of operations outside the United States, upon return to the home base.

[Docket No. 6258, 29 FR 19219, Dec. 31, 1964; as amended by Amdt. 121–253, 61 FR 2614, Jan. 26, 1996]

## §121.561 Reporting potentially hazardous meteorological conditions and irregularities of ground and navigation facilities.

**(a)** Whenever he encounters a meteorological condition or an irregularity in a ground or navigational facility, in flight, the knowledge of which he considers essential to the safety of other flights, the pilot in command shall notify an appropriate ground station as soon as practicable.

**(b)** The ground radio station that is notified under paragraph (a) of this section shall report the information to the agency directly responsible for operating the facility.

## §121.563 Reporting mechanical irregularities.

The pilot in command shall ensure that all mechanical irregularities occurring during flight time are entered in the maintenance log of the airplane at the end of that flight time. Before each flight the pilot in command shall ascertain the status of each irregularity entered in the log at the end of the preceding flight.

[Docket No. 17897, 45 FR 41594, June 19, 1980; as amended by Amdt. 121–179, 47 FR 33390, Aug. 2, 1982]

## §121.565 Engine inoperative: Landing; reporting.

**(a)** Except as provided in paragraph (b) of this section, whenever an engine of an airplane fails or whenever the rotation of an engine is stopped to prevent possible damage, the pilot in command shall land the airplane at the nearest suitable airport, in point of time, at which a safe landing can be made.

**(b)** If not more than one engine of an airplane that has three or more engines fails or its rotation is stopped, the pilot in command may proceed to an airport that he selects if, after considering the following, he decides that proceeding to that airport is as safe as landing at the nearest suitable airport:

(1) The nature of the malfunction and the possible mechanical difficulties that may occur if flight is continued.

(2) The altitude, weight, and usable fuel at the time of engine stoppage.

(3) The weather conditions en route and at possible landing points.

(4) The air traffic congestion.

(5) The kind of terrain.

(6) His familiarity with the airport to be used.

**(c)** The pilot in command shall report each stoppage of engine rotation in flight to the appropriate ground radio station as soon as practicable and shall keep that station fully informed of the progress of the flight.

**(d)** If the pilot in command lands at an airport other than the nearest suitable airport, in point of time, he or she shall (upon completing the trip) send a written report, in duplicate, to his or her director of operations stating the reasons for determining that the selection of an airport, other than the nearest airport, was as safe a course of action as landing at the nearest suitable airport. The director of operations shall, within 10 days after the pilot returns to his or her home base, send a copy of this report with the director of operation's comments to the certificate-holding district office.

[Docket No. 6258, 29 FR 19219, Dec. 31, 1964; as amended by Amdt. 121–207, 54 FR 39293, Sept. 25, 1989; Amdt. 121–253, 61 FR 2614, Jan. 26, 1996]

## §121.567 Instrument approach procedures and IFR landing minimums.

No person may make an instrument approach at an airport except in accordance with IFR weather minimums and instrument approach procedures set forth in the certificate holder's operations specifications.

## §121.569 Equipment interchange: Domestic and flag operations.

**(a)** Before operating under an interchange agreement, each certificate holder conducting domestic or flag operations shall show that—

(1) The procedures for the interchange operation conform with this chapter and with safe operating practices;

(2) Required crewmembers and dispatchers meet approved training requirements for the airplanes and equipment to be used and are familiar with the communications and dispatch procedures to be used;

(3) Maintenance personnel meet training requirements for the airplanes and equipment, and are familiar with the maintenance procedures to be used;

(4) Flight crewmembers and dispatchers meet appropriate route and airport qualifications; and

(5) The airplanes to be operated are essentially similar to the airplanes of the certificate holder with whom the interchange is effected with respect to the arrangement of flight instruments and the arrangement and motion of controls that are critical to safety unless the Administrator determines that the certificate holder has adequate training programs to insure that any potentially hazardous dissimilarities are safely overcome by flight crew familiarization.

**(b)** Each certificate holder conducting domestic or flag operations shall include the pertinent provisions and procedures involved in the equipment interchange agreement in its manuals.

[Docket No. 6258, 29 FR 19219, Dec. 31, 1964; as amended by Amdt. 121–253, 61 FR 2614, Jan. 26, 1996]

## §121.570 Airplane evacuation capability.

**(a)** No person may cause an airplane carrying passengers to be moved on the surface, take off or land unless each automatically deployable emergency evacuation assisting means installed pursuant to §121. 310(a), is ready for evacuation.

**(b)** Each certificate holder shall ensure that, at all times passengers are on board prior to airplane movement on the surface, at least one floor-level exit provides for the egress of passengers through normal or emergency means.

[Docket No. 26142, 57 FR 42674, Sept. 15, 1992]



## §121.571 Briefing passengers before take-off.

**(a)** Each certificate holder operating a passenger-carrying airplane shall insure that all passengers are orally briefed by the appropriate crewmember as follows:

(1) Before each takeoff, on each of the following:

(i) *Smoking.* Each passenger shall be briefed on when, where, and under what conditions smoking is prohibited (including, but not limited to, any applicable requirements of part 252 of this title). This briefing shall include a statement that the Federal Aviation Regulations require passenger compliance with the lighted passenger information signs, posted placards, areas designated for safety purposes as no smoking areas, and crewmember instructions with regard to these items. The briefing shall also include a statement that Federal law prohibits tampering with, disabling, or destroying any smoke detector in an airplane lavatory; smoking in passenger compartments.

(ii) The location of emergency exits.

(iii) The use of safety belts, including instructions on how to fasten and unfasten the safety belts. Each passenger shall be briefed on when, where, and under what condition the safety belt must be fastened about that passenger. This briefing shall include a statement that the Federal Aviation Regulations require passenger compliance with lighted passenger information signs and crewmembers instructions concerning the use of safety belts.

(iv) The location and use of any required emergency flotation means.

(v) On operations that do not use a flight attendant, the following additional information:

(A) The placement of seat backs in an upright position before takeoff and landing.

(B) Location of survival equipment.

(C) If the flight involves operations above 12,000 MSL, the normal and emergency use of oxygen.

(D) Location and operation of fire extinguisher.

(2) After each takeoff, immediately before or immediately after turning the seat belt sign off, an announcement shall be made that passengers should keep their seat belts fastened, while seated, even when the seat belt sign is off.

(3) Except as provided in paragraph (a)(4) of this section, before each takeoff a required crewmember assigned to the flight shall conduct an individual briefing of each person who may need the assistance of another person to move expeditiously to an exit in the event of an emergency. In the briefing the required crewmember shall—

(i) Brief the person and his attendant, if any, on the routes to each appropriate exit and on the most appropriate time to begin moving to an exit in the event of an emergency; and

(ii) Inquire of the person and his attendant, if any, as to the most appropriate manner of assisting the person so as to prevent pain and further injury.

(4) The requirements of paragraph (a)(3) of this section do not apply to a person who has been given a briefing before a previous leg of a flight in the same aircraft when the crewmembers on duty have been advised as to the most appropriate manner of assisting the person so as to prevent pain and further injury.

**(b)** Each certificate holder shall carry on each passenger-carrying airplane, in convenient locations for use of each passenger, printed cards supplementing the oral briefing and containing—

(1) Diagrams of, and methods of operating, the emergency exits; and

(2) Other instructions necessary for use of emergency equipment.

Each card required by this paragraph must contain information that is pertinent only to the type and model airplane used for that flight.

**(c)** The certificate holder shall describe in its manual the procedure to be followed in the briefing required by paragraph (a) of this section.

[Docket No. 2033, 30 FR 3206, March 9, 1965; as amended by Amdt. 121–30, 32 FR 13268, Sept. 20, 1967; Amdt. 121–84, 37 FR 3975, Feb. 24, 1972; Amdt. 121–133, 42 FR 18394, April 7, 1977; Amdt. 121–144, 43 FR 22648, May 25, 1978; Amdt. 121–146, 43 FR 28403, June 29, 1978; Amdt. 121–196, 53 FR 12362, April 13, 1988; Amdt. 121–230, 57 FR 42674, Sept. 15, 1992; Amdt. 121–251, 60 FR 65935, Dec. 20, 1995]

## §121.573 Briefing passengers: Extended overwater operations.

**(a)** In addition to the oral briefing required by §121.571(a), each certificate holder operating an airplane in extended overwater operations shall ensure that all passengers are orally briefed by the appropriate crewmember on the location and operation of life preservers, liferafts, and other flotation means, including a demonstration of the method of donning and inflating a life preserver.

**(b)** The certificate holder shall describe in its manual the procedure to be followed in the briefing required by paragraph (a) of this section.

**(c)** If the airplane proceeds directly over water after takeoff, the briefing required by paragraph (a) of this section must be done before takeoff.

**(d)** If the airplane does not proceed directly over water after takeoff, no part of the briefing required by paragraph (a) of this section has to be given before takeoff, but the entire briefing must be given before reaching the overwater part of the flight.

[Docket No. 2033, 30 FR 3206, March 9, 1965; as amended by Amdt. 121–144, 43 FR 22648, May 25, 1978; Amdt. 121–146, 43 FR 28403, June 29, 1978]

## §121.574 Oxygen for medical use by passengers.

**(a)** A certificate holder may allow a passenger to carry and operate equipment for the storage, generation, or dispensing of oxygen when the following conditions are met:

(1) The equipment is—

(i) Furnished by the certificate holder;

(ii) Of an approved type or is in conformity with the manufacturing, packaging, marking, labeling, and maintenance requirements of 49 CFR parts 171, 172, and 173, except §173.24(a)(1);

(iii) Maintained by the certificate holder in accordance with an approved maintenance program;

(iv) Free of flammable contaminants on all exterior surfaces;

(v) Capable of providing a minimum mass flow of oxygen to the user of four liters per minute;

(vi) Constructed so that all valves, fittings, and gauges are protected from damage; and

(vii) Appropriately secured.

(2) When the oxygen is stored in the form of a liquid, the equipment has been under the certificate holder's approved maintenance program since its purchase new or since the storage container was last purged.

(3) When the oxygen is stored in the form of a compressed gas as defined in 49 CFR 173.300(a)—

(i) The equipment has been under the certificate holder's approved maintenance program since its purchase new or since the last hydrostatic test of the storage cylinder; and

(ii) The pressure in any oxygen cylinder does not exceed the rated cylinder pressure.

(4) Each person using the equipment has a medical need to use it evidenced by a written statement to be kept in that person's possession, signed by a licensed physician which specifies the maximum quantity of oxygen needed each hour and the maximum flow rate needed for the pressure altitude corresponding to the pressure in the cabin of the airplane under normal operating conditions. This paragraph does not apply to the carriage of oxygen in an airplane in which the only passengers carried are persons who may have a medical need for oxygen during flight, no more than one relative or other interested person for each of those persons, and medical attendants.

(5) When a physician's statement is required by paragraph (a)(4) of this section, the total quantity of oxygen carried is equal to the maximum quantity of oxygen needed each hour, as specified in the physician's statement, multiplied by the number of hours used to compute the amount of airplane fuel required by this part.

(6) The pilot in command is advised when the equipment is on board, and when it is intended to be used.

(7) The equipment is stowed, and each person using the equipment is seated, so as not to restrict access to or use of any required emergency, or regular exit or of the aisle in the passenger compartment.

**(b)** No person may, and no certificate holder may allow any person to, smoke within 10 feet of oxygen storage and dispensing equipment carried in accordance with paragraph (a) of this section.

**(c)** No certificate holder may allow any person to connect or disconnect oxygen dispensing equipment, to or from a gaseous oxygen cylinder while any passenger is aboard the airplane.

**(d)** The requirements of this section do not apply to the carriage of supplemental or first-aid oxygen and related equipment required by this chapter.

[Docket No. 12169, 39 FR 42677, Dec. 6, 1974; as amended by Amdt. 121–159, 45 FR 41594, June 19, 1980]

## §121.575 Alcoholic beverages.

**(a)** No person may drink any alcoholic beverage aboard an aircraft unless the certificate holder operating the aircraft has served that beverage to him.

**(b)** No certificate holder may serve any alcoholic beverage to any person aboard any of its aircraft who—

(1) Appears to be intoxicated;

(2) Is escorting a person or being escorted in accordance with 49 CFR 1544.221; or

(3) Has a deadly or dangerous weapon accessible to him while aboard the aircraft in accordance with 49 CFR 1544.219, 1544.221, or 1544.223.

**(c)** No certificate holder may allow any person to board any of its aircraft if that person appears to be intoxicated.

**(d)** Each certificate holder shall, within five days after the incident, report to the Administrator the refusal of any person to comply with paragraph (a) of this section, or of any disturbance caused by a person who appears to be intoxicated aboard any of its aircraft.

[Docket No. 6258, 29 FR 19219, Dec. 31, 1964; as amended by Amdt. 121–118, 40 FR 17552, April 21, 1975; Amdt. 121–178, 47 FR 13316, March 29, 1982; Amdt. 121–275, 67 FR 31932, May 10, 2002]

## §121.576 Retention of items of mass in passenger and crew compartments.

The certificate holder must provide and use means to prevent each item of galley equipment and each serving cart, when not in use, and each item of crew baggage, which is carried in a passenger or crew compartment from becoming a hazard by shifting under the appropriate load factors corresponding to the emergency landing conditions under which the airplane was type certificated.

[Docket No. 16383, 43 FR 22648, May 25, 1978]

## §121.577 Stowage of food, beverage, and passenger service equipment during airplane movement on the surface, takeoff, and landing.

**(a)** No certificate holder may move an airplane on the surface, take off or land an airplane when any food, beverage, or tableware, furnished by the certificate holder is located at any passenger seat.

**(b)** No certificate holder may move on the surface, take off, or land unless each food and beverage tray and seat back tray table is secured in its stowed position.

**(c)** No certificate holder may permit an airplane to move on the surface, takeoff, or land unless each passenger serving cart is secured in its stowed position.

**(d)** No certificate holder may permit an airplane to move on the surface, takeoff, or land unless each movie screen that extends into an aisle is stowed.

**(e)** Each passenger shall comply with instructions given by a crewmember in compliance with this section.

[Docket No. 26142, 57 FR 42674, Sept. 15, 1992]

## §121.578 Cabin ozone concentration.

**(a)** For the purpose of this section, the following definitions apply:

(1) *Flight segment* means scheduled nonstop flight time between two airports.

(2) *Sea level equivalent* refers to conditions of 25°C and 760 millimeters of mercury pressure.

**(b)** Except as provided in paragraphs (d) and (e) of this section, no certificate holder may operate an airplane above the following flight levels unless it is successfully demonstrated to the Administrator that the concentration of ozone inside the cabin will not exceed—

(1) For flight above flight level 320, 0.25 parts per million by volume, sea level equivalent, at any time above that flight level; and

(2) For flight above flight level 270, 0.1 parts per million by volume, sea level equivalent, time-weighted average for each flight segment that exceeds 4 hours and includes flight above that flight level. (For this purpose, the amount of ozone below flight level 180 is considered to be zero.)

**(c)** Compliance with this section must be shown by analysis or tests, based on either airplane operational procedures and performance limitations or the certificate holder's operations. The analysis or tests must show either of the following:

(1) Atmospheric ozone statistics indicate, with a statistical confidence of at least 84%, that at the altitudes and locations at which the airplane will be operated cabin ozone concentrations will not exceed the limits prescribed by paragraph (b) of this section.

(2) The airplane ventilation system including any ozone control equipment, will maintain cabin ozone concentrations at or below the limits prescribed by paragraph (b) of this section.

**(d)** A certificate holder may obtain an authorization to deviate from the requirements of paragraph (b) of this section, by an amendment to its operations specifications, if—

(1) It shows that due to circumstances beyond its control or to unreasonable economic burden it cannot comply for a specified period of time; and

(2) It has submitted a plan acceptable to the Administrator to effect compliance to the extent possible.

**(e)** A certificate holder need not comply with the requirements of paragraph (b) of this section for an aircraft—

(1) When the only persons carried are flight crewmembers and persons listed in §121.583;

(2) If the aircraft is scheduled for retirement before January 1, 1985; or

(3) If the aircraft is scheduled for re-engining under the provisions of Subpart E of part 91, until it is re-engined.

[Docket No. 121–154, 45 FR 3883, Jan. 21, 1980. Redesignated by Amdt. 121–162, 45 FR 46739, July 10, 1980; Amdt. 121–181, 47 FR 58489, Dec. 30, 1982; Amdt. 121–251, 60 FR 65935, Dec. 20, 1995]

## §121.579 Minimum altitudes for use of autopilot.

**(a)** *En route operations.* Except as provided in paragraphs (b), (c), and (d) of this section, no person may use an autopilot en route, including climb and descent, at an altitude above the terrain that is less than twice the maximum altitude loss specified in the Airplane Flight Manual for a malfunction of the autopilot under cruise conditions, or less than 500 feet, whichever is higher.

**(b)** *Approaches.* When using an instrument approach facility, no person may use an autopilot at an altitude above the terrain that is less than twice the maximum altitude loss specified in the Airplane Flight Manual for a malfunction of the autopilot under approach conditions, or less than 50 feet below the approved minimum descent altitude or decision height for the facility, whichever is higher, except—

(1) When reported weather conditions are less than the basic VFR weather conditions in §91.155 of this chapter, no person may use an autopilot with an approach coupler for ILS approaches at an altitude above the terrain that is less than 50 feet higher than the maximum altitude loss specified in the Airplane Flight Manual for the malfunction of the autopilot with approach coupler under approach conditions; and

(2) When reported weather conditions are equal to or better than the basic VFR minimums in §91.155 of this chapter, no person may use an autopilot with an approach coupler for ILS approaches at an altitude above the terrain that is less than the maximum altitude loss specified in the Airplane Flight Manual for the malfunction of the autopilot with approach coupler under approach conditions, or 50 feet, whichever is higher.

**(c)** Notwithstanding paragraph (a) or (b) of this section, the Administrator issues operations specifications to allow the use, to touchdown, of an approved flight control guidance system with automatic capability, in any case in which—

(1) The system does not contain any altitude loss (above zero) specified in the Airplane Flight Manual for malfunction of the autopilot with approach coupler; and

(2) He finds that the use of the system to touchdown will not otherwise affect the safety standards required by this section.

**(d)** *Takeoffs.* Notwithstanding paragraph (a) of this section, the Administrator issues operations specifications to allow the use of an approved autopilot system with automatic capability below the altitude specified in paragraph (a) of this section during the takeoff and initial climb phase of flight provided:

(1) The Airplane Flight Manual specifies a minimum altitude engagement certification restriction;

(2) The system is not engaged prior to the minimum engagement certification restriction specified in the Airplane Flight Manual or an altitude specified by the Administrator, whichever is higher; and

(3) The Administrator finds that the use of the system will not otherwise affect the safety standards required by this section.

[Docket No. 6258, 29 FR 19219, Dec. 31, 1964; as amended by Amdt. 121–13, 30 FR 14781, Nov. 30, 1965; Amdt. 121–33, 32 FR 13912, Oct. 6, 1967; Amdt. 121–130, 41 FR 47229, Oct. 28, 1976; Amdt. 121–206, 54 FR 34331, Aug. 18, 1989; Amdt. 121–265, 62 FR 27922, May 21, 1997]

## §121.580 Prohibition on interference with crewmembers.

No person may assault, threaten, intimidate, or interfere with a crewmember in the performance of the crewmember's duties aboard an aircraft being operated under this part.

[Docket No. 6258, 29 FR 19219, Dec. 31, 1964; as amended by Amdt. 121–270, 64 FR 1080, Jan. 7, 1999]

## §121.581 Observer's seat: En route inspections.

**(a)** Except as provided in paragraph (c) of this section, each certificate holder shall make available a seat on the flight deck of each airplane, used by it in air commerce, for occupancy by the Administrator while conducting en route inspections. The location and equipment of the seat, with respect to its suitability for use in conducting en route inspections, is determined by the Administrator.

**(b)** In each airplane that has more than one observer's seat, in addition to the seats required for the crew complement for which the airplane was certificated, the forward observer's seat or the observer's seat selected by the Administrator must be made available when complying with paragraph (a) of this section.

**(c)** For any airplane type certificated before December 20, 1995, for not more than 30 passengers that does not have an observer seat on the flightdeck, the certificate holder must provide a forward passenger seat with headset or speaker for occupancy by the Administrator while conducting en route inspections.

[Docket No. 6258, 29 FR 19219, Dec. 31, 1964; as amended by Amdt. 121–144, 43 FR 22648, May 25, 1978; Amdt. 121–251, 60 FR 65935, Dec. 20, 1995; Amdt. 121–288, 67 FR 2128, Jan. 15, 2002]

## §121.583 Carriage of persons without compliance with the passenger-carrying requirements of this part.

**(a)** When authorized by the certificate holder, the following persons, but no others, may be carried aboard an airplane without complying with the passenger-carrying airplane requirements in §§121.309(f), 121.310, 121.391, 121.571, and 121.587; the passenger-carrying operation requirements in §§121.157(c), and 121.291; and the requirements pertaining to passengers in §§121.285, 121.313(f), 121.317, 121.547, and 121.573:

(1) A crewmember.

(2) A company employee.

(3) An FAA air carrier inspector, or an authorized representative of the National Transportation Safety Board, who is performing official duties.

(4) A person necessary for—

(i) The safety of the flight;

(ii) The safe handling of animals;

(iii) The safe handling of hazardous materials whose carriage is governed by regulations in 49 CFR part 175;

(iv) The security of valuable or confidential cargo;

(v) The preservation of fragile or perishable cargo;

(vi) Experiments on, or testing of, cargo containers or cargo handling devices;

(vii) The operation of special equipment for loading or unloading cargo; and

(viii) The loading or unloading of outsize cargo.

(5) A person described in paragraph (a)(4) of this section, when traveling to or from his assignment.

(6) A person performing duty as an honor guard accompanying a shipment made by or under the authority of the United States.

(7) A military courier, military route supervisor, military cargo contract coordinator, or a flight crewmember of another military cargo contract air carrier or commercial operator, carried by a mili-

tary cargo contract air carrier or commercial operator in operations under a military cargo contract, if that carriage is specifically authorized by the appropriate armed forces.

(8) A dependent of an employee of the certificate holder when traveling with the employee on company business to or from outlying stations not served by adequate regular passenger flights.

**(b)** No certificate holder may operate an airplane carrying a person covered by paragraph (a) of this section unless—

(1) Each person has unobstructed access from his seat to the pilot compartment or to a regular or emergency exit;

(2) The pilot in command has a means of notifying each person when smoking is prohibited and when safety belts must be fastened; and

(3) The airplane has an approved seat with an approved safety belt for each person. The seat must be located so that the occupant is not in any position to interfere with the flight crewmembers performing their duties.

**(c)** Before each takeoff, each certificate holder operating an airplane carrying persons covered by paragraph (a) of this section shall ensure that all such persons have been orally briefed by the appropriate crewmember on—

(1) Smoking;

(2) The use of seat belts;

(3) The location and operation of emergency exits;

(4) The use of oxygen and emergency oxygen equipment; and

(5) For extended overwater operations, the location of life rafts, and the location and operation of life preservers including a demonstration of the method of donning and inflating a life preserver.

**(d)** Each certificate holder operating an airplane carrying persons covered by paragraph (a) of this section shall incorporate procedures for the safe carriage of such persons into the certificate holder's operations manual.

**(e)** The pilot in command may authorize a person covered by paragraph (a) of this section to be admitted to the crew compartment of the airplane.

[Docket No. 10580, 35 FR 14612, Sept. 18, 1970; as amended by Amdt. 121–96, 37 FR 19608, Sept. 21, 1972; Amdt. 121–159, 45 FR 41594, June 19, 1980; Amdt. 121–232, 57 FR 48663, Oct. 27, 1992; Amdt. 121–251, 60 FR 65935, Dec. 20, 1995; Amdt. 121–253, 61 FR 2614, Jan. 26, 1996]

## §121.585 Exit seating.

**(a)**(1) Each certificate holder shall determine, to the extent necessary to perform the applicable functions of paragraph (d) of this section, the suitability of each person it permits to occupy an exit seat, in accordance with this section. For the purpose of this section—

(i) *Exit seat* means—

(A) Each seat having direct access to an exit; and,

(B) Each seat in a row of seats through which passengers would have to pass to gain access to an exit, from the first seat inboard of the exit to the first aisle inboard of the exit.

(ii) A passenger seat having direct access means a seat form which a passenger can proceed directly to an exit without entering an aisle or passing around an obstruction.

(2) Each certificate holder shall make the passenger exit seating determinations required by this paragraph in a non-discriminatory manner consistent with the requirements of this section, by person designated in the certificate holder's required operations manual.

(3) Each certificate holder shall designate the exit seats for each passenger seating configuration in its fleet in accordance with the definitions in this paragraph and submit those designations for approval as part of the procedures required to be submitted for approval under paragraphs (n) and (p) of this section.

**(b)** No certificate holder may seat a person in a seat affected by this section if the certificate holder determines that it is likely that the person would be unable to perform one or more of the applicable functions listed in paragraph (d) of this section because—

(1) The person lacks sufficient mobility, strength, or dexterity in both arms and hands, and both legs:

(i) To reach upward, sideways, and downward to the location of emergency exit and exit-slide operating mechanisms;

(ii) To grasp and push, pull, turn, or otherwise manipulate those mechanisms;

(iii) To push, shove, pull, or otherwise open emergency exits;

(iv) To lift out, hold, deposit on nearby seats, or maneuver over the seatbacks to the next row objects the size and weight of over-wing window exit doors;

(v) To remove obstructions similar in size and weight to over-wing exit doors;

(vi) To reach the emergency exit expeditiously;

(vii) To maintain balance while removing obstructions;

(viii) To exit expeditiously;

(ix) To stabilize an escape slide after deployment; or

(x) To assist others in getting off an escape slide;

(2) The person is less than 15 years of age or lacks the capacity to perform one or more of the applicable functions listed in paragraph (d) of this section without the assistance of an adult companion, parent, or other relative;

(3) The person lacks the ability to read and understand instructions required by this section and related to emergency evacuation provided by the certificate holder in printed or graphic form or the ability to understand oral crew commands.

(4) The person lacks sufficient visual capacity to perform one or more of the applicable functions in paragraph (d) of this section without the assistance of visual aids beyond contact lenses or eyeglasses;

(5) The person lacks sufficient aural capacity to hear and understand instructions shouted by flight attendants, without assistance beyond a hearing aid;

(6) The person lacks the ability adequately to impart information orally to other passengers; or,

(7) The person has:

(i) A condition or responsibilities, such as caring for small children, that might prevent the person from performing one or more of the applicable functions listed in paragraph (d) of this section; or

(ii) A condition that might cause the person harm if he or she performs one or more of the applicable functions listed in paragraph (d) of this section.

**(c)** Each passenger shall comply with instructions given by a crewmember or other authorized employee of the certificate holder implementing exit seating restrictions established in accordance with this section.

**(d)** Each certificate holder shall include on passenger information cards, presented in the language in which briefings and oral commands are given by the crew, at each exit seat affected by this section, information that, in the event of an emergency in which a crewmember is not available to assist, a passenger occupying an exit seat may use if called upon to perform the following functions:

(1) Locate the emergency exit;

(2) Recognize the emergency exit opening mechanism;

(3) Comprehend the instructions for operating the emergency exit;

(4) Operate the emergency exit;

(5) Assess whether opening the emergency exit will increase the hazards to which passengers may be exposed;

(6) Follow oral directions and hand signals given by a crewmember;

(7) Stow or secure the emergency exit door so that it will not impede use of the exit;

(8) Assess the condition of an escape slide, activate the slide, and stabilize the slide after deployment to assist others in getting off the slide;

(9) Pass expeditiously through the emergency exit; and

(10) Assess, select, and follow a safe path away from the emergency exit.

**(e)** Each certificate holder shall include on passenger information cards, at each exit seat—

(1) In the primary language in which emergency commands are given by the crew, the selection criteria set forth in paragraph (b) of this section, and a request that a passenger identify himself or herself to allow reseating if he or she:

(i) Cannot meet the selection criteria set forth in paragraph (b) of this section;

(ii) Has a nondiscernible condition that will prevent him or her from performing the applicable functions listed in paragraph (d) of this section;

(iii) May suffer bodily harm as the result of performing one or more of those functions; or,

(iv) Does not wish to perform those functions.

(2) In each language used by the certificate holder for passenger information cards, a request that a passenger identify himself or herself to allow reseating if he or she lacks the ability to read, speak, or understand the language or the graphic form in which instructions required by this section and related to emergency evacuation are provided by the certificate holder, or the ability to understand the specified language in which crew commands will be given in an emergency.

(3) May suffer bodily harm as the result of performing one or more of those functions; or,

(4) Does not wish to perform these functions.

A certificate holder shall not require the passenger to disclose his or her reason for needing reseating.

**(f)** Each certificate holder shall make available for inspection by the public at all passenger loading gates and ticket counters at each airport where it conducts passenger operations, written procedures established for making determinations in regard to exit row seating.

**(g)** No certificate holder may allow taxi or pushback unless at least one required crewmember has verified that no exit seat is occupied by a person the crewmember determines is likely to be unable to perform the applicable functions listed in paragraph (d) of this section.

**(h)** Each certificate holder shall include in its passenger briefings a reference to the passenger information cards, required by paragraphs (d) and (e), the selection criteria set forth in paragraph (b), and the functions to be performed, set forth in paragraph (d) of this section.

**(i)** Each certificate holder shall include in its passenger briefings a request that a passenger identify himself or herself to allow reseating if he or she—

(1) Cannot meet the selection criteria set forth in paragraph (b) of this section;

(2) Has a nondiscernible condition that will prevent him or her from performing the applicable functions listed in paragraph (d) of this section;

(3) May suffer bodily harm as the result of performing one or more of those functions listed in paragraph (d) of this section; or,

(4) Does not wish to perform those functions listed in paragraph (d) of this section.

A certificate holder shall not require the passenger to disclose his or her reason for needing reseating.

**(j)** [Reserved]

**(k)** In the event a certificate holder determines in accordance with this section that it is likely that a passenger assigned to an exit row seat would be unable to perform the functions listed in paragraph (d) of this section, or a passenger requests a non-exit seat, the certificate holder shall relocate the passenger to a non-exit seat.

**(l)** In the event of full booking in the non-exit seats, the certificate holder shall move a passenger, if necessary to accommodate a passenger being relocated from an exit row seat, who is willing and able to assume the evacuation functions that may be required, to an exit row seat.

**(m)** A certificate holder may deny transportation to any passenger under this section only because—

(1) the passenger refuses to comply with instructions given by a crewmember or other authorized employee of the certificate holder, implementing exit row seating restrictions established in accordance with this section, or

(2) the only seat that will physically accommodate the person's handicap is an exit seat.

**(n)** In order to comply with this section certificate holders shall—

(1) Establish procedures that address:

(i) The criteria listed in paragraph (b) of this section;

(ii) The functions listed in paragraph (d) of this section;

(iii) The requirements for airport information, passenger information cards, crewmember verification of appropriate seating in exit seats, passenger briefings, seat assignments, and denial of transportation as set forth in this section;

(iv) How to resolve disputes arising from implementation of this section, including identification of the certificate holder employee on the airport to whom complaints should be addressed for resolution; and,

(2) Submit their procedures for preliminary review and approval to the principal operations inspectors assigned to them at the certificate-holding district office.

**(o)** Certificate holders shall assign seats prior to boarding consistent with the criteria listed in paragraph (b) and the functions listed in paragraph (d) of this section, to the maximum extent feasible.

**(p)** The procedures required by paragraph (n) of this section will not become effective until final approval is granted by the Director, Flight Standards Service, Washington, DC. Approval will be based solely upon the safety aspects of the certificate holder's procedures.

[Docket No. 25821, 55 FR 8072, March 6, 1990; as amended by Amdt. 121–232, 57 FR 48663, Oct. 27, 1992; Amdt. 121–253, 61 FR 2614, Jan. 26, 1996]

## §121.586 Authority to refuse transportation.

**(a)** No certificate holder may refuse transportation to a passenger on the basis that, because the passenger may need the assistance of another person to move expeditiously to an exit in the event of an emergency, his transportation would or might be inimical to safety of flight unless—

(1) The certificate holder has established procedures (including reasonable notice requirements) for the carriage of passengers who may need the assistance of another person to move expeditiously to an exit in the event of an emergency; and

(2) At least one of the following conditions exist:

(i) The passenger fails to comply with the notice requirements in the certificate holder's procedures.

(ii) The passenger cannot be carried in accordance with the certificate holder's procedures.

**(b)** Each certificate holder shall provide the certificate-holding district office with a copy of each procedure it establishes in accordance with paragraph (a)(2) of this section.

**(c)** Whenever the Administrator finds that revisions in the procedures described in paragraph (a)(2) of this section are necessary in the interest of safety or in the public interest, the certificate holder, after notification by the Administrator, shall make those revisions in its procedures. Within 30 days after the certificate holder receives such notice, it may file a petition to reconsider the notice with the certificate-holding district office. The filing of a petition to reconsider stays the notice pending a decision by the Administrator. However, if the Administrator finds that there is an emergency that requires immediate action in the interest of safety in air commerce, he may, upon a statement of the reasons, require a change effective without stay.

**(d)** Each certificate holder shall make available to the public at each airport it serves a copy of each procedure it establishes in accordance with paragraph (a)(1) of this section.

[Docket No. 12881, 42 FR 18394, April 7, 1977; as amended by Amdt. 121–174, 46 FR 38051, July 23, 1981; Amdt. 121–207, 54 FR 39293, Sept. 25, 1989; Amdt. 121–253, 61 FR 2614, Jan. 26, 1996]

## §121.587 Closing and locking of flightcrew compartment door.

**(a)** Except as provided in paragraph (b) of this section, a pilot in command of an airplane that has a lockable flightcrew compartment door in accordance with §121.313 and that is carrying passengers shall ensure that the door separating the flightcrew compartment from the passenger compartment is closed and locked at all times when the aircraft is being operated.

**(b)** The provisions of paragraph (a) of this section do not apply at any time when it is necessary to permit access and egress by persons authorized in accordance with §121.547 and provided the part 119 operator complies with FAA approved procedures regarding the opening, closing and locking of the flightdeck doors.

[Docket No. FAA–2001–11032, 67 FR 2128, Jan. 15, 2002]

## §121.589 Carry-on baggage.

**(a)** No certificate holder may allow the boarding of carry-on baggage on an airplane unless each passenger's baggage has been scanned to control the size and amount carried on board in accordance with an approved carry-on baggage program in its operations specifications. In addition, no passenger may board an airplane if his/her carry-on baggage exceeds the baggage allowance prescribed in the carry-on baggage program in the certificate holder's operations specifications.

**(b)** No certificate holder may allow all passenger entry doors of an airplane to be closed in preparation for taxi or pushback unless at least one required crewmember has verified that each article of baggage is stowed in accordance with this section and §121.285 (c) and (d) of this part.

**(c)** No certificate holder may allow an airplane to take off or land unless each article of baggage is stowed:

(1) In a suitable closet or baggage or cargo stowage compartment placarded for its maximum weight and providing proper restraint for all baggage or cargo stowed within, and in a manner that does not hinder the possible use of any emergency equipment; or

(2) As provided in §121.285 (c) and (d) of this part; or

(3) Under a passenger seat.

**(d)** Baggage, other than articles of loose clothing, may not be placed in an overhead rack unless that rack is equipped with approved restraining devices or doors.

**(e)** Each passenger must comply with instructions given by crewmembers regarding compliance with paragraphs (a), (b), (c), (d), and (g) of this section.

**(f)** Each passenger seat under which baggage is allowed to be stowed shall be fitted with a means to prevent articles of baggage stowed under it from sliding forward. In addition, each aisle seat shall be fitted with a means to prevent articles of baggage stowed under it from sliding sideward into the aisle under crash impacts severe enough to induce the ultimate inertia forces specified in the emergency landing condition regulations under which the airplane was type certificated.

**(g)** In addition to the methods of stowage in paragraph (c) of this section, flexible travel canes carried by blind individuals may be stowed—

(1) Under any series of connected passenger seats in the same row, if the cane does not protrude into an aisle and if the cane is flat on the floor; or

(2) Between a nonemergency exit window seat and the fuselage, if the cane is flat on the floor; or

(3) Beneath any two nonemergency exit window seats, if the cane is flat on the floor; or

(4) In accordance with any other method approved by the Administrator.

[Docket No. 24996, 52 FR 21476, June 5, 1987; as amended by Amdt. 121–251, 60 FR 65935, Dec. 20, 1995]

## §121.590 Use of certificated land airports.

**(a)** Except as provided in paragraph (b) of this section or unless otherwise authorized by the Administrator, no air carrier, and no pilot being used by an air carrier may, in the conduct of operations governed by this part, operate an airplane designed for at least 31 passenger seats into a land airport in any State of the United States, the District of Columbia, or any territory or possession of the United States, unless that airport is certificated under part 139 of this chapter. However, an air carrier may designate and use as a required alternate airport for departure or destination an airport that is not certificated under part 139 of this chapter.

**(b)** Certificate holders conducting passenger-carrying operations with airplanes designed for less than 31 passenger seats may operate those airplanes into airports not certificated under part 139 of this chapter if the following conditions are met:

(1) The airport is adequate for the proposed operation, considering such items as size, surface, obstructions, and lighting.

(2) For an airplane carrying passengers at night, the pilot may not take off from, or land at, an airport unless—

(i) The pilot has determined the wind direction from an illuminated wind direction indicator or local ground communications or, in the case of takeoff, that pilot's personal observations; and

(ii) The limits of the area to be used for landing or takeoff are clearly shown by boundary or runway marker lights. If the area to be used for takeoff or landing is marked by flare pots or lanterns, their use must be approved by the Administrator.

[Docket No. 28154, 60 FR 65935, Dec. 20, 1995; as amended by Amdt. 121–262, 62 FR 13257, March 19, 1997]

# Subpart U—Dispatching and Flight Release Rules

**Source:** Docket No. 6258, 29 FR 19222, Dec. 31, 1964, unless otherwise noted.

## §121.591 Applicability.

This subpart prescribes dispatching rules for domestic and flag operations and flight release rules for supplemental operations.

[Docket No. 28154, 61 FR 2614, Jan. 26, 1996]

## §121.593 Dispatching authority: Domestic operations.

Except when an airplane lands at an intermediate airport specified in the original dispatch release and remains there for not more than one hour, no person may start a flight unless an aircraft dispatcher specifically authorizes that flight.

[Docket No. 6258, 29 FR 19222, Dec. 31, 1964; as amended by Amdt. 121–253, 61 FR 2614, Jan. 26, 1996]

## §121.595 Dispatching authority: Flag operations.

**(a)** No person may start a flight unless an aircraft dispatcher specifically authorizes that flight.

**(b)** No person may continue a flight from an intermediate airport without redispatch if the airplane has been on the ground more than six hours.

[Docket No. 6258, 29 FR 19222, Dec. 31, 1964; as amended by Amdt. 121–253, 61 FR 2614, Jan. 26, 1996]

## §121.597 Flight release authority: Supplemental operations.

**(a)** No person may start a flight under a flight following system without specific authority from the person authorized by the operator to exercise operational control over the flight.

**(b)** No person may start a flight unless the pilot in command or the person authorized by the operator to exercise operational control over the flight has executed a flight release setting forth the conditions under which the flights will be conducted. The pilot in command may sign the flight release only when he and the person authorized by the operator to exercise operational control believe that the flight can be made with safety.

**(c)** No person may continue a flight from an intermediate airport without a new flight release if the aircraft has been on the ground more than six hours.

[Docket No. 6258, 29 FR 19222, Dec. 31, 1964; as amended by Amdt. 121–3, 30 FR 3639, March 19, 1965; Amdt. 121–253, 61 FR 2614, Jan. 26, 1996]

## §121.599 Familiarity with weather conditions.

**(a)** *Domestic and flag operations.* No aircraft dispatcher may release a flight unless he is thoroughly familiar with reported and forecast weather conditions on the route to be flown.

**(b)** *Supplemental operations.* No pilot in command may begin a flight unless he is thoroughly familiar with reported and forecast weather conditions on the route to be flown.

[Docket No. 6258, 29 FR 19222, Dec. 31, 1964; as amended by Amdt. 121–253, 61 FR 2614, Jan. 26, 1996]

## §121.601 Aircraft dispatcher information to pilot in command: Domestic and flag operations.

**(a)** The aircraft dispatcher shall provide the pilot in command all available current reports or information on airport conditions and irregularities of navigation facilities that may affect the safety of the flight.

**(b)** Before beginning a flight, the aircraft dispatcher shall provide the pilot in command with all available weather reports and forecasts of weather phenomena that may affect the safety of flight, including adverse weather phenomena, such as clear air turbulence, thunderstorms, and low altitude wind shear, for each route to be flown and each airport to be used.

**(c)** During a flight, the aircraft dispatcher shall provide the pilot in command any additional available information of meteorological conditions (including adverse weather phenomena, such as clear air turbulence, thunderstorms, and low altitude wind shear), and irregularities of facilities and services that may affect the safety of the flight.

[Docket No. 6258, 29 FR 19222, Dec. 31, 1964; as amended by Amdt. 121–134, 42 FR 27573, May 31, 1977; Amdt. 121–144, 43 FR 22649, May 25, 1978; Amdt. 121–253, 61 FR 2614, Jan. 26, 1996]

## §121.603 Facilities and services: Supplemental operations.

**(a)** Before beginning a flight, each pilot in command shall obtain all available current reports or information on airport conditions and irregularities of navigation facilities that may affect the safety of the flight.

**(b)** During a flight, the pilot in command shall obtain any additional available information of meteorological conditions and irregularities of facilities and services that may affect the safety of the flight.

[Docket No. 6258, 29 FR 19222, Dec. 31, 1964; as amended by Amdt. 121–253, 61 FR 2614, Jan. 26, 1996]

## §121.605 Airplane equipment.

No person may dispatch or release an airplane unless it is airworthy and is equipped as prescribed in §121.303.

## §121.607 Communication and navigation facilities: Domestic and flag operations.

**(a)** Except as provided in paragraph (b) of this section for a certificate holder conducting flag operations, no person may dispatch an airplane over an approved route or route segment unless the communication and navigation facilities required by §§121.99 and 121.103 for the approval of that route or segment are in satisfactory operating condition.

**(b)** If, because of technical reasons or other reasons beyond the control of a certificate holder conducting flag operations, the facilities required by §§121.99 and 121.103 are not available over a route or route segment outside the United States, the certificate holder may dispatch an airplane over that route or route segment if the pilot in command and dispatcher find that communication and navigation facilities equal to those required are available and are in satisfactory operating condition.

[Docket No. 6258, 29 FR 19222, Dec. 31, 1964; as amended by Amdt. 121–253, 61 FR 2614, Jan. 26, 1996]

## §121.609 Communication and navigation facilities: Supplemental operations.

No person may release an aircraft over any route or route segment unless communication and navigation facilities equal to those required by §121.121 are in satisfactory operating condition.

[Docket No. 6258, 29 FR 19222, Dec. 31, 1964; as amended by Amdt. 121–253, 61 FR 2614, Jan. 26, 1996]

## §121.611 Dispatch or flight release under VFR.

No person may dispatch or release an aircraft for VFR operation unless the ceiling and visibility en route, as indicated by available weather reports or forecasts, or any combination thereof, are and will remain at or above applicable VFR minimums until the aircraft arrives at the airport or airports specified in the dispatch or flight release.

## §121.613 Dispatch or flight release under IFR or over the top.

Except as provided in §121.615, no person may dispatch or release an aircraft for operations under IFR or over-the-top, unless appropriate weather reports or forecasts, or any combination thereof, indicate that the weather conditions will be at or above the authorized minimums at the estimated time of arrival at the airport or airports to which dispatched or released.

[Docket No. 6258, 29 FR 19222, Dec. 31, 1964; as amended by Amdt. 121–33, 32 FR 13912, Oct. 6, 1967]

## §121.615 Dispatch or flight release over water: Flag and supplemental operations.

**(a)** No person may dispatch or release an aircraft for a flight that involves extended overwater operation unless appropriate weather reports or forecasts or any combination thereof, indicate that the weather conditions will be at or above the authorized minimums at the estimated time of arrival at any airport to which dispatched or released or to any required alternate airport.

**(b)** Each certificate holder conducting a flag or supplemental operation or a domestic operation within the State of Alaska shall conduct extended overwater operations under IFR unless it shows that operating under IFR is not necessary for safety.

**(c)** Each certificate holder conducting a flag or supplemental operation or a domestic operation within the State of Alaska shall conduct other overwater operations under IFR if the Administrator determines that operation under IFR is necessary for safety.

**(d)** Each authorization to conduct extended overwater operations under VFR and each requirement to conduct other overwater operations under IFR will be specified in the certificate holder's operations specifications.

[Docket No. 6258, 29 FR 19222, Dec. 31, 1964; as amended by Amdt. 121–33, 32 FR 13912, Oct. 6, 1967; Amdt. 121–253, 61 FR 2614, Jan. 26, 1996]

## §121.617 Alternate airport for departure.

**(a)** If the weather conditions at the airport of takeoff are below the landing minimums in the certificate holder's operations specifications for that airport, no person may dispatch or release an aircraft from that airport unless the dispatch or flight release specifies an alternate airport located within the following distances from the airport of takeoff:

(1) *Aircraft having two engines.* Not more than one hour from the departure airport at normal cruising speed in still air with one engine inoperative.

(2) *Aircraft having three or more engines.* Not more than two hours from the departure airport at normal cruising speed in still air with one engine inoperative.

**(b)** For the purpose of paragraph (a) of this section, the alternate airport weather conditions must meet the requirements of the certificate holder's operations specifications.

**(c)** No person may dispatch or release an aircraft from an airport unless he lists each required alternate airport in the dispatch or flight release.

### §121.619 Alternate airport for destination: IFR or over-the-top: Domestic operations.

**(a)** No person may dispatch an airplane under IFR or over-the-top unless he lists at least one alternate airport for each destination airport in the dispatch release. When the weather conditions forecast for the destination and first alternate airport are marginal at least one additional alternate must be designated. However, no alternate airport is required if for at least 1 hour before and 1 hour after the estimated time of arrival at the destination airport the appropriate weather reports or forecasts, or any combination of them, indicate—

(1) The ceiling will be at least 2,000 feet above the airport elevation; and

(2) Visibility will be at least 3 miles.

**(b)** For the purposes of paragraph (a) of this section, the weather conditions at the alternate airport must meet the requirements of §121.625.

**(c)** No person may dispatch a flight unless he lists each required alternate airport in the dispatch release.

[Docket No. 6258, 29 FR 19222, Dec. 31, 1964; as amended by Amdt. 121–159, 45 FR 41594, June 19, 1980; Amdt. 121–253, 61 FR 2614, Jan. 26, 1996]

### §121.621 Alternate airport for destination: Flag operations.

**(a)** No person may dispatch an airplane under IFR or over-the-top unless he lists at least one alternate airport for each destination airport in the dispatch release, unless—

(1) The flight is scheduled for not more than 6 hours and, for at least 1 hour before and 1 hour after the estimated time of arrival at the destination airport, the appropriate weather reports or forecasts, or any combination of them, indicate the ceiling will be:

(i) At least 1,500 feet above the lowest circling MDA, if a circling approach is required and authorized for that airport; or

(ii) At least 1,500 feet above the lowest published instrument approach minimum or 2,000 feet above the airport elevation, whichever is greater; and

(iii) The visibility at that airport will be at least 3 miles, or 2 miles more than the lowest applicable visibility minimums, whichever is greater, for the instrument approach procedures to be used at the destination airport; or

(2) The flight is over a route approved without an available alternate airport for a particular destination airport and the airplane has enough fuel to meet the requirements of §121.641(b) or §121.645(c).

**(b)** For the purposes of paragraph (a) of this section, the weather conditions at the alternate airport must meet the requirements of the certificate holder's operations specifications.

**(c)** No person may dispatch a flight unless he lists each required alternate airport in the dispatch release.

[Docket No. 6258, 29 FR 19222, Dec. 31, 1964; as amended by Amdt. 121–159, 45 FR 41594, June 19, 1980; Amdt. 121–253, 61 FR 2614, Jan. 26, 1996]

### §121.623 Alternate airport for destination; IFR or over-the-top: Supplemental operations.

**(a)** Except as provided in paragraph (b) of this section, each person releasing an aircraft for operation under IFR or over-the-top shall list at least one alternate airport for each destination airport in the flight release.

**(b)** An alternate airport need not be designated for IFR or over-the-top operations where the aircraft carries enough fuel to meet the requirements of §§121.643 and 121.645 for flights outside the 48 contiguous States and the District of Columbia over routes without an available alternate airport for a particular airport of destination.

**(c)** For the purposes of paragraph (a) of this section, the weather requirements at the alternate airport must meet the requirements of the certificate holder's operations specifications.

**(d)** No person may release a flight unless he lists each required alternate airport in the flight release.

[Docket No. 6258, 29 FR 19222, Dec. 31, 1964; as amended by Amdt. 121–253, 61 FR 2614, Jan. 26, 1996]

### §121.625 Alternate airport weather minimums.

No person may list an airport as an alternate airport in the dispatch or flight release unless the appropriate weather reports or forecasts, or any combination thereof, indicate that the weather conditions will be at or above the alternate weather minimums specified in the certificate holder's operations specifications for that airport when the flight arrives.

[Docket No. 6258, 29 FR 19222, Dec. 31, 1964; as amended by Amdt. 121–33, 32 FR 13912, Oct. 6, 1967]

## §121.627 Continuing flight in unsafe conditions.

**(a)** No pilot in command may allow a flight to continue toward any airport to which it has been dispatched or released if, in the opinion of the pilot in command or dispatcher (domestic and flag operations only), the flight cannot be completed safely; unless, in the opinion of the pilot in command, there is no safer procedure. In that event, continuation toward that airport is an emergency situation as set forth in §121.557.

**(b)** If any instrument or item of equipment required under this chapter for the particular operation becomes inoperative en route, the pilot in command shall comply with the approved procedures for such an occurrence as specified in the certificate holder's manual.

[Docket No. 6258, 29 FR 1922, Dec. 31, 1964; as amended by Amdt. 121–222, 56 FR 12310, March 22, 1991; Amdt. 121–253, 61 FR 2615, Jan. 26, 1996]

## §121.628 Inoperable instruments and equipment.

**(a)** No person may take off an airplane with inoperable instruments or equipment installed unless the following conditions are met:

(1) An approved Minimum Equipment List exists for that airplane.

(2) The certificate-holding district office has issued the certificate holder operations specifications authorizing operations in accordance with an approved Minimum Equipment List. The flight crew shall have direct access at all times prior to flight to all of the information contained in the approved Minimum Equipment List through printed or other means approved by the Administrator in the certificate holders operations specifications. An approved Minimum Equipment List, as authorized by the operations specifications, constitutes an approved change to the type design without requiring recertification.

(3) The approved Minimum Equipment List must:

(i) Be prepared in accordance with the limitations specified in paragraph (b) of this section.

(ii) Provide for the operation of the airplane with certain instruments and equipment in an inoperable condition.

(4) Records identifying the inoperable instruments and equipment and the information required by paragraph (a)(3)(ii) of this section must be available to the pilot.

(5) The airplane is operated under all applicable conditions and limitations contained in the Minimum Equipment List and the operations specifications authorizing use of the Minimum Equipment List.

**(b)** The following instruments and equipment may not be included in the Minimum Equipment List:

(1) Instruments and equipment that are either specifically or otherwise required by the airworthiness requirements under which the airplane is type certificated and which are essential for safe operations under all operating conditions.

(2) Instruments and equipment required by an airworthiness directive to be in operable condition unless the airworthiness directive provides otherwise.

(3) Instruments and equipment required for specific operations by this part.

**(c)** Notwithstanding paragraphs (b)(1) and (b)(3) of this section, an airplane with inoperable instruments or equipment may be operated under a special flight permit under §§21.197 and 21.199 of this chapter.

[Docket No. 25780, 56 FR 12310, March 22, 1991; as amended by Amdt. 121–222, 56 FR 14290, April 8, 1991; Amdt. 121–253, 61 FR 2615, Jan. 26, 1996]

## §121.629 Operation in icing conditions.

**(a)** No person may dispatch or release an aircraft, continue to operate an aircraft en route, or land an aircraft when in the opinion of the pilot in command or aircraft dispatcher (domestic and flag operations only), icing conditions are expected or met that might adversely affect the safety of the flight.

**(b)** No person may take off an aircraft when frost, snow, or ice is adhering to the wings, control surfaces, propellers, engine outlet or other critical surfaces of the aircraft or when the takeoff would not be in compliance with paragraph (c) of this section. Takeoffs with frost under the wing in the area of the fuel tanks may be authorized by the Administrator.

**(c)** Except as provided in paragraph (d) of this section, no person may dispatch, release, or take off an aircraft any time conditions are such that frost, ice, or snow may be reasonably expected to adhere to the aircraft, unless the certificate holder has an approved ground deicing/anti-icing program in its operations specifications and unless the dispatch, the release, and takeoff comply with that program. The approved ground deicing/anti-icing program must include at least the following items:

(1) A detailed description of—

(i) How the certificate holder determines that conditions are such that frost, ice, or snow may reasonably be expected to adhere to the aircraft and that ground deicing/anti-icing operational procedures must be in effect;

(ii) Who is responsible for deciding that ground deicing/anti-icing operational procedures are in effect;

(iii) The procedures for implementing ground deicing/anti-icing operational procedures;

(iv) The specific duties and responsibilities of each operational position or group responsible for getting the aircraft safely airborne while ground deicing/anti-icing operational procedures are in effect.

(2) Initial and annual recurrent ground training and testing for flight crewmembers and qualification for all other affected personnel (e.g. aircraft dispatchers, ground crews, contract personnel) concerning the specific requirements of the approved program and each person's responsibilities and duties under the approved program, specifically covering the following areas:

(i) The use of holdover times.

(ii) Aircraft deicing/anti-icing procedures, including inspection and check procedures and responsibilities.

(iii) Communications procedures.

(iv) Aircraft surface contamination (i.e., adherence of frost, ice, and snow) and critical area identification, and how contamination adversely affects aircraft performance and flight characteristics.

(v) Types and characteristics of deicing/anti-icing fluids.

(vi) Cold weather preflight inspection procedures

(vii) Techniques for recognizing contamination on the aircraft.

(3) The certificate holder's holdover time-tables and the procedures for the use of these tables by the certificate holder's personnel. Holdover time is the estimated time deicing/anti-icing fluid will prevent the formation of frost or ice and the accumulation of snow on the protected surfaces of an aircraft. Holdover time begins when the final application of deicing/anti-icing fluid commences and expires when the deicing/anti-icing fluid applied to the aircraft loses its effectiveness. The holdover times must be supported by data acceptable to the Administrator. The certificate holder's program must include procedures for flight crewmembers to increase or decrease the determined holdover time in changing conditions. The program must provide that takeoff after exceeding any maximum holdover time is permitted only when at least one of the following conditions exists:

(i) A pretakeoff contamination check, as defined in paragraph (c)(4) of this section, determines that the wings, control surfaces, and other critical surfaces, as defined in the certificate holder's program, are free of frost, ice, or snow.

(ii) It is otherwise determined by an alternate procedure approved by the Administrator in accordance with the certificate holder's approved program that the wings, control surfaces, and other critical surfaces, as defined in the certificate holder's program, are free of frost, ice, or snow

(iii) The wings, control surfaces, and other critical surface are redeiced and a new holdover time is determined.

(4) Aircraft deicing/anti-icing procedures and responsibilities, pretakeoff check procedures and responsibilities, and pretakeoff contamination check procedures and responsibilities. A pretakeoff check is a check of the aircraft's wings or representative aircraft surfaces for frost, ice, or snow within the aircraft's holdover time. A pre-takeoff contamination check is a check to make sure the wings, control surfaces, and other critical surfaces, as defined in the certificate holder's program, are free of frost, ice, and snow. It must be conducted within five minutes prior to beginning take off. This check must be accomplished from outside the aircraft unless the program specifies otherwise.

**(d)** A certificate holder may continue to operate under this section without a program as required in paragraph (c) of this section, if it includes in its operation specifications a requirement that, any time conditions are such that frost, ice, or snow may reasonably be expected to adhere to the aircraft, no aircraft will take off unless it has been checked to ensure that the wings, control surfaces, and other critical surfaces are free of frost, ice and snow. The check must occur within five minutes prior to beginning takeoff. This check must be accomplished from outside the aircraft.

[Docket No. 6258, 29 FR 19222, Dec. 31, 1964; as amended by Amdt. 121–231, 57 FR 44942, Sept. 29, 1992; Amdt. 121–253, 61 FR 2615, Jan. 26, 1996]

## §121.631 Original dispatch or flight release, redispatch or amendment of dispatch or flight release.

**(a)** A certificate holder may specify any regular, provisional, or refueling airport, authorized for the type of aircraft, as a destination for the purpose of original dispatch or release.

**(b)** No person may allow a flight to continue to an airport to which it has been dispatched or released unless the weather conditions at an alternate airport that was specified in the dispatch or flight release are forecast to be at or above the alternate minimums specified in the operations specifications for that airport at the time the aircraft would arrive at the alternate airport. However, the dispatch or flight release may be amended en route to include any alternate airport that is within the fuel range of the aircraft as specified in §§121.639 through 121.647.

**(c)** No person may change an original destination or alternate airport that is specified in the original dispatch or flight release to another airport while the aircraft is en route unless the other airport is authorized for that type of aircraft and the appropriate requirements of §§121.593 through 121.661 and 121.173 are met at the time of redispatch or amendment of the flight release.

**(d)** Each person who amends a dispatch or flight release en route shall record that amendment.

[Docket No. 6258, 29 FR 19222, Dec. 31, 1964; as amended by Amdt. 121–65, 35 FR 12709, Aug. 11, 1970]

## §121.633 [Reserved]

## §121.635 Dispatch to and from refueling or provisional airports: Domestic and flag operations.

No person may dispatch an airplane to or from a refueling or provisional airport except in accordance with the requirements of this part applicable to dispatch from regular airports and unless that airport meets the requirements of this part applicable to regular airports.

[Docket No. 16383, 43 FR 22649, May 25, 1978; as amended by Amdt. 121–253, 61 FR 2615, Jan. 26, 1996]

## §121.637 Takeoffs from unlisted and alternate airports: Domestic and flag operations.

**(a)** No pilot may takeoff an airplane from an airport that is not listed in the operations specifications unless—

(1) The airport and related facilities are adequate for the operation of the airplane;

(2) He can comply with the applicable airplane operating limitations;

(3) The airplane has been dispatched according to dispatching rules applicable to operation from an approved airport; and

(4) The weather conditions at that airport are equal to or better than the following:

(i) *Airports in the United States.* The weather minimums for takeoff prescribed in part 97 of this chapter; or where minimums are not prescribed for the airport, 800-2, 900-1-1/2, or 1,000-1.

(ii) *Airports outside the United States.* The weather minimums for takeoff prescribed or approved by the government of the country in which the airport is located; or where minimums are not prescribed or approved for the airport, 800-2, 900- 1-1/2, or 1,000-1.

**(b)** No pilot may take off from an alternate airport unless the weather conditions are at least equal to the minimums prescribed in the certificate holder's operations specifications for alternate airports.

[Docket No. 6258, 29 FR 19222, Dec. 31, 1964; as amended by Amdt. 121–33, 32 FR 13912, Oct. 6, 1967; Amdt. 121–253, 61 FR 2615, Jan. 26, 1996]

## §121.639 Fuel supply: All domestic operations.

No person may dispatch or take off an airplane unless it has enough fuel—

**(a)** To fly to the airport to which it is dispatched;

**(b)** Thereafter, to fly to and land at the most distant alternate airport (where required) for the airport to which dispatched; and

**(c)** Thereafter, to fly for 45 minutes at normal cruising fuel consumption or, for certificate holders who are authorized to conduct day VFR operations in their operations specifications and who are operating nontransport category airplanes type certificated after December 31, 1964, to fly for 30 minutes at normal cruising fuel consumption for day VFR operations.

[Docket No. 6258, 29 FR 19222, Dec. 31, 1964; as amended by Amdt. 121–251, 60 FR 65935, Dec. 20, 1995]

## §121.641 Fuel supply: nonturbine and turbo-propeller-powered airplanes: Flag operations.

**(a)** No person may dispatch or take off a nonturbine or turbo-propeller-powered airplane unless, considering the wind and other weather conditions expected, it has enough fuel—

(1) To fly to and land at the airport to which it is dispatched;

(2) Thereafter, to fly to and land at the most distant alternate airport specified in the dispatch release; and

(3) Thereafter, to fly for 30 minutes plus 15 percent of the total time required to fly at normal cruising fuel consumption to the airports specified in paragraphs (a)(1) and (2) of this section or to fly for 90 minutes at normal cruising fuel consumption, whichever is less.

**(b)** No person may dispatch a nonturbine or turbo-propeller-powered airplane to an airport for which an alternate is not specified under §121.621(a)(2), unless it has enough fuel, considering wind and forecast weather conditions, to fly to that airport and thereafter to fly for three hours at normal cruising fuel consumption.

[Docket No. 6258, 29 FR 19222, Dec. 31, 1964; as amended by Amdt. 121–253, 61 FR 2615, Jan. 26, 1996]

## §121.643 Fuel supply: Nonturbine and turbo-propeller-powered airplanes: Supplemental operations.

**(a)** Except as provided in paragraph (b) of this section, no person may release for flight or takeoff a nonturbine or turbo-propeller-powered airplane unless, considering the wind and other weather conditions expected, it has enough fuel—

(1) To fly to and land at the airport to which it is released;

(2) Thereafter, to fly to and land at the most distant alternate airport specified in the flight release; and

(3) Thereafter, to fly for 45 minutes at normal cruising fuel consumption or, for certificate holders who are authorized to conduct day VFR operations in their operations specifications and who are operating nontransport category airplanes type certificated after December 31, 1964, to fly for 30 minutes at normal cruising fuel consumption for day VFR operations.

**(b)** If the airplane is released for any flight other than from one point in the contiguous United States to another point in the contiguous United States, it must carry enough fuel to meet the requirements of paragraphs (a)(1) and (2) of this section and thereafter fly for 30 minutes plus 15 percent of the total time required to fly at normal cruising fuel consumption to the airports specified in paragraphs (a)(1) and (2) of this section, or to fly for 90 minutes at normal cruising fuel consumption, whichever is less.

**(c)** No person may release a nonturbine or turbo-propeller-powered airplane to an airport for which an alternate is not specified under §121.623(b), unless it has enough fuel, considering wind and other weather conditions expected, to fly to that airport and thereafter to fly for three hours at normal cruising fuel consumption.

[Docket No. 6258, 29 FR 19222, Dec. 31, 1964; as amended by Amdt. 121–10, 30 FR 10025, Aug. 12, 1965; Amdt. 121–251, 60 FR 65935, Dec. 20, 1995; Amdt. 121–253, 61 FR 2615, Jan. 26, 1996]

## §121.645 Fuel supply: Turbine-engine powered airplanes, other than turbo-propeller: Flag and supplemental operations.

**(a)** Any flag operation within the 48 contiguous United States and the District of Columbia may use the fuel requirements of §121.639.

**(b)** For any certificate holder conducting flag or supplemental operations outside the 48 contiguous United States and the District of Columbia, unless authorized by the Administrator in the operations specifications, no person may release for flight or takeoff a turbine-engine powered airplane (other than a turbo-propeller powered airplane) unless, considering wind and other weather conditions expected, it has enough fuel—

(1) To fly to and land at the airport to which it is released;

(2) After that, to fly for a period of 10 percent of the total time required to fly from the airport of departure to, and land at, the airport to which it was released;

(3) After that, to fly to and land at the most distant alternate airport specified in the flight release, if an alternate is required; and

(4) After that, to fly for 30 minutes at holding speed at 1,500 feet above the alternate airport (or the destination airport if no alternate is required) under standard temperature conditions.

**(c)** No person may release a turbine-engine powered airplane (other than a turbo-propeller airplane) to an airport for which an alternate is not specified under §121.621(a)(2) or §121.623(b) unless it has enough fuel, considering wind and other weather conditions expected, to fly to that airport and thereafter to fly for at least two hours at normal cruising fuel consumption.

**(d)** The Administrator may amend the operations specifications of a certificate holder conducting flag or supplemental operations to require more fuel than any of the minimums stated in paragraph (a) or (b) of this section if he finds that additional fuel is necessary on a particular route in the interest of safety.

**(e)** For a supplemental operation within the 48 contiguous States and the District of Columbia with a turbine engine powered airplane the fuel requirements of §121.643 apply.

[Docket No. 6258, 29 FR 19222, Dec. 31, 1964; as amended by Amdt. 121–10, 30 FR 10025, Aug. 12, 1965; Amdt. 121–144, 43 FR 22649, May 25, 1978. Amdt. 121–253, 61 FR 2615, Jan. 26, 1996]

## §121.647 Factors for computing fuel required.

Each person computing fuel required for the purposes of this subpart shall consider the following:

**(a)** Wind and other weather conditions forecast.

**(b)** Anticipated traffic delays.

**(c)** One instrument approach and possible missed approach at destination.

**(d)** Any other conditions that may delay landing of the aircraft.

For the purposes of this section, required fuel is in addition to unusable fuel.

## §121.649 Takeoff and landing weather minimums: VFR: Domestic operations.

**(a)** Except as provided in paragraph (b) of this section, regardless of any clearance from ATC, no pilot may takeoff or land an airplane under VFR when the reported ceiling or visibility is less than the following:

(1) For day operations—1,000-foot ceiling and one-mile visibility.

(2) For night operations—1,000-foot ceiling and two-mile visibility.

**(b)** Where a local surface restriction to visibility exists (e.g., smoke, dust, blowing snow or sand) the visibility for day and night operations may be reduced to $\frac{1}{2}$ mile, if all turns after takeoff and prior to landing, and all flight beyond one mile from the airport boundary can be accomplished above or outside the area of local surface visibility restriction.

**(c)** The weather minimums in this section do not apply to the VFR operation of fixed-wing aircraft at any of the locations where the special weather minimums of §91.157 of this chapter are not applicable (See part 91, Appendix D, section 3 of this chapter). The basic VFR weather minimums of §91.155 of this chapter apply at those locations.

[Docket No. 6258, 29 FR 19222, Dec. 31, 1964; as amended by Amdt. 121–39, 33 FR 4097, March 2, 1968; Amdt. 121–206, 54 FR 34331, Aug. 18, 1989; Amdt. 121–226, 56 FR 65663, Dec. 17, 1991; Amdt. 121–253, 61 FR 2615, Jan. 26, 1996]

## §121.651 Takeoff and landing weather minimums: IFR: All certificate holders.

**(a)** Notwithstanding any clearance from ATC, no pilot may begin a takeoff in an airplane under IFR when the weather conditions reported by the U.S. National Weather Service, a source approved by that Service, or a source approved by the Administrator, are less than those specified in—

(1) The certificate holder's operations specifications; or

(2) Parts 91 and 97 of this chapter, if the certificate holder's operations specifications do not specify takeoff minimums for the airport.

**(b)** Except as provided in paragraph (d) of this section, no pilot may continue an approach past the final approach fix, or where a final approach fix is not used, begin the final approach segment of an instrument approach procedure—

(1) At any airport, unless the U.S. National Weather Service, a source approved by that Service, or a source approved by the Administrator, issues a weather report for that airport; and

(2) At airports within the United States and its territories or at U.S. military airports, unless the latest weather report for that airport issued by the U.S. National Weather Service, a source approved by that Service, or a source approved by the Administrator, reports the visibility to be equal to or more than the visibility minimums prescribed for that procedure. For the purpose of this section, the term "U.S. military airports" means airports in foreign countries where flight operations are under the control of U.S. military authority.

**(c)** If a pilot has begun the final approach segment of an instrument approach procedure in accordance with paragraph (b) of this section and after that receives a later weather report indicating below-minimum conditions, the pilot may continue the approach to DH or MDA. Upon reaching DH or at MDA, and at any time before the missed approach point, the pilot may continue the approach below DH or MDA and touch down if—

(1) The aircraft is continuously in a position from which a descent to a landing on the intended runway can be made at a normal rate of descent using normal maneuvers, and where that descent rate will allow touchdown to occur within the touchdown zone of the runway of intended landing;

(2) The flight visibility is not less than the visibility prescribed in the standard instrument approach procedure being used;

(3) Except for Category II or Category III approaches where any necessary visual reference requirements are specified by authorization of the Administrator, at least one of the following visual references for the intended runway is distinctly visible and identifiable to the pilot:

(i) The approach light system, except that the pilot may not descend below 100 feet above the touchdown zone elevation using the approach lights as a reference unless the red terminating bars or the red side row bars are also distinctly visible and identifiable.

(ii) The threshold.

(iii) The threshold markings.

(iv) The threshold lights.

(v) The runway end identifier lights.

(vi) The visual approach slope indicator.

(vii) The touchdown zone or touchdown zone markings.

(viii) The touchdown zone lights.

(ix) The runway or runway markings.

(x) The runway lights; and

(4) When the aircraft is on a straight-in nonprecision approach procedure which incorporates a visual descent point, the aircraft has reached the visual descent point, except where the aircraft is not equipped for or capable of establishing that point, or a descent to the runway cannot be made using normal procedures or rates of descent if descent is delayed until reaching that point.

**(d)** A pilot may begin the final approach segment of an instrument approach procedure other than a Category II or Category III procedure at an airport when the visibility is less than the visibility minimums prescribed for that procedure if that airport is served by a operative ILS and an operative PAR, and both are used by the pilot. However, no pilot may operate an aircraft below the authorized

MDA, or continue an approach below the authorized DH, unless—

(1) The aircraft is continuously in a position from which a descent to a landing on the intended runway can be made at a normal rate of descent using normal maneuvers and where such a descent rate will allow touchdown to occur within the touchdown zone of the runway of intended landing;

(2) The flight visibility is not less than the visibility prescribed in the standard instrument approach procedure being used; and

(3) Except for Category II or Category III approaches where any necessary visual reference requirements are specified by the authorization of the Administrator, at least one of the following visual references for the intended runway is distinctly visible and identifiable to the pilot:

(i) The approach light system, except that the pilot may not descend below 100 feet above the touchdown zone elevation using the approach lights as a reference unless the red terminating bars or the red side row bars are also distinctly visible and identifiable.

(ii) The threshold.

(iii) The threshold markings.

(iv) The threshold lights.

(v) The runway end identifier lights.

(vi) The visual approach slope indicator.

(vii) The touchdown zone or touchdown zone markings.

(viii) The touchdown zone lights.

(ix) The runway or runway markings.

(x) The runway lights.

**(e)** For the purpose of this section, the final approach segment begins at the final approach fix or facility prescribed in the instrument approach procedure. When a final approach fix is not prescribed for a procedure that includes a procedure turn, the final approach segment begins at the point where the procedure turn is completed and the aircraft is established inbound toward the airport on the final approach course within the distance prescribed in the procedure.

**(f)** Unless otherwise authorized in the certificate holder's operations specifications, each pilot making an IFR takeoff, approach, or landing at a foreign airport shall comply with the applicable instrument approach procedures and weather minimums prescribed by the authority having jurisdiction over the airport.

[Docket No. 20060, 46 FR 2291, Jan. 8, 1981]

## §121.652 Landing weather minimums: IFR: All certificate holders.

**(a)** If the pilot in command of an airplane has not served 100 hours as pilot in command in operations under this part in the type of airplane he is operating, the MDA or DH and visibility landing minimums in the certificate holder's operations specification for regular, provisional, or refueling airports are increased by 100 feet and one-half mile (or the RVR equivalent). The MDA or DH and visibility minimums need not be increased above those applicable to the airport when used as an alternate airport, but in no event may the landing minimums be less than 300 and 1. However, a Pilot in command employed by an a certificate holder conducting operations in large aircraft under Part 135 of this chapter, may credit flight time acquired in operations conducted for that operator under part 91 in the same type airplane for up to 50 percent of the 100 hours of pilot in command experience required by this paragraph.

**(b)** The 100 hours of pilot in command experience required by paragraph (a) of this section may be reduced (not to exceed 50 percent) by substituting one landing in operations under this part in the type of airplane for 1 required hour of pilot in command experience, if the pilot has at least 100 hours as pilot in command of another type airplane in operations under this part.

**(c)** Category II minimums and the sliding scale when authorized in the certificate holder's operations specifications do not apply until the pilot in command subject to paragraph (a) of this section meets the requirements of that paragraph in the type of airplane he is operating.

[Docket No. 7594, 33 FR 10843, July 31, 1968; as amended by Amdt. 121–143, 43 FR 22642, May 25, 1978; Amdt. 121–253, 61 FR 2615, Jan. 26, 1996]

## §121.653 [Reserved]

## §121.655 Applicability of reported weather minimums.

In conducting operations under §§121.649 through 121.653, the ceiling and visibility values in the main body of the latest weather report control for VFR and IFR takeoffs and landings and for instrument approach procedures on all runways of an airport. However, if the latest weather report, including an oral report from the control tower, contains a visibility value specified as runway visibility or runway visual range for a particular runway of an airport, that specified value controls for VFR and IFR landings and takeoffs and straight-in instrument approaches for that runway.

## §121.657 Flight altitude rules.

**(a)** *General.* Notwithstanding §91.119 or any rule applicable outside the United States, no person may operate an aircraft below the minimums set forth in paragraphs (b) and (c) of this section, except when necessary for takeoff or landing, or except when, after considering the character of the terrain, the quality and quantity of meteorological services, the navigational facilities available, and other flight conditions, the Administrator prescribes other minimums for any route or part of a route where he finds that the safe conduct of the flight requires other altitudes. Outside of the United States the minimums prescribed in this section are controlling unless higher minimums are prescribed in the certificate holder's operations specifications or by the foreign country over which the aircraft is operating.

**(b)** *Day VFR operations.* No certificate holder conducting domestic operations may operate a passenger-carrying aircraft and no certificate holder conducting flag or supplemental operations may operate any aircraft under VFR during the day at an altitude less than 1,000 feet above the surface or less than 1,000 feet from any mountain, hill, or other obstruction to flight.

**(c)** *Night VFR, IFR, and over-the-top operations.* No person may operate an aircraft under IFR including over-the-top or at night under VFR at an altitude less than 1,000 feet above the highest obstacle within a horizontal distance of five miles from the center of the intended course, or, in designated mountainous areas, less than 2,000 feet above the highest obstacle within a horizontal distance of five miles from the center of the intended course.

**(d)** *Day over-the-top operations below minimum en route altitudes.* A person may conduct day over-the-top operations in an airplane at flight altitudes lower than the minimum en route IFR altitudes if—

(1) The operation is conducted at least 1,000 feet above the top of lower broken or overcast cloud cover;

(2) The top of the lower cloud cover is generally uniform and level;

(3) Flight visibility is at least five miles; and

(4) The base of any higher broken or overcast cloud cover is generally uniform and level and is at least 1,000 feet above the minimum en route IFR altitude for that route segment.

[Docket No. 6258, 29 FR 19222, Dec. 31, 1964; as amended by Amdt. 121–144, 43 FR 22649, May 25, 1978; Amdt. 121–206, 54 FR 34331, Aug. 18, 1989; Amdt. 121–253, 61 FR 2615, Jan. 26, 1996]

## §121.659 Initial approach altitude: Domestic and supplemental operations.

**(a)** Except as provided in paragraph (b) of this section, when making an initial approach to a radio navigation facility under IFR, no person may descend an aircraft below the pertinent minimum altitude for initial approach (as specified in the instrument approach procedure for that facility) until his arrival over that facility has been definitely established.

**(b)** When making an initial approach on a flight being conducted under §121.657(d), no pilot may commence an instrument approach until his arrival over the radio facility has definitely been established. In making an instrument approach under these circumstances no person may descend an aircraft lower than 1,000 feet above the top of the lower cloud or the minimum altitude determined by the Administrator for that part of the IFR approach, whichever is lower.

[Docket No. 6258, 29 FR 19222, Dec. 31, 1964; as amended by Amdt. 121–253, 61 FR 2615, Jan. 26, 1996]

## §121.661 Initial approach altitude: Flag operations.

When making an initial approach to a radio navigation facility under IFR, no person may descend below the pertinent minimum altitude for initial approach (as specified in the instrument approach procedure for that facility) until his arrival over that facility has been definitely established.

[Docket No. 6258, 29 FR 19222, Dec. 31, 1964; as amended by Amdt. 121–253, 61 FR 2615, Jan. 26, 1996]

## §121.663 Responsibility for dispatch release: Domestic and flag operations.

Each certificate holder conducting domestic or flag operations shall prepare a dispatch release for each flight between specified points, based on information furnished by an authorized aircraft dispatcher. The pilot in command and an authorized aircraft dispatcher shall sign the release only if they both believe that the flight can be made with safety. The aircraft dispatcher may delegate authority to sign a release for a particular flight, but he may not delegate his authority to dispatch.

[Docket No. 28154, 61 FR 2615, Jan. 26, 1996]

## §121.665 Load manifest.

Each certificate holder is responsible for the preparation and accuracy of a load manifest form before each takeoff. The form must be prepared and signed for each flight by employees of the certificate holder who have the duty of supervising the loading of aircraft and preparing the load manifest forms or by other qualified persons authorized by the certificate holder.

## §121.667 Flight plan: VFR and IFR: Supplemental operations.

**(a)** No person may take off an aircraft unless the pilot in command has filed a flight plan, containing the appropriate information required by part 91, with the nearest FAA communication station or appropriate military station or, when operating outside the United States, with other appropriate authority. However, if communications facilities are not readily available, the pilot in command shall file the flight plan as soon as practicable after the aircraft is airborne. A flight plan must continue in effect for all parts of the flight.

**(b)** When flights are operated into military airports, the arrival or completion notice required by §§91.153 and 91.169 may be filed with the appropriate airport control tower or aeronautical communication facility used for that airport.

[Docket No. 6258, 29 FR 19222, Dec. 31, 1964; as amended by Amdt. 121–206, 54 FR 34331, Aug. 18, 1989; Amdt. 121–253, 61 FR 2615, Jan. 26, 1996]

# Subpart V— Records and Reports

SOURCE: Docket No. 6258, 29 FR 19226, Dec. 31, 1964, unless otherwise noted.

## §121.681 Applicability.

This subpart prescribes requirements for the preparation and maintenance of records and reports for all certificate holders.

## §121.683 Crewmember and dispatcher record.

**(a)** Each certificate holder shall—

(1) Maintain current records of each crewmember and each aircraft dispatcher (domestic and flag operations only) that show whether the crewmember or aircraft dispatcher complies with the applicable section of this chapter, including, but not limited to, proficiency and route checks, airplane and route checks, airplane and route qualifications, training, any required physical examinations, flight, duty, and rest time records; and

(2) Record each action taken concerning the release from employment or physical or professional disqualification of any flight crewmember or aircraft dispatcher (domestic and flag operations only) and keep the record for at least six months thereafter.

**(b)** Each certificate holder conducting supplemental operations shall maintain the records required by paragraph (a) of this section at its principal base of operations, or at another location used by it and approved by the Administrator.

**(c)** Computer record systems approved by the Administrator may be used in complying with the requirements of paragraph (a) of this section.

[Docket No. 6258, 29 FR 19226, Dec. 31, 1964; as amended by Amdt. 121–144, 43 FR 22649, May 25, 1978; Amdt. 121–241, 59 FR 42993, Aug. 19, 1994; Amdt. 121–253, 61 FR 2615, Jan. 26, 1996]

## §121.685 Aircraft record: Domestic and flag operations.

Each certificate holder conducting domestic or flag operations shall maintain a current list of each aircraft that it operates in scheduled air transportation and shall send a copy of the record and each change to the certificate-holding district office. Airplanes of another certificate holder operated under an interchange agreement may be incorporated by reference.

[Docket No. 28154, 61 FR 2615, Jan. 26, 1996]

## §121.687 Dispatch release: Flag and domestic operations.

**(a)** The dispatch release may be in any form but must contain at least the following information concerning each flight:

(1) Identification number of the aircraft.

(2) Trip number.

(3) Departure airport, intermediate stops, destination airports, and alternate airports.

(4) A statement of the type of operation (e.g., IFR, VFR).

(5) Minimum fuel supply.

**(b)** The dispatch release must contain, or have attached to it, weather reports, available weather forecasts, or a combination thereof, for the destination airport, intermediate stops, and alternate airports, that are the latest available at the time the release is signed by the pilot in command and dispatcher. It may include any additional available weather reports or forecasts that the pilot in command or the aircraft dispatcher considers necessary or desirable.

[Docket No. 6258, 29 FR 19222, Dec. 31, 1964; as amended by Amdt. 121–253, 61 FR 2615, Jan. 26, 1996]

## §121.689 Flight release form: Supplemental operations.

**(a)** Except as provided in paragraph (c) of this section, the flight release may be in any form but must contain at least the following information concerning each flight:

(1) Company or organization name.

(2) Make, model, and registration number of the aircraft being used.

(3) Flight or trip number, and date of flight.

(4) Name of each flight crewmember, flight attendant, and pilot designated as pilot in command.

(5) Departure airport, destination airports, alternate airports, and route.

(6) Minimum fuel supply (in gallons or pounds).

(7) A statement of the type of operation (e.g., IFR, VFR).

**(b)** The aircraft flight release must contain, or have attached to it, weather reports, available weather forecasts, or a combination thereof, for the destination airport, and alternate airports, that are the latest available at the time the release is signed. It may include any additional available weather reports or forecasts that the pilot in command considers necessary or desirable.

**(c)** Each certificate holder conducting domestic or flag operations under the rules of this part applicable to supplemental operations shall comply with the dispatch or flight release forms required for scheduled operations under this subpart.

[Docket No. 6258, 29 FR 19222, Dec. 31, 1964; as amended by Amdt. 121–253, 61 FR 2615, Jan. 26, 1996]

## §121.691 [Reserved]

## §121.693 Load manifest: All certificate holders.

The load manifest must contain the following information concerning the loading of the airplane at takeoff time:

**(a)** The weight of the aircraft, fuel and oil, cargo and baggage, passengers and crewmembers.

**(b)** The maximum allowable weight for that flight that must not exceed the least of the following weights:

(1) Maximum allowable takeoff weight for the runway intended to be used (including corrections for altitude and gradient, and wind and temperature conditions existing at the takeoff time).

(2) Maximum takeoff weight considering anticipated fuel and oil consumption that allows compliance with applicable en route performance limitations.

(3) Maximum takeoff weight considering anticipated fuel and oil consumption that allows compliance with the maximum authorized design landing weight limitations on arrival at the destination airport.

(4) Maximum takeoff weight considering anticipated fuel and oil consumption that allows compliance with landing distance limitations on arrival at the destination and alternate airports.

**(c)** The total weight computed under approved procedures.

**(d)** Evidence that the aircraft is loaded according to an approved schedule that insures that the center of gravity is within approved limits.

**(e)** Names of passengers, unless such information is maintained by other means by the certificate holder.

[Docket No. 6258, 29 FR 19226, Dec. 31, 1964; as amended by Amdt. 121–159, 45 FR 41595, June 19, 1980; Amdt. 121–253, 61 FR, Jan. 26, 1996]

## §121.695 Disposition of load manifest, dispatch release, and flight plans: Domestic and flag operations.

**(a)** The pilot in command of an airplane shall carry in the airplane to its destination—

(1) A copy of the completed load manifest (or information from it, except information concerning cargo and passenger distribution);

(2) A copy of the dispatch release; and

(3) A copy of the flight plan.

**(b)** The certificate holder shall keep copies of the records required in this section for at least three months.

[Docket No. 6258, 29 FR 19226, Dec. 31, 1964; as amended by Amdt. 121–178, 47 FR 13316, March 29, 1982; Amdt. 121–253, 61 FR 2616, Jan. 26, 1996]

## §121.697 Disposition of load manifest, flight release, and flight plans: Supplemental operations.

**(a)** The pilot in command of an airplane shall carry in the airplane to its destination the original or a signed copy of the—

(1) Load manifest;

(2) Flight release;

(3) Airworthiness release;

(4) Pilot route certification; and

(5) Flight plan.

**(b)** If a flight originates at the certificate holder's principal base of operations, it shall retain at that base a signed copy of each document listed in paragraph (a) of this section.

**(c)** Except as provided in paragraph (d) of this section, if a flight originates at a place other than the certificate holder's principal base of operations, the pilot in command (or another person not aboard the airplane who is authorized by the certificate holder) shall, before or immediately after departure of the flight, mail signed copies of the documents listed in paragraph (a) of this section, to the principal base of operations.

**(d)** If a flight originates at a place other than the certificate holder's principal base of operations, and there is at that place a person to manage the flight departure for the certificate holder who does not himself or herself depart on the airplane, signed copies of the documents listed in paragraph (a) of this section may be retained at that place for not more than 30 days before being sent to the certificate holder's principal base of operations. However, the documents for a particular flight need not be further retained at that place or be sent to the principal base of operations, if the originals or other copies of them have been previously returned to the principal base of operations.

**(e)** The certificate holder conducting supplemental operations shall:

(1) Identify in its operations manual the person having custody of the copies of documents retained in accordance with paragraph (d) of this section; and

(2) Retain at its principal base of operations either an original or a copy of the records required by this section for at least three months.

[Docket No. 6258, 29 FR 19226, Dec. 31, 1964; as amended by Amdt. 121–123, 40 FR 44541, Sept. 29, 1975; Amdt. 121–143, 43 FR 22642, May 25, 1978; Amdt. 121–178, 47 FR 13316, March 29, 1982; Amdt. 121–253, 61 FR 2616, Jan. 26, 1996]

## §121.698–121.699 [Reserved]

## §121.701 Maintenance log: Aircraft.

**(a)** Each person who takes action in the case of a reported or observed failure or malfunction of an airframe, engine, propeller, or appliance that is critical to the safety of flight shall make, or have made, a record of that action in the airplane's maintenance log.

**(b)** Each certificate holder shall have an approved procedure for keeping adequate copies of the record required in paragraph (a) of this section in the airplane in a place readily accessible to each flight crewmember and shall put that procedure in the certificate holder's manual.

## §121.703 Service difficulty reports (operational).

**(a)** Each certificate holder shall report the occurrence or detection of each failure, malfunction, or defect concerning—

(1) Any fire and, when monitored by a related fire-warning system, whether the fire-warning system functioned properly;

(2) Any false warning of fire or smoke;

(3) An engine exhaust system that causes damage to the engine, adjacent structure, equipment, or components;

(4) An aircraft component that causes the accumulation or circulation of smoke, vapor, or toxic or noxious fumes;

(5) Any engine flameout or shutdown during flight or ground operations;

(6) A propeller feathering system or ability of the system to control overspeed;

(7) A fuel or fuel-dumping system that affects fuel flow or causes hazardous leakage;

(8) A landing gear extension or retraction, or the opening or closing of landing gear doors during flight;

(9) Any brake system component that results in any detectable loss of brake actuating force when the aircraft is in motion on the ground;

(10) Any aircraft component or system that results in a rejected takeoff after initiation of the takeoff roll or the taking of emergency actions, as defined by the Aircraft Flight Manual or Pilot's Operating Handbook;

(11) Any emergency evacuation system or component including any exit door, passenger emergency evacuation lighting system, or evacuation equipment found to be defective or that fails to perform the intended function during an actual emergency or during training, testing, maintenance, demonstrations, or inadvertent deployments; and

(12) Autothrottle, autoflight, or flight control systems or components of these systems.

**(b)** For the purpose of this section *during flight* means the period from the moment the aircraft leaves the surface of the earth on takeoff until it touches down on landing.

**(c)** In addition to the reports required by paragraph (a) of this section, each certificate holder shall report any other failure, malfunction, or defect in an aircraft, system, component, or powerplant that occurs or is detected at any time if that failure, malfunction, or defect has endangered or may endanger the safe operation of an aircraft.

**(d)** Each certificate holder shall submit each report required by this section, covering each 24-hour period beginning at 0900 local time of each day and ending at 0900 local time on the next day, to a centralized collection point as specified by the Administrator. Each report of occurrences during a 24-hour period shall be submitted to the FAA within the next 96 hours. However, a report due on Saturday or Sunday may be submitted on the following Monday, and a report due on a holiday may be submitted on the next workday. Each certificate holder also shall make the report data available for 30 days for examination by the certificate-holding district office in a form and manner acceptable to the Administrator.

**(e)** The certificate holder shall submit the reports required by this section on a form or in another format acceptable to the Administrator. The reports shall include the following information:

(1) The manufacturer, model, and serial number of the aircraft, engine, or propeller;

(2) The registration number of the aircraft;

(3) The operator designator;

(4) The date on which the failure, malfunction, or defect was discovered;

(5) The stage of flight or ground operation during which the failure, malfunction, or defect was discovered;

(6) The nature of the failure, malfunction, or defect;

(7) The applicable Joint Aircraft System/Component Code;

(8) The total cycles, if applicable, and total time of the aircraft, aircraft engine, propeller, or component;

(9) The manufacturer, manufacturer part number, part name, serial number, and location of the component that failed, malfunctioned, or was defective, if applicable;

(10) The manufacturer, manufacturer part number, part name, serial number, and location of the part that failed, malfunctioned, or was defective, if applicable;

(11) The precautionary or emergency action taken;

(12) Other information necessary for a more complete analysis of the cause of the failure, malfunction, or defect, including available information pertaining to type designation of the major component and the time since the last maintenance overhaul, repair, or inspection; and

(13) A unique control number for the occurrence, in a form acceptable to the Administrator.

**(f)** A certificate holder that also is the holder of a Type Certificate (including a Supplemental Type Certificate), a Parts Manufacturer Approval, or a Technical Standard Order authorization, or that is a licensee of a Type Certificate holder, need not report a failure, malfunction, or defect under this section if the failure, malfunction, or defect has been reported by that certificate holder under §21.3 of this chapter or under the accident reporting provisions of 49 CFR part 830.

**(g)** A report required by this section may be submitted by a certificated repair station when the reporting task has been assigned to that repair station by a part 121 certificate holder. However, the part 121 certificate holder remains primarily responsible for ensuring compliance with the provisions of this section. The part 121 certificate holder shall receive a copy of each report submitted by the repair station.

**(h)** No person may withhold a report required by this section although all information required by this section is not available.

**(i)** When a certificate holder gets supplemental information to complete the report required by this section, the certificate holder shall expeditiously submit that information as a supplement to the original report and use the unique control number from the original report.

***Editorial Note:*** *Per the November 23, 2001 Federal Register, mandatory compliance was changed to January 16, 2003. To view the rule currently in effect, go to ASA's website and click on "Product Updates":*

www.asa2fly.com

[Docket No. 6258, 29 FR 19226, Dec. 31, 1964; as amended by Docket No. 8084, 32 FR 5770, April 11, 1967; Amdt. 121–72, 35 FR 18188, Nov. 28, 1970; Amdt. 121–143, 43 FR 22642, May 25, 1978; Amdt. 121–178, 47 FR 13316, March 29, 1982; Amdt. 121–187, 50 FR 32375, Aug. 9, 1985; Amdt. 121–195, 53 FR 8728, March 16, 1988; Amdt. 121–251, 60 FR 65935, Dec. 20, 1995; Amdt. 121–279, 65 FR 56201, Sept. 15, 2000; Amdt. 121–284, 66 FR 21626, April 30, 2001; Docket No. FAA–2000–7952, 66 FR 58912, Nov. 23, 2001]

## §121.704 Service difficulty reports (structural).

**(a)** Each certificate holder shall report the occurrence or detection of each failure or defect related to—

(1) Corrosion, cracks, or disbonding that requires replacement of the affected part;

(2) Corrosion, cracks, or disbonding that requires rework or blendout because the corrosion, cracks, or disbonding exceeds the manufacturer's established allowable damage limits;

(3) Cracks, fractures, or disbonding in a composite structure that the equipment manufacturer has designated as a primary structure or a principal structural element; or

(4) Repairs made in accordance with approved data not contained in the manufacturer's maintenance manual.

**(b)** In addition to the reports required by paragraph (a) of this section, each certificate holder shall report any other failure or defect in aircraft structure that occurs or is detected at any time if that failure or defect has endangered or may endanger the safe operation of an aircraft.

**(c)** Each certificate holder shall submit each report required by this section, covering each 24-hour period beginning at 0900 local time of each day and ending at 0900 local time on the next day, to a centralized collection point as specified by the Administrator. Each report of occurrences during a 24-hour period shall be submitted to the FAA within the next 96 hours. However, a report due on Saturday or Sunday may be submitted on the following Monday, and a report due on a holiday may be submitted on the next workday. Each certificate holder also shall make the report data available for 30 days for examination by the certificate-holding district office in a form and manner acceptable to the Administrator.

**(d)** The certificate holder shall submit the reports required by this section on a form or in another format acceptable to the Administrator. The reports shall include the following information:

(1) The manufacturer, model, serial number, and registration number of the aircraft;

(2) The operator designator;

(3) The date on which the failure or defect was discovered;

(4) The stage of ground operation during which the failure or defect was discovered;

(5) The part name, part condition, and location of the failure or defect;

(6) The applicable Joint Aircraft System/Component Code;

121

(7) The total cycles, if applicable, and total time of the aircraft;

(8) Other information necessary for a more complete analysis of the cause of the failure or defect, including corrosion classification, if applicable, or crack length and available information pertaining to type designation of the major component and the time since the last maintenance overhaul, repair, or inspection; and

(9) A unique control number for the occurrence, in a form acceptable to the Administrator.

**(e)** A certificate holder that also is the holder of a Type Certificate (including a Supplemental Type Certificate), a Parts Manufacturer Approval, or a Technical Standard Order authorization, or that is a licensee of a Type Certificate holder, need not report a failure or defect under this section if the failure or defect has been reported by that certificate holder under §21.3 of this chapter or under the accident reporting provisions of 49 CFR part 830.

**(f)** A report required by this section may be submitted by a certificated repair station when the reporting task has been assigned to that repair station by the part 121 certificate holder. However, the part 121 certificate holder remains primarily responsible for ensuring compliance with the provisions of this section. The part 121 certificate holder shall receive a copy of each report submitted by the repair station.

**(g)** No person may withhold a report required by this section although all information required by this section is not available.

**(h)** When a certificate holder gets supplemental information to complete the report required by this section, the certificate holder shall expeditiously submit that information as a supplement to the original report and use the unique control number from the original report.

***Editorial Note:*** *Per the November 23, 2001 Federal Register, mandatory compliance was changed to January 16, 2003. To view the rule currently in effect, go to ASA's website and click on "Product Updates":*

www.asa2fly.com

[Docket No. 28293, 65 FR 56202, Sept. 15, 2000; as amended by Amdt. 121–284, 66 FR 21626, April 30, 2001; Docket No. FAA–2000–7952, 66 FR 58912, Nov. 23, 2001]

## §121.705 Mechanical interruption summary report.

Each certificate holder shall submit to the Administrator, before the end of the 10th day of the following month, a summary report for the previous month of each interruption to a flight, unscheduled change of aircraft en route, unscheduled stop or diversion from a route, or unscheduled engine removal caused by known or suspected mechanical difficulties or malfunctions that are not required to be reported under §121.703 or §121.704 of this part.

***Editorial Note:*** *Per the November 23, 2001 Federal Register, mandatory compliance was changed to January 16, 2003. To view the rule currently in effect, go to ASA's website and click on "Product Updates":*

www.asa2fly.com

[Docket No. 28293, 65 FR 56203, Sept. 15, 2000; as amended by Amdt. 121–284, 66 FR 21626, April 30, 2001; Docket No. FAA–2000–7952, 66 FR 58912, Nov. 23, 2001]

## §121.707 Alteration and repair reports.

**(a)** Each certificate holder shall, promptly upon its completion, prepare a report of each major alteration or major repair of an airframe, aircraft engine, propeller, or appliance of an aircraft operated by it.

**(b)** The certificate holder shall submit a copy of each report of a major alteration to, and shall keep a copy of each report of a major repair available for inspection by, the representative of the Administrator who is assigned to it.

## §121.709 Airworthiness release or aircraft log entry.

**(a)** No certificate holder may operate an aircraft after maintenance, preventive maintenance or alterations are performed on the aircraft unless the certificate holder, or the person with whom the certificate holder arranges for the performance of the maintenance, preventive maintenance, or alterations, prepares or causes to be prepared—

(1) An airworthiness release; or

(2) An appropriate entry in the aircraft log.

**(b)** The airworthiness release or log entry required by paragraph (a) of this section must—

(1) Be prepared in accordance with the procedures set forth in the certificate holder's manual;

(2) Include a certification that—

(i) The work was performed in accordance with the requirements of the certificate holder's manual;

(ii) All items required to be inspected were inspected by an authorized person who determined that the work was satisfactorily completed;

(iii) No known condition exists that would make the airplane unairworthy; and

(iv) So far as the work performed is concerned, the aircraft is in condition for safe operation; and

(3) Be signed by an authorized certificated mechanic or repairman except that a certificated repairman may sign the release or entry only for the work for which he is employed and certificated.

**(c)** Notwithstanding paragraph (b)(3) of this section, after maintenance, preventive maintenance, or alterations performed by a repair station that is located outside the United States, the airworthiness release or log entry required by paragraph (a) of this section may be signed by a person authorized by that repair station.

**(d)** When an airworthiness release form is prepared the certificate holder must give a copy to the pilot in command and must keep a record thereof for at least 2 months.

**(e)** Instead of restating each of the conditions of the certification required by paragraph (b) of this section, the air carrier may state in its manual that the signature of an authorized certificated mechanic or repairman constitutes that certification.

[Docket No. 6258, 29 FR 19226, Dec. 31, 1964; as amended by Amdt. 121–6, 30 FR 6432, May 8, 1965; Amdt. 121–21, 31 FR 10613, Aug. 9, 1966; Amdt. 121–286, 66 FR 41116, Aug. 6, 2001]

### §121.711 Communication records: Domestic and flag operations.

Each certificate holder conducting domestic or flag operations shall record each en route radio contact between the certificate holder and its pilots and shall keep that record for at least 30 days.

[Docket No. 28154, 61 FR 2616, Jan. 26, 1996]

### §121.713 Retention of contracts and amendments: Commercial operators who conduct intrastate operations for compensation or hire.

**(a)** Each commercial operator who conducts intrastate operations for compensation or hire shall keep a copy of each written contract under which it provides services as a commercial operator for a period of at least 1 year after the date of execution of the contract. In the case of an oral contract, it shall keep a memorandum stating its elements, and of any amendments to it, for a period of at least one year after the execution of that contract or change.

**(b)** Each commercial operator who conducts intrastate operations for compensation or hire shall submit a financial report for the first 6 months of each fiscal year and another financial report for each complete fiscal year. If that person's operating certificate is suspended for more than 29 days, that person shall submit a financial report as of the last day of the month in which the suspension is terminated. The report required to be submitted by this section shall be submitted within 60 days of the last day of the period covered by the report and must include—

(1) A balance sheet that shows assets, liabilities, and net worth on the last day of the reporting period;

(2) The information required by §119.36(e)(2), (e)(7), and (e)(8) of this chapter.

(3) An itemization of claims in litigation against the applicant, if any, as of the last day of the period covered by the report;

(4) A profit and loss statement with the separation of items relating to the applicant's commercial operator activities from his other business activities, if any; and

(5) A list of each contract that gave rise to operating income on the profit and loss statement, including the names and addresses of the contracting parties and the nature, scope, date, and duration of each contract.

[Docket No. 28154, 60 FR 65936, Dec. 20, 1995; as amended by Amdt. 121–262, 62 FR 13257, March 19, 1997]

## Subpart W— Crewmember Certificate: International

### §121.721 Applicability.

This section describes the certificates that were issued to United States citizens who were employed by air carriers at the time of issuance as flight crewmembers on United States registered aircraft engaged in international air commerce. The purpose of the certificate is to facilitate the entry and clearance of those crewmembers into ICAO contracting states. They were issued under Annex 9; as amended, to the Convention on International Civil Aviation.

[Docket No. 28154, 61 FR 30435, June 14, 1996]

### §121.723 Surrender of international crewmember certificate.

The holder of a certificate issued under this section, or the air carrier by whom the holder is employed, shall surrender the certificate for cancellation at the nearest FAA Flight Standards District Office at the termination of the holder's employment with that air carrier.

[Docket No. 28154, 61 FR 30435, June 14, 1996]

# Subpart X— Emergency Medical Equipment and Training

**Source:** Docket No. FAA–2000–7119, 66 FR 19028, April 12, 2001 unless otherwise noted.

**Sec.**



## § 121.801 Applicability.

This subpart prescribes the emergency medical equipment and training requirements applicable to all certificate holders operating passenger-carrying airplanes under this part. Nothing in this subpart is intended to require certificate holders or its agents to provide emergency medical care or to establish a standard of care for the provision of emergency medical care.

## § 121.803 Emergency medical equipment.

**(a)** No person may operate a passenger-carrying airplane under this part unless it is equipped with the emergency medical equipment listed in this section.

**(b)** Each equipment item listed in this section—

(1) Must be inspected regularly in accordance with inspection periods established in the operations specifications to ensure its condition for continued serviceability and immediate readiness to perform its intended emergency purposes;

(2) Must be readily accessible to the crew and, with regard to equipment located in the passenger compartment, to passengers;

(3) Must be clearly identified and clearly marked to indicate its method of operation; and

(4) When carried in a compartment or container, must be carried in a compartment or container marked as to contents and the compartment or container, or the item itself, must be marked as to date of last inspection.

**(c)** For treatment of injuries, medical events, or minor accidents that might occur during flight time each airplane must have the following equipment that meets the specifications and requirements of appendix A of this part:

(1) Approved first-aid kits.

(2) In airplanes for which a flight attendant is required, an approved emergency medical kit.

(3) In airplanes for which a flight attendant is required, an approved emergency medical kit as modified effective April 12, 2004.

(4) In airplanes for which a flight attendant is required and with a maximum payload capacity of more than 7,500 pounds, an approved automated external defibrillator as of April 12, 2004.

## § 121.805 Crewmember training for in-flight medical events.

**(a)** Each training program must provide the instruction set forth in this section with respect to each airplane type, model, and configuration, each required crewmember, and each kind of operation conducted, insofar as appropriate for each crewmember and the certificate holder.

**(b)** Training must provide the following:

(1) Instruction in emergency medical event procedures, including coordination among crewmembers.

(2) Instruction in the location, function, and intended operation of emergency medical equipment.

(3) Instruction to familiarize crewmembers with the content of the emergency medical kit.

(4) Instruction to familiarize crewmembers with the content of the emergency medical kit as modified on April 12, 2004.

(5) For each flight attendant—

(i) Instruction, to include performance drills, in the proper use of automated external defibrillators.

(ii) Instruction, to include performance drills, in cardiopulmonary resuscitation.

(iii) Recurrent training, to include performance drills, in the proper use of an automated external defibrillators and in cardiopulmonary resuscitation at least once every 24 months.

**(c)** The crewmember instruction, performance drills, and recurrent training required under this section are not required to be equivalent to the expert level of proficiency attained by professional emergency medical personnel.

## Appendix A to Part 121

### First-Aid Kits and Emergency Medical Kits

#### *First-Aid Kits*

Approved first-aid kits, at least one approved emergency medical kit, and at least one approved automated external defibrillator required under §121.803 of this part must be readily accessible to the crew, stored securely, and kept free from dust, moisture, and damaging temperatures.

**(1)** The minimum number of first-aid kits required is set forth in the following table:

| No. of passenger seats | No. of first-aid kits |
|---|---|
| 0–50 | 1 |
| 51–150 | 2 |
| 151–250 | 3 |
| More than 250 | 4 |

**(2)** Except as provided in paragraph (3), each approved first-aid kit must contain at least the following appropriately maintained contents in the specified quantities:

| Contents | Quantity |
|---|---|
| Adhesive bandage compresses, 1-inch | 16 |
| Antiseptic swabs | 20 |
| Ammonia inhalants | 10 |
| Bandage compresses, 4-inch | 8 |
| Triangular bandage compresses, 40-inch | 5 |
| Arm splint, noninflatable | 1 |
| Leg splint, noninflatable | 1 |
| Roller bandage, 4-inch | 4 |
| Adhesive tape, 1-inch standard roll | 2 |
| Bandage scissors | 1 |

**(3)** Arm and leg splints which do not fit within a first-aid kit may be stowed in a readily accessible location that is as near as practicable to the kit.

### *Emergency Medical Kits*

**(1)** Until April 12, 2004, at least one approved emergency medical kit that must contain at least the following appropriately maintained contents in the specified quantities:

| Contents | Quantity |
|---|---|
| Sphygmomanometer | 1 |
| Stethoscope | 1 |
| Airways, oropharyngeal (3 sizes) | 3 |
| Syringes (sizes necessary to administer required drugs) | 4 |
| Needles (sizes necessary to administer required drugs) | 6 |
| 50% Dextrose injection, 50cc | 1 |
| Epinephrine 1:1000, single dose ampule or equivalent | 2 |
| Diphenhydramine HCl injection, single dose ampule or equivalent | 2 |
| Nitroglycerin tablets | 10 |
| Basic instructions for use of the drugs in the kit | 1 |
| Protective nonpermeable gloves or equivalent | 1 pair |

**(2)** As of April 12, 2004, at least one approved emergency medical kit that must contain at least the following appropriately maintained contents in the specified quantities:

| Contents | Quantity |
|---|---|
| Sphygmonanometer | 1 |
| Stethoscope | 1 |
| Airways, oropharyngeal (3 sizes): 1 pediatric, 1 small adult, 1 large adult or equivalent | 3 |
| Self-inflating manual resuscitation device with 3 masks (1 pediatric, 1 small adult, 1 large adult or equivalent) | 1:3 masks |
| CPR mask (3 sizes), 1 pediatric, 1 small adult, 1 large adult, or equivalent | 3 |
| IV Admin Set: Tubing w/ 2 Y connectors<br>Alcohol sponges<br>Adhesive tape, 1-inch standard roll adhesive<br>Tape scissors<br>Tourniquet | 1<br>2<br>1<br>1 pair<br>1 |
| Saline solution, 500 cc | 1 |
| Protective nonpermeable gloves or equivalent | 1 pair |
| Needles (2-18 ga., 2-20 ga., 2-22 ga., or sizes necessary to administer required medications) | 6 |
| Syringes (1-5 cc, 2-10 cc, or sizes necessary to administer required medications) | 4 |
| Analgesic, non-narcotic, tablets, 325 mg | 4 |
| Antihistamine tablets, 25 mg | 4 |
| Antihistamine injectable, 50 mg, (single dose ampule or equivalent) | 2 |
| Atropine, 0.5 mg, 5 cc (single dose ampule or equivalent) | 2 |
| Aspirin tablets, 325 mg | 4 |
| Bronchodilator, inhaled (metered dose inhaler or equivalent) | 1 |
| Dextrose, 50%/50 cc injectable, (single dose ampule or equivalent) | 1 |
| Epinephrine 1:1000, 1 cc, injectable, (single dose ampule or equivalent) | 2 |
| Epinephrine 1:10,000, 2 cc, injectable, (single dose ampule or equivalent) | 2 |
| Lidocaine, 5 cc, 20 mg/ml, injectable (single dose ampule or equivalent) | 2 |
| Nitroglycerin tablets, 0.4 mg | 10 |
| Basic instructions for use of the drugs in the kit | 1 |

**(3)** If all of the above-listed items do not fit into one container, more than one container may be used.

*AUTOMATED EXTERNAL DEFIBRILLATORS*

At least one approved automated external defibrillator, legally marketed in the United States in accordance with Food and Drug Administration requirements, that must:

**(1)** Be stored in the passenger cabin.

**(2)** Meet FAA Technical Standard Order requirements for power sources for electronic devices used in aviation as approved by the Administrator.

**(3)** Be maintained in accordance with the manufacturer's specifications.

[Docket No. 12384, 38 FR 35234, Dec. 26, 1973; as amended by Amdt. 121–115, 40 FR 1039, Jan. 6, 1975; Amdt. 121–188, 51 FR 1223, Jan. 9, 1986; Amdt. 121–236, 59 FR 1781, Jan. 12, 1994; Amdt. 121–242, 59 FR 52642, October 18, 1994; Amdt. 121–243, 59 FR 62276, Dec. 2, 1994; Amdt. 121–281, 66 FR 19044, April 12, 2001]

## Appendix B to Part 121
### Airplane Flight Recorder Specifications

| Parameters | Range | Accuracy sensor input to DFDR readout | Sampling interval (per second) | Resolution[4] readout |
|---|---|---|---|---|
| Time (GMT or Frame Counter) (range 0 to 4095, sampled 1 per frame) | 24 Hrs | ±0.125% per hour | 0.25 (1 per 4 seconds) | 1 sec. |
| Altitude | –1,000 ft to max certificated altitude of aircraft | ±100 to ±700 ft (See Table 1, TSO–C51a) | 1 | 5' to 35[1] |
| Airspeed | 50 KIAS to $V_{S0}$, and $V_{S0}$ to 1.2 $V_D$ | ±5%, ±3% | 1 | 1 kt. |
| Heading | 360° | ±2° | 1 | 0.5° |
| Normal Acceleration (Vertical) | –3g to +6g | ±1% of max range excluding datum error of ±5% | 8 | 0.01g |
| Pitch Attitude | ±75° | ±2° | 1 | 0.5° |
| Roll Attitude | ±180° | ±2° | 1 | 0.5° |
| Radio Transmitter Keying | On-Off (Discrete) | ±2° | ±2% | |
| Thrust / Power on Each Engine | Full Range Forward | ±2° | 1 (per engine) | 0.2%[2] |
| Trailing Edge Flap or Cockpit Control Selection | Full Range or Each Discrete Position | ±3° or as Pilot's Indicator | 0.5 | 0.5%[2] |
| Leading Edge Flap or Cockpit Control Selection | Full Range or Each Discrete Position | ±3° or as Pilot's Indicator | 0.5 | 0.5%[2] |
| Thrust Reverser Position | Stowed, in Transit, and Reverse (Discrete) | | 1 (per 4 seconds per engine) | |
| Ground Spoiler Position / Speed Brake Selection | Full Range or Each Discrete Position | ±2% Unless Higher Accuracy Uniquely Required | 1 | 0.2%[2] |
| Marker Beacon Passage | Discrete | | 1 | |
| Autopilot Engagement | Discrete | | 1 | |
| Longitudinal Acceleration | ±1g | ±1.5% max range excluding datum error of ±5% | 4 | 0.01g |
| Pilot Input and/or Surface Position — Primary Controls (Pitch, Roll, Yaw)[3] | Full Range | ±2° Unless Higher Accuracy Uniquely Required | 1 | 0.2%[2] |
| Lateral Acceleration | ±1g | ±1.5% max range excluding datum error of ±5% | 4 | 0.01g |
| Pitch Trim Position | Full Range | ±3% Unless Higher Accuracy Uniquely Required | 1 | 0.3%[2] |
| Glideslope Deviation | ±400 Microamps | ±3% | 1 | 0.3%[2] |
| Localizer Deviation | ±400 Microamps | ±3% | 1 | 0.3%[2] |
| AFCS Mode and Engagement Status | Discrete | | 1 | |
| Radio Altitude | –20 ft to 2,500 ft | ±2 Ft or ±3% Whichever is Greater Below 500 Ft and ±5% Above 500 Ft | 1 | 1 ft + 5%[2] above 500' |
| Master Warning | Discrete | | 1 | |
| Main Gear Squat Switch Status | Discrete | | 1 | |
| Angle of Attack (if recorded directly) | As installed | As installed | 2 | 0.3%[2] |
| Outside Air Temperature or Total Air Temperature | –50°C to +90°C | ±2°C | 0.5 | 0.3°C |
| Hydraulics, Each System Low Pressure | Discrete | | 0.5 | or 0.5%[2] |
| Groundspeed | As installed | Most Accurate Systems Installed (IMS Equipped Aircraft Only) | 1 | 0.2%[2] |

121

## APPENDIX B TO PART 121

AIRPLANE FLIGHT RECORDER SPECIFICATION (*CONTINUED*)

If additional recording capacity is available, recording of the following parameters is recommended. The parameters are listed in order of significance:

| Parameters | Range | Accuracy sensor input to DFDR readout | Sampling interval (per second) | Resolution[4] readout |
|---|---|---|---|---|
| Drift Angle | When available, As installed | As installed | 4 | |
| Wind Speed and Direction | When available, As installed | As installed | 4 | |
| Latitude and Longitude | When available, As installed | As installed | 4 | |
| Brake pressure / Brake pedal position | As installed | As installed | 1 | |
| Additional engine parameters:<br>EPR<br>N1<br>N2<br>EGT | <br>As installed<br>As installed<br>As installed<br>As installed | <br>As installed<br>As installed<br>As installed<br>As installed | <br>1 (per engine)<br>1 (per engine<br>1 (per engine)<br>1 (per engine) | |
| Throttle Lever Position | As installed | As installed | 1 (per engine) | |
| Fuel Flow | As installed | As installed | 1 (per engine) | |
| TCAS:<br>TA<br>RA<br>Sensitivity level (as selected by crew) | <br>As installed<br>As installed<br>As installed | <br>As installed<br>As installed<br>As installed | <br>1<br>1<br>2 | |
| GPWS (ground proximity warning system) | Discrete | | 1 | |
| Landing gear or gear selector position | Discrete | | 0.25 (1 per 4 seconds) | |
| DME 1 and 2 Distance | 0–200 NM | As installed | 0.25 | 1 mi. |
| Nav 1 and 2 Frequency Selection | Full range | As installed | 0.25 | |

[1] When altitude rate is recorded. Altitude rate must have sufficient resolution and sampling to permit the derivation of altitude to 5 feet.

[2] Percent of full range

[3] For airplanes that can demonstrate the capability of deriving either the control input on control movement (one from the other) for all modes of operation and flight regimes, the "or" applies. For airplanes with non-mechanical control systems (fly-by-wire) the "and" applies. In airplanes with split surfaces, suitable combination of inputs is acceptable in lieu of recording each surface separately.

[4] This column applies to aircraft manufactured after October 11, 1991.

[Docket No. 25530, 53 FR 26147, July 11, 1988; 53 FR 30906, August 16, 1988]

## Appendix C to Part 121

### C-46 Nontransport Category Airplanes

#### *Cargo Operations*

1. *Required engines.*

(a) Except as provided in paragraph (b) of this section, the engines specified in subparagraphs (1) or (2) of this section must be installed in C-46 nontransport category airplanes operated at gross weights exceeding 45,000 pounds:

(1) Pratt and Whitney R2800-51-M1 or R2800-75-M1 engines (engines converted from basic model R2800-51 or R2800-75 engines in accordance with FAA approved data) that—

(i) Conform to Engine Specification 5E–8;

(ii) Conform to the applicable portions of the operator's manual;

(iii) Comply with all the applicable airworthiness directives; and

(iv) Are equipped with high capacity oil pump drive gears in accordance with FAA approved data.

(2) Other engines found acceptable by the FAA Regional Flight Standards Division having type certification responsibility for the C–46 airplane.

(b) Upon application by an operator conducting cargo operations with nontransport category C-46 airplanes between points within the State of Alaska, the appropriate FAA Flight Standards District Office, Alaskan Region, may authorize the operation of such airplanes, between points within the State of Alaska; without compliance with paragraph (a) of this section if the operator shows that, in its area of operation, installation of the modified engines is not necessary to provide adequate cooling for single-engine operations. Such authorization and any conditions or limitations therefor is made a part of the Operations Specifications of the operator.

**2.** *Minimum acceptable means of complying with the special airworthiness requirements.* Unless otherwise authorized under §121.213, the data set forth in sections 3 through 34 of this appendix, as correlated to the C-46 nontransport category airplane, is the minimum means of compliance with the special airworthiness requirements of §§121.215 through 121.281.

**3.** *Susceptibility of material to fire.* [Deleted as unnecessary]

**4.** *Cabin interiors.* C-46 crew compartments must meet all the requirements of §121.215, and, as required in §121.221, the door between the crew compartment and main cabin (cargo) compartment must be flame resistant.

**5.** *Internal doors.* Internal doors, including the crew to main cabin door, must meet all the requirements of §121.217.

**6.** *Ventilation.* Standard C-46 crew compartments meet the ventilation requirements of §121.219 if a means of ventilation for controlling the flow of air is available between the crew compartment and main cabin. The ventilation requirement may be met by use of a door between the crew compartment and main cabin. The door need not have louvers installed; however, if louvers are installed, they must be controllable.

**7.** *Fire precautions.* Compliance is required with all the provisions of §121.221.

(a) In establishing compliance with this section, the C-46 main cabin is considered as a Class A compartment if—

(1) The operator utilizes a standard system of cargo loading and tiedown that allows easy access in flight to all cargo in such compartment, and, such system is included in the appropriate portion of the operator's manual; and

(2) A cargo barrier is installed in the forward end of the main cabin cargo compartment. The barrier must—

(i) Establish the most forward location beyond which cargo cannot be carried;

(ii) Protect the components and systems of the airplane that are essential to its safe operation from cargo damage; and

(iii) Permit easy access, in flight, to cargo in the main cabin cargo compartment.

The barrier may be a cargo net or a network of steel cables or other means acceptable to the Administrator which would provide equivalent protection to that of a cargo net. The Barrier need not meet crash load requirements of FAR §25.561; however, it must be attached to the cargo retention fittings and provide the degree of cargo retention that is required by the operators' standard system of cargo loading and tiedown.

(b) C-46 forward and aft baggage compartments must meet, as a minimum, Class B requirements of this section or be placarded in a manner to preclude their use as cargo or baggage compartments.

**8.** *Proof of compliance.* The demonstration of compliance required by §121.223 is not required for C-46 airplanes in which—

(1) The main cabin conforms to Class A cargo compartment requirements of §121.219; and

(2) Forward and aft baggage compartments conform to Class B requirements of §121.221, or are placarded to preclude their use as cargo or baggage compartments.

**9.** *Propeller deicing fluid.* No change from the requirements of §121.225. Isopropyl alcohol is a combustible fluid within the meaning of this section.

**10.** *Pressure cross-feed arrangements, location of fuel tanks, and fuel system lines and fittings.* C-46 fuel systems which conform to all applicable Curtiss design specifications and which comply with the FAA type certification requirements are in compliance with the provisions of §§121.227 through 121.231.

**11.** *Fuel lines and fittings in designated fire zones.* No change from the requirements of §121.233.

**12.** *Fuel valves.* Compliance is required with all the provisions of §121.235. Compliance can be established by showing that the fuel system conforms to all the applicable Curtiss design specifications, the FAA type certification requirements, and, in addition, has explosion-proof fuel booster pump electrical selector switches installed in lieu of the open contact type used originally.

**13.** *Oil lines and fittings in designated fire zones.* No change from the requirements of §121.237.

**14.** *Oil valves.* C-46 oil shutoff valves must conform to the requirements of §121.239. In addition, C-46 airplanes using Hamilton Standard propellers must provide, by use of stand pipes in the engine oil tanks or other approved means, a positive source of oil for feathering each propeller.

**15.** *Oil system drains.* The standard C-46 "Y" drains installed in the main oil inlet line for each engine meet the requirements of §121.241.

**16.** *Engine breather line.* The standard C-46 engine breather line installation meets the requirements of §121.243 if the lower breather lines actually extend to the trailing edge of the oil cooler air exit duct.

**17.** *Firewalls and firewall construction.* Compliance is required with all of the provisions of §§121.245 and 121.247. The following requirements must be met in showing compliance with these sections:

(a) *Engine compartment.* The engine firewalls of the C-46 airplane must—

(1) Conform to type design, and all applicable airworthiness directives;

(2) Be constructed of stainless steel or approved equivalent; and

(3) Have fireproof shields over the fairleads used for the engine control cables that pass through each firewall.

(b) *Combustion heater compartment.* C-46 airplanes must have a combustion heater fire extinguishing system which complies with AD-49-18-1 or an FAA approved equivalent.

**18.** *Cowling.* Standard C-46 engine cowling (cowling of aluminum construction employing stainless steel exhaust shrouds) which conforms to the type design and cowling configurations which conform to the C-46 transport category requirements meet the requirements of §121.249.

**19.** *Engine accessory section diaphragm.* C-46 engine nacelles which conform to the C-46 transport category requirements meet the requirements of §121.251. As provided for in that section, a means of equivalent protection which does not require provision of a diaphragm to isolate the engine power section and exhaust system from the engine accessory compartment is the designation of the entire engine compartment forward of and including the firewall as a designated fire zone, and the installation of adequate fire detection and fire extinguishing systems which meet the requirements of §121.263 and §121.273, respectively, in such zone.

**20.** *Powerplant fire protection.* C-46 engine compartments and combustion heater compartments are considered as designated fire zones within the meaning of §121.253.

**21.** *Flammable fluids*—

(a) *Engine compartment.* C-46 engine compartments which conform to the type design and which comply with all applicable airworthiness directives meet the requirements of §121.255.

(b) *Combustion heater compartment.* C-46 combustion heater compartments which conform to type design and which meet all the requirements of AD-49-18-1 or an FAA approved equivalent meet the requirements of §121.255.

**22.** *Shutoff* means—

(a) *Engine compartment.* C-46 engine compartments which comply with AD-62-10-2 or FAA approved equivalent meet the requirements of §121.257 applicable to engine compartments, if, in addition, a means satisfactory to the Administrator is provided to shut off the flow of hydraulic fluid to the cowl flap cylinder in each engine nacelle. The shutoff means must be located aft of the engine firewall. The operator's manual must include, in the emergency portion, adequate instructions for proper operation of the additional shutoff means to assure correct sequential positioning of engine cowl flaps under emergency conditions. In accordance with §121.315, this positioning must also be incorporated in the emergency section of the pilot's checklist.

(b) *Combustion heater compartment.* C-46 heater compartments which comply with paragraph (5) of AD-49-18-1 or FAA approved equivalent meet the requirements of §121.257 applicable to heater compartments if, in addition, a shutoff valve located above the main cabin floor level is installed in the alcohol supply line or lines between the alcohol supply tank and those alcohol pumps located under the main cabin floor. If all of the alcohol pumps are located above the main cabin floor, the alcohol shutoff valve need not be installed. In complying with paragraph (5) of AD-49-18-1, a fail-safe electric fuel shutoff valve may be used in lieu of the manually operated valve.

**23.** *Lines and fittings*—

(a) *Engine compartment.* C-46 engine compartments which comply with all applicable airworthiness directives, including AD-62-10-2, by using FAA approved fire-resistant lines, hoses, and end fittings, and engine compartments which meet the C-46 transport category requirements, meet the requirements of §121.259.

(b) *Combustion heater compartments.* All lines, hoses, and end fittings, and couplings which carry fuel to the heaters and heater controls, must be of FAA approved fire-resistant construction.

**24.** *Vent and drain lines.*

(a) *Engine compartment.* C-46 engine compartments meet the requirements of §121.261 if—

(1) The compartments conform to type design and comply with all applicable airworthiness directives or FAA approved equivalent; and

(2) Drain lines from supercharger case, engine-driven fuel pump, and engine-driven hydraulic pump reach into the scupper drain located in the lower cowling segment.

(b) *Combustion heater compartment.* C-46 heater compartments meet the requirements of §121.261 if they conform to AD-49-18-1 or FAA approved equivalent.

**25.** *Fire-extinguishing system.*

(a) To meet the requirements of §121.263, C-46 airplanes must have installed fire extinguishing systems to serve all designated fire zones. The fire-extinguishing systems, the quantity of extinguishing agent, and the rate of discharge shall be such as to provide a minimum of one adequate discharge for each designated fire zone. Compliance with this provision requires the installation of a separate fire extinguisher for each engine compartment. Insofar as the engine compartment is concerned, the system shall be capable of protecting the entire compartment against the various types of fires likely to occur in the compartment.

(b) Fire-extinguishing systems which conform to the C-46 transport category requirements meet the requirements set forth in paragraph (a). Furthermore, fire-extinguishing systems for combustion heater compartments which conform to the requirements of AD-49-18-1 or an FAA approved equivalent also meet the requirements in paragraph (a).

In addition, a fire-extinguishing system for C-46 airplanes meets the adequacy requirement of paragraph (a) if it provides the same or equivalent protection to that demonstrated by the CAA in tests conducted in 1941 and 1942, using a CW-20 type engine nacelle (without diaphragm). These tests were conducted at the Bureau of Standards facilities in Washington, DC, and copies of the test reports are available through the FAA Regional Engineering Offices. In this connection, the flow rates and distribution of extinguishing agent substantiated in American Airmotive Report No. 128-52-d, FAA approved February 9, 1953, provides protection equivalent to that demonstrated by the CAA in the CW-20 tests. In evaluating any C-46 fire-extinguishing system with respect to the aforementioned CW-20 tests, the Administration would require data in a narrative form, utilizing drawings or photographs to show at least the following:

Installation of containers; installation and routing of plumbing; type, number, and location of outlets or nozzles; type, total volume, and distribution of extinguishing agent; length of time required for discharging; means for thermal relief, including type and location of discharge indicators; means of discharging, e.g., mechanical cutterheads, electric cartridge, or other method; and whether a one- or two-shot system is used; and if the latter is used, means of cross-feeding or otherwise selecting distribution of extinguishing agent; and types of materials used in makeup of plumbing.

High rate discharge (HRD) systems using agents such as bromotrifluoromethane, dibrodifluoromethane and chlorobromomethane (CB), may also meet the requirements of paragraph (a).

**26.** *Fire-extinguishing agents, Extinguishing agent container pressure relief, Extinguishing agent container compartment temperatures, and Fire-extinguishing system materials.* No change from the requirements of §§121.265 through 121.271.

**27.** *Fire-detector system.* Compliance with the requirements of §121.273 requires that C-46 fire detector systems conform to:

(a) AD-62-10-2 or FAA approved equivalent for engine compartments; and

(b) AD-49-18-1 or FAA approved equivalent for combustion heater compartments

**28.** *Fire detectors.* No change from the requirements of §121.275.

**29.** *Protection of other airplane components against fire.* To meet the requirements of §121.277, C-46 airplanes must—

(a) Conform to the type design and all applicable airworthiness directives; and

(b) Be modified or have operational procedures established to provide additional fire protection for the wheel well door aft of each engine compartment. Modifications may consist of improvements in sealing of the main landing gear wheel well doors. An operational procedure which is acceptable to the Agency is one requiring the landing gear control to be placed in the up position in case of in-flight engine fire. In accordance with §121.315, such procedure must be set forth in the emergency portion of the operator's emergency checklist pertaining to in-flight engine fire.

**30.** *Control of engine rotation.* C-46 propeller feathering systems which conform to the type design and all applicable airworthiness directives meet the requirements of §121.279.

**31.** *Fuel system independence.* C-46 fuel systems which conform to the type design and all applicable airworthiness directives meet the requirements of §121.281.

**32.** *Induction system ice prevention.* The C-46 carburetor anti-icing system which conforms to the type design and all applicable airworthiness directives meets the requirements of §121.283.

**33.** *Carriage of cargo in passenger compartments.* Section 121.285 is not applicable to non-transport category C-46 cargo airplanes.

**34.** *Carriage of cargo in cargo compartments.* A standard cargo loading and tiedown arrangement set forth in the operator's manual and found acceptable to the Administrator must be used in complying with §121.287.

**35.** *Performance data.* Performance data on Curtiss model C-46 airplane certificated for maximum weight of 45,000 and 48,000 pounds for cargo-only operations.

1. The following performance limitation data, applicable to the Curtiss model C-46 airplane for cargo-only operation, must be used in determining compliance with §§121.199 through 121.205. These data are presented in the tables and figures of this appendix.

*TABLE 1—TAKEOFF LIMITATIONS*

(a) Curtiss C-46 certificated for maximum weight of 45,000 pounds.

(1) *Effective length* of runway required when effective length is determined in accordance with §121.171 (distance to accelerate to 93 knots TIAS and stop, with zero wind and zero gradient). (Factor=1.00)

[DISTANCE IN FEET]

| Standard altitude in feet | Airplane weight in pounds | | |
|---|---|---|---|
| | 39,000 | 42,000 | 45,000[1] |
| S.L. | 4,110 | 4,290 | 4,570 |
| 1,000 | 4,250 | 4,440 | 4,720 |
| 2,000 | 4,400 | 4,600 | 4,880 |
| 3,000 | 4,650 | 4,880 | 5,190 |
| 4,000 | 4,910 | 5,170 | 5,500 |
| 5,000 | 5,160 | 5,450 | 5,810 |
| 6,000 | 5,420 | 5,730 | 6,120 |
| 7,000 | 5,680 | 6,000 | 6,440 |
| 8,000 | 5,940 | 6,280 | ([1]) |

[1] Ref. Fig. 1(a)(1) for weight and distance for altitudes above 7,000'.

(2) Actual length of runway required when *effective length*, considering obstacles, is not determined (distance to accelerate to 93 knots TIAS and stop, divided by the factor 0.85).

[DISTANCE IN FEET]

| Standard altitude in feet | Airplane weight in pounds | | |
|---|---|---|---|
| | 39,000 | 42,000 | 45,000[1] |
| S.L. | 4,830 | 5,050 | 5,370 |
| 1,000 | 5,000 | 5,230 | 5,550 |
| 2,000 | 5,170 | 5,410 | 5,740 |
| 3,000 | 5,470 | 5,740 | 6,100 |
| 4,000 | 5,770 | 6,080 | 6,470 |
| 5,000 | 6,070 | 6,410 | 6,830 |
| 6,000 | 6,380 | 6,740 | 7,200 |
| 7,000 | 6,680 | 7,070 | 7,570 |
| 8,000 | 6,990 | 7,410 | ([1]) |

[1] Ref. Fig. 1(a)(2) for weight and distance for altitudes above 7,000'.

(b) Curtiss C-46 certificated for maximum weight 48,000 pounds.

(1) *Effective length* of runway required when effective length is determined in accordance with §121.171 (distance to accelerate to 93 knots TIAS and stop, with zero wind and zero gradient). (Factor=1.00)

[DISTANCE IN FEET]

| Standard altitude in feet | Airplane weight in pounds | | | |
|---|---|---|---|---|
| | 39,000 | 42,000 | 45,000 | 48,000[1] |
| S.L. | 4,110 | 4,290 | 4,570 | 4,950 |
| 1,000 | 4,250 | 4,440 | 4,720 | 5,130 |
| 2,000 | 4,400 | 4,600 | 4,880 | 5,300 |
| 3,000 | 4,650 | 4,880 | 5,190 | 5,670 |
| 4,000 | 4,910 | 5,170 | 5,500 | 6,050 |
| 5,000 | 5,160 | 5,450 | 5,810 | 6,420 |
| 6,000 | 5,420 | 5,730 | 6,120 | 6,800 |
| 7,000 | 5,680 | 6,000 | 6,440 | ([1]) |
| 8,000 | 5,940 | 6,280 | 6,750 | ([1]) |

[1] Ref. Fig. 1(b)(1) for weight and distance for altitudes above 6,000'.

(2) Actual length of runway required when *effective length*, considering obstacles, is not determined (distance to accelerate to 93 knots TIAS and stop, divided by the factor 0.85).

[DISTANCE IN FEET]

| Standard altitude in feet | Airplane weight in pounds | | | |
|---|---|---|---|---|
| | 39,000 | 42,000 | 45,000 | 48,000[1] |
| S.L. | 4,830 | 5,050 | 5,370 | 5,830 |
| 1,000 | 5,000 | 5,230 | 5,550 | 6,030 |
| 2,000 | 5,170 | 5,410 | 5,740 | 6,230 |
| 3,000 | 5,470 | 5,740 | 6,100 | 6,670 |
| 4,000 | 5,770 | 6,080 | 6,470 | 7,120 |
| 5,000 | 6,070 | 6,410 | 6,830 | 7,560 |
| 6,000 | 6,380 | 6,740 | 7,200 | 8,010 |
| 7,000 | 6,680 | 7,070 | 7,570 | ([1]) |
| 8,000 | 6,990 | 7,410 | 7,940 | ([1]) |

[1] Ref. Fig. 1(b)(2) for weight and distance for altitudes above 6,000'.

### *TABLE 2— EN ROUTE LIMITATIONS*

(a) Curtiss model C-46 certificated for maximum weight of 45,000 pounds (based on a climb speed of 113 knots (TIAS)).

| Weight (pounds) | Terrain clearance (feet)[1] | Blower setting |
|---|---|---|
| 45,000 | 6,450 | Low. |
| 44,000 | 7,000 | Do. |
| 43,000 | 7,500 | Do. |
| 42,200 | 8,000 | High. |
| 41,000 | 9,600 | Do. |
| 40,000 | 11,000 | Do. |
| 39,000 | 12,300 | Do. |

[1] Highest altitude of terrain over which airplanes may be operated in compliance with §121.201.
Ref. Fig. 2(a).

(b) Curtiss model C-46 certificated for maximum weight of 48,000 pounds or with engine installation approved for 2,550 revolutions per minute (1,700 brake horsepower). Maximum continuous power in low blower (based on a climb speed of 113 knots (TIAS)).

| Weight (pounds) | Terrain clearance (feet)[1] | Blower setting |
|---|---|---|
| 48,000 | 5,850 | Low. |
| 47,000 | 6,300 | Do. |
| 46,000 | 6,700 | Do. |
| 45,000 | 7,200 | Do. |
| 44,500 | 7,450 | Do. |
| 44,250 | 8,000 | High. |
| 44,000 | 8,550 | Do. |
| 43,000 | 10,800 | Do. |
| 42,000 | 12,500 | Do. |
| 41,000 | 13,000 | Do. |

[1] Highest altitude of terrain over which airplanes may be operated in compliance with §121.201.
Ref. Fig. 2(b).

*TABLE 3 — LANDING LIMITATIONS*

(a) Intended Destination.

*Effective length* of runway required for intended destination when effective length is determined in accordance with §121.171 with zero wind and zero gradient.

(1) Curtiss model C-46 certificated for maximum weight of 45,000 pounds. (0.60 factor)

DISTANCE IN FEET

| Standard altitude in feet | Airplane weight in pounds and approach speeds[1] in knots | | | | | | | |
|---|---|---|---|---|---|---|---|---|
| | 40,000 | $V_{50}$ | 42,000 | $V_{50}$ | 44,000 | $V_{50}$ | 45,000 | $V_{50}$ |
| S.L. | 4,320 | 86 | 4,500 | 88 | 4,700 | 90 | 4,800 | 91 |
| 1,000 | 4,440 | 86 | 4,620 | 88 | 4,830 | 90 | 4,930 | 91 |
| 2,000 | 4,550 | 86 | 4,750 | 88 | 4,960 | 90 | 5,050 | 91 |
| 3,000 | 4,670 | 86 | 4,880 | 88 | 5,090 | 90 | 5,190 | 91 |
| 4,000 | 4,800 | 86 | 5,000 | 88 | 5,220 | 90 | 5,320 | 91 |
| 5,000 | 4,920 | 86 | 5,140 | 88 | 5,360 | 90 | 5,460 | 91 |
| 6,000 | 5,040 | 86 | 5,270 | 88 | 5,550 | 90 | 5,600 | 91 |
| 7,000 | 5,170 | 86 | 5,410 | 88 | 5,650 | 90 | 5,750 | 91 |
| 8,000 | 5,310 | 86 | 5,550 | 88 | 5,800 | 90 | 5,900 | 91 |

[1] Steady approach speed through 50-foot height TIAS denoted by symbol $V_{50}$.
Ref. Fig. 3(a)(1).

(2) Curtiss model C-46 certificated for maximum weight of 48,000 pounds.[1] (0.60 factor)

DISTANCE IN FEET

| Standard altitude in feet | Airplane weight in pounds and approach speeds[2] in knots | | | | | | | |
|---|---|---|---|---|---|---|---|---|
| | 42,000 | $V_{50}$ | 44,000 | $V_{50}$ | 46,000 | $V_{50}$ | 43,000 | $V_{50}$ |
| S.L. | 3,370 | 80 | 3,490 | 82 | 3,620 | 84 | 3,740 | 86 |
| 1,000 | 3,460 | 80 | 3,580 | 82 | 3,710 | 84 | 3,830 | 86 |
| 2,000 | 3,540 | 80 | 3,670 | 82 | 3,800 | 84 | 3,920 | 86 |
| 3,000 | 3,630 | 80 | 3,760 | 82 | 3,890 | 84 | 4,020 | 86 |
| 4,000 | 3,720 | 80 | 3,850 | 82 | 3,980 | 84 | 4,110 | 86 |
| 5,000 | 3,800 | 80 | 3,940 | 82 | 4,080 | 84 | 4,220 | 86 |
| 6,000 | 3,890 | 80 | 4,040 | 82 | 4,180 | 84 | 4,320 | 86 |
| 7,000 | 3,980 | 80 | 4,140 | 82 | 4,280 | 84 | 4,440 | 86 |
| 8,000 | 4,080 | 80 | 4,240 | 82 | 4,390 | 84 | 4,550 | 86 |

[1] For use with Curtiss model C-46 airplanes when approved for this weight.
[2] Steady approach speed through 50 height knots TIAS denoted by the symbol $V_{50}$[3].
Ref. Fig. 3(a)(2).

(b) Alternate Airports.

*Effective length* of runway required when effective length is determined in accordance with §121.171 with zero wind and zero gradient.

(1) Curtiss model C-46 certificated for maximum weight of 45,000 pounds. (0.70 factor.)

DISTANCE IN FEET

| Standard altitude in feet | Airplane weight in pounds and approach speeds[1] in knots | | | | | | | |
|---|---|---|---|---|---|---|---|---|
| | 40,000 | $V_{50}$ | 42,000 | $V_{50}$ | 44,000 | $V_{50}$ | 45,000 | $V_{50}$ |
| S.L. | 3,700 | 86 | 3,860 | 88 | 4,030 | 90 | 4,110 | 91 |
| 1,000 | 3,800 | 86 | 3,960 | 88 | 4,140 | 90 | 4,220 | 91 |
| 2,000 | 3,900 | 86 | 4,070 | 88 | 4,250 | 90 | 4,340 | 91 |
| 3,000 | 4,000 | 86 | 4,180 | 88 | 4,360 | 90 | 4,450 | 91 |
| 4,000 | 4,110 | 86 | 4,290 | 88 | 4,470 | 90 | 4,560 | 91 |
| 5,000 | 4,210 | 86 | 4,400 | 88 | 4,590 | 90 | 4,680 | 91 |
| 6,000 | 4,330 | 86 | 4,510 | 88 | 4,710 | 90 | 4,800 | 91 |
| 7,000 | 4,430 | 86 | 4,630 | 88 | 4,840 | 90 | 4,930 | 91 |
| 8,000 | 4,550 | 86 | 4,750 | 88 | 4,970 | 90 | 5,060 | 91 |

[1] Steady approach speed through 50 foot-height-knots TIAS denoted by symbol $V_{50}$.

Ref. Fig. 3(b)(1).

(2) Curtiss model C-46 certificated for maximum weight of 48,000 pounds.[1] (0.70 factor.)

DISTANCE IN FEET

| Standard altitude in feet | Airplane weight in pounds and approach speeds[2] in knots | | | | | | | |
|---|---|---|---|---|---|---|---|---|
| | 42,000 | $V_{50}$ | 44,000 | $V_{50}$ | 46,000 | $V_{50}$ | 48,000 | $V_{50}$ |
| S.L. | 2,890 | 80 | 3,000 | 82 | 3,110 | 84 | 3,220 | 86 |
| 1,000 | 2,960 | 80 | 3,070 | 82 | 3,180 | 84 | 3,280 | 86 |
| 2,000 | 3,040 | 80 | 3,150 | 82 | 3,260 | 84 | 3,360 | 86 |
| 3,000 | 3,110 | 80 | 3,220 | 82 | 3,340 | 84 | 3,440 | 86 |
| 4,000 | 3,180 | 80 | 3,300 | 82 | 3,410 | 84 | 3,520 | 86 |
| 5,000 | 3,260 | 80 | 3,380 | 82 | 3,500 | 84 | 3,610 | 86 |
| 6,000 | 3,330 | 80 | 3,460 | 82 | 3,580 | 84 | 3,700 | 86 |
| 7,000 | 3,420 | 80 | 3,540 | 82 | 3,670 | 84 | 3,800 | 86 |
| 8,000 | 3,500 | 80 | 3,630 | 82 | 3,760 | 84 | 3,900 | 86 |

[1] For use with Curtiss model C-46 airplanes when approved for this weight.

[2] Steady approach speed through 50 foot-height-knots TIAS denoted by the symbol $V_{50}$.

Ref. Fig. 3(b)(2).

(c) Actual length of runway required when effective length, considering obstacles, is not determined in accordance with §121.171.

(1) Curtiss model C-46 certificated for maximum weight of 45,000 pounds. (0.55 factor.)

DISTANCE IN FEET

| Standard altitude in feet | Airplane weight in pounds and approach speeds[1] in knots | | | | | | | |
|---|---|---|---|---|---|---|---|---|
| | 40,000 | $V_{50}$ | 42,000 | $V_{50}$ | 44,000 | $V_{50}$ | 45,000 | $V_{50}$ |
| S.L. | 4,710 | 86 | 4,910 | 88 | 5,130 | 90 | 5,230 | 91 |
| 1,000 | 4,840 | 86 | 5,050 | 88 | 5,270 | 90 | 5,370 | 91 |
| 2,000 | 4,960 | 86 | 5,180 | 88 | 5,410 | 90 | 5,510 | 91 |
| 3,000 | 5,090 | 86 | 5,320 | 88 | 5,550 | 90 | 5,660 | 91 |
| 4,000 | 5,230 | 86 | 5,460 | 88 | 5,700 | 90 | 5,810 | 91 |
| 5,000 | 5,360 | 86 | 5,600 | 88 | 5,850 | 90 | 5,960 | 91 |
| 6,000 | 5,500 | 86 | 5,740 | 88 | 6,000 | 90 | 6,110 | 91 |
| 7,000 | 5,640 | 86 | 5,900 | 88 | 6,170 | 90 | 6,280 | 91 |
| 8,000 | 5,790 | 86 | 6,050 | 88 | 6,340 | 90 | 6,450 | 91 |

[1] Steady approach speed through 50 foot-height-knots TIAS denoted by symbol $V_{50}$.
Ref. Fig. 3(c)(1).

(2) Curtiss C-46 certificated for maximum weight of 48,000 pounds.[1] (0.55 factor.)

DISTANCE IN FEET

| Standard altitude in feet | Airplane weight in pounds and approach speeds[2] in knots | | | | | | | |
|---|---|---|---|---|---|---|---|---|
| | 42,000 | $V_{50}$ | 44,000 | $V_{50}$ | 46,000 | $V_{50}$ | 48,000 | $V_{50}$ |
| S.L. | 3,680 | 80 | 3,820 | 82 | 3,960 | 84 | 4,090 | 86 |
| 1,000 | 3,770 | 80 | 3,910 | 82 | 4,050 | 84 | 4,180 | 86 |
| 2,000 | 3,860 | 80 | 4,000 | 82 | 4,140 | 84 | 4,280 | 86 |
| 3,000 | 3,960 | 80 | 4,090 | 82 | 4,240 | 84 | 4,380 | 86 |
| 4,000 | 4,050 | 80 | 4,190 | 82 | 4,340 | 84 | 4,490 | 86 |
| 5,000 | 4,150 | 80 | 4,290 | 82 | 4,450 | 84 | 4,600 | 86 |
| 6,000 | 4,240 | 80 | 4,400 | 82 | 4,560 | 84 | 4,710 | 86 |
| 7,000 | 4,350 | 80 | 4,510 | 82 | 4,670 | 84 | 4,840 | 86 |
| 8,000 | 4,450 | 80 | 4,620 | 82 | 4,790 | 84 | 4,960 | 86 |

[1] For use with Curtiss model C-46 airplanes when approved for this weight.
[2] Steady approach speed through 50 foot-height-knots TIAS denoted by the symbol $V_{50}$.
Ref. Fig. 3(c)(2).

CURTISS C-46 MODELS
CERTIFICATED FOR MAX. WEIGHT OF 45,000 LBS.

TAKEOFF LIMITATION.
ZERO WIND AND ZERO GRADIENT.

BASED ON EFFECTIVE TAKEOFF LENGTH. (1.00 FACTOR)

FAR 121.199

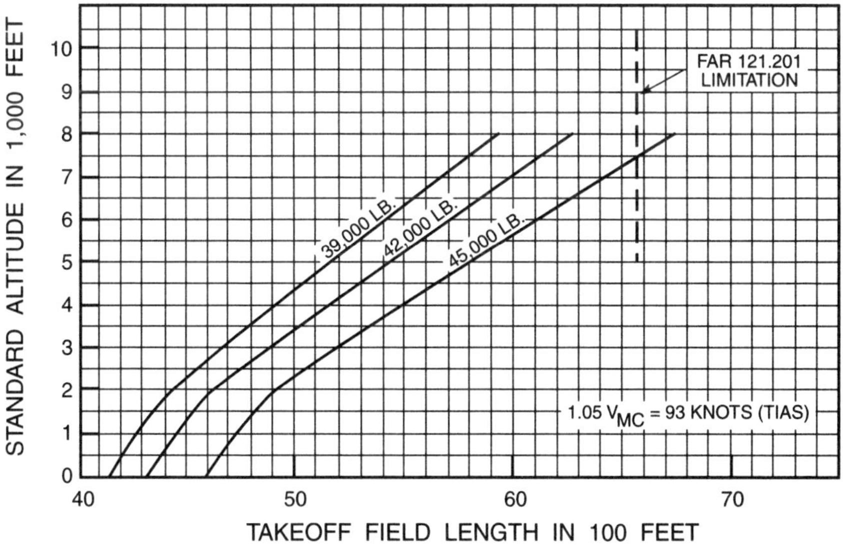


**Reference Table 1(a)(1)** **Fig. 1(a)(1)**

CURTISS C-46 MODELS
CERTIFICATED FOR MAX. WEIGHT OF 45,000 LBS.

TAKEOFF LIMITATION
ZERO WIND AND ZERO GRADIENT

BASED ON ACTUAL TAKEOFF LENGTH
WHEN EFFECTIVE LENGTH IS NOT
DETERMINED. (0.85 FACTOR)

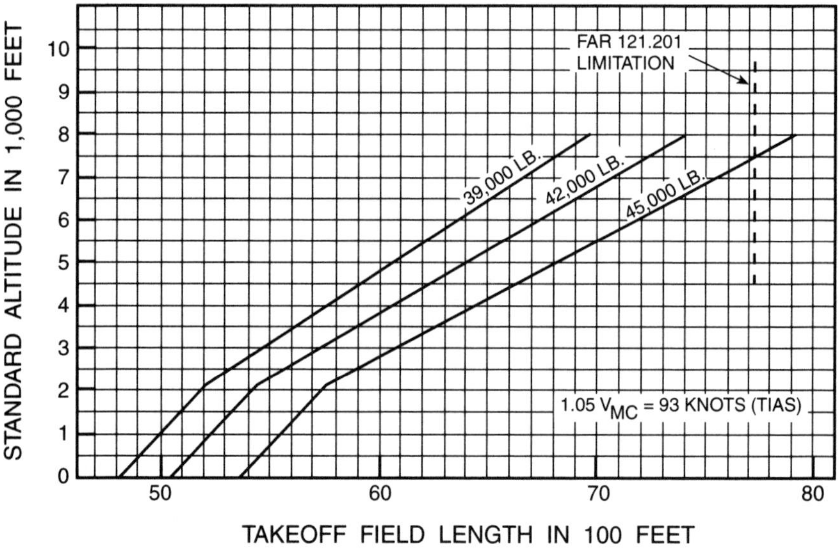


Reference Table 1(a)(2) Fig. 1(a)(2)

CURTISS C-46 MODELS
CERTIFICATED FOR MAX. WEIGHT OF 48,000 LBS.

TAKEOFF LIMITATION
ZERO WIND AND ZERO GRADIENT

BASED ON EFFECTIVE TAKEOFF
LENGTH. (1.00 FACTOR)

FAR 121.199

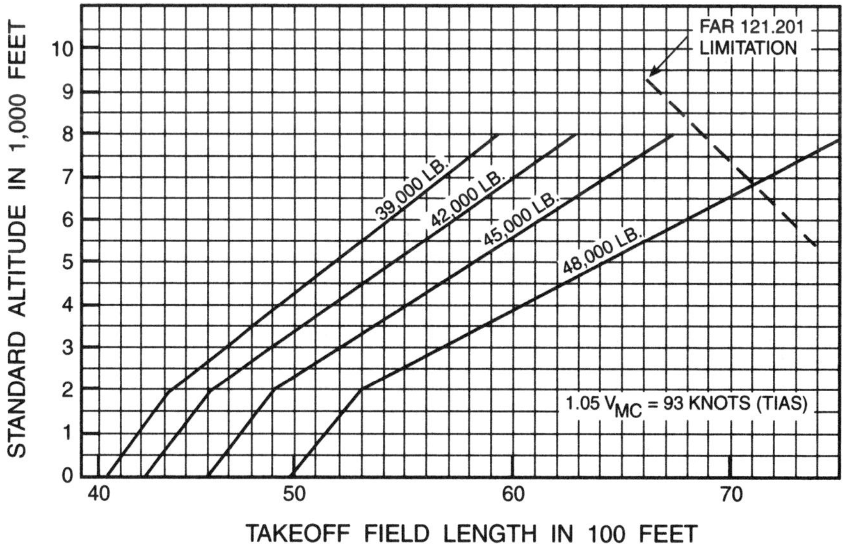


Reference Table 1(b)(1)

Fig. 1(b)(1)

CURTISS C-46 MODELS
CERTIFICATED FOR MAX. WEIGHT OF 48,000 LBS.

TAKEOFF LIMITATION
ZERO WIND AND ZERO GRADIENT

BASED ON ACTUAL TAKEOFF LENGTH
WHEN EFFECTIVE LENGTH IS NOT
DETERMINED. (0.85 FACTOR)

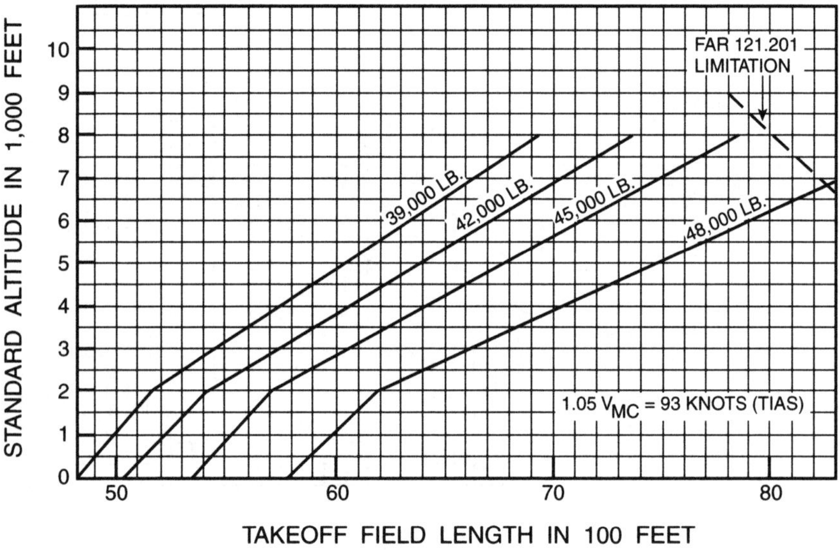


Reference Table 1(b)(2)

Fig. 1(b)(2)

RUNWAY GRADIENT CORRECTION
FOR ACCELERATE-STOP DISTANCE

FOR C-46 AIRPLANES UNDER FAR 121.199

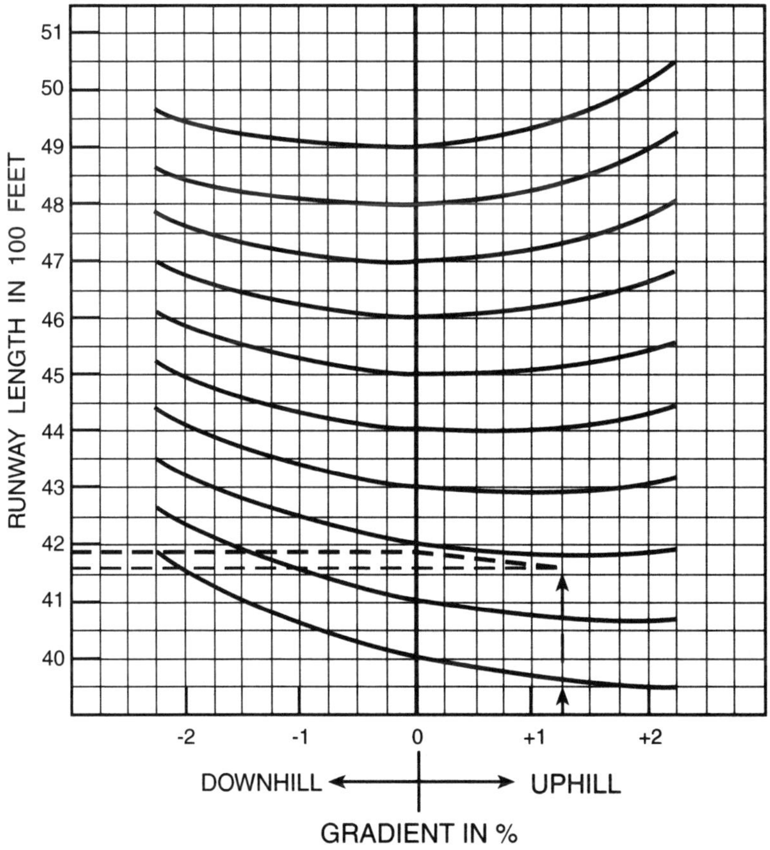


Fig. 1(c)

## CURTISS C-46 MODELS
## ENROUTE LIMITATIONS—ONE ENGINE INOPERATIVE

FAR 121.201

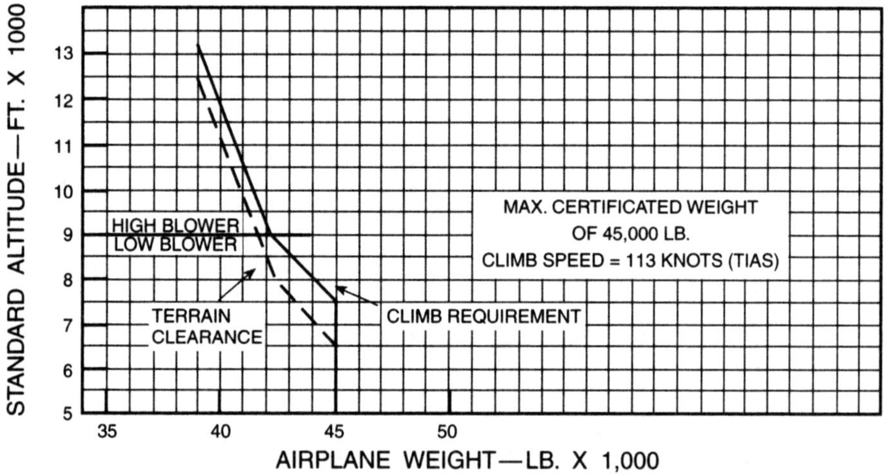


**Reference Table 2(a)** **Fig. 2(a)**

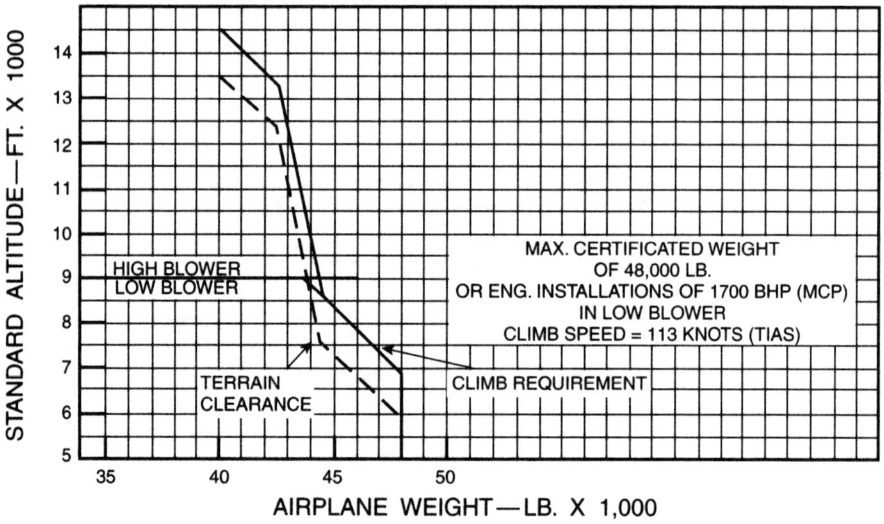


**Reference Table 2(b)** **Fig. 2(b)**

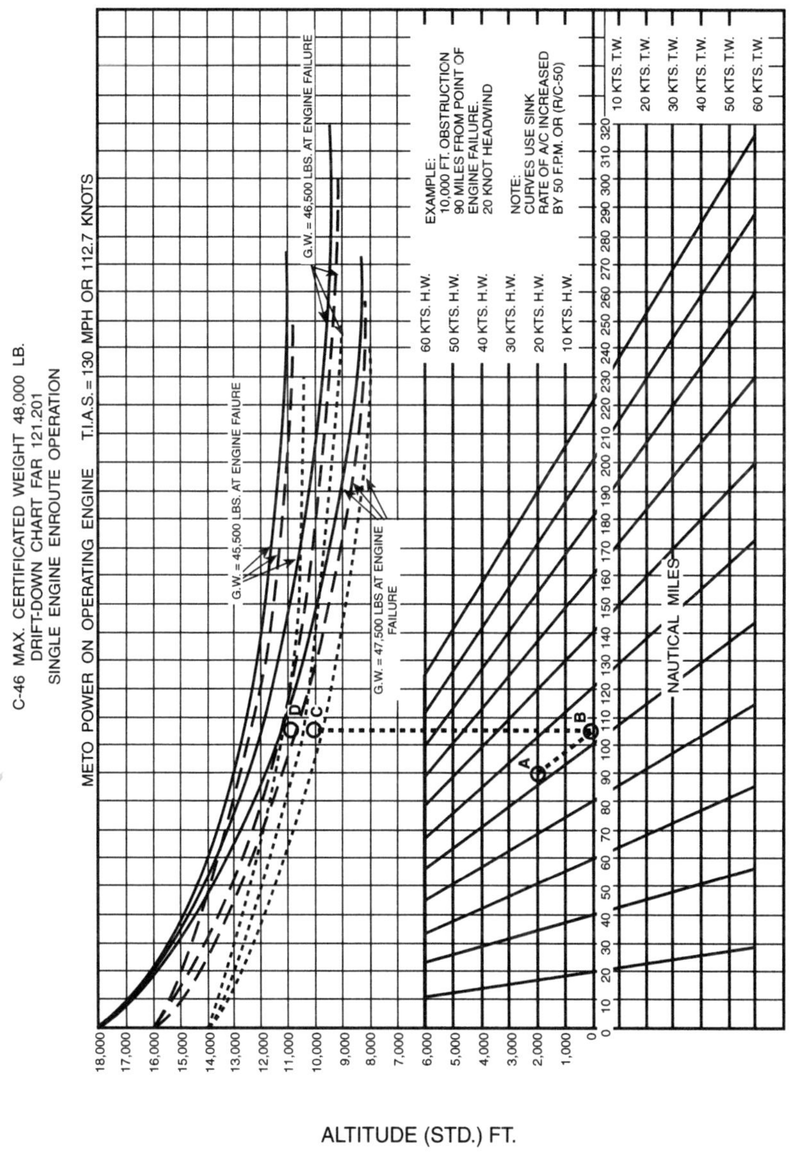
C-46 MAX. CERTIFICATED WEIGHT 48,000 LB.
DRIFT-DOWN CHART FAR 121.201
SINGLE ENGINE ENROUTE OPERATION
METO POWER ON OPERATING ENGINE T.I.A.S. = 130 MPH OR 112.7 KNOTS
G.W. = 46,500 LBS. AT ENGINE FAILURE
G.W. = 45,500 LBS. AT ENGINE FAILURE
G.W. = 47,500 LBS AT ENGINE FAILURE
EXAMPLE:
10,000 FT. OBSTRUCTION
90 MILES FROM POINT OF
ENGINE FAILURE.
20 KNOT HEADWIND
NOTE:
CURVES USE SINK
RATE OF A/C INCREASED
BY 50 F.P.M. OR (R/C-50)
60 KTS. H.W.
50 KTS. H.W.
40 KTS. H.W.
30 KTS. H.W.
20 KTS. H.W.
10 KTS. H.W.
10 KTS. T.W.
20 KTS. T.W.
30 KTS. T.W.
40 KTS. T.W.
50 KTS. T.W.
60 KTS. T.W.
NAUTICAL MILES
A
B
C
D
ALTITUDE (STD.) FT.

C-46 MAX. CERTIFICATED WEIGHT 48,000 LBS.
ENROUTE CLIMB SUMMARY

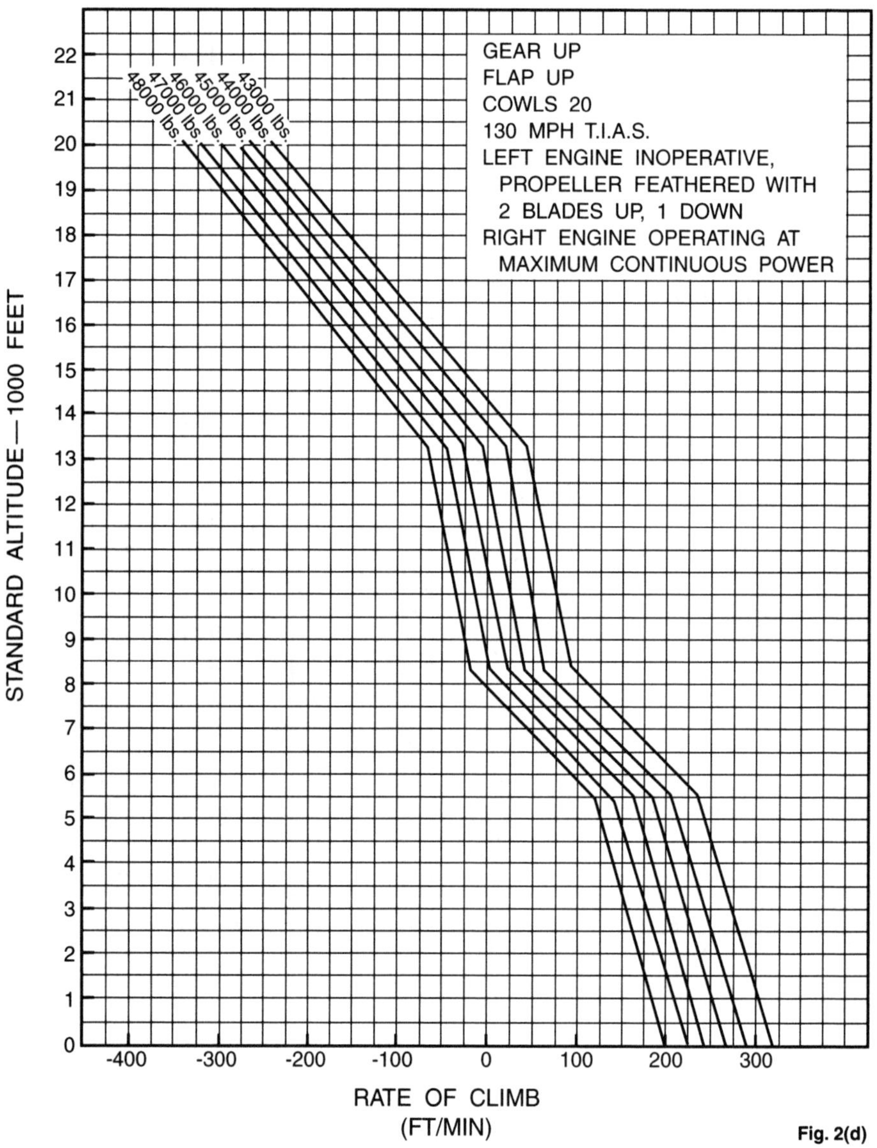


Fig. 2(d)

CURTISS C-46 MODELS
CERTIFICATED FOR MAX. WEIGHT OF 45,000 LBS.

LANDING LIMITATIONS.
ZERO WIND AND ZERO GRADIENT.

BASED ON EFFECTIVE LANDING LENGTH
AT INTENDED DESTINATION. (0.60 FACTOR)

FAR 121.203

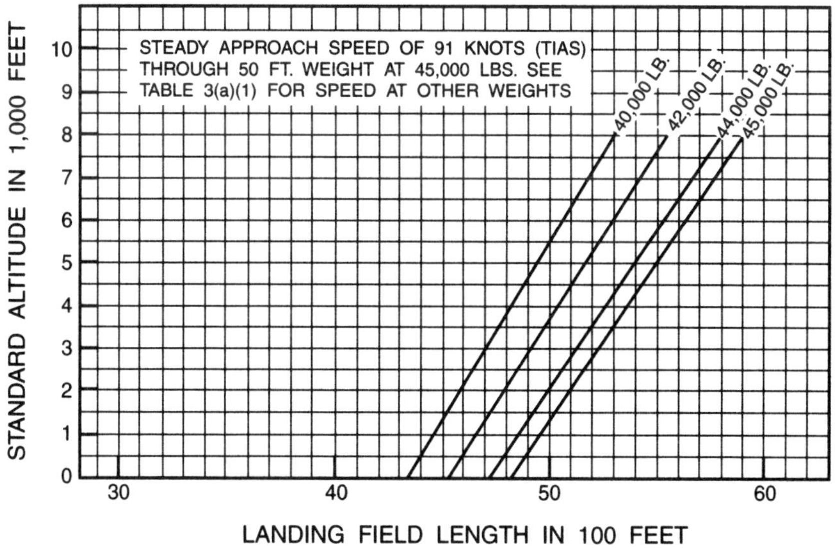


Fig. 3(a)(1)

121

CURTISS C-46 MODELS
CERTIFICATED FOR MAX. WEIGHT OF 48,000 LBS.

LANDING LIMITATIONS.
ZERO WIND AND ZERO GRADIENT.

BASED ON EFFECTIVE LANDING LENGTH
AT INTENDED DESTINATION. (0.60 FACTOR)

FAR 121.203

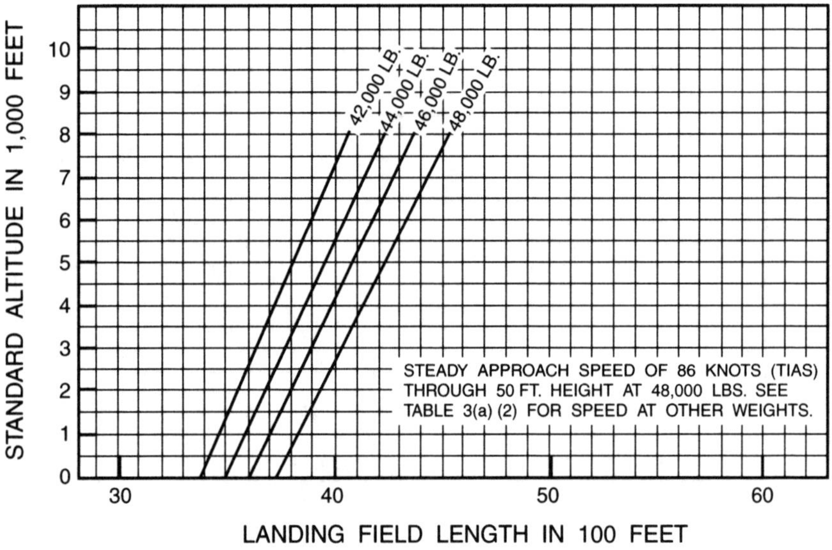


**Fig. 3(a)(2)**

CURTISS C-46 MODELS
CERTIFICATED FOR MAX. WEIGHT OF 45,000 LBS.

LANDING LIMITATIONS,
ZERO WIND AND ZERO GRADIENT.

BASED ON EFFECTIVE LANDING LENGTH
AT ALTERNATE AIRPORTS. (0.70 FACTOR).

FAR 121.205

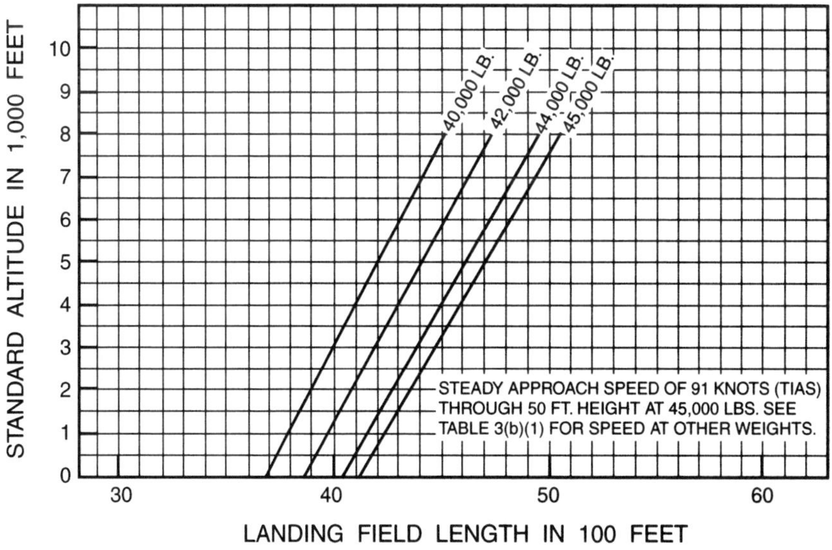


Fig. 3(b)(1)

CURTISS C-46 MODELS
CERTIFICATED FOR MAX. WEIGHT OF 48,000 LBS.

LANDING LIMITATIONS.
ZERO WIND AND ZERO GRADIENT.

BASED ON EFFECTIVE LANDING LENGTH
AT ALTERNATE AIRPORTS. (0.70 FACTOR).

FAR 121.205

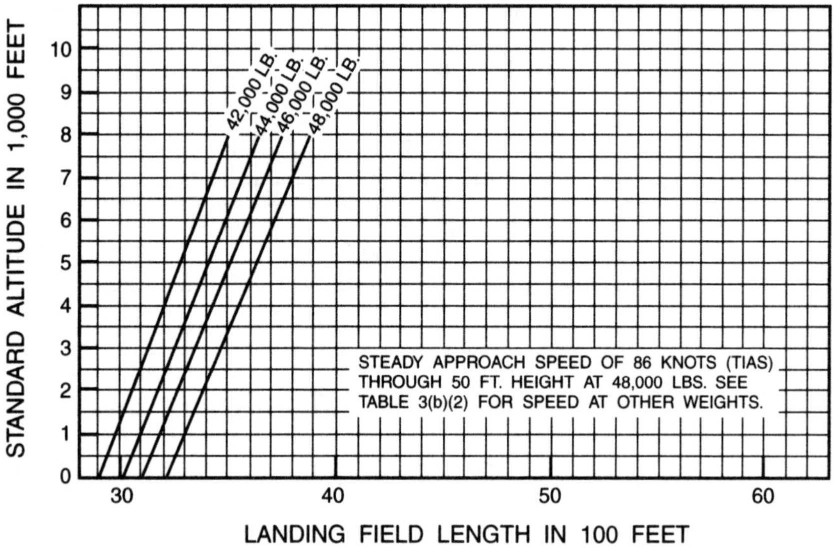


Fig. 3(b)(2)

CURTISS C-46 MODELS
CERTIFICATED FOR MAX. WEIGHT OF 45,000 LBS.

LANDING LIMITATIONS.
ZERO WIND AND ZERO GRADIENT.

BASED ON ACTUAL LANDING LENGTH
WHEN EFFECTIVE LENGTH IS NOT
DETERMINED. (0.55 FACTOR).

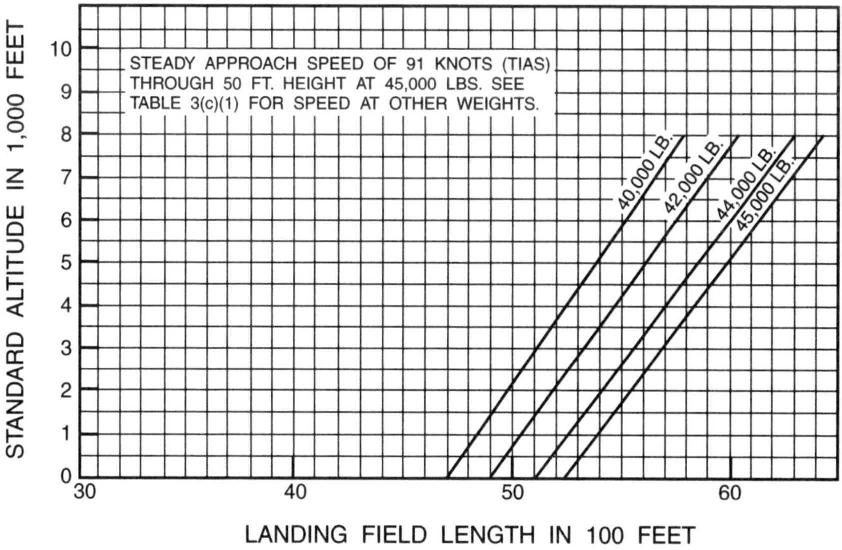


Fig. 3(c)(1)

121

CURTISS C-46 MODELS
CERTIFICATED FOR MAX. WEIGHT OF 48,000 LBS.

LANDING LIMITATIONS.
ZERO WIND AND ZERO GRADIENT.

BASED ON ACTUAL LANDING LENGTH
WHEN EFFECTIVE LENGTH IS NOT
DETERMINED. (0.55 FACTOR).

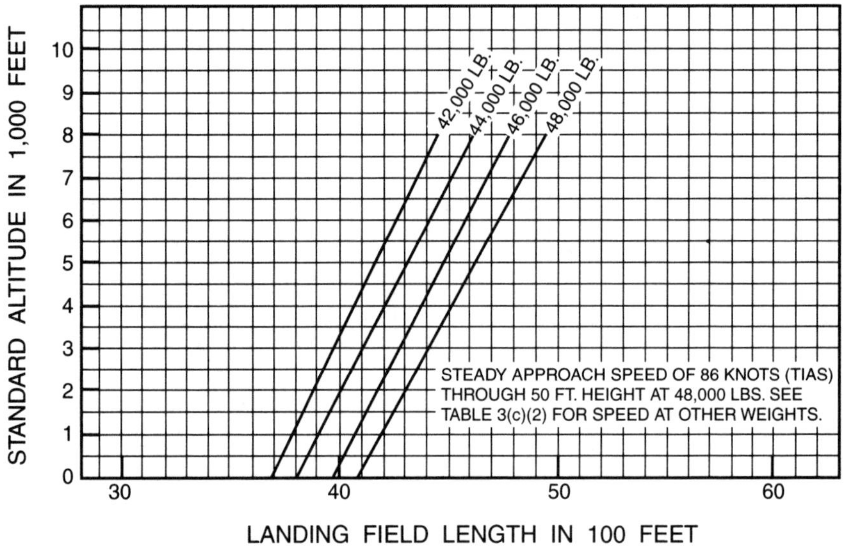


**Fig. 3(c)(2)**

[Docket No. 4080, 30 FR 258, Jan. 3, 1965; 30 FR 481, Jan. 14, 1965; as amended by Amdt. 121–207, 54 FR 39293, Sept. 25, 1989]

## APPENDIX D TO PART 121

### CRITERIA FOR DEMONSTRATION OF EMERGENCY EVACUATION PROCEDURES UNDER § 121.291

**(a)** *Aborted takeoff demonstration.*

(1) The demonstration must be conducted either during the dark of the night or during daylight with the dark of the night simulated. If the demonstration is conducted indoors during daylight hours, it must be conducted with each window covered and each door closed to minimize the daylight effect. Illumination on the floor or ground may be used, but it must be kept low and shielded against shining into the airplane's windows or doors.

(2) The airplane must be a normal ground attitude with landing gear extended.

(3) Unless the airplane is equipped with an off-wing descent means, stands or ramps may be used for descent from the wing to the ground. Safety equipment such as mats or inverted life rafts may be placed on the floor or ground to protect participants. No other equipment that is not part of the emergency evacuation equipment of the airplane may be used to aid the participants in reaching the ground.

(4) The airplane's normal electrical power sources must be deenergized.

(5) All emergency equipment for the type of passenger-carrying operation involved must be installed in accordance with the certificate holder's manual.

(6) Each external door and exit, and each internal door or curtain must be in position to simulate a normal takeoff.

(7) A representative passenger load of persons in normal health must be used. At least 40 percent of the passenger load must be females. At least 35 percent of the passenger load must be over 50 years of age. At least 15 percent of the passenger load must be female and over 50 years of age. Three life-size dolls, not included as part of the total passenger load, must be carried by passengers to simulate live infants 2 years old or younger. Crewmembers, mechanics, and training personnel, who maintain or operate the airplane in the normal course of their duties, may not be used as passengers.

(8) No passenger may be assigned a specific seat except as the Administrator may require. Except as required by item (12) of this paragraph, no employee of the certificate holder may be seated next to an emergency exit.

(9) Seat belts and shoulder harnesses (as required) must be fastened.

(10) Before the start of the demonstration, approximately one-half of the total average amount of carry-on baggage, blankets, pillows, and other similar articles must be distributed at several locations in the aisles and emergency exit access ways to create minor obstructions.

(11) The seating density and arrangement of the airplane must be representative of the highest capacity passenger version of that airplane the certificate holder operates or proposes to operate.

(12) Each crewmember must be a member of a regularly scheduled line crew, except that flight crewmembers need not be members of a regularly scheduled line crew, provided they have knowledge of the airplane. Each crewmember must be seated in the seat the crewmember is normally assigned for takeoff, and must remain in that seat until the signal for commencement of the demonstration is received.

(13) No crewmember or passenger may be given prior knowledge of the emergency exits available for the demonstration.

(14) The certificate holder may not practice, rehearse, or describe the demonstration for the participants nor may any participant have taken part in this type of demonstration within the preceding 6 months.

(15) The pretakeoff passenger briefing required by §121.571 may be given in accordance with the certificate holder's manual. The passengers may also be warned to follow directions of crewmembers, but may not be instructed on the procedures to be followed in the demonstration.

(16) If safety equipment as allowed by item (3) of this section is provided, either all passenger and cockpit windows must be blacked out or all of the emergency exits must have safety equipment in order to prevent disclosure of the available emergency exits.

(17) Not more than 50 percent of the emergency exits in the sides of the fuselage of an airplane that meet all of the requirements applicable to the required emergency exits for that airplane may be used for the demonstration. Exits that are not to be used in the demonstration must have the exit handle deactivated or must be indicated by red lights, red tape, or other acceptable means, placed outside the exits to indicate fire or other reason that they are unusable. The exits to be used must be representative of all of the emergency exits on the airplane and must be designated by the certificate holder, subject to approval by the Administrator. At least one floor level exit must be used.

(18) Except as provided in paragraph (a)(3) of this appendix, all evacuees must leave the airplane by a means provided a part of the airplane's equipment.

(19) The certificate holder's approved procedures and all of the emergency equipment that is normally available, including slides, ropes, lights, and megaphones, must be fully utilized during the

demonstration, except that the flightcrew must take no active role in assisting others inside the cabin during the demonstration.

(20) The evacuation time period is completed when the last occupant has evacuated the airplane and is on the ground. Evacuees using stands or ramps allowed by item (3) above are considered to be on the ground when they are on the stand or ramp: *Provided,* That the acceptance rate of the stand or ramp is no greater than the acceptance rate of the means available on the airplane for descent from the wing during an actual crash situation.

**(b)** *Ditching demonstration.* The demonstration must assume that daylight hours exist outside the airplane, and that all required crewmembers are available for the demonstration.

(1) If the certificate holder's manual requires the use of passengers to assist in the launching of liferafts, the needed passengers must be aboard the airplane and participate in the demonstration according to the manual.

(2) A stand must be placed at each emergency exit and wing, with the top of the platform at a height simulating the water level of the airplane following a ditching.

(3) After the ditching signal has been received, each evacuee must don a life vest according to the certificate holder's manual.

(4) Each liferaft must be launched and inflated, according to the certificate holder's manual, and all other required emergency equipment must be placed in rafts.

(5) Each evacuee must enter a liferaft, and the crewmembers assigned to each liferaft must indicate the location of emergency equipment aboard the raft and describe its use.

(6) Either the airplane, a mockup of the airplane or a floating device simulating a passenger compartment must be used.

(i) If a mockup of the airplane is used, it must be a life-size mockup of the interior and representative of the airplane currently used by or proposed to be used by the certificate holder, and must contain adequate seats for use of the evacuees. Operation of the emergency exits and the doors must closely simulate those on the airplane. Sufficient wing area must be installed outside the over-the-wing exits to demonstrate the evacuation.

(ii) If a floating device simulating a passenger compartment is used, it must be representative, to the extent possible, of the passenger compartment of the airplane used in operations. Operation of the emergency exits and the doors must closely simulate operation on that airplane. Sufficient wing area must be installed outside the over-the-wing exits to demonstrate the evacuation. The device must be equipped with the same survival equipment as is installed on the airplane, to accommodate all persons participating in the demonstration.

[Docket No. 2033, 30 FR 3206, March 9, 1965; as amended by Amdt. 121–30, 32 FR 13268, Sept. 20, 1967; Amdt. 121–41, 33 FR 9067, June 20, 1968; Amdt. 121–46, 34 FR 5545, March 22, 1969; Amdt. 121–47, 34 FR 11489, July 11, 1969; Amdt. 121–233, 58 FR 45230, Aug. 26, 1993]

## Appendix E to Part 121

### Flight Training Requirements

The maneuvers and procedures required by §121.424 of this part for pilot initial, transition, and upgrade flight training are set forth in the certificate holder's approved low-altitude windshear flight training program and in this appendix and must be performed inflight except that windshear maneuvers and procedures must be performed in an airplane simulator in which the maneuvers and procedures are specifically authorized to be accomplished and except to the extent that certain other maneuvers and procedures may be performed in an airplane simulator with a visual system (visual simulator), an airplane simulator without a visual system (nonvisual simulator), a training device, or a static airplane as indicated by the appropriate symbol in the respective column opposite the maneuver or procedure.

Whenever a maneuver or procedure is authorized to be performed in a nonvisual simulator, it may be performed in a visual simulator; when authorized in a training device, it may be performed in a visual or nonvisual simulator, and in some cases, a static airplane. Whenever the requirement may be performed in either a training device or a static airplane, the appropriate symbols are entered in the respective columns.

For the purpose of this appendix, the following symbols mean—

P = Pilot in Command (PIC).

S = Second in Command (SIC).

B = PIC and SIC.

F = Flight Engineer.

PJ = PIC transition Jet to Jet.

PP = PIC transition Prop. to Prop.

SJ = SIC transition Jet to Jet.

SP = SIC transition Prop. to Prop.

AT = All transition categories (PJ, PP, SJ, SP).

PS = SIC upgrading to PIC (same airplane).

SF = Flight Engineer upgrading to SIC (same airplane).

BU = Both SIC and Flight Engineer upgrading (same airplane).

## Flight Training Requirements

| Maneuvers / Procedures | Initial training | | | | | Transition training | | | | | Upgrade training | | | | |
|---|---|---|---|---|---|---|---|---|---|---|---|---|---|---|---|
| | A/P | | Simulator | | | A/P | | Simulator | | | A/P | | Simulator | | |
| | Inflight | Static | Visual simulator | Nonvisual simulator | Training device | Inflight | Static | Visual simulator | Nonvisual simulator | Training device | Inflight | Static | Visual simulator | Nonvisual simulator | Training device |
| As appropriate to the airplane and the operation involved, flight training for pilots must include the following maneuvers and procedures. | | | | | | | | | | | | | | | |
| **I. Preflight** | | | | | | | | | | | | | | | |
| **(a)** Visual inspection of the exterior and interior of the airplane, the location of each item to be inspected, and the purpose for inspecting it. If a flight engineer is a required crewmember for the particular type of airplane, the visual inspection may be replaced by using an approved pictorial means that realistically portrays the location and detain of preflight inspection items. | | B | | | | | AT | | | | | BU | | | |
| **(b)** Use of the prestart check list, appropriate control system checks, starting procedures, radio and electronic equipment checks, and the selection of proper navigation and communications radio facilities and frequencies prior to flight. | | | | B | | | | | AT | | | | | BU | |
| **(c)** Taxiing, sailing, and docking procedures in compliance with instructions issued by the appropriate Traffic Control Authority or by the person conducting the training. | B | | | | | AT | | | | | BU | | | | |
| **(d)** Pretakeoff checks that include powerplant checks. | | | | B | | | | | AT | | | | | BU | |
| **II. Takeoffs:** | | | | | | | | | | | | | | | |
| **(a)** Normal takeoffs which, for the purpose of this maneuver, begin when the airplane is taxied into position on the runway to be used. | B | | | | | AT | | | | | BU | | | | |
| **(b)** Takeoffs with instrument conditions simulated at or before reaching an altitude of 100' above the airport elevation. | | | B | | | | | AT | | | | | BU | | |
| **(c)** Crosswind takeoffs | B | | | | | AT | | | | | BU | | | | |
| **(d)** Takeoffs with a simulated failure of the most critical powerplant —<br>(1) At a point after $V_1$ and before $V_2$ that in the judgment of the person conducting the training is appropriate to the airplane type under the prevailing conditions; or<br>(2) At a point as close as possible after $V_1$ when $V_1$ and $V_2$ or $V_1$ and $V_R$ are identical; or<br>(3) At the appropriate speed for nontransport category airplanes. | | | B | | | | | AT | | | | | BU | | |
| For transition training in an airplane group with engines mounted in similar positions, or from wing-mounted engines to aft fuselage-mounted engines, the maneuver may be performed in a nonvisual simulator. | | | | | | | | | | | | | | | |

121

## Flight Training Requirements (Continued)

| Maneuvers / Procedures | Initial training | | | | | Transition training | | | | | Upgrade training | | | | |
|---|---|---|---|---|---|---|---|---|---|---|---|---|---|---|---|
| | A/P | | Simulator | | | A/P | | Simulator | | | A/P | | Simulator | | |
| | Inflight | Static | Visual simulator | Nonvisual simulator | Training device | Inflight | Static | Visual simulator | Nonvisual simulator | Training device | Inflight | Static | Visual simulator | Nonvisual simulator | Training device |
| **(e)** Rejected takeoffs accomplished during a normal takeoff run after reaching a reasonable speed determined by giving due consideration to aircraft characteristics, runway length, surface conditions, wind directions and velocity, brake heat energy, and any other pertinent factors that may adversely affect safety or the airplane. | | | | B | | | | | AT | | | | | BU | |
| Training in at least one of the above takeoffs must be accomplished at night. For transitioning pilots this requirement may be met during the operating experience required under § 121.434 of this part by performing a normal takeoff at night when a check airman serving as pilot-in-command is occupying a pilot station. | | | | | | | | | | | | | | | |
| **III. Flight Maneuvers and Procedures:** | | | | | | | | | | | | | | | |
| **(a)** Turns with and without spoilers | | | | B | | | | | AT | | | | | BU | |
| **(b)** Tuck and Mach buffet | | | | B | | | | | AT | | | | | BU | |
| **(c)** Maximum endurance and maximum range procedures | | | | B | | | | | AT | | | | | BU | |
| **(d)** Operation of systems and controls at the flight engineer station | | | | B | | | | | AT | | | | | PS | |
| **(e)** Runway and jammed stabilizer | | | | B | | | | | AT | | | | | BU | |
| **(f)** Normal and abnormal or alternate operation of the following systems and procedures: | | | | | | | | | | | | | | | |
| (1) Pressurization | | | | | B | | | | | AT | | | | | BU |
| (2) Pneumatic | | | | | B | | | | | AT | | | | | BU |
| (3) Air conditioning | | | | | B | | | | | AT | | | | | BU |
| (4) Fuel and oil | | B | | | B | | AT | | | AT | | BU | | | BU |
| (5) Electrical | | B | | | B | | AT | | | AT | | BU | | | BU |
| (6) Hydraulic | | B | | | B | | AT | | | AT | | BU | | | BU |
| (7) Flight control | | B | | | B | | AT | | | | | BU | | | BU |
| (8) Anti-icing and deicing | | | | B | | | | | AT | | | | | BU | |
| (9) Auto-pilot | | | | B | | | | | AT | | | | | BU | |
| (10) Automatic or other approach aids | B | | | B | | | | | AT | | SF | | | BU | |
| (11) Stall warning devices, stall avoidance devices, and stability augmentation devices | B | | | B | | | | | AT | | SF | | | BU | |
| (12) Airborne radar devices | | | | B | | | | | AT | | | | | BU | |
| (13) Any other systems, devices, or aids available | | | | B | | | | | AT | | | | | BU | |
| (14) Electrical, hydraulic, flight control, and flight instrument system malfunctioning or failure | | B | | | B | | AT | | | AT | | BU | | | BU |

## Flight Training Requirements (Continued)

| Maneuvers / Procedures | Initial training | | | | | Transition training | | | | | Upgrade training | | | | |
|---|---|---|---|---|---|---|---|---|---|---|---|---|---|---|---|
| | A/P | | Simulator | | | A/P | | Simulator | | | A/P | | Simulator | | |
| | Inflight | Static | Visual simulator | Nonvisual simulator | Training device | Inflight | Static | Visual simulator | Nonvisual simulator | Training device | Inflight | Static | Visual simulator | Nonvisual simulator | Training device |
| (15) Landing gear and flap systems failure or malfunction | | B | | | B | | AT | | | AT | | BU | | | BU |
| (16) Failure of navigation or communications equipment | | | | B | | | | | AT | | | | | BU | |
| **(g)** Flight emergency procedures that include at least the following: | | | | | | | | | | | | | | | |
| (1) Powerplant, heater, cargo compartment, cabin, flight deck, wing, and electrical fires | | B | | | B | | AT | | | AT | | BU | | | BU |
| (2) Smoke control | | B | | | B | | AT | | | AT | | BU | | BU | BU |
| (3) Powerplant failures | | | | B | | | | | AT | | | | | | BU |
| (4) Fuel jettisoning | | B | | | B | | B | | | B | | BU | | | BU |
| (5) Any other emergency procedures outlined in the appropriate flight manual | | | | B | | | | | AT | | | | | BU | |
| **(h)** Steep turns in each direction. Each steep turn must involve a bank angle of 45° with a heading change of at least 180° but not more than 360°. | | | | P | | | | | PJ | | | | | PS | |
| **(i)** Approaches to stalls in the takeoff configuration (except where the airplane uses only a zero-flap configuration), in the clean configuration, and in the landing configuration. | | | | B | | | | | AT | | | | | BU | |
| Training in at least one of the above configurations must be accomplished while in a turn with a bank angle between 15° and 30°. | | | | | | | | | | | | | | | |
| **(j)** Recovery from specific flight characteristics that are peculiar to the airplane type. | | | | B | | | | | AT | | | | | BU | |
| **(k)** Instrument procedures that include the following: | | | | | | | | | | | | | | | |
| (1) Area departure and arrival | | | | B | | | | | AT | | | | | BU | |
| (2) Use of navigation systems including adherence to assigned radials | | | | B | | | | | AT | | | | | BU | |
| (3) Holding | | | | B | | | | | AT | | | | | BU | |
| **(l)** ILS instrument approaches that include the following: | | | | | | | | | | | | | | | |
| (1) Normal ILS approaches | B | | | | | AT | | | | | BU | | | | |
| (2) Manually controlled ILS approaches with a simulated failure of one powerplant which occurs before initiating the final approach course and continues to touchdown or through the missed approach procedure. | B | | | | | | | AT | | | | | BU | | |
| **(m)** Instrument approaches and missed approaches other than ILS which include the following: | | | | | | | | | | | | | | | |
| (1) Nonprecision approaches that the trainee is likely to use. | | | | | B | | | | | AT | | | BU | | |
| (2) In addition to subparagraph (1) of this paragraph, at least one other nonprecision approach and missed approach procedure that the trainee is likely to use. | | | B | | | | | AT | | | | | BU | | |

## Flight Training Requirements (Continued)

| Maneuvers / Procedures | Initial training | | | | | Transition training | | | | | Upgrade training | | | | |
|---|---|---|---|---|---|---|---|---|---|---|---|---|---|---|---|
| | A/P | | Simulator | | | A/P | | Simulator | | | A/P | | Simulator | | |
| | Inflight | Static | Visual simulator | Nonvisual simulator | Training device | Inflight | Static | Visual simulator | Nonvisual simulator | Training device | Inflight | Static | Visual simulator | Nonvisual simulator | Training device |
| In connection with paragraph III(k) and III(l), each instrument approach must be performed according to any procedures and limitations approved for the approach facility used. The instrument approach begins when the airplane is over the initial approach fix for the approach procedure being used (or turned over to the final approach controller in the case of GCA approach) and ends when the airplane touches down on runway or when transition to a missed approach configuration is completed. | | | | | | | | | | | | | | | |
| **(n)** Circling approaches which include the following: | B | | | | | AT | | | | | BU | | | | |
| (1) That portion of the circling approach to the authorized minimum altitude for the procedure being used must be made under simulate instrument conditions. | | | | | | | | | | | | | | | |
| (2) The circling approach must be made to the authorized minimum circling approach altitude followed by a change in heading and the necessary maneuvering (by visual reference) to maintain a flight path that permits a normal landing on a runway at least 90° from the final approach course of the simulated instrument portion of the approach. | | | | | | | | | | | | | | | |
| (3) The circling approach must be performed without excessive maneuvering, and without exceeding the normal operating limits of the airplane. The angle of bank should not exceed 30°. | | | | | | | | | | | | | | | |
| Training in the circling approach maneuver is not required for a pilot employed by a certificate holder subject to the operating rules of Part 121 of this chapter if the certificate holder's manual prohibits a circling approach in weather conditions below 1000–3 (ceiling and visibility); for a SIC if the certificate holder's manual prohibits the SIC from performing a circling approach in operations under this part. | | | | | | | | | | | | | | | |
| **(o)** Zero-flap approaches. Training in this maneuver is not required for a particular airplane type if the Administrator had determined that the probability of flap extension failure on that type airplane is extreme remote due to system design. In making this determination, the Administrator determines whether training on slats only and partial flap approaches is necessary. | P | | | | | | | PP, PJ | | | | | PS | | |
| **(p)** Missed approaches which include the following: | | | | | | | | | | | | | | | |
| (1) Missed approaches from ILS approaches | | | B | | | | | AT | | | | | BU | | |
| (2) Other missed approaches | | | | | B | | | | | AT | | | | | BU |
| (3) Missed approaches that include a complete approved missed approach procedure | | | | | B | | | | | AT | | | | | BU |
| (4) Missed approaches that include a powerplant failure | | | B | | | | | AT | | | | | BU | | |

## Flight Training Requirements (Continued)

| Maneuvers / Procedures | Initial training | | | | | Transition training | | | | | Upgrade training | | | | |
|---|---|---|---|---|---|---|---|---|---|---|---|---|---|---|---|
| | A/P | | Simulator | | | A/P | | Simulator | | | A/P | | Simulator | | |
| | Inflight | Static | Visual simulator | Nonvisual simulator | Training device | Inflight | Static | Visual simulator | Nonvisual simulator | Training device | Inflight | Static | Visual simulator | Nonvisual simulator | Training device |
| **IV. Landings and Approaches to Landings:** | | | | | | | | | | | | | | | |
| **(a)** Normal landings | B | | | | | AT | | | | | BU | | | | |
| **(b)** Landing and go around with the horizontal stabilizer out of trim | P | | | | | | | PJ, PP | | | | | | | PS |
| **(c)** Landing in sequence from an ILS instrument approach | B | | | | | AT | | AT | | | | | BU | | |
| **(d)** Cross wind landing | B | | | | | AT | | | | | BU | | | | |
| **(e)** Maneuvering to a landing with simulated powerplant failure, as follows: | | | | | | | | | | | | | | | |
| (1) Except as provided in subparagraph (3) of this paragraph in the case of 3-engine airplanes, maneuvering to a landing with an approved procedure that approximates the loss of two powerplants (center and one outboard engine). | P | | | | | | | PJ, PP | | | | | PS | | |
| (2) Except as provided in subparagraph (3) of this paragraph, in the case of other multiengine airplanes, maneuvering to a landing with a simulated failure of 50 percent of available powerplants with the simulated loss of power on one side of the airplane. | P | | | | | | | PJ, PP | | | | | PS | | |
| (3) Notwithstanding the requirements of subparagraphs (1) and (2) of this paragraph, flight crewmembers who satisfy those requirements in a visual simulator must also: | | | | | | | | | | | | | | | |
| (*i*) Take inflight training in one-engine inoperative landings; and | | | | | | | | | | | | | | | |
| (*ii*) In the case of a second-in-command upgrading to a pilot-in-command and who has not previously performed the maneuvers required by this paragraph in flight, meet the requirements of this paragraph applicable to initial training for pilots-in-command. | | | | | | | | | | | | | | | |
| (4) In the case of flight crewmembers other than the pilot-in-command, perform the maneuver with the simulated loss of power of the most critical powerplant only. | | | | | | | | | | | | | | | |
| **(f)** Landing under simulated circling approach conditions (exceptions under III(n) applicable to this requirement. | B | | | | | | | AT | | | | | BU | | |
| **(g)** Rejected landings that include a normal missed approach procedure after the landing is rejected. For the purpose of this maneuver the landing should be rejected at approximately 50 feet and approximately over the runway threshold. | B | | | | | | | AT | | | | | BU | | |
| **(h)** Zero-flap landings if the Administrator finds that maneuver appropriate for training in the airplane. | P | | | | | | | PP, PJ | | | | | PS | | |

121

## Flight Training Requirements (Continued)

| Maneuvers / Procedures | Initial training | | | | | Transition training | | | | | Upgrade training | | | | |
|---|---|---|---|---|---|---|---|---|---|---|---|---|---|---|---|
| | A/P | | Simulator | | | A/P | | Simulator | | | A/P | | Simulator | | |
| | Inflight | Static | Visual simulator | Nonvisual simulator | Training device | Inflight | Static | Visual simulator | Nonvisual simulator | Training device | Inflight | Static | Visual simulator | Nonvisual simulator | Training device |
| **(i)** Manual reversion (if appropriate) | | | B | | | | | AT | | | | | BU | | |
| Training in landings and approaches to landings must include the types and conditions provided in IV(a) through (i) but more than one type may be combined where appropriate. | | | | | | | | | | | | | | | |
| Training in one of the above landings must be accomplished at night. For transitioning pilots, this requirement may be met during the operating experience required under § 121.434 of this part by performing a normal landing when a check pilot serving as pilot-in-command is occupying a pilot station. | B | | | | | AT | | | | | BU | | | | |

[Docket No. 9509, 35 FR 97, Jan. 3, 1970; as amended by Amdt. 121–91, 37 FR 10730, May 27, 1972; Amdt. 121–108, 38 FR 35446, Dec. 28, 1973; Amdt. 121–159, 45 FR 41595, June 19, 1980; Amdt. 121–199, 53 FR 37697, Sept. 27, 1988]

P = Pilot in Command (PIC)
S = Second in Command (SIC)
B = PIC and SIC
F = Flight Engineer
PJ = PIC transition Jet to Jet
PP = PIC transition Prop. to Prop
SJ = SIC transition Jet to Jet
SP = SIC transition Prop. to Prop
AT = All transition categories (PJ, PP, SJ, SP)
PS = SIC upgrading to PIC (same airplane)
SF = Flight Engineer upgrading to SIC (same airplane)
BU = Both SIC and Flight Engineer upgrading (same airplane)

# Appendix F to Part 121

## Proficiency Check Requirements

The maneuvers and procedures required by §121.441 for pilot proficiency checks are set forth in this appendix and must be performed inflight except to the extent that certain maneuvers and procedures may be performed in an airplane simulator with a visual system (visual simulator), an airplane simulator without a visual system (nonvisual simulator), or a training device as indicated by the appropriate symbol in the respective column opposite the maneuver or procedure.

Whenever a maneuver or procedure is authorized to be performed in a nonvisual simulator, it may also be performed in a visual simulator; when authorized in a training device, it may be performed in a visual or nonvisual simulator. For the purpose of this appendix, the following symbols mean—

P = Pilot in Command.

B = Both Pilot in Command and Second in Command.

* = A symbol and asterisk (B*) indicates that a particular condition is specified in the maneuvers and procedures column.

# = When a maneuver is preceded by this symbol it indicates the maneuver may be required in the airplane at the discretion of the person conducting the check.

Throughout the maneuvers prescribed in this appendix, good judgment commensurate with a high level of safety must be demonstrated. In determining whether such judgment has been shown, the person conducting the check considers adherence to approved procedures, actions based on analysis of situations for which there is no prescribed procedure or recommended practice, and qualities of prudence and care in selecting a course of action.

| Maneuvers / Procedures | Required | | Permitted | | | |
|---|---|---|---|---|---|---|
| | Simulated instrument conditions | Inflight | Visual simulator | Nonvisual simulator | Training device | Waiver provisions of § 121.441(d) |
| The procedures and maneuvers set forth in this appendix must be performed in a manner that satisfactorily demonstrates knowledge and skill with respect to—<br>(1) The airplane, its systems and components;<br>(2) Proper control of airspeed, configuration, direction, altitude, and attitude in accordance with procedures and limitations contained in the approved Airplane Flight Manual, the certificate holder's operations Manual, check lists, or other approved material appropriate to the airplane type; and<br>(3) Compliance with approach, ATC, or other applicable procedures. | | | | | | |
| **I. Preflight:** | | | | | | |
| **(a)** *Equipment examination (oral or written).* As part of the practical test, the equipment examination must be closely coordinated with, and related to, the flight maneuvers portion but may not be given during the flight maneuvers portion. The equipment examination must cover—<br>(1) Subjects requiring a practical knowledge of the airplane, its powerplants, systems, components, operational and performance factors;<br>(2) Normal, abnormal, and emergency procedures, and the operations and limitations relating thereto; and<br>(3) The appropriate provisions of the approved Airplane Flight Manual. | | | | | B | |
| The person conducting the check may accept, as equal to this equipment test, an equipment test given to the pilot in the certificate holder's ground school within the preceding 6 calendar months. | | | | | B | B* |
| **(b)** *Preflight inspection.* The pilot must—<br>(1) Conduct an actual visual inspection of the exterior and interior of the airplane, locating each item and explaining briefly the purpose for inspecting it; and<br>(2) Demonstrate the use of the prestart check list, appropriate control system checks, starting procedures, radio and electronic equipment checks, and the selection of proper navigation and communications radio facilities and frequencies prior to flight. | | | | | | |

| Maneuvers / Procedures | Required: Simulated instrument conditions | Required: Inflight | Permitted: Visual simulator | Permitted: Nonvisual simulator | Permitted: Training device | Permitted: Waiver provisions of § 121.441(d) |
|---|---|---|---|---|---|---|
| Except for flight checks required by § 121.424(d)(2), an approved pictorial means that realistically portrays the location and detail of preflight inspection items and provides for the portrayal of abnormal conditions may be substituted for the preflight inspection. If a flight engineer is a required flight crewmember for the particular type airplane, the visual inspection may be waived under § 121.441(d). | | | | | | |
| **(c)** *Taxiing.* This maneuver includes taxiing (in the case of a second in command proficiency check to the extent practical from the second in command crew position), sailing, or docking procedures in compliance with instructions issued by the appropriate traffic control authority or by the person conducting the checks. | | B | | | | |
| # **(d)** *Powerplant checks.* As appropriate to the airplane type. | | | | B | | |
| **II. Takeoff:** **(a)** *Normal.* One normal takeoff which, for the purpose of this maneuver, begins when the airplane is taxied into position on the runway to be used. | | B* | | | | |
| **(b)** *Instrument.* One takeoff with instrument conditions simulated at or before reaching an altitude of 100' above the airport elevation. | B | | B* | | | |
| **(c)** *Crosswind.* One crosswind takeoff, if practicable, under the existing meteorological, airport, and traffic conditions. | | B* | | | | |
| Requirements (a) and (c) may be combined, and requirements (a), (b), and (c) may be combined if (b) is performed inflight. | | | | | | |
| **(d)** *Powerplant failure.* One takeoff with a simulated failure of the most critical powerplant— | | | B | | | |
| (1) At a point after $V_1$ and before $V_2$ that in the judgment of the person conducting the check is appropriate to the airplane type under the prevailing conditions; | | | | | | |
| (2) At a point as close as possible after $V_1$ when $V_1$ and $V_2$ or $V_1$ and $V_r$ are identical; or | | | | | | |
| (3) At the appropriate speed for non-transport category airplanes. | | | | | | |
| In an airplane group with aft fuselage-mounted engines, this maneuver may be performed in a non-visual simulator. | | | | | | |
| **(e)** *Rejected.* A rejected takeoff may be performed in an airplane during a normal takeoff run after reaching a reasonable speed determined by giving due consideration to aircraft characteristics, runway length, surface conditions, wind direction and velocity, brake heat energy, and any other pertinent factors that may adversely affect safety or the airplane. | | | | B* | | B |
| **III. Instrument procedures:** (a) *Area departure and area arrival.* During each of these maneuvers the applicant must— | B | | | B | | B* |
| (1) Adhere to actual or simulated ATC clearances (including assigned radials); and | | | | | | |
| (2) Properly use available navigation facilities. | | | | | | |
| Either area arrival or area departure, but not both, may be waived under § 121.441(d). | | | | | | |
| **(b)** *Holding.* This maneuver includes entering, maintaining, and leaving holding patterns. It may be performed in connection with either area departure or area arrival. | B | | | B | | B |
| **(c)** *ILS and other instrument approaches.* There must be the following: | | | | | | |
| (1) At least one normal ILS approach. | B | | B | | | |
| (2) At least one manually-controlled ILS approach with a simulated failure of one powerplant. The simulated failure should occur before initiating the final approach course and must continue to touchdown or through the missed approach procedure. | B | | | | | |
| (3) At least one nonprecision approach procedure that is representative of the nonprecision approach procedures that the certificate holder is likely to use. | B | | B | | | |

| Maneuvers / Procedures | Required | | Permitted | | | |
|---|---|---|---|---|---|---|
| | Simulated instrument conditions | Inflight | Visual simulator | Nonvisual simulator | Training device | Waiver provisions of § 121.441(d) |
| (4) Demonstration of at least one nonprecision approach procedure on a letdown aid other than the approach procedure performed under subparagraph (3) of this paragraph that the certificate holder is approved to use. If performed in a training device, the procedures must be observed by a check pilot or an approved instructor. | B | | | | B | |
| Each instrument approach must be performed according to any procedures and limitations approved for the approach facility used. The instrument approach begins when the airplane is over the initial approach fix for the approach procedure being used (or turned over to the final approach controller in the case of GCA approach), and ends when the airplane touches down on the runway or when transition to a missed approach configuration is completed. Instrument conditions need not be simulated below 100' above touchdown zone elevation. | | | | | | |
| **(d)** *Circling approaches.* If the certificate holder is approved for circling minimums below 1000–3, at least one circling approach must be made under the following conditions— | | | B* | | | B* |
| (1) The portion of the approach to the authorized minimum circling approach altitude must be made under simulated instrument conditions. | B | | | | | |
| (2) The approach must be made to the authorized minimum circling approach altitude followed by a change in heading and the necessary maneuvering (by visual reference) to maintain a flight path that permits a normal landing on a runway at least 90° from the final approach course of the simulated instrument portion of the approach. | | | | | | |
| (3) The circling approach must be performed without excessive maneuvering, and without exceeding the normal operating limits of the airplane. The angle of bank should not exceed 30°. | | | | | | |
| If local conditions beyond the control of the pilot prohibit the maneuver or prevent it from being performed as required, it may be waived as provided in § 121.441(d): Provided, however, that the maneuver may not be waived under this provision for two successive proficiency checks. The circling approach maneuver is not required for a second-in-command if the certificate holder's manual prohibits a second-in-command from performing a circling approach in operations under this part. | | | | | | |
| **(e)** *Missed approach.* | | | | | | |
| (1) Each pilot must perform at least one missed approach from an ILS approach. | | | B* | | | |
| (2) Each pilot in command must perform at least one additional missed approach. | | | P* | | | |
| A complete approved missed approach procedure must be accomplished at least once. At the discretion of the person conducting the check, a simulated powerplant failure may be required during any of the missed approaches. These maneuvers may be performed either independently or in conjunction with maneuvers required under Sections III or V of this appendix. At least one missed approach must be performed in flight. | | | | | | |
| **IV. Inflight Maneuvers:** | | | | | | |
| **(a)** *Steep turns.* At least one steep turn in each direction must be performed. Each steep turn must involve a bank angle of 45° with a heading change of at least 180° but not more than 360°. | P | | | P | | P |
| **(b)** *Approaches to stalls.* For the purpose of this maneuver the required approach to a stall is reached when there is a perceptible buffet or other response to the initial stall entry. Except as provided below, there must be at least three approaches to stalls as follows: | B | | | B | | B* |
| (1) One must be in the takeoff configuration (except where the airplane used only a zero-flap takeoff configuration). | | | | | | |
| (2) One in a clean configuration. | | | | | | |
| (3) One in a landing configuration. | | | | | | |

| Maneuvers / Procedures | Required | | Permitted | | | |
|---|---|---|---|---|---|---|
| | Simulated instrument conditions | Inflight | Visual simulator | Nonvisual simulator | Training device | Waiver provisions of § 121.441(d) |
| At the discretion of the person conducting the check, one approach to a stall must be performed in one of the above configurations while in a turn with the bank angle between 15° and 30°. Two out of the three approaches required by this paragraph may be waived. | | | | | | |
| If the certificate holder is authorized to dispatch or flight release the airplane with a stall warning device inoperative, the device may not be used during this maneuver. | | | | | | |
| **(c)** *Specific flight characteristics.* Recovery from specific flight characteristics that are peculiar to the airplane type. | | | | B | | B |
| **(d)** *Powerplant failures.* In addition to specific requirements fro maneuvers with simulated powerplant failures, the person conducting the check may require a simulated powerplant failure at any time during the check. | | | | B | | |
| **V. Landings and Approaches to Landings:** Notwithstanding the authorizations for combining and waiving maneuvers and for the use of a simulator, at least two actual landings (one to a full stop) must be made for all pilot-in-command and initial second-in-command proficiency checks. Landings, and approaches to landings, must include the following, but more than one type may be combined where appropriate: | | | | | | |
| Landings and approaches to landings must include the types listed below, but more than one type may be combined where appropriate: | | | | | | |
| **(a)** *Normal landing.* | | B | | | | |
| **(b)** Landing in sequence from an ILS instrument approach except that if circumstances beyond the control of the pilot prevent an actual landing, the person conducting the check may accept an approach to a point where in his judgment a landing to a full stop could have been made. | | B* | | | | |
| **(c)** Crosswind landing, if practical under existing meteorological, airport, and traffic conditions. | | B* | | | | |
| **(d)** *Maneuvering to a landing with simulated powerplant failure as follows:* | | | | | | |
| (1) In the case of 3-engine airplanes, maneuvering to a landing with an approved procedure that approximates the loss of two powerplants (center and one outboard engine); or | | | B* | | | |
| (2) In the case of other multiengine airplanes, maneuvering to a landing with a simulated failure of 50 percent of available powerplants, with the simulated loss of power on one side of the airplane. | | | B* | | | |
| Notwithstanding the requirements of subparagraphs (d)(1) and (2) of this paragraph, in a proficiency check for other than a pilot-in-command, the simulated loss of power may be only the most critical powerplant. However, if a pilot satisfies the requirements of subparagraphs (d)(1) or (2) of this paragraph in a visual simulator, he also must maneuver in flight to a landing with a simulated failure of the most critical powerplant. In addition, a pilot-in-command may omit the maneuver required by subparagraph (d)(1) or (d)(2) of this paragraph during a required proficiency check or simulator course of training if he satisfactorily performed that maneuver during the preceding proficiency check, or during the preceding approved simulator course of training under the observation of a check airman, whichever was completed later. | | | | | | |
| **(e)** Except as provided in paragraph (f) of this section, if the certificate holder is approved for circling minimums below 1000–3, a landing under simulated circling approach conditions. However, when performed in an airplane, if circumstances beyond the control of the pilot prevent a landing, the person conducting the check may accept an approach to a point where, in his judgment, a landing to a full stop could have been made. | | | B* | | | |
| # **(f)** A rejected landing, including a normal missed approach procedure, that is rejected approximately 50' over the runway and approximately over the runway threshold. This maneuver may be combined with instrument, circling, or missed approach procedures, but instrument conditions need not be simulated below 100 feet above the runway. | | | B | | | |

| Maneuvers / Procedures | Required: Simulated instrument conditions | Required: Inflight | Permitted: Visual simulator | Permitted: Nonvisual simulator | Permitted: Training device | Permitted: Waiver provisions of § 121.441(d) |
|---|---|---|---|---|---|---|
| **VI. Normal and Abnormal Procedures:** Each applicant must demonstrate the proper use of as many of the systems and devices listed below as the person conducting the check finds are necessary to determine that the person being checked has a practical knowledge of the use of the systems and devices appropriate to the airplane type: | | | | | | |
| **(a)** Anti-icing and de-icing systems | | | | B | | |
| **(b)** Auto-pilot systems | | | | B | | |
| **(c)** Automatic or other approach aid systems | | | | B | | |
| **(d)** Stall warning devices, stall avoidance devices, and stability augmentation devices | | | | B | | |
| **(e)** Airborne radar devices | | | | B | | |
| **(f)** Any other systems, devices, or aids available | | | | B | | |
| **(g)** Hydraulic and electrical system failures and malfunctions | | | | | B | |
| **(h)** Landing gear and flap systems failure or malfunction | | | | | B | |
| **(i)** Failure of navigation or communications equipment | | | | B | | |
| **VII. Emergency Procedures:** Each applicant must demonstrate the proper emergency procedures for as many of the emergency situations listed below as the person conducting the check finds are necessary to determine that the person being checked has an adequate knowledge of, and ability to perform, such procedure: | | | | | | |
| **(a)** Fire in flight | | | | B | | |
| **(b)** Smoke control | | | | B | | |
| **(c)** Rapid decompression | | | | B | | |
| **(d)** Emergency descent | | | | B | | |
| **(e)** Any other emergency procedures outlined in the appropriate approved Airplane Flight Manual. | | | | B | | |

[Docket No. 9509, 35 FR 99, Jan. 3, 1970; as amended by Amdt. 121–80, 36 FR 19362, Oct. 5, 1971; Amdt. 121–91, 37 FR 10730, May 27, 1972; Amdt. 121–92, 37 FR 12717, June 28, 1972; Amdt. 121–108, 38 FR 35448, Dec. 28, 1973; Amdt. 121–136, 42 FR 43389, Aug. 29, 1977]

# Appendix G to Part 121

## Doppler Radar and Inertial Navigation System (INS): Request for Evaluation; Equipment and Equipment Installation; Training Program; Equipment Accuracy and Reliability; Evaluation Program

**1.** *Application authority.*

(a) An applicant for authority to use a Doppler Radar or Inertial Navigation System must submit a request for evaluation of the system to the Flight Standards District Office or International Field Office charged with the overall inspection of its operations 30 days prior to the start of evaluation flights.

(b) The application must contain:

(1) A summary of experience with the system showing to the satisfaction of the Administrator a history of the accuracy and reliability of the system proposed to be used.

(2) A training program curriculum for initial approval under §121.405.

(3) A maintenance program for compliance with Subpart L of this part.

(4) A description of equipment installation.

(5) Proposed revisions to the Operations Manual outlining all normal and emergency procedures relative to use of the proposed system, including detailed methods for continuing the navigational function with partial or complete equipment failure, and methods for determining the most accurate system when an unusually large divergence between systems occurs. For the purpose of this appendix, a large divergence is a divergence that results in a track that falls beyond clearance limits.

(6) Any proposed revisions to the minimum equipment list with adequate justification therefor.

(7) A list of operations to be conducted using the system, containing an analysis of each with respect to length, magnetic compass reliability, availability of en route aids, and adequacy of gateway and terminal radio facilities to support the system. For the purpose of this appendix, a gateway is a specific navigational fix where use of long range navigation commences or terminates.

**2.** *Equipment and equipment installation—Inertial Navigation Systems (INS) or Doppler Radar System.*

(a) Inertial Navigation and Doppler Radar Systems must be installed in accordance with applicable airworthiness requirements.

(b) Cockpit arrangement must be visible and usable by either pilot seated at his duty station.

(c) The equipment must provide, by visual, mechanical, or electrical output signals, indications of the invalidity of output data upon the occurrence of probable failures or malfunctions within the system.

(d) A probable failure or malfunction within the system must not result in loss of the aircraft's required navigation capability.

(e) The alignment, updating, and navigation computer functions of the system must not be invalidated by normal aircraft power interruptions and transients.

(f) The system must not be the source of cause of objectionable radio frequency interference, and must not be adversely affected by radio frequency interference from other aircraft systems.

(g) The FAA-approved airplane flight manual, or supplement thereto, must include pertinent material as required to define the normal and emergency operating procedures and applicable operating limitations associated with INS and Doppler performance (such as maximum latitude at which ground alignment capability is provided, or deviations between systems).

**3.** *Equipment and equipment installation—Inertial Navigation Systems (INS).*

(a) If an applicant elects to use an Inertial Navigation System it must be at least a dual system (including navigational computers and reference units). At least two systems must be operational at takeoff. The dual system may consist of either two INS units, or one INS unit and one Doppler Radar unit.

(b) Each Inertial Navigation System must incorporate the following:

(1) Valid ground alignment capability at all latitudes appropriate for intended use of the installation.

(2) A display of alignment status or a ready to navigate light showing completed alignment to the flight crew.

(3) The present position of the airplane in suitable coordinates.

(4) Information relative to destinations or waypoint positions:

(i) The information needed to gain and maintain a desired track and to determine deviations from the desired track.

(ii) The information needed to determine distance and time to go to the next waypoint or destination.

(c) For INS installations that do not have memory or other inflight alignment means, a separate electrical power source (independent of the main propulsion system) must be provided which can supply, for at least 5 minutes, enough power (as shown by analysis or as demonstrated in the airplane) to maintain the INS in such condition that its full capability is restored upon the reactivation of the normal electrical supply.

(d) The equipment must provide such visual, mechanical, or electrical output signals as may be required to permit the flight crew to detect probable failures or malfunctions in the system.

**4.** *Equipment and equipment installation — Doppler Radar Systems.*

(a) If an applicant elects to use a Doppler Radar System it must be at least a dual system (including dual antennas or a combined antenna designed for multiple operation), except that:

(1) A single operating transmitter with a standby capable of operation may be used in lieu of two operating transmitters.

(2) Single heading source information to all installations may be utilized, provided a compass comparator system is installed and operational procedures call for frequent cross-checks of all compass heading indicators by crewmembers.

The dual system may consist of either two Doppler Radar units or one Doppler Radar unit and one INS unit.

(b) At least two systems must be operational at takeoff.

(c) As determined by the Administrator and specified in the certificate holder's operations specifications, other navigational aids may be required to update the Doppler Radar for a particular operation. These may include Loran, Consol, DME, VOR, ADF, ground-based radar, and airborne weather radar. When these aids are required, the cockpit arrangement must be such that all controls are accessible to each pilot seated at his duty station.

**5.** *Training programs.* The initial training program for Doppler Radar and Inertial Navigation Systems must include the following:

(a) Duties and responsibilities of flight crewmembers, dispatchers, and maintenance personnel.

(b) For pilots, instruction in the following:

(1) Theory and procedures, limitations, detection of malfunctions, preflight and inflight testing, and cross-checking methods.

(2) The use of computers, an explanation of all systems, compass limitations at high latitudes, a review of navigation, flight planning, and applicable meteorology.

(3) The methods for updating by means of reliable fixes.

(4) The actual plotting of fixes.

(c) Abnormal and emergency procedures.

**6.** *Equipment accuracy and reliability.*

(a) Each Inertial Navigation System must meet the following accuracy requirements, as appropriate:

(1) For flights up to 10 hours' duration, no greater than 2 nautical miles per hour of circular error on 95 percent of system flights completed is permitted.

(2) For flights over 10 hours' duration, a tolerance of ±20 miles cross-track and ±25 miles along-track on 95 percent of system flights completed is permitted.

(b) Compass heading information to the Doppler Radar must be maintained to an accuracy of ±1° and total system deviations must not exceed 2. When free gyro techniques are used, procedures shall be utilized to insure that an equivalent level of heading accuracy and total system deviation is attained.

(c) Each Doppler Radar System must meet accuracy requirements of ±20 miles cross-track and ±25 miles along-track for 95 percent of the system flights completed. Updating is permitted.

A system that does not meet the requirements of this section will be considered a failed system.

**7.** *Evaluation program.*

(a) Approval by evaluation must be requested as a part of the application for operational approval of a Doppler Radar or Inertial Navigation System.

(b) The applicant must provide sufficient flights which show to the satisfaction of the Administrator the applicant's ability to use cockpit navigation in his operation.

(c) The Administrator bases his evaluation on the following:

(1) Adequacy of operational procedures.

(2) Operational accuracy and reliability of equipment and feasibility of the system with regard to proposed operations.

(3) Availability of terminal, gateway, area, and en route ground-based aids, if required, to support the self-contained system.

(4) Acceptability of cockpit workload.

(5) Adequacy of flight crew qualifications.

(6) Adequacy of maintenance training and availability of spare parts.

After successful completion of evaluation demonstrations, FAA approval is indicated by issuance of amended operations specifications and en route flight procedures defining the new operation. Approval is limited to those operations for which the adequacy of the equipment and the feasibility of cockpit navigation has been satisfactorily demonstrated.

[Docket No. 10204, 37 FR 6464, March 30, 1972; as amended by Amdt. 121–207, 54 FR 39293, Sept. 25, 1989]

## Appendix H to Part 121

### Advanced Simulation

This appendix provides guidelines and a means for achieving flightcrew training in advanced airplane simulators. This appendix describes the simulator and visual system requirements which must be achieved to obtain approval of certain types of training in the simulator. The requirements in this appendix are in addition to the simulator approval requirements in §121.407. Each simulator which is used under this appendix must be approved as a Level B, C, or D simulator, as appropriate.

To obtain FAA approval of the simulator for a specific level, the following must be demonstrated to the satisfaction of the Administrator:

**1.** Documented proof of compliance with the appropriate simulator, visual system, and additional training requirements of this appendix for the level for which approval is requested.

**2.** An evaluation of the simulator to ensure that its ground, flight, and landing performance matches the type of airplane simulated.

**3.** An evaluation of the appropriate simulator and visual system requirements of the level for which approval is requested.

*Changes to Simulator Programming*

While a need exists for some flexibility in making changes in the software program, strict scrutiny of these changes is essential to ensure that the simulator retains its ability to duplicate the airplane's flight and ground characteristics. Therefore, the following procedure must be followed to allow these changes without affecting the approval of an Appendix H simulator:

**1.** Twenty-one calendar days before making changes to the software program which might impact flight or ground dynamics of an Appendix H simulator, a complete list of these planned changes, including dynamics related to the motion and visual systems, must be provided in writing to the FAA office responsible for conducting the recurrent evaluation of that simulator.

**2.** If the FAA does not object to the planned change within 21 calendar days, the operator may make the change.

**3.** Changes which might affect the approved simulator Level B test guide must be tested by the operator in the simulator to determine the impact of the change before submission to the FAA.

**4.** Software changes actually installed must be summarized and provided to the FAA. When the operator's test shows a difference in simulator performance due to a change, an amended copy of the test guide page which includes the new simulator test results will also be provided to update the FAA's copy of the test guide.

**5.** The FAA may examine supporting data or flight check the simulator, or both, to ensure that the aerodynamic quality of the simulator has not been degraded by any change in software programming.

**6.** All requests for changes are evaluated on the basis of the same criteria used in the initial approval of the simulator for Level B, C, or D.

*Simulator Minimum Equipment List (MEL)*

Because of the strict tolerances and other approval requirements of Appendix H simulators, the simulator can provide realistic training with certain nonessential items inoperative. Therefore, an operator may operate its simulator under an MEL which has been approved by the Administrator for that simulator. The MEL includes simulator components and indicates the type of training or checking that is authorized if the component becomes inoperative. To accomplish this, the component is placed in one of the following categories along with any remarks applicable to the component's use in the training program:

**1.** No training or checking.

**2.** Training in specific maneuvers.

**3.** Certification and checking.

**4.** Line Oriented Flight Training (LOFT).

*Advanced Simulation Training Program*

For an operator to conduct Level B, C, or D training under this appendix all required simulator instruction and checks must be conducted under an advanced simulation training program which is approved by the Administrator for the operator. This program must also ensure that all instructors and check airmen used in Appendix H training and checking are highly qualified to provide the training required in the training program. The advanced simulation training program shall include the following:

**1.** The operator's initial, transition, upgrade, and recurrent simulator training programs and its procedures for re-establishing recency of experience in the simulator.

**2.** How the training program will integrate Level B, C, or D simulators with other simulators and training devices to maximize the total training, checking, and certification functions.

**3.** Documentation that each instructor and check airman has served for at least 1 year in that capacity in a certificate holder's approved program or has served for at least 1 year as a pilot in command or second in command in an airplane of the group in which that pilot is instructing or checking.

**4.** A procedure to ensure that each instructor and check airman actively participates in either an approved regularly scheduled line flying program as a flight crewmember or an approved line observation program in the same airplane type for which that person is instructing or checking.

**5.** A procedure to ensure that each instructor and check airman is given a minimum of 4 hours of training each year to become familiar with the operator's advanced simulation training program, or changes to it, and to emphasize their respective roles in the program. Training for simulator instructors and check airmen shall include training policies and procedures, instruction methods and techniques, operation of simulator controls (including environmental and trouble panels), limitations of the simulator, and minimum equipment required for each course of training.

**6.** A special Line Oriented Flight Training (LOFT) program to facilitate the transition from the simulator to line flying. This LOFT program consists of at least a 4-hour course of training for each flightcrew. It also contains at least two representative flight segments of the operator's route. One of the flight segments contains strictly normal operating procedures from push back at one airport to arrival at another. Another flight segment contains training in appropriate abnormal and emergency flight operations.

## Level B

### *Training and Checking Permitted*

**1.** Recency of experience (§121.439).

**2.** Night takeoffs and landings (part 121, Appendix E).

**3.** Landings in a proficiency check without the landing on the line requirements (§121.441).

### *Simulator Requirements*

**1.** Aerodynamic programming to include:

a. Ground effect—for example, roundout, flare, and touchdown. This requires data on lift, drag, and pitching moment in ground effect.

b. Ground reaction—Reaction of the airplane upon contact with the runway during landing to include strut deflections, tire friction, and side forces.

c. Ground handling characteristics—steering inputs to include crosswind, braking, thrust reversing, deceleration, and turning radius.

**2.** Minimum of 3-axis freedom of motion systems.

**3.** Level B landing maneuver test guide to verify simulator data with actual airplane flight test data, and provide simulator performance tests for Level B initial approval.

**4.** Multichannel recorders capable of recording Level B performance tests.

### *Visual Requirements*

**1.** Visual system compatibility with aerodynamic programming.

**2.** Visual system response time from pilot control input to visual system output shall not exceed 300 milliseconds more than the movement of the airplane to a similar input. Visual system response time is defined as the completion of the visual display scan of the first video field containing different information resulting from an abrupt control input.

**3.** A means of recording the visual response time for comparison with airplane data.

**4.** Visual cues to assess sink rate and depth perception during landings.

**5.** Visual scene to instrument correlation to preclude perceptible lags.

## Level C

### *Training and Checking Permitted*

**1.** For all pilots, transition training between airplanes in the same group, and for a pilot in command the certification check required by §61.153(g) of this chapter.

**2.** Upgrade to pilot-in-command training and the certification check when the pilot—

a. Has previously qualified as second in command in the equipment to which the pilot is upgrading;

b. Has at least 500 hours of actual flight time while serving as second in command in an airplane of the same group; and

c. Is currently serving as second in command in an airplane in this same group.

**3.** Initial pilot-in-command training and the certification check when the pilot—

a. Is currently serving as second in command in an airplane of the same group;

b. Has a minimum of 2,500 flight hours as second in command in an airplane of the same group; and

c. Has served as second in command on at least two airplanes of the same group.

**4.** For all second-in-command pilot applicants who meet the aeronautical experience requirements of §61.159 of this chapter in the airplane, the initial and upgrade training and checking required by this part, and the certification check requirements of §61.153 of this chapter.

*Simulator Requirements*

**1.** Representative crosswind and three-dimensional windshear dynamics based on airplane related data.

**2.** Representative stopping and directional control forces for at least the following runway conditions based on airplane related data:

a. Dry.
b. Wet.
c. Icy.
d. Patchy wet.
e. Patchy icy.
f. Wet on rubber residue in touchdown zone.

**3.** Representative brake and tire failure dynamics (including antiskid) and decreased brake efficiency due to high brake temperatures based on airplane related data.

**4.** A motion system which provides motion cues equal to or better than those provided by a six-axis freedom of motion system.

**5.** Operational principal navigation systems, including electronic flight instrument systems, INS, and OMEGA, if applicable.

**6.** Means for quickly and effectively testing simulator programming and hardware.

**7.** Expanded simulator computer capacity, accuracy, resolution, and dynamic response to meet Level C demands. Resolution equivalent to that of at least a 32-bit word length computer is required for critical aerodynamic programs.

**8.** Timely permanent update of simulator hardware and programming subsequent to airplane modification.

**9.** Sound of precipitation and significant airplane noises perceptible to the pilot during normal operations and the sound of a crash when the simulator is landed in excess of landing gear limitations.

**10.** Aircraft control feel dynamics shall duplicate the airplane simulated. This shall be determined by comparing a recording of the control feel dynamics of the simulator to airplane measurements in the takeoff, cruise, and landing configuration.

**11.** Relative responses of the motion system, visual system, and cockpit instruments shall be coupled closely to provide integrated sensory cues. These systems shall respond to abrupt pitch, roll, and yaw inputs at the pilot's position within 150 milliseconds of the time, but not before the time, when the airplane would respond under the same conditions. Visual scene changes from steady state disturbance shall not occur before the resultant motion onset but within the system dynamic response tolerance of 150 milliseconds. The test to determine compliance with these requirements shall include simultaneously recording the analog output from the pilot's control column and rudders, the output from an accelerometer attached to the motion system platform located at an acceptable location near the pilots' seats, the output signal to the visual system display (including visual system analog delays), and the output signal to the pilot's attitude indicator or an equivalent test approved by the Administrator. The test results in a comparison of a recording of the simulator's response to actual airplane response data in the takeoff, cruise, and landing configuration.

*Visual Requirements*

**1.** Dusk and night visual scenes with at least three specific airport representations, including a capability of at least 10 levels of occulting, general terrain characteristics, and significant landmarks.

**2.** Radio navigation aids properly oriented to the airport runway layout.

**3.** Test procedures to quickly confirm visual system color, RVR, focus, intensity, level horizon, and attitude as compared to the simulator attitude indicator.

**4.** For the approach and landing phase of flight, at and below an altitude of 2,000 feet height above the airport (HAA) and within a radius of 10 miles from the airport, weather representations including the following:

a. Variable cloud density.
b. Partial obscuration of ground scenes; that is, the effect of a scattered to broken cloud deck.
c. Gradual break out.
d. Patchy fog.
e. The effect of fog on airport lighting.
f. Category II and III weather conditions.

**5.** Continuous minimum visual field of view of 75° horizontal and 30° vertical per pilot seat. Visual gaps shall occur only as they would in the airplane simulated or as required by visual system hardware. Both pilot seat visual systems shall be able to be operated simultaneously.

**6.** Capability to present ground and air hazards such as another airplane crossing the active runway or converging airborne traffic.

## Level D

*Training and Checking Permitted*

Except for the requirements listed in the next sentence, all pilot flight training and checking required by this part and the certification check requirements of §61.153(g) of this chapter. The line check required by §121.440 of this part, the static airplane requirements of Appendix E to this part, and the operating experience requirements of §121.434 of this part must still be performed in the airplane.

*SIMULATOR REQUIREMENTS*

**1.** Characteristic buffet motions that result from operation of the airplane (for example, high-speed buffet, extended landing gear, flaps, nose-wheel scuffing, stall) which can be sensed at the flight deck. The simulator must be programmed and instrumented in such a manner that the characteristic buffet modes can be measured and compared to airplane data. Airplane data are also required to define flight deck motions when the airplane is subjected to atmospheric disturbances such as rough air and cobblestone turbulence. General purpose disturbance models that approximate demonstrable flight test data are acceptable.

**2.** Aerodynamic modeling for aircraft for which an original type certificate is issued after June 1, 1980, including low-altitude, level-flight ground effect, mach effect at high altitude, effects of airframe icing, normal and reverse dynamic thrust effect on control surfaces, aero-elastic representations, and representations of nonlinearities due to side slip based on airplane flight test data provided by the manufacturer.

**3.** Realistic amplitude and frequency of cockpit noises and sounds, including precipitation static and engine and airframe sounds. The sounds shall be coordinated with the weather representations required in visual requirement No. 3.

**4.** Self-testing for simulator hardware and programming to determine compliance with Level B, C, or D simulator requirements.

**5.** Diagnostic analysis printout of simulator malfunctions sufficient to determine MEL compliance. These printouts shall be retained by the operator between recurring FAA simulator evaluations as part of the daily discrepancy log required under §121.407(a)(5).

*VISUAL REQUIREMENTS*

**1.** Daylight, dusk, and night visual scenes with sufficient scene content to recognize a specific airport, the terrain, and major landmarks around that airport and to successfully accomplish a visual landing. The daylight visual scene must be part of a total daylight cockpit environment which at least represents the amount of light in the cockpit on an overcast day. For the purpose of this rule, daylight visual system is defined as a visual system capable of producing, as a minimum, full color presentations, scene content comparable in detail to that produced by 4,000 edges or 1,000 surfaces for daylight and 4,000 light points for night and dusk scenes, 6-foot lamberts of light at the pilot's eye (highlight brightness), 3-arc minutes resolution for the field of view at the pilot's eye, and a display which is free of apparent quantization and other distracting visual effects while the simulator is in motion. The simulation of cockpit ambient lighting shall be dynamically consistent with the visual scene displayed. For daylight scenes, such ambient lighting shall neither "washout" the displayed visual scene nor fall below 5-foot lamberts of light as reflected from an approach plate at knee height at the pilot's station and/or 2-foot lamberts of light as reflected from the pilot's face.

**2.** Visual scenes portraying representative physical relationships which are known to cause landing illusions in some pilots, including short runway, landing over water, runway gradient, visual topographic features, and rising terrain.

**3.** Special weather representations which include the sound, visual, and motion effects of entering light, medium, and heavy precipitation near a thunderstorm on takeoff, approach, and landings at and below an altitude of 2,000 feet HAA and within a radius of 10 miles from the airport.

**4.** Level C visual requirements in daylight as well as dusk and night representations.

**5.** Wet and, if appropriate for the operator, snow-covered runway representations, including runway lighting effects.

**6.** Realistic color and directionality of airport lighting.

**7.** Weather radar presentations in aircraft where radar information is presented on the pilot's navigation instruments.

(Secs. 313, 601, 603, 604, Federal Aviation Act of 1958; as amended (49 U.S.C. 1354, 1421, 1423, 1424); sec. 6(c), Department of Transportation Act (49 U.S.C. 1655(c)))

[Docket No. 19758, 45 FR 44183, June 30, 1980; 45 FR 48599, July 31, 1980; as amended by Amdt. 121–258, 61 FR 30732, June 17, 1996; Amdt. 121–258, 61 FR 39859, July 31, 1996; Amdt. 121–267, 62 FR 68137, Dec. 30, 1997]

## APPENDIX I TO PART 121

### DRUG TESTING PROGRAM

This appendix contains the standards and components that must be included in an anti-drug program required by this chapter.

**I.** *General*

A. *Purpose.* The purpose of this appendix is to establish a program designed to help prevent accidents and injuries resulting from the use of prohibited drugs by employees who perform safety-sensitive functions.

B. *DOT Procedures.* Each employer shall ensure that drug testing programs conducted pursuant to 14 CFR parts 65, 121, and 135 comply with the requirements of this appendix and the "Procedures for Transportation Workplace Drug Testing Programs" published by the Department of Transportation (DOT) (49 CFR part 40). An employer may not use or contract with any drug testing laboratory that is not certified by the Department of

Health and Human Services (HHS) under the National Laboratory Certification Program.

C. Employer Responsibility. As an employer, you are responsible for all actions of your officials, representatives, and service agents in carrying out the requirements of this appendix and 49 CFR part 40.

**II.** *Definitions.* For the purpose of this appendix, the following definitions apply:

*Accident* means an occurrence associated with the operation of an aircraft which takes place between the time any person boards the aircraft with the intention of flight and all such persons have disembarked, and in which any person suffers death or serious injury, or in which the aircraft receives substantial damage.

*Annualized rate* for the purposes of unannounced testing of employees based on random selection means the percentage of specimen collection and testing of employees performing a safety-sensitive function during a calendar year. The employer shall determine the annualized rate by referring to the total number of employees performing a sensitive safety function for the employer at the beginning of the calendar year.

*Contractor company* means a company that has employees who perform safety-sensitive functions by contract for an employer.

*DOT agency* means an agency (or "operating administration") of the United States Department of Transportation administering regulations requiring drug testing (14 CFR part 61 et al,; 46 CFR part 16; 49 CFR parts 199, 219, and 382) in accordance with 49 CFR part 40.

*Employee* is a person who performs, either directly or by contract, a safety-sensitive function for an employer, as defined below. Provided, however, that an employee who works for an employer who holds a part 135 certificate and who holds a part 121 certificate is considered to be an employee of the part 121 certificate holder for the purposes of this appendix.

*Employer* is a part 121 certificate holder, a part 135 certificate holder, an operator as defined in §135.1(c) of this chapter, or an air traffic control facility not operated by the FAA or by or under contract to the U.S. military. Provided, however, that an employer may use a person who is not included under that employer's drug program to perform a safety-sensitive function, if that person is subject to the requirements of another employer's FAA-approved antidrug program.

*Performing* (a safety-sensitive function): an employee is considered to be performing a safety-sensitive function during any period in which he or she is actually performing, ready to perform, or immediately available to perform such function.

*Positive rate* means the number of positive results for random drug tests conducted under this appendix plus the number of refusals to take random tests required by this appendix, divided by the total number of random drug tests conducted under this appendix plus the number of refusals to take random tests required by this appendix.

*Prohibited drug* means marijuana, cocaine, opiates, phencyclidine (PCP), and amphetamines, as specified in 49 CFR 40.85.

*Refusal to submit* means that a covered employee engages in conduct specified in 49 CFR 40.191.

*Safety-sensitive function* means a function listed in section III of this appendix.

*Verified negative drug test result* means a drug test result from an HHS-certified laboratory that has undergone review by an MRO and has been determined by the MRO to be a negative result.

*Verified positive drug test result* means a drug test result from an HHS-certified laboratory that has undergone review by an MRO and has been determined by the MRO to be a positive result.

**III.** *Employees Who Must Be Tested.* Each person who performs a safety-sensitive function directly or by contract for an employer must be tested pursuant to an FAA-approved antidrug program conducted in accordance with this appendix:

A. Flight crewmember duties.
B. Flight attendant duties.
C. Flight instruction duties.
D. Aircraft dispatcher duties.
E. Aircraft maintenance or preventive maintenance duties.
F. Ground security coordinator duties.
G. Aviation screening duties.
H. Air traffic control duties.

**IV.** *Substances For Which Testing Must Be Conducted.* Each employer shall test each employee who performs a safety-sensitive function for evidence of marijuana, cocaine, opiates, phencyclidine (PCP), and amphetamines during each test required by section V. of this appendix.

**V.** *Types of Drug Testing Required.* Each employer shall conduct the following types of testing in accordance with the procedures set forth in this appendix and the DOT "Procedures for Transportation Workplace Drug Testing Programs" (49 CFR part 40):

**A.** *Pre-employment testing.*

1. Prior to the first time an individual performs a safety-sensitive function for an employer, the employer shall require the individual to undergo testing for prohibited drug use.

2. An employer is permitted to require pre-employment testing of an individual if the following criteria are met:

(a) The individual previously performed a covered function for the employer;

(b) The employer removed the individual from the employer's random testing program conducted under this appendix for reasons other than a verified positive test result on an FAA-mandated drug test or a refusal to submit to such testing; and

(c) The individual will be returning to the performance of a safety-sensitive function.

3. No employer shall allow an individual required to undergo pre-employment testing under section V, paragraph A.1 or A.2 of this appendix to perform a safety-sensitive function unless the employer has received a verified negative drug test result for the individual.

4. The employer shall advise each individual applying to perform a safety-sensitive function at the time of application that the individual will be required to undergo pre-employment testing to determine the presence of marijuana, cocaine, opiates, phencyclidine (PCP), and amphetamines, or a metabolite of those drugs in the individual's system. The employer shall provide this same notification to each individual required by the employer to undergo pre-employment testing under section V, paragraph A.(2) of this appendix.

**B.** *Periodic testing.* Each employee who performs a safety-sensitive function for an employer and who is required to undergo a medical examination under part 67 of this chapter shall submit to a periodic drug test. The employee shall be tested for the presence of marijuana, cocaine, opiates, phencyclidine (PCP), and amphetamines or a metabolite of those drugs during the first calendar year of implementation of the employer's antidrug program. The tests shall be conducted in conjunction with the first medical evaluation of the employee or in accordance with an alternative method for collecting periodic test specimens detailed in an employer's approved antidrug program. An employer may discontinue periodic testing of its employees after the first calendar year of implementation of the employer's antidrug program when the employer has implemented an unannounced testing program based on random selection of employees.

**C.** *Random testing.*

1. Except as provided in paragraphs 2–4 of this section, the minimum annual percentage rate for random drug testing shall be 50 percent of covered employees.

2. The Administrator's decision to increase or decrease the minimum annual percentage rate for random drug testing is based on the reported positive rate for the entire industry. All information used for this determination is drawn from the statistical reports required by section X of this appendix. In order to ensure reliability of the data, the Administrator considers the quality and completeness of the reported data, may obtain additional information or reports from employers, and may make appropriate modifications in calculating the industry positive rate. Each year, the Administrator will publish in the **Federal Register** the minimum annual percentage rate for random drug testing of covered employees. The new minimum annual percentage rate for random drug testing will be applicable starting January 1 of the calendar year following publication.

3. When the minimum annual percentage rate for random drug testing is 50 percent, the Administrator may lower this rate to 25 percent of all covered employees if the Administrator determines that the data received under the reporting requirements of this appendix for two consecutive calendar years indicate that the reported positive rate is less than 1.0 percent.

4. When the minimum annual percentage rate for random drug testing is 25 percent, and the data received under the reporting requirements of this appendix for any calendar year indicate that the reported positive rate is equal to or greater than 1.0 percent, the Administrator will increase the minimum annual percentage rate for random drug testing to 50 percent of all covered employees.

5. The selection of employees for random drug testing shall be made by a scientifically valid method, such as a random-number table or a computer-based random number generator that is matched with employees' Social Security numbers, payroll identification numbers, or other comparable identifying numbers. Under the selection process used, each covered employee shall have an equal chance of being tested each time selections are made.

6. The employer shall randomly select a sufficient number of covered employees for testing during each calendar year to equal an annual rate not less than the minimum annual percentage rate for random drug testing determined by the Administrator. If the employer conducts random drug testing through a Consortium/Third-party administrator (C/TPA), the number of employees to be tested may be calculated for each individual employer or may be based on the total number of covered employees covered by the C/TPA who are subject to random drug testing at the same minimum annual percentage rate under this part or any DOT drug testing rule.

7. Each employer shall ensure that random drug tests conducted under this appendix are unannounced and that the dates for administering random tests are spread reasonably throughout the calendar year.

8. If a given covered employee is subject to random drug testing under the drug testing rules of more than one DOT agency, the employee shall

be subject to random drug testing at the percentage rate established for the calendar year by the DOT agency regulating more than 50 percent of the employee's function.

9. If an employer is required to conduct random drug testing under the drug testing rules of more than one DOT agency, the employer may—

(a) Establish separate pools for random selection, with each pool containing the covered employees who are subject to testing at the same required rate; or

(b) Randomly select covered employees for testing at the highest percentage rate established for the calendar year by any DOT agency to which the employer is subject.

10. An employer required to conduct random drug testing under the anti drug rules of more than one DOT agency shall provide each such agency access to the employer's records of random drug testing, as determined to be necessary by the agency to ensure the employer's compliance with the rule.

**D.** *Post-accident testing.* Each employer shall test each employee who performs a safety-sensitive function for the presence of marijuana, cocaine, opiates, phencyclidine (PCP), and amphetamines or a metabolite of those drugs in the employee's system if that employee's performance either contributed to an accident or can not be completely discounted as a contributing factor to the accident. The employee shall be tested as soon as possible but not later than 32 hours after the accident. The decision not to administer a test under this section must be based on a determination, using the best information available at the time of the determination, that the employee's performance could not have contributed to the accident. The employee shall submit to post-accident testing under this section.

**E.** *Testing based on reasonable cause.* Each employer shall test each employee who performs a safety-sensitive function and who is reasonably suspected of using a prohibited drug. Each employer shall test an employee's specimen for the presence of marijuana, cocaine, opiates, phencyclidine (PCP), and amphetamines or a metabolite of those drugs. An employer may test an employee's specimen for the presence of other prohibited drugs or drug metabolites only in accordance with this appendix and the DOT "Procedures for Transportation Workplace Drug Testing Programs" (49 CFR part 40). At least two of the employee's supervisors, one of whom is trained in detection of the symptoms of possible drug use, shall substantiate and concur in the decision to test an employee who is reasonably suspected of drug use; provided, however, that in the case of an employer other than a part 121 certificate holder who employs 50 or fewer employees who perform safety-sensitive functions, one supervisor who is trained in detection of symptoms of possible drug use shall substantiate the decision to test an employee who is reasonably suspected of drug use. The decision to test must be based on a reasonable and articulable belief that the employee is using a prohibited drug on the basis of specific contemporaneous physical, behavioral, or performance indicators of probable drug use.

**F.** *Return to Duty Testing.* Each employer shall ensure that before an individual is returned to duty to perform a safety-sensitive function after refusing to submit to a drug test required by this appendix or receiving a verified positive drug test result on a test conducted under this appendix the individual shall undergo a return to duty drug test. No employer shall allow an individual required to undergo return to duty testing to perform a safety-sensitive function unless the employer has received a verified negative drug test result for the individual. The test cannot occur until after the SAP has determined that the employee has successfully complied with the prescribed education and/or treatment.

**G.** *Follow-up Testing.*

1. Each employer shall implement a reasonable program of unannounced testing of each individual who has be hired to perform or who has be returned to the performance of a safety-sensitive function after refusing to submit to a drug test required by this appendix or receiving a verified positive drug test result on a test conducted under this appendix.

2. The number and frequency of such testing shall be determined by the employer's Substance Abuse Professional conducted in accordance with the provisions of 49 CFR part 40, but shall consist of at least six tests in the first 12 months following the employee's return to duty.

3. The employer must direct the employee to undergo testing for alcohol in accordance with appendix J of this part, in addition to drugs, if the Substance Abuse Professional determines that alcohol testing is necessary for the particular employee. Any such alcohol testing shall be conducted in accordance with the provisions of 49 CFR part 40.

4. Follow-up testing shall not exceed 60 months after the date the individual begins to perform or returns to the performance of a safety-sensitive function. The Substance Abuse Professional may terminate the requirement for follow-up testing at any time after the first six tests have been conducted, if the Substance Abuse Professional determines that such testing is no longer necessary.

**VI.** *Administrative and Other Matters.*

**A.** *MRO Record Retention Requirements.*

1. Records concerning drug tests confirmed positive by the laboratory shall be maintained by the MRO for 5 years. Such records include the MRO copies of the custody and control form, medical interviews, documentation of the basis for verifying as negative test results confirmed as positive by the laboratory, any other documentation concerning the MRO's verification process.

2. Should the employer change MROs for any reason, the employer shall ensure that the former MRO forwards all records maintained pursuant to this rule to the new MRO within ten working days of receiving notice from the employer of the new MRO's name and address.

3. Any employer obtaining MRO services by contract, including a contract through a C/TPA, shall ensure that the contract includes a recordkeeping provision that is consistent with this paragraph, including requirements for transferring records to a new MRO.

**B.** *Access to Records.* The employer and the MRO shall permit the Administrator or the Administrator's representative to examine records required to be kept under this appendix and 49 CFR part 40. The Administrator or the Administrator's representative may require that all records maintained by the service agent for the employer must be produced at the employer's place of business.

**C.** *Release of Drug Testing Information.* An employer shall release information regarding an employee's drug testing results, evaluation, or rehabilitation to a third party in accordance with 49 CFR part 40. Except as required by law, this appendix, or 49 CFR part 40, no employer shall release employee information.

**D.** *Refusal To Submit to Testing.*

1. Each employer shall notify the FAA within 5 working days of any employee who holds a certificate issued under part 61, part 63, or part 65 of this chapter who has refused to submit to a drug test required under this appendix. Notification should be sent to: Federal Aviation Administration, Office of Aviation Medicine, Drug Abatement Division (AAM–800), 800 Independence Avenue SW, Washington, DC 20591.

2. Employers are not required to notify the above office of refusals to submit to pre-employment or return to duty testing.

**E.** *Permanent Disqualification From Service.* An employee who has verified positive drug test results on two drug tests required by Appendix I to part 121 of this chapter and conducted after September 19, 1994, is permanently precluded from performing for an employer the safety-sensitive duties the employee performed prior to the second drug test.

2. An employee who has engaged in prohibited drug use during the performance of a safety-sensitive function after September 19, 1994 is permanently precluded from performing that safety-sensitive function for an employer.

**VII.** *Medical Review Officer, Substance Abuse Professional, and Employer Responsibilities.* The employer shall designate or appoint a Medical Review Officer (MRO) who shall be qualified in accordance with 49 CFR part 40 and shall perform the functions set forth in 49 CFR part 40 and this appendix. If the employer does not have a qualified individual on staff to serve as MRO, the employer may contract for the provision of MRO services as part of its drug testing program.

**A.** *Medical Review Officer (MRO).* The MRO must perform the functions set forth in 49 CFR part 40, Subpart G, and this appendix. The MRO shall not delay verification of the primary test result following a request for a split specimen test unless such delay is based on reasons other than the fact that the split specimen test result is pending. If the primary test result is verified as positive, actions required under this rule (e.g., notification to the Federal Air Surgeon, removal from safety-sensitive position) are not stayed during the 72-hour request period or pending receipt of the split specimen test result.

**B.** *Substance Abuse Professional (SAP).* The SAP must perform the functions set forth in 49 CFR part 40, Subpart O.

**C.** *Additional Medical Review Officer, Substance Abuse Professional, and Employer Responsibilities Regarding 14 CFR part 67 Airman Medical Certificate Holders.*

1. As part of verifying a confirmed positive test result, the MRO shall inquire, and the individual shall disclose, whether the individual is or would be required to hold a medical certificate issued under 14 CFR part 67 to perform a safety-sensitive function for the employer. If the individual answers in the negative, the MRO shall then inquire, and the individual shall disclose whether the individual currently holds a medical certificate issued under 14 CFR part 67. If the individual answers in the affirmative to either question, in addition to notifying the employer in accordance with 49 CFR part 40, the MRO must forward to the Federal Air Surgeon, at the address listed in paragraph 5, the name of the individual, along with identifying information and supporting documentation, within 12 working days after verifying a positive drug test result.

2. The SAP shall inquire, and the individual shall disclose, whether the individual is or would be required to hold a medical certificate issued under 14 CFR part 67 of this chapter to perform a safety sensitive function for the employer. If the individual answers in the affirmative, the SAP cannot recommend that the individual be returned to a safety-sensitive function that requires the individual to hold a 14 CFR part 67 medical certificate unless and until such individual has received a medical certificate or a special issuance medical certificate from the Federal Air Surgeon. The receipt of a medical certificate or a special issuance medical certificate does not alter any obligations otherwise required by 49 CFR part 40 or this appendix.

3. The employer must forward to the Federal Air Surgeon a copy of any report provided by the SAP, if available, regarding an individual for whom the MRO has provided a report to the Federal Air Surgeon under section VII.C.1 of this appendix, within 12 working days of the employer's receipt of the report.

4. The employer cannot permit an employee who is required to hold a medical certificate under part 67 of this chapter to perform a safety-sensitive duty to resume that duty until the employee has received a medical certificate or a special issuance medical certificate from the Federal Air Surgeon unless and until the employer has ensured that the employee meets the return-to-duty requirements in accordance with 49 CFR part 40.

5. Reports required under this section shall be forwarded to the Federal Air Surgeon, Federal Aviation Administration, Attn: Drug Abatement Division (AAM-800), 800 Independence Avenue SW, Washington, DC 20591.

**VIII.** *Employee Assistance Program (EAP).*

The employer shall provide an EAP for employees. The employer may establish the EAP as a part of its internal personnel services or the employer may contract with an entity that will provide EAP services to an employee. Each EAP must include education and training on drug use for employees and training for supervisors making determinations for testing of employees based on reasonable cause.

**A.** *EAP Education Program.* Each EAP education program must include at least the following elements: display and distribution of informational material; display and distribution of a community service hot-line telephone number for employee assistance; and display and distribution of the employer's policy regarding drug use in the workplace. The employer's policy shall include information regarding the consequences under the rule of using drugs while performing safety-sensitive functions, receiving a verified positive drug test result, or refusing to submit to a drug test required under the rule.

**B.** *EAP Training Program.* Each employer shall implement a reasonable program of initial training for employees. The employee training program must include at least the following elements: The effects and consequences of drug use on personal health, safety, and work environment; the manifestations and behavioral cues that may indicate drug use and abuse; and documentation of training given to employees and employer's supervisory personnel. The employer's supervisory personnel who will determine when an employee is subject to testing based on reasonable cause shall receive specific training on specific, contemporaneous physical, behavioral, and performance indicators of probable drug use in addition to the training specified above. The employer shall ensure that supervisors who will make reasonable cause determinations receive at least 60 minutes of initial training. The employer shall implement a reasonable recurrent training program for supervisory personnel making reasonable cause determinations during subsequent years. The employer shall identify the employee and supervisor EAP training in the employer's drug testing plan submitted to the FAA for approval.

**IX.** *Employer's Antidrug Program Plan.*

**A.** *Schedule for Submission of Plans and Implementation.*

1. Each employer shall submit an antidrug program plan to the Federal Aviation Administration, Office of Aviation Medicine, Drug Abatement Division (AAM–800), 800 Independence Avenue SW, Washington DC, 20591.

2.(a) Any person who applies for a certificate under the provisions of part 121 or part 135 of this chapter after September 19, 1994 shall submit an antidrug program plan to the FAA for approval and must obtain such approval prior to beginning operations under the certificate. The program shall be implemented not later than the date of inception of operations. Contractor employees to a new certificate holder must be subject to an FAA-approved antidrug program within 60 days of the implementation of the employer's program.

(b) Any person who intends to begin sightseeing operations as an operator under 14 CFR 135.1(c) after September 19, 1994 shall, not later than 60 days prior to the proposed initiation of such operations, submit an antidrug program plan to the FAA for approval. No operator may begin conducting sightseeing flights prior to receipt of approval; the program shall be implemented concurrently with the inception of operations. Contractor employees to a new operator must be subject to an FAA-approved program within 60 days of the implementation of the employer's program.

(c) any person who intends to begin air traffic control operations as an employer as defined in 14 CFR 65.46(a)(2) (air traffic control facilities not operated by the FAA or by or under contract to the U.S. military) after September 19, 1994 shall, not later than 60 days prior to the proposed initiation of such operations, submit an antidrug program plan to the FAA for approval. No air traffic control facility may begin conducting air traffic control operations prior to receipt of approval; the program shall be implemented concurrently with the inception of operations. Contractor employees to a new air traffic control facility must be subject to an FAA-approved program within 60 days of the implementation of the facility's program.

3. In accordance with this appendix, an entity or individual that holds a repair station certificate issued by the FAA pursuant to part 145 of this chapter and employs individuals who perform a safety-sensitive function pursuant to a primary or direct contract with an employer or an operator may submit an antidrug program plan (specifying the procedures for complying with this appendix) to the FAA for approval. Each certificated repair station shall implement its approved antidrug program in accordance with its terms.

4. Any entity or individual whose employees perform safety-sensitive functions pursuant to a contract with an employer (as defined in section II of this appendix), may submit an antidrug program plan to the FAA for approval on a form and in a manner prescribed by the Administrator.

(a) The plan shall specify the procedures that will be used to comply with the requirements of this appendix.

(b) Each contractor shall implement its antidrug program in accordance with the terms of its approved plan.

5. Each air traffic control facility operating under contract to the FAA shall submit an antidrug program plan to the FAA (specifying the procedures for all testing required by this appendix) not later than November 17, 1994. Each facility shall implement its antidrug program not later than 60 days after approval of the program by the FAA. Employees performing air traffic control duties by contract for the air traffic control facility (i.e., not directly employed by the facility) must be subject to an FAA-approved antidrug program within 60 days of implementation of the air traffic control facility's program.

6. Each employer, or contractor company that has submitted an antidrug plan directly to the FAA, shall obtain appropriate approval from the FAA prior to changing programs.

**B.** An employer's antidrug plan must specify the methods by which the employer will comply with the testing requirements of this appendix. The plan must provide the name and address of the laboratory which has been selected by the employer for analysis of the specimens collected during the employer's antidrug testing program.

**C.** An employer's antidrug plan must specify the procedures and personnel the employer will use to ensure that a determination is made as to the veracity of test results and possible legitimate explanations for an employee receiving a verified positive drug test result.

**D.** The employer shall consider its antidrug program to be approved by the Administrator, unless notified to the contrary by the FAA, within 60 days after submission of the plan to the FAA.

**X.** *Reporting of Antidrug Program Results.*

**A.** Annual reports of antidrug program results shall be submitted to the FAA in the form and manner prescribed by the Administrator by March 15 of the succeeding calendar year for the prior calendar year (January 1 through December 31) in accordance with the provisions below.

1. Each part 121 certificate holder shall submit an annual report each year.

2. Each entity conducting an antidrug program under an FAA-approved antidrug plan, other than a part 121 certificate holder, that has 50 or more employees performing a safety-sensitive function January 1 of any calendar year shall submit an annual report to the FAA for that calendar year.

3. The Administrator reserves the right to require that aviation employers not otherwise required to submit annual reports prepare and submit such reports to the FAA. Employers that will be required to submit annual reports under this provision will be notified in writing by the FAA.

**B.** Each report shall be submitted in the form and manner prescribed by the Administrator. No other form, including another DOT Operating Administration's form, is acceptable for submission to the FAA.

**C.** Each report shall be signed by the employer's antidrug program manager or other designated representative.

**D.** Each report with verified positive drug test results shall include all of the following informational elements:

1. Number of covered employees by employee category.

2. Number of covered employees affected by the antidrug rule of another operating administration identified and reported by number and employee category.

3. Number of specimens collected by type of test and employee category.

4. Number of positive drug test results verified by a Medical Review Officer (MRO) by type of test, type of drug, and employee category.

5. Number of negative drug test results reported by an MRO by type of test and employee category.

6. Number of persons denied a safety-sensitive position based on a verified positive pre-employment drug test result reported by an MRO.

7. Action taken following a verified positive drug test result(s), by type of action.

8. Number of employees returned to duty during the reporting period after having received a verified positive drug test result on or refused to submit to a drug test required under the FAA rule.

9. Number of employees by employee category with tests verified positive for multiple drugs by an MRO.

10. Number of employees who refused to submit to a drug test and the action taken in response to the refusal(s).

11. Number of covered employees who have received required initial training.

12. Number of supervisory personnel who have received required initial training.

13. Number of supervisors who have received required recurrent training.

**E.** Each report with only negative test results shall include all of the following informational elements. (This report may *only* be submitted by employers with *no* verified positive drug test results during the reporting year.)

1. Number of covered employees by employee category.

2. Number of covered employees affected by the antidrug rule of another operating administration identified and reported by number and employee category.

3. Number of specimens collected by type of test and employee category.

4. Number of negative tests reported by an MRO by type of test and employee category

5. Number of employees who refused to submit to a drug test and the action taken in response to the refusal(s).

6. Number of employees returned to duty during the reporting period after having received a verified positive drug test result on or refused to submit to a drug test required under the FAA rule.

7. Number of covered employees who have received required initial training.

8. Number of supervisory personnel who have received required initial training.

9. Number of supervisors who have received required recurrent training.

**F.** A C/TPA may prepare reports on behalf of individual aviation employers for purposes of compliance with this reporting requirement. However, the aviation employer shall sign and submit such a report and shall remain responsible for ensuring the accuracy and timeliness of each report prepared on its behalf by a C/TPA. A C/TPA must not sign the form.

**XI.** *Preemption.*

**A.** The issuance of 14 CFR parts 65, 121, and 135 by the FAA preempts any state or local law, rule, regulation, order, or standard covering the subject matter of 14 CFR parts 65, 121, and 135, including but not limited to, drug testing of aviation personnel performing safety-sensitive functions.

**B.** The issuance of 14 CFR parts 65, 121, and 135 does not preempt provisions of state criminal law that impose sanctions for reckless conduct of an individual that leads to actual loss of life, injury, or damage to property whether such provisions apply specifically to aviation employees or generally to the public.

**XII.** *Testing Outside the Territory of the United States.*

**A.** No part of the testing process (including specimen collection, laboratory processing, and MRO actions) shall be conducted outside the territory of the United States.

1. Each employee who is assigned to perform safety-sensitive functions solely outside the territory of the United States shall be removed from the random testing pool upon the inception of such assignment.

2. Each covered employee who is removed from the random testing pool under this paragraph A shall be returned to the random testing pool when the employee resumes the performance of safety-sensitive functions wholly or partially within the territory of the United States.

**B.** The provisions of this appendix shall not apply to any person who performs a function listed in section III of this appendix by contract for an employer outside the territory of the United States.

**XIII.** *Waivers from 49 CFR 40.21.* An employer subject to this part may petition the Drug Abatement Division, Office of Aviation Medicine, for a waiver allowing the employer to stand down an employee following a report of a laboratory confirmed positive drug test or refusal, pending the outcome of the verification process.

**A.** Each petition for a waiver must be in writing and include substantial facts and justification to support the waiver. Each petition must satisfy the substantive requirements for obtaining a waiver, as provided in 49 CFR 40.21.

**B.** Each petition for a waiver must be submitted to the Federal Aviation Administration, Office of Aviation Medicine, Drug Abatement Division (AAM-800), 800 Independence Avenue, SW, Washington, DC 20591.

**C.** The Administrator may grant a waiver subject to 49 CFR 40.21(d).

[Docket No. 25148, 53 FR 47057, Nov. 21, 1988; as amended by Amdt. 121–240, 59 FR 42933, Aug. 19, 1994; Amdt. 121–240, 59 FR 53869, Oct. 26, 1994; 59 FR 62226, Dec. 2, 1994; Amdt. 121–240, 59 FR 66672, Dec. 28, 1994; 61 FR 37224, July 17, 1996; Amdt. 121–276, 65 FR 18887, April 10, 2000; Amdt. 121–285, 66 FR 41966, Aug. 9, 2001; Amdt. 121–287, 66 FR 57866, Nov. 19, 2001]

## Appendix J to Part 121

### Alcohol Misuse Prevention Program

This appendix contains the standards and components that must be included in an alcohol misuse prevention program required by this chapter.

### *I. General*

**A.** *Purpose.* The purpose of this appendix is to establish programs designed to help prevent accidents and injuries resulting from the misuse of alcohol by employees who perform safety-sensitive functions in aviation.

**B.** *Alcohol testing procedures.* Each employer shall ensure that all alcohol testing conducted pursuant to this appendix complies with the procedures set forth in 49 CFR part 40. The provisions of 49 CFR part 40 that address alcohol testing are made applicable to employers by this appendix.

**C.** *Employer Responsibility.* As an employer, you are responsible for all actions of your officials, representatives, and service agents in carrying out the requirements of the DOT agency regulations.

**D.** *Definitions.*

As used in this appendix—

*Accident* means an occurrence associated with the operation of an aircraft which takes place between the time any person boards the aircraft with the intention of flight and the time all such person have disembarked, and in which any person suffers death or serious injury or in which the aircraft receives substantial damage.

*Administrator* means the Administrator of the Federal Aviation Administration or his or her designated representative.

*Alcohol* means the intoxicating agent in beverage alcohol, ethyl alcohol, or other low molecular weight alcohols, including methyl or isopropyl alcohol.

*Alcohol concentration (or content)* means the alcohol in a volume of breath expressed in terms of grams of alcohol per 210 liters of breath as indicated by an evidential breath test under this appendix.

*Alcohol use* means the consumption of any beverage, mixture, or preparation, including any medication, containing alcohol.

*Contractor company* means a company that has employees who perform safety-sensitive functions by contract for an employer.

*Covered employee* means a person who performs, either directly or by contract, a safety-sensitive function listed in section II of this appendix for an employer (as defined below). For purposes of pre-employment testing only, the term "covered employee" includes a person applying to perform a safety-sensitive function.

*DOT agency* means an agency (or "operating administration") of the United States Department of Transportation administering regulations requiring alcohol testing (14 CFR parts 65, 121, and 135; 49 CFR parts 199, 219, and 382) in accordance with 49 CFR part 40.

*Employer* means a part 121 certificate holder; a part 135 certificate holder; an air traffic control facility not operated by the FAA or by or under contract to the U.S. military; and an operator as defined in 14 CFR 135.1(c).

*Performing* (a safety-sensitive function); an employee is considered to be performing a safety-sensitive function during any period in which he or she is actually performing, ready to perform, or immediately available to perform such functions.

*Refusal to submit* means that a covered employee engages in conduct specified in 49 CFR 40.261.

*Safety-sensitive function* means a function listed in section II of this appendix.

*Violation rate* means the number of covered employees (as reported under section IV of this appendix) found during random tests given under this appendix to have an alcohol concentration of 0.04 or greater plus the number of employees who refused a random test required by this appendix, divided by the total reported number of employees in the industry given random alcohol tests under this appendix plus the total reported number of employees in the industry who refuse a random test required by this appendix.

**E.** *Preemption of State and local laws.*

1. Except as provided in subparagraph 2 of this paragraph, these regulations preempt any State or local law, rule, regulation, or order to the extent that:

(a) Compliance with both State or local requirement and this appendix is not possible; or

(b) Compliance with the State or local requirement is an obstacle to the accomplishment and execution of any requirement in this appendix.

2. The alcohol misuse requirements of this title shall not be construed to preempt provisions of State criminal law that impose sanctions for reckless conduct leading to actual loss of life, injury, or

damage to property, whether the provisions apply specifically to transportation employees or employers or to the general public.

**F.** *Other requirements imposed by employers.*

Except as expressly provided in these alcohol misuse requirements, nothing in these requirements shall be construed to affect the authority of employers, or the rights of employees, with respect to the use or possession of alcohol, including any authority and rights with respect to alcohol testing and rehabilitation.

**G.** *Requirement for notice.*

Before performing an alcohol test under this appendix, each employer shall notify a covered employee that the alcohol test is required by this appendix. No employer shall falsely represent that a test is administered under this appendix.

## *II. Covered Employees*

Each employee who performs a function listed in this section directly or by contract for an employer as defined in this appendix must be subject to alcohol testing under an FAA-approved alcohol misuse prevention program implemented in accordance with this appendix. The covered safety-sensitive functions are:

**1.** Flight crewmember duties.
**2.** Flight attendant duties.
**3.** Flight instruction duties.
**4.** Aircraft dispatcher duties.
**5.** Aircraft maintenance or preventive maintenance duties.
**6.** Ground security coordinator duties.
**7.** Aviation screening duties.
**8.** Air traffic control duties.

## *III. Tests Required*

**A.** *Pre-employment testing*

As an employer, you may, but are not required to, conduct pre-employment alcohol testing under this part. If you choose to conduct pre-employment alcohol testing, you must comply with the following requirements:

1. You must conduct a pre-employment alcohol test before the first performance of safety-sensitive functions by every covered employee (whether a new employee or someone who has transferred to a position involving the performance of safety-sensitive functions).

2. You must treat all safety-sensitive employees performing safety-sensitive functions the same for the purpose of pre-employment alcohol testing (i.e., you must not test some covered employees and not others).

3. You must conduct the pre-employment tests after making a contingent offer of employment or transfer, subject to the employee passing the pre-employment alcohol test.

4. You must conduct all pre-employment alcohol tests using the alcohol testing procedures of 49 CFR part 40.

5. You must not allow a covered employee to begin performing safety-sensitive functions unless the result of the employee's test indicates an alcohol concentration of less than 0.04. If a pre-employment test result under this paragraph indicates an alcohol concentration of 0.02 or greater but less than 0.04, the provisions of paragraph F of section V of this appendix apply.

**B.** *Post-Accident Testing*

1. As soon as practicable following an accident, each employer shall test each surviving covered employee for alcohol if that employee's performance of a safety-sensitive function either contributed to the accident or cannot be completely discounted as a contributing factor to the accident. The decision not to administer a test under this section shall be based on the employer's determination, using the best available information at the time of the determination, that the covered employee's performance could not have contributed to the accident.

2. If a test required by this section is not administered within 2 hours following the accident, the employer shall prepare and maintain on file a record stating the reasons the test was not promptly administered. If a test required by this section is not administered within 8 hours following the accident, the employer shall cease attempts to administer an alcohol test and shall prepare and maintain the same record. Records shall be submitted to the FAA upon request of the Administrator or his or her designee.

(b) For the years stated in this paragraph, employers who submit MIS reports shall submit to the FAA each record of a test required by this section that is not completed within 8 hours. The employer's records of tests that are not completed within 8 hours shall be submitted to the FAA by March 15, 1996; March 15, 1997; and March 15, 1998; for calendar years 1995, 1996, and 1997, respectively. Employers shall append these records to their MIS submissions. Each record shall include the following information:

(i) Type of test (reasonable suspicion/post-accident);

(ii) Triggering event (including date, time, and location);

(iii) Employee category (do not include employee name or other identifying information);

(iv) Reason(s) test could not be completed within 8 hours; and

(v) If blood alcohol testing could have been completed within eight hours, the name, address, and telephone number of the testing site where blood testing could have occurred.

3. A covered employee who is subject to post-accident testing shall remain readily available for such testing or may be deemed by the employer to have refused to submit to testing. Nothing in this section shall be construed to require the delay of necessary medical attention for injured people following an accident or to prohibit a covered employee from leaving the scene of an accident for the period necessary to obtain assistance in responding to the accident or to obtain necessary emergency medical care.

C. *Random Testing*

1. Except as provided in paragraphs 2–4 of this section, the minimum annual percentage rate for random alcohol testing will be 25 percent of the covered employees.

2. The Administrator's decision to increase or decrease the minimum annual percentage rate for random alcohol testing is based on the violation rate for the entire industry. All information used for this determination is drawn from alcohol MIS reports required by this appendix. In order to ensure reliability of the data, the Administrator considers the quality and completeness of the reported data, may obtain additional information or reports from employers, and may make appropriate modifications in calculating the industry violation rate. Each year, the Administrator will publish in the **Federal Register** the minimum annual percentage rate for random alcohol testing of covered employees. The new minimum annual percentage rate for random alcohol testing will be applicable starting January 1 of the calendar year following publication.

3.(a) When the minimum annual percentage rate for random alcohol testing is 25 percent or more, the Administrator may lower this rate to 10 percent of all covered employees if the Administrator determines that the data received under the reporting requirements of this appendix for two consecutive calendar years indicate that the violation rate is less than 0.5 percent.

(b) When the minimum annual percentage rate for random alcohol testing is 50 percent, the Administrator may lower this rate to 25 percent of all covered employees if the Administrator determines that the data received under the reporting requirements of this appendix for two consecutive calendar years indicate that the violation rate is less than 1.0 percent but equal to or greater than 0.5 percent.

4.(a) When the minimum annual percentage rate for random alcohol testing is 10 percent, and the data received under the reporting requirements of this appendix for that calendar year indicate that the violation rate is equal to or greater than 0.5 percent but less than 1.0 percent, the Administrator will increase the minimum annual percentage rate for random alcohol testing to 25 percent of all covered employees.

(b) When the minimum annual percentage rate for random alcohol testing is 25 percent or less, and the data received under the reporting requirements of this appendix for that calendar year indicate that the violation rate is equal to or greater than 1.0 percent, the Administrator will increase the minimum annual percentage rate for random alcohol testing to 50 percent of all covered employees.

5. The selection of employees for random alcohol testing shall be made by a scientifically valid method, such as a random-number table or a computer-based random number generator that is matched with employee's Social Security numbers, payroll identification numbers, or other comparable identifying numbers. Under the selection process used, each covered employee shall have an equal chance of being tested each time selections are made.

6. The employer shall randomly select a sufficient number of covered employees for testing during each calendar year to equal an annual rate not less than the minimum annual percentage rate for random alcohol testing determined by the Administrator. If the employer conducts random testing through a Consortium/Third-party administrator (C/TPA), the number of employees to be tested may be calculated for each individual employer or may be based on the total number of covered employees who are subject to random alcohol testing at the same minimum annual percentage rate under this appendix or any DOT alcohol testing rule.

7. Each employer shall ensure that random alcohol tests conducted under this appendix are unannounced and that the dates for administering random tests are spread reasonably throughout the calendar year.

8. Each employer shall require that each covered employee who is notified of selection for random testing proceeds to the testing site immediately; provided, however, that if the employee is performing a safety-sensitive function at the time of the notification, the employer shall instead ensure that the employee ceases to perform the safety-sensitive function and proceeds to the testing site as soon as possible.

9. A covered employee shall only be randomly tested while the employee is performing safety-sensitive functions; just before the employee is to perform safety-sensitive functions; or just after the employee has ceased performing such functions.

10. If a given covered employee is subject to random alcohol testing under the alcohol testing rules of more than one DOT agency, the employee shall be subject to random alcohol testing at the percentage rate established for the calen-

dar year by the DOT agency regulating more than 50 percent of the employee's functions.

11. If an employer is required to conduct random alcohol testing under the alcohol testing rules of more than one DOT agency, the employer may—

(a) Establish separate pools for random selection, with each pool containing the covered employees who are subject to testing at the same required rate; or

(b) Randomly select such employees for testing at the highest percentage rate established for the calendar year by any DOT agency to which the employer is subject.

**D.** *Reasonable Suspicion Testing*

1. An employer shall require a covered employee to submit to an alcohol test when the employer has reasonable suspicion to believe that the employee has violated the alcohol misuse prohibitions in §65.46a, 121.458, or 135.253 of this chapter.

2. The employer's determination that reasonable suspicion exists to require the covered employee to undergo an alcohol test shall be based on specific, contemporaneous, articulable observations concerning the appearance, behavior, speech or body odors of the employee. The required observations shall be made by a supervisor who is trained in detecting the symptoms of alcohol misuse. The supervisor who makes the determination that reasonable suspicion exists shall not conduct the breath alcohol test on that employee.

3. Alcohol testing is authorized by this section only if the observations required by paragraph 2 are made during, just preceding, or just after the period of the work day that the covered employee is required to be in compliance with this rule. An employee may be directed by the employer to undergo reasonable suspicion testing for alcohol only while the employee is performing safety-sensitive functions; just before the employee is to perform safety-sensitive functions; or just after the employee has ceased performing such functions.

4.(a) If a test required by this section is not administered within 2 hours following the determination made under paragraph 2 of this section, the employer shall prepare and maintain on file a record stating the reasons the test was not promptly administered. If a test required by this section is not administered within 8 hours following the determination made under paragraph 2 of this section, the employer shall cease attempts to administer an alcohol test and shall state in the record the reasons for not administering the test.

(b) Notwithstanding the absence of a reasonable suspicion alcohol test under this section, no covered employee shall report for duty or remain on duty requiring the performance of safety-sensitive functions while the employee is under the influence of or impaired by alcohol, as shown by the behavioral, speech, or performance indicators of alcohol misuse, nor shall an employer permit the covered employee to perform or continue to perform safety-sensitive functions until:

(1) An alcohol test is administered and the employee's alcohol concentration measures less than 0.02; or

(2) The start of the employee's next regularly scheduled duty period, but not less than 8 hours following the determination made under paragraph 2 of this section that there is reasonable suspicion that the employee has violated the alcohol misuse provisions in §65.46a, 121.458, 135.253 of this chapter.

(c) No employer shall take any action under this appendix against a covered employee based solely on the employee's behavior and appearance in the absence of an alcohol test. This does not prohibit an employer with authority independent of this appendix from taking any action otherwise consistent with law.

**E.** *Return to Duty Testing*

Each employer shall ensure that before a covered employee returns to duty requiring the performance of a safety-sensitive function after engaging in conduct prohibited in §65.46a, §121.458, or §135.253 of this chapter, the employee shall undergo a return to duty alcohol test with a result indicating an alcohol concentration of less than 0.02. The test cannot occur until after the SAP has determined that the employee has successfully complied with the prescribed education and/or treatment.

**F.** *Follow-up Testing*

1. Each employer shall ensure that the employee who engages in conduct prohibited by §65.46a, §121.458, or §135.253 of this chapter is subject to unannounced follow-up alcohol testing as directed by a SAP.

2. The number and frequency of such testing shall be determined by the employer's SAP, but must consist of at least six tests in the first 12 months following the employee's return to duty.

3. The employer must direct the employee to undergo testing for drugs in accordance with appendix I of this part, in addition to alcohol, if the SAP determines that drug testing is necessary for the particular employee. Any such drug testing shall be conducted in accordance with the provisions of 49 CFR part 40.

4. Follow-up testing shall not exceed 60 months after the date the individual begins to perform or returns to the performance of a safety-sensitive function. The SAP may terminate the requirement for follow-up testing at any time after

the first six tests have been conducted, if the SAP determines that such testing is no longer necessary.

5. A covered employee shall be tested for alcohol under this paragraph only while the employee is performing safety-sensitive functions, just before the employee is to perform safety-sensitive functions, or just after the employee has ceased performing such functions.

**G.** *Retesting of Covered Employees With an Alcohol Concentration of 0.02 or Greater but Less Than 0.04*

Each employer shall retest a covered employee to ensure compliance with the provisions of section V, paragraph F of this appendix, if the employer chooses to permit the employee to perform a safety-sensitive function within 8 hours following the administration of an alcohol test indicating an alcohol concentration of 0.02 or greater but less than 0.04.

## *IV. Handling of Test Results, Record Retention, and Confidentiality*

**A.** *Retention of Records*

1. General Requirement. In addition to the records required to be maintained under 49 CFR part 40, employers must maintain records required by this appendix in a secure location with controlled access.

2. Period of retention.

(a) Five years.

(1) Copies of any annual reports submitted to the FAA under this appendix for a minimum of 5 years.

(2) Records of notifications to the Federal Air Surgeon of violations of the alcohol misuse prohibitions in this chapter by covered employees who hold medical certificates issued under part 67 of this chapter.

(3) Documents presented by a covered employee to dispute the result of an alcohol test administered under this appendix.

(4) Records related to other violations of §65.46a, §121.458, or §135.253 of this chapter.

(b) Two years. Records related to the testing process and training required under this appendix.

(1) Documents related to the random selection process.

(2) Documents generated in connection with decisions to administer reasonable suspicion alcohol tests.

(3) Documents generated in connection with decisions on post-accident tests.

(4) Documents verifying existence of a medical explanation of the inability of a covered employee to provide adequate breath for testing.

(5) Materials on alcohol misuse awareness, including a copy of the employer's policy on alcohol misuse.

(6) Documentation of compliance with the requirements of section VI, paragraph A of this appendix.

(7) Documentation of training provided to supervisors for the purpose of qualifying the supervisors to make a determination concerning the need for alcohol testing based on reasonable suspicion.

(8) Certification that any training conducted under this appendix complies with the requirements for such training.

**B.** *Reporting of Results in a Management Information System*

1. Annual reports summarizing the results of alcohol misuse prevention programs shall be submitted to the FAA in the form and manner prescribed by the Administrator by March 15 of each year covering the previous calendar year (January 1 through December 31) in accordance with the provisions below.

(a) Each part 121 certificate holder shall submit an annual report each year.

(b) Each entity conducting an alcohol misuse prevention program under the provisions of this appendix, other than a part 121 certificate holder, that has 50 or more covered employees on January 1 of any calendar year shall submit an annual report to the FAA for that calendar year.

(c) The administrator reserves the right to require employers not otherwise required to submit annual reports to prepare and submit such reports to the FAA. Employers that will be required to submit annual reports under this provision will notified in writing by the FAA.

2. Each employer that is subject to more than one DOT agency alcohol rule shall identify each employee covered by the regulations of more than one DOT agency. The identification will be by the total number and category of covered function. Prior to conducting any alcohol test on a covered employee subject to the rules of more than one DOT agency, the employer shall determine which DOT agency rule or rules authorizes or requires the test. The test result information shall be directed to the appropriate DOT agency or agencies.

3. Each employer shall ensure the accuracy and timeliness of each report submitted.

4. Each report shall be submitted in the form and manner prescribed by the Administrator.

5. Each report shall be signed by the employer's alcohol misuse prevention program manager or other designated representative.

6. Each report that contains information on an alcohol screening test result of 0.02 or greater or a violation of the alcohol misuse provisions of

§65.46a, 121.458, or 135.253 of this chapter shall include the following informational elements:

(a) Number of covered employees by employee category.

(b) Number of covered employees in each category subject to alcohol testing under the alcohol misuse rule of another DOT agency, identified under each agency.

(c)(1) Number of screening tests by type of test and employee category.

(2) Number of confirmation tests, by type of test and employee category.

(d) Number of confirmation alcohol tests indicating an alcohol concentration of 0.02 or greater but less than 0.04 by type of test and employee category.

(e) Number of confirmation alcohol tests indicating an alcohol concentration of 0.04 or greater, by type of test and employee category.

(f) Number of persons denied a position as a covered employee following a pre-employment alcohol test indicating an alcohol concentration of 0.04 or greater.

(g) Number of covered employees with a confirmation alcohol test indicating an alcohol concentration of 0.04 or greater who were returned to duty in covered positions (having complied with the recommendations of a substance abuse professional as described 49 CFR Part 40).

(h) Number of covered employees who were administered alcohol and drug tests at the same time, with both a positive drug test result and an alcohol test result indicating an alcohol concentration of 0.04 or greater.

(i) Number of covered employees who were found to have violated other alcohol misuse provisions of §§65.46a, 121.458, or 135.253 of this chapter, and the action taken in response to the violation.

(j) Number of covered employees who refused to submit to an alcohol test required under this appendix, the number of such refusals that were for random tests, and the action taken in response to each refusal.

(k) Number of supervisors who have received required training during the reporting period in determining the existence of reasonable suspicion of alcohol misuse.

7. Each report with no screening test results of 0.02 or greater or violations of the alcohol misuse provisions of §§65.46a, 121.458, or 135.253 of this chapter shall include the following informational elements. (This report may only be submitted if the program results meet these criteria.)

(a) Number of covered employees by employee category.

(b) Number of covered employees in each category subject to alcohol testing under the alcohol misuse rule of another DOT agency, identified by each agency.

(c) Number of screening tests by type of test and employee category.

(d) Number of covered employees who engaged in alcohol misuse who were returned to duty in covered positions (having complied with the recommendations of a substance abuse professional as described in 49 CFR Part 40).

(e) Number of covered employees who refused to submit to an alcohol test required under this appendix, and the action taken in response to each refusal.

(f) Number of supervisors who have received required training during the reporting period in determining the existence of reasonable suspicion of alcohol misuse.

8. A C/TPA may prepare reports on behalf of individual aviation employers for purposes of compliance with this reporting requirement. However, the aviation employer shall sign and submit such a report and shall remain responsible for ensuring the accuracy and timeliness of each report prepared on its behalf by a C/TPA. A C/TPA must not sign the form.

**C.** *Access to Records and Facilities*

1. Except as required by law or expressly authorized or required in this appendix, no employer shall release covered employee information that is contained in records required to be maintained under this appendix.

2. A covered employee is entitled, upon written request, to obtain copies of any records pertaining to the employee's use of alcohol, including any records pertaining to his or her alcohol tests in accordance with 49 CFR part 40. The employer shall promptly provide the records requested by the employee. Access to an employee's records shall not be contingent upon payment for records other than those specifically requested.

3. Each employer shall permit access to all facilities utilized in complying with the requirements of this appendix to the Secretary of Transportation or any DOT agency with regulatory authority over the employer or any of its covered employees.

## *V. Consequences for Employees Engaging in Alcohol-Related Conduct*

**A.** *Removal From Safety-sensitive Function*

1. Except as provided in 49 CFR Part 40, no covered employee shall perform safety-sensitive functions if the employee has engaged in conduct prohibited by §§65.46a, 121.458, or 135.253 of this chapter or an alcohol misuse rule of another DOT agency.

2. No employer shall permit any covered employee to perform safety-sensitive functions if the employer has determined that the employee has violated this paragraph.

**B.** *Permanent Disqualification From Service*

An employee who violates §§65.46a(c), 121.458(c), or 135.253(c) of this chapter, or who engages in alcohol use that violates another alcohol misuse provision of §§65.46a, 121.458, or 135.253 of this chapter and had previously engaged in alcohol use that violated the provisions of §§65.46a, 121.458, or 135.253 of this chapter after becoming subject to such prohibitions is permanently precluded from performing for an employer the safety-sensitive duties the employee performed before such violation.

**C.** *Notice to the Federal Air Surgeon*

1. An employer who determines that a covered employee who holds an airman medical certificate issued under part 67 of this chapter has engaged in alcohol use that violated the alcohol misuse provisions of §§65.46a, 121.458, or 135.253 of this chapter shall notify the Federal Air Surgeon within 2 working days.

2. Each such employer shall forward to the Federal Air Surgeon a copy of the report of any evaluation performed under the provisions of section VI.C. of this appendix within 2 working days of the employer's receipt of the report.

3. All documents shall be sent to the Federal Air Surgeon, Office of Aviation Medicine, Federal Aviation Administration, Attn: Drug Abatement Division (AAM–800), 800 Independence Avenue SW, Washington, DC 20591.

4. No covered employee who is required to hold a medical certificate under part 67 of this chapter to perform a safety-sensitive duty shall perform that duty following a violation of this appendix until and unless the Federal Air Surgeon has recommended that the employee be permitted to perform such duties.

5. Once the Federal Air Surgeon has recommended under paragraph C.4. of this section that the employee be permitted to perform safety-sensitive duties, the employer cannot permit the employee to perform those safety-sensitive duties until the employer has ensured that the employee meets the return to duty requirements in accordance with 49 CFR part 40.

**D.** *Notice of Refusals*

1. Except as provided in subparagraph 2 of this paragraph D, each employer shall notify the FAA within 5 working days of any covered employee who holds a certificate issued under 14 CFR part 61, part 63, or part 65 who has refused to submit to an alcohol test required under this appendix. Notifications should be sent to: Federal Aviation Administration, Office of Aviation Medicine, Drug Abatement Division (AAM–800), 800 Independence Avenue SW, Washington, DC 20591.

2. An employer is not required to notify the above office of refusals to submit to pre-employment alcohol tests or refusals to submit to return to duty tests.

**E.** *Required Evaluation and Testing*

No covered employee who has engaged in conduct prohibited by §§65.46a, 121.458, or 135.253 of this chapter shall perform safety-sensitive functions unless the employee has met the requirements of 49 CFR part 40. No employer shall permit a covered employee who has engaged in such conduct to perform safety-sensitive functions unless the employee has met the requirements of 49 CFR part 40.

**F.** *Other Alcohol-Related Conduct*

1. No covered employee tested under the provisions of section III of this appendix who is found to have an alcohol concentration of 0.02 or greater but less than 0.04 shall perform or continue to perform safety-sensitive functions for an employer, nor shall an employer permit the employee to perform or continue to perform safety-sensitive functions, until:

(a) The employee's alcohol concentration measures less than 0.02; or

(b) The start of the employee's next regularly scheduled duty period, but not less than 8 hours following administration of the test.

2. Except as provided in subparagraph 1 of this paragraph, no employer shall take any action under this rule against an employee based solely on test results showing an alcohol concentration less than 0.04. This does not prohibit an employer with authority independent of this rule from taking any action otherwise consistent with law.

## *VI. Alcohol Misuse Information, Training, and Substance Abuse Professional*

**A.** *Employer Obligation to Promulgate a Policy on the Misuse of Alcohol*

1. *General requirements.* Each employer shall provide educational materials that explain these alcohol misuse requirements and the employer's policies and procedures with respect to meeting those requirements.

(a) The employer shall ensure that a copy of these materials is distributed to each covered employee prior to the start of alcohol testing under the employer's FAA-mandated alcohol misuse prevention program and to each person subsequently hired for or transferred to a covered position.

(b) Each employer shall provide written notice to representatives of employee organizations of the availability of this information.

2. *Required content.* The materials to be made available to employees shall include detailed discussion of at least the following:

(a) The identity of the person designated by the employer to answer employee questions about the materials.

(b) The categories of employees who are subject to the provisions of these alcohol misuse requirements.

(c) Sufficient information about the safety-sensitive functions performed by those employees to make clear what period of the work day the covered employee is required to be in compliance with these alcohol misuse requirements.

(d) Specific information concerning employee conduct that is prohibited by this chapter.

(e) The circumstances under which a covered employee will be tested for alcohol under this appendix.

(f) The procedures that will be used to test for the presence of alcohol, protect the employee and the integrity of the breath testing process, safeguard the validity of the test results, and ensure that those results are attributed to the correct employee.

(g) The requirement that a covered employee submit to alcohol tests administered in accordance with this appendix.

(h) An explanation of what constitutes a refusal to submit to an alcohol test and the attendant consequences.

(i) The consequences for covered employees found to have violated the prohibitions in this chapter, including the requirement that the employee be removed immediately from performing safety-sensitive functions, and the process in 49 CFR Part 40, subpart O.

(j) The consequences for covered employees found to have an alcohol concentration of 0.02 or greater but less than 0.04.

(k) Information concerning the effects of alcohol misuse on an individual's health, work, and personal life; signs and symptoms of an alcohol problem; and available methods of evaluating and resolving problems associated with the misuse of alcohol; and intervening when an alcohol problem is suspected, including confrontation, referral to any available employee assistance program, and/or referral to management.

(l) *Optional provisions.* The materials supplied to covered employees may also include information on additional employer policies with respect to the use or possession of alcohol, including any consequences for an employee found to have a specified alcohol level, that are based on the employer's authority independent of this appendix. Any such additional policies or consequences must be clearly and obviously described as being based on independent authority.

**B.** *Training for Supervisors*

Each employer shall ensure that persons designated to determine whether reasonable suspicion exists to require a covered employee to undergo alcohol testing under section II of this appendix receive at least 60 minutes of training on the physical, behavioral, speech, and performance indicators of probable alcohol misuse.

**C.** *Substance Abuse Professional (SAP) Duties*

The SAP must perform the functions set forth in 49 CFR part 40, Subpart O, and this appendix.

### *VII. Employer's Alcohol Misuse Prevention Program*

**A.** *Schedule for Submission of Certification Statements and Implementation*

1. Each employer shall submit an alcohol misuse prevention program (AMPP) certification statement as prescribed in paragraph B of section VII of this appendix, in duplicate, to the Federal Aviation Administration, Office of Aviation Medicine, Drug Abatement Division (AAM–800), 800 Independence Avenue SW, Washington, DC 20591, in accordance with the schedule below.

(a) Each employer that holds a part 121 certificate, each employer that holds a part 135 certificate and directly employs more than 50 covered employees, and each air traffic control facility affected by this rule shall submit a certification statement to the FAA by July 1, 1994. Each employer must implement an AMPP meeting the requirements of this appendix on January 1, 1995. Contractor employees to these employers must be subject to an AMPP meeting the requirements of this appendix by July 1, 1995.

(b) Each employer that holds a part 135 certificate and directly employs from 11 to 50 covered employees shall submit a certification statement to the FAA by January 1, 1995. Each employer must implement an AMPP meeting the requirements of this appendix on July 1, 1995. Contractor employees to these employers must be subject to an AMPP meeting the requirements of this appendix by January 1, 1996.

(c) Each employer that holds a part 135 certificate and directly employs ten or fewer covered employees, and each operator as defined in 14 CFR 135.1(c) shall submit a certification statement to the FAA by July 1, 1995. Each employer must implement an AMPP meeting the requirements of this appendix on January 1, 1996. Contractor employees to these employers must be subject to an AMPP meeting the requirements of this appendix by July 1, 1996.

2. A company providing covered employees by contract to employers may be authorized by the FAA to establish an AMPP under the auspices of

this appendix by submitting a certification statement meeting the requirements of paragraph B of section VII of this appendix directly to the FAA. Each contractor company that establishes an AMPP shall implement its AMPP in accordance with the provisions of this appendix.

(a) The FAA may revoke its authorization in the case of any contractor company that fails to properly implement its AMPP.

(b) No employer shall use a contractor company's employee who is not subject to the employer's AMPP unless the employer has first determined that the employee is subject to the contractor company's FAA-mandated AMPP.

3. Any person who applies for a certificate under the provisions of parts 121 or 135 of this chapter after the effective date of the final rule shall submit an alcohol misuse prevention program (AMPP) certification statement to the FAA prior to beginning operations pursuant to the certificate. The AMPP shall be implemented concurrently with beginning such operations or on the date specified in paragraph A.1. of this section, whichever is later. Contractor employees to a new certificate holder must be subject to an FAA-mandated AMPP within 180 days of the implementation of the employer's AMPP.

4. Any person who intends to begin air traffic control operations as an employer as defined in 14 CFR 65.46(a)(2) (air traffic control facilities not operated by the FAA or by or under contract to the U.S. military) after March 18, 1994 shall, not later than 60 days prior to the proposed initiations of such operations, submit an alcohol misuse prevention program certification statement to the FAA. The AMPP shall be implemented concurrently with the inception of operations or on the date specified in paragraph A.1 of this section, whichever is later. Contractor employees to a new air traffic control facility must be subject to an FAA-approved program within 180 days of the implementation of the facility's program.

5. Any person who intends to begin sightseeing operations as an operator under 14 CFR 135.1(c) after March 18, 1994 shall, not later than 60 days prior to the proposed initiation of such operations, submit an alcohol misuse prevention program (AMPP) certification statement to the FAA. The AMPP shall be implemented concurrently with the inception of operations or on the date specified in paragraph A.1 of this section, whichever is later. Contractor employees to a new operator must be subject to an FAA-mandated AMPP within 180 days of the implementation of the employer's AMPP.

6. The duplicate certification statement shall be annotated indicating receipt by the FAA and returned to the employer or contractor company.

7. Each employer, and each contractor company that submits a certification statement directly to the FAA, shall notify the FAA of any proposed change in status, (e.g., join another carrier's program) prior to the effective date of such change. The employer or contractor company must ensure that it is continuously covered by an FAA-mandated alcohol misuse prevention program.

**B.** *Required Content of AMPP Certification Statements*

1. Each AMPP certification statement submitted by an employer or a contractor company shall provide the following information:

(a) The name, address, and telephone number of the employer/contractor company and for the employer/contractor company AMPP manager.

(b) FAA operating certificate number (if applicable);

(c) The date on which the employer or contractor company will implement its AMPP;

(d) A statement signed by an authorized representative of the employer or contractor company certifying an understanding of and agreement to comply with the provisions of the FAA's alcohol misuse prevention regulations.

## *VIII. Employees Located Outside the U.S.*

**A.** No covered employees shall be tested for alcohol misuse while located outside the territory of the United States.

1. Each covered employee who is assigned to perform safety-sensitive functions solely outside the territory of the United States shall be removed from the random testing pool upon the inception of such assignment.

2. Each covered employee who is removed from the random testing pool under this paragraph shall be returned to the random testing pool when the employee resumes the performance of safety-sensitive functions wholly or partially within the territory of the United States.

**B.** The provisions of this appendix shall not apply to any person who performs a safety-sensitive function by contract for an employer outside the territory of the United States.

[Docket No. 27065, 59 FR 7390, Feb. 15, 1994; as amended by Amdt. 121–237, 59 FR 53086, Oct. 21, 1994; 59 FR 62238, Dec. 2, 1994; Amdt. 121–237, 60 FR 24766, May 10, 1995; 61 FR 37224, July 17, 1996; Amdt. 121–276, 65 FR 18887, April 10, 2000; Amdt. 121–285, 66 FR 41968, Aug. 9, 2001; Amdt. 121–287, 66 FR 57866, Nov. 19, 2001]

# Appendix K to Part 121

## Performance Requirements for Certain Turbopropeller Powered Airplanes

**1.** Applicability. This appendix specifies requirements for the following turbopropeller powered airplanes that must comply with the Airplane Performance Operating Limitations in §§121.189 through 121.197:

a. After December 20, 2010, each airplane manufactured before March 20, 1997, and type certificated in the:

i. Normal category before July 1, 1970, and meets special conditions issued by the Administrator for airplanes intended for use in operations under part 135 of this chapter.

ii. Normal category before July 19, 1970, and meets the additional airworthiness standards in SFAR No. 23 of 14 CFR part 23.

iii. Normal category, and complies with the additional airworthiness standards in Appendix A of part 135 of this chapter.

iv. Normal category, and complies with section 1.(a) or 1.(b) of SFAR No. 41 of 14 CFR part 21.

b. After March 20, 1997, each airplane:

i. Type certificated prior to March 29, 1995, in the commuter category.

ii. Manufactured on or after March 20, 1997, and that was type certificated in the normal category, and complies with the requirements described in paragraphs 1.a.i through iii of this appendix.

**2.** Background. Sections 121.157 and 121.173(b) require that the airplanes operated under this part and described in paragraph 1 of this appendix, comply with the Airplane Performance Operating Limitations in §§121.189 through 121.197. Airplanes described in §121.157(f) and paragraph 1.a of this appendix must comply on and after December 20, 2010. Airplanes described in §121.157(e) and paragraph 1.b of this appendix must comply on and after March 20, 1997. (Airplanes type certificated in the normal category, and in accordance with SFAR No. 41 of 14 CFR part 21, as described in paragraph 1.a.iv of this appendix, may not be produced after October 17, 1991.)

**3.** References. Unless otherwise specified, references in this appendix to sections of part 23 of this chapter are to those sections of 14 CFR part 23; as amended by Amendment No. 23–45 (August 6, 1993, 58 FR 42156).

### *Performance*

**4.** Interim Airplane Performance Operating Limitations.

a. Until December 20, 2010, airplanes described in paragraph 1.a of this appendix may continue to comply with the requirements in Subpart I of part 135 and §135.181(a)(2) of this chapter that apply to small, nontransport category airplanes.

b. Until March 20, 1997, airplanes described in paragraph 1.b.i of this appendix may continue to comply with the requirements in Subpart I of part 135 of this chapter that apply to commuter category airplanes.

**5.** Final Airplane Performance Operating Limitations.

a. Through an amended type certification program or a supplemental type certification program, each airplane described in paragraph 1.a and 1.b.ii of this appendix must be shown to comply with the commuter category performance requirements specified in this appendix, which are included in part 23 of this chapter. Each new revision to a current airplane performance operating limitation for an airplane that is or has been demonstrated to comply, must also be approved by the Administrator. An airplane approved to the requirements of section 1.(b) of SFAR No. 41 of 14 CFR part 21, as described in paragraph 1.a.iv of this appendix, and that has been demonstrated to comply with the additional requirements of section 4.(c) of SFAR No. 41 of 14 CFR part 21 and International Civil Aviation Organization Annex 8 (available from the FAA, 800 Independence Avenue SW, Washington, DC 20591), will be considered to be in compliance with the commuter category performance requirements.

b. Each turbopropeller powered airplane subject to this appendix must be demonstrated to comply with the airplane performance operating limitation requirements of this chapter specified as follows:

i. Section 23.45 Performance General.

ii. Section 23.51 Takeoff.

iii. Section 23.53 Takeoff speeds.

iv. Section 23.55 Accelerate stop distance.

v. Section 23.57 Takeoff path.

vi. Section 23.59 Takeoff distance and takeoff run.

vii. Section 23.61 Takeoff flight path.

viii. Section 23.65 Climb: All engines operating.

ix. Section 23.67 Climb: one engine inoperative.

x. Section 23.75 Landing.

xi. Section 23.77 Balked landing.

xii. Sections 23.1581 through 23.1589 Airplane flight manual and approved manual material.

**6.** Operation. After compliance with the final airplane performance operating limitations requirements has been demonstrated and added to the Airplane Flight Manual performance data of the affected airplane, that airplane must be operated in accordance with the performance limitations of §§121.189 through 121.197.

[Docket No. 28154, 60 FR 65936, Dec. 20, 1995]

## Appendix L to Part 121

### Type Certification Regulations Made Previously Effective

Appendix L lists regulations in this part that require compliance with standards contained in superseded type certification regulations that continue to apply to certain transport category airplanes. The tables set out citations to current CFR section, applicable aircraft, superseded type certification regulation and applicable time periods, and the CFR edition and Federal Register documents where the regulation having prior effect is found. Copies of all superseded regulations may be obtained at the Federal Aviation Administration Law Library, Room 924, 800 Independence Avenue SW, Washington, D.C.

*Editorial Note: Table begins on the next page*

| Part 121 section | Applicable aircraft | Provisions: CFR/FR references |
|---|---|---|
| §121.312(a)(1)(i) | Transport category; or nontransport category type certificated before January 1, 1965; passenger capacity of 20 or more; manufactured prior to August 20, 1990. | Heat release rate testing. 14 CFR 25.853(d) in effect March 6, 1995: 14 CFR parts 1 to 59, Revised as of January 1, 1995, and amended by Amdt 25–83, 60 FR 6623, February 2, 1995.<br>Formerly 14 CFR 25.853(a-1) in effect August 20, 1986: 14 CFR parts 1 to 59, Revised as of January 1, 1986. |
| §121.312(a)(1)(ii) | Transport category; or nontransport category type certificated before January 1, 1965; passenger capacity of 20 or more; manufactured after August 19, 1990. | Heat release rate and smoke testing. 14 CFR 25.853(d) in effect March 6, 1995: 14 CFR parts 1 to 59, Revised as of January 1, 1995, and amended by Amdt 25–83, 60 FR 6623, February 2, 1995.<br>Formerly 14 CFR 25.853(a-1) in effect September 26, 1988: 14 CFR parts 1 to 59, Revised as of January 1, 1988, and amended by Amdt 25–66, 53 FR 32584, August 25, 1988. |
| §121.312(a)(2)(i) | Transport category; or nontransport category type certificate before January 1, 1965; application for type certificate filed prior to May 1, 1972; substantially complete replacement of cabin interior on or after May 1, 1972. | Provisions of 14 CFR 25.853 in effect on April 30, 1972: 14 CFR parts 1 to 59, Revised as of January 1, 1972. |
| §121.312(a)(3)(i) | Transport category type certificated after January 1, 1958; nontransport category type certificated after January 1, 1958, but before January 1, 1965; passenger capacity of 20 or more; substantially complete replacement of the cabin interior on or after March 6, 1995. | Heat release rate testing. 14 CFR 25.853(d) in effect March 6, 1995: 14 CFR parts 1 to 59, Revised as of January 1, 1995; and amended by Amdt 25–83, 60 FR 6623, February 2, 1995.<br>Formerly 14 CFR 25.853(a-1) in effect August 20, 1986: 14 CFR parts 1 to 59, Revised as of January 1, 1986. |
| §121.312(a)(3)(ii) | Transport category type certificated after January 1, 1958; nontransport category type certificated after January 1, 1958, but before January 1, 1965; passenger capacity of 20 or more; substantially complete replacement of the cabin interior on or after August 20, 1990. | Heat release rate and smoke testing. 14 CFR 25.853(d) in effect March 6, 1995; 14 CFR parts 1 to 59, Revised as of January 1, 1995; and amended by Amdt 25–83, 60 FR 6623, February 2, 1995.<br>Formerly 14 CFR 25.853(a-1) in effect September 26, 1988: CFR, Title 14, Parts 1 to 59, Revised as of January 1, 1988, and amended by Amdt 25–66, 53 FR 32584, August 25, 1988. |
| §121.312(b) (1) and (2) | Transport category airplane type certificated after January 1, 1958; Nontransport category airplane type certificated after December 31, 1964. | Seat cushions. 14 CFR 25.853(c) effective on November 26, 1984: 14 CFR parts 1 to 59, Revised as of January 1, 1984, and amended by Amdt 25–59, 49 FR 43188, October 26, 1984. |
| §121.312(c) | Airplane type certificated in accordance with SFAR No. 41; maximum certificated takeoff weight in excess of 12,500 pounds. | Compartment interior requirements. 14 CFR 25.853(a) in effect March 6, 1995: 14 CFR parts 1 to 59, Revised as of January 1, 1995, and amended by Amdt 25–83, 60 FR 6623, February 2, 1995.<br>Formerly 14 CFR 25.853(a), (b-1), (b-2), and (b-3) in effect on September 26, 1978: 14 CFR parts 1 to 59, Revised as of January 1, 1978. |
| §121.314(a) | Transport category airplanes type certificated after January 1, 1958. | Class C or D cargo or baggage compartment definition, 14 CFR 25.857 in effect on June 16, 1986, 14 CFR parts 1 to 59, Revised 1/1/97, and amended by Amendment 25–60, 51 FR 18243, May 16, 1986. |

[Docket No. 28154, 60 FR 65936, Dec. 20, 1995; as amended by Amdt. 121–269, 63 FR 8049, Feb. 17, 1998]

# Appendix M to Part 121
## Airplane Flight Recorder Specifications

The recorded values must meet the designated range, resolution, and accuracy requirements during dynamic and static conditions. All data recorded must be correlated in time to within one second.

| Parameters | Range | Accuracy (sensor input) | Seconds per sampling interval | Resolution | Remarks |
|---|---|---|---|---|---|
| 1. Time or Relative Times Counts[1] | 24 Hrs, 0 to 4095 | ± 0.125% Per Hour | 4 | 1 sec | UTC time preferred when available. Count increments each 4 seconds of system operation. |
| 2. Pressure Altitude | –1000 ft to max certificated altitude of aircraft. +5000 ft. | ± 100 to ± 700 ft (see table, TSO C124a or TSO C51a) | 1 | 5' to 35' | Data should be obtained from the air data computer when practicable. |
| 3. Indicated airspeed or Calibrated airspeed | 50 KIAS or minimum value to Max $V_{S0}$ to 1.2 $V_D$ | ± 5% and ± 3% | 1 | 1 kt | Data should be obtained from the air data computer when practicable. |
| 4. Heading (Primary flight crew reference) | 0–360° and Discrete "true" or "mag" | ± 2° | 1 | 0.5° | When true or magnetic heading can be selected as the primary heading reference, a discrete indicating selection must be recorded. |
| 5. Normal Acceleration (Vertical) | –3g to +6g | ± 1% of max range excluding datum error of ± 5 | 0.125 | 0.004g | — |
| 6. Pitch Attitude | ± 75° | ± 2° | 1 or 0.25 for airplanes operated under §121.344(f) | 0.5° | A sampling rate of 0.25 is recommended. |
| 7. Roll Attitude[2] | ± 180° | ± 2° | 1 or 0.5 for airplanes operated under §121.344(f) | 0.5° | A sampling rate of 0.5 is recommended. |
| 8. Manual Radio Transmitter Keying or CVR/DFDR synchronization reference | On-Off (Discrete) None | — | 1 | — | Preferably each crewmember but one discrete acceptable for all transmission provided the CVR/FDR system complies with TSO C124a CVR synchronization requirements (paragraph 4.2.1 ED-55). |
| 9. Thrust/Power on Each Engine — primary flight crew reference[14] | Full Range Forward | ± 2% | 1 (per engine) | 0.2% of full range | Sufficient parameters (e.g. EPR, N1 or Torque, NP) as appropriate to the particular engine be recorded to determine power in forward and reverse thrust, including potential overspeed conditions. |
| 10. Autopilot Engagement | Discrete "on" or "off" | — | 1 | — | — |
| 11. Longitudinal Acceleration | ± 1g | ± 1.5% max. range excluding datum error of ± 5% | 0.25 | 0.004g | — |

121

## Appendix M to Part 121
## Airplane Flight Recorder Specifications (Continued)

The recorded values must meet the designated range, resolution, and accuracy requirements during dynamic and static conditions. All data recorded must be correlated in time to within one second.

| Parameters | Range | Accuracy (sensor input) | Seconds per sampling interval | Resolution | Remarks |
|---|---|---|---|---|---|
| 12a. Pitch Control(s) position (non-fly-by-wire systems) | Full Range | ± 2% Unless Higher Accuracy Uniquely Required | 0.5 or 0.25 for airplanes operated under §121.344(f) | 0.2% of full range | For airplanes that have a flight control break away capability that allows either pilot to operate the controls independently, record both control inputs. The control inputs may be sampled alternately once per second to produce the sampling interval of 0.5 or 0.25, as applicable. |
| 12b. Pitch Control(s) position (fly-by-wire systems) [3] | Full Range | ± 2° Unless Higher Accuracy Uniquely Required | 0.5 or 0.25 for airplanes operated under §121.344(f) | 0.2% of full range | — |
| 13a. Lateral Control position(s) (non-fly-by-wire) | Full Range | ± 2° Unless Higher Accuracy Uniquely Required | 0.5 or 0.25 for airplanes operated under §121.344(f) | 0.2% of full range | For airplanes that have a flight control break away capability that allows either pilot to operate the controls independently, record both control inputs. The control inputs may be sampled alternately once per second to produce the sampling interval of 0.5 or 0.25, as applicable. |
| 13b. Lateral Control position(s) (fly-by-wire) [4] | Full Range | ± 2° Unless Higher Accuracy Uniquely Required | 0.5 or 0.25 for airplanes operated under §121.344(f) | 0.2% of full range | — |
| 14a. Yaw Control position(s) (non-fly-by-wire) [5] | Full Range | ± 2° Unless Higher Accuracy Uniquely Required | 0.5 | 0.2% of full range | For airplanes that have a flight control break away capability that allows either pilot to operate the controls independently, record both control inputs. The control inputs may be sampled alternately once per second to produce the sampling interval of 0.5. |
| 14b. Yaw Control position(s) (fly-by-wire) | Full Range | ± 2° Unless Higher Accuracy Uniquely Required | 0.5 | 0.2% of full range | — |
| 15. Pitch Control Surface(s) Position [6] | Full Range | ± 2° Unless Higher Accuracy Uniquely Required | 0.5 or 0.25 for airplanes operated under §121.344(f) | 0.2% of full range | For airplanes fitted with multiple or split surfaces, a suitable combination of inputs is acceptable in lieu of recording each surface separately. The control surfaces may be sampled alternately to produce the sampling interval of 0.5 or 0.25. |

## Appendix M to Part 121
## Airplane Flight Recorder Specifications (Continued)

The recorded values must meet the designated range, resolution, and accuracy requirements during dynamic and static conditions. All data recorded must be correlated in time to within one second.

| Parameters | Range | Accuracy (sensor input) | Seconds per sampling interval | Resolution | Remarks |
|---|---|---|---|---|---|
| 16. Lateral Control Surface(s) Position [7] | Full Range | ± 2° Unless Higher Accuracy Uniquely Required | 0.5 or 0.25 for airplanes operated under §121.344(f) | 0.2% of full range | A suitable combination of surface position sensors is acceptable in lieu of recording each surface separately. The control surfaces may be sampled alternately to produce the sampling interval of 0.5 or 0.25. |
| 17. Yaw Control Surface(s) Position [8] | Full Range | ± 2° Unless Higher Accuracy Uniquely Required | 0.5 [8] | 0.2% of full range | For airplanes with multiple or split surfaces, a suitable combination of surface position sensors is acceptable in lieu of recording each surface separately. The control surfaces may be sampled alternately to produce the sampling interval of 0.5. |
| 18. Lateral Acceleration | ± 1g | ± 1.5% max. range excluding datum error of ± 5% | 0.25 | 0.004g | — |
| 19. Pitch Trim Surface Position [9] | Full Range | ± 3% Unless Higher Accuracy Uniquely Required | 1 | 0.3% of full range | — |
| 20. Trailing Edge Flap or Cockpit Control Selection [10] | Full Range or Each Position (discrete) | ± 3° or as Pilot's indicator | 2 | 0.5% of full range | Flap position and cockpit control may each be sampled alternately at 4 second intervals, to give a data point every 2 seconds. |
| 21. Leading Edge Flap or Cockpit Control Selection [11] | Full Range or Each Discrete Position | ± 3° or as Pilot's indicator and sufficient to determine each discrete position | 2 | 0.5% of full range | Left and right sides, or flap position and cockpit control may each be sampled at 4 second intervals, so as to give a data point every 2 seconds. |
| 22. Each Thrust Reverser Position (or equivalent for propeller airplane) | Stowed, In Transit, and Reverse (Discrete) | — | 1 (per engine) | — | Turbo-jet – 2 discretes enable the 3 states to be determined.<br>Turbo-prop – discrete. |
| 23. Ground Spoiler Position or Speed Brake Selection [12] | Full Range or Each Position (discrete) | ± 2° Unless Higher Accuracy Uniquely Required | 1 or 0.5 for airplanes operated under §121.344(f) | 0.2% of full range | — |
| 24. Outside Air Temperature or Total Air Temperature [13] | –50°C to +90°C | ± 2°C | 2 | 0.3°C | — |
| 25. Autopilot/Auto-throttle/AFCS Mode and Engagement Status | A suitable combination of discretes | — | 1 | — | Discretes should show which systems are engaged and which primary modes are controlling the flight path and speed of the aircraft. |

## Appendix M to Part 121
## Airplane Flight Recorder Specifications (Continued)

The recorded values must meet the designated range, resolution, and accuracy requirements during dynamic and static conditions. All data recorded must be correlated in time to within one second.

| Parameters | Range | Accuracy (sensor input) | Seconds per sampling interval | Resolution | Remarks |
|---|---|---|---|---|---|
| 26. Radio Altitude | –20 ft to 2,500 ft | ±2 ft or ±3% Whichever is Greater Below 500 ft and ± 5% Above 500 ft | 1 | 1 ft + 5% above 500 ft | For autoland/category 3 operations. Each radio altimeter should be recorded, but arranged so that at least one is recorded each second. |
| 27. Localizer Deviation, MLS Azimuth, or GPS Latitude Deviation | ± 400 Microamps or available sensor range as installed ± 62° | As installed ± 3% recommended | 1 | 0.3% of full range | For autoland/category 3 operations. Each system should be recorded, but arranged so that at least one is recorded each second. It is not necessary to record ILS and MLS at the same time, only the approach aid in use need be recorded. |
| 28. Glideslope Deviation, MLS Elevation, or GPS Vertical Deviation | ± 400 Microamps or available sensor range as installed 0.9 to +30° | As installed ± 3% recommended | 1 | 0.3% of full range | For autoland/category 3 operations. Each system should be recorded, but arranged so that at least one is recorded each second. It is not necessary to record ILS and MLS at the same time, only the approach aid in use need be recorded. |
| 29. Marker Beacon Passage | Discrete "on" or "off" | | 1 | — | A single discrete is acceptable for all markers. |
| 30. Master Warning | Discrete | — | 1 | — | Record the master warning and record each "red" warning that cannot be determined from other parameters or from the cockpit voice recorder. |
| 31. Air/ground sensor (primary airplane system reference nose or main gear) | Discrete "air" or "ground" | — | 1 (0.25 recommended) | — | — |
| 32. Angle of Attack (if measured directly) | As installed | As installed | 2 or 0.5 for airplanes operated under §121.344(f) | 0.3% of full range | If left and right sensors are available, each may be recorded at 4 or 1 second intervals, as appropriate, so as to give a data point at 2 seconds or 0.5 second, as required. |
| 33. Hydraulic Pressure Low, Each System | Discrete or available sensor range, "low" or "normal" | ± 5% | 2 | 0.5% of full range | — |
| 34. Groundspeed | As installed | Most Accurate Systems Installed | 1 | 0.2% of full range | — |
| 35. GPWS (ground proximity warning system) | Discrete "warning" or "off" | — | 1 | — | A suitable combination of discretes unless recorder capacity is limited in which case a single discrete for all modes is acceptable. |

## Appendix M to Part 121
## Airplane Flight Recorder Specifications (Continued)

The recorded values must meet the designated range, resolution, and accuracy requirements during dynamic and static conditions. All data recorded must be correlated in time to within one second.

| Parameters | Range | Accuracy (sensor input) | Seconds per sampling interval | Resolution | Remarks |
|---|---|---|---|---|---|
| 36. Landing Gear Position or Landing gear cockpit control selection | Discrete | — | 4 | — | A suitable combination of discretes should be recorded. |
| 37. Drift Angle[15] | As installed | As installed | 4 | 0.1° | — |
| 38. Wind Speed and Direction | As installed | As installed | 4 | 1 knot, and 1.0° | — |
| 39. Latitude and Longitude | As installed | As installed | 4 | 0.002°, or as installed | Provided by the Primary Navigation System Reference. Where capacity permits Latitude/longitude resolution should be 0.0002°. |
| 40. Stick shaker and pusher activation | Discrete(s) "on" or "off" | — | 1 | — | A suitable combination of discretes to determine activation. |
| 41. Windshear Detection | Discrete "warning" or "off" | — | 1 | — | — |
| 42. Throttle/power lever position[16] | Full Range | ± 2% | 1 for each lever | 2% of full range | For airplanes with non-mechanically linked cockpit engine controls. |
| 43. Additional Engine Parameters | As installed | As installed | Each engine, each second | 2% of full range | Where capacity permits, the preferred priority is indicated vibration level, N2, EGT, Fuel Flow, Fuel Cut-off lever position and N3, unless engine manufacturer recommends otherwise. |
| 44. Traffic Alert and Collision Avoidance System (TCAS) | Discretes | As installed | 1 | — | A suitable combination of discretes should be recorded to determine the status of – Combined Control, Vertical Control, Up Advisory, and Down Advisory. (ref. ARINC Characteristic 735 Attachment 6E, TCAS VERTICAL RA DATA OUTPUT WORD.) |
| 45. DME 1 and 2 Distance | 0–200 NM | As installed | 4 | 1 NM | 1 mile |
| 46. Nav 1 and 2 Selected Frequency | Full Range | As installed | 4 | — | Sufficient to determine selected frequency. |
| 47. Selected barometric setting | Full Range | ± 5% | (1 per 64 sec.) | 0.2% of full range | — |
| 48. Selected Altitude | Full Range | ± 5% | 1 | 100 ft | — |

121

## Appendix M to Part 121
## Airplane Flight Recorder Specifications (Continued)

The recorded values must meet the designated range, resolution, and accuracy requirements during dynamic and static conditions. All data recorded must be correlated in time to within one second.

| Parameters | Range | Accuracy (sensor input) | Seconds per sampling interval | Resolution | Remarks |
|---|---|---|---|---|---|
| 49. Selected speed | Full Range | ± 5% | 1 | 1 knot | — |
| 50. Selected Mach | Full Range | ± 5% | 1 | .01 | — |
| 51. Selected vertical speed | Full Range | ± 5% | 1 | 100 ft/min | — |
| 52. Selected heading | Full Range | ± 5% | 1 | 1° | — |
| 53. Selected flight path | Full Range | ± 5% | 1 | 1° | — |
| 54. Selected decision height | Full Range | ± 5% | 64 | 1 ft | — |
| 55. EFIS display format | Discrete(s) | — | 4 | — | Discretes should show the display system status (e.g., off, normal, fail, composite, sector, plan, nav aids, weather radar, range, copy). |
| 56. Multi-function/Engine Alerts Display format | Discrete(s) | — | 4 | — | Discretes should show the display system status (e.g., off, normal, fail, and the identity of display pages for emergency procedures, need not be recorded.) |
| 57. Thrust command [17] | Full Range | ± 2% | 2 | 2% of full range | — |
| 58. Thrust target | Full Range | ± 2% | 4 | 2% of full range | — |
| 59. Fuel quantity in CG trim tank | Full Range | ± 5% | (1 per 64 sec.) | 1% of full range | — |
| 60. Primary Navigation System Reference | Discrete GPS, INS, VOR/DME, MLS, Loran C, Omega, Localizer Glideslope | — | 4 | — | A suitable combination of discretes to determine the Primary Navigation System reference. |
| 61. Ice Detection | Discrete "ice" or "no ice" | — | 4 | — | — |
| 62. Engine warning each engine vibration | Discrete | — | 1 | — | — |
| 63. Engine warning each engine over temp | Discrete | — | 1 | — | — |
| 64. Engine warning each engine oil pressure low | Discrete | — | 1 | — | — |

## Appendix M to Part 121
## Airplane Flight Recorder Specifications (Continued)

The recorded values must meet the designated range, resolution, and accuracy requirements during dynamic and static conditions. All data recorded must be correlated in time to within one second.

| Parameters | Range | Accuracy (sensor input) | Seconds per sampling interval | Resolution | Remarks |
|---|---|---|---|---|---|
| 65. Engine warning each engine over speed | Discrete | — | 1 | — | — |
| 66. Yaw Trim Surface Position | Full Range | ± 3% Unless Higher Accuracy Uniquely Required | 2 | 0.3% of full range | — |
| 67. Roll Trim Surface Position | Full Range | ± 3% Unless Higher Accuracy Uniquely Required | 2 | 0.3% of full range | — |
| 68. Brake Pressure (left and right) | As installed | ± 5% | 1 | — | To determine braking effort applied by pilots or by autobrakes. |
| 69. Brake Pedal Application (left and right) | Discrete or Analog "applied" or "off" | ± 5% (Analog) | 1 | — | To determine braking applied by pilots. |
| 70. Yaw or sideslip angle | Full Range | ± 5% | 1 | 0.5° | — |
| 71. Engine bleed valve position | Discrete "open" or "closed" | — | 4 | — | — |
| 72. De-icing or anti-icing system selection | Discrete "on" or "off" | — | 4 | — | — |
| 73. Computed center of gravity | Full Range | ± 5% | (1 per 64 sec.) | 1% of full range | — |
| 74. AC electrical bus status | Discrete "power" or "off" | — | 4 | — | Each bus. |
| 75. DC electrical bus status | Discrete "power" or "off" | — | 4 | — | Each bus. |
| 76. APU bleed valve position | Discrete "open" or "closed" | — | 4 | — | — |
| 77. Hydraulic Pressure (each system) | Full Range | ± 5% | 2 | 100 psi | — |
| 78. Loss of cabin pressure | Discrete "loss" or "normal" | — | 1 | — | — |
| 79. Computer failure (critical flight and engine control systems) | Discrete "fail" or "normal" | — | 4 | — | — |
| 80. Heads-up display (when an information source is installed) | Discrete(s) "on" or "off" | — | 4 | — | — |
| 81. Para-visual display (when an information source is installed) | Discrete(s) "on" or "off" | — | — | — | — |

121

## Appendix M to Part 121
## Airplane Flight Recorder Specifications (Continued)

The recorded values must meet the designated range, resolution, and accuracy requirements during dynamic and static conditions. All data recorded must be correlated in time to within one second.

| Parameters | Range | Accuracy (sensor input) | Seconds per sampling interval | Resolution | Remarks |
|---|---|---|---|---|---|
| 82. Cockpit trim control input position – pitch | Full Range | ± 5% | 1 | 0.2% of full range | Where mechanical means for control inputs are not available, cockpit display trim positions should be recorded. |
| 83. Cockpit trim control input position – roll | Full Range | ± 5% | 1 | 0.7% of full range | Where mechanical means for control inputs are not available, cockpit display trim position should be recorded. |
| 84. Cockpit trim control input position – yaw | Full Range | ± 5% | 1 | 0.3% of full range | Where mechanical means for control input are not available, cockpit display trim positions should be recorded. |
| 85. Trailing edge flap and cockpit flap control position | Full Range | ± 5% | 2 | 0.5% of full range | Trailing edge flaps and cockpit flap control position may each be sampled alternately at 4 second intervals to provide a sample each 0.5 second. |
| 86. Leading edge flap and cockpit flap control position | Full Range or Discrete | ± 5% | 1 | 0.5% of full range | — |
| 87. Ground spoiler position and speed brake selection | Full Range or Discrete | ± 5% | 0.5 | 0.2% of full range | — |
| 88. All cockpit flight control input forces (control wheel, control column, rudder pedal) | Full Range<br>Control wheel ±70 lbs<br>Control column ±85 lbs<br>Rudder pedal ±165 lbs | ± 5% | 1 | 0.2% of full range | For fly-by-wire flight control systems, where flight control surface position is a function of the displacement of the control input device only, it is not necessary to record this parameter. For airplanes that have a flight control break away capability that allows either pilot to operate the control independently, record both control force inputs. The control force inputs may be sampled alternately once per 2 seconds to produce the sampling interval of 1. |

*See Appendix M Footnotes on the next page.*

[1] For A300 B2/B4 airplanes, resolution = 6 seconds.

[2] For A330/A340 series airplanes, resolution = 0.703°.

[3] For A318/A319/A320/A321 series airplanes, resolution = 0.275% (0.088° > 0.064°).
For A330/A340 series airplanes, resolution = 2.20% (0.703° > 0.064°).

[4] For A318/A319/A320/A321 series airplanes, resolution = 0.22% (0.088° > 0.080°).
For A330/A340 series airplanes, resolution = 1.76% (0.703° > 0.080°).

[5] For A318/A319/A320/A321 series airplanes, resolution = 0.21% (0.088° > 0.084°).
For A330/A340 series airplanes, resolution = 1.18% (0.703° > 0.120°).

[6] For A330/A340 series airplanes, resolution = 0.783% (0.352° > 0.090°).

[7] For A330/A340 series airplanes, aileron resolution = 0.704% (0.352° > 0.100°).
For A330/A340 series airplanes, spoiler resolution = 1.406% (0.703° > 0.100°).

[8] For A330/A340 series airplanes, resolution = 0.30% (0.176° > 0.12°).
For A330/A340 series airplanes, seconds per sampling interval = 1.

[9] For all Airbus airplanes, resolution = 0.518% (0.088° > 0.051°).

[10] For A330/A340 series airplanes, resolution = 1.05% (0.250° > 0.120°).

[11] For A330/A340 series airplanes, resolution = 1.05% (0.250° > 0.120°).
For A300 B2/B4 series airplanes, resolution = 0.92% (0.230° > 0.125°).

[12] For A300-600/A310 series airplanes, speed brake resolution = 0.224% (0.112° > 0.100°).
For A330/A340 series airplanes, spoiler resolution = 1.406% (0.703° > 0.100°).

[13] For A330/A340 series airplanes, resolution = 0.5°C.

[14] For A330 airplanes with PW or RR engines, resolution = .29%.

[15] For A330/A340 series airplanes, resolution = 0.352°.

[16] For A318/A319/A320/A321 series airplanes, resolution = 4.32%.
For A330/A340 series airplanes, resolution is 3.27% of full range for throttle lever angle (TLA); for reverse thrust, reverse throttle lever angle (RLA) resolution is nonlinear over the active reverse thrust range, which is 51.54° to 96.14°. The resolved element is 2.8° uniformly over the entire active reverse thrust range, or 2.9% of the full range value of 96.14°.

[17] For A318/A319/A320/A321 series airplanes, with IAE engines, resolution = 2.58%.

[Docket No. 28109, 62 FR 38382, July 17, 1997; as amended by Amdt. 121–266, 62 FR 48135, Sept. 12, 1997; Amdt. 121–271, 64 FR 46120, Aug. 24, 1999; Amdt. 121–278, 65 FR 51745, Aug. 24, 2000; Amdt. 121–292, 67 FR 54322, Aug. 21, 2002]

## A

## B



## C

## G



## H

## I



## K



## L



## M

FAR INDEX

## U



## V



## W

# Reader Response

Your feedback is invaluable to us. Please take a moment to provide us with your comments and suggestions. Thank you — *Aviation Supplies & Academics, Inc.*

*Please print clearly.*

Name______________________________

Title *(Student, Instructor, Pilot, other)* ______________________________

Business/School ______________________________

Address______________________________

City ______________________________

State ____________________ Zip/Postal Code ____________

Country ______________________________

Telephone ______________________________

E-mail ______________________________

Where did you purchase this book? ______________________________

Was the text recommended to you? ______________________________

By whom? ______________________________

Please tell us what you liked or disliked about the text: Does this book meet your needs? (if no, please explain) Would you like to see certain regulations added or removed? (list each Part number) Is there anything you'd like to see different regarding the book content or organization?

______________________________

______________________________

______________________________

______________________________

______________________________

______________________________

______________________________

General comments and suggestions: ______________________________

______________________________

______________________________

______________________________

______________________________

❑ **Please send me an ASA Catalog**

**Mail or fax this form to:**

Aviation Supplies & Academics, Inc.
7005 132nd Place SE
Newcastle, WA 98059-3153
Fax: 425.235.0128

**Visit ASA's website: www.asa2fly.com**

03-FAR-FC